Allium Crop Science: Recent Advances

Haim D. Rabinowitch would like to dedicate this book to the memory of his mother, Sara Rabinowitch.

Lesley Currah dedicates this book to the memory of Allan Jackson (Wye College staff, 1946–1973), an inspiring teacher of Vegetable Science for many generations of students.

To my wife Shoshie
HDR

To my husband Ian
LC

Allium Crop Science: Recent Advances

Edited by

H.D. Rabinowitch

Faculty of Agricultural, Food and Environment Quality Sciences
The Hebrew University of Jerusalem
Israel

and

L. Currah

Currah Consultancy
Stratford-upon-Avon
UK

CABI *Publishing*

CABI *Publishing* is a division of CAB *International*

CABI Publishing
CAB International
Wallingford
Oxon OX10 8DE
UK

Tel: +44 (0)1491 832111
Fax: +44 (0)1491 833508
Email: cabi@cabi.org
Web site: www.cabi-publishing.org

CABI Publishing
10 E 40th Street
Suite 3203
New York, NY 10016
USA

Tel: +1 212 481 7018
Fax: +1 212 686 7993
Email: cabi-nao@cabi.org

A catalogue record for this book is available from the British Library, London, UK.

Library of Congress Cataloging-in-Publication Data

Allium crop science: recent advances/edited by H.D. Rabinowitch and L. Currah.
p. cm.
Includes bibliographical references (p.).
ISBN 0-85199-510-1 (alk. paper)
1. Allium I. Rabinowitch, Haim D. II. Currah, Lesley
SB413.A45 A44 2002
635′.25--dc21

2002025904

ISBN 0 85199 510 1

Typeset by Columns Design Ltd, Reading.
Printed and bound in the UK by Biddles Ltd, Guildford and King's Lynn.

Contents

The colour section can be found following page 4.

Contributors

B. Bohanec, Biotechnical Faculty, Centre for Plant Biotechnology and Breeding, University of Ljubljana, Jamnikarjeva 101, 1000 Ljubljana, Slovenia
A.-D. Bosch Serra, Departament de Medi Ambient i Ciències del Sòl, Universitat de Lleida, Av. Alcalde Rovira Roure 177, E-25198 Lleida, Spain
L. Currah, Currah Consultancy, 14 Eton Road, Stratford-upon-Avon CV37 7EJ, UK
H. De Clercq, Department of Plant Genetics and Breeding (DvP), Centre for Agricultural Research-Ghent (CLO-Gent), Caritasstraat 21, 9090 Melle, Belgium
C.C. Eady, New Zealand Institute for Crop & Food Research Limited, Private Bag 4704, Christchurch, New Zealand
T. Etoh, Laboratory of Vegetable Crops, Faculty of Agriculture, Kagoshima University, 21–24 Korimoto 1, Kagoshima 890-0065, Japan
N. Friesen, Botanical Garden of the University of Osnabrück, Albrechtstraße 29, D-49076, Osnabrück, Germany
R.M. Fritsch, Institut für Pflanzengenetik und Kulturpflanzenforschung, D-06466 Gatersleben, Germany
R.D. Gitaitis, Department of Plant Pathology, University of Georgia, Coastal Plain Experiment Station, Tifton, GA 31793-0748, USA
I.R. Gubb, Fresh Produce Consultancy, Mulberry Lodge, Culmstock, Cullompton, Devon EX15 3JB, UK
M.J. Havey, Agricultural Research Service – USDA, Department of Horticulture, 1575 Linden Drive, University of Wisconsin, Madison, WI 53706, USA
M.P. Hoffmann, Department of Entomology, Cornell University, Ithaca, NY 14853, USA
R. Kamenetsky, Department of Ornamental Horticulture, The Volcani Center, Bet Dagan 50250, Israel
M. Keusgen, Institute for Pharmaceutical Biology, University of Bonn, Nußallee 6, D-53115 Bonn, Germany
C. Kik, Plant Research International, Wageningen University and Research Center, PO Box 16, 6700 AA Wageningen, The Netherlands
M. Klaas, Gotthard Müller Straße 57, D-70794 Filderstadt-Bernhausen, Germany
T.P. Kuhar, Department of Entomology, Cornell University, Ithaca, NY 14853, USA
J.E. Lancaster, AgriFood Solutions Ltd., Voss Road, RD4, Christchurch, New Zealand
J.W. Lorbeer, Department of Plant Pathology, Cornell University, Ithaca, NY 14853, USA
H.S. MacTavish, ADAS Arthur Rickwood, Mepal, Ely CB6 2AB, UK

G.L. Mark, Department of Plant Pathology, Cornell University, Ithaca, NY 14853, USA

H.D. Rabinowitch, Institute of Plant Science and Genetics in Agriculture, The Hebrew University of Jerusalem, Faculty of Agricultural, Food and Environmental Quality Sciences, PO Box 12, Rehovot 76100, Israel

W.M. Randle, Department of Horticulture, University of Georgia, 1111 Plant Sciences Building, Athens, GA 30602-7273, USA

R. Salomon, Agricultural Research Organization, The Volcani Center, Department of Virology, PO Box 6, Bet Dagan 50250, Israel

P.W. Simon, USDA/ARS, Department of Horticulture, 1575 Linden Drive, University of Wisconsin, Madison, WI 53706, USA

E. Van Bockstaele, Department of Plant Genetics and Breeding (DvP), Centre for Agricultural Research-Ghent (CLO-Gent), Caritasstraat 21, 9090 Melle, Belgium

Abbreviations

A	*Azotobacter*
ABA	Abscisic acid
ACSO	*S*-alk(en)yl-L-cysteine sulphoxide
ADP	Adenosine diphosphate
AFLP	Amplified fragment length polymorphism
6-AG	6-Azoguanine
AMP	Adenosine monophosphate
AP	Ammonium phosphate
API	Analytical profile index
APS	Adenosine phosphosulphate
Asn	Asparagine
AT	Adenine–thymine (base pair)
ATP	Adenosine triphosphate
AVRDC	Asian Vegetable Research and Development Center (Taiwan)
BAP	Benzylaminopurine
BLASTN	Software used for sequence resemblance analysis
BLASTP	Software used for sequence resemblance analysis
bp	Base pair
BSA	Bulked segregant analysis
Bt	*Bacillus thuringiensis*
CA	Controlled-atmosphere (storage)
cAMP	Cyclic AMP
CAPS	Cleaved amplified polymorphic sequence
CDL	Critical disease level
cDNA	Complementary DNA
CI	Consistency index
cM	Centimorgan
CMS	Cytoplasmic male sterility
CoA	Coenzyme A
CP	Coat protein
cpDNA	Chloroplast DNA
CPE	Cumulative Class A pan evaporation
CULTAN	Controlled-uptake long-term ammonia nutrition
2,4-D	2,4-Dichlorophenoxyacetic acid
Da	Dalton

DCPA	Dimethyl-2,3,5,6-tetrachloro-1,4-benzenedicarboxylate
DD	Day-degrees/degree-days
DH	Doubled haploid
DM	Dry matter
DNA	Deoxyribonucleic acid
DP	Degree of polymerization
DRIS	Diagnosis and recommendation integrated system
dS	Decisiemens
EBDC	Ethylene bis-dithiocarbamate
EC	Electrical conductivity
ELISA	Enzyme-linked immunosorbent assay
EM	Electron microscope
EMB	Eosin methylene blue
EMBL	European Molecular Biology Laboratory
EPSPS	5-Enolpyruvylshikimate-3-phosphate synthase
EPY	Enzymatically determined pyruvic acid
EST	Esterase
ETc	Crop evapotranspiration
FAO	Food and Agriculture Organization of the United Nations
FFT	Fructan : fructan-fructosyl transferase
FISH	Fluorescent *in situ* hybridization
5-FU	5 Fluorouracil
FYM	Farmyard manure
GA, GA_3	Gibberellic acid
GAL	Galactosidase
Gar V	Garlic virus V
GC	Guanine–cytosine (base pair)
GCLV	Garlic common latent virus
GFP	Green fluorescent protein
GISH	Genomic *in situ* hybridization
GLV	Garlic latent virus
GM	Genetically modified
GMS	Genic male sterility
GMV	Garlic mosaic virus
Gna	*Galanthus nivalis* (snowdrop) agglutinin
GPS	Global positioning (satellite) system
GV2	Garlic virus 2
GVA	Garlic virus A
GVC	Garlic virus C
HMG-CoA	Hydroxymethylglutaryl coenzyme A (reductase)
HPLC	High-performance liquid chromatography
IBA	Indolebutyric acid
ICM	Integrated crop management
IC-RT-PCR	Immunocapture-reverse transcriptase PCR
ID	Intermediate-day (type onion)
IEF	Isoelectric focusing
IGS	Intergenic spacer
IIHR	Indian Institute of Horticultural Research
INRA	Institut National de la Recherche Agronomique (France)
INTA	Instituto Nacional de Tecnología Agropecuaria (Argentina)
2iP	*N*-6-(2-isopentenyl)-adenine
IPA	Empresa Pernambucana de Pesquisa Agropecuária (Brazil)
IPM	Integrated pest management
IS	Insertion sequence
ITS	Intergenic transcribed spacer or Internal transcribed spacer

kb	Kilobase
kc	Crop coefficient
kDa	Kilodalton
KDF	Light extinction coefficient
kGy	Kilogray
kPa	Kilopascal
LAI	Leaf-area index
LD	Long-day (type onion)
LDL	Low-density lipoprotein fraction
LF	Lachrymatory factor
LSC	Large single copy
LWH	Leaf-wetness hours
Lys	Lysine
LYSV	Leek yellow-stripe virus
MAB	Marker-assisted breeding
MAFF	Ministry of Agriculture, Fisheries and Food (UK)
Mb	Megabase
MCP	Methyl cyclopropene
MCSO	(+)-*S*-methyl-L-cysteine sulphoxide
MDH	Malate dehydrogenase
MH	Maleic hydrazide
MIC	Minimum inhibitory concentration
MJ	Megajoule
MPa	Megapascal
mRNA	Messenger RNA
MS, Ms	Male-sterile
mtDNA	Mitochondrial DNA
N	Normal cytoplasm
NAA	Naphthalene acetic acid
NADPH	Nicotinamide adenine dinucleotide phosphate (reduced)
NAFTA	North American Free Trade Agreement
Nc	Organic nitrogen in the plant
nDNA	Nuclear DNA
NHRDF	National Horticultural Research and Development Foundation (India)
NRI	Natural Resources Institute (UK)
nuDNA	Nuclear DNA
OP	Open-pollinated
OWR	Onion white rot (*Sclerotium cepivorum*)
OYDV	Onion yellow dwarf virus
PAF	Platelet activating factor
PAPS	Phosphoadenosine phosphosulphate
PAR	Photosynthetically active radiation
PCA	Principal-component analysis
PCR	Polymerase chain reaction
PCSO	(+)-*S*-propyl-L-cysteine sulphoxide
1-PECSO	*trans*-(+)-*S*-(1-propenyl)-L-cysteine sulphoxide
2-PECSO	(+)-*S*-(2-propenyl)-L-cysteine sulphoxide, alliin
PEG	Polyethylene glycol
PG	Phosphogypsum
PG	Endopolygalacturonase
pg	Picogram
PGI	Phosphoglucoisomerase
PGM	Phosphoglucomutase
PMC	Pollen mother cell
ppm	Parts per million

PPT	Phosphinothricin
PR	Pathogenesis-related
PRR	Pink-root-resistant (= tolerant)
PTM	Primary thickening meristem
PVP	Plant variety protected
QDG	Quercetin-3,4′-*O*-diglucoside
QMG	Quercetin-4′-*O*-monoglucoside
QTL	Quantitative-trait loci
RAPD	Randomly amplified polymorphic DNA
rbcL	Ribulose-1,5-biphosphate carboxylase
rDNA	Ribosomal DNA
RFLP	Restriction fragment length polymorphism
RH	Relative humidity
RI	Retention index
RL	Root length
RN	Recombination nodule
RNA	Ribonucleic acid
rRNA	Ribosomal RNA
RT-PCR	Reverse-transcription polymerase chain reaction
S	One type of sterile cytoplasm
s, S	Svedberg constant (see Chapter 3)
SAT	Satellite chromosome genetic material
SC	Synaptonemal complex
SCAR	Sequence-characterized amplified region
SD	Short-day (type onion)
SDS-PAGE	Sodium dodecyl sulphate – polyacrylamide gel electrophoresis
SDW	Shoot dry weight
SLV	Shallot latent virus
SSC	Small single copy (see Chapter 3)
SSC	Soluble-solids content
SSD	Single-seed descent
SST	Sucrose : sucrose-fructosyl transferase
SVX	Shallot virus X
SWP	Soil water potential
SYSV	Shallot yellow-stripe virus
T	One type of sterile cytoplasm
Ti	Tumour-inducing
TMV	Tobacco mosaic virus
TOFMS	Time-of-flight mass spectroscopy
tRNA	Transfer RNA
TRV	Tobacco rattle virus
TuMV	Turnip mosaic virus
UAN	Urea ammonium nitrate
UPGMA	Unweighted pair-group method using arithmetic averages
VAM	Vesicular-arbuscular mycorrhiza
VLD	Very-long-day (type onion)
W	Plant mass
WHO	World Health Organization of the United Nations
WSC	Water-soluble carbohydrates
WSMV	Wheat streak mosaic virus
WUE	Water-use efficiency

Note: not included: chemical symbols, culture media, primers, genes.

Introduction

Onion, Japanese bunching onion, leek and garlic are the most important edible *Allium* crops. Onion, the principal *Allium*, ranks second in value after tomatoes on the list of cultivated vegetable crops worldwide (FAO, 2001). In addition, for generations, over 20 other *Allium* species have been consumed by humans (van der Meer, 1997): the most popular alliums include garlic, chives and several Oriental species which are both cultivated and collected from the wild. Lately, old and new alliums, both edible and ornamental, have started to become popular worldwide. They include culinary species such as Chinese chives (*A. tuberosum*) on the one hand, and beautiful flowering bulbous plants such as *A. aflatunense* on the other (Colour Plate 1A). Consumers and researchers alike have also become more aware of the health benefits and medicinal properties of alliums in recent years (Keusgen, Chapter 15, this volume).

Research on the physiological, biochemical and genetic traits of alliums is gaining momentum, but good accounts of modern advances in the biology of alliums have been lacking. In their 1928 book, *Truck Crop Plants*, Jones and Rosa devoted 26 pages to alliums. The chapter focused mainly on agronomy and varietal maintenance. Thirty-seven years later, Jones and Mann (1963) published their classic book *Onions and their Allies*. The authors reviewed the state-of-the-art of agronomy and physiology with some emphasis on genetics, based on the pioneering work of Henry A. Jones from the 1930s up to the early 1960s. At that time, topics such as tissue culture, sulphur and carbohydrate biochemistry and the biology of seed development were not yet a significant part of *Allium* science, and the initial steps of molecular biology did not include any *Allium* species. Twenty-two years later, Fenwick and Hanley (1985) published a comprehensive review of various physiological and biochemical aspects of the genus *Allium* from the point of view of food science and uses of the crops. Since then *Allium* research has diversified significantly and a single person or a small number of authors can no longer put together an expert review of all the biological aspects of *Allium* research. In 1990, Rabinowitch and Brewster published their three-volume multi-authored book *Onions and Allied Crops*. These works provided comprehensive coverage of *Allium* science in the late 1980s. Pollination biology, seed development, genetic resources, anatomy, tissue culture, weed competition and herbicides, mycorrhizal associations and their significance, carbohydrate and sulphur biochemistry, and therapeutic and medicinal values of alliums were among the important topics reviewed for the first time. Brewster's 1994 book *Onions and other Vegetable Alliums* was a condensed and updated summary of the 1990 volumes, aiming 'to introduce the scientific principles that underline production practices'. It provided a valuable, concise textbook for students, with particularly good coverage of physiological topics; these were updated again by Brewster in 1997.

More specialized *Allium* topics were also the subject of publications during the last two decades. *Pest Control in Tropical Onions* (Anon., 1986) was a compendium of advice on current practice in the use of pesticides at a time before Integrated Pest Management (IPM) had yet made much impact. *Onions in Tropical Regions* (Currah and Proctor, 1990) summarized work from many countries, bringing together survey results and research literature from a tropical perspective. Brice *et al.* (1997) summarized knowledge of the factors affecting onion storage in the tropics, aiming to assist growers in determining which storage methods to adopt. Other recent specialist publications we recommend are those by van Deven (1992), Diekmann (1997) and Gregory *et al.* (1998).

In 2000, we felt that the new and striking developments in *Allium* science over the past decade had reached the point where an advanced comprehensive picture should be drawn for the benefit of *Allium* scientists and for students new to the topics. We agreed that the book would focus on topics developed in recent years and not yet reviewed earlier. Hence, in this book we aim to cover the subjects on which significant new knowledge has accumulated, newly emerged topics or those that have gained a marked momentum in the last quarter of the 20th century. These include genome organization in alliums; exploitation of wild and cultivated relatives for the breeding of *Allium* crops; diversity, fertility and seed production of garlic; genetic transformation of onions; doubled-haploid onions; molecular markers in alliums; and ornamental alliums. We include reviews of shallots, and onions in the tropics, as these were not yet treated in detail in mainstream literature available in English. For leeks, we are fortunate to include a review by scientists from Belgium, where the crop is being intensively researched. The topic of postharvest of onions is also thoroughly reviewed from the biological point of view. In plant pathology, reviews cover the detection of garlic viruses and the propagation of virus-free crops; bacterial diseases of the alliums, including descriptions of diseases that have become significant recently; and the important topic of forecasting and monitoring pests and diseases in connection with IPM methods of control.

The lengthy chapter on agronomy of onions may need an explanation. We have tried to provide an overview of recent technical work, including seed priming, modelling of onion growth, irrigation and weed control studies at the 'high tech' end of agronomy, while also taking account of the tendency towards lower-input and environmentally friendly production methods: these are becoming relevant for producers worldwide. Hence we present some examples of organic production methods and research topics (such as weed control without herbicides), which are being actively pursued at the present time.

'The science of alliums involves knowledge ranging from the level of the molecule to that of the agroecosystem' (Brewster, 1994). Brewster stated that his book was 'concerned with processes in the "upper middle" part of this spectrum'. The primary aim of the present book is to bring together, in a single volume, up-to-date knowledge obtained by a variety of scientific disciplines – from the basic level of the molecule to application in the field in *Allium* crops. We hope that this book will help to bridge disciplinary barriers in *Allium* research, that it will be of value to workers interested in all the biological aspects of alliums and that it will facilitate discussions and interactions between scientists and field experts in the study of bulbous plants and horticultural plant sciences. We also hope that it will be enjoyable to read and provide an introduction to some unfamiliar aspects of *Allium* science for specialists and generalists alike.

We do not claim to have covered all the research topics which are currently being investigated and we are aware that some new areas receive detailed attention here while others may be omitted since they are not currently attracting much research interest. As a more commercially orientated companion to this volume, we would like to draw readers' attention to the recent appearance of the Proceedings of the Second International Symposium on Edible *Alliaceae*, held in Adelaide, Australia in 1997, which has just appeared in the *Acta Horticulturae* series (Armstrong, 2001). This follows the proceedings of two earlier international meetings on alliums held in 1993 and 1994 (Midmore, 1994; Burba and Galmarini, 1997). The new volume gives good coverage of marketing and of problems connected to the export of *Allium* crops, as well as highlighting research work from Australia and New Zealand.

The experts who we approached generously agreed to share their knowledge through this book project. We thank our authors for their willingness to contribute, for their time and expertise and for their patience with our editorial demands. We are also particularly grateful to our two principal helpers in the production of the manuscripts: Janine Harpaz, at the Faculty of Agricultural, Food and Environmental Quality Sciences, for her friendly and wholehearted assistance and for her valuable and meticulous work throughout the compilation of this book; and Ian Currah in the UK, for his valuable and timely help with computing and in maintaining electronic communications. We thank the publishers at CABI, especially Tim Hardwick and Claire Gwilt, for their patience and for the professional job they have done on the combined intellectual creation of 26 authors and in expeditiously seeing the book through press. Lesley Currah would like to thank the staff of the library at Horticulture Research International (HRI), Wellesbourne for allowing her to use the literature collection. We also thank the many individuals who helped us to trace references, reviewed chapters and provided collections of reprints on specialist topics. In particular, we are glad to acknowledge the help of Brian Smith, Ian Puddephat, Helen Robinson and Tijs Gilles at HRI, Wellesbourne; Ray Fordham, Charles Wright and James Brewster, UK; Florence Esnault in Brittany and S.R. Bhonde in India. We are most grateful to the Vegetable Research Trust at HRI, Wellesbourne and to the Production and Marketing Board of Ornamental Plants of Israel for their generous support for the inclusion of colour plates in this book.

References

Anon. (eds) (1986) *Pest Control in Tropical Onions*. Tropical Development and Research Institute, London, UK, 109 pp.

Armstrong, J. (ed.) (2001) Proceedings of the Second International Symposium on Edible *Alliaceae*, Adelaide, South Australia, 10–13 November 1997. *Acta Horticulturae* 555, 304 pp.

Brewster, J.L. (1994) *Onions and Other Vegetable Alliums*. CAB International, Wallingford, UK, 236 pp.

Brewster, J.L. (1997) Onions and garlic. In: Wien, H.C. (ed.) *The Physiology of Vegetable Crops*. CAB International, Wallingford, UK, pp. 581–619.

Brewster, J.L. and Rabinowitch, H.D. (eds) (1990) *Onions and Allied Crops, III. Biochemistry, Food Science, and Minor Crops*. CRC Press, Boca Raton, Florida, 265 pp.

Brice, J., Currah, L., Malins, A. and Bancroft, R. (1997). *Onion Storage in the Tropics: A Short Practical Guide to Methods of Storage and their Selection*. Natural Resources Institute, The University of Greenwich, Chatham, UK, 120 pp.

Burba, J.L. and Galmarini, C.R. (1997) Proceedings of the First International Symposium on Edible *Alliaceae*, 14–18 March 1994, Mendoza, Argentina. *Acta Horticulturae* 433, 652 pp.

Currah, L. and Proctor, F.J. (1990) *Onions in Tropical Regions*. Bulletin 35, Natural Resources Institute, Chatham, UK, 232 pp.

Diekmann, M. (ed.) (1997) *FAO/IPGRI Technical Guidelines for the Safe Movement of Germplasm. No. 18,* Allium *spp*. Food and Agriculture Organization of the United Nations, Rome/International Plant Genetic Resources Institute, Rome, Italy, 60 pp.

FAO (2001) Agrostat database, updated annually: http://apps.fao.org/

Fenwick, G.R. and Hanley, A.B. (1985) The genus *Allium*, Part 1. *CRC Critical Reviews on Food Science and Nutrition* 22, 199–271.

Gregory, M., Fritsch, R.M., Friesen, N., Khassanov, F.O. and McNeal, D.W. (1998) Nomenclator Alliorum. *Allium Names and Synonyms – A World Guide*. The Trustees, Royal Botanic Garden, Kew, Richmond, UK, 83 pp.

Jones, H.A. and Mann, L.K. (1963) *Onions and their Allies*. InterScience, New York, 286 pp.

Jones, H.A. and Rosa, J.T. (1928) *Allium*. In: *Truck Crop Plants*. McGraw-Hill, New York, pp. 37–63.

Midmore, D.J. (ed.) (1994) Proceedings of an International Symposium on Alliums for the Tropics, 15–19 February 1993, Bangkok and Chiang Mai, Thailand. *Acta Horticulturae* 358, 431 pp.

Rabinowitch, H.D. and Brewster, J.L. (eds) (1990) *Onions and Allied Crops, I. Botany, Physiology, and Genetics*. CRC Press, Boca Raton, Florida, 273 pp.

Rabinowitch, H.D. and Brewster, J.L. (eds) (1990) *Onions and Allied Crops, II. Agronomy, Biotic Interactions, Pathology, and Crop Protection*. CRC Press, Boca Raton, Florida, 320 pp.

van der Meer, Q.P. (1997) Old and new crops within edible alliums. *Acta Horticulturae* 433, 17–31.

van Deven, L. (1992). *Onions and Garlic Forever*. Louis van Deven, 608 North Main, PO Box 72, Carrollton, Illinois, 114 pp.

Plate 1. Topsets in the inflorescence of *A. aflatunense*, promoted by poor storage conditions. **A.** Normal inflorescence. **B.** Topsets form in the periphery of the inflorescence, the centre consists of normal flowers. **C.** Topsets develop all over the inflorescence.

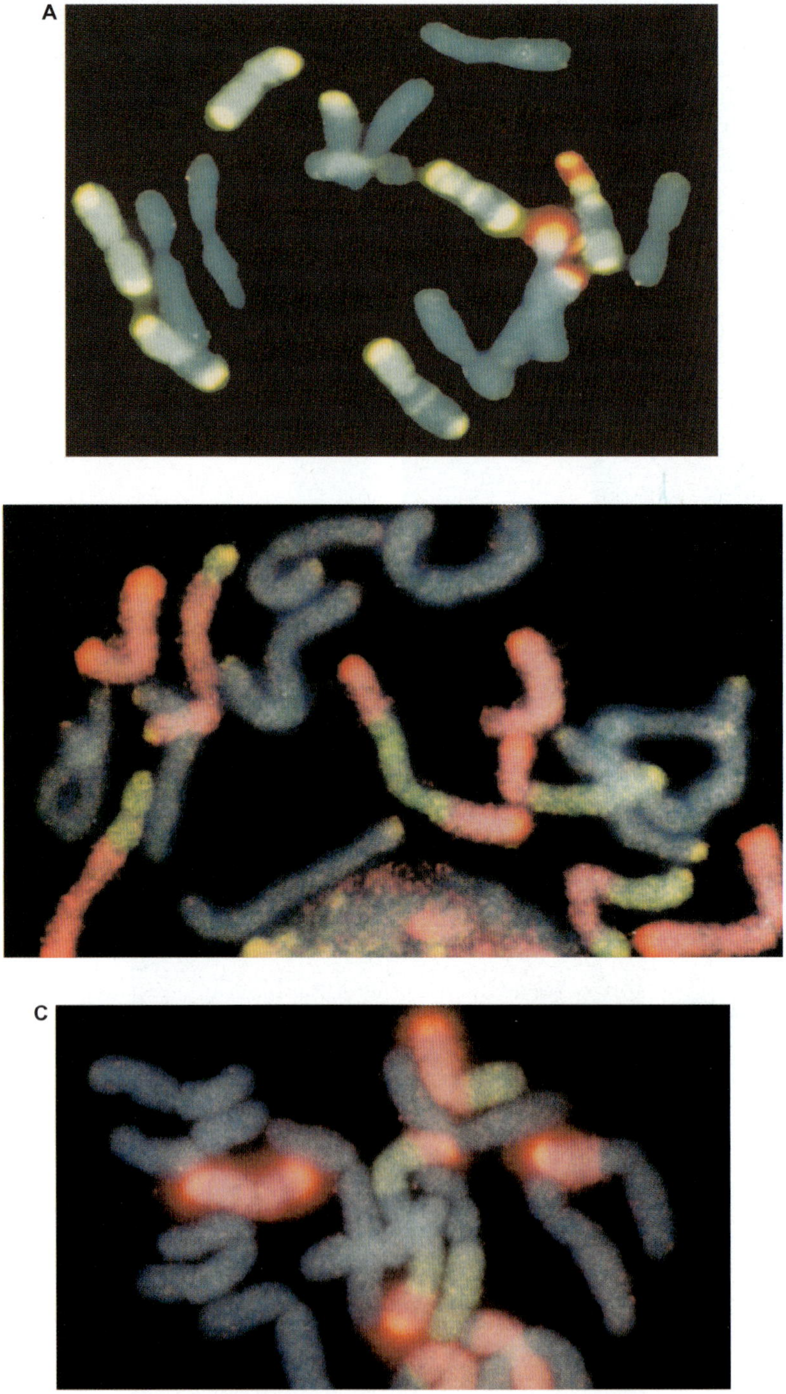

Plate 2. A. GISH and FISH in the backcross (*A. fistulosum* x *A. cepa*) x *A. cepa*. The *A. cepa* chromatin fluoresces blue (block, DAPI); the *A. fistulosum* chromatin fluoresces yellow (Dig, FITC), with a pronounced signal at the telomeres; the 45S rDNA probe fluoresces red (Biotin, CY3). **B.** Multicolour GISH in the first generation bridge-cross *A. cepa* x (*A. fistulosum* x *A. roylei*). *A. cepa*, blue (block, DAPI); *A. fistulosum*, red (Biotin, CY3); *A. roylei*, green (Dig, FITC). (Khrustaleva and Kik, 1998, with permission.) **C.** Multicolour GISH in the second generation bridge-cross *A. cepa* x [*A. cepa* x (*A. fistulosum* x *A. roylei*)]. *A. cepa*, blue (block, DAPI); *A. fistulosum*, red (Biotin, CY3); *A. roylei*, green (Dig, FITC). (Khrustaleva and Kik, 2000, with permission.) Dig, digoxygenin-11-2'-deoxy-uridine-5'-triphosphate.

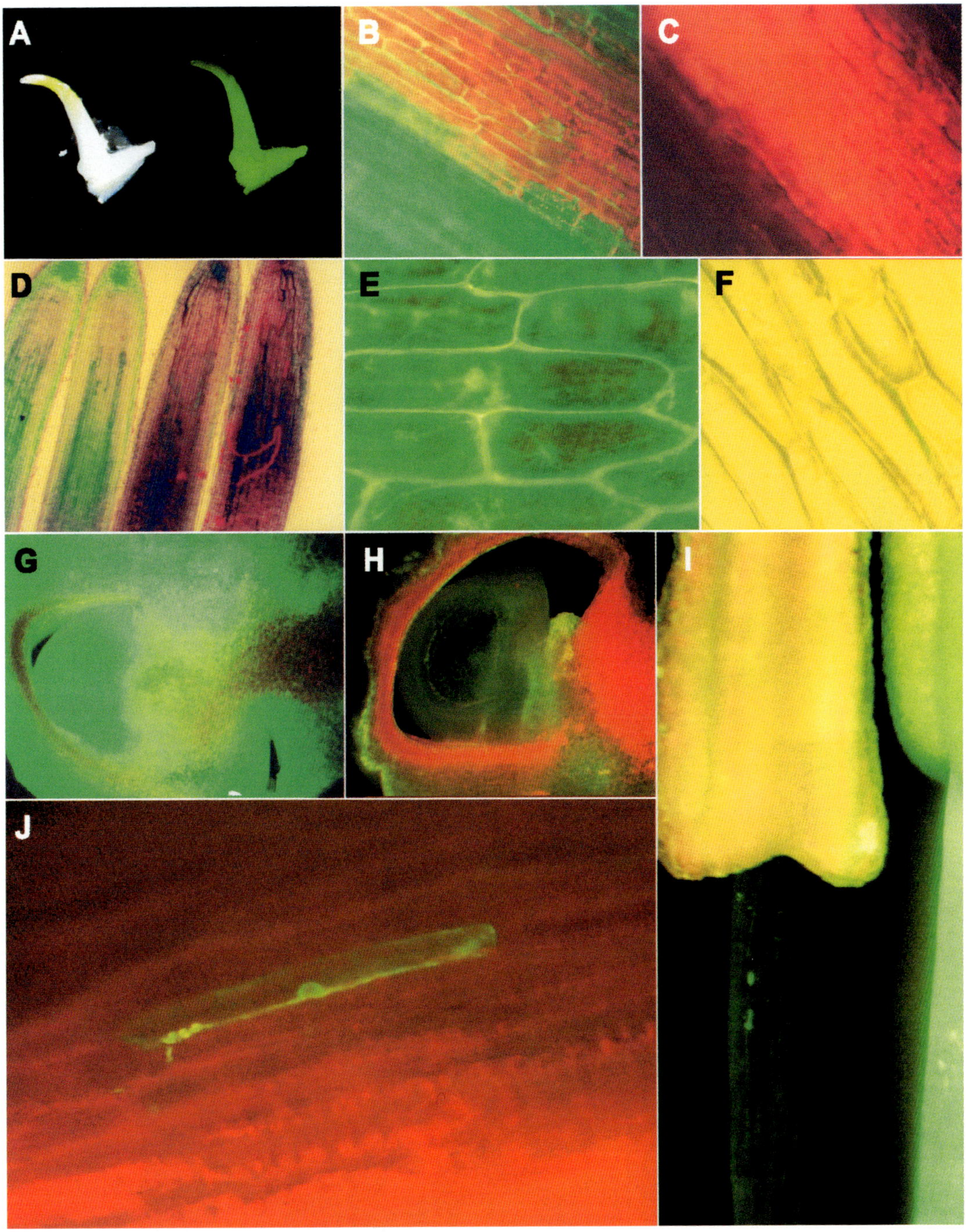

Plate 3. Transgenic onion tissues expressing the *gfp* gene and non-transgenic control tissues under blue light, except where indicated. **A.** Transgenic shoot culture under white light (left) and blue light (right). **B.** Transgenic petal. **C.** Non-transgenic petal. **D.** Transgenic (left) and non-transgenic (right) roots (some additional background lighting). **E.** Transgenic bulb epidermal skin cells. **F.** Non-transgenic bulb epidermal skin cells. **G.** Section through a transgenic ovary. **H.** Section through a non-transgenic ovary. **I.** Transgenic (right) and non-transgenic (left) stamen. **J.** Transgenic petal from a line exhibiting a punctate expression profile.

A

B

C

D

E

Plate 4. A. Sprouting haploid embryo forms a typical loop structure. **B.** Extracted haploids form a complete bipolar embryo with already-developed roots. Some embryos develop into abnormal structures. **C.** Differential development of embryos from a single harvest (some from a single ovary) cultured in the same Petri dish. Some are considerably smaller than others; some are deformed. **D.** Haploid and doubled haploid plants grow vigorously and form bulbs similar to heterozygous plants. **E.** Normal inflorescence of a doubled haploid.

A

B

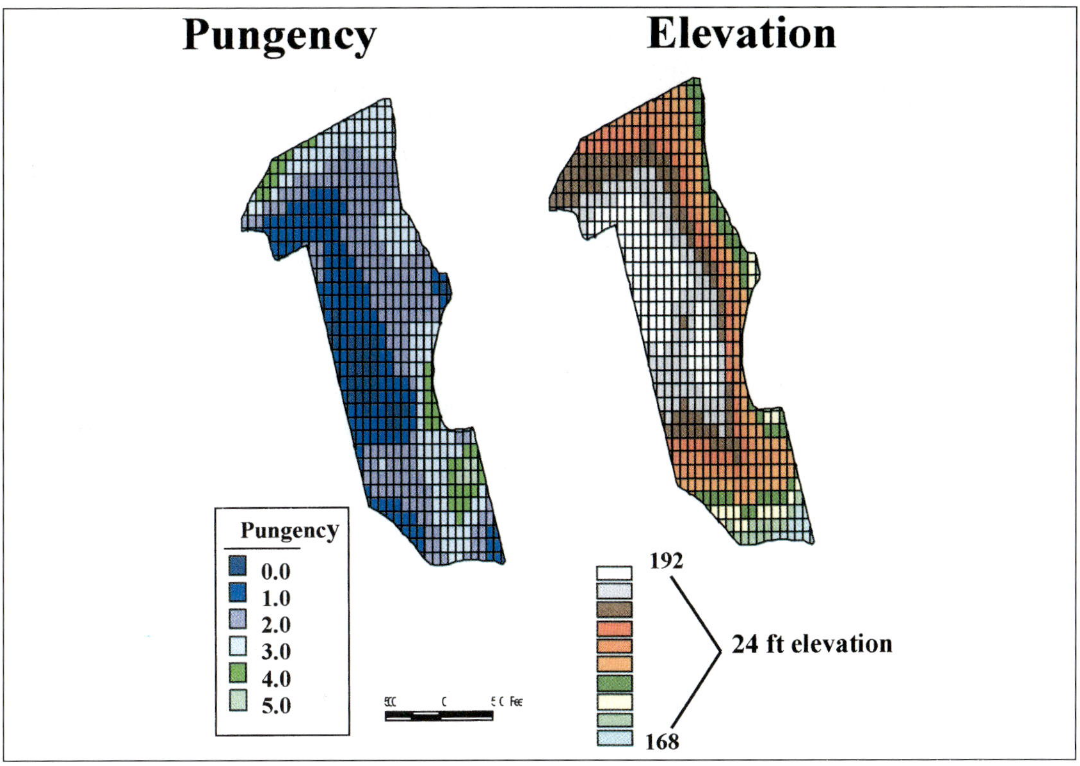

Plate 5 (upper plate). A. Leaves of garlic (*A. sativum*) (left) and great-headed garlic (*A. ampeloprasum* Great-headed Garlic Group) (right) exhibiting mosaic symptoms typical of mixed virus infection. **B.** Leaves of great-headed garlic infected with LYSV, showing typical mosaic symptoms.

Plate 6 (lower plate). Bulb pungency distribution and elevation changes in a commercial onion field mapped using global positioning satellites and measured by bulb pyruvate content. (Courtesy of Vidalia Labs International, Collins, Georgia, USA.)

Plate 7. A. A bunch of inflorescences for consumption as green vegetables. (Picture taken in 1994 in a market in Bangkok, Thailand, Haim Rabinowitch.) **B.** Transverse sections of shallot (top) and bulb onion (bottom). Note the subdivisions of the doubling bulbs. **C.** Shallot storage in the field (Indonesia, 2000). The stored shallot bulbs are well protected from direct sun irradiation and rain (Ada Harazy, FAHU, Israel). **D.** Shallot storage in an open, well-ventilated shed (Indonesia, 2000) (Ada Harazy, FAHU, Israel).

Plate 8. Species of subgenus *Melanocrommyum*. **A.** Cv. 'Globus' (*A. karataviense* x *A. stipitatum*), general view. **B.** *A. nevskianum*, general view. **C.** Inflorescence of *A. macleanii*. **D.** *A. lipskyanum*, general view.

Plate 9. Species of subgenus *Melanocrommyum*. **A.** *A. giganteum*, general view. **B.** Inflorescences of *A. regelii*. **C.** Inflorescence of *A. oreophilum*. **D.** *A. karataviense*, general view. **E.** Inflorescence of *A. cupuliferum*.

Plate 10. Species of subgenus *Rhizirideum s. lat*. **A.** *A. galanthum*, general view. **B.** Inflorescences of *A. platyspathum*. **C.** Inflorescences of *A. lusitanicum*. **D.** Inflorescence of *A. cyaneum*. **E.** Inflorescences of *A. obliquum*. **F.** *A. cyathophorum* var. *farreri*, general view.

Plate 11. Species of subgenus *Amerallium*. **A.** *A. moly*, general view. **B.** Inflorescences of *A. amplectens*. **C.** Inflorescence of *A. unifolium*. **D.** Inflorescences of *A. insubricum*. **E.** Inflorescence of *A. siculum*.

Plate 12. Species of subgenus *Allium*. **A.** Inflorescences of *A. atroviolaceum*, general view. **B.** *A. flavum*, general view. **C.** Inflorescences of *A. cyaneum*. **D.** Inflorescence of *A. sphaerocephalon*. **E.** Inflorescences of *A. carinatum* subsp. *pulchellum*.

Plate 13. Species with pleasant odour or suitable for forcing. **A.** Inflorescence of *A. stipitatum*. **B.** Inflorescences of *A. triquetrum*. **C.** Inflorescence of *A. nigrum*. **D.** Inflorescence of *A. darwasicum*. **E.** Inflorescence of *A. aschersonianum*. **F.** Inflorescence of *A. schubertii*.

1 Evolution, Domestication and Taxonomy

R.M. Fritsch[1] and N. Friesen[2]
[1]*Institut für Pflanzengenetik und Kulturpflanzenforschung, D-06466 Gatersleben, Germany;* [2]*Botanischer Garten der Universität, D-49076 Osnabrück, Germany*

1. The Genus *Allium* L.

1.1 General characteristics

The taxonomic position of *Allium* and related genera has long been a matter of controversy. In early classifications of the angiosperms (Melchior, 1964), they were placed in the *Liliaceae*. Later, they were more often included in the *Amaryllidaceae*, on the basis of inflorescence structure. Recently, molecular data have favoured a division into a larger number of small monophyletic families. In the most recent and competent taxonomic treatment of the monocotyledons, *Allium* and its close relatives were recognized as a distinct family, the *Alliaceae*, close to the *Amaryllidaceae*. The following hierarchy has been adopted (Takhtajan, 1997):

1. Class *Liliopsida*.
2. Subclass *Liliidae*.
3. Superorder *Liliianae*.
4. Order *Amaryllidales*.
5. Family *Alliaceae*.
6. Subfamily *Allioideae*.
7. Tribe *Allieae*.
8. Genus *Allium*.

However, other classifications still have their proponents and are still used in some literature.

There is more agreement about the delimitation of the genus *Allium* itself. It is a large genus of perennial, mostly bulbous plants sharing as characteristics:

- Underground storage organs: bulbs, rhizomes or swollen roots.
- Bulbs: often on rhizomes; true bulbs (one or two extremely thickened prophylls) or false bulbs (thickened basal sheaths plus thickened prophylls (bladeless 'true scales')); several tunics, membranous, fibrous or coriaceous; annual or perennial roots.
- Rhizomes: condensed or elongated; rarely runner-like; with very diverse branching patterns.
- Leaves: basally arranged, frequently covering the flower scape and thus appearing cauline.
- Bracts: two to several, often fused into an involucre ('spathe').
- Inflorescence: fasciculate to often umbel- or head-like, (one-) few- to many-flowered, loose to dense.
- Flowers: pedicelled, actinomorphic, hypogynous, trimerous.
- Tepals: in two slightly differentiated whorls, free.
- Stamens: in two whorls, sometimes basally connected, the inner ones often widened and/or toothed.
- Ovary: trilocular, three septal nectaries of various shape, two or more curved (campylotropous) ovules per locule, sometimes diverse apical appendages (crests and horns); developing into a loculicidal capsule dehiscing along the midrib of the carpels.
- Style: single, with slender, capitate or, more rarely, trilobate stigma.
- Seeds: angular to globular, black (epidermal layer contains phytomelan), ornamentation of the cells extremely variable.
- Chemical characters: reserve compounds consist of sugars, mainly fructans, and no starch; enzymatic decomposition products of several cysteine sulphoxides (see Randle and Lancaster, Chapter 14, and Keusgen, Chapter 15, this volume) cause the species- and group-specific (though sometimes missing) characteristic odour.
- Karyology: predominant basic chromosome numbers x = 8 and x = 7 with polyploids in both series; chromosome morphology and banding pattern different between taxonomic groups.

Shape, size, colour and texture of rhizomes, bulbs, roots, leaves (e.g. flat, channelled, terete or fistulose, sheath/lamina ratio), scapes, spathes, inflorescences, tepals (mostly white or rose to violet, rarely blue or yellow), stamens, ovaries and seeds may vary considerably and in very different manners. The same is true for the anatomy, cross-sections and internal structure of all the listed plant parts.

Basal bulblets and bulbils (topsets) are important in vegetative propagation. As far as known, most *Allium* species are allogamous. Spontaneous interspecific hybridization is not as rare as formerly believed, but strong crossing barriers exist in some groups, even between morphologically similar species.

1.2 Distribution, ecology and domestication

The genus *Allium* is widely distributed over the holarctic region from the dry subtropics to the boreal zone (Fig. 1.1). One or two

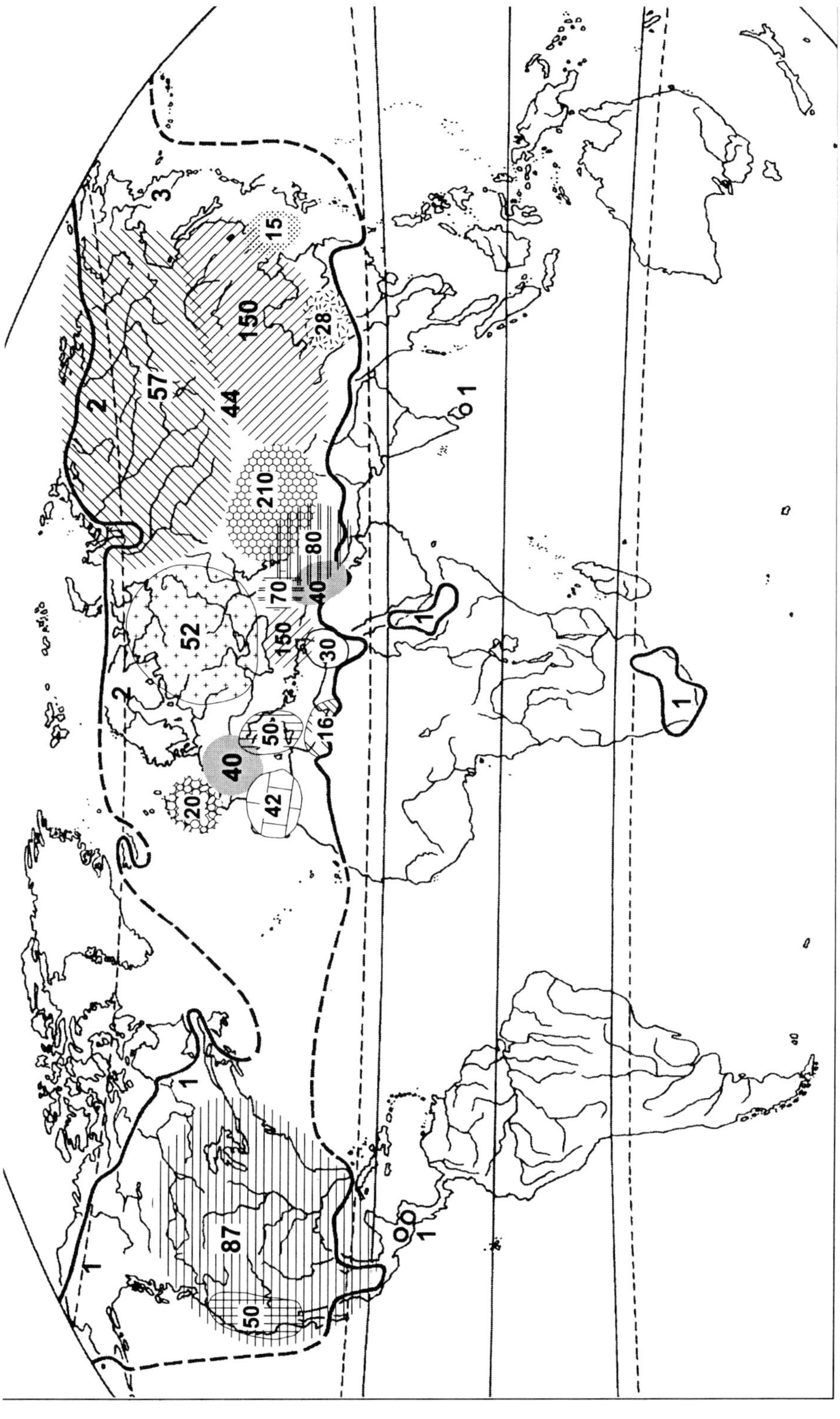

Fig. 1.1. World distribution of wild species of the genus *Allium*. The numbers on the map indicate the number of species found in each region.

species even occur in the subarctic belt, e.g. *A. schoenoprasum* L., and a few alliums are scattered in mountains or highlands within the subtropics and tropics. Only *A. dregeanum* Kth. has been described from the southern hemisphere (South Africa) (de Sarker *et al.*, 1997).

A region of especially high species diversity stretches from the Mediterranean basin to Central Asia and Pakistan (Fig. 1.1). A second, less pronounced centre of species diversity occurs in western North America. These centres of diversity possess differing percentages of the several subgroups of the genus and are thus clearly distinguishable in taxonomic terms.

Evolution of the genus has been accompanied by ecological diversification. The majority of species grow in open, sunny, rather dry sites in arid and moderately humid climates. However, *Allium* species have adapted to many other ecological niches. Different types of forests, European subalpine pastures and moist subalpine and alpine grasslands of the Himalayan and Central Asian high mountains all contain some *Allium* species, and gravelly places along river-banks do as well. Even saline and alkaline environments are tolerated by some taxa.

Allium species from these diverse habitats exhibit a parallel diversity in their rhythms of growth (phenology). Spring-, summer- and autumn-flowering taxa exist. There are short- and long-living perennials, species with one or several annual cycles of leaf formation, and even continuously leafing ones. Species may show summer or winter dormancy. For many species (named 'ephemeroids'), annual growth is limited to a very short period in spring and early summer when the cycle from leaf sprouting to seed maturation is completed in 2 or 3 months.

Conditions suitable for seed germination vary between species. Seed dormancy is variable between wild species. For most species the germinability of the seeds seems to be limited to a few years, unless the seed is stored under cold and very dry conditions, when its life can be greatly extended.

The genus is of great economic significance because it includes several important vegetable crops and ornamental species. However, in contrast, some *Allium* species are noxious weeds of cultivated ground. The cultivated *Allium* crop species are listed in Table 1.1.

Generally, all plant parts of alliums may be consumed by humans (except perhaps the seeds), and many wild species are exploited by the local inhabitants. These natural resources are often improperly managed at the present time (see Section 2.3.4), and overcollecting caused severe decline of wild sources in the past. Very probably, both protection and the rational use of wild plants growing close to settlements, as well as the transfer of plants into existing garden plots (as explained below under *A. cepa*) (Hanelt, 1990), may all have been important at the initial stages of domestication. Further human and natural selection then led to the development of the different plant types present in several cultivated species.

Domestication did not change the ploidy status of onion, shallot, garlic and many other diploid species, and introgression of other species only rarely played a role during the selection processes. The same seems to be true for the cultivated taxa of *A. ampeloprasum*, which apparently arose from ancestors of different ploidy levels (see Section 4.2).

However, cultivated strains of *A. ramosum* and *A. chinense* include diploids, triploids and tetraploids. Because diploid and tetraploid wild strains exist, polytopic, i.e. at different places (and at several times), domestication of *A. ramosum* seems probable. The history of domestication of *A. chinense* is still being disputed. Either the existence of wild strains in Central and East China is accepted, or cultivars are traced back to the closely related wild species *A. komarovianum* Vved. Participation of other wild species, such as *A. thunbergii* G. Don, seems possible (Hanelt, 2001).

Domestication of wild plants is still continuing. *A. komarovianum* was reportedly taken again into cultivation as a vegetable in North Korea quite recently (Hanelt, 2001), and the case of *A. pskemense* is described below (Section 3.4). Some species listed below (Section 4) have also been recently taken into cultivation, but usually exact data are lacking.

Table 1.1. Cultivated *Allium* species and their areas of cultivation.

Botanical names of the crop groups	Other names used in the literature	Area of cultivation	English names
A. altaicum Pall.	*A. microbulbum* Prokh.	South Siberia	Altai onion
A. ampeloprasum L.			
Leek group	*A. porrum* L., *A. ampeloprasum* var. *porrum* (L.) J. Gay	Mainly Europe, North America	Leek
Kurrat group	*A. kurrat* Schweinf. ex Krause	Egypt and adjacent areas	Kurrat, salad leek
Great-headed garlic group	*A. ampeloprasum* var. *holmense* (Mill.) Aschers. et Graebn.	Eastern Mediterranean, California	Great-headed garlic
Pearl-onion group	*A. ampeloprasum* var. *sectivum* Lued.	Atlantic and temperate Europe	Pearl onion
Tarée group		Iran	Tarée irani
A. canadense L.		Cuba	Canada onion
A. cepa L.			
Common onion group	*A. cepa* ssp. *cepa*/var. *cepa*, *A. cepa* ssp. *australe* Kazakova	Worldwide	Onion, common onion
Ever-ready onions	*A. cepa* var. *perutile* Stearn	Great Britain	Ever-ready onion
Aggregatum group	*A. ascalonicum* auct. hort., *A. cepa* var. *aggregatum* G. Don, var. *ascalonicum* Backer, ssp. *orientalis* Kazakova	Nearly worldwide	Shallot, potato onion, multiplier onion
A. consanguineum Kunth		North-East India	
A. × *cornutum* Clem. ex Vis.*	*A. cepa* var. *viviparum* auct.*	Locally in South Asia, Europe, Canada, Antilles	
A. chinense G. Don	*A. bakeri* Regel	China, Korea, Japan, South-East Asia	Rakkyo, Japanese scallions
A. fistulosum L.		East Asia, temperate Europe and America	Japanese bunching onion, Welsh onion
A. hookeri Thw.		Bhutan, Yunnan, North-West Thailand	
A. kunthii G. Don	*A. longifolium* (Kunth) Humb.	Mexico	
A. macrostemon Bunge	*A. uratense* Franch., *A. grayi* Regel	China, Korea, Japan	Chinese garlic, Japanese garlic
A. neapolitanum Cyr.	*A. cowanii* Lindl.	Central Mexico	Naples garlic
A. nutans L.		West and South Siberia, Russia, Ukraine	
A. obliquum L.		West Siberia, East Europe	Oblique onion
A. oschaninii O. Fedtsch.		France, Italy	French shallot*
A. × *proliferum* (Moench) Schrader			
East Asian group	*A. aobanum* Araki, *A. wakegi* Araki	China, Japan, South-East Asia	Wakegi onion
Eurasian group	*A. cepa* var. *viviparum* (Metzg.) Alef., *A. cepa* var. *proliferum* (Moench) Alef.	North America, Europe, North-East Asia	Top onion, tree onion, Egyptian onion, Catawissa onion
A. pskemense B. Fedtsch.		Uzbekistan, Kyrgyzstan, Kazakhstan	

Continued.

Table 1.1. *Continued.*

Botanical names of the crop groups	Other names used in the literature	Area of cultivation	English names
A. ramosum L.	*A. odorum* L., *A. tuberosum* Rottl. ex Sprengel	China and Japan, worldwide now	Chinese chive, Chinese leek
A. rotundum L.	*A. scorodoprasum* ssp. *rotundum* (L.) Stearn	Turkey	
A. sativum L.			Garlic
Common garlic group	*A. sativum* var. *sativum*, *A. sativum* var. *typicum* Regel	Mediterranean area, also worldwide	
Longicuspis group	*A. longicuspis* Regel	Central to South and East Asia	
Ophioscorodon group	*A. sativum* var. *ophioscorodon* (Link) Döll	Europe, also worldwide	
A. schoenoprasum L.	*A. sibiricum* L.	Worldwide in temperate areas	Chive
A. ursinum L.		Central and North Europe	Ramsons
A. victorialis L.	*A. microdictyon* Prokh., *A. ochotense* Prokh.	Caucasus, Japan, Korea, Europe (formerly)	Long-root onion, long-rooted garlic
A. wallichii Kunth	*A. platyphyllum* Diels, *A. lancifolium* Stearn	East Tibet	

* See Friesen and Klaas (1998).

As with many ancient cultivated plants, only a limited amount of circumstantial evidence and no hard facts are available on the evolutionary history of cultivated alliums. Sculptural and painted representations from ancient Egypt support the assumption that onion, garlic and leek were already cultivated at that time. However, it is impossible to pursue these traces during antiquity because many plant names of that era cannot with certainty be assigned to particular species of plants. Unfortunately, a great part of the recent and historical diversity of onion, garlic and several other *Allium* crops, such as chives, was developed during that time and therefore will remain obscure.

1.3 Phylogeny and classification

Recent estimations accept about 750 species in the genus *Allium* (Stearn, 1992), and 650 more synonymous species names exist (Gregory *et al.*, 1998). It is important to divide this large number of species into smaller units or groups for practical purposes. This is also theoretically justified because the genus consists of groups differing in phylogenetic history, in geographical affinity and in evolutionary state and age.

The early monographer of *Allium*, Regel (1875, 1887), grouped the 285 species he accepted into six sections, which trace back to informal groups established by Don (1832). A more recent classification was proposed by Hanelt *et al.* (1992), including six subgenera, 57 sections and subsections. In this scheme, the authors combined some essential ideas from earlier classifications and our own research data as a landmark at the beginning of the molecular research era.

Later, regional revisions on Mediterranean section *Allium* (Mathew, 1996), Central Asia (Khassanov, 1997), China (Xu and Kamelin, 2000) and North America (McNeal and Jacobsen, 2002) supplemented the partly outdated older ones available for Europe (Stearn, 1980; Pastor and Valdes, 1983), most parts of Asia (Vvedensky and Kovalevskaya, 1971; Wendelbo, 1971; Matin, 1978; Kollmann, 1984; Friesen, 1988) and Africa (Wilde-Duyfjes, 1976). The latest compilation of *Allium* names (Gregory *et al.*, 1998) allows us to trace information across the different species concepts, the complicated classifications and the nomenclatural incongruities presented in earlier classifications.

1.3.1 Evolutionary lineages

The genus *Allium* is generally adapted to arid conditions. This makes it difficult to select natural evolutionary lineages using easily discernible characteristics. Phylogenetically different structures, e.g. leaf blades with one or two rows of vascular bundles, are often hidden by morphological similarities forced by functional reasons. Therefore, the traditional infrageneric classifications include homoplasies, i.e. excess changes resulting from parallel or convergent evolution, and do not necessarily represent evolutionary lineages.

- In the past, detailed investigations using modern methods have contributed more supportive data to evaluate and establish evolutionary lineages, and have resulted in more elaborate classifications with more and necessarily smaller groups. However, many facts remain open to interpretation, and neither the phylogenetically most basic *Allium* group nor the evolutionary lineages could be precisely determined (Hanelt *et al.*, 1992). Thus, the unknown phylogenetic connections between the taxonomic groups remain the most prominent problem of all *Allium* classification studies.
- Most recently, molecular studies have resulted in independent data on the evolutionary history of the genus (see Fig. 8.1, Klaas and Friesen, Chapter 8, this volume). Three main evolutionary lines were detected: (i) subgenus *Amerallium sensu* Hanelt *et al.* (1992), subgenus *Nectaroscordum*, subgenus *Microscordum*; (ii) subgenus *Melanocrommyum sensu* Hanelt *et al.* (1992), subgenus *Caloscordum*, subgenus *Anguinum*; and (iii) subgenus *Rhizirideum sensu stricto*, subgenus *Butomissa*, subgenus *Cepa*, subgenus *Allium s. str.*, subgenus *Reticulatobulbosa s. str.*

The taxonomically unclear subgenus *Bromatorrhiza* (Hanelt *et al.*, 1992) was an artificial assemblage (Samoylov *et al.*, 1999). Its members are now considered to belong to the subgenera *Amerallium* (sect. *Bromatorrhiza*) and *Rhizirideum* (sects *Cyathophora* and *Coleoblastus*).

Based on molecular data, the phylogenetic information now available allows us to conclude that the bulbous subgenera *Amerallium* and *Melanocrommyum* represent more ancient lines. The development of elongated rhizomes and of false bulbs are advanced character states (synapomorphies), as are fistulose leaves in the sections *Cepa* and *Schoenoprasum*. This new classification mainly uses well-known taxonomic groups and names, but several sections have been given another rank or another formal circumscription. The accepted subgenera are characterized as follows.

1.3.2 Subgenera with a basal chromosome number of x = 7

Subgenus Amerallium. Subgenus *Amerallium* is not exclusively a New World group, although its name may seem to indicate this. Several sections are Eurasian (European, Mediterranean, Himalayan). Nevertheless, molecular data have verified the monophyly of this subgenus as well as the distinctness of both geographical subgroups (Samoylov *et al.*, 1999). Most species of subgenus *Amerallium* produce true bulbs but others have bulbs on rhizomes. Vegetative anatomy and other characters, including molecular data, strongly support its separate status. The basic chromosome number x = 7 dominates, and yet x = 8, 9 and 11 also occur in several morphologically derived groups.

Subgenus Microscordum. The monotypic East Asian section *Microscordum* shares anatomical and morphological characters with the species of subgenus *Amerallium*, although the plants are tetraploids ($2n$ = 32) on the basic number x = 8. Molecular data have verified the systematic position close to the subgenera *Amerallium* and *Nectaroscordum*.

Subgenus Nectaroscordum. A basic chromosome number of x = 9 and the special and unique characters of most flower parts and of other morphological traits were the main arguments for separating this oligotypic group at generic level. However, leaf anatomical characters and molecular data suggest a close relationship to subgenus *Amerallium*.

1.3.3 Subgenera with a basal chromosome number of x = 8

RHIZOMATOUS PLANTS. All rhizomatous species with x = 8 chromosomes share many characters and have been included in the classical subgenus *Rhizirideum s. lato* (Hanelt *et al.*, 1992). Rhizomes have always been considered an indication of primitive or ancestral origin, irrespective of the existing morphological diversity (Cheremushkina, 1992). However, dendrograms based on molecular data (Mes *et al.*, 1997; Friesen *et al.*, 1999a; Fig. 1.2) showed several clades with rhizomatous species being 'dislocated' between clades of the bulbous subgenera *Melanocrommyum* and *Allium*. This fact provides evidence that rhizomes are not necessarily ancestral, and may have evolved and developed independently several times.

Irrespective of the different phylogenetic status, rhizomatous alliums are adapted to similar ecological conditions and have much in common in their horticultural traits. For practical reasons, the '*Rhizirideum* group' will remain a handy and workable unit for a long time.

Subgenus Rhizirideum s. str. This small subgenus comprises several oligotypic sections to which Eurasian steppe species belong, as well as others which show the most diversity in South Siberia and Mongolia. A few species, which would perhaps best be separated as subgenus *Cyathophora*, formerly incorrectly included in the subgenus *Bromatorrhiza*, are distributed in Tibet and the Himalayas.

Subgenus Anguinum. The morphologically well characterized section *Anguinum* is disjunctively distributed in high mountains from south-western Europe to East Asia, and also in north-eastern North America. The plants possess well-developed rhizomes and show a distinct and unique type of simple seed testa sculpture (Kruse, 1988). According to molecular studies, the subgenus is more closely related to the bulbous subgenus *Melanocrommyum* than to any other *Allium* lineage.

Subgenus Butomissa. This small and unique subgenus includes only a few species, which partly inhabit the Siberian–Mongolian–North Chinese steppes, while other species are distributed in the mountains from East Asia to Central Asia and up to the eastern Mediterranean area.

Subgenus Cepa. Species with fistulose leaves, often well-developed bulbs and short vertical rhizomes dominate. Several species of the well-known sections *Cepa* and *Schoenoprasum* occupy most of the Eurasian continent, but most species are distributed in the mountain belt from the Alps and Caucasus to East Asia.

Subgenus Reticulatobulbosa. This is the largest segregate from subgenus *Rhizirideum sensu lato* (*s. lato*), characterized by narrow linear leaves and reticulate bulb tunics. The centre of diversity of the different species-rich sections is located in South Siberia and Central Asia, with wide extensions into adjacent regions of Asia, Europe, Tibet and the Himalayas. Species from section *Scorodon s. str.* (*A. moschatum*) are bulbous but with a well-developed small rhizome. Molecular data support their inclusion in this subgenus.

BULBOUS PLANTS

Subgenus Allium. The subgenus *Allium* is the largest one of the genus and originates exclusively from the Old World. The section *Allium* shows the strongest species diversity: it mainly ranges from the Mediterranean to Central Asia. The section *Codonoprasum* has a centre of diversity in the Mediterranean area. The section *Scorodon* in the broad sense was an artificial assemblage, and its reclassification into several sections, mainly distributed in the Irano-Turanian floristic region (Khassanov, 1997), is supported by molecular data.

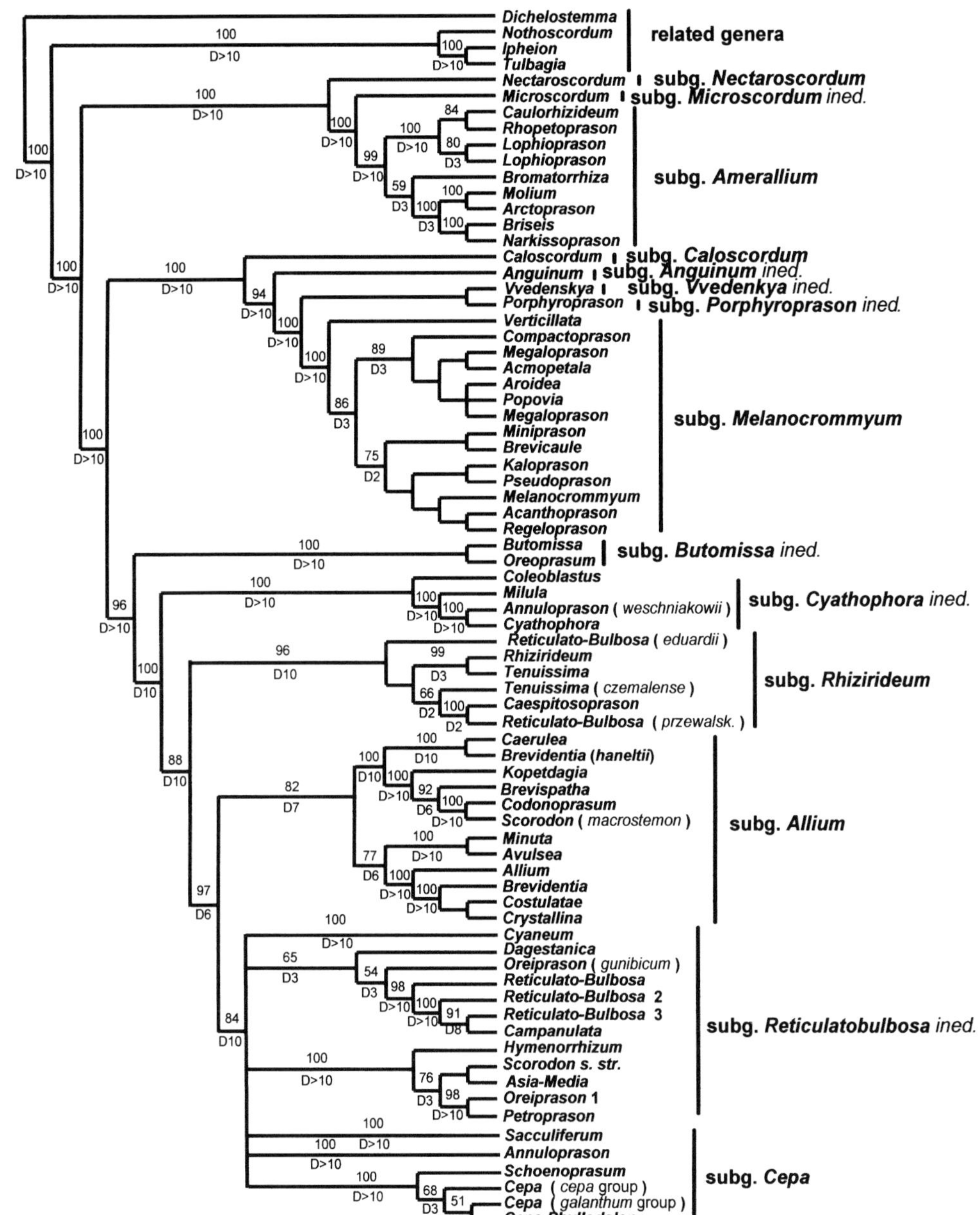

Fig. 1.2. Dendrogram of the genus *Allium* based on molecular markers (strict consensus tree, internal transcribed spacer (ITS) sequences; some group names are provisional). The less advanced groups are close to the related genera (above), the most advanced ones on the opposite side (below).

Subgenus Melanocrommyum. The phenotypically extremely variable subgenus *Melanocrommyum* is well delimited and thus occupies a special evolutionary branch of the genus. For instance, all hitherto investigated species contain only a few cysteine sulphoxides and inactive alliinase, and many plants of this taxon are therefore odourless (Keusgen, 1999). Apparently, rather recently the number and diversity of taxa rapidly increased in the very arid climates of the Near and Middle East to Central Asia. Its recent geographical speciation centre in Central Asia (*c.* 36–40°N, 66–70°E) was identified and confirmed by molecular markers (Mes *et al.*, 1999). The reticulate phylogenies of several groups explain the existence of small but polyphyletic groups, which conflict with the conventional use of taxonomic categories. A pragmatic taxonomic classification of the subgenus is still awaited.

Subgenus Caloscordum. Only three species distributed in East Asia belong to this small but well-characterized group. Morphological reasons to separate it at subgeneric level rather close to the subgenus *Melanocrommyum* are supported by molecular data. The distantly related sections *Vvedenskya* and *Porphyroprason* would also best be raised to subgeneric rank.

2. The Section *Cepa* (Mill.) Prokh.

This small group includes the two economically important cultivated species, *A. cepa* L. and *A. fistulosum* L. The section shares several morphological and molecular characters with the section *Schoenoprasum*, and is only distantly related to most of the other rhizomatous species.

2.1 Morphology, distribution and ecology

The species are characterized by cylindrical, fistulose, distichous leaves. The cylindrical to globose bulbs are composed of several leaf-bases and are covered by membranous skins. The sheath part of the leaves forms a pseudostem, which hides a great part of the above-ground scape. The inflated scape is fistulose and terminates with a multi-flowered head-like inflorescence. Bracteoles are present at the bases of the pedicels. The spathe is short and the flowers are campanulate or with spreading tepals. The inner stamens are strongly widened at the base, where they may possess short teeth. The stigma is capitate. The triloculate ovary has septal nectaries with distinct nectariferous pores, and two ovules per locule, which develop into angular seeds. Usually, axillary daughter bulbs are developed on short rhizomes, building up rather large tufts. A gradual reduction of the rhizome can be seen within the section, leading finally to the flat, disc-like corm or basal plate of the common onion, *A. cepa*.

The wild species of the section *Cepa* occur within the Irano-Turanian floristic region, mainly in the mountainous areas of the Tien-Shan and Pamir-Alai. Occurrences in neighbouring floristic provinces are marginal extensions of the main area. The exceptions are *A. altaicum* and *A. rhabdotum*, which grow in the mountains of southern Siberia and Mongolia and in the eastern Himalayas, respectively (Hanelt, 1985; Friesen *et al.*, 1999b). For details, see Fig. 1.3.

The wild taxa of section *Cepa* are petrophytes, which always grow in open plant formations, such as rocks, rock crevices, stony slopes, river-banks, gravelly deposits and similar sites with a shallow soil layer. Their occurrence is not strongly correlated either to the mineral content or pH of the soil or to particular plant-sociological associations or vegetation types. This distribution pattern often results in groups of small populations (Levichev and Krassovskaja, 1981; Hanelt, 1985). However, the occurrence of large populations has also been reported (Hanelt, 1990).

Unlike some other *Allium* species from the same area, taxa of the section *Cepa* have a fairly long annual growth period and are not ephemeroids. Leaf growth begins after the frost has ceased in the spring, and may be next limited by low temperatures in the following autumn and winter. Species growing in arid areas have a weak, drought-induced summer dormancy but this is easily broken by summer rainfall. Therefore, they commonly lack leaf blades during bloom in summer. All the wild taxa of this section have a prolonged juvenile phase, lasting 3–10 years, before the first flowers are produced (Hanelt, 1985).

These species have long been gathered by local people, who use the bulbs and leaves for food or preserve them for winter use. Often, large-scale collection for commercial or semi-commercial purposes still continues. This has resulted in the disappearance of species from many localities, and a shrinking of their population sizes (Hanelt, 1990). Taxa of more local distribution are seriously endangered or threatened by the rapidly decreasing number of localities at which they occur. Therefore, they were listed in the 'Red Books' of the former Soviet Union and of all Central Asian republics. This situation is serious, because all wild species of the section *Cepa* are the secondary gene pool of *A. cepa* and *A. fistulosum*. The evaluation and exploitation of these genetic resources could contribute significantly to the improvement of these two cultivated species (see Kik, Chapter 4, this volume).

2.2 Cytological limitations

The species of the section *Cepa* are diploid ($2n = 16$), although the occasional occurrence of individual tetraploid bulbs has been reported. Contrary to what is found in some other *Allium* groups, the chromosomes are metacentric or submetacentric and differ only somewhat in length. Only the satellite chromosome pair is subtelocentric (subacrocentric), the satellites being attached to the short arms. Most species of the section *Cepa* have very small dotlike satellites, as in other subgroups of the genus, apart from *A. fistulosum* and *A. altaicum*, which both possess significantly larger satellites. Similar fluorochrome and Giemsa-stained chromosome banding patterns occur in the whole section. However, marker chromosomes with specific intercalary bands on some chromosomes, as well as differences in total length of the chromosome complement were detected (Ohle, 1992; van Raamsdonk and de Vries, 1992b). In spite of the morphological and cytological similarities between the species of section *Cepa*, there are strong crossing barriers between them, which prevent interspecific gene flow even where sympatric distribution of two species occurs.

2.3 Grouping of the species

Section *Cepa* belongs to the morphologically, karyologically and biochemically well-circumscribed *Allium* groups, whose coherence has additionally been demonstrated by molecular data (Pich *et al.*, 1996; Klaas, 1998). The main morphological species-specific characters were presented by van Raamsdonk and de Vries (1992a, b).

The taxa of the section fall into three groups on the basis of morphological and geographical differences (Hanelt, 1985). However, the results of crossing experiments (van Raamsdonk and de Vries, 1992a) and of recent molecular studies show the isolated position of *A. oschaninii* as a sister group to the *A. cepa*/*A. vavilovii* evolutionary lineage (Friesen and Klaas, 1998). Therefore, the *Cepa* alliance is proposed as a fourth informal group.

1. *Galanthum* alliance. White flowers with spreading tepals and filaments above the adnation to the tepals, coalescent into a narrow ring, are characteristic. Nectariferous tubes end in a tangentially widened pocket. Flowering plants have only about two to four assimilating leaves per shoot. Scapes are evenly inflated. The species show a disjunctive distribution in the Irano-Turanian region.

2. *Oschaninii* alliance. White flowers with spreading tepals and filaments without the above-mentioned ring are characteristic. Nectariferous pores are also pocket-like. There are greater numbers of cylindrical leaves, usually four to nine, and a bubble-like swelling in the lower half of the scape. Distribution is concentrated in the Turkestanian province.

3. *Cepa* alliance. The taxa share most characters with the *Oschaninii* alliance but the flowers may also be greenish and the leaves are initially flat or semi-cylindrical. Distribution is mainly Turkmenian–Iranian.

4. *Altaicum* alliance. These species have campanulate to broadly tubular flowers of a whitish–transparent colour. Filaments are distinctly longer than in the other alliances and do not coalesce into a ring. Nectariferous tubes end in a simple lateral hole. Few leaves are present, and the scapes are evenly inflated. Main distribution is in South Siberia and Mongolia and possibly in Himalaya.

2.4 Enumeration of the species

2.4.1 Galanthum *alliance*

Allium galanthum *Kar. et Kir.* This *Allium* is widely distributed in north-east Kazakhstan to the northern Tien-Shan chains, with isolated occurrences east and south of that area. It has the most continental distribution of all species of the section and occurs mainly within the desert zone.

Allium farctum *Wendelbo.* This is a recently described species from the mountains of West Pakistan, East Afghanistan and the marginal area of West Himalaya. The distribution is not yet fully known. Although morphologically similar to *A. galanthum*, the seed-coat structure is as in the *Oschaninii* alliance (Kruse, 1988). Morphological reasons exclude this species as a possible progenitor of the common onion (Hanelt, 1990).

Allium pskemense *B. Fedtsch.* This is an endangered local species from the western Tien-Shan range, where the borders of Kyrgyzstan, Uzbekistan and Kazakhstan meet. Inhabitants of this area collect the bulbs and sometimes transplant the species and cultivate it in their gardens (Levichev and Krassovskaja, 1981). It has rather large bulbs with a very pungent taste.

2.4.2 Oschaninii *alliance*

Allium oschaninii *O. Fedtsch.* This species is distributed in the transitional area from Central to South-West Asia (Fig. 1.3), with isolated occurrences in north-eastern Iran (Hanelt, 1985). It is often found only in inaccessible places, because the leaves are eaten by livestock and its large bulbs are collected by local inhabitants. The plants are morphologically very variable and sometimes resemble *A. cepa*. It was formerly thought to be conspecific with it (*A. cepa* var. *sylvestre* Regel), but recent molecular studies show it to be a sister group to the *A. cepa*/*A. vavilovii* evolutionary lineage (Friesen and Klaas, 1998).

Unexpectedly, the latter report gave convincing molecular evidence that the 'French grey shallot' is a domesticate of *A. oschaninii*. This divergent form is highly esteemed for its excellent taste, and has been cultivated in southern France and Italy for a long time (Messiaen *et al.*, 1993; D'Antuono, 1998; Rabinowitch and Kamenetsky, Chapter 17, this volume).

Allium praemixtum *Vved.* This recently described species is endemic in the south-western marginal chains of the Tien-Shan range, on both sides of the border between Tajikistan and Uzbekistan. Its classification is still in doubt because it differs from *A. oschaninii* only by some minor morphological characters.

2.4.3 Cepa *alliance*

Allium vavilovii *M. Pop. et Vved.* This is an endangered local species of the central Kopetdag range in Turkmenia (Fig. 1.4) and North-East Iran. Its bubble-like hollow stem is similar to that of *A. oschaninii* but the leaves are completely flat and falcate. Molecular analysis revealed that it is the closest known relative of the common onion (Friesen and Klaas, 1998; Fritsch *et al.*, 2001).

Allium asarense *R.M. Fritsch et Matin.* Only very recently this species was identified at a single place in the Elburz range west of Tehran, where it grows on very steep scree and rocky slopes. The plants have semi-cylindrical, falcate, not inflated leaves, a stem with a bubble-like inflation (Fig. 1.5) and small semi-globose umbels with small greenish, brown-flushed flowers. Initially it was believed to represent another subspecies of *A. vavilovii*, but molecular studies assigned it to be a basal group of the *A. cepa*/*A. vavilovii* evolutionary lineage, which deserves species status (Friesen and Klaas, 1998; Fritsch *et al.*, 2001).

Allium cepa *L.* A variable plant cultivated worldwide. Unknown in the wild, although sometimes naturalized (see Section 3).

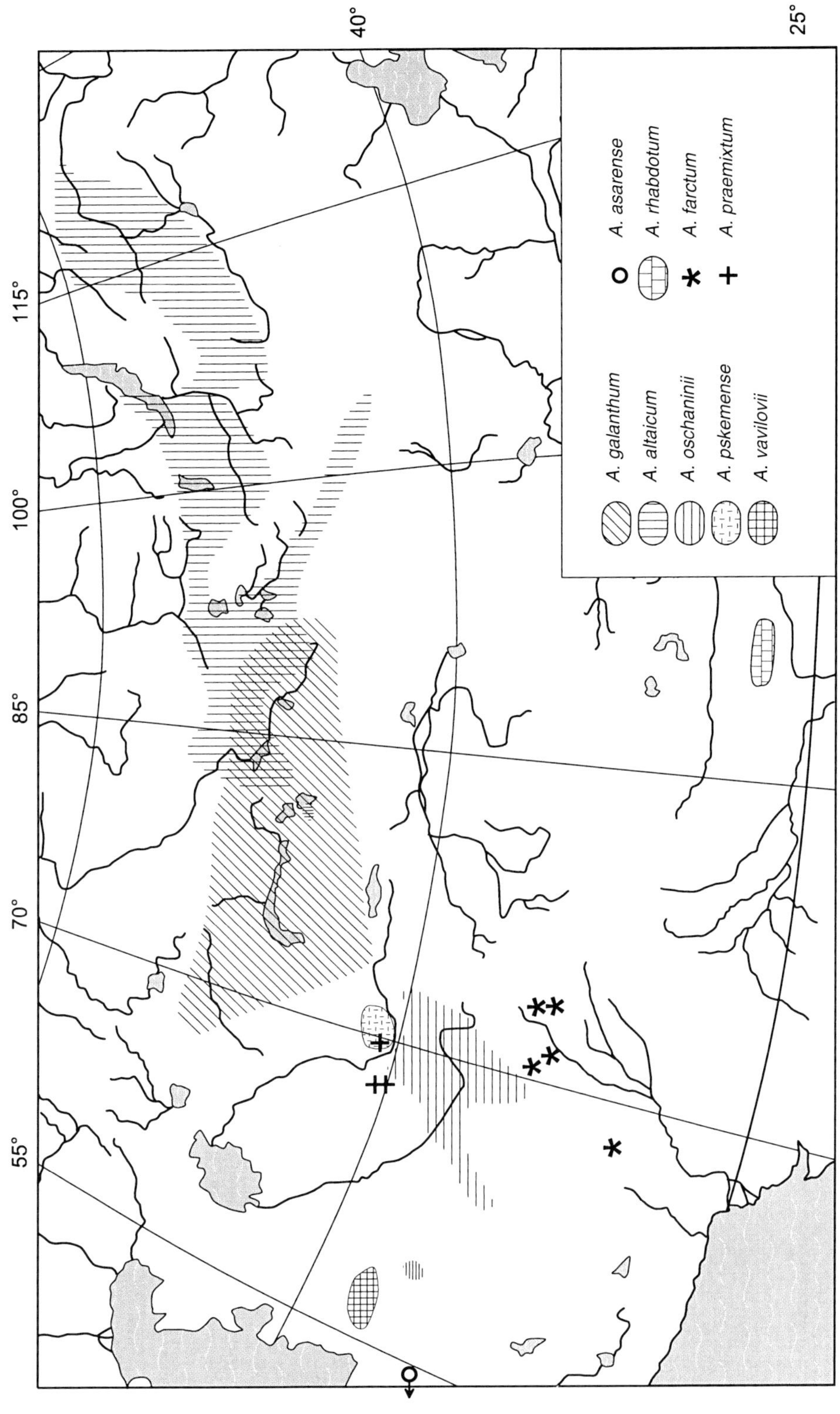

Fig. 1.3. Natural distribution of wild species of section *Cepa*.

Fig. 1.4. *Allium vavilovii* on a scree slope, Kopetdag range, Turkmenia.

2.4.4 Altaicum *alliance*

Allium altaicum *Pall.* This is the most widely distributed species of the section. It occurs in the mountains of southern Siberia, North and Central Mongolia to the Trans-Baikal and in the upper Amur region. The bulbs are extremely frost-resistant. Populations are often threatened by mass collection for food. Occasionally plants are transplanted into backyard gardens (N. Friesen, personal observations). *Allium microbulbum* Prokh., which was described decades ago as a cultivated plant in the Trans-Baikal area, may refer to such casual domesticates.

Allium altaicum is a variable species, having at least two phylogenetically distinct morphotypes. It is the wild progenitor of *A. fistulosum*, which was most probably selected from populations near the southernmost border of its natural area (Friesen *et al.*, 1999b), confirming earlier assumptions about its domestication in North China. Literature sources refer to domestication more than 2000 years ago (cited in Maaß, 1997a).

Allium fistulosum *L.* This is a variable cultivated species, of primary importance in China, Korea and Japan (Inden and Asahira, 1990). It is grown mainly for the slender bulbs and basal parts of the pseudostem, which are much esteemed as fresh or cooked vegetables. In the West it is more rarely grown, mainly for the fresh green leaves, and is eaten as a salad onion (scallion).

2.4.5 Insufficiently known and hybrid taxa

Allium rhabdotum *Stearn.* A recently described species, known so far only from herbarium collections made in Bhutan in the eastern Himalayas (Stearn, 1960). It possibly belongs to the *Altaicum* alliance (Hanelt, 1985) but needs more thorough study from living plants.

Allium roylei *Baker.* Formerly only known as a very rare species from north-west India. One *A. roylei* strain was introduced into the European research scene in the 1960s. All living plants investigated in Europe trace back to this single fertile strain. It crosses

Fig. 1.5. *Allium asarense* under cultivation at Gatersleben, Germany.

easily with *A. cepa* and *A. fistulosum*, and shares a high degree of genetic similarity with other taxa of section *Cepa*. However, most morphological characters differ remarkably from others in this section and are much more similar to those of section *Oreiprason*. The study of other wild populations is essential (Klaas, 1998). Recent evidence indicates that *A. roylei* might have a hybrid origin, as its nuclear DNA profile is related to species of the section *Cepa* but its chloroplast DNA profile to the section *Schoenoprasum* (van Raamsdonk *et al.*, 1997, 2000).

Allium × proliferum *(Moench) Schrad.* It has been shown recently that some minor cultivated taxa, formerly thought to be varieties of *A. cepa* or *A. fistulosum*, or which were described as distinct species, are in fact hybrids of these two species. Analysis of the karyotypes (Schubert *et al.*, 1983), biochemical and molecular data (Havey, 1991; Friesen and Klaas, 1998) and isozyme analysis (Maaß, 1997a) have univocally confirmed the hybrid nature of the plants in question. Top onion and the Wakegi onion are two diploid hybrid types, both having the same parentage. Therefore, they should be combined into one (hybridogenic) nothospecies, according to the rules of botanical nomenclature.

It should be noted that there exist topset-producing forms of *A. cepa* (Jones and Mann, 1963) and *A. fistulosum* (Havey, 1992), which have originated by minor genetic changes and not by species hybridization.

Top onion, tree onion, Egyptian onion, Catawissa onion. These plants are hybrids between *A. fistulosum* and the common onion type of *A. cepa*, and were named *A.* × *proliferum* in its narrow sense. Most or all of the flowers in an inflorescence do not develop, but some bulbils (topsets) grow instead. These may sprout while still on the mother plant. Flowers, if developed, are completely sterile. The plants are widely cultivated in home gardens in North America, Europe and north-eastern Asia for their topsets and young sprout leaves. A seed-fertile tetraploid strain having the same parental species is known and consumed as scallions ('Beltsville Bunching') (McCollum, 1976).

The origin and place of domestication remain unsolved. Chinese scripts and the overlapping areas of both *A. cepa* and *A. fistulosum* in north-western China suggest a Chinese origin (Hanelt, 1990) but comparison of isozyme patterns supports a possible polytopic origin (Maaß, 1997a).

Wakegi onion. The Wakegi onion is used as a green salad onion and has been cultivated for centuries in China, Japan and South-East Asia. It is completely sterile (although the inflorescence is normal, if developed) and is therefore reproduced only vegetatively. It is a hybrid between shallot (the *Aggregatum* type of *A. cepa*) and *A. fistulosum* as maternal parent (Tashiro *et al.*, 1995). Arifin *et al.* (2000), using material from Indonesia, concluded from restriction fragment length polymorphism (RFLP) analysis of amplified *mat*K gene from chloroplast DNA (cpDNA) that *A.* × *wakegi* originated from shallot as maternal parent and Japanese bunching onion as paternal parent, as well as from the reciprocal cross.

Triploid viviparous onions Allium × cornutum *Clem. ex Vis.* Another type of sterile viviparous onions with a more slender stature and pinkish-flushed flowers is locally cultivated in Tibet, Jammu, Croatia, Central and West Europe, Canada and the Antilles. The plants are triploids. Unanimously, *A. cepa* is accepted as donor of two chromosome sets. The source of the third chromosome set is still disputed. However, *A. fistulosum* is rejected as the second parent (Havey, 1991; Friesen and Klaas, 1998). Puizina *et al.* (1999) proposed *A. roylei*, which was not accepted by Maaß (1997b) and Friesen and Klaas (1998).

3. *Allium cepa* L.

3.1 Description and variability

Allium cepa is cultivated mainly as a biennial, but some types are treated as perennials. It is propagated by seeds, bulbs or sets (small bulbs). Bulbs have a reduced disc-like rhizome at the base. Scapes are up to 1.8 m tall and gradually tapering from an expanded lower part. The leaves have rather short sheaths and differ in size and are near circular in cross-section but somewhat flattened on the adaxial side. The umbel is subglobose, dense, many-flowered (50 to several hundred) and with a short persistent spathe. Pedicels are equal and much longer than the white and star-like flowers with spreading tepals. Stamens are somewhat exserted, and the inner ones bear short teeth on both sides of the broadened base. The fruit is a capsule approximately 5 mm long.

The wide variation in bulb characteristics indicates intensive selection. Bulb weight may be up to 1 kg in some southern European cultivars, and the shape covers a wide range from globose to bottle-like and to flattened-disciform. The colour of the membranous skins may be white, silvery, buff, yellowish, bronze, rose red, purple or violet. The colour of the fleshy scales can vary from white to bluish-red. There is also much variation in flavour, the keeping ability of the bulbs and the ability to produce daughter bulbs in the first season. Great variability in ecophysiological growth pattern has developed. There exist varieties adapted to bulbing in a wide range of photoperiodic and temperature conditions (see Bosch Serra and Currah, Chapter 9, and Currah, Chapter 16, this volume). Similarly, adaptation exists for bolting and flowering in a broad range of climates, but non-bolting strains are found in many shallots (Hanelt, 1986a; Kamenetsky and Rabinowitch, Chapter 2, and Rabinowitch and Kamenetsky, Chapter 17, this volume). Organs not selected for by humans, e.g. the flower and the capsule, have been very little affected by domestication and exhibit no striking variations.

3.2 Infraspecific classification

The great variability within the species has led to different proposals for infraspecific groupings, whose historical development has been discussed in detail by Hanelt (1990). Kazakova (1978) presented the most recent version of a classical system which held shallots apart at species level and recognized three formal subspecies, eight formal varieties and 17 cultivar groups (named conculta) based exclusively on quantitative characters. This rather cumbersome classification of *A. cepa* involves statistical methods. The characteristics used are affected strongly by environment and need to be tested in a range of climates. Also, in modern breeding, many 'classical' cultivar groups have been crossed and the boundaries between the different taxa are becoming blurred, making it difficult to place material within the scheme.

The broadly accepted concept of the species *A. cepa* used here includes races with many lateral bulbs and/or shoots, which rarely bolt, and which are partly seed-sterile, namely shallots and potato onions. Other morphological and karyological characters, isozyme and molecular-marker patterns are almost identical to those of *A. cepa* (Hanelt, 1990; Maaß, 1997a, b; Klaas, 1998). Here a simple informal classification will be applied, similar to that of Jones and Mann (1963), accepting two large and one small horticultural groups. The advantages of flexibility and the lack of nomenclature constraints have been discussed in detail elsewhere (Hanelt, 1986b). This approach is convenient for both breeders and horticulturists.

3.2.1 Common onion group

The variability of the species, as discussed above, occurs mainly in this group, economically the most important *Allium* crop. It includes hundreds of open-pollinated traditional and modern cultivars, F_1 hybrids and local races, cultivated in most regions of the world. The bulbs are large and normally single, and plants reproduce from seeds or from seed-grown sets. The majority of cultivars grown for dry bulbs belong to this group, as do salad or pickling onions. In many countries, gene erosion has recently accelerated with the widespread introduction of high-quality, high-yielding F_1 hybrids. However, great diversity still exists in North India and Pakistan, in the former Soviet Union, European and Middle Asian republics, in the Middle East and in the eastern and south-eastern parts of the Mediterranean area (Astley *et al.*, 1982; Bosch Serra and Currah, Chapter 9, and Currah, Chapter 16, this volume).

3.2.2 Aggregatum group

The bulbs are smaller than in common onions, and several to many form an aggregated cluster. Traditional reproduction is almost exclusively vegetative via daughter bulbs, though recently lines of seed-reproduced shallots have been developed (see Rabinowitch and Kamenetsky, Chapter 17, this volume).

The group is of minor economic importance. Locally adapted clones and cultivars are grown mainly in home gardens in Europe, America and Asia for dry bulbs and, more rarely, for green leaves. Cultivation on a larger scale takes place in France, Holland, England and Scandinavia, in Argentina and in some tropical regions, e.g. West Africa, Thailand, Sri Lanka and other South-East Asian countries, and the Caribbean area. In France and other European countries, as well as in the USA, shallots are favoured for their special flavour. In tropical areas, shallots are used as onion substitutes because of their ability to propagate vegetatively and their short growth cycle, and perhaps because they are resistant to local diseases. The variability within this group is poorly represented in gene-bank collections, where the capacity for carrying latent viruses formerly made them a dubious asset. This problem can be solved by meristem culture, followed by *in vitro* propagation (Keller *et al.*, 2000), or by establishing seed-propagated cultivars (Rabinowitch and Kamenetsky, Chapter 17, this volume).

Shallots are the most important subgroup of the Aggregatum group and the only ones grown commercially to any extent. They produce aggregations of many small, narrowly ovoid to pear-shaped bulbs, which often have red-brown (coppery) skins. The plants have narrow leaves and short scapes (see Rabinowitch and Kamenetsky, Chapter 17, this volume).

Not easily distinguishable from shallots are the potato or multiplier onions. They differ from shallots (though many intermediate forms exist) by their larger bulb size, by fewer daughter bulbs, which remain enclosed by the skin of the mother bulb for longer than in the shallots, and often by their somewhat flattened shape. They are cultivated in home gardens in Europe, North America, the Caucasus, Kazakhstan and the south-east of European Russia (Kazakova, 1978), and commercially in Brazil and southern India (Currah, Chapter 16, this volume).

3.2.3 Ever-ready onion group

This third group of *A. cepa* may be distinguished from the other two by its prolific vegetative growth and by the lack of a dormant period. Bulbs or leaves can be gathered at all times of the year. It is used mainly as a salad onion and was commonly cultivated in British gardens in the mid-20th century. Detailed descriptions were given by Stearn (1943) and Jones and Mann (1963). Isozyme (Maaß, 1997a) and molecular-marker patterns (Friesen and Klaas, 1998) fall inside the variability of the common onion group.

3.3 Evolutionary lineages

Only a few hard facts plus some circumstantial evidence are available to help us to trace the evolutionary history of *A. cepa*. The ancestral group from which *A. cepa* must have originated includes only the wild taxa of the *Oschaninii* and *Cepa* alliances (see Section 3.4). They share with *A. cepa* many morphological characters and have in common the special sculpturing of the seed-coat (Kruse, 1988). The current natural distribution of this alliance indicates that domestication of *A. cepa* probably started in the Middle East (Hanelt, 1990).

Recent molecular data support the conclusion of Hanelt (1990), who assigned only *A. vavilovii* as the closest wild relative of *A. cepa* (Friesen and Klaas, 1998; Fritsch *et al.*, 2001). However, the immediate ancestor remains as yet unknown. The recent discovery of *A. asarense* in northern Iran (see

Section 3.4) nurtures once more the scientists' hope of discovering the direct wild ancestor of the onion, perhaps in a very restricted refugial area.

Abandonment of *A. oschaninii* as a possible ancestor will shift the probable area of domestication of the common onion in a south-westerly direction, approximating to the ancient advanced civilizations of the Near East, where the earliest evidence of common onions and garlic comes from. Therefore, we concur with Hanelt (1990), who proposed that the South-West Asian gene centre of *A. cepa* should be acknowledged as the primary centre of domestication and variability. Other regions, such as the Mediterranean basin, where onions exhibit a great variability, are secondary centres.

3.4 History of domestication and cultivation

Prehistoric remains of cultivated plants are often extremely helpful for reconstructing their evolution and history. This is especially true for long-living seed crops, such as cereals, but much less so for species like the bulb onion, which have little chance of long-term preservation. Therefore, one has to rely mostly upon written records, carvings and paintings. Hence, the picture one obtains of the history of such species is fragmentary, at least for the earlier epochs. The conventional wisdom on the history of cultivation of the common onion has been summarized by Helm (1956), Jones and Mann (1963), Kazakova (1978) and Havey (1995) and was briefly discussed by Hanelt (1990). Hence, only a very short review is given here.

Allium cepa is one of the oldest cultivated vegetables, recorded for over 4000 years. The earliest records come from Egypt, where it was cultivated at the time of the Old Kingdom. Onions appear as carvings on pyramid walls and in tombs from the third and fourth dynasties (2700 BC), indicating their importance in the daily diet of many people. The biblical records of the Exodus (1500 BC) are also well known. From Mesopotamia there is evidence of cultivation in Sumer at the end of the third millennium BC. This, together with the records from Egypt, indicates that the initial domestication began earlier than 4000 years ago.

The current exploitation of *A. pskemense* can be used as an illustration of how early cultivation of the onion might have started. This species is consumed by inhabitants of the Pskem and Chatkal valleys, who frequently transplant it from the wild to their gardens, where it is cultivated and propagated (Levichev and Krassovskaja, 1981). Perhaps, thousands of years ago, overcollecting made bulbs of the onion's ancestor scarce, thus stimulating their transfer into gardens and so initiating domestication (Hanelt, 1986a). Further human and natural selection probably favoured a change in allometric growth pattern towards bulbs, a shortening of the life cycle of the plants to bienniality and adaptation to many environments (Hanelt, 1990).

In India there are reports of onion in writings from the 6th century BC. In the Greek and Roman Empires, it was a common cultivated garden plant. Its medicinal properties and details on cultivation and recognition of different cultivars were described. It is thought that the Romans, who cultivated onions in special gardens (*cepinae*), took onions north of the Alps, as all the names for onion in West and Central European languages are derived from Latin. Different cultivars of onion are listed in garden catalogues from the 9th century AD, but the onion became widespread as a crop in Europe only during the Middle Ages and was probably introduced into Russia in the 12th or 13th century.

The onion was among the first cultivated plants taken to the Americas from Europe, beginning with Columbus in the Caribbean. Later it was imported several times and established in the early 17th century in what is now the northern USA. Europeans took the species to East Asia during the 19th century. The indigenous cultivated species of this region, especially *A. fistulosum*, are still more widespread and popular for culinary uses there.

This history of cultivation applies solely to the common onion group. The Aggregatum group is poorly documented in historical records. Most probably, the 'Ascalonian onions' of the authors of antiquity were not shallots. The first reliable records are from the 12th and 13th centuries in France and 16th and 17th centuries in England and Germany. In the herbals of that time, there are good illustrations of this group (Helm, 1956).

4. Other Economic Species

4.1 Garlic and garlic-like forms

4.1.1 Allium sativum *L.*

Garlic is the second most important *Allium* species. It is grown worldwide in all temperate to subtropical (and mountainous tropical) areas as an important spice and medicinal plant. The bulb, composed of few to many densely packed elongated side bulbs ('cloves'), is the main economic organ, and the fresh leaves, pseudostems and bulbils (topsets) are also consumed by humans. Enzymatic decomposition products of alliin, present in all plant parts, have antibacterial and antifungal activity (see Keusgen, Chapter 15, this volume) and cause the intense and specific odour.

Like onion, garlic has been used by humans from very ancient times, when the historical traces fade away and cannot be followed either to a wild ancestor or even to the exact area of domestication. For taxonomic reasons, its wild ancestor (if still extant, or its close relatives) should grow anywhere in an area from the Mediterranean to southern Central Asia. Wild-growing and profusely flowering garlic with long protruding anthers has been described as *Allium longicuspis* Regel from Central Asia. However, such long filaments are developed in all investigated garlic groups if flower development is artificially forced by removing the bulbils in the umbel at a very early stage (Maaß, 1996; Kamenetsky and Rabinowitch, Chapter 2, this volume). Vegetative descendants of 'wild' garlic resemble common bolting garlic types, which have long been cultivated (R.M. Fritsch, personal observation). Thus, no reliable character remains to maintain *A. longicuspis* at species level, but proponents continue to regard it as the truly wild ancestor of garlic (Lallemand *et al.*, 1997). More recently, a remarkable similarity to garlic of the Turkish wild species *A. tuncelianum* was detected, denoting this taxon as another candidate for the wild ancestor (Mathew, 1996; Etoh and Simon, Chapter 5, this volume).

Unlike the case of the seed-bearing onion, the lost ability for generative multiplication has led to a much more restricted morphological and genetic variation in garlic, irrespective of the large area where it is in cultivation. Contrary to former formal infraspecific classifications, recent proposals classify the many existing selections into informal cultivar groups (Maaß and Klaas, 1995; Lallemand *et al.*, 1997). Most garlic from Central Asia belongs to the rather diverse *Longicuspis* group (large bolting plants, many small topsets, to some extent still fertile cultivars). They might have been the genetic pool from which the other cultivar groups developed – the *subtropical* and *Pekinense* subgroups (smaller plants, few large topsets) – which possibly developed under the special climatic conditions of South, South-East and East Asia; the Mediterranean *Sativum* group (bolting and non-bolting types, large topsets); and the *Ophioscorodon* group from Central and East Europe (long coiling scapes, few large topsets).

4.1.2 Allium ampeloprasum *L., great-headed garlic group*

This hexaploid seed-sterile domesticate of *A. ampeloprasum* is locally cultivated in Asia Minor to Iran and Caucasus, and sporadically in California and in other regions of America and Europe. These plants appear to be 'siblings' of garlic with somewhat less intensive odour and taste. They develop large cloves, which are used for both consumption and multiplication. The new sprouts bulb and flower in the first year (in subtropical Israel and California) of cultivation from autumn to spring (H.D. Rabinowitch, Israel, 2000, personal communication) or the second year (in the temperate zone) as a summer crop (van der Meer, 1997; Hanelt, 2001).

4.1.3 Allium macrostemon *Bunge*

Native in the northern central parts of China and Mongolia, this species is grown for the garlic-like taste of its leaves and bulbs. Some strains flower normally and produce fertile seeds (*A. uratense*; in Korea and Japan the synonym *A. grayi* is still sometimes in use), but others develop only bulbils (topsets) (*A. macrostemon s. str.*). Apparently it is a local domesticate of China that reached Korea and Japan earlier than true garlic. In recent times it has become a neglected crop because of its low yield (Hanelt, 2001).

4.2 Taxa of Asiatic origin

4.2.1 Allium ampeloprasum *alliance*

Allium ampeloprasum s. lato is a very variable species (or a group of closely related taxa) widely distributed in the Mediterranean basin. In ancient times, tetraploid populations from the eastern part of its area of distribution were domesticated as vegetables and spice plants. The plants multiply by seeds, apart from pearl onions and great-headed garlic, which are mainly propagated by bulbs/cloves. Formerly named at species level (see Table 1.1), informal classification into cultivar groups is proposed (Hanelt, 2001).

KURRAT GROUP. A leek-like vegetable, used mainly in Egypt and some neighbouring Arab countries, where the rather narrow leaves are used fresh as salad or as a condiment in special dishes (Mathew, 1996; van der Meer, 1997; Hanelt, 2001). The fertile plant freely crosses with leek to produce fertile hybrids, which were utilized in a leek-breeding programme for resistance to leek yellow-stripe virus in Holland by the late Q.P. van der Meer (H.D. Rabinowitch, personal communication).

TARÉE GROUP. A similar use as a condiment is reported for narrow-leafed Caucasian strains of leek and for Tarée cultivated in northern Iran (van der Meer, 1997), which are sometimes included in the Kurrat group (Hanelt, 2001).

LEEK GROUP. Although probably already cultivated in ancient Egypt, in recent times this annual crop has mainly been commercially produced in West and Central Europe, being less important in other European countries, North America and temperate Asia, and is sporadically grown elsewhere. The plants are broad-leaved and stocky. Pseudostems and the basal leaf parts of juvenile plants are mainly consumed as cooked vegetables or condiments (van der Meer and Hanelt, 1990; van der Meer, 1997; Hanelt, 2001; De Clercq and Van Bockstaele, Chapter 18, this volume). When grown as a biennial, leek develops basal bulbs in the second year (van der Meer and Hanelt, 1990; van der Meer, 1997).

PEARL-ONION GROUP. Currently only under small-scale cultivation in house gardens in Central and South Europe, the rather small and slender plants develop large numbers of small subglobular daughter bulbs, which are pickled as a spice (van der Meer, 1997; Hanelt, 2001).

4.2.2 East Asian onions

ALLIUM HOOKERI THWAIT. Naturally distributed in Tibet and North-West China, this species is also cultivated by several non-Chinese tribes in mountainous regions from Bhutan to Yunnan and North-West Thailand. Mainly the fleshy roots but also the leaves are used as vegetables and for soups, fried or pickled (Hanelt, 2001).

ALLIUM RAMOSUM L. (INCLUDING *A. TUBEROSUM* ROTTL. EX SPRENGEL). In East Asia (*A. tuberosum*; local name: Nira) and Central Asia (*A. ramosum*; local name: Djusai) are widely cultivated for the leaves and the flowering umbels, which combine garlic and sweet flavours and are used for soups, salads and other traditional Chinese and Japanese dishes. The plants were taken by immigrants to many other countries. In recent times this species has started to become more popular in Central and West Europe where the leaves are said to have therapeutic effects on tumours (van der Meer, 1997). Its culture and uses in the Orient were described by Saito (1990).

A. tuberosum is usually accepted as the crop species. However, *A. ramosum* (early-flowering, large tepals) and *A. tuberosum* (late-flowering, small tepals) are related by all kinds of transitional forms. Most cultivated strains are tetraploids or triploids; they often develop seeds apomictically (facultative apomicts). Recent molecular data (N. Friesen, unpublished) clearly segregate all cultivated strains as a sister group to the wild species.

ALLIUM CHINENSE G. DON. This kind of oriental garlic, also called rakkyo, is cultivated in China, Korea, Japan, Vietnam, Indonesia and other countries of South-East Asia as a minor or moderately important crop. It is an ancient crop in China, from where it spread to Japan, probably at the end of the first millennium AD (Hanelt, 2001). The domestication history of rakkyo is still being disputed (see Section 1.2). Immigrants from East Asia introduced it into the Americas.

The bulbs are mostly used for pickles and, more rarely, boiled or used as a medicine. The uses and cultivation methods of rakkyo were described by Toyama and Wakamiya (1990).

ALLIUM WALLICHII KUNTH. This species grows wild in the East Himalayas and Tibet to south-west, south and central China. In eastern Tibet, it is grown as a vegetable in traditional home gardens (Hanelt, 2001).

ALLIUM CONSANGUINEUM KUNTH. In its area of natural distribution in West and Central Himalayas, this species is collected from the wild as a vegetable and spice plant. Minor cultivation for the edible leaves was reported from north-eastern India (Hanelt, 2001).

ALLIUM OBLIQUUM L. This tall species grows wild from East Europe to Central Siberia and north-western China, where it is often collected as a substitute for garlic. For a long time it has traditionally been grown for the bulbs in home gardens in West Siberia. Recently it has also become attractive as a medicinal plant in Europe (Hanelt, 2001).

4.3 Chives and locally important onions from other areas

4.3.1 Allium schoenoprasum *L.*

Chives are naturally distributed in most parts of the northern hemisphere (they are the most widely distributed *Allium* of all). In Europe, the young leaves are appreciated as an early vitamin source in spring and are used as a condiment for salads, sauces and special dishes (Poulsen, 1990; van der Meer, 1997; Hanelt, 2001). The species is extremely polymorphous and is being developed by commercial breeders as both a vegetable and an ornamental. Cultivation probably began in Italy, from where it was distributed to Central and West Europe in the early Middle Ages (Helm, 1956), but independent beginnings of cultivation are assumed for Japan and perhaps elsewhere (Hanelt, 2001).

4.3.2 Allium nutans *L.*

In its natural area of distribution from West Siberia to the Yenisei area, it has been collected as a wild vegetable since ancient times. It is transplanted and grown for that reason in home gardens of West Siberia and the Altai mountains. Its cultivation has spread during recent decades to other parts of Russia and the Ukraine (van der Meer, 1997; Hanelt, 2001).

4.3.3 Allium canadense *L.*

This variable species is naturally widespread in North America east of the 103rd meridian. Formerly much collected by native American tribes and later by European settlers, it was introduced to Cuba, where it is locally grown in home gardens as a vegetable (Hanelt, 2001).

4.3.4 Allium kunthii *G. Don*

Wild growing in Mexico and Texas, this species is (semi-)cultivated for its bulbs by the Tarahumara and Tzeltal tribes of Mexico (Hanelt, 2001).

4.3.5 Allium ursinum *L.*

A species which is naturally widespread in temperate Europe to the Caucasus, the leaves and bulbs are sometimes collected for their garlic-like flavour. In earlier centuries, this species was cultivated as a vegetable, medicinal and spice plant in Central and North Europe. Cultivation trials have also been started in recent times. In Germany and mountainous regions of Caucasus it is sometimes transplanted into home gardens (Hanelt, 2001).

4.3.6 Allium neapolitanum *Cyr.*

A common species in the Mediterranean region, which in the past has escaped from cultivation as an ornamental in other warmer countries. It is currently cultivated in Central Mexico, where bulbs and leaves are salted or fried as condiments for several dishes (Hanelt, 2001).

4.3.7 Allium victorialis *L.*

In Europe and Caucasus this polymorphous species grows wild at high altitudes, but in East Asia it usually grows in the forest belt. In former centuries in several European mountain areas, it was cultivated as a medicinal and fetish plant. In Caucasus it is occasionally sown or transplanted in home gardens as a vegetable (Hanelt, 2001). The leaves are often collected in Siberia and the Russian Far East for fresh use, or the basal parts are preserved with salt for the winter period. Recently, it has been offered as a vegetable in catalogues of Japanese seed firms, and it was also introduced in Korea (Hanelt, 2001).

4.3.8 Species of uncertain cultivation status

About two dozen more alliums than mentioned above are collected as wild vegetables and medicinal and spice plants. Several of them were also sporadically cultivated, but the attempts were usually unsuccessful (e.g. *A. triquetrum* (Hanelt, 2001)) or were abandoned (e.g. *A. stipitatum*). Former cultivation is assumed for topset-bearing forms of *A. ampeloprasum* L. and *A. scorodoprasum* L. (Stearn, 1980), but the incomplete old records do not permit exact determination as to the nature of the tested plants (Helm, 1956). Certainly, more species than mentioned in this chapter are potential crops of local importance (van der Meer, 1997).

5. Conclusions

Allium is a species-rich and taxonomically complicated genus. Modern classifications accept more than 750 species and about 60 taxonomic groups at subgeneric, sectional and subsectional ranks.

Recent molecular data provide evidence for three main evolutionary lines. The most ancient line contains bulbous plants, with only rarely a notably elongated rhizome, while the other two lines contain both rhizomatous and bulbous taxa. Thus, the presence of elongated rhizomes is an advanced character state, which developed several times independently. However, probably most sections with rhizomatous species will be retained provisionally together in one subgenus for practical reasons.

Further progress in compiling a phylogenetically based natural *Allium* classification will mainly depend on the accessibility of living material from the hitherto under-investigated arid areas of South-West, southern Central and western East Asia.

Common onion and garlic are species of worldwide economic importance and they consist of several infraspecific groups. Their cultivation traces back to very ancient times, and thus their direct wild ancestors and places of domestication remain unknown. Other *Allium* species of minor economic importance, such as leek, chives, etc., as well as about two dozen species and hybrids grown sporadically or in restricted regions only, have been mostly taken into cultivation in the historical period.

In this time of increasing general mobility and easy contact between peoples and continents, not only formerly unknown fruits and vegetables but also condiments, such as *A. tuberosum*, have been recently introduced, especially into Europe and

North America. New data about the beneficial effects of the fresh greens of these and other alliums will further accelerate their acceptance as part of a healthy daily diet and support their use as phytopharmaceuticals. Therefore, in the future cultivation of minor species, as well as cultivation trials of hitherto uncultivated species, will be enhanced without changing the dominant position of common onion and garlic, and locally of rakkyo and other traditional species. Domestication of other interesting wild *Allium* taxa will be necessary in the future in order to protect their natural resources from overexploitation.

Acknowledgements

We are grateful for stimulating discussions with our colleagues from Gatersleben and we would like to thank especially Dr P. Hanelt and Prof. Dr K. Bachmann. The drawings are by Mrs A. Kilian.

References

Arifin, N.S., Ozaki, Y. and Okubo, H. (2000) Genetic diversity in Indonesian shallot (*Allium cepa* var. *ascalonicum*) and *Allium × wakegi* revealed by RAPD markers and origin of *A. × wakegi* identified by RFLP analyses of amplified chloroplast genes. *Euphytica* 111, 23–31.

Astley, D., Innes, N.L. and van der Meer, Q.P. (1982) *Genetic Resources of* Allium *Species – a Global Report.* IBPGR, Rome, 38 pp.

Cheremushkina, V.A. (1992) Evolution of life forms of species in subgenus *Rhizirideum* (Koch) Wendelbo, genus *Allium* L. In: Hanelt, P., Hammer, K. and Knüpffer, H. (eds) *The Genus* Allium *– Taxonomic Problems and Genetic Resources. Proceedings of an International Symposium, Gatersleben, 11–13 June 1991.* IPK, Gatersleben, Germany, pp. 27–34.

D'Antuono, L.F. (1998) A new taxon among vegetable crops? *Allium Improvement Newsletter* 8, 1–3.

de Sarker, D., Johnson, M.A.T., Reynolds, A. and Brandham, P.E. (1997) Cytology of the highly polyploid disjunct species, *Allium dregeanum* (*Alliaceae*), and of some Eurasian relatives. *Botanical Journal of the Linnean Society* 124, 361–373.

Don, G. (1832) *A Monograph of the Genus* Allium. *Memoirs of the Wernerian Natural History Society*. Adam Black, Edinburgh, 102 pp.

Friesen, N. (1988) *Lukovye Sibiri*. Nauka, Novosibirsk, USSR, 185 pp.

Friesen, N. and Klaas, M. (1998) Origin of some minor vegetatively propagated *Allium* crops studied with RAPD and GISH. *Genetic Resources and Crop Evolution* 45, 511–523.

Friesen, N., Blattner, F.R., Klaas, M. and Bachmann, K. (1999a) Phylogeny of *Allium* L. (*Alliaceae*) based on ITS sequences. In: *Abstracts, XVI International Botanical Congress, St Louis, USA, 2–8 August 1999.* Abstract 674, Missouri Botanical Garden, St Louis, Missouri, p. 405.

Friesen, N., Pollner, S., Bachmann, K. and Blattner, F.R. (1999b) RAPDs and non-coding chloroplast DNA reveal a single origin of the cultivated *Allium fistulosum* from *A. altaicum* (*Alliaceae*). *American Journal of Botany* 86, 554–562.

Fritsch, R.M., Matin, F. and Klaas, M. (2001) *Allium vavilovii* M. Popov et Vved. and a new Iranian species are the closest among the known relatives of the common onion *A. cepa* L. (Alliaceae). *Genetic Resources and Crop Evolution* 48, 401–408.

Gregory, M., Fritsch, R.M., Friesen, N.W., Khassanov, F.O. and McNeal, D.W. (1998) *Nomenclator Alliorum.* Allium *Names and Synonyms – a World Guide*. Royal Botanic Gardens, Kew, UK, 83 pp.

Hanelt, P. (1985) Zur Taxonomie, Chorologie und Ökologie der Wildarten von *Allium* L. sect. *Cepa* (Mill.) Prokh. *Flora* 176, 99–116.

Hanelt, P. (1986a) Pathway of domestication with regard to crop types (grain legumes, vegetables). In: Barrigozzi, C. (ed.) *The Origin and Domestication of Cultivated Plants.* Elsevier, Amsterdam, pp. 179–199.

Hanelt, P. (1986b) Formal and informal classifications of the infraspecific variability of cultivated plants – advantages and limitations. In: Styles, B.T. (ed.) *Infraspecific Classification of Wild and Cultivated Plants*. Clarendon Press, Oxford, pp. 139–156.

Hanelt, P. (1990) Taxonomy, evolution, and history. In: Rabinowitch, H.D. and Brewster, J.L. (eds) *Onions and Allied Crops*, Vol. I. *Botany, Physiology, and Genetics*. CRC Press, Boca Raton, Florida, pp. 1–26.

Hanelt, P. (2001) *Alliaceae*. In: Hanelt, P. (ed.) *Mansfeld's Encyclopedia of Agricultural and Horticultural Crops*, Vol. 4, 3rd edn. Springer-Verlag, Vienna, pp. 2250–2269.

Hanelt, P., Schultze-Motel, J., Fritsch, R., Kruse, J., Maaß, H.I., Ohle, H. and Pistrick, K. (1992) Infrageneric grouping of *Allium* – the Gatersleben approach. In: Hanelt, P., Hammer, K. and Knüpffer, H. (eds) *The Genus* Allium *– Taxonomic Problems and Genetic Resources. Proceedings of an International Symposium, Gatersleben, 11–13 June 1991*. IPK, Gatersleben, Germany, pp. 107–123.

Havey, M.J. (1991) Molecular characterization of the interspecific origin of viviparous onion. *Journal of Heredity* 82, 501–502.

Havey, M.J. (1992) A viviparous *Allium fistulosum*. *Allium Improvement Newsletter* 2, 13–14.

Havey, M.J. (1995) Onions and other cultivated alliums. In: Smartt, J. and Simmonds, N.W. (eds) *Evolution of Crop Plants*, 2nd edn. Longman Scientific and Technical, Burnt Mill, UK, pp. 344–350.

Helm, J. (1956) Die zu Würz- und Speisezwecken kultivierten Arten der Gattung *Allium* L. *Kulturpflanze* 4, 130–180.

Inden, H. and Asahira, T. (1990) Japanese bunching onion (*Allium fistulosum* L.). In: Brewster, J.L. and Rabinowitch, H.D. (eds) *Onions and Allied Crops*, Vol. III. *Biochemistry, Food Science, and Minor Crops*. CRC Press, Boca Raton, Florida, pp. 159–178.

Jones, H.A. and Mann, L.K. (1963) *Onions and Their Allies: Botany, Cultivation and Utilization*. Leonard Hill, London and Interscience, New York, 285 pp.

Kazakova, A.A. (1978) *Luk*. Kul'turnaja Flora SSSR, X, Kolos, Leningrad, USSR, 264 pp.

Keller, E.R.J., Senula, A. and Lesemann, D.E. (2000) Elimination of viruses through meristem culture and thermotherapy for the establishment of an *in vitro* collection of garlic (*Allium sativum*) in the genebank of the IPK Gatersleben. In: Doyle, B.M., Curry, R.F. and Cassells, A.C. (eds) *Methods and Markers for Quality Assurance in Micropropagation*. ISHS Working Group 'Quality Management in Micropropagation'. University of Cork, Republic of Ireland, 24–27 August 1999. *Acta Horticulturae* 530, 121–127.

Keusgen, M. (1999) *Biosensorische Methoden zur qualitativen Bestimmung von Cysteinsulfoxiden*. Berichte aus der Pharmazie, Shaker Verlag, Aachen, Germany, 152 pp.

Khassanov, F.O. (1997) Conspectus of the wild growing *Allium* species of Middle Asia. In: Öztürk, M., Seçmen, Ö. and Görk, G. (eds) *Plant Life in Southwest and Central Asia*. Ege University Press, Izmir, Turkey, pp. 141–159.

Klaas, M. (1998) Applications and impact of molecular markers on evolutionary and diversity studies in the genus *Allium*. *Plant Breeding* 117, 297–308.

Kollmann, F. (1984) *Allium*. In: Davies, P.H. (ed.) *Flora of Turkey and the East Aegean Islands*, Vol. 8. Edinburgh University Press, Edinburgh, pp. 98–208.

Kruse, J. (1988) Rasterelektronenmikroskopische Untersuchungen an Samen der Gattung *Allium* L. III. *Kulturpflanze* 36, 355–368.

Lallemand, J., Messiaen, C.M., Briand, F. and Etoh, T. (1997) Delimitation of varietal groups in garlic (*Allium sativum* L.) by morphological, physiological and biochemical characters. *Acta Horticulturae* 433, 123–132.

Levichev, I.G. and Krassovskaja, L.S. (1981) The Pskemski onion *Allium pskemense* B. Fedtsch. in the southern part of its range. *Bjulletin Moskovskogo Obshchestva Ispytatelej Prirody, Otdel Biologicheskij* 86, 105–112 (in Russian).

Maaß, H.I. (1996) Morphologische Beobachtungen an Knoblauch. *Palmengarten* 60, 65–69.

Maaß, H.I. (1997a) Genetic diversity in the top onion, *Allium* × *proliferum* (*Alliaceae*), analysed by isozymes. *Plant Systematics and Evolution* 208, 35–44.

Maaß, H.I. (1997b) Studies on triploid viviparous onions and their origin. *Genetic Resources and Crop Evolution* 44, 95–99.

Maaß, H.I. and Klaas, M. (1995) Infraspecific differentiation of garlic (*Allium sativum* L.) by isozyme and RAPD markers. *Theoretical and Applied Genetics* 91, 89–97.

McCollum, G.D. (1976) Onion and allies *Allium* (*Liliaceae*). In: Simmonds, N.W. (ed.) *Evolution of Crop Plants*. Longman, London, pp. 186–190.

McNeal, D.W., Jr and Jacobsen, T.D. (2002) XX. *Allium* onion, garlic, leek, chives. In: Kiger, E. (ed.) *Flora of North America*, Vol. 26. Oxford University Press, Oxford (in press).

Mathew, B. (1996) *A Review of* Allium *Section* Allium. Royal Botanic Gardens, Kew, UK, 176 pp.

Matin, F. (1978) *Study of the family* Alliaceae *in Iran*. Département Botanique No. 6, Institut de Recherches Entomologiques et Phytopathologiques d'Evine, Tehran, 74 pp.

Melchior, H. (1964) 3. Reihe *Liliiflorae* (*Liliales*). In: Melchior, H. (ed.) *A. Engler's Syllabus der Pflanzenfamilien. 12. Auflage*. Gebrüder Borntraeger, Berlin-Nikolassee, pp. 513–543.

Mes, T.H.M., Friesen, N., Fritsch, R.M., Klaas, M. and Bachmann, K. (1997) Criteria for sampling in *Allium* based on chloroplast DNA PCR-RFLPs. *Systematic Botany* 22, 701–712.

Mes, T.H.M., Fritsch, R.M., Pollner, S. and Bachmann, K. (1999) Evolution of the chloroplast genome and polymorphic ITS regions in *Allium* subg. *Melanocrommyum*. *Genome* 42, 237–247.

Messiaen, C.-M., Cohat, J., Leroux, J.P., Pichon, M. and Beyries, A. (1993) *Les Allium Alimentaires Reproduits par Voie Végétative*. Institut National de Recherche Agronomique, Paris, 228 pp.

Ohle, H. (1992) Karyotype analysis using Giemsa C-banding technique in *Allium* species of six sections of the subgenus *Rhizirideum*. In: Hanelt, P., Hammer, K. and Knüpffer, H. (eds) *The Genus* Allium *– Taxonomic Problems and Genetic Resources. Proceedings of an International Symposium, Gatersleben, 11–13 June 1991*. IPK, Gatersleben, Germany, pp. 221–232.

Pastor, J. and Valdes, B. (1983) *Revision del genero* Allium (Liliaceae) *en la peninsula Iberica e islas Baleares*. Publicationes de la Universidad de Sevilla, Serie Ciencias: Otras Publicaciones 3, Sevilla, 182 pp.

Pich, U., Fritsch, R. and Schubert, I. (1996) Closely related *Allium* species (*Alliaceae*) share a very similar satellite sequence. *Plant Systematics and Evolution* 202, 255–264.

Poulsen, N. (1990) Chives, *Allium schoenoprasum* L. In: Brewster, J.L. and Rabinowitch, H.D. (eds) *Onions and Allied Crops*, Vol. III. *Biochemistry, Food Science, and Minor Crops*. CRC Press, Boca Raton, Florida, pp. 231–250.

Puizina, J., Javornik, B., Bohanec, B., Schweizer, D., Maluszynska, J. and Papes, D. (1999) Random amplified polymorphic DNA analysis, genome size, and genomic *in situ* hybridization of triploid viviparous onions. *Genome* 42, 1208–1216.

Regel, E. (1875) Alliorum adhuc cognitorum monographia. *Acta Horti Petropolitani* 3, 1–266.

Regel, E. (1887) Allii species Asiae Centralis in Asia Media a Turcomania desertisque Araliensibus et Caspicis usque ad Mongoliam crescentes. *Acta Horti Petropolitani* 10, 278–362.

Saito, S. (1990) Chinese chives, *Allium tuberosum* Rottl. In: Brewster, J.L. and Rabinowitch, H.D. (eds) *Onions and Allied Crops*, Vol. III. *Biochemistry, Food Science, and Minor Crops*. CRC Press, Boca Raton, Florida, pp. 219–230.

Samoylov, A., Friesen, N., Pollner, S. and Hanelt, P. (1999) Use of chloroplast DNA polymorphisms for the phylogenetic study of *Allium* subgenus *Amerallium* and subgenus *Bromatorrhiza* (*Alliaceae*) II. *Feddes Repertorium* 110, 103–109.

Schubert, I., Ohle, H. and Hanelt, P. (1983) Phylogenetic conclusions from Giemsa banding and NOR staining in top onions (*Liliaceae*). *Plant Systematics and Evolution* 143, 245–256.

Stearn, W.T. (1943) The Welsh onion and the Ever-ready onion. *Gardeners Chronicle* 143, 86–88.

Stearn, W.T. (1960) *Allium* and *Milula* in the Central and Eastern Himalaya. *Bulletin of the British Museum of Natural History (Botany)* B2, 159–191.

Stearn, W.T. (1980) *Allium* L. In: Tutin, T.G., Heywood, V.H., Burges, N.A., Moore, D.M., Valentine, D.H., Walters, S.M. and Webb, D.A. (eds) *Flora Europaea*, Vol. 5. Cambridge University Press, Cambridge, pp. 49–69.

Stearn, W.T. (1992) How many species of *Allium* are known? *Kew Magazine* 9, 180–182.

Takhtajan, A. (1997) *Diversity and Classification of Flowering Plants*. Columbia University Press, New York, 643 pp.

Tashiro, Y., Oyama, T., Iwamoto, Y., Noda, R. and Miyazaki, S. (1995) Identification of maternal and paternal plants of *Allium wakegi* Araki by RFLP analysis of chloroplast DNA. *Journal of the Japanese Society for Horticultural Science* 63, 819–824.

Toyama, S. and Wakamiya, I. (1990) Rakkyo *Allium chinense* G. Don. In: Brewster, J.L. and Rabinowitch, H.D. (eds) *Onions and Allied Crops*, Vol. III. *Biochemistry, Food Science, and Minor Crops*. CRC Press, Boca Raton, Florida, pp. 197–218.

van der Meer, Q.P. (1997) Old and new crops within edible *Allium*. *Acta Horticulturae* 433, 17–31.

van der Meer, Q.P. and Hanelt, P. (1990) Leek (*Allium ampeloprasum* var. *porrum*). In: Brewster, J.L. and Rabinowitch, H.D. (eds) *Onions and Allied Crops*, Vol. III. *Biochemistry, Food Science, and Minor Crops*. CRC Press, Boca Raton, Florida, pp. 179–196.

van Raamsdonk, L.W.D. and de Vries, T. (1992a) Biosystematic studies in *Allium* L. section *Cepa*. *Botanical Journal of the Linnean Society* 109, 131–143.

van Raamsdonk, L.W.D. and de Vries, T. (1992b) Systematics and phylogeny of *Allium cepa* L. and allies. In: Hanelt, P., Hammer, K. and Knüpffer, H. (eds) *The Genus* Allium *– Taxonomic Problems and Genetic Resources. Proceedings of an International Symposium, Gatersleben, 11–13 June 1991*. IPK, Gatersleben, Germany, pp. 257–263.

van Raamsdonk, L.W.D., Smiech, M.P. and Sandbrink, J.M. (1997) Introgression explains incongruence between nuclear and chloroplast DNA-based phylogenies in *Allium* section *Cepa*. *Botanical Journal of the Linnean Society* 123, 91–108.

van Raamsdonk, L.W.D., Vrielink-Van Ginkel, M. and Kik, C. (2000) Phylogeny reconstruction and hybrid analysis in *Allium* subgenus *Rhizirideum*. *Theoretical and Applied Genetics* 100, 1000–1009.

Vvedensky, A.I. and Kovalevskaya, S.S. (1971) Rod 151, (7) *Allium* L. – Luk zhua (kaz.) piez (tadzh.). In: Vvedensky, A.I. and Kovalevskaya, S.S. (eds) *Opredelitel rastenij Srednej Azii. Kriticheskij konspekt flory*, Vol. 2. Izdatel'stvo 'FAN' Uzbekskoj SSR, Tashkent, pp. 39–89, 311–328.

Wendelbo, P. (1971) *Alliaceae*. In: Rechinger, K.H. (ed.) *Flora Iranica*, Vol. 76. Akademische Druck- und Verlagsanstalt, Graz, Austria, 100 pp.

Wilde-Duyfjes, B.E.E. (1976) *A Revision of the Genus* Allium *L. (*Liliaceae*) in Africa*. 76–11, Mededelingen Landbouwhogeschool, Wageningen, 237 pp.

Xu, J. and Kamelin, R.V. (2000) 32. *Allium* Linnaeus, Sp. Pl. 1: 294. 1753. In: Wu, Z. and Raven, P.H. (eds) *Flora of China*, Vol. 24. Science Press and Missouri Botanical Garden Press, Beijing and St Louis, Missouri, pp. 165–202.

2 Florogenesis

R. Kamenetsky[1] and H.D. Rabinowitch[2]
[1]*Department of Ornamental Horticulture, The Volcani Center, Bet Dagan 50250, Israel;* [2]*Institute of Plant Science and Genetics in Agriculture, The Hebrew University of Jerusalem, Faculty of Agricultural, Food and Environmental Quality Sciences, PO Box 12, Rehovot 76100, Israel*

1. Introduction

Flowering is one of the most fascinating and yet complicated processes in nature, involving a variety of strategies and physiological processes to guarantee the development of the generative organs for optimal production of seeds and to ensure continuation of the species. Flowering of various taxa within the genus *Allium* is extremely diverse with regard to morphology, developmental biology, genetic control and response to the environment. Until now, florogenesis has only been studied in a few species of the large *Allium* genus, mainly those of current economic importance.

We shall review the transition of *Allium* plants from the vegetative to the generative phase, the development of the *Allium* inflorescence from initiation to anthesis and its regulation by internal and external factors. We shall also discuss the factors involved in the differentiation of floral parts and inflorescence structure, with special attention to differences between biomorphological groups. Pollination and seed development in edible *Alliums* have been reviewed comparatively recently (Rabinowitch, 1985, 1990a, b; Currah, 1990; Brewster, 1994) and will only be mentioned when appropriate.

2. Morphological Structure and Differences among Biomorphological Groups

The complex process of flowering varies among members of the genus *Allium*. The various biomorphological groups respond differently to inductive conditions and develop from initiation to bloom in different ways. They also vary significantly in the morphological organization of the storage organs and in life cycle.

Wild *Allium* species have been divided into three main biomorphological groups (Pastor and Valdes, 1985; Hanelt *et al.*, 1992; Kamenetsky, 1992, 1996a; Fritsch and Friesen, Chapter 1, and Kamenetsky and Fritsch, Chapter 19, this volume).

2.1 The rhizomatous group

This group includes members of the subgenera *Rhizirideum* and *Amerallium*, which, in the wild, are confined mainly to mesoxerophytic habitats: meadows, forests and high mountain zones (Hanelt *et al.*, 1992). The fleshy rhizomes are built up through successive concrescence of the basal plates over several generations and function primarily as underground storage organs. Bulbs of these species are composed of leaf sheaths of different thickness. Wild rhizomatous species grow continuously all year round with no apparent dormant stage, and low winter temperatures only slow this down (Cheremushkina, 1985, 1992; Pistrick, 1992). The juvenile period lasts 1–2 years. In post-juvenile plants, flowering occurs late in the spring or in the summer. Differentiation of the inflorescence occurs at the base of the youngest leaf, and the number of flowering cycles ranges from one to three per season in different species (Kruse, 1992; Fig. 2.1).

2.2 The bulbous group

This group includes members of the subgenera *Allium* and *Melanocrommyum* and some members of the subgenus *Amerallium*. Wild plants of these taxa inhabit mainly steppes, semi-desert and desert areas. The storage organs are completely or partially subterranean and consist of a compressed

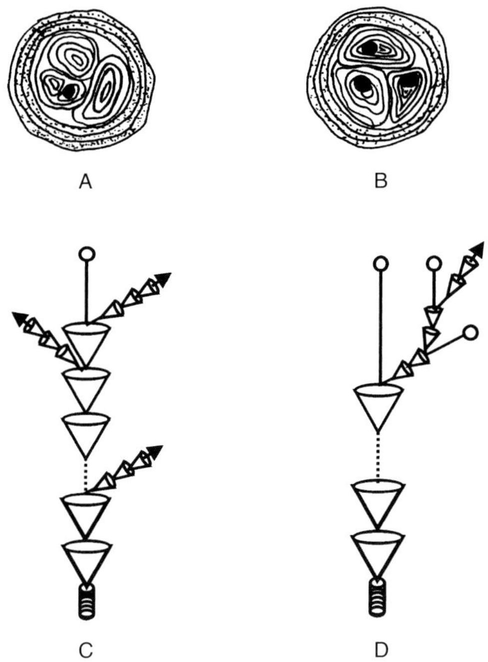

Fig. 2.1. Diagrammatic representation of morphological structure of rhizomatous *Allium* species. A, C. Cross-section and diagram of *A. victorialis*, showing a single terminal generative shoot and several vegetative shoots. B, D. Cross-section and diagram of *A. tuberosum*, showing several flowering cycles during one season. ▽, Foliage leaf; ♀, inflorescence; ▮, rhizome; ▲, vegetative growing point or shoot apex.

and flattened stem – the basal plate – together with the fleshy, succulent leaf-bases and/or specialized true scales, which assume the storage functions (De Mason, 1990; Kamenetsky, 1996a). In the summer, the bulbs enter a rest period, and sprouting recommences either in the autumn or in the spring (Pistrick, 1992). The juvenile period lasts 2–5 years and post-juvenile plants flower in the spring. Differentiation of the inflorescence occurs at the base of the youngest leaf during summer/autumn of the previous year (Kamenetsky, 1997; Fig. 2.2).

2.3 Edible *Allium* species

These are probably best considered as a separate group. For several millennia, these plant species have been selected by humans for specific morphological and physiological traits (Hanelt, 1990). Today, the domesticated onion, *A. cepa* of the subgenus *Rhizirideum*, behaves very much like a true bulbous plant. Its bulb consists of specialized leaf sheaths ('false scales') and modified bladeless leaves ('true scales'), which swell to form a bulb, the storage organ (Brewster, 1990, 1994; De Mason, 1990). In contrast, leek, a selection from the bulbous *A. ampeloprasum*, forms a long false stem, consisting of leaf sheaths and within them, folded immature leaf blades (the storage organ) (van der Meer and Hanelt, 1990; De Clercq and Van Bockstaele, Chapter 18, this volume).

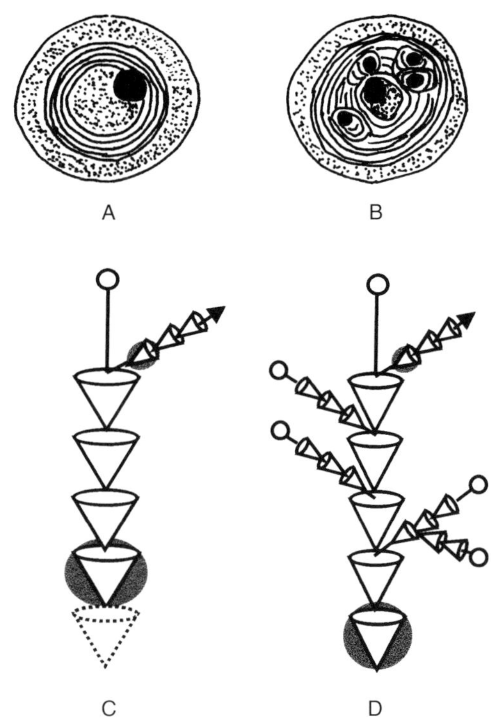

Fig. 2.2. Diagrammatic representation of morphological structure of bulbous species. A, C. Cross-section and diagram of *A. nigrum*, showing development of a single terminal inflorescence and one renewal bud. B, D. Cross-section and diagram of *A. moly*, showing the main flowering shoot and several lateral shoots and secondary inflorescences. ▽, Foliage leaf; ♀, inflorescence; ⬤, storage leaf (scale); ▲, vegetative growing point or shoot apex.

Taxonomically, many economically important species belong to the subgenus *Rhizirideum*, e.g. *A. cepa* (onion, shallot), *A. fistulosum* (Japanese bunching onion) and *A. schoenoprasum* (chives), while *A. sativum* (garlic) and *A. ampeloprasum* (leek, elephant garlic, kurrat and pearl onion) belong to the subgenus *Allium* (Hanelt, 1990; Fritsch and Friesen, Chapter 1, this volume). The two groups differ markedly in both morphological organization and life cycle. Moreover, significant physiological differences occur even within one botanical species (e.g. *A. cepa*) (Rabinowitch, 1990a; Krontal *et al.*, 1998).

3. Transition from the Vegetative to the Generative Stage

In many geophytes, florogenesis can be divided into five consecutive steps, comprising induction, initiation, differentiation (organogenesis), maturation and growth of floral parts and anthesis (Le Nard and De Hertogh, 1993).

The induction and initiation of flowering are greatly affected both by the genetic make-up of the individual plant and by environmental factors; their interactions affect a series of molecular and biochemical processes, leading to the transition from vegetative to reproductive development (Halevy, 1990; Bernier *et al.*, 1993).

3.1 Genetic effects

There is a significant genetic variation in the response of *Allium* genotypes to the environment. Differences in the length of the juvenile phase (physiological age), the responses to photoperiod and to the optimum, minimum and maximum temperatures for floral induction have been recorded within the gene pools of bulb onion (Rabinowitch, 1985, 1990a), shallot (Messiaen *et al.*, 1993; Krontal *et al.*, 1998; Rabinowitch and Kamenetsky, Chapter 17, this volume), Japanese bunching onion (Tindall, 1983) and leek (van der Meer and Hanelt, 1990; De Clercq and Van Bockstaele, Chapter 18, this volume). Garlic clones differ significantly in their ability to form a floral stem and inflorescence (Takagi, 1990; Section 3.4.6 below; Etoh and Simon, Chapter 5, this volume). Some garlic clones develop normal flower primordia and long scapes and go on to bloom, but topsets (bulbils), widely varying in numbers, develop in the inflorescence concurrently with flowers. Plants of other clones initiate a flower scape but the inflorescence degenerates prematurely. A third group comprises non-bolting clones.

3.2 Physiological age

When propagated from seeds, all *Allium* plants need to reach a certain physiological age (or critical mass) before being capable of florogenesis and blooming. The length of the juvenile phase ranges from a few months, e.g. bulb onion (Rabinowitch, 1990a), chives *A. schoenoprasum* (Poulsen, 1990), Japanese bunching onion (Inden and Asahira, 1990), leek (van der Meer and Hanelt, 1990; De Clercq and Van Bockstaele, Chapter 18, this volume) and shallot (Messiaen *et al.*, 1993; Krontal *et al.*, 1998; Rabinowitch and Kamenetsky, Chapter 17, this volume), to 5–6 years, e.g. *A. giganteum* and *A. karataviense* (De Hertogh and Zimmer, 1993). The length of the juvenile phase depends on the genetic make-up of the plant and the growth environment, e.g. bulb onion (Heath and Mathur, 1944; Ito, 1956; Shishido and Saito, 1976; Brewster, 1985; Rabinowitch, 1990a) and Japanese bunching onion (Inden and Asahira, 1990). Both factors control the amount of accumulated reserves necessary for successful blooming. It has been suggested, however, that the ability to flower depends not only on the amount of available reserves but also on the size of the apical meristem (Halevy, 1990; Le Nard and De Hertogh, 1993).

With a few exceptions, in seedlings of bulb onion (Rabinowitch, 1990a) and of Japanese bunching onion (Inden and Asahira, 1990), the transition to the reproductive stage normally occurs in the first or second growing season after the formation of 10–14 leaves (including leaf buds). Under inductive conditions, floral initiation in shallot (Krontal *et al.*, 1998) and in leek (van der Meer and Hanelt, 1990; De Clercq and Van Bockstaele, Chapter 18, this volume) is already possible after formation of the first six and seven true leaves (including leaf primordia), respectively.

In nature, seedlings of the rhizomatous *A. senescens* branch after emergence to form a primary clump. Growth and branching continue for 3–5 years before the vegetative plant reaches the required physiological age (or critical mass) for blooming; then all shoots become reproductive simultaneously (Cheremushkina, 1985).

In ornamental bulbous *Allium* species, the ability to flower depends on the amount of reserves (the critical mass of the bulb). The minimum bulb circumference needed for flowering varies between 3 and 5 cm for *A. caeruleum*, *A. neapolitanum* and *A. unifolium*, between 12 and 14 cm for *A. aflatunense* (= *A. hollandicum*), *A. cristophii* and *A. karataviense* and between 20 and 22 cm for *A. giganteum* (De Hertogh and Zimmer, 1993). In general, seedlings of ornamental species with small bulbs flower in the second year of development (e.g. *A. neapolitanum*, *A. caeruleum*; R.M. Fritsch, Gatersleben, 1999, personal communication), whereas those of plants with large bulbs (e.g. members of the subgenus *Melanocrommyum*) require 3–5 years of growth before they reach the blooming phase (De Hertogh and Zimmer, 1993; Kamenetsky, 1994). In *A. aschersonianum* (subgenus *Melanocrommyum*) the transition of the apical meristem to the reproductive stage occurred in bulbs as young as 2 years. However, these plants were too small to support a normal bloom and therefore the young reproductive bud aborted inside the bulb (Kamenetsky *et al.*, 2000).

3.3 Morphological changes during floral initiation

Juvenile *Allium* plants exhibit a monopodial growth habit, and only become sympodial after the formation of the first generative meristem. Thereafter, *Allium* plants produce renewal bulbs and flower every year. During the vegetative stage, the apical meristem is flat and leaf primordia initiate from the periphery towards the centre (Fig. 2.3A, B). On the transition of the apical meristem from vegetative to generative, the meristem swells to form a dome shape, a spathe is formed in the apex and leaf initiation ceases. The spathe arises as a nearly uniform ring, elongates quickly and envelops the reproductive meristem (Fig. 2.3C, D).

3.4 Environmental control of flower induction and initiation

Cold exposure is required for floral induction in the major cultivated *Allium* crops, including bulb onion (Rabinowitch, 1985, 1990a), chives (Poulsen, 1990), shallot (Krontal *et al.*, 2000), garlic (Takagi, 1990; Section 3.4.6 below) and Japanese bunching onion (Inden and Asahira, 1990). In addition, some *Allium* crops require a long photoperiod for inflorescence initiation and further differentiation; they include Chinese chives (*A. tuberosum*) (Saito, 1990), leek (van der Meer and Hanelt, 1990; De Clercq and Van Bockstaele, Chapter 18, this volume) and rakkyo (*A. chinense*) (Toyama and Wakamiya, 1990).

The ornamental species of the subgenus *Melanocrommyum* show a different physiology, as the transition from the vegetative to the reproductive phase occurs at the end of the growth period or during the 'rest' period without cold induction.

3.4.1 Bulb onion

In the bulb onion, floral initiation in the post-juvenile plant requires cold induction. This thermophase occurs in plants with a minimum leaf and leaf primordia number estimated variously as 7–10 (Brewster, 1985, 1994), 11–12 (Heath and Mathur, 1944; Ito, 1956) or 12–14 (Heath and Mathur, 1944; P.B. Mathur, unpublished). For many onion cultivars, optimum temperatures in the thermophase range between 8 and 12°C, and response is markedly slower at temperatures below 6 and above 17°C (Brewster,

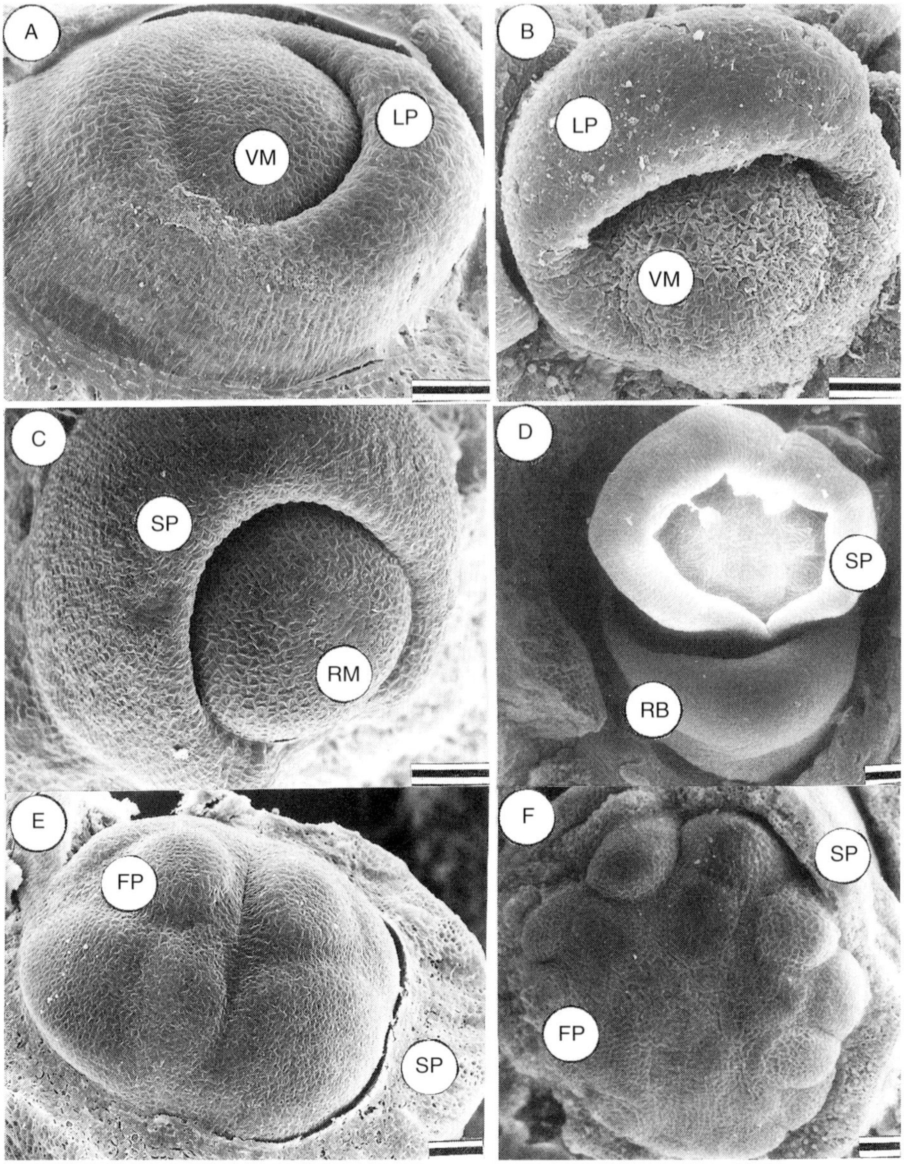

Fig. 2.3. Scanning electron photomicrographs of *Allium* spp. vegetative apex and initial stages of floral development. Bar = 0.1 mm. A. Initiation of leaf primordia (LP) in the vegetative apical meristem (VM) of *A. aschersonianum*. Older leaf primordia removed. B. Further development of leaf primordium (LP) in the vegetative apical meristem (VM) of *A. aschersonianum*. Older leaf primordia removed. C. Floral initiation in *A. aschersonianum*. Spathe (SP) surrounds the swollen reproductive meristem (RM). D. Spathe (SP) of *A. nigrum* envelops the reproductive meristem. Renewal bulb (RB) is initiated in the axis of the last leaf (removed), near the base of the floral stalk. E. Differentiation of four centres of development in a reproductive meristem of shallot, with spathe (SP) removed. First flower primordia (FP) are visible. F. Initiation of the first flower primordia (FP) in the periphery of the reproductive meristem of *A. nigrum*. Spathe removed.

1985). However, the West African onion cv. 'Bawku' is optimally induced to flower between 15 and 21°C (Sinnadurai, 1970a, b) and some landraces from northern Russia have an optimum of 3–4°C (for reviews, see Rabinowitch, 1985, 1990a). Relatively little is yet known of the responses of tropical onions and shallots in this respect (for more details, see Currah and Proctor, 1990; Currah, Chapter 16, this volume).

Post-juvenile onion plants respond to cold induction both at rest and during active growth in the field, and their sensitivity to cold induction increases with age, i.e. older plants require less cold induction (Gregory, 1936; Thompson and Smith, 1938; Heath and Mathur, 1944). In growing seedlings, the critical dry weight of shoot (basal plate plus leaves) for inflorescence induction ranges from 60 to 450 mg (Brewster, 1985). The minimum critical dry weight required by dry bulbs to initiate flowering during storage is much higher than that in growing plants and, in both cases, the threshold is determined by the genetics of the plant (Brewster, 1994).

High temperatures of 28–30°C throughout storage not only inhibited inflorescence initiation in onion, but also exerted a marked after-effect during the subsequent growing season, expressed as delayed flowering (Heath and Mathur, 1944; Aoba, 1960), or led to greatly reduced flowering (Jones, 1927; Jones and Emsweller, 1936; Heath, 1943a, 1945; van Beekom, 1953; Lachman and Michelson, 1960; van Kampen, 1970).

3.4.2 Shallot

In shallot (*A. cepa* Aggregatum group) seedlings of tropical origin, floral initiation becomes evident after formation of the sixth true leaf (Krontal *et al.*, 1998). Unlike the bulb onion, in which, after floral initiation, the lateral meristems form dormant adventitious buds, shallot leaf formation continues at the axillary meristems, simultaneously with floral development at the main apex. As in bulb onion, low temperatures induce flowering of shallots, with the optimum between 5 and 10°C, either in storage or during growth, whereas high or intermediate growing or storage temperatures delay or prevent inflorescence development (Krontal *et al.*, 2000; Rabinowitch and Kamenetsky, Chapter 17, this volume). Some shallot genotypes, however, are very resistant to flowering, possibly due to a long history of selection against this trait, as suggested with garlic.

3.4.3 Japanese bunching onion

Like the bulb onion, Japanese bunching onion varies according to the cultivar in both juvenile age and cold requirement (Table 2.1; Watanabe, 1955; Yakura and Okimizu, 1969; Lin and Chang, 1980; Inden and Asahira, 1990; Yamasaki *et al.*, 2000a). Genotypic differences exist in the interaction between low temperature and photoperiod: two mid-season flowering cultivars exhibited a similar response to temperature in flower initiation and bolting, but

Table 2.1. Effect of genotype, physiological age, day length and temperature on floral induction in Japanese bunching onion.

		Physiological age		Induction requirements	
Cultivar	Origin	Leaf number	Pseudostem diameter (mm)	Temperature (°C)	Duration (days)
Kaga[1,2]	Japan	11–12	5–7	<13	30
Gao Jiao[3]	China	–	3	5	30
Pei Chang[4]	Taiwan	–	4.5	5	5
			4.5	20	10

[1]Yakura and Okimizu (1969).
[2]Watanabe (1955).
[3]Lin and Chang (1980).
[4]Inden and Asahira (1990).

they differed markedly in their photoperiodic response. The primary requirement in cv. 'Kincho' was low temperature, while in 'Asagi-kujo' it was a short day (Yamasaki *et al.*, 2000a).

3.4.4 Wild members of the section Cepa *(subgenus* Rhizirideum*)*

In their natural habitats, *A. altaicum*, *A. oschaninii* and *A. pskemense* have a short summer 'rest' period. Sprouting begins in the autumn, but the low winter temperatures retard or completely inhibit leaf development and elongation (Pistrick, 1992). Only plants with more than 10 or 11 leaves (including leaf primordia) progress to the reproductive stage, which occurs in the autumn, when temperatures decrease and day length becomes short (Cheremushkina, 1985).

3.4.5 Wild rhizomatous species

Little is known about florogenesis in this group of plants. Under natural conditions, the renewal bulbs of nine Siberian species formed in the leaf axils of the parent plants, which remained vegetative during the first and second growing seasons. In the third season, and following the development of 7–10 (*A. senescens*) or 16–20 (*A. nutans*) leaves, the renewal bulbs became reproductive (Cheremushkina, 1985). Initiation of flowering occurs either in the spring, when it is followed by instant scape elongation and bloom (*A. nutans*, *A. senescens*, *A. galanthum*), or in the autumn, before the harsh winter (*A. obliquum*) (Cheremushkina, 1985).

In Israel, where winters are mild, rhizomatous species such as *A. trachyscordum*, *A. petraeum*, *A. platyspathum* and *A. nutans* from Siberia and Kazakhstan bloom in the spring and summer, between May and July, without any additional cold treatment (Kamenetsky, 1996b).

3.4.6 Garlic

All current commercial clones of *A. sativum* (subgenus *Allium*) are completely sterile (Etoh and Simon, Chapter 5, this volume). Possible reasons for this include competition for nutrients between generative and vegetative buds (topsets) within the developing inflorescence (Koul and Gohil, 1970), premature degeneration of the tapetum (Novak, 1972), or infection with degenerative-like diseases (Konvicka, 1973, 1984). Etoh (1985) suggested that garlic is in transition from a sexual to an asexual reproductive state and that farmers have accelerated the process through numerous generations of selection.

Garlic clones vary in their ability to bolt and have been classified accordingly (Gvaladze, 1961; Takagi, 1990; Etoh and Simon, Chapter 5, this volume), as follows:

1. Complete bolting – plants produce a long thick flower stalk, with many topsets and a variable number of flowers.
2. Incomplete bolting – plants produce a thin short flower stalk, with a few large topsets; usually no flowers are formed.
3. Non-bolting – plants do not normally form a flower stalk; instead, only cloves are produced inside the pseudostem (Takagi, 1990).

When grown under the appropriate environmental conditions, plants of the first two groups, but not those of the third group, produce inflorescences and floral buds. Genotypes from the temperate zone require stronger cold induction for inflorescence formation than those from subtropical and tropical regions. The inductive temperatures vary significantly with cultivar and range between −2 and 10°C (Takagi, 1990; R. Kamenetsky, personal observations). Long storage at low temperatures resulted in the blooming of plants with smaller numbers of leaves and in earlier flowering than in bulbs stored for a shorter period. However, a very long cold treatment (2°C for 5 months) reduced blooming of garlic cv. 'Yamagata' (Takagi, 1990).

Transition of the apical meristem from the vegetative to the reproductive state occurs only in growing plants with a minimum of six to eight leaves and leaf primordia, but not during cold or ambient storage (Kamenetsky and Rabinowitch, 2001), and low field temperatures promote inflorescence induction

(Takagi, 1990). However, after storage at low temperatures, garlic plants from the complete-bolting group (Israeli Gene Bank, Rehovot, plant introduction no. 2091) were able to initiate flowers at relatively high growth temperatures (23/15°C, day/night, respectively) (Kamenetsky and Rabinowitch, 2001).

3.4.7 Ornamental species (subgenus Allium*)*

In this group, inflorescence initiation occurs only in growing plants, following the formation of seven to nine green leaves. Growth temperatures of 17–20°C and long days are essential for floral initiation and scape elongation, whereas high field temperatures and short days are not inductive and plants remain vegetative (Berghoef and Zevenbergen, 1992; Kamenetsky, 1996b).

3.4.8 Ornamental species (subgenus Amerallium *= former section* Molium*)*

Plants originating in Mediterranean climates (Mediterranean basin, California) remain vegetative during a summer 'rest', when soil temperatures are high. A visible transition of the apex to the reproductive state occurs only in the autumn, when temperatures decrease. Thus, an optimum temperature range of 9–17°C has been recorded for floral initiation in members of the subgenus *Amerallium*, including *A. unifolium* (Kodaira *et al.*, 1996), *A. neapolitanum* and *A. roseum* (Maeda *et al.*, 1994; van Leeuwen and van der Weijden, 1994).

3.4.9 Ornamental species (subgenus Melanocrommyum*)*

Plants from the Irano-Turanian region (Central Asia, Iran, Afghanistan) (e.g. *A. aflatunense* = *A. hollandicum*, *A. altissimum*, *A. karataviense*) initiate leaf primordia in the renewal bulb during the flowering of the mother plant. Following the differentiation of five to seven leaf primordia, the apical meristems of *A. altissimum* and *A. karataviense* become latent. No detectable changes occur for 6–10 weeks, and then floral initiation becomes visible at the apex within the bulb (Kamenetsky, 1997; Kamenetsky and Japarova, 1997).

In *A. aflatunense* (= *A. hollandicum*) the transition from the vegetative to the reproductive phase occurs at the end of the growth period, immediately after the cessation of leaf initiation. The differentiation of the floral meristem has been observed in plants grown at all temperatures from 4 to 26°C (Zemah *et al.*, 2001).

In *A. aschersonianum*, *A. nigrum* and *A. rothii* of the Israeli flora, flowering of the mother plant in February–March is followed by the high-temperature induction of a 12–15-week latent period of the apical meristem within the bulb. During July–October, five to seven leaf primordia form and the meristem becomes reproductive without cold induction. When the plants are stored at 20–25°C, the summer rest becomes considerably shorter and floral initiation occurs in August. In such cases, plants can be forced into flower 2–3 months earlier than under ambient Israeli summer conditions (Kamenetsky, 1994, 1997; Kamenetsky *et al.*, 2000).

To the best of our knowledge, there are no data on the photoperiod effect on floral induction in wild species of the subgenus *Melanocrommyum*.

4. Floral Differentiation (Organogenesis) and Inflorescence Structure

The *Allium* inflorescence appears to be simple. In reality, however, it is very complex. For many years, botanists referred to it as a monopodial umbel. However, as early as 1837, Louis and Auguste Bravais described the inflorescence of *A. moly* (subgenus *Amerallium*) as having two sequential layers of sympodial flower clusters (cited by Mann, 1959). Later, Weber (1929) reported that the inflorescence of *A. odorum* (= *A. ramosum*) consists of a terminal flower, which bears two bracts on its pedicel, each with an axillary flower. This dichasial branching continues, and thus each flower gives rise to two lateral flowers.

4.1 Bulb onion

Jones and Emsweller (1936) made an analysis of the structure and development of the onion inflorescence and of the individual flower. Over the broad surface of the stem tip, which is situated within the developing spathe, numerous membranous bracts develop, which cover the cluster of young flowers in their first stages. De Mason (1990) notes that the generative meristem of onion subdivides into multiple centres, each of which gives rise to a group of flowers, a cyme (= bostryx). The flower buds in each cyme are arranged in a spiral order. Thus, the bulb-onion inflorescence, often with 400–600 flowers, comprises many flower clusters, each consisting of several flowers.

4.2 Shallot

Krontal *et al.* (1998) reported that differentiation of shallot flowers begins with subdivision of the apical meristem into four centres (Fig. 2.3E). The floral initials occur in one of these centres only after the scape reaches 5–7 mm in length above the basal plate. In each of the four centres of differentiation, floral primordia develop unevenly in a helical order. Each centre of development is covered by thin membranous bracts and contains six or seven developing flower clusters (Fig. 2.4A). Initiation and differentiation of additional new primordia continue simultaneously with the sequential differentiation, growth and development of older flowers. Thus, the shallot inflorescence consists of clusters, each containing five to ten flower buds arranged in a spiral order: it can therefore be described as an umbel-like flower arrangement, the branches (flower clusters) of which arise from a common meristem (Rabinowitch and Kamenetsky, Chapter 17, this volume).

4.3 Garlic

Morphological events in the flower development of garlic are of special interest because of its inherited sterility (Konvicka, 1984; Etoh *et al.*, 1988; Etoh and Simon, Chapter 5, this volume). Floral development has been described in Japan for the bolting-garlic cv. 'Shanhai-wase' (Etoh, 1985) and in Israel for accession no. 2091, introduced from Russia (Kamenetsky and Rabinowitch, 2001). The differentiation of floral initials begins only after the scape has reached 5–7 mm in length and the apex diameter exceeds 0.5 mm. Later, the apical meristem subdivides into several swellings, each of which gives rise to a number of individual flower primordia (Fig. 2.4B). When the floral stalk reaches 15 cm in length, the pedicels elongate and the inflorescence becomes spherical (Fig. 2.4C).

Long leaf-like bracts develop both at the periphery and in the centre of the inflorescence, thus separating the developing umbel into distinct floral clusters (Fig. 2.4D). Further inflorescence growth and development include both initiation and differentiation of new flower primordia, and sequential differentiation, growth and development of older flowers. At this time, new undifferentiated domes, 0.15 mm in diameter, form at the base of the inflorescence. These swellings quickly differentiate into vegetative buds and grow to form small inflorescence bulbils: the topsets (Fig. 2.4E, F).

Topset differentiation begins in the periphery of the apical surface; their number, size and rate of development are determined by the genotype and show great variability. After differentiating, the topsets develop quickly, a process followed by degeneration and abortion of many of the developing flowers. Similar observations by Etoh (1985) led to the conclusion that garlic is in a transitional state from sexual to asexual reproduction.

When the spathe breaks open, differentiated flower buds of garlic become visible to the naked eye, but the fast-growing topsets stifle them and the flower buds quickly degenerate. Therefore, in some garlic clones, continuous removal of the developing topsets can result in normal flowering, pollination and seed production (Koul and Gohil, 1970; Konvicka, 1984; Etoh *et al.*, 1988; Pooler and Simon, 1994; Jenderek and Hannan, 2001; Etoh and Simon, Chapter 5, this volume).

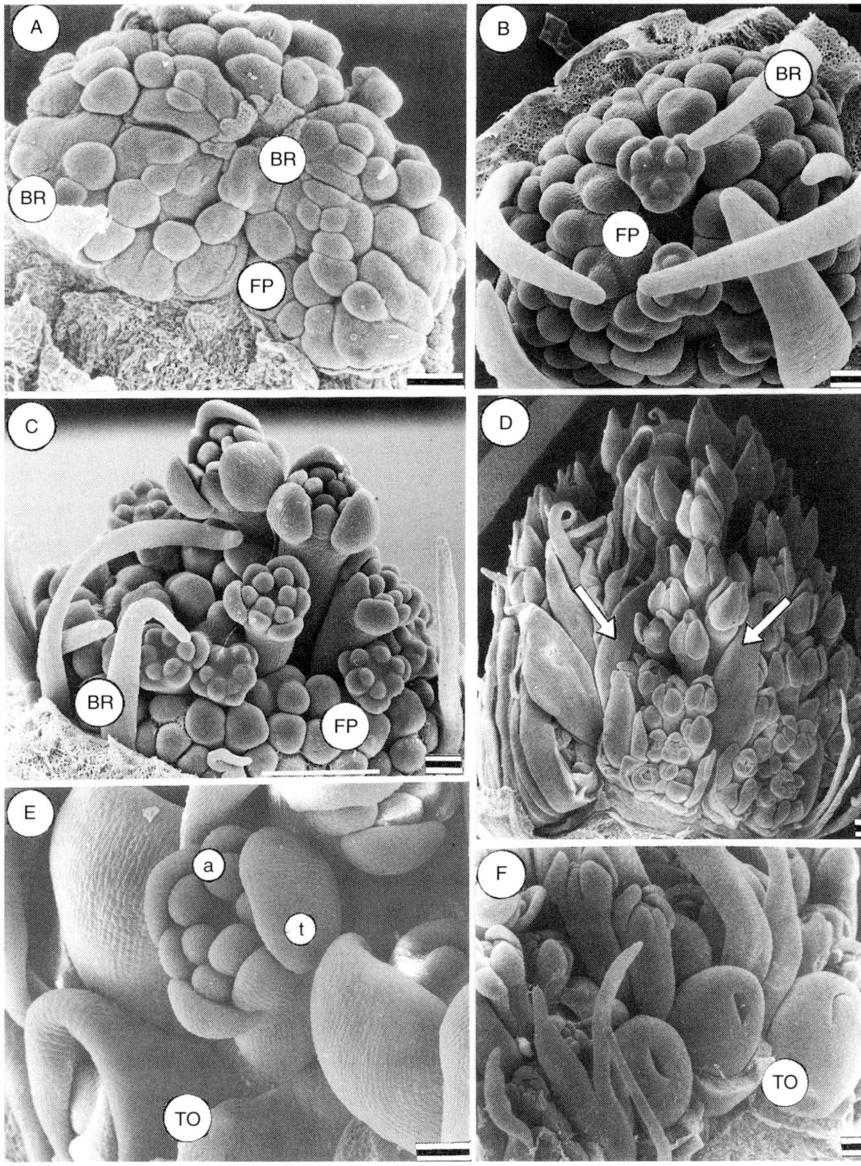

Fig. 2.4. Scanning electron photomicrographs of *Allium* spp. floral development. Bar = 0.1 mm. A. Flower primordia (FP) are visible in shallot inflorescence. Four centres of differentiation are separated by the bracts (BR). Spathe and peripheral and central bracts removed. B. Early stages of garlic floral development. Floral differentiation is visible in older flower primordia (FP), while younger flowers still appear as meristematic domes. Leaflike bracts (BR) form at the periphery of the inflorescence. Spathe removed. C. The inflorescence of garlic becomes hemispherical in shape and consists of numerous floral primordia (FP). Differentiation of floral primordia is uneven: floral parts occur in the oldest floral primordia, while younger ones still appear as undifferentiated meristematic domes. Spathe removed. D. Floral pedicels and leaf-like bracts elongate. In individual garlic flower clusters (arrows), which are separated by leaf-like bracts, floral primordia develop unevenly in a helical order. New flower primordia continue to appear at the base of the inflorescence. E. Magnification of the basal part of garlic inflorescence. Newly developed meristems appear and rapidly differentiate to form small inflorescence bulbs: topsets (TO). In the individual flowers, tepals (t) and anthers (a) are visible. F. Topsets (TO) in garlic inflorescence.

4.4 Japanese bunching onion

After flower initiation, the early stage of flower development is day-neutral and, after floret formation stage, a long-day photoperiod promotes flower development and elongation of the seed-stalk (Yamasaki *et al.*, 2000b). In Israel, cultivated *A. fistulosum* plants bloom in the spring and early summer, just before entering summer dormancy (H.D. Rabinowitch, personal observation), thus ending the production season. In Japan, this crop is of high economic value. Work is in progress by Yamasaki and colleagues to exploit the genetic variability within existing cultivars for day-length response (e.g. stronger requirement for a short day (SD) in cv. 'Asagi-kujo' compared with cv. 'Kincho' (Yamasaki *et al.*, 2000b)) in order to control or delay flowering so as to extend the harvest season, which is normally curtailed when the plants start to flower in Japan.

Recently greenhouse culture and plug-seedling transplanting of Japanese bunching onion have increased in Japan, where a new method of bolting control using long-day treatment is easily applicable.

4.5 Ornamental species (subgenus *Amerallium* = former section *Molium*)

A detailed description of the developing inflorescences of six Mediterranean species (Mann, 1959) indicated that the single spathe consists of four bracts, each of which bears in its axil a flower cluster (a helicoid cyme or bostryx) of three to seven flowers. Several smaller cymes differentiate later in the centre of the inflorescence; they contain smaller numbers of flowers. The first peripheral cyme is formed opposite to the uppermost foliage leaf; the others follow in alternating positions. The four peripheral cymes flower first and the central ones flower last. Within each cyme, the flowers open in a strict sequence from oldest to youngest (Fig. 2.5).

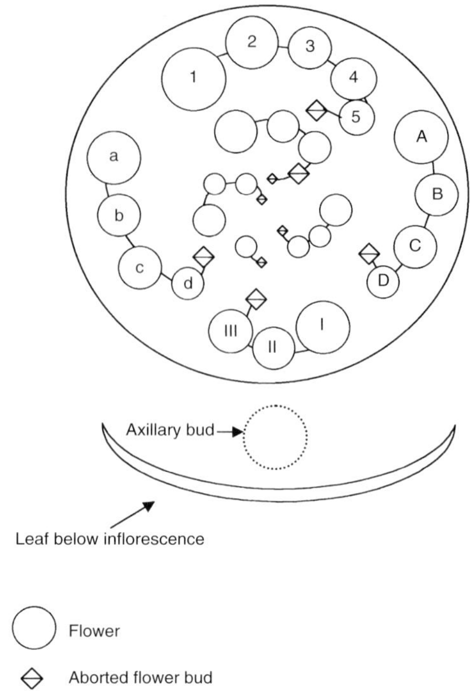

Fig. 2.5. Cross-section of the inflorescence of *A. neapolitanum*, showing its flower arrangement. The first four helicoid cymes (bostryces) are located around the periphery of the inflorescence and are designated by different signs, in the following order: 1,2,3…; I,II,III…; A,B,C…; a,b,c…. The first cyme develops opposite the uppermost foliage leaf (last leaf) and axillary bud. (From Mann, 1959, with permission.)

4.6 Ornamental species (subgenus *Melanocrommyum*)

Flower differentiation of the Central Asian species *A. aflatunense* (= *A. hollandicum*), *A. altissimum* and *A. karataviense*, as well as the species from the Mediterranean area *A. nigrum*, *A. rothii* and *A. tel-avivense*, begins during the rest period of the bulb (Kamenetsky, 1994, 1997; Kamenetsky and Japarova, 1997). Despite the significant variation in their life cycle and pace of floral development, they all have a similar inflorescence structure within a spathe, which is shaped at first as a nearly uniform ring.

Following the cessation of leaf formation and the initiation of a spathe, the apical meristem grows markedly in size, and several peripheral swellings differentiate to produce a row of flower primordia (Fig. 2.3F). Within each peripheral swelling, the flat meristematic surface protrudes to become round and smooth; it later divides into many centres, each of which gives rise to a flower cluster (Fig. 2.6A, B). As flower primordia continue their development, the circular spathe grows upward to envelop the developing inflorescence.

Flower number per umbel and per flower cluster vary with species, plant age and size, and probably with growth conditions (Table 2.2). Differentiation and development of flowers within each cluster proceed in a spiral order to form a complex monochasium, the cyme (or bostryx). New flower primordia continue to form within each cyme while older flowers already have differentiated floral parts (Fig. 2.6C, D). The youngest primordium in each cyme sometimes aborts.

The sequence of differentiation affects the inflorescence structure and is maintained throughout from flower anthesis to seed maturation (Kamenetsky, 1997). The strict developmental sequence from the oldest to the youngest flower was observed within each flower cluster (Fig. 2.6E, F). As flowers open, the pedicels reach similar lengths, so that, in a fully developed inflorescence, cymes can no longer be recognized.

Based on the sequence of inflorescence differentiation, we hereby propose the following classifications of *Allium* inflorescence structures:

1. The apical meristem divides initially into several (usually four) centres, separated by leaf-like bracts. Each centre gives rise to a number of flower clusters (cymes). Floral differentiation and organogenesis occur simultaneously with both scape elongation and vegetative growth and development. This type of florogenetic process has been reported for onion (Jones and Emsweller, 1936; De Mason, 1990), garlic (Etoh, 1985; Kamenetsky and Rabinowitch, 2001) and shallot (Krontal *et al.*, 1998).

2. The inflorescence is composed of monopodially arranged clusters, of which the first one is formed opposite to the uppermost foliage leaf and others follow in alternating positions. Within each cyme, the flowers differentiate and open in a strict sequence from oldest to youngest. Floral differentiation and organogenesis occur both during storage and during active growth and development (Mann, 1959). This type of florogenetic process has been reported for species from the subgenus *Amerallium*, e.g. *A. neapolitanum* and *A. roseum*.

3. All flower clusters (cymes) arise from a common meristem. Differentiation of clusters commences in the periphery of the apical meristem and continues towards its centre. Within each cluster, flowers are formed in a helical order. Floral differentiation and organogenesis take place during the summer rest period (Kamenetsky, 1994, 1997; Zemah *et al.*, 2001). This type of florogenetic process has been reported for the subgenus *Melanocrommyum*, e.g. *A. aflatunense* (= *A. hollandicum*), *A. altissimum*, *A. karataviense*, *A. nigrum* and *A. rothii*.

5. Differentiation of the Individual Flower

All *Alliums* produce flowers with six perianth lobes, six stamens and a tricarpellary pistil, situated in the centre of the flower. Ovaries of differentiated *Allium* flowers include the

Table 2.2. Number of flowers within umbels of *Allium* spp. of subgenus *Melanocrommyum* (adapted from Kamenetsky, 1997).

Species and stage of development	Number of flowers		
	Umbel	Peripheral cymes	Central cymes
A. karataviense, old*	450–600	12–15	6–10
A. nigrum, old	140–160	6–7	3–5
A. nigrum, young	60–75	4–5	2–3

*Young and old = the plants have undergone the first flowering cycles or experienced a number of cycles, respectively.

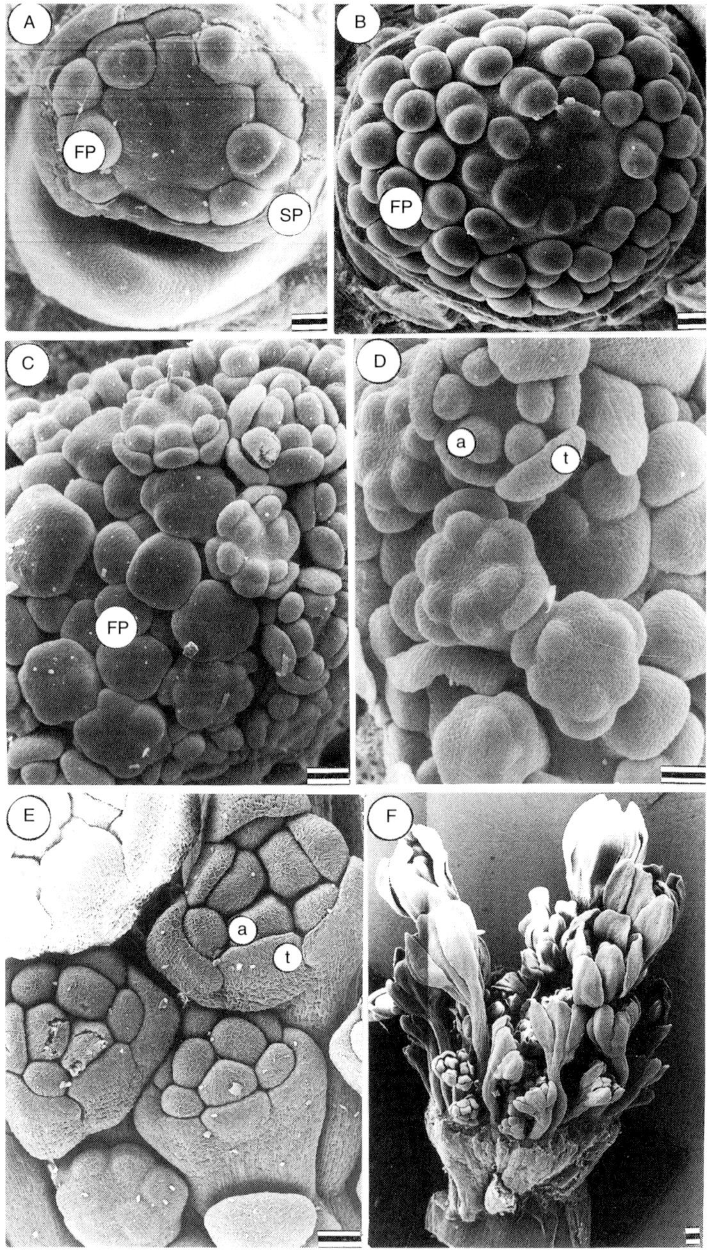

Fig. 2.6. Scanning electron photomicrographs of *Allium* floral development. Bar = 0.1 mm. A. Flower differentiation in *A. nigrum*. Individual peripheral flower primordia (FP) are visible. Spathe (SP) removed. The development of the renewal bulb is visible. B. Progress in flower differentiation in *A. nigrum*. Within the peripheral floral primordia (FP) differentiation occurs centripetally and the flat meristematic surface subdivides into several central cymes. Spathe removed. C. Differentiation of flower parts in older (first-formed) flower primordia of *A. aflatunense*. Younger primordia are not yet differentiated. D. Magnification of individual cluster in the inflorescence of *A. aflatunense*. Floral primordia develop unevenly in a helical order. In older flowers, tepals (t) and anthers (a) are visible. E. Flower differentiation in individual inflorescence cluster of *A. karataviense*. Young flower buds are undifferentiated. F. Fully developed inflorescence of *A. karataviense* at the end of September.

nectaries, which consist of secretory cells situated on the outer ovary walls (Fritsch, 1992). Shapes and positions of the nectaries and their canals differ between taxonomic groups of the genus. At anthesis, nectar secretion begins through a spurlike prolonged part of the ovary or special canal. The nectar accumulates in the gap between the ovary and the bases of the filaments and tepals.

In the bulb onion, anthers shed their pollen at anthesis or 1–2 days later. The delicate style of the protandrous flower reaches full length and becomes receptive (develops a sticky surface to retain pollen) 2–3 days after anthesis, when the flower's own pollen has already been shed (Jones and Rosa, 1928; Jones and Emsweller, 1933; Moll, 1954; Chang and Struckmeyer, 1976; Currah and Ockendon, 1978; Ali *et al.*, 1984; Currah, 1990; De Mason, 1990). Colour of the tepals varies with species, from white or yellow to pink, red, purple and blue (Brewster, 1994; Kamenetsky and Fritsch, Chapter 19, this volume). The number of flowers per umbel varies within and between species and is greatly affected by environment, age and the position within the plant – e.g. a primary inflorescence consists of more flowers than a secondary umbel. In the bulb onion, there are commonly 200–600 flowers per umbel (Currah and Ockendon, 1978; Ali *et al.*, 1984), and similar numbers were reported for leek and Japanese bunching onion. Shallot inflorescences are smaller, while chives, Chinese chives and rakkyo produce between a few and 30–40 flowers per umbel (De Mason, 1990; Brewster, 1994). The ornamental value of the most popular species is based on their multiflowered inflorescences, which include 400–500 flowers (e.g. *A. aflatunense*, *A. giganteum*, *A. karataviense*). However, some ornamental alliums have only a few large flowers per umbel (e.g. *A. insubricum*, *A. moly*, *A. oreophilum*) (Kamenetsky and Fritsch, Chapter 19, this volume).

5.1 Bulb onion

In *A. cepa*, the outer three tepals arise first, each simultaneously with its respective stamen in its axil. These outer tepals and their associated stamens occur in a clockwise succession, whereas the inner tepals also arise together with their subtended stamens, but in an anticlockwise direction. The carpels develop as three protruding areas within the inner stamens and meet at the heart of the flower to form the trilocular ovary (Jones and Emsweller, 1936; Esau, 1965; De Mason, 1990). Each flower has three nectaries located between the broad bases of the filaments of the inner stamens and the lower ovarian walls. The nectaries open to the surface through a pore (De Mason, 1990).

5.2 Shallot

The floral morphology in shallot is very similar to that of bulb onion, but no clear direction of primordia differentiation in individual shallot flowers has been observed (Krontal *et al.*, 1998; Rabinowitch and Kamenetsky, Chapter 17, this volume).

5.3 Garlic

As in onion and shallot, during the differentiation of flower primordia, each perianth lobe and the subtended stamen arise simultaneously from a single primordium (Kamenetsky and Rabinowitch, 2001).

5.4 Ornamental species (subgenus *Melanocrommyum*)

The outer perianth lobes and stamens of *A. aflatunense* (= *A. hollandicum*), *A. altissimum*, *A. aschersonianum*, *A. karataviense* and *A. nigrum* usually arise first, followed by the differentiation of the inner whorl (Fig. 2.7A–D). The carpels initiate last, when the outer perianth lobes overarch the stamens (Fig. 2.7E, F; Kamenetsky, 1994, 1997; Kamenetsky and Japarova, 1997). When the stigma becomes receptive, the tepals spread widely to expose the accumulated nectar to potential pollinators. Nectar attractiveness depends on its aroma as well as on its fluorescence in the ultraviolet (UV) range,

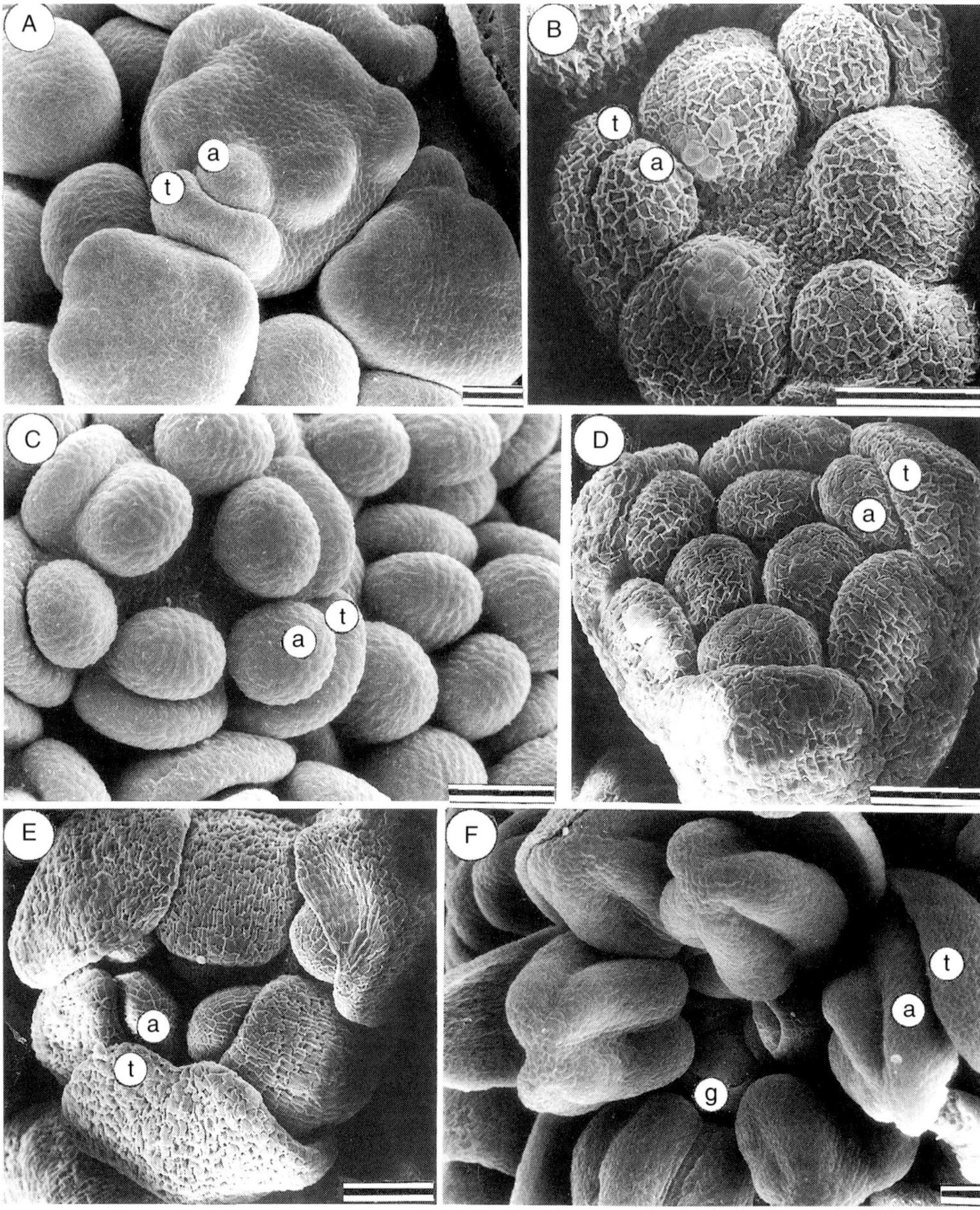

Fig. 2.7. Scanning electron photomicrographs of differentiation of individual flowers in *Allium* spp. Bar = 0.1 mm. A. Initial stages in differentiation of individual flower of *A. aflatunense*. Tepals and their respective anthers form from common primordia. One outer tepal (t) and its respective anther (a) form first, then two adjacent inner tepals with stamens are differentiated. B. Differentiation of individual flower of *A. altissimum*. Outer whorl of three tepals (t) and their related anthers (a) are visible.Three undifferentiated common primordia are formed in the inner whorl. C. Differentiation of flower parts of *A. karataviense*. Tepals (t) and anthers (a) of outer and inner whorls form simultaneously. D. Advanced development of individual flower of *A. karataviense*. Tepals (t) and anthers (a) increase in size. E. Further development of individual flower of *A. karataviense*. Tepals (t) elongate and overarch the anthers (a). F. Final stage in flower differentiation of *A. karataviense*. Gynaecium segments (g) form in the centre of the flower, and the anthers (a) reach their characteristic form.

which is visible to insects (Waller and Martin, 1978), but a high potassium level in the nectar may discourage honey-bees from visiting onion flowers (Waller *et al.*, 1972).

6. Floral Malformations and Topset Formation

Aberrations in floral initiation and differentiation may lead to modifications in inflorescence formation. In many *Allium* species, floral malformations occur as a direct consequence of adverse conditions during floral initiation and differentiation, but sometimes, as in garlic, the major factor is genetic.

Abnormal floral development may negatively affect seed production or, in the case of ornamental species, reduce the decorative value of the plant.

6.1 Bulb onion

High temperatures during storage of bulbs with inflorescence initials or in the field may cause reversion from the floral to the vegetative phase. The more advanced the reproductive bud, the longer the treatment required to cause such a reversion (Heath and Mathur, 1944; Sinnadurai, 1970a). When exposed to high temperatures, flower primordia in bulbs that had been stored at 21–27°C shrank, withered and turned brown (Woodbury, 1950). After the emergence of the scape, injury to the spathe of the developing inflorescence promotes the development of topsets (Rabinowitch, 1990a), probably because of a significant change in the endogenous hormonal balance. Cytokinin applications can be used to promote higher rates of bulbil formation in umbels from which the flower buds have been trimmed (Thomas, 1972).

Male sterility has been known in the bulb onion since 1925 (Jones and Clarke, 1943; Berninger, 1965). Male sterility in 'Italian Red 13–53' was conditioned by the interaction of a particular form of cytoplasm (S cytoplasm) with a homozygous recessive form (*ms*) of the single nuclear restorer (Ms) locus. In plants carrying S cytoplasm, fertility is restored by a dominant nuclear allele (*Ms*) at this restorer locus. Additional cytoplasmic and genetic mechanisms were described later (Berninger, 1965; Schweisguth, 1973). The latter, however, were hardly used in hybrid seed production (Dowker, 1990; Rabinowitch, 1990a; Havey, 1995, 2000; Havey, Chapter 3, and Eady, Chapter 6, this volume).

6.2 Shallot

Malformed flowers and topsets have been observed in tropical shallots grown from seeds under high temperatures of 26/18°C, day/night, respectively (Rabinowitch and Kamenetsky, Chapter 17, this volume). Storage of bulbs at 30°C caused a delay in the emergence of scapes as compared with plants from low and intermediate storage temperatures, but did not induce floral malformations (Krontal *et al.*, 2000). Male sterility is common in shallots grown in Israel and elsewhere (H.D. Rabinowitch, personal observation). Male-sterile shallots are readily fertilized by pollen from shallot and/or bulb onion to form viable seeds. The inherited characteristics of shallot enable male-sterile plants to be easily maintained and multiplied by vegetative propagation. To the best of our knowledge, no information is available on the heredity of male sterility in shallot; however, we can speculate that it will be much the same as that of bulb onion.

6.3 Garlic

Gustafsson (1946/47, cited by Etoh, 1985) assigned garlic and some other *Allium* species (*A. caeruleum*, *A. carinatum*, *A. proliferum*, *A. scorodoprasum*, *A. vineale*) to the group of viviparous plants, in which topsets (bulbils) develop instead of flowers or intermingle with flowers in the inflorescence. Topsets differentiate in between the flower initials, at the base of the inflorescence (Fig. 2.4E, F; Kamenetsky and Rabinowitch, 2001). As a result of the strong competition with the developing topsets, the garlic flowers wither and die.

Flowers in the Japanese garlic cv. 'Shanhai-wase' exhibited floral malformations and abnormal development of the embryo sacs, possibly because of an unfavourable environment during floral differentiation (Etoh, 1985), but perhaps more probably due to the numerous generations of selection by humans for larger bulbs and cloves and against flowering.

6.4 Chives, Japanese bunching onion and leek

For a detailed review on these crops, see Havey, Chapter 3, this volume. In chives, male sterility is conditioned by genic male sterility (GMS), which is controlled by a single nuclear gene *wi*, with recessive inheritance (Engelke and Tatlioglu, 2000a). An alternative cytoplasmic male sterility (CMS) depends on the interaction between the cytoplasm (S) and a single nuclear fertility-restoration locus (*X*) (Tatlioglu, 1982). There is a high degree of variability of the mitochondrial genome in chives (Engelke and Tatlioglu, 2000b) and consequently two CMS systems were described (Engelke and Tatlioglu, 2000c). Fertility of some male-sterile plants, however, can be regained under favourable environmental conditions. Hence, exposure to a constant temperature of 24°C resulted in production of viable pollen (Tatlioglu, 1985). This temperature sensitivity is controlled by a single dominant allele (*T*) (Tatlioglu, 1987). A third gene, *a*, restores fertility in combination with tetracycline treatment (Tatlioglu and Wricke, 1988).

In Japanese bunching onion, male sterility is controlled by the interaction of a cytoplasmic factor (S) with two nuclear genes: ms_1 and ms_2 (Moue and Uehara, 1985).

In leek, a genic male-sterility system has been described (Schweisguth, 1970; De Clercq and Van Bockstaele, Chapter 18, this volume) and naturally occurring male-sterile plants reproduced clonally now provide the basis for hybrid leek production (Smith, 1994; Smith and Crowther, 1995). The appearance of male-sterile leek flowers is described by De Clercq and Van Bockstaele (Chapter 18, this volume), who also illustrate how removing or wounding young flower buds can induce topset formation in the leek umbel (see Fig. 18.3a, b).

6.5 Ornamental species (subgenus *Melanocrommyum*)

These plants develop topsets in response to adverse storage conditions. High temperatures at the time of differentiation promoted floral malformations in *A. aflatunense* (= *A. hollandicum*) (Fig. 2.8A–D; Colour Plate 1A–C) (H. Zemah, Israel, 2000, personal communication).

Preplanting exposure of *A. aschersonianum*, from the Mediterranean semi-desert, to relatively low temperatures of 9–13°C during floral initiation and differentiation, affected apical meristem division and led to the formation of two or three short scapes with small and partly malformed flowers (Z. Gilad, Israel, 2000, personal communication). However, exposure of the bulbs of *A. aschersonianum* to 48–50°C for 4–6 h in September–October, during within-bulb flower differentiation, resulted in a small number of flowers in the inflorescence, with the simultaneous formation of topsets and/or lateral bulbs (Kamenetsky *et al.*, 2000; E. Hovav, Israel, 2000, personal communication).

7. Maturation and Growth of Floral Parts and Floral Stalk Elongation

Interactions between storage and growth temperatures play the most important role in normal scape elongation and flowering of *Allium* species, although light conditions can markedly affect this process.

As with leaf-blade structure, there is a great variation in the morphological structure of the inflorescence axis (also named the floral stem, scape or stalk) among *Allium* species (Jones and Mann, 1963; Vvedensky, 1968). In all cases, the stalk represents a single internode, elongating out of the innermost ensheathing leaf base. In the bulb onion, the scape is hollow and its anatomy reveals more similarity to that of onion leaves than to that of the vegetative stem.

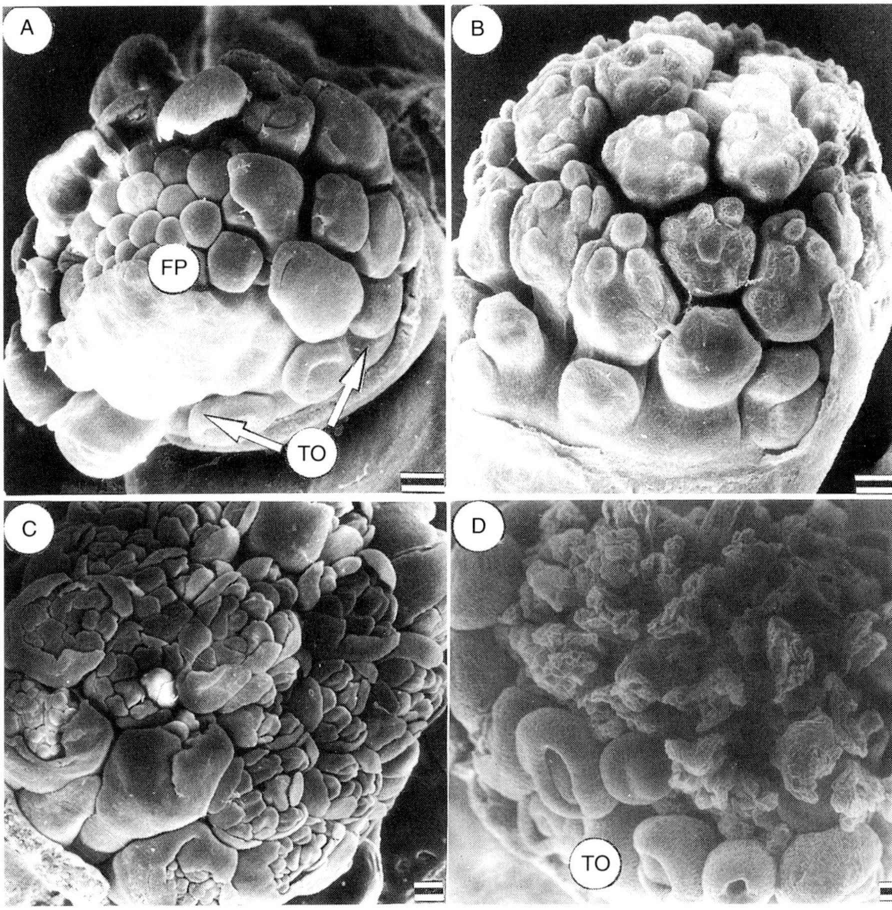

Fig. 2.8. Scanning electron photomicrographs of floral malformations in *A. aflatunense*. Bar = 0.1 mm. A. Leaf-like bracts are formed in the centre of the reproductive meristem. Topsets (TO) are formed in the periphery of the inflorescence. FP, flower primordia. B. Irregular development of individual flowers leads to flower abortion. C. Development of numerous anthers and tepals. D. Topset (TO) formation in the periphery of the inflorescence.

The epidermis is heavily cutinized and contains stomata, and the mesophyll has palisade cells on the outside and spongy cells on the inside (De Mason, 1990). In most of the *Melanocrommyum* species used as ornamentals, as well as in garlic, leek and chives, the scape is round and solid (Jones and Mann, 1963; De Mason, 1990; Fritsch, 1993). Others (e.g. *A. neapolitanum*, *A. triquetrum*) produce solid triangular scapes, whereas *A. fistulosum*, *A. proliferum* and species from the section *Cepa* produce cylindrical fistulose scapes (Jones and Mann, 1963; R. Kamenetsky, personal observations). The distribution of several anatomical characters of floral scapes broadly corresponds to taxonomic relationships within the genus *Allium* (Fritsch, 1993).

7.1 Bulb onion and shallot

Cool temperatures of around 17°C (Thompson and Smith, 1938; Holdsworth and Heath, 1950) or 10–16°C in the greenhouse enhanced scape elongation in onion (Woodbury, 1950) and shallot (Krontal *et al.*, 2000), while high temperatures of 25–30°C suppressed the emergence of inflorescences already initiated (Heath, 1943a, b; Heath and Mathur, 1944; Holdsworth and Heath, 1950; Rabinowitch, 1985, 1990a; Krontal *et al.*, 2000).

In tropical shallots grown at high temperatures (29/21°C, day/night), normal bloom was evident only in plants from bulbs stored at 5°C, while those from bulbs stored at 10, 20 and 30°C had shrivelled scapes. When grown at 17/9°C, the first to bloom were plants from bulbs stored at 10°C, followed by those stored at 5, 20 and 30°C (Krontal *et al.*, 2000; Rabinowitch and Kamenetsky, Chapter 17, this volume).

7.2 Garlic

Storage at low temperatures (from –2 to 9°C) and growth at mild temperatures (from 17 to 23°C during the day and from 9 to 15°C at night) promote early scape emergence and elongation. In bolting types, day length in the field plays a dominant role in the promotion of scape elongation (Takagi, 1990; Kamenetsky and Rabinowitch, 2001).

7.3 Ornamental species (subgenus *Allium*)

During growth and development, *A. ampeloprasum* (a domesticated long-scape cut flower, selected from plants growing wild in Israel) and *A. sphaerocephalon* require intermediate temperatures (17–20°C) and long days for normal scape elongation and flowering (Berghoef and Zevenbergen, 1992; De Hertogh and Zimmer, 1993; Maeda *et al.*, 1994). Under high growth temperatures and short days, the plants remain vegetative and do not bloom (*A. sphaerocephalon*) (Berghoef and Zevenbergen, 1992). Storage temperatures affect floral initiation and flowering percentage but do not influence scape emergence and bloom. Autumn storage at 2, 5 or 9°C reduced the percentage of flowering plants and resulted in inferior flower quality (*A. caeruleum*) (van Leeuwen and van der Weijden, 1994).

7.4 Ornamental species (subgenus *Amerallium*)

Storage temperatures of 9–17°C, followed by mild temperatures of 10–20°C during growth, enhance stem elongation. Storage at lower temperatures (2–5°C) or growth at temperatures higher than 20°C accelerated flowering but also resulted in a low percentage of flowering plants and short scapes (Maeda *et al.*, 1994; van Leeuwen and van der Weijden, 1994; Kodaira *et al.*, 1996).

7.5 Ornamental species (subgenus *Melanocrommyum*)

As in other geophytes from the Irano-Turanian region (e.g. tulip), *Allium* species require a long cold exposure for stem elongation, normal flowering and initiation of the renewal bud(s). Moderate growth temperatures (17–23°C during the day and 9–15°C at night) also promote scape elongation (Dosser, 1980; Zimmer and Renken, 1984; De Hertogh and Zimmer, 1993; Zemah *et al.*, 1999, 2001). However, day length has no effect on scape elongation in *A. aflatunense* (= *A. hollandicum*) (Zemah *et al.*, 2001).

A few exceptions are the *Melanocrommyum* species from the Mediterranean basin, such as *A. rothii*, *A. aschersonianum* and *A. nigrum*, which flower without post-differentiation cold treatment, possibly due to adaptation to local climatic conditions.

8. Concluding Remarks

In most plants, flowering plays an essential role in the perpetuation of the species, including the majority of the *Allium* spp. This is particularly evident in ornamentals, where flowers are the final product, but it is also true in edible crops. Understanding of the flowering processes, including the developmental biology, physiology and genetics of the reproductive organs, improves our knowledge of one of the most important processes in nature. The added value from studies of this topic results from increased efficiency in forcing, blooming and shelf-life of ornamental species, in induction of flowering for breeding and seed production and/or the prevention of undesired bolting in all crops. However, regardless of numerous works on the flowering of geophytes (for reviews, see Hartsema, 1961; Halevy, 1985, 1990; Rabinowitch, 1985, 1990a; Rees, 1992; Le Nard and De Hertogh, 1993), we know little of the basic chain of processes which, if successful, ends in normal flowering.

For ornamentals and edible species, florogenesis studies focus on two major objectives: (i) timing of flowering; and (ii) prevention of flowering. When put into practice, manipulation of earliness and lateness allows for year-round production, while the prevention of flowering (including flower bud/scape abortion) facilitates vegetative propagation and bulb production, which may be essential for clonal production.

A gene coding for flowering in *Arabidopsis* has recently been identified (Samech *et al.*, 2000). This discovery may stimulate similar studies in other plant species and in alliums. However, little is known about the endogenous changes during flower induction and initiation, including hormonal balance and hormone functions, from dormancy release to anthesis, as well as gene and protein expression (genomics and proteomics). Molecular markers for the various developmental phases are urgently needed (Le Nard and De Hertogh, 2000).

The role of physiological age and that of the size of critical mass in relation to flowering are of paramount importance for the ornamental industry and for seed production. The wide range of critical sizes found in *Allium* species (Brewster, 1994; Kamenetsky *et al.*, 2000) indicates that, while energy balance may provide one explanation for the plant's state of readiness for floral induction, it may not be the only one. Better understanding of the role of juvenile phase/plant age in flowering should eventually enable us to shorten breeding cycles and reduce production costs.

Apomixis has been demonstrated in *A. odorum* (= *A. ramosum*) (Modilewski, 1930; Hakanson and Levan, 1957) and in *A. tuberosum* (Kojima and Kawaguchi, 1989; Kojima *et al.*, 1991; Bohanec, Chapter 7, this volume). The trait is of high value for clonal propagation of new selections, especially of ornamentals with low rates of vegetative propagation, as well as for maintaining male-sterile lines. On the other hand, it greatly interferes with genetic studies and breeding. Hence, intimate knowledge of the apomixis mechanism and the means of switching it on and off will have a great importance in the future. Likewise, the production of topsets is common in alliaceous crops, such as garlic and great-headed garlic (Jones and Mann, 1963), and occurs infrequently in other alliums, such as bulb onion and leek. Better understanding of the control mechanism leading to the conversion of the umbel from generative to vegetative and vice versa could serve similar ends, though the generative process is to be preferred, due to the biotic cleansing that is associated with the production of true seed.

Male sterility is important for hybrid seed production and for extended vase-life of ornamentals. Identification of male-sterile genotypes in many other alliums could be of high importance to both industries (seed production and floriculture). The understanding of genetic make-up or the introgression of simply controlled mechanisms encoding for male sterility could improve our capabilities in breeding, production and product handling. Hence, cytoplasmic male sterility, if introduced in ornamentals, could facilitate the production of hybrids with sterile flowers and long flower life.

In *Allium* spp., the genetics of most important flowering traits is unknown. With regard to flowering, male sterility (*Ms*, *T*) (for details see Havey, Chapter 3, this volume) and *dw* are the only known genes in bulb onion (Rabinowitch *et al.*, 1984; Horobin, 1986; Friedlander, 1988). There is no information on genetic regulation of umbel size, flower colour, length of bloom and odour, just to name a few traits, in any *Allium* spp., nor do we have any knowledge of the genetic control of the five stages of flower development or of the genetic × environment interactions.

Undoubtedly, of the biological sciences, genetics is emerging as the leading discipline in the 21st century. It is expected that, with the new molecular (Havey, Chapter 3, and Kik, Chapter 4, this volume) and physiological (Kik, Chapter 4, and Bohanec, Chapter 7, this volume) tools, breakthroughs in the genetics of *Allium* spp. will enable the crossing of species barriers, the managing of economically important traits and the improvement of our capabilities in controlling blooming in alliums and other plant species.

References

Ali, M., Dowker, B.D., Currah, L. and Mumford, P.M. (1984) Floral biology and pollen viability of parental lines of onion. *Annals of Applied Biology* 104, 167–174.

Aoba, T. (1960) The influence of the storage temperature for onion bulbs on their seed production. *Journal of the Japanese Society for Horticultural Science* 29, 135–141 (in Japanese).

Berghoef, J. and Zevenbergen, A.P. (1992) Effects of environmental conditions on flower initiation and development of *Allium sphaerocephalon* L. *Acta Horticulturae* 325, 91–96.

Bernier, G., Havelange, A., Houssa, C., Petitjean, A. and Lejeune, P. (1993) Physiological signals that induce flowering. *The Plant Cell* 5, 1147–1155.

Berninger, E. (1965) Contribution à l'étude de la stérilité mâle de l'oignon (*Allium cepa* L.). *Annales de l'Amélioration des Plantes* 15, 183–199.

Brewster, J.L. (1985) The influence of seedling size and carbohydrate status of photon flux density during vernalization on inflorescence initiation in onion (*Allium cepa* L.). *Annals of Applied Botany* 55, 403–414.

Brewster, J.L. (1990) Physiology of crop growth and bulbing. In: Rabinowitch, H.D. and Brewster, J.L. (eds) *Onions and Allied Crops, I. Botany, Physiology, and Genetics*. CRC Press, Boca Raton, Florida, pp. 53–88.

Brewster, J.L. (1994) *Onions and Other Vegetable Alliums*. CAB International, Wallingford, UK, 236 pp.

Chang, W.N. and Struckmeyer, B.E. (1976) The influence of temperature, time of day, and flower age on pollen germination, stigma receptivity, pollen tube growth, and fruit set of *Allium cepa* L. *Journal of the American Society for Horticultural Science* 101, 81–83.

Cheremushkina, V.A. (1985) Osobennosty ritma sezonnogo razvitija i varianty malogo jiznennogo zikla kornevishnikh lukov (Seasonal development rhythm and variants of the minor life cycle in rhizomatous onions). *Bjulleten Moskovskogo Obshchestva Ispitatelei Prirody* 90(4), 96–106.

Cheremushkina, V.A. (1992) Evolution of life forms of species in subgenus *Rhizirideum* (Koch) Wendelbo, genus *Allium* L. In: Hanelt, P., Hammer, K. and Knüpffer, H. (eds) *The Genus* Allium *– Taxonomic Problems and Genetic Resources. Proceedings of an International Symposium held at Gatersleben, 11–13 June 1991*. IPK, Gatersleben, Germany, pp. 27–34.

Currah, L. (1990) Pollination biology. In: Rabinowitch, H.D. and Brewster, J.L. (eds) *Onions and Allied Crops, I. Botany, Physiology, and Genetics*. CRC Press, Boca Raton, Florida, pp. 135–149.

Currah, L. and Ockendon, D.J. (1978) Protandry and the sequence of flower opening in the onion (*Allium cepa* L.). *New Phytologist* 81, 419–428.

Currah, L. and Proctor, F.J. (1990) *Onions in Tropical Regions*. Bulletin 35, Natural Resources Institute, Chatham, UK, 232 pp.

De Hertogh, A.A. and Zimmer, K. (1993) *Allium* – ornamental species. In: De Hertogh, A.A. and Le Nard, M. (eds) *The Physiology of Flower Bulbs*. Elsevier, Amsterdam, pp. 187–200.

De Mason, D.A. (1990) Morphology and anatomy of *Allium*. In: Rabinowitch, H.D. and Brewster, J.L. (eds) *Onions and Allied Crops, I. Botany, Physiology, and Genetics*. CRC Press, Boca Raton, Florida, pp. 27–51.

Dosser, A. (1980) *Allium karataviense* – A lonely little onion in a petunia patch. *North Carolina Flower Growers Bulletin* 24, 11–12.

Dowker, B.D. (1990) Onion breeding. In: Rabinowitch, H.D. and Brewster, J.L. (eds) *Onions and Allied Crops, I. Botany, Physiology, and Genetics*. CRC Press, Boca Raton, Florida, pp. 215–232.

Engelke, T. and Tatlioglu, T. (2000a) The *wi* gene causes genic male sterility in *Allium schoenoprasum*. *Plant Breeding* 119, 325–328.

Engelke, T. and Tatlioglu, T. (2000b) Mitochondrial genome diversity in connection with male sterility in *Allium schoenoprasum* L. *Theoretical and Applied Genetics* 100, 942–948.

Engelke, T. and Tatlioglu, T. (2000c) Genetic analysis supported by molecular methods provide evidence of a new genic (*st1*) and new cytoplasmic (*st2*) male sterility in *Allium schoenoprasum* L. *Theoretical and Applied Genetics* 101, 478–486.

Esau, K. (1965) *Plant Anatomy*, 2nd edn. John Wiley & Sons, New York, 767 pp.

Etoh, T. (1985) Studies on the sterility in garlic, *Allium sativum* L. *Memoirs of the Faculty of Agriculture, Kagoshima University* 21, 77–132.

Etoh, T., Noma, Y., Nishitarumizu, Y. and Wakomoto, T. (1988) Seed productivity and germinability of various garlic clones collected in Soviet Central Asia. *Memoirs of the Faculty of Agriculture, Kagoshima University* 24, 129–139.

Friedlander, B. (1988) Characterization of dwarf scape in onion (*Allium cepa* L.): physiology, anatomy and genetics. MSc thesis, Faculty of Agriculture, The Hebrew University of Jerusalem, Rehovot, Israel.

Fritsch, R. (1992) Septal nectaries in the genus *Allium* – shape, position and excretory canals. In: Hanelt, P., Hammer, K. and Knüpffer, H. (eds) *The Genus* Allium *– Taxonomic Problems and Genetic Resources. Proceedings of an International Symposium held at Gatersleben, 11–13 June 1991.* IPK, Gatersleben, Germany, pp. 77–85.

Fritsch, R. (1993) Anatomische Merkmale des Blütenschaftes in der Gattung *Allium* L. und ihre systematische Bedeutung. *Botanische Jahrbücher* 115, 97–131.

Gregory, F.G. (1936) The effect of length of day on the flowering of plants. *Scientific Horticulture* 4, 143–154.

Gvaladze, G.E. (1961) The embryology of the genus *Allium* L. *Bulletin of the Academy of Sciences of the Georgian SSR* 26, 193–200 (in Russian).

Hakanson, A. and Levan, A. (1957) Endo-duplicational meiosis in *Allium odorum*. *Hereditas* 43, 179–200.

Halevy, A.H. (ed.) (1985) *Handbook of Flowering*. CRC Press, Boca Raton, Florida, six vols.

Halevy, A.H. (1990) Recent advances in control of flowering and growth habit of geophytes. *Acta Horticulturae* 266, 35–42.

Hanelt, P. (1990) Taxonomy, evolution and history. In: Rabinowitch, H.D. and Brewster, J.L. (eds) *Onions and Allied Crops, I. Botany, Physiology, and Genetics*. CRC Press, Boca Raton, Florida, pp. 1–26.

Hanelt, P., Schultze-Motel, J., Fritsch, R., Kruse, J., Maaß, H.I., Ohle, H. and Pistrick, K. (1992) Infrageneric grouping of *Allium* – the Gatersleben approach. In: Hanelt, P., Hammer, K. and Knüpffer, H. (eds) *The Genus* Allium *– Taxonomic Problems and Genetic Resources. Proceedings of an International Symposium held at Gatersleben, 11–13 June 1991.* IPK, Gatersleben, Germany, pp. 107–123.

Hartsema, A.M. (1961) Influence of temperatures on flower formation and flowering of bulbous and tuberous plants. In: Ruhland, W. (ed.) *Handbuch der Pflanzenphysiologie*, Vol.16. Springer-Verlag, Berlin, pp. 123–167.

Havey, M.J. (1995) Cytoplasmic determinations using the polymerase chain reaction to aid in the extraction of maintainer lines from open-pollinated populations of onion. *Theoretical and Applied Genetics* 90, 263–268.

Havey, M.J. (2000) Diversity among male-sterility-inducing and male-fertile cytoplasms of onion. *Theoretical and Applied Genetics* 101, 778–782.

Heath, O.V.S. (1943a) Studies in the physiology of the onion plant. I. An investigation of factors concerned in the flowering ('bolting') of onion grown from sets and its prevention. Part I. Production and storage of onion sets, and field results. *Annals of Applied Biology* 30, 208–220.

Heath, O.V.S. (1943b) Studies in the physiology of the onion plant. I. An investigation of factors concerned in the flowering ('bolting') of onions grown from sets and its prevention. Part II. Effects of day length and temperature on onion grown from sets, and general discussion. *Annals of Applied Biology* 30, 308–319.

Heath, O.V.S. (1945) Formative effect of environmental factors as exemplified in the development of the onion plant. *Nature (London)* 155, 623–626.

Heath, O.V.S. and Mathur, P.B. (1944) Studies in the physiology of the onion plant. II. Inflorescence initiation and development, and other changes in the internal morphology of onion sets, as influenced by temperature and day length. *Annals of Applied Biology* 31, 173–187.

Holdsworth, M. and Heath, O.V.S. (1950) Studies in the physiology of the onion plant. IV. The influence of day-length and temperature on the flowering of the onion plant. *Journal of Experimental Botany* 1, 353–375.

Horobin, J.F. (1986) Inheritance of a dwarf seed stalk character in onion. *Annals of Applied Biology* 108, 199–204.

Inden, H. and Asahira, T. (1990) Japanese bunching onion (*Allium fistulosum* L.). In: Brewster, J.L and Rabinowitch, H.D. (eds) *Onions and Allied Crops, III. Biochemistry, Food Science, and Minor Crops*. CRC Press, Boca Raton, Florida, pp. 159–178.

Ito, K. (1956) Studies on the bolting of the onion. I. Relation between flower-bud formation and bulb division. *Journal of the Japanese Society for Horticultural Science* 25, 187–193.

Jenderek, M.M. and Hannan, R.M. (2001) Seed producing ability of garlic (*Allium sativum*) clones from two public US collections. In: *Proceedings of the Third International Symposium on Edible Alliaceae, University of Georgia, Athens, Georgia, USA, 29 October–3 November 2000*. University of Georgia, Athens, Georgia, pp. 73–75.

Jones, H.A. (1927) The influence of storage temperature on seed production in the Ebenezer onion. *Proceedings of the American Society for Horticultural Science* 24, 62–63.

Jones, H.A. and Clarke, A.E. (1943) Inheritance of male sterility in the onion and the production of hybrid seeds. *Proceedings of the American Society for Horticultural Science* 43, 189–194.

Jones, H.A. and Emsweller, S.L. (1933) Methods of breeding onions. *Hilgardia* 7, 625–642.

Jones, H.A. and Emsweller, S.L. (1936) Development of the flower and macrogametophyte of *Allium cepa*. *Hilgardia* 10, 415–428.

Jones, H.A. and Mann, L.K. (1963) *Onions and Their Allies. Botany, Cultivation and Utilization*. Interscience Publishers, New York, 286 pp.

Jones, H.A. and Rosa, J.T. (1928) *Allium*. In: *Truck Crop Plants*. McGraw-Hill, New York, pp. 37–63.

Kamenetsky, R. (1992) Morphological types and root system as indicators of evolutionary pathways in the genus *Allium*. In: Hanelt, P., Hammer, K. and Knüpffer, H. (eds) *The Genus* Allium *– Taxonomic Problems and Genetic Resources. Proceedings of an International Symposium held at Gatersleben, 11–13 June 1991*. IPK, Gatersleben, Germany, pp. 129–135.

Kamenetsky, R. (1994) Life cycle, flower initiation and propagation of the desert geophyte *Allium rothii*. *International Journal of Plant Science* 155, 597–605.

Kamenetsky, R. (1996a) Life cycle and morphological features of *Allium* L. species in connection with geographical distribution. *Bocconea* 5, 251–257.

Kamenetsky, R. (1996b) *Creation of a Living Collection of the Genus* Allium, *their Development and Possible Uses as New Ornamental Plants*. Final Report, Chief Scientist of the Ministry of Agriculture, Grant No. 256–0413 (1994–1996), Bet Dagan, Israel, 41 pp.

Kamenetsky, R. (1997) Inflorescence of *Allium* species (subgenus *Melanocrommyum*): structure and development. *Acta Horticulturae* 430, 141–146.

Kamenetsky, R. and Japarova, N. (1997) Relationship between annual cycle and floral development of three *Allium* species from subgenus *Melanocrommyum*. *Journal of Arid Environments* 35, 473–485.

Kamenetsky, R. and Rabinowitch, H.D. (2001) Floral development in bolting garlic. *Sexual Plant Reproduction* 13, 235–241.

Kamenetsky, R., Gilad, Z. and Rabinowitch, E. (2000) *Development of* A. aschersonianum *from Israeli Flora as New Ornamental Crop for Cut Flower and Bulb Production*. Final Report, The Foundation of the Chief Scientist of the Ministry of Agriculture of Israel, Bet Dagan, 22 pp. (in Hebrew, with English summary).

Kodaira, E., Mori., G., Takeuchi, M. and Imanishi, H. (1996) Effects of temperature on the growth and flowering of *Allium unifolium* Kellogg. *Journal of the Japanese Society for Horticultural Science* 65, 373–380.

Kojima, A. and Kawaguchi, T. (1989) Apomictic nature of Chinese chives (*Allium tuberosum* Rottl.) detected in unpollinated ovule culture. *Japanese Journal of Breeding* 41, 73–83.

Kojima, A., Nagato, Y. and Hinata, K. (1991) Degree of apomixis in Chinese chives (*Allium tuberosum*) estimated by esterase isozyme analysis. *Plant Breeding* 104, 177–183.

Konvicka, O. (1973) The causes of sterility in *Allium sativum* L. *Biologia Plantarum (Praha)* 15, 144–149 (in Czech).

Konvicka, O. (1984) Generative reproduction of garlic (*Allium sativum*). *Allium Newsletter* 1, 28–37.

Koul, A.K. and Gohil, R.N. (1970) Causes averting sexual reproduction in *Allium sativum*. Linn. *Cytologia* 35, 197–202.

Krontal, Y., Kamenetsky, R. and Rabinowitch, H.D. (1998) Lateral development and florogenesis of a tropical shallot – a comparison with bulb onion. *International Journal of Plant Science* 159, 57–64.

Krontal, Y., Kamenetsky, R. and Rabinowitch, H.D. (2000) Flowering physiology and some vegetative traits of short-day shallot – a comparison with bulb onion. *Journal of Horticultural Science and Biotechnology* 75, 35–41.

Kruse, J. (1992) Growth form characters and their variation in *Allium* L. In: Hanelt, P., Hammer, K. and Knüpffer, H. (eds) *The Genus* Allium *– Taxonomic Problems and Genetic Resources. Proceedings of an International Symposium held at Gatersleben, 11–13 June 1991*. IPK, Gatersleben, Germany, pp. 173–179.

Lachman, W.H. and Michelson, L.F. (1960) Effect of warm storage on the bolting of onions grown from sets. *Proceedings of the American Society for Horticultural Science* 75, 495–499.

Le Nard, M. and De Hertogh, A.A. (1993) Bulb growth and development and flowering. In: De Hertogh, A.A. and Le Nard, M. (eds) *The Physiology of Flower Bulbs*. Elsevier, Amsterdam, pp. 29–44.

Le Nard, M. and De Hertogh, A.A. (2000) Growth, development and flowering: research needs for flower bulbs (geophytes). In: *Abstracts of the VIIIth International Symposium on Flower Bulbs, 28–31 August 2000, Kirstenbosch, Cape Town, South Africa*, p. 21.

Lin, M.W. and Chang, W.N. (1980) Interspecific hybridization in the genus *Allium*. I. Effect of different temperatures on bolting of Japanese bunching onion (*Allium fistulosum* L.) *Chinese Horticulture* 26, 173–179 (in Chinese).

Maeda, M., Dubouzet, J.G., Arisumi, K.I., Etoh, T. and Sakata, Y. (1994) Effects of cold storage and staggered planting in forcing culture of spring-flowering *Allium* species. *Journal of the Japanese Society for Horticultural Science* 63, 629–638.

Mann, L.K. (1959) The *Allium* inflorescence: some species of the section *Molium*. *American Journal of Botany* 46, 730–739.

Messiaen, C.M., Cohat, J., Leroux, J.P., Pichon, M. and Beyries, A. (1993) *Les Allium Alimentaires Reproduits par Voie Végétative*. INRA, Paris, 228 pp.

Modilewski, J. (1930) Neue Beitrage zur Polyembryonie von *Allium odorum*. *Berichte der Deutschen Botanischen Gesellschaft* 48, 285–294.

Moll, R.H. (1954) Receptivity of the individual onion flower and some factors affecting its duration. *Proceedings of the American Society for Horticultural Science* 64, 399–404.

Moue, T. and Uehara, T. (1985) Inheritance of cytoplasmic male sterility in *Allium fistulosum* L. (Welsh onion). *Journal of the Japanese Society for Horticultural Science* 53, 432–437.

Novak, F.J. (1972) Tapetal development in the anthesis of *Allium sativum* L. and *Allium longicuspis* Regel. *Experientia* 28, 1380–1381.

Pastor, J. and Valdes, B. (1985) Bulb structure in some species of *Allium* (*Liliaceae*) of the Iberian Peninsula. *Annales Musei Goulandris* 7, 249–261.

Pistrick, K. (1992) Phenological variability in the genus *Allium* L. In: Hanelt, P., Hammer, K. and Knüpffer, H. (eds) *The Genus* Allium *– Taxonomic Problems and Genetic Resources. Proceedings of an International Symposium held at Gatersleben, 11–13 June 1991*. IPK, Gatersleben, Germany, pp. 243–249.

Pooler, M.R. and Simon, P.W. (1994) True seed production in garlic. *Sexual Plant Reproduction* 7, 282–286.

Poulsen, N. (1990) Chives *Allium schoenoprasum* L. In: Brewster, J.L. and Rabinowitch, H.D. (eds) *Onions and Allied Crops, III. Biochemistry, Food Science, and Minor Crops*. CRC Press, Boca Raton, Florida, pp. 231–250.

Rabinowitch, H.D. (1985) Onions and other edible *Alliums*. In: Halevy, A.H. (ed.) *Handbook of Flowering*. CRC Press, Boca Raton, Florida, pp. 398–409.

Rabinowitch, H.D. (1990a) Physiology of flowering. In: Rabinowitch, H.D. and Brewster, J.L. (eds) *Onions and Allied Crops, I. Botany, Physiology, and Genetics*. CRC Press, Boca Raton, Florida, pp. 113–134.

Rabinowitch, H.D. (1990b) Seed development. In: Rabinowitch, H.D. and Brewster, J.L. (eds) *Onions and Allied Crops, I. Botany, Physiology, and Genetics*. CRC Press, Boca Raton, Florida, pp. 151–159.

Rabinowitch, H.D., Friedlander, B. and Peters, R. (1984) Inheritance and characterization of dwarf scape in onions: a progress report. In: *Proceedings of 3rd Eucarpia* Allium *Symposium*, Wageningen, The Netherlands, pp. 83–89.

Rees, A.R. (1992) *Ornamental Bulbs, Corms and Tubers*. CAB International, Wallingford, UK, 220 pp.

Saito, S. (1990) Chinese chives *Allium tuberosum* Rottl. In: Brewster, J.L. and Rabinowitch, H.D. (eds) *Onions and Allied Crops, III. Biochemistry, Food Science, and Minor Crops*. CRC Press, Boca Raton, Florida, pp. 219–230.

Samech, A., Onouchi, H., Gold, S.E., Ditta, G.S., Schwartz-Sommer, Z., Yankofsky, M. and Coupland, G. (2000) Distinct roles of CONSTANS target genes in reproductive development of *Arabidopsis*. *Science* 288, 1613–1616.

Schweisguth, B. (1970) Études préliminaires à l'amélioration du poireau *A. porrum* L. Proposition d'une méthode d'amélioration. *Annales de l'Amélioration des Plantes* 20, 215–231.

Schweisguth, B. (1973) Étude d'un nouveau type de stérilité mâle chez l'oignon, *Allium cepa* L. *Annales de l'Amélioration des Plantes* 23, 221–233.

Shishido, Y. and Saito, T. (1976) Studies on the flower bud formation in onion plants. II. Effect of physiological conditions on the low temperature induction of flower buds on green plants. *Journal of the Japanese Society for Horticultural Science* 45, 160–167.

Sinnadurai, S. (1970a) The effect of light and temperature on onions. *Ghana Journal of Agricultural Science* 3, 13–15.

Sinnadurai, S. (1970b) A note on the bulbing and flowering habit of the Bawku onion. *Tropical Agriculture (Trinidad)* 47, 77–79.

Smith, B.M. (1994) *Annual Report for 1993–1994*. Horticulture Research International, Wellesbourne, UK, p. 13.

Smith, B.M. and Crowther, T.C. (1995) Inbreeding depression and single cross hybrids in leeks (*Allium ampeloprasum* ssp. *porrum*). *Euphytica* 86, 87–94.

Takagi, H. (1990) Garlic *Allium sativum* L. In: Brewster, J.L. and Rabinowitch, H.D. (eds) *Onions and Allied Crops, III. Biochemistry, Food Science, and Minor Crops*. CRC Press, Boca Raton, Florida, pp. 109–146.

Tatlioglu, T. (1982) Cytoplasmic male sterility in chives (*Allium schoenoprasum* L.). *Zeitschrift für Pflanzenzüchtung* 89, 251–262.

Tatlioglu, T. (1985) Influence of temperature on the expression of male sterility in chives (*Allium schoenoprasum* L.). *Zeitschrift für Pflazenzüchtung* 94, 156–161.

Tatlioglu, T. (1987) Genetic control of temperature-sensitivity of cytoplasmic male sterility (*cms*) sterility in chives (*Allium schoenoprasum* L.). *Plant Breeding* 99, 65–76.

Tatlioglu, T. and Wricke, G. (1988) Genetic control of tetracycline sensitivity of cytoplasmic male sterility (*cms*) in chives (*Allium schoenoprasum* L.). *Plant Breeding* 100, 34–40.

Thomas, T.H. (1972) Stimulation of onion bulblet production by N^6-benzyladenine. *Horticultural Research* 12, 77–79.

Thompson, H.C. and Smith, O. (1938) *Seedstalk and Bulb Development in the Onion (*Allium cepa *L.)*. Bulletin of Cornell University Agricultural Experiment Station, No. 708, Ithaca, New York, 21 pp.

Tindall, H.D. (1983) *Vegetables in the Tropics*. Macmillan, London, pp. 14–35.

Toyama, M. and Wakamiya, I. (1990) Rakkyo (*Allium chinense*) G. Don. In: Brewster, J.L. and Rabinowitch, H.D. (eds) *Onions and Allied Crops, III. Biochemistry, Food Science, and Minor Crops*. CRC Press, Boca Raton, Florida, pp. 197–218.

van Beekom, C.W.C. (1953) *Uien en Sjaloten*. Mededelingen Tuinbouw Voorlichting, Vol. 49, 132 pp. (in Dutch, with English summary).

van Kampen, J. (1970) Shortening the breeding cycle in onions. PhD thesis, Wageningen Agricultural University, Wageningen, The Netherlands, 69 pp. (in Dutch).

van Leeuwen, P.J. and van der Weijden, J.A. (1994) Vervroegen beperkt mogelijk voor enkele soorten. *Vakblad voor de Bloemisterij* 29, 28–29.

van der Meer, Q.P. and Hanelt, P. (1990) Leek (*Allium ampeloprasum*). In: Brewster, J.L. and Rabinowitch, H.D. (eds) *Onions and Allied Crops, III. Biochemistry, Food Science, and Minor Crops*. CRC Press, Boca Raton, Florida, pp. 179–196.

Vvedensky, A. (1968) Genus *Allium* L. In: Komarov, V.L. (ed.) *Flora of the USSR*, Vol. IV. Translation from Russian. Israel Program for Scientific Translations, Jerusalem, pp. 141–280.

Waller, G.D. and Martin, J.H. (1978) Fluorescence for identification of onion nectar in foraging honeybees. *Environmental Entomology* 7, 766–768.

Waller, G.D., Carpenter, E.W. and Ziehl, O.A. (1972) Potassium in onion nectar and its probable effect on attractiveness of onion flowers to honeybees. *Journal of the American Society for Horticultural Science* 97, 535–539.

Watanabe, H. (1955) Studies on the flower bud differentiation and bolting of Welsh onion varieties. *Studies of the Institute of Horticulture, Kyoto University* 7, 101–108.

Weber, E. (1929) Entwicklungsgeschichtliche untersuchungen über die Gattung *Allium*. *Botanisches Archiv* 25, 1–44.

Woodbury, G.W. (1950) *A Study of Factors Influencing Floral Initiation and Seedstalk Development in Onion (*Allium cepa *Linn.)*. Agricultural Experiment Station of the University of Idaho, Moscow, Idaho, Research Bulletin 18, 27 pp.

Yakura, T. and Okimizu, S. (1969) Studies on flowering in the genus *Allium*. I. Relationship between temperature and photoperiod, and flower bud initiation, bolting and flowering. *Agriculture and Horticulture* 44, 1131–1132 (in Japanese).
Yamasaki, A., Tanaka, K., Yoshida, M. and Miura, H. (2000a) Effects of day and night temperatures on flower-bud formation and bolting of Japanese bunching onion (*Allium fistulosum* L.). *Journal of the Japanese Society for Horticultural Science* 69, 40–46.
Yamasaki, A., Miura, H. and Tanaka, K. (2000b) Effect of photoperiods before, during and after vernalization on flower initiation and development and its varietal difference in Japanese bunching onion (*Allium fistulosum* L.). *Journal of Horticultural Science and Biotechnology* 75, 645–650.
Zemah, H., Bendel, P., Rabinowitch, H.D. and Kamenetsky, R. (1999) Visualization of morphological structure and water status during storage of *Allium aflatunense* bulbs by NMR imaging. *Plant Science* 147, 65–73.
Zemah, H., Rabinowitch, H.D. and Kamenetsky, R. (2001) Florogenesis and the effect of temperatures on the development of *Allium aflatunense*. *Journal of Horticultural Science and Biotechnology* 76, 507–513.
Zimmer, K. and Renken, M. (1984) Untersuchungen an *Allium aflatunense*. *Deutscher Gartenbau* 38, 2004–2008.

3 Genome Organization in *Allium*

M.J. Havey
Agricultural Research Service – USDA, Department of Horticulture, 1575 Linden Drive, University of Wisconsin, Madison, WI 53706, USA

1. Introduction

Our understanding of the structure, transmission and diversity among the plant genomes has steadily increased over the last 100 years. The beginning of the 20th century saw the rediscovery of Mendel's work in pea (*Pisum sativum* L.) and his laws of inheritance. Since that time, plants have been useful model organisms for studies on chromosome morphologies, aneuploidy, polyploidy, maternal transmission of phenotypes and transposable elements. With the advent of molecular biology in the 1970s and 1980s, the ability to directly analyse DNA and to clone specific genes substantially increased our understanding of gene function and regulation. Representative higher-plant chloroplast (Shinozaki *et al.*, 1986) and mitochondrial (Unseld *et al.*, 1997) genomes were completely sequenced during the 1980s and 1990s. The 20th century closed with the publishing of the complete sequence of the smaller nuclear chromosomes of *Arabidopsis thaliana* L. (Lin *et al.*, 1999). Coinciding with the steady increase in knowledge about the plant genomes throughout the 20th century, research on species within the genus *Allium* has revealed much about the structure of their chloroplast, mitochondrial and nuclear genomes. The goal of this chapter is to review the literature describing these genomes and to recognize their commonalities, as well as their uniqueness, as compared with other angiosperms.

2. The Nuclear Genome

2.1 Genetic architecture

Most of the cultivated alliums, e.g. bulb onion (*A. cepa* L.), Japanese bunching onion (*A. fistulosum* L.), leek (*A. ampeloprasum* L.), chives (*A. schoenoprasum* L.) and Chinese chives (*A. tuberosum* L.), are seed-propagated. Outcrossing is encouraged by both protandry (Currah and Ockendon, 1978) and the natural occurrence of cytoplasmic male sterility (Jones and Clarke, 1943; Berninger, 1965). Rates of self-pollination in seed fields of the cultivated alliums have been estimated at 5–25% (Berninger and Buret, 1967). Many generations of random outcrossing have probably had two major effects on the *Allium* nuclear genome. The first is that the relatively high heterozygosity, sustained by outcrossing, has allowed deleterious recessive alleles to be maintained in populations. The second is that populations should be at or near linkage equilibrium.

Regarding the frequency of deleterious alleles in *Allium* populations, Berninger and Buret (1967) scored the frequencies of chlorophyll deficiencies among self- and open-pollinated plants of diploid ($2n = 2x = 16$) bulb onion and tetraploid ($2n = 4x = 32$) leek. For onion, 20–30% of the tested plants were scored as heterozygous at a chlorophyll-deficiency locus. The authors estimated that approximately 20 chlorophyll-deficiency loci were polymorphic in the scored populations. The frequency of the deleterious recessive allele at any specific locus was low (0.01–0.04), but the numbers of chlorophyll-deficiency loci were high enough for the homozygous recessive genotype to appear frequently. In contrast, autotetraploid leek had fewer chlorophyll-deficiency loci (7–14), but frequencies of the deleterious alleles at these loci were over ten times those of onion. This high frequency would be expected because self-pollination would reveal recessive alleles only for plants simplex or duplex at the chlorophyll-deficiency loci.

Regarding the second effect of outcrossing, I know of no reports estimating linkage equilibrium in outcrossing *Allium* populations. However, unpublished data from my laboratory have revealed that two linked loci in onion, a restriction fragment length polymorphism (RFLP) located 0.9 centimorgans (cM) from the male-sterile (*Ms*) locus, are in linkage equilibrium in the open-pollinated onion populations, cvs 'Brigham Yellow Globe', 'Mountain Danvers' and 'Sapporo-Ki' (Gökçe, 2001).

Individual diploid plants have a maximum of two alleles per locus. However, greater numbers of alleles can be maintained within or among populations of diploid plants. Classical genetic studies revealed only two alleles at all morphological loci in onion (Cramer and Havey, 1999), except possibly at the loci conditioning red bulb colour (El-Shafie and Davis, 1967). Biochemical and molecular markers can reveal greater allelic diversity than morphological markers. Isozyme markers are relatively rare within cultivated populations of the bulb onion (Peffley and Orozco-Castillo, 1987), but occur more frequently among closely related *Allium* species (Peffley *et al.*, 1985; Peffley, 1986). Eickmeyer *et al.* (1990) studied segregations at 16 isozyme loci in chives, two of which had more than two alleles. Using 43 non-duplicated codominant RFLPs from a genetic map of bulb onion, King *et al.* (1998b) scored putative allelic diversity among a set of 14 inbred lines from diverse bulb-onion breeding programmes.

Segregation analyses had previously established known alleles at these loci. When both of these characterized alleles were absent and another uniquely sized fragment was present in an inbred line derived from a single S_0 plant (i.e. a maximum of two alleles can exist at each locus in a specific individual), this uniquely sized fragment was scored as an additional allele. This survey revealed that there were more than two alleles at 46% of the 43 studied RFLP loci (King *et al.*, 1998b). Hence, molecular markers can reveal greater allelic diversity than classical techniques in bulb onion.

Genetic maps are powerful tools for plant breeding and studies on plant-genome evolution (Tanksley, 1993). Two genetic maps of bulb onion have been developed. The first was from a cross within *A. cepa* and was primarily composed of RFLPs (King *et al.*, 1998a). The second was from an interspecific cross (*A. cepa* × *A. roylei* Stearn) and was composed primarily of amplified fragment length polymorphisms (AFLPs) (van Heusden *et al.*, 2000a).

Recent research has revealed synteny (conservation of genetic linkages) among the chromosomes of related crop plants (Devos and Gale, 1997). Sequence and linkage conservation are extensive within the *Poaceae* (Ahn *et al.*, 1993; Devos *et al.*, 1994; Dunford *et al.*, 1995) and the *Solanaceae* (Bonierbale *et al.*, 1988; Tanksley *et al.*, 1992). These studies demonstrate that speciation may be associated with chromosome rearrangements that shift blocks of linked loci (Bonierbale *et al.*, 1988; Tanksley *et al.*, 1988; Bennetzen and Freeling, 1993; Moore, 1995). Colinearity of linkages among evolutionarily distant species can aid in the genetic analyses and cloning of economically important loci. Nothing is known about the synteny among the cultivated alliums or among the alliums and other monocots, such as the grasses or asparagus. The genetic map of King *et al.* (1998a) is primarily based on RFLPs and may be useful for syntenic studies. Bradeen and Havey (1995) demonstrated that the complementary DNAs (cDNAs) (i.e. DNA synthesized from a messenger RNA (mRNA) molecule) revealing RFLPs in bulb onion cross-hybridize well to DNAs of other species in *Allium* section *Cepa*. It remains to be determined if hybridizations of bulb-onion cDNA can be used to reveal and map RFLPs in other cultivated alliums.

The large numbers of polymorphisms revealed by AFLPs makes map development possible for most crosses. AFLP maps will be less useful for syntenic studies among the cultivated alliums, but will be useful for the assignment of linkage groups to chromosomes and for map integration. Shigyo *et al.* (1996) developed a set of alien addition lines of *A. fistulosum* carrying a single bulb-onion chromosome that have proved useful for assigning morphological traits (Shigyo *et al.*, 1997a, b) and molecular markers (Shigyo *et al.*, 1997c; van Heusden *et al.*, 2000b) to chromosomes. Other markers based on the polymerase chain reaction (PCR), such as simple sequence repeats, will provide additional polymorphisms to the genetic map of the alliums (Fischer and Bachmann, 2000). Because seed propagation of garlic (*A. sativum* L.) is becoming a reality (Etoh, 1983; Pooler and Simon, 1994; Etoh and Simon, Chapter 5, this volume), the study of segregating families and development of a genetic map of this important species will become possible. A fascinating study would be to determine the syntenic relationships among the most economically important *Allium* species, such as bulb and Japanese bunching onions, garlic and leek.

2.2 Chromosome numbers and karyotypes

The majority of *Allium* species are indigenous to Eurasia and the Mediterranean basin, and over 90% of species from these areas have a basic chromosome number of eight (Ved Brat, 1965a). More than 95% of the North American *Allium* species have a basic chromosome number of seven; a few *Allium* species from Eurasia possess a basic chromosome number of nine (Ved Brat, 1965a). Most *Allium* species are diploid ($2n = 2x = 14$, 16 or 18). Polyploidy is less common, but occurs among botanical varieties of the cultivated forms *A. ampeloprasum* ($2n =$

4x = 32 or 6x = 48), *A. schoenoprasum* (2*n* = 4x = 32), *A. chinense* (2*n* = 4x = 32), and *A. tuberosum* (2*n* = 4x = 32), as well as in wild species such as *A. babingtonii* or *A. oreoprasum* (both with 48 chromosomes) (Ved Brat, 1965a). Triploids (2*n* = 3x= 24) exist in *A. schoenoprasum*, *A. chinense* (Jones, 1990) and *A. trifoliatum* Cyr. var. *sterile* Kollm., and both triploids and pentaploids (2*n* = 5x = 40) in *A. ampeloprasum* (Kollmann, 1971, 1972). Supernumerary (B) chromosomes have been documented in *A. ampeloprasum* (Khazanehdari and Jones, 1996), *A. schoenoprasum* (Bougourd and Parker, 1976), *A. paniculatum*, *A. cernuum* and *A. canadense* (Ved Brat, 1965a).

The formation of multivalents during meiosis among large metacentric chromosomes of *A. ampeloprasum* leek group is avoided by localization of chiasmata near the centromere (Koul and Gohil, 1970; Kollmann, 1972; Stack and Roelofs, 1996). Localized chiasmata also occur among diploid species, such as *A. fistulosum*, *A. kochii* and *A. cyathophorum* (Maeda, 1937; Ved Brat, 1965b). Chiasmata localized near the centromere of a diploid species could be explained if this region were largely euchromatic and therefore gene-rich. Fiskesjö (1975) observed that the terminal ends of *A. fistulosum* chromosomes were largely heterochromatic, and Villanueva-Mosqueda (1999) used genomic *in situ* hybridization (GISH) to reveal strong signal intensities at the ends of the chromosomes. If genes conditioning desirable traits in *A. fistulosum* (van der Meer and van Bennekom, 1978) are located near the centromere, transfer of these genes to another species, such as bulb onion, may be difficult. Usually interspecific hybrids show more terminal (rather than near the centromere) chiasmata (Maeda, 1937) – hence some of the difficulties in transferring euchromatic regions from Japanese bunching onion to bulb onion.

Almost all *Allium* species possess symmetrical median- to submedian-centromeric chromosomes with relatively small size differences, although some telocentric chromosomes are present in a few species (Ved Brat, 1965a). Karotypic analyses have revealed little variability among the chromosomes within a species (Ved Brat, 1965a); however, the size of the telomorphic heterochromatin was variable (El-Gadi and Elkington, 1975). The species in *Allium* section *Cepa* have the best-studied karyotypes. Saini and Davis (1970) observed that these species have very similar sizes, centromere locations and absence of knobs. *Allium cepa* and *A. fistulosum* differ for Giemsa C-banding (El-Gadi and Elkington, 1975; Fiskesjö, 1975), show non-typical bivalent or multivalent pairing during meiosis (Emsweller and Jones, 1935, 1945; Maeda, 1937), have chromosome rearrangements (Emsweller and Jones, 1938; Peffley, 1986; Peffley and Mangum, 1990; Cryder *et al.*, 1991) and differ by about 28% in amounts of DNA (Jones and Rees, 1968).

Telomeres are located at the ends of chromosomes and are required for stable maintenance and transmission of chromosomes. Almost all plants possess the *Arabidopsis*-type telomere as multimeric repeats of TTTAGGG (Fuchs *et al.*, 1995). However, some species of the *Alliaceae* and *Liliaceae* are unique because they do not possess the widely conserved *Arabidopsis*-type telomeric repeat (Fuchs *et al.*, 1995; Pich *et al.*, 1996a, b). Pich *et al.* (1996a) reported that a previously identified 375-base-pair (bp) guanosine and cytidine (GC)-rich satellite DNA (Barnes *et al.*, 1985) replaced the *Arabidopsis*-type telomere to stabilize the chromosome ends. The replacement of original telomeres by different repetitive sequences has also been documented in insects (reviewed by Pich *et al.*, 1996b).

3. DNA

3.1 DNA amounts

Plants possess nuclear genomes of hugely different sizes (Bennett and Smith, 1976). Polyploidy is common among angiosperms and can explain some genome size differences. Gymnosperms tend to have the largest nuclear genomes among diploid plants (Price, 1976). However, among diploid angiosperms, the huge differences in the amounts of nuclear DNA cannot be explained solely by polyploidy (Fig. 3.1). Great differences in genome size exist within the genus *Allium*. Ohri *et al.* (1998) documented these differences among the major *Allium* subgenera and presented an excellent treatise on the congruence of genome size with other taxonomic data. Bennett (1972, 1976) and Ohri *et al.* (1998) proposed that perennial species with long generation times and indigenous to temperate regions, typical of most alliums, tended to have larger genomes.

Onion is often used in the classroom for cytogenetic analyses, because it possesses relatively few, very large chromosomes, which directly reflect an enormous amount of nuclear DNA. The nuclear genome of onion contains 17.9 pg (Labani and Elkington, 1987) or 15,290 megabase pairs (Arumuganathan and Earle, 1991) of DNA per 1C nucleus, making it one of the largest genomes among cultivated plants (6, 16 and 107 times greater than maize (*Zea mays*), tomato (*Lycopersicon esculentum*) and *Arabidopsis thaliana*, respectively) (Fig. 3.1). Diploid onion contains as much DNA as hexaploid wheat (*Triticum aestivum* L.) and, on average, each onion chromosome carries an amount of DNA equal to 75% of the 1C content of the maize nuclear genome (Bennett and Smith, 1976). Molecular studies have revealed important characteristics of this extremely large genome. The GC content of onion DNA is 32%, the lowest known for any angiosperm (Kirk *et al.*, 1970; Stack and Comings, 1979). CsCl and Cs_2SO_4-Ag^+ density-gradient centrifugation revealed no significant satellite DNA bands, except for a 375-bp telomeric sequence representing 4% of the genome (Barnes *et al.*, 1985). Stack and Comings (1979) used reassociation kinetics to reveal three repetitive fractions in the bulb-onion genome. The first fraction represents 41% of the genome and is repeated approximately 21,600 times, fraction two comprises 36% of the genome and is repeated approximately 225 times and fraction three comprises 6% of the genome and

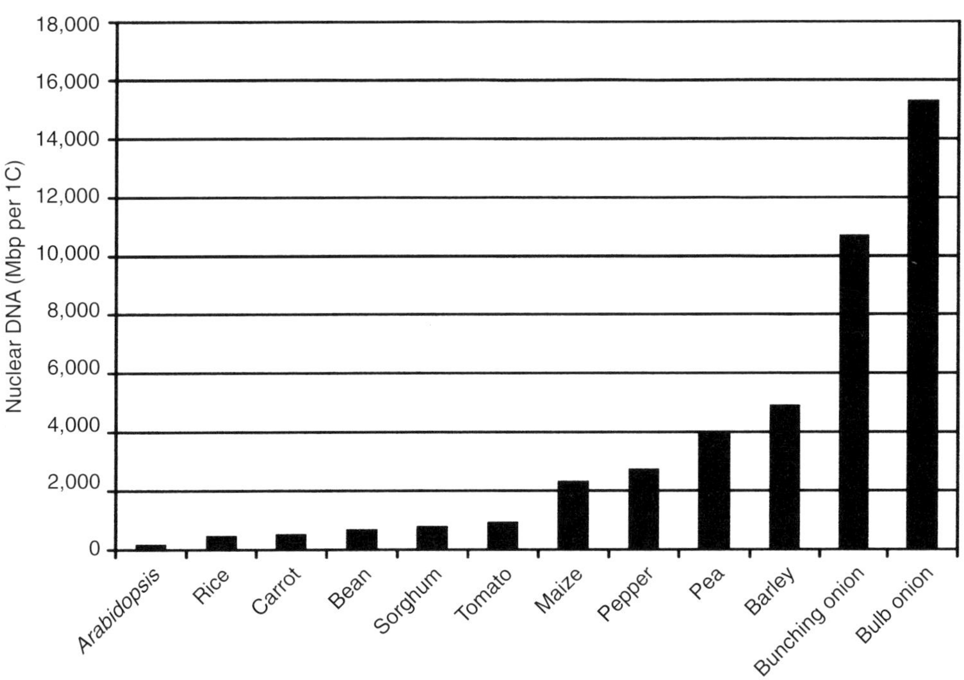

Fig. 3.1. Histogram showing the relative amounts of nuclear DNA (megabasepairs per 1C nucleus) for some major diploid ($2n = 2x$) crop species (DNA amounts were estimated by Arumuganathan and Earle, 1991).

consists of single-copy DNA. Approximately 10% of the DNA is not detectable by reassociation kinetics (Stack and Comings, 1979). The results of these studies indicate that the onion genome consists of middle-repetitive sequences occurring in short-period interspersions among single-copy regions (Stack and Comings, 1979).

Significant differences in chromosome sizes and nuclear DNA content have evolved among closely related *Allium* species. For example, a close phylogenetic relationship between bulb onion and Japanese bunching onion is supported by karyotype and heterochromatic banding (Vosa, 1976; Narayan, 1988), crossability (van Raamsdonk *et al.*, 1992) and shared mutations in the chloroplast and nuclear 45s ribosomal DNAs (Havey, 1992a). However, bulb onion has approximately 28% more nuclear DNA than *A. fistulosum* (Jones and Rees, 1968; Labani and Elkington, 1987). This difference of 5.4 pg per 1C nucleus is approximately equal to the total 1C DNA content of barley (*Hordeum vulgare*), pepper (*Capsicum annuum*) or radish (*Raphanus sativus*) (Bennett and Smith, 1976). This increase in DNA content cannot be attributed to duplication of one or a few chromosomes. Jones and Rees (1968) and Narayan (1988) studied interspecific hybrids between *A. cepa* and *A. fistulosum* and observed that all eight bivalents were asymmetric, indicating that DNA differences were spread across all eight chromosomes. However, size differences among individual bivalents varied from a maximum of 60% to a minimum of 20% (Jones and Rees, 1968). Pairing at pachytene revealed loops and overlaps, which are evidence for accumulation of repetitive sequences or tandem duplication of chromosome segments (Jones and Rees, 1968). Intrachromosomal transposition is not likely to be a mechanism contributing to larger chromosome sizes because multiple loops per bivalent were not observed.

Increased genome sizes could result from ancient polyploidization event(s), termed palaeopolyploidy. This occurrence would have increased chromosome numbers in the past by duplicating individual chromosomes or chromosome sets. Centric fusions among duplicated telocentric chromosomes would produce fewer and larger metacentric chromosomes (Ohno, 1970) and result in diploidization of the duplicated genome. Palaeopolyploidy can be identified by conserved linkage relationships among duplicated genomic regions (Helentjaris *et al.*, 1988; Slocum *et al.*, 1990; Shoemaker *et al.*, 1996) or, in the case of maize, the existence of putative progenitors with lower base chromosome numbers (Anderson, 1945; Celarier, 1956). It is unlikely that onion has undergone a relatively recent polyploidization event, because there is no evidence of duplicated linkage blocks (King *et al.*, 1998a) or related species with a chromosome number of four.

Jones and Rees (1968) and Ranjekar *et al.* (1978) proposed that intrachromosomal duplications contributed to increased chromosome sizes in onion. This model of genome evolution would increase chromosome sizes, not chromosome numbers. Mechanisms producing intrachromosomal duplications include transposition events involving duplication of DNA fragments or RNA-mediated retrotransposition (Vanin, 1985) or tandem duplications by unequal crossing over (Smith, 1976). Tandem duplication of specific genes by unequal crossing over at meiosis has been proposed as the duplicating mechanism for the *R* (Robbins *et al.*, 1991), *Rp1* (Hulbert and Bennetzen, 1991) and *Kn1* (Veit *et al.*, 1990) loci of maize. Transposition of DNA (Pichersky, 1990) and RNA-mediated retrotransposition (Vanin, 1985) are known to duplicate coding (Matters and Goodenough, 1992; Kvarnheden *et al.*, 1995) and non-coding (Smith, 1976; San Miguel *et al.*, 1996) regions of plant genomes.

A low-density genetic map of onion based primarily on RFLPs (King *et al.*, 1998a) provided support for intrachromosomal duplications contributing to increased chromosome sizes in onion, as previously proposed by Jones and Rees (1968) and Ranjekar *et al.* (1978). Duplicated RFLPs were revealed by 20% of cDNA clones, a frequency lower than that found in palaeopolyploid species, but higher than in diploids (King *et al.*, 1998a). Sequencing of cDNAs detecting duplicated RFLP loci revealed that two-thirds showed homology to known gene families showing linkage in other plants (King *et al.*, 1998a). However, the remainder of the cDNA clones had no matches in the sequence databases at that time or showed high homologies to low-copy genes in other organisms. The distributions of multiple loci detected by single clones were 42% tightly linked (< 10 cM), 5% loosely linked (10–30 cM) and 53% unlinked (> 30 cM) (King *et al.*, 1998a). Forty per cent of RFLP loci were dominant, the highest reported for any plant species (King *et al.*, 1998a). Among duplicated loci detected by single clones, 19% segregated as two loci each with two codominant alleles, 52% segregated as one locus with codominant alleles and one locus with only a dominant fragment, and 29% segregated as two loci with only dominant fragments. These dominant RFLPs could be due to hemizygous duplications (present in only one parent of the mapping population) or comigration of duplicated fragments.

The linked nature of many duplicated RFLP loci, the prevalence of dominant RFLPs and the absence of conserved, duplicated linkage blocks in onion are features that differentiate it from most palaeopolyploids. Tandem duplication of DNA by unequal crossing over would increase DNA content without increasing chromosome numbers, produce closely linked duplicated loci and account for the loops observed during pachytene in interspecific hybrids between *A. cepa* and *A. fistulosum* (Jones and Rees, 1968). Meiotic pairing and unequal crossing over at homologous middle-repetitive regions flanking single-copy sequences (Stack and Comings, 1979) could duplicate the single-copy regions. This event would produce gametes with tandemly duplicated and deficient regions. Union of a wild-type gamete with the gamete carrying the tandemly duplicated region would produce a viable progeny with linked codominant and dominant loci. Presumably, the deficient gamete would be detrimental and selected against. Continued unequal crossing over within the middle-repetitive region could separate the tandemly duplicated single-copy regions, allowing for occasional recombinants.

A second scenario consistent with the RFLP-mapping results is retrotransposition. A DNA molecule would be synthesized from an mRNA intermediate by an indigenous reverse transcriptase (Hirochika and Hirochika, 1993). The DNA molecule would then be reinserted into the genome. Retrotransposed sequences are duplicated and tend to insert randomly into the genome (Vanin, 1985). Retrotransposition would explain the unlinked duplications in the onion genome. Detection of multiple loci and numerous restriction fragments with cDNA probes could reveal retrotransposed pseudogenes.

4. Gene Content

There is no evidence of increased numbers of coding regions among diploid *Allium* species with significantly larger nuclear genomes. Hybridization of random cDNA clones to DNA-gel blots of related species in *Allium* section *Cepa* revealed no significant differences in the numbers of fragments (Bradeen and Havey, 1995). The amounts of nuclear DNA in this group ranged from 17.9 (bulb onion) to 12.5 (Japanese bunching onion) pg per 1C (Bennett and Smith, 1976), indicating that palaeopolyploidy has probably not occurred in the recent evolution of species in *Allium* section *Cepa*.

4.1 Retroviral sequences

Retroviral sequences can contribute to huge increases in the sizes of plant nuclear genomes. For example, repetitive cycles of duplication and insertion of retroviral sequences significantly added to the sizes of non-coding regions flanking the *Adh*-1 region in the maize nuclear genome (San Miguel *et al.*, 1996). Retroviral sequences can be classified into four types: retroviruses, *Ty*3/*gypsy* retrotransposon, *Ty*1/*copia* retrotransposon and LINE elements (Bennetzen, 1996). These classes differ for the presence of specific coding regions or structural formats. *Ty*1-*copia*-like sequences are highly abundant and distributed over all chromosomes in large-genome gymnosperms (Kamm *et al.*, 1996). *Ty*1-*copia*-like retrotransposons are present throughout the bulb-onion genome, although they are concentrated in terminal heterochromatic regions (Pearce *et al.*, 1996). Significant homology to a retroviral reverse transcriptase has been identified in bulb onion (Hirochika and Hirochika, 1993). Hybridization of *del*2, an abundant non-long-terminal-repeat retrotransposon from *Lilium speciosum*, to *Bam*HI-digested onion DNA detected a prominent band at 6.6 kilobases (kb) (Leeton and Smyth, 1993). These studies document retrotransposon-like sequences in the bulb-onion genome; however, their specific role in the evolution of the enormous nuclear genome of onion remains to be established.

4.2 Ribosomal DNA

The structure and coding sequences of the nuclear ribosomal (r) DNA are highly conserved among plants. The rDNA region consists of highly repeated 45S monomeric units (S = the Svedberg coefficient, a unit that measures sedimentation rates in density centrifugation, thus providing a relative measure of density, used to differentiate between molecules). Each unit consists of three conserved regions encoding the 5.8S, 18S and 26S rRNAs (Appels and Honeycutt, 1986). Two internal transcribed spacers (ITS) separate the three rRNA-coding regions, and an intergenic spacer (IGS) separates the unit comprised of rRNA-coding regions and the ITS. Variation at or between restriction-enzyme sites in the nuclear 45S rDNA has been observed between genera, species and occasionally individuals within a population (Jorgensen and Cluster, 1988). Sequence variation among related *Allium* species is concentrated in the ITS or IGS regions and polymorphisms in the length of the 45S repeat are primarily due to length differences in the IGS (Havey, 1992b). The sizes of the 45S rDNA repeats are 11.8, 13.1, 11.5, 10.4 and 12.3 kb for bulb onion, Japanese bunching onion, garlic, leek and chives, respectively (Havey, 1992b). Intra- and interspecific differences were reported for the relative proportion of rDNA and number of nucleolus organizer regions (NOR) per cell in *Allium* (Maggini and Garbari, 1977; Maggini *et al.*, 1978). Maggini and Carmona (1981) mapped sites for *Bam*HI, *Eco*RI and *Hin*dIII in the 45S rDNA of bulb onion and reported sequence heterogeneity in the IGS within a single cultivar. Schubert *et al.* (1983) and Havey (1991b) used differences in the NOR and at restriction-enzyme sites in the 45S rDNA, respectively, between *A. cepa* and *A. fistulosum* to establish the interspecific origin of the top-setting (viviparous) onion (*Allium* × *proliferum* (Moench) Schrad. syn. *Allium cepa* L. var. *viviparum* (Metzger) Alefeld).

Radioactive and fluorescence *in situ* hybridization of the nuclear 45S rDNA revealed the locations and numbers of NOR in bulb and bunching onions (Schubert and Wobus, 1985; Ricroch *et al.*, 1992). In bulb onion, NOR were localized to chromosomes 6 and 8; bunching onion possessed NOR on chromosomes 5 and 8. For both species, the smallest chromosome (no. 8) carried one NOR. A fascinating anomaly, NOR jumping, was reported by Schubert and Wobus (1985). These researchers observed that recombination can occur between chromosomes carrying NOR regions, resulting in exchanges of terminal regions.

5. The Mitochondrial Genome

5.1 Basic structure

The gene content of mitochondrial genomes is generally conserved among all plants, encoding rRNAs, transfer RNAs (tRNAs) and some of the enzymatic subunits responsible for oxidative phosphorylation (Lonsdale, 1988). However, the mitochondrial genome synthesizes relatively few of the subunits required for the energy-producing machinery. Many mitochondrial enzymatic subunits are nuclear-encoded, cytoplasmically translated and imported into the mitochondria (Newton, 1988). The sizes and structures of the plant mitochondrial genomes differ significantly from those of animals or fungi. In these latter organisms, the mitochondrial genomes are relatively small (approximately 17 kb in humans and yeast) and highly conserved in structure (Gillham, 1994). Angiosperms, on the other hand, show huge size variations, from the relatively small mitochondrial genome of approximately 210 kb in the brassicas to the enormous genomes (up to 2300 kb) in the genus *Cucumis* (Ward *et al.*, 1981). Sequence variation in the plant mitochondrial genome accumulates more slowly than in the nuclear genome (Palmer and Herbon, 1988). However, the structure of the plant mitochondrial genome changes relatively quickly by recombination among direct repeats to produce smaller circular molecules and gross rearrangements (Stern and Palmer, 1984; Palmer, 1985, 1990). For this reason, the plant mitochondrial genome is not as useful as the chloroplast genome for phylogenetic studies.

The sizes and structures of the *Allium* mitochondrial genomes are unknown. Restriction-enzyme analyses of the *Allium* mitochondrial genomes have concentrated on differences among cytoplasmic male-sterile (CMS) versus normal (N) male-fertile cytoplasms. These analyses demonstrated that the *Allium* mitochondrial genome possesses regions homologous to genes found in most, if not all, angiosperms, such as apocytochrome B; subunits α, 6 and 9 of the F_0–F_1 adenosine triphosphatase (ATPase) complex; subunits 1, 2 and 3 of the cytochrome oxidase complex; and subunits 1, 3 and 5 of the nicotinamide adenine dinucleotide (NAD):Q_1 complex (Holford *et al.*, 1991a; Satoh *et al.*, 1993; Havey, 1995, 1997). Sato (1998) identified sequences in the mitochondrial genome of S-cytoplasmic onion showing high homology to the chloroplast genome. The promiscuous transfer of chloroplast sequences to the plant mitochondrial genome is well established in many other species (Moon *et al.*, 1988; Nugent and Palmer, 1988; Nakazono and Hirai, 1993).

5.2 Cytoplasmic male-sterile vs. normal male-fertile cytoplasm

CMS is known in many crops and is commonly used to produce hybrid seed (Jones and Clarke, 1943; Hanson and Conde, 1985). Hybrid allium crop cultivars have numerous advantages over open-pollinated cultivars, including significant heterosis over the inbred parent or open-pollinated source of the inbred parent, greater uniformity, the possibility of combining dominantly inherited disease resistances and the protection of breeders' rights (Jones and Davis, 1944; Currah, 1986). Pollination control used to be a major obstacle to the production of hybrid alliums. The *Allium* umbel contains hundreds of perfect flowers and, although outcrossing is encouraged by protandry (Currah and Ockendon, 1978), mature pollen and receptive stigmas are present at the same time in the densely packed inflorescence. Large-scale emasculation is not practical. At present, CMS is used commercially to produce hybrid seed of bulb onion, Japanese bunching onion and chives.

The most widely used source of CMS in bulb onion was discovered in Davis, California, in 1925. Dr H.A. Jones and colleagues were inbreeding plants of the cultivar 'Italian Red' to develop a red storage onion. One plant (13–53) did not set seed after self-pollination and was saved by the presence of bulbils (topsets) in the inflorescence (Jones and Emsweller, 1936). Male sterility in 'Italian Red 13–53' was conditioned by the interaction of a particular form of cytoplasm (S cytoplasm) with recessive alleles at a single nuclear male-fertility restorer (*Ms*) locus in the homozygous recessive form (*ms/ms*). In plants carrying S cytoplasm, fertility is restored by a dominant allele at this restorer locus (Jones and Clarke, 1943). Although Rhoades (1931) had previously recognized a maternal effect conditioning male sterility in maize, Jones and Clarke (1943) were the first to demonstrate that CMS is conditioned by the interaction of cytoplasmic (maternal) and nuclear factors. S cytoplasm has been widely used to produce hybrid-onion seed for most of the world's major onion-producing areas. This source of CMS is widely used because of the relatively simple inheritance of nuclear male-fertility restoration (making the extraction of maintainer lines easier), stable expression of male sterility across a range of temperatures (Barham and Munger, 1950) and the fact that the first suite of male-sterile inbred lines developed in the USA exclusively used this source of CMS (Havey, 1991a).

The morphology of N- and S-cytoplasmic onions has been well studied. The development of anthers is identical in both cytoplasms until the time of pollen shedding (Holford *et al.*, 1991b). No pollen is released from anthers of male-sterile plants possessing S cytoplasm, due to premature breakdown of the tapetum at the tetrad stage, hypertrophy of the tapetum at the dyad stage or abnormally long retention of the tapetum (Monosmith, 1925; Tatabe, 1952; Holford *et al.*, 1991b). Unlike T cytoplasm of maize, there were no differences in number or structure of mitochondria in the tapetum of N- and S-cytoplasmic onions (Holford *et al.*, 1991b).

A second source of CMS (T cytoplasm) in bulb onion was discovered by Berninger (1965) in the French cultivar 'Jaune paille des Vertus'. Schweisguth (1973) demonstrated that male fertility in T-cytoplasmic plants is restored by dominant alleles at one locus (*A*–) or at both of two complementary loci (*B*–*C*–). To my knowledge, there have been no cytological analyses of male sterility in T-cytoplasmic onion. This source or similar sources of male-sterile cytoplasm have been used to produce hybrid-onion seed in France, Holland and Japan (Havey, 2000).

Some confusion has resulted from early studies on sources of CMS extracted from indigenous open-pollinated onion populations. Male sterility has been identified and studied in onion plants from the USA (Peterson and Foskett, 1953), Germany (Kobabe, 1958), Turkey (Davis, 1958), New Zealand (Yen, 1959), Holland (van der Meer and van Bennekom, 1969) and India (Pathak and Gowda, 1994). Researchers initially assumed that all sources of CMS were S-cytoplasmic. The genetics of male-fertility restoration (Jones and Clarke, 1943; Schweisguth, 1973; Havey, 2000) and molecular analyses of S and T cytoplasms (de Courcel *et al.*, 1989; Holford *et al.*, 1991a; Havey, 1993, 2000) have clearly demonstrated that independent sources of CMS exist. The male sterility observed in cv. 'Pukekohe Longkeeper' in New Zealand by Yen (1959) was probably conditioned by S cytoplasm. We (Havey, 1993) demonstrated that this open-pollinated population exclusively possessed S cytoplasm. CMS extracted in India from a population of cv. 'Nasik White' (Pathak and Gowda, 1994) is identical to S cytoplasm (Havey 2000). However sources of male sterility extracted from the Dutch cultivar 'Rijnsburger' were probably T-cytoplasmic (Havey, 2000), although the authors at the time assumed this to be S cytoplasm (van der Meer and van Bennekom, 1969). This would explain why van der Meer and van Bennekom (1969) observed that male sterility from cv. 'Rijnsburger' broke down at high temperatures. Hybrid-onion seed is routinely produced in the USA under extremely high summer temperatures in the Treasure Valley of Idaho or the Central Valley of California, without breakdown of male sterility.

Restriction-enzyme analyses have revealed differences in the mitochondrial DNAs among normal (N) male-fertile and male-sterile cytoplasms in crops such as maize (Levings and Pring, 1976; Kemble *et al.*, 1980), sorghum (Pring *et al.*, 1982) and sunflower (Brown *et al.*, 1986). Mitochondrial polymorphisms have been identified among mitochondrial DNAs (mtDNAs) of N, S and T cytoplasms of onion (de Courcel *et al.*, 1989; Holford *et al.*, 1991a; Havey, 1993, 1995, 2000; Satoh *et al.*, 1993). De Courcel *et al.* (1989) and Holford *et al.* (1991a) isolated mtDNA, digested with restriction enzymes, and identified polymorphisms on ethidium-bromide-stained agarose gels between N and S cytoplasms. De Courcel *et al.* (1989) proposed that two main groups of cytoplasms exist. M cytoplasm was the most common. *Bam*HI digests of the mtDNA distinguished four subgroups (M_1–M_4) and included N (M_2) and T cytoplasms (M_4). S cytoplasm was easily distinguished from the M-cytoplasmic group using mtDNA. Holford *et al.* (1991a) were able to distinguish among N, S and T cytoplasms with *Bam*HI and *Hin*dIII digests of mtDNA. They observed no plasmid-like molecules on gels of undigested mitochondrial DNA, indicating that S cytoplasm in onion does not possess small circular DNA molecules (episomes) like the S-cytoplasmic source of CMS in maize. Additionally, the male sterility of S cytoplasm was not transmissible by grafting (van der Meer and van Bennekom, 1970) and virus-like particles were not found (Holford *et al.*, 1991a).

A problem with scoring polymorphisms in the mitochondrial genomes using restriction-enzyme digests and visualization on gels is that a single structural change can be scored as a polymorphism for more than one enzyme. Hybridization of mitochondrial probes to DNA-gel blots is a better way to characterize polymorphisms and this technique has revealed many polymorphisms between N- and S-cytoplasmic onions (Holford *et al.*, 1991a; Satoh *et al.*, 1993; Havey, 1995; Sato, 1998). Few polymorphisms have been revealed between N and T cytoplasms (Holford *et al.*, 1991a; Havey, 1995, 2000). These studies are in agreement with the original proposal of de Courcel *et al.* (1989), in which T cytoplasm is a member of the M-cytoplasmic class. The S cytoplasm is different from that of the M-cytoplasmic onions. Holford *et al.* (1991a) proposed that S is an alien cytoplasm introduced into onion; Havey (1993) proposed that the transfer from an unknown donor species occurred via the viviparous triploid cultivar 'Pran'.

CMS has been well characterized in chives by Dr T. Tatlioglu and his students. In chives, CMS is conditioned by the interaction of the cytoplasm (S) and a single nuclear fertility-restoration locus (*X*) (Tatlioglu, 1982). Microsporogenesis is similar for male-fertile and sterile plants until the tetrad stage, when the microspores die in male-sterile plants (Ruge *et al.*, 1993). Polymorphisms in the mitochondrial genome between CMS and normal male-fertile chive plants have been identified and a unique 18 kilodalton protein was specifically associated with the male-sterile phenotype (Potz and Tatlioglu, 1993). CMS in chives is sensitive to chemical and environmental factors. Tetracycline treatments restored male fertility for CMS of chives (Tatlioglu, 1986) and sensitivity was conditioned by recessive alleles at a single locus (*aa*) (Tatlioglu and Wricke, 1988). CMS in chives also showed temperature sensitivity; some genetically male-sterile plants grown at a constant temperature of 24°C produce viable pollen (Tatlioglu, 1985). Crosses among temperature-sensitive and insensitive CMS chive plants revealed a dominant allele at one locus (*T*) conditioning temperature sensitivity of the CMS (Tatlioglu, 1987).

CMS has been described in Japanese bunching onion (*A. fistulosum*) and is used commercially to produce hybrid seed in Japan. Male-fertility restoration is inherited in a more complex manner than in either bulb onion or chives. CMS is controlled by the interaction of the cytoplasm (S) with two nuclear restorer loci (MS_1 and MS_2) (Moue and Uehara, 1985). Male sterility occurs when both of these nuclear fertility-restoration loci are homozygous recessive.

CMS in leek has not been described. Male sterility has been observed in leek, but subsequent genetic studies revealed a genic male-sterility system (Schweisguth, 1970). Asexual propagation of genic male-sterile plants is currently used to produce hybrid-leek seed (Smith and Crowther, 1995; De Clercq and Van Bockstaele, Chapter 18, this volume). In order to develop a CMS system for leek, Peterka *et al.* (1997) generated an interspecific hybrid between CMS onion and leek as an initial step to transfer CMS from onion to leek. At least two generations of backcrossing to leek have been completed. Kik *et al.* (1997) purified mtDNA from leek populations and, after digesting with restriction enzymes, identified two mitochondrial types. We (Havey and Lopes Leite, 1999) used DNA-gel blots to evaluate for polymorphisms in the mtDNA among cultivated populations of *A. ampeloprasum*. We observed only five polymorphisms among cultivated leek accessions and kurrat (*A. ampeloprasum* kurrat group), agreeing with Kik *et al.* (1997) that little variability exists in the mtDNA of leek and kurrat. Both Kik *et al.* (1997) and my laboratory (Havey and Lopes Leite, 1999) observed that the mitochondrial genome of great-headed garlic (*A. ampeloprasum* var. *holmense*) showed polymorphic differences from that of leek and kurrat.

Alien cytoplasms are known to condition male sterility (Hanson and Conde, 1985).

Potentially useful sources of CMS have been recently developed by transfer to bulb onion of the cytoplasm of *A. galanthum* Kar. et Kir. (Havey, 1999), shallot (*A. cepa* Aggregatum group) (Yamashita and Tashiro, 1999) and Japanese bunching onion (Yamashita *et al.*, 1999a). Alleles known to restore male fertility for onion plants possessing S cytoplasm showed no male-fertility restoration for the galanthum-CMS bulb-onion lines (Havey, 1999). Yamashita *et al.* (1999a) observed a dominant allele at a single nuclear locus (*Rf*) which restored male fertility for galanthum-CMS lines of Japanese bunching onion. The *Rf* locus probably originated from the *A. galanthum* parent used in the original interspecific cross. Subsequently, Yamashita *et al.* (1999b) identified isozyme and randomly amplified polymorphic DNA (RAPD) markers tagging the *Rf* locus from *A. galanthum*.

6. The Chloroplast Genome

6.1 Basic structure

The linear array of genes in the chloroplast DNA is highly conserved among evolutionarily distant species and is generally a circular DNA molecule of approximately 150 kb (Palmer and Stein, 1986). The chloroplast DNA usually possesses two inverted repeats carrying the rRNA-coding regions (Palmer and Stein, 1986). Unlike the direct repeats found in the mitochondrial genome, recombination between the inverted repeats of the chloroplast DNA does not produce subgenomic circular molecules. Flanking the inverted repeats are large (LSC) and small single-copy (SSC) regions carrying conserved linear arrays of genes.

The structure of the *Allium* chloroplast genome was first studied by Chase and Palmer (1989). They demonstrated that the bulb onion possesses a chloroplast genome of standard size, gene order and structure. Chase and Palmer (1989) and Katayama *et al.* (1991) observed that the bulb-onion chloroplast DNA is more similar to that of tobacco (as a representative species of the dicotyledons) than to that of members of the monocotyledonous *Poaceae*. Katayama *et al.* (1991) estimated the size of the bulb-onion chloroplast DNA at 155 kb, with the two inverted repeats of 26 kb each and the LSC and SSC regions of 86 and 16 kb, respectively. Chase and Palmer (1989) and Havey (1991c) made slightly smaller size estimates of the bulb-onion chloroplast DNA at 145 and 140 kb, respectively. These size differences can be attributed to the errors associated with estimations of restriction fragment sizes from agarose gels or autoradiograms.

6.2 Variability among species

RFLPs in the chloroplast DNA can be revealed using heterologous probes and can result from mutations at restriction-enzyme sites, inversions, insertions or deletions. Analyses of polymorphisms at or between restriction-enzyme sites in the chloroplast DNA are often used to estimate phylogenetic relationships among species, which may be more difficult to elucidate by using morphologies or crossabilities. Phylogenetic analyses based on the chloroplast genome have been widely used in *Allium* (Klaas, 1998; Klaas and Friesen, Chapter 8, this volume). Havey (1991c) was the first to use polymorphic restriction-enzyme sites in the chloroplast DNA to estimate phylogenies. In this initial study, chloroplast-based phylogenetic relationships among the major cultivated alliums were consistent, although not in complete agreement, with previous classifications. Nevertheless, this preliminary study established the basis for more detailed phylogenetic estimates among more closely related species (Havey, 1992a). Phylogenetic estimates based on polymorphic restriction-enzyme sites in the chloroplast genome among species within *Allium* section *Cepa* agreed with crossabilities (van Raamsdonk *et al.*, 1992). The research group at the Institut für Pflanzengenetik und Kulturpflanzenforschung (IPK), Gatersleben, Germany, has undertaken larger phylogenetic studies using chloroplast polymorphisms. This group has a long and distinguished research record on the taxonomy and phylogeny of *Allium* (Hanelt *et al.*, 1992). The most wide-ranging phylogenetic study using restriction-enzyme analysis of the chloroplast DNA involved 49 *Allium* species from the major subgenera, sections and subsections (Linne von Berg *et al.*, 1996). The resulting phenogram agreed with previous morphological-based classifications proposed by the Gatersleben group (Hanelt *et al.*, 1992). Subsequent studies based on cladistic analyses of chloroplast polymorphisms supported the subdivision of *Allium* into two main subgeneric groups corresponding to basic chromosome numbers of seven and eight (Samoylov *et al.*, 1995, 1999).

Phylogenetic estimates among closely related species using the chloroplast DNA are difficult because of its conserved nature (Sandbrink *et al.*, 1990). Introns and intergenic spacers may accumulate nucleotide differences or structural rearrangements more quickly, as compared with coding regions (Wolfe *et al.*, 1987; Kelchner and Wendel, 1996). Phylogenetic estimates based on characters in these faster-evolving regions may (Mes *et al.*, 1997) or may not (Goldenberg *et al.*, 1993; Morton and Clegg, 1993) produce phylogenetically informative characters, as compared with the chloroplast genome as a whole or other specific chloroplast regions. One of the main problems with phylogenetic estimates based on polymorphisms in non-coding regions of the chloroplast DNA is that adenosine–thymidine (AT) slippage during replication can generate similarly sized (homoplasious) fragments derived from independent events. Alcala *et al.* (1999) reported polymorphisms in a non-coding chloroplast region within single bulb-onion plants. Nevertheless, non-coding chloroplast regions are useful. Mes *et al.* (1997) identified numerous intergenic regions useful for phylogenetic studies in *Allium*. Friesen *et al.* (1999) amplified five non-coding regions in the chloroplast DNA, digested with a battery of restriction enzymes, and demonstrated a single origin of cultivated *A. fistulosum* from the wild species *A. altaicum* Pall. Yamashita *et al.* (1998, 1999a) used PCR to amplify the ribulose-1,5-biphosphate carboxylase (rbcL)-ORF106 region and identified polymorphic restriction-enzyme sites useful in the classification of cytoplasms.

6.3 Cytoplasmic male-sterile vs. normal male-fertile cytoplasm

CMS is always associated with mutations or chimeric genes in the mitochondrial genome (Hanson, 1991). However, polymorphisms have occasionally been identified in the chloroplast DNA between male-fertile and male-sterile cytoplasms. Examples include beet (Saumitou-Laprade *et al.*, 1993), sorghum (Chen *et al.*, 1990) and bulb onion (de Courcel *et al.*, 1989; Holford *et al.*, 1991a; Havey, 1993). This does not necessarily mean that CMS is encoded by the chloroplast DNA, but more probably means that the chloroplast polymorphisms reveal an alien or divergent cytoplasm. Strict maternal inheritance of the organellar genome would guarantee that both genomes are maintained together. In bulb onion, de Courcel *et al.* (1989) and Holford *et al.* (1991a) isolated chloroplast DNA from N- and S-cytoplasmic onions, digested with restriction enzymes, and identified polymorphisms on ethidium-bromide-stained agarose gels. Havey (1993) used DNA-gel blots to reveal five polymorphisms between N and S cytoplasms of onion. These chloroplast polymorphisms were found both in S cytoplasm from bulb onions and in its putative donor 'Pran' (Havey, 1993). No differences in the chloroplast DNA have been identified between N- and T-cytoplasmic onions (Havey, 1993).

Using classical crosses, the cytoplasm of an individual onion plant can be established after 4–8 years (Havey, 1995). Molecular markers distinguishing male-fertile and male-sterile cytoplasms offer great advantages by significantly reducing the time required to establish the type of a cytoplasm. Isolation of organellar DNA or DNA-gel blot analyses are significantly quicker than crossing, but are still relatively time-consuming and labour-intensive. PCR is a significantly quicker and cheaper method of evaluating for DNA polymorphisms. The chloroplast genome of onion possesses a useful PCR-based polymorphism to distinguish N and S cytoplasms. Because both the chloroplast and mitochondrial DNAs in bulb onion show maternal inheritance (Tatabe, 1968; Havey, 1995), polymorphisms in either genome should be useful for classifying cytoplasms. We (Havey, 1995) identified oligonucleotide primers that preferentially amplify by PCR a 100-bp insertion in the chloroplast DNA of N cytoplasm (Havey, 1993). This region was chosen because the polymorphism can be scored directly after gel electrophoresis of the PCR reaction and requires no further manipulation, such as digestion with a restriction enzyme. Sato (1998) developed a similar PCR-based mitochondrial marker to distinguish N and S cytoplasm of onion. Lilly and Havey (2001) developed oligonucleotide primers that specifically amplify from the chloroplast genome of N cytoplasm to reveal cytoplasmic mixtures in hybrid-onion seed lots. These organellar markers are widely used by the onion-breeding community to quickly and cheaply classify cytoplasms for maintainer or male-sterile line development and for quality control in hybrid-onion seed production.

7. Conclusions and Future Developments

The plant genome is composed of DNA carried in the nucleus, mitochondrion and chloroplast. The *Allium* organellar genomes are similar in structure and gene content to those of other angiosperms. However, the nuclear genome is unique. The enormous accumulation of DNA, without recent polyploidization, is similar to that of the large genomes of lilies and gymnosperms. The unique aspect of the *Allium* nuclear genome is the uniform accumulation of huge amounts of DNA, and therefore the large chromosome size, across all chromosomes among closely related species. Cytological (Jones and Rees, 1968; Ranjekar *et al.*, 1978) and genetic-mapping (King *et al.*, 1998a) experiments support a role for tandem duplication in the evolution of the bulb-onion nuclear genome. Sequencing of *Arabidopsis* chromosome 2 revealed a plethora of tandem duplications (Lin *et al.*, 1999), in spite of the relatively small size of this nuclear genome. The *Allium* nuclear genome, as typified by the bulb onion, may have unique evolutionary mechanisms allowing tandem duplication and diversification of coding regions. As our understanding of plant genomes continues to grow, future research will provide insights about forces contributing to the huge nuclear genome of the alliums.

References

Ahn, S., Anderson, J.A., Sorrells, M.E. and Tanksley, S.D. (1993) Homoeologous relationships of rice, wheat and maize chromosomes. *Molecular and General Genetics* 241, 483–490.

Alcala, J., Pike, L. and Giovannoni, J. (1999) Identification of plastome variants useful for cytoplasmic selection and cultivar identification in onion. *Journal of the American Society for Horticultural Science* 124, 122–127.

Anderson, E. (1945) What is *Zea mays*? A report of progress. *Chronicles of Botany* 9, 88–92.

Appels, R. and Honeycutt, R. (1986) rDNA: evolution over a billion years. In: Dutta, S. (ed.) *DNA Systematics*, Vol. II. *Plants*. CRC Press, Boca Raton, Florida, pp. 81–135.

Arumuganathan, K. and Earle, E.D. (1991) Nuclear DNA content of some important plant species. *Plant Molecular Biology Reporter* 9, 208–218.

Barham, W.S. and Munger, H.M. (1950) The stability of male sterility in onion. *Proceedings of the American Society for Horticultural Science* 56, 401–409.

Barnes, S.R., James, A.M. and Jamieson, G. (1985) The organization, nucleotide sequence, and chromosomal distribution of a satellite DNA from *Allium cepa*. *Chromosoma* 92, 185–192.

Bennett, M.D. (1972) Nuclear DNA content and minimum generation time in herbaceous plants. *Philosophical Transactions of the Royal Society of London Series B* 181, 109–135.

Bennett, M.D. (1976) DNA amount, latitude, and crop plant distribution. *Environmental and Experimental Botany* 16, 93–108.

Bennett, M.D. and Smith, J.B. (1976) Nuclear DNA amounts in angiosperms. *Philosphical Transactions of the Royal Society of London Series B* 274, 227–274.

Bennetzen, J.L. (1996) The contributions of retroelements to plant genome organization, function, and evolution. *Trends in Microbiology* 4, 347–353.

Bennetzen, J.L. and Freeling, M. (1993) Grasses as a single genetic system: genome composition, colinearity and compatibility. *Trends in Genetics* 9, 259–261.

Berninger, E. (1965) Contribution à l'étude de la stérilité mâle de l'oignon (*Allium cepa* L.). *Annales de l'Amélioration des Plantes* 15, 183–199.

Berninger, E. and Buret, P. (1967) Étude des déficients chlorophylliens chez deux espèces cultivées du genre *Allium*: l'oignon (*A. cepa* L.) et le poireau (*A. porrum* L.). *Annales de l'Amélioration des Plantes* 17, 175–194.

Bonierbale, M.W., Plaisted, R.L. and Tanksley, S.D. (1988) RFLP maps based on a common set of clones reveal modes of chromosomal evolution in potato and tomato. *Genetics* 120, 1095–1103.

Bougourd, S.M. and Parker, J.S. (1976) The B-chromosome system of *Allium schoenoprasum*. II. Stability, inheritance and phenotypic effects. *Chromosoma* 75, 369–383.

Bradeen, J.M. and Havey, M.J. (1995) Randomly amplified polymorphic DNA in bulb onion and its use to assess inbred integrity. *Journal of the American Society for Horticultural Science* 120, 752–758.

Brown, G., Bussey, H. and Des Rosiers, L. (1986) Analysis of mitochondrial DNA, chloroplast DNA, and double-stranded RNA in fertile and cytoplasmic male-sterile sunflower. *Canadian Journal of Genetics and Cytology* 28, 121–129.

Celarier, R.P. (1956) Additional evidence for five as the basic chromosome number of the *Andropogoneae*. *Rhodora* 58, 135–143.

Chase, M. and Palmer, J. (1989) Chloroplast DNA systematics of lilioid monocots, feasibility, resources, and an example from the Orchidaceae. *American Journal of Botany* 76, 1720–1730.

Chen, Z., Liang, G., Muthukrishnan, S. and Kofoid, K. (1990) Chloroplast DNA polymorphism in fertile and male-sterile cytoplasms of sorghum. *Theoretical and Applied Genetics* 80, 727–731.

Cramer, C.S. and Havey, M.J. (1999) Morphological, biochemical, and molecular markers in onion. *HortScience* 34, 589–593.

Cryder, C.M., Corgan, J.N., Urquhart, N.S. and Clason, D. (1991) Isozyme analysis of progeny derived from (*Allium fistulosum* × *Allium cepa*) × *Allium cepa*. *Theoretical and Applied Genetics* 82, 337–345.

Currah, L. (1986) Leek breeding: a review. *Journal of Horticultural Science* 61, 407–415.

Currah, L. and Ockendon, D. (1978) Protandry and the sequence of flower opening in the onion. *New Phytologist* 81, 419–428.

Davis, E. (1958) Male sterility in onion plants from Turkey. *Journal of Heredity* 49, 31–32.

de Courcel, A., Veder, F. and Boussac, J. (1989) DNA polymorphism in *Allium cepa* cytoplasms and its implications concerning the origin of onions. *Theoretical and Applied Genetics* 77, 793–798.

Devos, K.M. and Gale, M.D. (1997) Comparative genetics in the grasses. *Plant Molecular Biology* 35, 3–15.

Devos, K.M., Chao, S. and Li, Q.Y. (1994) Relationship between chromosome 9 of maize and wheat homeologous group 7 chromosomes. *Genetics* 138, 1287–1292.

Dunford, R.P., Kurata, N., Laurie, D.A., Money, T.A., Minobe, Y. and Moore, G. (1995) Conservation of fine-scale DNA marker order in the genomes of rice and the *Triticeae*. *Nucleic Acids Research* 23, 2724–2728.

Eickmeyer, E., Primus-Kopfer, S. and Wricke, G. (1990) Genetic analysis of isozyme loci in chives (*Allium schoenoprasum* L.). *Plant Breeding* 105, 327–331.

El-Gadi, A. and Elkington, T. (1975) Comparison of the Giemsa c-band karyotypes and the relationships of *Allium cepa*, *A. fistulosum* and *A. galanthum*. *Chromosoma* 51, 19–23.

El-Shafie, M. and Davis, G. (1967) Inheritance of bulb color in *Allium cepa*. *Hilgardia* 38, 607–622.

Emsweller, S.L. and Jones, H.A. (1935) An interspecific hybrid in *Allium*. *Hilgardia* 9, 265–273.

Emsweller, S.L. and Jones, H.A. (1938) Crossing-over, fragmentation, and formation of new chromosomes in *Allium* species hybrid. *Botanical Gazette* 99, 729–772.

Emsweller, S.L. and Jones, H.A. (1945) Further studies on the chiasmata of the *Allium cepa* × *Allium fistulosum* hybrid and its derivatives. *American Journal of Botany* 32, 370–379.

Etoh, T. (1983) Germination of seeds obtained from a clone of garlic (*Allium sativum* L.). *Proceedings of the Japanese Academy* 59(B), 83–87.

Fischer, D. and Bachmann, K. (2000) Onion microsatellites for germplasm analysis and their use in assessing intra- and interspecific relatedness within the subgenus *Rhizirideum*. *Theoretical and Applied Genetics* 101, 153–164.

Fiskesjö, G. (1975) Chromosomal relationships between three species of *Allium* as revealed by C-banding. *Hereditas* 81, 23–32.

Friesen, N., Pollner, S., Blattner, F. and Bachmann, K. (1999) RAPDs and non-coding chloroplast DNA reveal a single origin of the cultivated *Allium fistulosum* from *A. altaicum*. *American Journal of Botany* 86, 554–562.

Fuchs, J., Brandes, A. and Schubert, I. (1995) Telomere sequences localization and karyotype evolution in higher plants. *Plant Systematics and Evolution* 196, 227–241.

Gillham, N.W. (1994) *Organelle Genes and Genomes*. Oxford University Press, New York, 424 pp.

Gökçe, A.F. (2001) Marker facilitated selection of maintainer lines in onion (*Allium cepa* L.). PhD thesis, University of Wisconsin-Madison, Wisconsin, USA.

Goldenberg, E., Clegg, M., Durbin, M., Doebley, J. and Ma, D. (1993) Evolution of a non-coding region of the chloroplast genome. *Molecular Phylogenetics and Evolution* 2, 52–64.

Hanelt, P., Schultze-Motel, J., Fritsch, R., Kruse, J., Maaß, H.I., Ohle, H. and Pistrick, K. (1992) Infrageneric grouping of *Allium* – the Gatersleben approach. In: Hanelt, P., Hammer, K. and Knüpffer, H. (eds) *The Genus Allium – Taxonomic Problems and Genetic Resources*. Institut für Pflanzengenetik und Kulturpflanzenzuchtung (IPK), Gatersleben, Germany, pp. 107–123.

Hanson, M. (1991) Plant mitochondrial mutations and male sterility. *Annual Review of Genetics* 25, 461–486.

Hanson, M. and Conde, M. (1985) Functioning and variation of cytoplasmic genomes: lessons from cytoplasmic–nuclear interactions affecting male fertility in plants. *International Review of Cytology* 94, 213–267.

Havey, M.J. (1991a) History of releases from the onion breeding program of the United States Department of Agriculture. *Allium Improvement Newsletter* 1, 50–54.

Havey, M.J. (1991b) Molecular characterization of the interspecific origin of viviparous onion. *Journal of Heredity* 82, 501–503.

Havey, M.J. (1991c) Phylogenetic relationships among cultivated *Allium* species from restriction enzyme analysis of the chloroplast genome. *Theoretical and Applied Genetics* 81, 752–757.

Havey, M.J. (1992a) Restriction enzyme analysis of the chloroplast and nuclear 45s ribosomal DNA of *Allium* sections *Cepa* and *Phyllodolon*. *Plant Systematics and Evolution* 183, 17–31.

Havey, M.J. (1992b) Restriction enzyme analysis of the nuclear 45s ribosomal DNA of six cultivated Alliums. *Plant Systematics and Evolution* 181, 45–55.

Havey, M.J. (1993) A putative donor of S cytoplasm and its distribution among open-pollinated populations of onion. *Theoretical and Applied Genetics* 86, 128–134.

Havey, M.J. (1995) Cytoplasmic determinations using the polymerase chain reaction to aid in the extraction of maintainer lines from open-pollinated populations of onion. *Theoretical and Applied Genetics* 90, 263–268.

Havey, M.J. (1997) On the origin and distribution of normal cytoplasm of onion. *Genetic Resources and Crop Evolution* 44, 307–313.

Havey, M.J. (1999) Seed yield, floral morphology, and lack of male-fertility restoration of male-sterile onion (*Allium cepa*) populations possessing the cytoplasm of *Allium galanthum*. *Journal of the American Society for Horticultural Science* 124, 626–629.

Havey, M.J. (2000) Diversity among male-sterility-inducing and male-fertile cytoplasms of onion. *Theoretical and Applied Genetics* 101, 778–782.

Havey, M.J. and Lopes Leite, D. (1999) Organellar DNA diversity among cultivated accessions of *Allium ampeloprasum*. *Journal of the American Society for Horticultural Science* 124, 163–165.

Helentjaris, T., Weber, D. and Wright, S. (1988) Identification of the genomic locations of duplicate nucleotide sequences in maize by analysis of restriction fragment length polymorphisms. *Genetics* 118, 353–363.

Hirochika, H. and Hirochika, R. (1993) *Ty*1-*copia* group retrotransposons as ubiquitous components of plant genomes. *Japanese Journal of Genetics* 68, 35–46.

Holford, P., Croft, J. and Newbury, H. (1991a) Differences between, and possible origins of, the cytoplasms found in fertile and male-sterile onions (*Allium cepa* L.). *Theoretical and Applied Genetics* 82, 737–744.

Holford, P., Croft, J. and Newbury, H. (1991b) Structural studies of microsporogenesis in fertile and male-sterile onions (*Allium cepa* L.) containing the cms-S cytoplasm. *Theoretical and Applied Genetics* 82, 745–755.

Hulbert, S.H. and Bennetzen, J.L. (1991) Recombination at the *Rp*1 locus of maize. *Molecular and General Genetics* 226, 377–382.

Jones, H.A. and Clarke, A. (1943) Inheritance of male sterility in the onion and the production of hybrid seed. *Proceedings of the American Society for Horticultural Science* 43, 189–194.

Jones, H.A. and Davis, G. (1944) *Inbreeding and Heterosis and their Relation to the Development of New Varieties of Onions*. Technical Bulletin Number 874, USDA, Washington, DC, 28 pp.

Jones, H.A. and Emsweller, S. (1936) A male-sterile onion. *Proceedings of the American Society for Horticultural Science* 34, 582–585.

Jones, R.N. (1990) Cytogenetics. In: Rabinowitch, H.D. and Brewster, J.L. (eds) *Onions and Allied Crops*, Vol. I. *Botany, Physiology, and Genetics*. CRC Press, Boca Raton, Florida, pp. 199–214.

Jones, R. and Rees, H. (1968) Nuclear DNA variation in *Allium*. *Heredity* 23, 591–605.

Jorgensen, R. and Cluster, P. (1988) Modes and tempos in the evolution of nuclear ribosomal DNA: new characters for evolutionary studies and new markers for genetic and population studies. *Annals of the Missouri Botanical Garden* 75, 1238–1247.

Kamm, A., Doudrick, R.L., Heslop-Harrison, J.S. and Schmidt, T. (1996) The genomic and physical organization of *Ty*1-*copia*-like sequences as a component of large genomes in *Pinus elliottii* var. *elliottii* and other gymnosperms. *Proceedings of the National Academy of Sciences of the USA* 93, 2708–2713.

Katayama, H., Sasakuma, T. and Ogihara, Y. (1991) Physical map of chloroplast DNA of the onion, *Allium cepa* L., showing the location of photosynthesis-related genes. *Japanese Journal of Genetics* 66, 421–431.

Kelchner, S. and Wendel, J. (1996) Hairpins create minor inversions in non-coding regions of chloroplast DNA. *Current Genetics* 30, 259–262.

Kemble, B., Grun, R. and Flavell, R. (1980) Classification of normal and male-sterile cytoplasms in maize. II: Electrophoretic analysis of DNA species in mitochondria. *Genetics* 95, 451–458.

Khazanehdari, K.A. and Jones, G.H. (1996) Meiotic synapsis of the *Allium porrum* B chromosome: evidence for a derived isochromosome origin. *Genome* 39, 1199–1204.

Kik, C., Samoylov, A.M., Verbeek, W.H.J. and van Raamsdonk, L.W.D. (1997) Mitochondrial DNA variation and crossability of leek (*Allium porrum*) and its wild relatives from the *Allium ampeloprasum* complex. *Theoretical and Applied Genetics* 94, 465–471.

King, J.J., Bradeen, J.M., Bark, O., McCallum, J.A. and Havey, M.J. (1998a) A low-density genetic map of onion reveals a role for tandem duplication in the evolution of an extremely large diploid genome. *Theoretical and Applied Genetics* 96, 52–62.

King, J.J., Bradeen, J.M. and Havey, M.J. (1998b) Variability for RFLPs and relationships among elite commercial inbred and virtual hybrid populations of onion. *Journal of the American Society for Horticultural Science* 123, 1034–1037.

Kirk, J.T.O., Rees, H. and Evans, G. (1970) Base composition of nuclear DNA within the genus *Allium*. *Heredity* 25, 507–512.

Klaas, M. (1998) Applications and impact of molecular markers on evolutionary and diversity studies in the genus *Allium*. *Plant Breeding* 117, 297–308.

Kobabe, G. (1958) Ontogenetical and genetical investigations on the new male-sterile mutants of the common onion (*Allium cepa* L.). *Zeitschrift für Pflanzenzüchtung* 40, 353–384.

Kollmann, F. (1971) *Allium ampeloprasum* – a polyploid complex. I. Ploidy levels. *Israel Journal of Botany* 20, 13–20.

Kollmann, F. (1972) *Allium ampeloprasum* – a polyploid complex. II. Meiosis and relationships between the ploidy types. *Caryologia* 25, 295–312.

Koul, A.K. and Gohil, R.N. (1970) Cytology of the tetraploid *Allium ampeloprasum* with chiasma localization. *Chromosoma* 29, 12–19.

Kvarnheden, A., Tandre, K. and Engstrom, P. (1995) A *cdc*2 homologue and closely related processed retropseudogenes from Norway spruce. *Plant Molecular Biology* 27, 391–403.

Labani, R. and Elkington, T. (1987) Nuclear DNA variation in the genus *Allium* L. (*Liliaceae*). *Heredity* 59, 119–128.

Leeton, P.R.J. and Smyth, D.R. (1993) An abundant LINE-like element amplified in the genome of *Lilium speciosum*. *Molecular and General Genetics* 237, 97–104.

Levings, C. and Pring, D. (1976) Restriction enzyme analysis of mitochondrial DNA from normal and Texas cytoplasmic male-sterile maize. *Science* 193, 158–160.

Lilly, J.W. and Havey, M.J. (2001) Sequence analysis of a chloroplast intergenic spacer for phylogenetic estimates in *Allium* section *Cepa* and a PCR-based polymorphism detecting mixtures of male-fertile and male-sterile cytoplasmic onion. *Theoretical and Applied Genetics* 102, 78–82.

Lin, X., Kaul, S., Rounsley, S., Shea, T., Benito, M., Town, C., Fujii, C., Mason, T., Bowman, C., Barnstead, M., Feldblyum, T., Buell, C., Ketchum, K., Lee, J., Ronning, C., Koo, H., Moffat, K., Cronin, L., Shen, M., Pai, G., van Aken, S., Umayama, L., Tallon, L., Gill, J., Adams, M., Carrera, A., Creasy, T., Goodman, H., Somerville, C., Copenhaver, G., Preuss, D., Nierman, W., White, O., Eisen, K., Salzberg, J.S., Fraser, C. and Venter, C. (1999) Sequence and analysis of chromosome 2 of the plant *Arabidopsis thaliana*. *Nature* 402, 761–768.

Linne von Berg, G., Samoylov, A., Klaas, M. and Hanelt, P. (1996) Chloroplast DNA restriction analysis and the infrageneric grouping in *Allium* L. *Plant Systematics and Evolution* 200, 253–261.

Lonsdale, D.M. (1988) Plant mitochondrial genes and sequences. *Plant Molecular Biology Reporter* 6, 266–273.

Maeda, T. (1937) Chiasma studies in *Allium fistulosum*, *Allium cepa*, and their F_1, F_2 and backcross hybrids. *Japanese Journal of Genetics* 13, 146–159.

Maggini, F. and Carmona, M. (1981) Sequence heterogeneity of the ribosomal DNA in *Allium cepa* (*Liliaceae*). *Protoplasma* 108, 163–171.

Maggini, F. and Garbari, F. (1977) Amounts of ribosomal DNA in *Allium* (*Liliaceae*). *Plant Systematics and Evolution* 128, 201–208.

Maggini, F., Barsanti, P. and Marazia, T. (1978) Individual variation of the nucleolus organizer regions in *Allium cepa* and *A. sativum*. *Chromosoma* 66, 173–183.

Matters, G.L. and Goodenough, U.W. (1992) A gene/pseudogene tandem duplication encodes a cysteine-rich protein expressed during zygote development in *Chlamydomonas reinhardtii*. *Molecular and General Genetics* 232, 81–88.

Mes, T., Friesen, N., Fritsch, R., Klaas, M. and Bachmann, K. (1997) Criteria for sampling in *Allium* based on chloroplast DNA PCR-RFLPs. *Systematic Botany* 22, 701–712.

Monosmith, H. (1925) Male sterility in *Allium cepa* L. PhD thesis, University of California, Davis, California, USA.

Moon, E., Kao, T.H. and Wu, R. (1988) Rice mitochondrial genome contains a rearranged chloroplast gene cluster. *Molecular and General Genetics* 213, 247–253.

Moore, G. (1995) Cereal genome evolution: pastoral pursuits with 'Lego' genomes. *Current Opinions in Genetics and Development* 5, 717–724.

Morton, B. and Clegg, M. (1993) A chloroplast DNA mutational hotspot and gene conversion in a noncoding region near *rbcL* in the grass family (*Poaceae*). *Current Genetics* 24, 357–365.

Moue, T. and Uehara, T. (1985) Inheritance of cytoplasmic male sterility in *Allium fistulosum* L. (Welsh onion). *Journal of the Japanese Society for Horticultural Science* 53, 432–437.

Nakazono, M. and Hirai, A. (1993) Identification of the entire set of transferred chloroplast DNA sequences in the mitochondrial genome of rice. *Molecular and General Genetics* 236, 341–346.

Narayan, R.K.J. (1988) Constraints upon the organization and evolution of chromosomes in *Allium*. *Theoretical and Applied Genetics* 75, 319–329.

Newton, K.J. (1988) Plant mitochondrial genomes: organization, expression, and variation. *Annual Review of Plant Physiology* 39, 503–532.

Nugent, J. and Palmer, J.D. (1988) Location, identity, amount and serial entry of chloroplast DNA sequences in crucifer mitochondrial DNAs. *Current Genetics* 14, 501–509.

Ohno, S. (1970) *Evolution by Gene Duplication*. Springer-Verlag, New York, 160 pp.

Ohri, D., Fritsch, R.M. and Hanelt, P. (1998) Evolution of genome size in *Allium* (*Alliaceae*). *Plant Systematics and Evolution* 210, 57–86.

Palmer, J. (1985) Evolution of chloroplast and mitochondrial DNA in plants and algae. In: MacIntyre, R. (ed.) *Molecular Evolutionary Genetics*. Plenum Press, New York, pp. 131–240.

Palmer, J. (1990) Contrasting modes and tempos of genome evolution in land plant organelles. *Trends in Genetics* 6, 115–120.

Palmer, J. and Stein, D. (1986) Conservation of chloroplast genome structure among vascular plants. *Current Genetics* 10, 823–833.

Palmer, J.D. and Herbon, L.A. (1988) Plant mitochondrial DNA evolves rapidly in structure, but slowly in sequence. *Journal of Molecular Evolution* 28, 87–97.

Pathak, C. and Gowda, R. (1994) Breeding for the development of onion hybrids in India: problems and prospects. *Acta Horticulturae* 358, 239–242.

Pearce, S.R., Pich, U., Harrison, G., Flavell, A.J., Heslop-Harrison, J.S., Schubert, I. and Kumar, A. (1996) The *Ty1-copia* group retrotransposons of *Allium cepa* are distributed throughout the chromosomes but are enriched in the terminal heterochromatin. *Chromosome Research* 4, 357–364.

Peffley, E.B. (1986) Evidence for chromosomal differentiation of *Allium fistulosum* and *A. cepa*. *Journal of the American Society for Horticultural Science* 111, 126–129.

Peffley, E.B. and Mangum, P.D. (1990) Introgression of *Allium fistulosum* L. into *Allium cepa* L.: cytogenetic evidence. *Theoretical and Applied Genetics* 79, 113–118.

Peffley, E.B. and Orozco-Castillo, C. (1987) Polymorphism of isozymes within plant introductions of *Allium cepa* L. and *A. fistulosum* L. *HortScience* 22, 956–957.

Peffley, E.B., Corgan, J., Horak, K. and Tanksley, S.D. (1985) Electrophoretic analysis of *Allium* alien addition lines. *Theoretical and Applied Genetics* 71, 176–184.

Peterka, H., Budahn, H. and Schrader, O. (1997) Interspecific hybrids between onion (*Allium cepa* L.) with S cytoplasm and leek (*Allium ampeloprasum* L.). *Theoretical and Applied Genetics* 94, 383–389.

Peterson, C.E. and Foskett, R.L. (1953) Occurrence of pollen sterility in seed fields of Scott County Globe onions. *Proceedings of the American Society for Horticultural Science* 62, 443–448.

Pich, U., Fritsch, R. and Schubert, I. (1996a) Closely related *Allium* species (*Alliaceae*) share a very similar satellite sequence. *Plant Systematics and Evolution* 202, 255–264.

Pich, U., Fuchs, J. and Schubert, I. (1996b) How do *Alliaceae* stabilize their chromosome ends in the absence of TTTAGGG sequences? *Chromosome Research* 4, 207–213.

Pichersky, E. (1990) Nomad DNA – a model for movement and duplication of DNA sequences in plant genomes. *Plant Molecular Biology* 15, 437–448.

Pooler, M.R. and Simon, P.W. (1994) True seed production in garlic. *Sexual Plant Reproduction* 7, 282–286.

Potz, H. and Tatlioglu, T. (1993) Molecular analysis of cytoplasmic male sterility in chives (*Allium schoenoprasum* L.). *Theoretical and Applied Genetics* 87, 439–445.

Price, H.J. (1976) Evolution of DNA content in higher plants. *The Botanical Review* 42, 27–52.

Pring, D., Conde, M., Schertz, K. and Levings, C. (1982) Plasmid-like DNAs associated with mitochondria of cytoplasmic male-sterile sorghum. *Molecular and General Genetics* 186, 180–184.

Ranjekar, P.K., Pallotta, D. and Lafontaine, J.G. (1978) Analysis of plant genomes. V. Comparative study of molecular properties of DNAs of seven *Allium* species. *Biochemical Genetics* 16, 957–970.

Rhoades, M. (1931) The cytoplasmic inheritance of male sterility in *Zea mays*. *Science* 73, 340–341.

Ricroch, A., Peffley, E.B. and Baker, R.J. (1992) Chromosomal location of rDNA in *Allium*: *in situ* hybridization using biotin- and fluorescin-labelled probe. *Theoretical and Applied Genetics* 83, 413–418.

Robbins, T.P., Walker, E.L., Kermicle, J.L., Alleman, M. and Dellaporta, S.L. (1991) Meiotic instability of the R–r complex arising from displaced intragenic exchange in intrachromosomal rearrangement. *Genetics* 129, 271–283.

Ruge, B., Potz, H. and Tatlioglu, T. (1993) Influence of different cytoplasms and nuclear genes involved in the cms system of chives (*Allium schoenoprasum* L.) on microsporogenesis. *Plant Breeding* 110, 24–28.

Saini, S. and Davis, G. (1970) Karyotypic analysis of some *Allium* species. *Journal of the American Society for Horticultural Science* 95, 102–105.

Samoylov, A., Klaas, M. and Hanelt, P. (1995) Use of chloroplast DNA polymorphisms for the phylogenetic study of the subgenera *Amerallium* and *Bromatorrhiza* (genus *Allium*). *Feddes Repertorium* 106, 161–167.

Samoylov, A., Klaas, M. and Hanelt, P. (1999) Use of chloroplast DNA polymorphisms for the phylogenetic study of *Allium* subgenus *Amerallium* and subgenus *Bromatorrhiza* (*Alliaceae*) II. *Feddes Repertorium* 110, 103–109.

Sandbrink, J., van Bruggen, A. and van Brederode, J. (1990) Patterns of infraspecific chloroplast DNA variation in species of *Silene*. *Biochemical and Systematic Ecology* 18, 233–238.

San Miguel, P., Tikhonov, A., Jin, Y.K., Motchoulskaia, N., Zakharov, D., Melake-Berhan, A., Springer, P.S., Edwards, K.J., Lee, M., Avramova, Z. and Bennetzen, J.L. (1996). Nested retrotransposons in the intergenic regions of the maize genome. *Science* 274, 765–768.

Sato, Y. (1998) PCR amplification of CMS-specific mitochondrial nucleotide sequences to identify cytoplasmic genotypes of onion (*Allium cepa* L.). *Theoretical and Applied Genetics* 96, 367–370.

Satoh, Y., Nagai, M., Mikami, T. and Kinoshita, T. (1993) The use of mitochondrial DNA polymorphism in the classification of individual plants by cytoplasmic genotypes. *Theoretical and Applied Genetics* 86, 345–348.

Saumitou-Laprade, P., Rouwendal, G., Cuguen, J., Krens, F. and Michaelis, G. (1993) Different CMS sources found in *Beta vulgaris* ssp. *maritima*: mitochondrial variability in wild populations revealed by a rapid screening procedure. *Theoretical and Applied Genetics* 85, 529–535.

Schubert, I. and Wobus, U. (1985) *In situ* hybridization confirms jumping nucleolus organizing regions in *Allium*. *Chromosoma* 92, 143–148.

Schubert, I., Ohle, H. and Hanelt, P. (1983) Phylogenetic conclusions from Giemsa banding and NOR staining in top onions (*Liliaceae*). *Plant Systematics and Evolution* 143, 245–256.

Schweisguth, B. (1970) Études préliminaires à l'amélioration du poireau: proposition d'une méthode d'amélioration. *Annales de l'Amélioration des Plantes* 20, 215–231.

Schweisguth, B. (1973) Étude d'un nouveau type de stérilité mâle chez l'oignon, *Allium cepa* L. *Annales de l'Amélioration des Plantes* 23, 221–233.

Shigyo, M., Tashiro, Y., Isshiki, S. and Miyazaki, S. (1996) Establishment of a series of monosomic addition lines of Japanese bunching onion (*Allium fistulosum* L.) with extra chromosomes from shallot (*A. cepa* L. Aggregatum group). *Genes and Genetic Systems* 71, 363–371.

Shigyo, M., Tashiro, Y., Iino, M., Terahara, N., Ishimaru, K. and Isshiki, S. (1997a) Chromosomal locations of genes related to flavonoid and anthocyanin production in leaf sheath of shallot (*A. cepa* L. Aggregatum group). *Genes and Genetic Systems* 72, 149–152.

Shigyo, M., Iino, M., Isshiki, S. and Tashiro, Y. (1997b) Morphological characteristics of a series of monosomic addition lines of Japanese bunching onion (*Allium fistulosum* L.) with extra chromosomes from shallot (*A. cepa* L. Aggregatum group). *Genes and Genetic Systems* 72, 181–186.

Shigyo, M., Miyazaki, S., Isshiki, S. and Tashiro, Y. (1997c) Assignment of randomly amplified polymorphic DNA markers to all chromosomes of shallot (*A. cepa* L. Aggregatum group). *Genes and Genetic Systems* 72, 249–252.

Shinozaki, K., Ohme, M., Tanaka, M., Wakasugi, T., Hayashida, N., Matsubayashi, T., Zaita, N., Chunwongse, J., Obokata, J., Yamaguchi-Shinozaki, K., Ohto, C., Torazawa, K., Meng, B.Y., Sugita, M., Deno, H., Kamogashira, T., Yamada, K., Kusuda, J., Takaiwa, F., Kato, A., Tohdoh, N., Shimada, H. and Sugiura, M. (1986) The complete nucleotide sequence of tobacco chloroplast genome: its gene organization and expression. *EMBO Journal* 5, 2043–2049.

Shoemaker, R.C., Polzin, K., Labate, J., Specht, J., Brummer, E.C., Olson, T., Young, N., Concibido, V., Wilcox, J., Tamulonis, J., Kochert, G. and Boersma, H. (1996) Genome duplication in soybean (*Glycine* subgenus *soja*). *Genetics* 144, 329–338.

Slocum, M.K., Figdore, S.S., Kennard, W.C., Suzuki, J.Y. and Osborn, T.C. (1990) Linkage arrangement of restriction fragment length polymorphism loci in *Brassica oleracea*. *Theoretical and Applied Genetics* 80, 57–64.

Smith, B. and Crowther, T. (1995) Inbreeding depression and single cross hybrids in leeks (*Allium ampeloprasum* ssp. *porrum*). *Euphytica* 86, 87–94.

Smith, G.P. (1976) Evolution of repeated DNA sequences by unequal crossover. *Science* 191, 528–535.

Stack, S.M. and Comings, D.E. (1979) The chromosomes and DNA of *Allium cepa*. *Chromosoma* 70, 161–181.

Stack, S.M. and Roelofs, D. (1996) Localized chiasmata and meiotic nodules in the tetraploid *Allium porrum*. *Genome* 39, 770–773.

Stern, D.B. and Palmer, J.D. (1984) Recombinational sequences in plant mitochondrial genomes: diversity and homologies to known mitochondrial genes. *Nucleic Acids Research* 12, 6141–6157.

Tanksley, S. (1993) Mapping polygenes. *Annual Review of Genetics* 27, 205–233.

Tanksley, S.D., Bernatzky, R., Lapitan, N.L. and Prince, J.P. (1988) Conservation of gene repertoire but not gene order in pepper and tomato. *Proceedings of the National Academy of Sciences of the USA* 85, 6419–6423.

Tanksley, S.D., Ganal, M.W., Prince, J.P., de Vicente, M.C., Bonierbale, M.W., Brown, P., Fulton, T., Giovannoni, J., Grandillo, S., Martin, G., Messeguer, R., Miller, J., Miller, L., Paterson, A., Pineda, O., Roder, M., Wing, R., Wu, W. and Young, N. (1992) High density molecular linkage maps of the tomato and potato genomes. *Genetics* 132, 1141–1160.

Tatabe, T. (1952) Cytological studies on pollen degeneration in male-sterile onion. *Journal of the Japanese Society for Horticultural Science* 21, 73–75.

Tatabe, T. (1968) Genetic studies on the leaf variegation of *Allium cepa* L. *Journal of the Japanese Society for Horticultural Science* 37, 345–348.

Tatlioglu, T. (1982) Cytoplasmic male sterility in chives (*Allium schoenoprasum* L.). *Zeitschrift für Pflanzenzüchtung* 89, 251–262.

Tatlioglu, T. (1985) Influence of temperature on the expression of cytoplasmic male sterility in chives (*Allium schoenoprasum* L.). *Zeitschrift für Pflanzenzüchtung* 94, 156–161.

Tatlioglu, T. (1986) Influence of tetracycline on the expression of cytoplasmic male sterility (cms) in chives (*Allium schoenoprasum* L.). *Plant Breeding* 97, 46–55.

Tatlioglu, T. (1987) Genetic control of temperature-sensitivity of cytoplasmic male sterility (cms) in chives (*Allium schoenoprasum* L.). *Plant Breeding* 99, 65–76.

Tatlioglu, T. and Wricke, G. (1988) Genetic control of tetracycline-sensitivity of cytoplasmic male sterility (cms) in chives (*Allium schoenoprasum* L.). *Plant Breeding* 100, 34–40.

Unseld, M., Marienfeld, J.R. and Brennicke, A. (1997) The mitochondrial genome of *Arabidopsis thaliana* contains 57 genes in 366,924 nucleotides. *Nature Genetics* 15, 57–61.

van der Meer, Q.P. and van Bennekom, J.L. (1969) Effect of temperature on the occurrence of male sterility in onion. *Euphytica* 18, 389–394.

van der Meer, Q.P. and van Bennekom, J.L. (1970) Failure of graft transmission of male-sterilizing substance in onion (*Allium cepa* L.). *Euphytica* 19, 430–432.

van der Meer, Q.P. and van Bennekom, J.L. (1978) Improving the onion crop (*Allium cepa* L.) by transfer of characters from *Allium fistulosum* L. *Biuletyn Warzywniczy* 22, 87–91.

van Heusden, A.W., van Ooijen, J.W., Vrielink-van Ginkel, R., Verbeek, W.H.J., Wietsma, W.A. and Kik, C. (2000a) A genetic map of an interspecific cross in *Allium* based on amplified fragment length polymorphism (AFLP) markers. *Theoretical and Applied Genetics* 100, 118–126.

van Heusden, A.W., Shigyo, M., Tashiro, Y., Vrielink-van Ginkel, R. and Kik, C. (2000b) AFLP linkage group assignment to the chromosomes of *Allium cepa* L. via monosomic addition lines. *Theoretical and Applied Genetics* 100, 480–486.

Vanin, E.F. (1985) Processed pseudogenes: characteristics and evolution. *Annual Review of Genetics* 19, 253–272.

van Raamsdonk, L., Wietsma, W. and de Vries, J. (1992) Crossing experiments in *Allium* L. section *Cepa*. *Botanical Journal of the Linnean Society* 109, 293–303.

Ved Brat, S. (1965a) Genetic systems in *Allium*. I. Chromosome variation. *Chromosoma* 16, 486–499.

Ved Brat, S. (1965b) Genetic systems in *Allium*. III. Meiosis and breeding systems. *Heredity* 20, 325–339.

Veit, B., Vollbrecht, E., Mathern, J. and Hake, S. (1990) A tandem duplication causes the *Kn1-0* allele of *Knotted*, a dominant morphological mutant of maize. *Genetics* 125, 623–631.

Villanueva-Mosqueda, E. (1999) GISH analysis of advanced backcross plants from an interspecific hybrid between *Allium cepa* and *A. fistulosum* and genetics of seed yield in onion. PhD thesis, University of Wisconsin-Madison, Wisconsin, USA.

Vosa, C. (1976) Heterochromatic patterns in *Allium*. I. The relationship between the species of the *cepa* group and its allies. *Heredity* 36, 383–392.

Ward, B.L., Anderson, R.S. and Bendich, A.J. (1981) The mitochondrial genome is large and variable in a family of plants (*Cucurbitaceae*). *Cell* 25, 793–803.

Wolfe, K., Li, W. and Sharp, P. (1987) Rates of nucleotide substitution vary greatly among plant mitochondrial, chloroplast, and nuclear DNAs. *Proceedings of the National Academy of Sciences of the USA* 84, 9054–9058.

Yamashita, K. and Tashiro, Y. (1999) Possibility of developing male-sterile line of shallot (*Allium cepa* L. Aggregatum group) with cytoplasm from *A. galanthum* Kar. et Kir. *Journal of the Japanese Society for Horticultural Science* 68, 256–262.

Yamashita, K., Oyama, T., Nora, R., Miyazaki, T. and Tashiro, Y. (1998) Comparative study on methods for identification of chloroplast DNA of cultivated and wild species in section *Cepa* of *Allium*. *Bulletin of the Faculty of Agriculture, Saga University (Japan)* 83, 111–120.

Yamashita, K., Arita, H. and Tashiro, Y. (1999a) Cytoplasm of a wild species *Allium galanthum* Kar. et Kir. is useful for developing male-sterile line of *A. fistulosum* L. *Journal of the Japanese Society for Horticultural Science* 68, 788–797.

Yamashita, K., Arita, H. and Tashiro, Y. (1999b) Isozyme and RAPD markers linked to fertility restoring gene for cytoplasmic male-sterile *Allium fistulosum* L. with cytoplasm of *A. galanthum* Kar. et Kir. *Journal of the Japanese Society for Horticultural Science* 68, 954–959.

Yen, D.E. (1959) Pollen sterility in Pukekohe Longkeeper onions. *New Zealand Journal of Agricultural Research* 2, 605–612.

4 Exploitation of Wild Relatives for the Breeding of Cultivated *Allium* Species*

C. Kik

Plant Research International, Wageningen University and Research Center, PO Box 16, 6700 AA Wageningen, The Netherlands

1. Introduction

Interspecific hybridization has attracted considerable attention throughout the centuries. The study of this phenomenon was initiated by Linnaeus, who suggested that new species originated via hybridization (Roberts, 1929). In the first half of the 20th century, there was speculation that hybridization may play a major role in adaptive evolution (Anderson, 1949; Stebbins, 1950). At that time, the importance of hybridization in evolution was difficult to assess because the tools used to study plant hybridization and related phenomena were relatively undeveloped (Grant, 1971; Heiser, 1973; Stace, 1975). Since 1980, the application of molecular-marker technology has made it clear that species hybridization has been greatly underestimated and that this phenomenon plays an important role in evolution (Rieseberg and Wendel, 1993; Rieseberg *et al.*, 1996; Rieseberg, 1998).

In plant-improvement programmes, species hybridization has always been an important tool for the introduction of genetic variation in the breeding of new cultivars, as wild relatives of cultivated species contain gene reservoirs for agronomically useful traits (Zeven and van Harten, 1978; Kalloo and Chowdhury, 1992). The classical route to enriching domesticated plants with 'wild' genes is via recurrent back-crossing, in which 'wild' donor genes are introgressed

*This chapter is dedicated to the memory of my father, Adriaan Kik.

into a recipient crop genome. However, several pre- and post-fertilization hybridization barriers are difficult to overcome (Hadley and Openshaw, 1980; Khush and Brar, 1992; van Tuyl, 1997). Hence, new routes were developed in the 1970s and 1980s to circumvent these barriers, e.g. somatic hy(cy)bridization (Glimelius *et al.*, 1991) and genetic transformation (*Agrobacterium*-mediated: Zupan and Zambryski, 1995; particle gun: Christou, 1993). The potential of these recently developed techniques for plant breeding and *Allium* breeding in particular is discussed by Eady (Chapter 6, this volume). This chapter will focus on the current state-of-the-art of sexual hybridization in the major cultivated edible and ornamental *Allium* species.

2. Alliums

The role of wild relatives in crop improvement in the economically important crops worldwide is impressive (for wheat: Jiang *et al.*, 1994; Sharma, 1995; Fedak, 1999; for rice: Brar and Khush, 1997; for cotton: Wendel *et al.*, 1989; for maize: Williams *et al.*, 1995; for sugarbeet: van Geyt *et al.*, 1990). Improvement of cultivated *Alliums* with wild relatives by introgression breeding, however, has not yet progressed very far. This is partly due to the prolonged juvenile phase of most economically important *Alliums*, which makes breeding a time-consuming process, and to the comparatively low economic importance of many alliaceous crops. Onion, the most important *Allium* crop, ranks second in value after tomatoes on the list of cultivated vegetable crops worldwide. However, compared with the leading economic crops, such as wheat, soybean, tobacco, maize and rice, it occupies a relatively modest position (FAO, 2001).

2.1 Edible alliums

Onion (*Allium cepa* L.), Japanese bunching onion (syn. Welsh onion: *A. fistulosum* L.), leek (*A. ampeloprasum* L. leek group) and garlic (*A. sativum* L.) are the most important cultivated edible *Allium* crops. Alliums are mostly used as condiments for a wide variety of dishes; however, since ancient times, their medicinal value has also been recognized (see Keusgen, Chapter 15, this volume). Nowadays, garlic preparations are commonly used in the prevention of cardiovascular diseases and specific types of cancer (Koch and Lawson, 1996).

Onion and garlic are grown worldwide, whereas leek is predominantly cultivated in Europe and Japanese bunching onion in East Asia. The productivity of these crops is affected by several factors, both biotic (diseases and pests: Rabinowitch, 1997) and abiotic (unfavourable soil, temperature and water conditions: e.g. Wannamaker and Pike, 1987). The genetic variability within the four crops is limited; therefore, there is a need to broaden these genomes with genes from diverse sources. Tissue-culture techniques have made the hybridization of distant species possible (Ohsumi *et al.*, 1993; Buiteveld *et al.*, 1998), and genetic transformation facilitates the introduction of alien genes into the species of interest (Myers and Simon, 1998; Eady *et al.*, 2000; Zheng *et al.*, 2001; Eady, Chapter 6, this volume). Moreover, recent advances in *Allium* molecular-marker (King *et al.*, 1998; Klaas, 1998; van Heusden *et al.*, 2000a, b; Klaas and Friesen, Chapter 8, this volume) and *in situ* hybridization technology (Ricroch *et al.*, 1992; Hizume, 1994; Khrustaleva and Kik, 2000) have enabled precise detection of introgressed chromosome segments from wild plants into cultivated *Allium* species.

Onion (Jones, 1990), Japanese bunching onion (Inden and Asahira, 1990) and garlic (Etoh, 1986; Etoh and Simon, Chapter 5, this volume) are diploid species ($2n = 2x = 16$), and leek is a tetraploid ($2n = 4x = 32$) (Currah, 1986; van der Meer and Hanelt, 1990; De Clercq and Van Bockstaele, Chapter 18, this volume). Onion, Japanese bunching onion and leek are biennials, and the reproductive system of these crops is predominantly cross-fertilization, although selfing is possible (onion: Pike, 1986; Dowker, 1990; bunching onion: Inden and Asahira, 1990; leek: Currah, 1986; van der Meer and Hanelt, 1990; De Clercq and Van

Bockstaele, Chapter 18, this volume; garlic: Etoh, 1997; Etoh and Simon, Chapter 5, this volume). In garlic, the development of flowers is severely suppressed by the development of bulbils (topsets) in the umbel (Pooler and Simon, 1994; Etoh and Simon, Chapter 5, and Kamenetsky and Rabinowitch, Chapter 2, this volume). This competition in the umbel between generative and vegetative meristems leads in practice to flower and flower bud degeneration (Kamenetsky and Rabinowitch, 2001; Kamenetsky and Rabinowitch, Chapter 2, this volume) and consequently to complete sterility. Therefore clonal selection has, for many centuries, been the only method for improving garlic (Etoh and Simon, Chapter 5, this volume). In onion, Japanese bunching onion and leek, the breeding methods employed are predominantly hybridization between remote genotypes, to increase genetic variability. This is followed by selfing and mass or family selection within segregating populations. Cross-pollination and hybrid-seed production is facilitated by male sterility, both genic and cytoplasmic, as in the bulb onion (Kaul, 1988; Dowker, 1990; Havey, Chapter 3, this volume), or by genic male sterility, as in leek (Smith and Crowther, 1995; De Clercq and Van Bockstaele, Chapter 18, this volume).

2.1.1 Onion

The bulb onion is a cultigen, which is not found in the wild. It has recently become clear that *A. vavilovii* is its closest known relative, because the two species are completely interfertile and morphologically they are quite similar (Hanelt, 1990; Fritsch and Friesen, Chapter 1, this volume). *Allium vavilovii* and many other relatives of onion grow wild in the Tien-Shan and Pamir-altai mountainous ranges, which form the border between Kazahkstan and China.

Onion belongs to the subgenus *Rhizirideum*, section *Cepa*. Based on the taxonomy of Hanelt (1990), Fritsch and Friesen (Chapter 1, this volume) have subdivided this section into four groups: (i) the *Galanthum* alliance with *A. galanthum*, *A. farctum* and *A. pskemense*; (ii) the *Oschaninii* alliance, with *A. oschaninii* and *A. praemixtum*; (iii) the informal *Cepa* alliance, with *A. cepa*, *A. asarense* and *A. vavilovii*; and (iv) the *Altaicum* alliance, with *A. altaicum* and *A. fistulosum*. Using the main representatives of these four alliances in an extensive phylogenetic analysis, this subdivision into four groups has been largely confirmed by van Raamsdonk *et al.* (1997, 2000; Fig. 4.1).

Crossability analysis of onion with its wild relatives showed that *A. vavilovii* is completely interfertile with onion, that *A. oschaninii* is completely intersterile, and that *A. fistulosum*, *A. altaicum*, *A. galanthum* and *A. pskemense* show low levels of interfertility, due to severe crossing barriers (Saini and Davis, 1969; McCollum, 1971; Gonzalez and Ford-Lloyd, 1987; van Raamsdonk *et al.*, 1992).

Crosses of onion with species from the other sections of the subgenus are possible. Peterka and Budahn (1996) presented evidence that onion can be crossed with chives (*A. schoenoprasum*) and Nomura and Makura (1996) crossed onion with rakkyo (*A. chinense*).

Keller *et al.* (1996) analysed in detail the hybrid status of a number of intersectional hybrids. The most notable intersectional hybrid, from a breeding point of view, is the hybrid between onion and *A. roylei* (van der Meer and de Vries, 1990). The taxonomic position of *A. roylei* is unclear: Hooker (cited by Stearn, 1946) placed it in the section *Schoenoprasum*, Wendelbo (1971) in the section *Rhizirideum* and Labani and Elkington (1987) in the section *Cepa*. Recently, van Raamsdonk *et al.* (1997, 2000) showed that the species probably has a hybrid origin, as its nuclear (nu) DNA profile is related to members of the section *Cepa* and its chloroplast (cp) DNA profile to the section *Schoenoprasum* (see also Fritsch and Friesen, Chapter 1, and Havey, Chapter 3, this volume).

Successful crosses of onion with species from subgenus *Allium* have also been carried out using the embryo-rescue technique: onion was crossed with leek (Peterka *et al.*, 1997), with *A. sphaerocephalon* (Keller *et al.*, 1996) and with garlic (Ohsumi *et al.*, 1993). However, all the interspecific hybrids were completely sterile.

It is evident that a number of wild species can be successfully crossed with onion, but

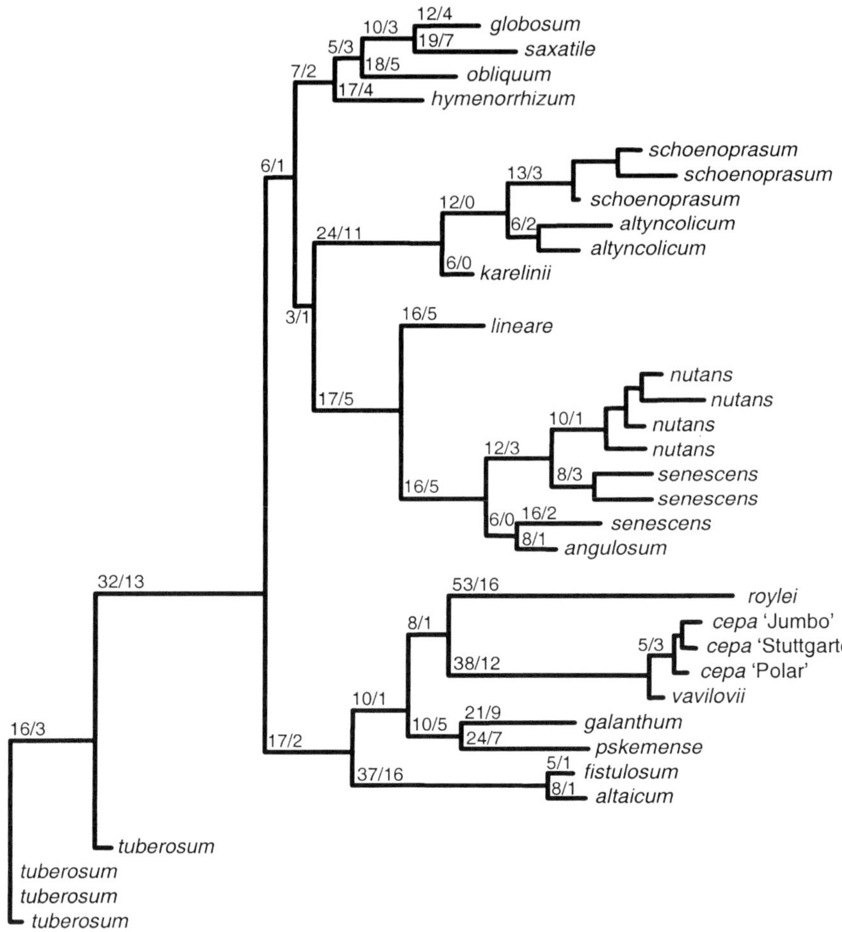

Fig. 4.1. A most parsimonious tree after phylogenetic analysis of 355 AFLP characters scored for a number of accessions of *Allium* subgenus *Rhizirideum.* Indices (x/y) at each branch indicate the total number of characters supporting the branch (x) and the number of unique characters (y). (From van Raamsdonk *et al.*, 2000, with permission.)

the actual exploitation of the genome of wild species depends on the fertility of the offspring and the presence of agronomically beneficial traits. Until now, most attention has been focused on disease and pest resistance (Rabinowitch, 1997). In view of the increasing demand for product diversification and the focus on health issues, it can be envisaged that in the near future attention will also be paid to the onion's sulphur-containing compounds (Block, 1992; Randle *et al.*, 1995; van Raamsdonk and Kik, 1997; Randle and Lancaster, Chapter 14, this volume), fructans (Simon, 1995; Vijn *et al.*, 1997; van Raamsdonk and Kik, 1997) and flavonoid (Patil *et al.*, 1995) metabolism. In the following sections, the three most important introgression cases for onion will be discussed, namely introgression from *A. roylei*, *A. fistulosum* and *A. galanthum*.

INTROGRESSION FROM *ALLIUM ROYLEI.* Researchers became interested in the potential gene reservoir of *A. roylei* for the bulb onion when the wild species proved to be completely resistant to downy mildew (*Peronospora destructor*) (Kofoet *et al.*, 1990) and partially resistant to onion-leaf blight

(*Botrytis squamosa*) (de Vries *et al.*, 1992a). The first successful sexual cross between a male-sterile onion and *A. roylei* was reported by van der Meer and de Vries (1990). They obtained a partially fertile interspecific hybrid, which was subsequently backcrossed to onion and generated a morphologically variable backcross (BC_1) population (Fig. 4.2). In the pollen meiosis of the interspecific hybrid, no multivalents were observed but only bivalents and a very limited number of univalents (92.2% bound bivalent arms; de Vries *et al.*, 1992c). Furthermore, two chiasmata per bivalent were usually present during diakinesis, and no indications were present, in metaphase I of the pollen mother cells' (PMCs') meiosis, of cytoplasmic effects on chiasma formation (de Vries *et al.*, 1992b, c). However, de Vries and Jongerius (1992) found a pericentric inversion in the first meiotic metaphase in 14% of the PMCs of the interspecific hybrid. This inversion appeared to be present in ten out of 20 BC_1 *A. cepa* × (*A. cepa* × *A. roylei*) plants analysed. De Vries and Jongerius (1992) concluded that the consequences of this inversion for introgression breeding are not important, because repeated backcrossing will automatically eliminate the plants with this genotype.

When analysing a BC_1 *A. cepa* × (*A. cepa* × *A. roylei*) and an interspecific F_1 population, Kofoet *et al.* (1990) found that the resistance to downy mildew present in *A. roylei* segregated in 1 : 1 and 1 : 0 (resistant : susceptible) ratios, respectively. This led to the conclusion that a single dominant gene controls this trait. When analysing two selfed populations of the interspecific hybrid (termed interspecific F_2 populations), however, de Vries *et al.* (1992b, c) concluded that the resistance was based on two weakly linked nuclear genes (recombination frequency 0.32). Using an advanced fluorescent *in situ* hybridization (FISH) technique, L.I. Khrustaleva (Wageningen, The Netherlands, 2000, personal communication) observed that the position of the *Pd* marker (*Pd* is the locus for downy mildew resistance) was on the distal end of *A. roylei* chromosome 2. Van Heusden *et al.* (2000b) reported that there was a considerably skewed segregation towards the wild *A. roylei* alleles in the F_2 population. Skewed

Fig. 4.2. Morphological variation in the backcross *A. cepa* × (*A. cepa* × *A. roylei*).

segregation is well known in interspecific crosses (Zamir and Tadmor, 1986), although the causes for this phenomenon are not completely clear. Xu *et al.* (1997) and Virk *et al.* (1998) analysed skewed segregation in rice and concluded that one-third of the skewed segregating loci were affected by gametophytic/sterility genes and another third by the association with the indica–japonica subspecies differentiation. Based on a segregating F_2 population, van Heusden *et al.* (2000a) constructed a high density amplified fragment length polymorphism (AFLP) molecular-marker map and finally showed that resistance to downy mildew is determined by a single locus that is located on the distal end of linkage group 2 of *A. roylei* (Fig. 4.3).

Bulked segregant analysis (BSA) (Michelmore *et al.*, 1991) revealed a linkage between the downy-mildew-resistance gene and a randomly amplified polymorphic DNA (RAPD) marker (Williams *et al.*, 1990). The most closely linked marker was at 2.7 centimorgans (cM) distance from the *Pd* resistance gene (de Vries *et al.*, 1992d; Kik *et al.*, 1997a). This RAPD marker has been cloned and 20-base-pair primers have been designed in order to develop a user-friendly sequence-characterized amplified region (SCAR) marker. Currently, several breeding companies are developing downy-mildew-resistant onion cultivars based on the resistance introgressed from *A. roylei*, using the *Pd*-SCAR marker in the selection process.

In the near future, the time span for introgressing alien genes into the cultivated gene pool will probably be reduced due to the use of molecular-marker maps. These maps will allow the selection of individual genotypes that predominantly contain the genome of the desired crop, but have retained the alien alleles of interest. Through this marker-assisted breeding (MAB) approach, the number of backcrosses for the introgression of 'wild' genes into a cultivated species can be reduced from the six or seven generations commonly employed today to two or three generations (Patterson, 1996). For the biennial-breeding onion and other alliaceous crops, this will have a major impact.

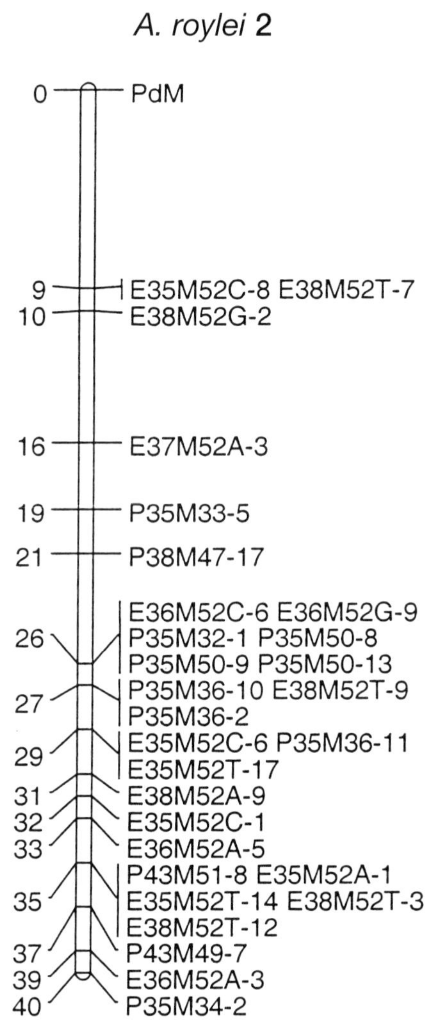

Fig. 4.3. A detail of *A. roylei* linkage group 2 involving the position of the SCAR marker (PdM) for downy mildew (*Peronospora destructor*) resistance. Map distances (in cM) are to the left, and AFLP locus designations to the right of the linkage group.

INTROGRESSION FROM *ALLIUM FISTULOSUM* (JAPANESE BUNCHING ONION). Among all the interspecific crosses in the genus *Allium*, the cross between *A. fistulosum* and onion has been studied the most extensively, because Japanese bunching onion harbours several agronomically desired traits. The species carries resistance genes against onion leaf blight (*B. squamosa*) (Currah and Maude, 1984), pink root (*Pyrenochaeta terrestris*)

(Netzer *et al.*, 1985), anthracnose (*Colletotrichum gloeosporioides*) (Galvan *et al.*, 1997), smut (*Urocystis cepulae*) (Felix, 1933) and onion yellow-dwarf virus (OYDV) (Rabinowitch, 1997). In addition, *A. fistulosum* has a higher dry-matter content, is more pungent and winter-hardy, flowers earlier, has a shorter flowering period and has a higher attractiveness for pollinators compared with the bulb onion (van der Meer and van Bennekom, 1978). Although strictly speaking *A. fistulosum* is not a wild species, we shall deal here with interspecific crosses involving this species and bulb onion.

The bulb onion and the non-bulbing *A. fistulosum* differ in scape morphology, leaf cross-section, bulbing degree and perianth colour and shape (Vvedensky, 1944). However, the two species appear to be closely related because of their equal chromosome numbers, similar karyotypes (Vosa, 1976) and similarities in their cpDNA restriction patterns (Havey, 1991). The first interspecific hybrid between *A. cepa* and *A. fistulosum* was obtained as early as 1931. It proved to be almost sterile, although occasionally a few seeds were obtained from self-pollination of the interspecific hybrid (Emsweller and Jones, 1935a, b). A number of interspecific hybrids between onion and *A. fistulosum* are cultivated commercially, the most important ones being 'Beltsville Bunching', 'Delta Giant', 'Top Onion' and 'Wakegi Onion'.

The fertile hybrid 'Beltsville Bunching' is an amphidiploid species, possessing two chromosome complements from *A. cepa* and two from *A. fistulosum* ($2n = 2x = 32$; CCFF). It is grown from seed, and cultivated as a minor crop in the USA. The triploid hybrid ($2n = 3x = 24$; CCF) 'Delta Giant' is cultivated on a small scale in the USA and is propagated vegetatively (for more details, see Rabinowitch and Kamenetsky, Chapter 17, this volume). 'Top Onion' (*A.* × *proliferum* (Moench) Schrad.) is a diploid interspecific hybrid ($2n = 2x = 16$; CF) between onion and *A. fistulosum*, as has been determined by biochemical and molecular methods (Havey, 1991; Maaß, 1997) and genomic *in situ* hybridization (GISH) (Friesen and Klaas, 1998). McCollum (1974) showed that the selfed progeny of 'Top Onion' either resemble *A. fistulosum* or the viviparous 'Top Onion', but not the bulb onion. Havey (1991) demonstrated that the cpDNA restriction pattern of 'Top Onion' closely resembles that of *A. fistulosum*, which might explain the type of segregation encountered, assuming nucleocytoplasmic incompatibility, i.e. the cytoplasm of one species does not allow expression of the nuclear genes of the other. 'Top Onion' is propagated vegetatively and used as a garden crop in temperate zones. 'Wakegi Onion' (*Allium wakegi* Araki) is a diploid interspecific hybrid ($2n = 2x = 16$; CF) between shallot and *A. fistulosum*. It is propagated vegetatively and is frequently confused with true shallots (Arifin *et al.*, 2000). The plant is cultivated predominantly in tropical and subtropical regions in Asia, for example the well-known cultivar 'Sumenep' in Indonesia. Using GISH, Hizume (1994) unequivocally established the hybrid origin of this crop. Arifin *et al.* (2000) found from restriction fragment length polymorphism (RFLP) analysis that shallot was the maternal parent and Japanese bunching onion the paternal parent of 'Wakegi Onion' and that reciprocal crosses also existed.

Electron-microscopy analysis of the synaptonemal complex (SC) of the interspecific hybrid between onion and *A. fistulosum* showed that heteromorphic bivalents are present, that the male chiasma frequency was reduced compared with that of both parents and that chiasmata were predominantly interstitial and distal (Albini and Jones, 1990). Furthermore, Albini and Jones (1990) found that synapsis in the centromeric region of the interspecific hybrid is disturbed and that irregularities occurred in the SC. In the interspecific hybrid between onion and *A. fistulosum*, Stevenson *et al.* (1998) observed a 20% deficit of chiasmata in metaphase I compared with GISH-based labelled exchanges in anaphase I, thus confirming the above conclusion.

The BC_1 of the interspecific hybrid with *A. cepa* usually results in a limited number of progeny and an irregular segregation (Cryder *et al.*, 1991; van der Valk *et al.*, 1991b; Bark *et al.*, 1994), probably due to a prefertilization barrier as the growth of the '*cepa*' pollen tubes is heavily disturbed in the style of the interspecific hybrid (van der Valk *et al.*, 1991a). Cytogenetic analysis of the backcross showed that at least three paracentric inversions and one translocation are present (Peffley and Mangum, 1990; Ulloa *et al.*, 1994, 1995). Peffley and Mangum (1990) and Cryder *et al.* (1991) provided evidence that limited recombination between the two genomes is possible. Analysis of a BC_2 population by Ulloa *et al.* (1995) showed that the majority of the plants resembled *A. cepa*. However, the reproductive organs of the *A. cepa*-type plants were morphologically abnormal, resulting in a negligible seed set of the BC_2 plants. This led Ulloa *et al.* (1995) to conclude that nucleocytoplasmic incompatibility might be the cause underlying the species barrier between *A. cepa* and *A. fistulosum*. Contrary to Ulloa *et al.* (1995), Peffley and Hou (2000) found in F_1BC_3 populations that introgression of *A. fistulosum* into the genome of onion is possible. They suggested that this was due to the fact that the cytoplasm of their backcross populations originated from onion, and not, as in previous studies, from *A. fistulosum*. Using GISH, L.I. Khrustaleva and C. Kik (unpublished results; Colour Plate 2A) showed that, in a BC_1 population ((*A. fistulosum* × *A. cepa*) × *A. cepa*), homologous recombination takes place between both genomes. Villanueva-Mosqueda and Havey (1998) reported that in a sixth back-cross population of *A. cepa* × *A. fistulosum* to *A. cepa*, *A. fistulosum* segments could be observed via GISH in an onion genetic background. Using GISH, this was not observed in monosomic addition lines between shallot and *A. fistulosum* (Shigyo *et al.*, 1998).

To circumvent the sterility problem and to investigate whether the two important gene reservoirs for onion could be exploited simultaneously, de Vries *et al.* (1992e) and Khrustaleva and Kik (1998) used *A. roylei* for bridge-crossing to introgress genes from *A. fistulosum* into onion. *A. roylei* is a good candidate for functioning as a bridge between the two species: it crosses readily with both *A. cepa* (van der Meer and de Vries, 1990) and *A. fistulosum* (McCollum, 1982) and has an amount of DNA (28.5 pg DNA per 2C) intermediate between that of *A. cepa* (33.5 pg DNA per 2C) and *A. fistulosum* (22.5 pg DNA per 2C) (Labani and Elkington, 1987). All three species are diploids with identical chromosome numbers ($2n = 2x = 16$).

Using a multicolour GISH method, Khrustaleva and Kik (1998) showed that the three parental genomes in the first generation bridge cross *A. cepa* × (*A. fistulosum* × *A. roylei*) could be distinguished from each other, indicating significant differences in repetitive DNA composition among the three species (Colour Plate 2B). A meiotic analysis of the first-generation bridge cross revealed a high percentage of bound bivalent arms (82.6%) at metaphase I of meiosis. However, some degree of genome instability existed, indicated by the presence of occasional univalents in meiosis. Pollen fertility in the first-generation bridge cross was average. In a more detailed study, Khrustaleva and Kik (2000) analysed the meiotic anaphase I and prophase II of the first-generation bridge-cross individuals, and showed a large number of recombinations between the three genomes. Occasional translocations were observed in the second-generation bridge cross. Irregularities in the SC might also have occurred, as the number of observed recombination points in anaphase I and prophase II greatly exceeded the value expected from chiasma frequency in metaphase I. Recombination points were randomly distributed over the chromosomes, suggesting that the *A. cepa* or *A. roylei* type of random chiasma distribution prevails over the *A. fistulosum* type of proximally localized chiasma distribution (Colour Plate 2C). Variation in pollen fertility occurred in the second-generation bridge-cross population.

From a breeding point of view this is fortunate, because genotypes with a high fertility can be selected for further breeding. The variation in pollen fertility could be due to the use of a cytoplasmic male-sterile *A. cepa* mother plant in the bridge cross, in combination with the presence or absence of restorer genes in the second-generation bridge-cross individuals (de Vries and Wietsma, 1992). However, incongruity between the three species could also be involved. Furthermore, it is not clear if all the 'wild' chromatin from both *A. fistulosum* and *A. roylei* can be introgressed into onion. As this bridge-cross approach represents for the first time a real possibility for simultaneously exploiting two important gene reservoirs for onion, more research in this direction is clearly warranted.

INTROGRESSION FROM *ALLIUM GALANTHUM*. The development of interspecific hybrids between the bulb onion and *A. galanthum* was attempted in a number of studies (Saini and Davis, 1967, 1969; McCollum, 1971; van Raamsdonk *et al.*, 1992). The interspecific hybrid proved to be highly sterile, although some seed set was observed. Yamashita and Tashiro (1999) reported that, although fertility was very low in the interspecific hybrid and early back-cross generations, it was eventually restored. They obtained a similar result when an interspecific hybrid between *A. galanthum* and *A. fistulosum* was repeatedly backcrossed with *A. fistulosum* (Yamashita *et al.*, 1999). Interestingly, they found that the cytoplasm of *A. galanthum* induced male sterility in *A. cepa* and also in *A. fistulosum*. They observed that microsporogenesis proceeded normally until the tetrad stage. Later, degeneration of the protoplasm in the tetrads took place, resulting in empty pollen grains. A similar course of events takes place in S-type cytoplasmic male sterility (CMS) of onion (Holford *et al.*, 1991), whereas T-type CMS has an abnormal pollen meiosis (Berninger, 1965; Schweisguth, 1973). Yamashita *et al.* (1999) also reported that fertility restoration in onion with *A. galanthum* cytoplasm is probably determined by one locus with two alleles. Havey (1999) reported that the galanthum-CMS type of male sterility yielded comparable seed quantities to those obtained from S, N and T types of CMS. Furthermore, he found that the restorer gene for S-type CMS did not restore galanthum-CMS. The development of a new source of CMS would be of value for the breeding of onions and shallots. The risk of using only a small cytoplasm gene pool for breeding has been clearly shown in the southern corn-blight incident, which was due to the ubiquitous usage of a single source of CMS-T maize during the 1960s and 1970s in the USA (Levings, 1990).

2.1.2 Leek

Leek, like onion, is not found in nature. It is thought that the leek and its cultivated relatives originate from *A. ampeloprasum* (Stearn, 1978; van der Meer and Hanelt, 1990; Fritsch and Friesen, Chapter 1, and De Clercq and Van Bockstaele, Chapter 18, this volume), which is common all over the Mediterranean basin (Feinbrun, 1943, 1948; Kollmann, 1971, 1972). In its gene centre, *A. ampeloprasum* forms the *A. ampeloprasum* complex (*sensu latu*) (von Bothmer, 1974) together with *A. commutatum*, *A. bourgeaui* and *A. atroviolaceum*. The existence of this complex led Mathew (1996) to suggest that species other than *A. ampeloprasum* could be the progenitor of leek.

The species complex comprises a polyploid series, and leek is at the tetraploid level ($2n = 4x = 32$) in this series. Mathew (1996) included 115 species in the subgenus *Allium*, and hypothesized an informal classification of this subgenus into six groups; the *ampeloprasum* group included both leek and garlic.

A few phylogenetic studies have been carried out to establish the evolutionary relationships between leek and its wild relatives. Kik *et al.* (1997b) analysed the mitochondrial (mt) DNA variation within and between the various cultivated relatives of leek and their wild relatives and concluded that the relationship between them is quite close, because the majority of the species are clustered within one group. Furthermore, Kik and co-workers observed that mtDNA variation in leek is very limited compared with that of its wild relatives (Fig. 4.4). Havey and Lopes Leite (1999) later confirmed this.

Scarcely any crossing experiments have been carried out between leek and its relatives. Van der Meer (1984) reported successful crosses between leek and kurrat, and subsequent back-crosses to leek were made in the context of introducing yellow-stripe virus resistance into leek. In Crete, where

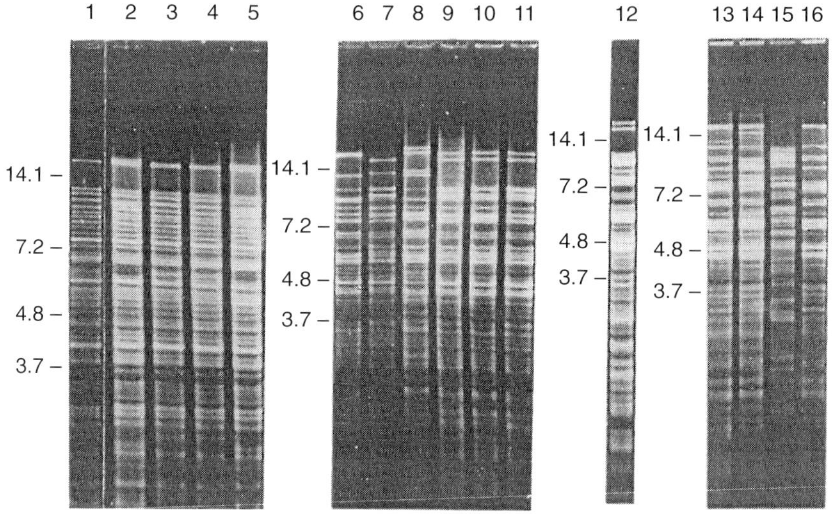

Fig. 4.4. *Eco*RI restriction patterns of mtDNAs from five individuals of leek cv. Porino (lanes 1–5) and various accessions of *A. commutatum* (lanes 6–14, 16) and *A. ampeloprasum* (lane 15). Fragment length sizes (kb) are indicated on the left. (From Kik *et al.*, 1997b, with permission.)

the three species of the *A. ampeloprasum* complex, namely *A. ampeloprasum*, *A. commutatum* and *A. bourgeaui*, grow sympatrically, von Bothmer (1974) observed plants that exhibited traits from more than one species. He concluded that genetic exchange most probably occurs within the *A. ampeloprasum* complex. Kik *et al.* (1997b) successfully crossed leek with its wild relatives of the *A. ampeloprasum* complex and suggested that these wild relatives can be exploited for the improvement of leek, especially to increase cytoplasmic variation. Peterka and Budahn (1996) and Peterka *et al.* (1997) also studied the possibilities for increasing cytoplasmic variation in leek and they successfully crossed onion, *A. fistulosum* and *A. schoenoprasum* with leek, though the triploid progenies were sterile.

The improvement of leek via sexual hybridization presents a potential problem because of the predominant occurrence in this species of proximal chiasmata, with only 0.03–2% of the chiasmata non-proximal (Levan, 1940; Jones *et al.*, 1996). These localized chiasmata are predominantly located in the pericentromeric one-third of metaphase I chromosomes (Stack and Roelofs, 1996). Stack (1993) and Khazanehdari and Jones (1997) hypothesized that this strong chiasma localization may have a survival value as a bivalentizing mechanism, which reduces the frequency of tetravalents and unbalanced gametes. A consequence of this chiasma localization could be a lack of recombination in the distal two-third ends of the chromosomes. It has therefore been speculated that genes in leek are inherited in tightly linked complexes named supergenes (Ved Brat, 1965; Gohil, 1984). On the other hand, the fact that leek chromosomes pair along their whole length in prophase I suggests that recombination points are distributed at random (Khazanehdari *et al.*, 1995). Smilde *et al.* (1999) analysed a population of 70 plants from a cross between two leek genotypes and found, on the basis of 97 segregating AFLP (Vos *et al.*, 1995) markers, no indications for the presence of large linkage blocks. However, their marker map spanned only 405 of the expected 6400 cM and consequently their results are not conclusive. The challenge for the future will be to develop a high-density genetic-linkage map combined with a physical map, obtained via FISH mapping of single-copy sequences on the leek chromosomes. This combined map will provide a deeper insight into the genome organization of leek and clearly show what consequences the occurrence of proximal chiasmata has for the breeding of leek.

2.1.3 Garlic

The evolutionary relationships between garlic and its wild relatives have been little investigated (Mathew, 1996). It has been proposed that *A. longicuspis* is the progenitor species of garlic, because the two species are morphologically similar and because *A. longicuspis* can be found in the centre of evolution of garlic, namely on the western side of the Tien-Shan mountains (Vvedensky, 1944; Fritsch and Friesen, Chapter 1, and Etoh and Simon, Chapter 5, this volume). Contrary to the situation at the interspecific level, variation at the intraspecific level has been studied in detail (Messiaen *et al.*, 1993; Pooler and Simon, 1993; Maaß and Klaas, 1995; Etoh and Simon, Chapter 5, this volume). In a most comprehensive study, Maaß and Klaas (1995) analysed 300 garlic clones from various locations on the Eurasian continent for polymorphism of 12 isozymes and 125 RAPD markers. The results were combined with those of the two other studies to give an integrated picture of the structure of the garlic germplasm and the domestication of garlic. A subdivision of the world's garlic germplasm into four groups – *sativum*, *ophioscorodon*, *longicuspis* and subtropical – was proposed by Maaß and Klaas (1995). They considered the heterogeneous *longicuspis* group as the most primitive, from which the other three groups were derived, and proposed that the *sativum* group was domesticated in the Mediterranean basin, the *ophioscorodon* group in Central-Eastern Europe and the Caucasus, and the subtropical group in the region encompassing India, Vietnam, Myanmar (Burma) and Malaysia. They also distinguished a subgroup *pekinense*, grown in China, which originated from the *longicuspis* group (Fig. 4.5).

Garlic has long been known only as a sterile species, but in 1953 Kononkov already reported the existence of fertile garlic plants. The Kononkov results were lost in obscurity for several decades, and it was Etoh (1983a, b, 1984) who brought the issue of fertile garlic back on to the scientific agenda. Etoh (1986) and Hong and Etoh (1996) collected garlic from Soviet Central Asia. They found a number of fertile, semi-fertile or male-sterile plants which, following the removal of the topsets and self- or cross-pollination, produced viable seeds. On average, 12% of the seeds germinated (Etoh, 1997). Hong *et al.* (1997) screened a pool of 12 fertile and 12 male-sterile genotypes, and found two RAPD markers linked to male sterility. Testing another group of 30 fertile and 30 male-sterile clones revealed the presence of both markers in the fertile clones but not in the male-sterile ones. There was, however, one male-sterile clone in which both fragments were amplified.

Garlic breeding via sexual hybridization is still in its infancy. However, one can envisage that, with the increasing occurrence of restored fertility, in the near future garlic improvement will be carried out as it is in potato: by crossing two highly heterozygous clones and the subsequent selection among the offspring to establish the best individuals for vegetative propagation. The use of wild relatives to increase the diversity of the garlic gene pool with agronomically important traits will be an obvious next step in modern breeding schemes.

2.2 Ornamental alliums

Ornamental alliums are found in a number of subgenera, but mostly in subgenus *Melanocrommyum* (de Hertogh and Zimmer, 1993). About 20 *Allium* species are commercially used as ornamentals, primarily as cut flowers and also as ornamentals in gardens (see Kamenetsky and Fritsch, Chapter 19, this volume). Their commercial value is low at present compared with that of the edible alliums, although they have great economic potential.

Based on analyses of nuDNA variation (internal transcribed spacer (ITS) region) of a restricted set of species, Dubouzet and Shinoda (1998) concluded that the subgenus *Melanocrommyum* has a monophyletic origin.

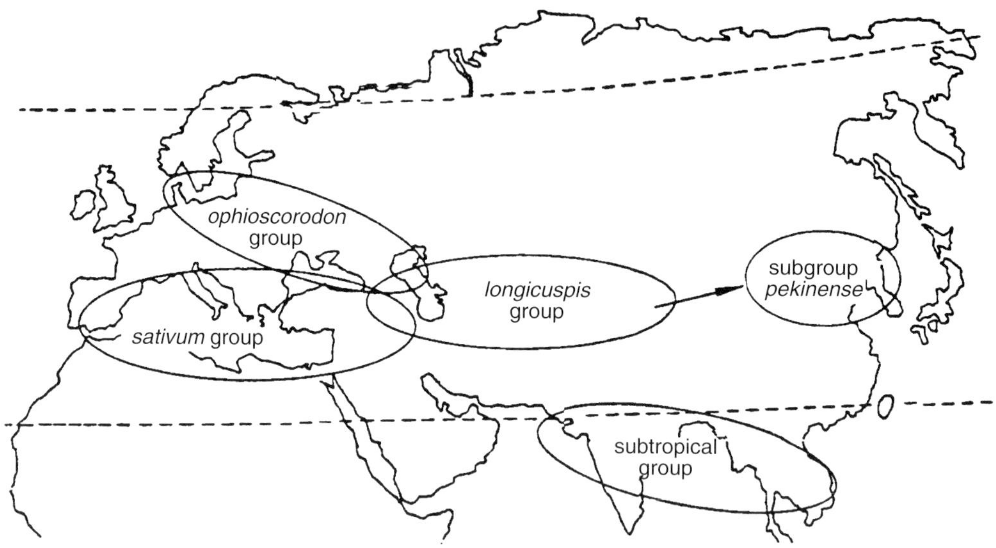

Fig. 4.5. Geographical distribution of the various garlic groups (from Maaß and Klaas, 1995, with permission).

Mes *et al.* (1999) reached the same conclusion when analysing cpDNA and nuDNA variation on a large number of species from various subgenera of the genus *Allium*.

2.2.1 *Subgenus* Melanocrommyum

Interspecific hybridization within the subgenus *Melanocrommyum* has been carried out to some extent with the main purpose of enhancing the attractiveness of the flowers. The most notable is the interspecific cross between *A. macleanii* (= *A. elatum*) and *A. cristophii*, which resulted in the commercially important cultivar 'Globemaster' (Bijl van Duyvenbode, 1990). The hybrid origin of this cultivar has been confirmed by GISH (Friesen *et al.*, 1997). Furthermore, using the same technique, Friesen *et al.* (1997) showed that the cultivar 'Globus' originated from a cross between *A. karataviensis* and *A. stipitatum* and not between *A. cristophii* and *A. giganteum*, as had been proposed on morphological grounds. They also suggested that the cultivars 'Lucy Ball' and 'Gladiator' are of hybrid origin. However, they have identified only one parent, namely *A. aflatunense* (= *A. hollandicum*) and not the other. Dubouzet *et al.* (1998) tried to cross *A. giganteum* with a number of other species, but obtained hybrid plantlets only from a cross with *A. schubertii*, using embryo rescue. Furthermore, Dubouzet *et al.* (1993, 1994) reported on the successful development of new ornamental species for cultivation in southern Japan, from crosses between the interspecific hybrids *A. chinense* × *A. thunbergii* as a female parent and *A. tuberosum* (subgenus *Rhizirideum*), *A. cowanii* (subgenus *Amerallium*) or *A. giganteum* (subgenus *Melanocrommyum*) as pollinators. However, the only proof for the successful production of an interspecific trihybrid cross was provided for *A. chinense* × *A. thunbergii* and *A. tuberosum* (all members of subgenus *Rhizirideum*) (Dubouzet *et al.*, 1996).

3. *Allium* Alien Introgression: Conclusions and Future Directions

Cultivated *Allium* species can be severely affected by various biotic and abiotic factors. The introduction of traits like CMS and the demand for new or modified metabolites (e.g. sulphur-containing compounds, fructans, carbohydrates, flavonoids) for health purposes are steadily increasing. There is a need to extend the gene pools of the various *Allium* crops for such traits from diverse sources. Wild species, such as *A. roylei* and *A. fistulosum*/*A. altaicum*, are important reservoirs of useful genes, and offer great potential for the incorporation of such genes into commercial cultivars. Therefore, it is expected that alien introgression in *Allium* will become an integral part of the breeding of new cultivars in the near future. The possibility of applying molecular-marker and *in situ* hybridization technology in breeding programmes should considerably speed up the process of breeding new cultivars.

For the bulb onion, this change in breeding strategy has already been partly implemented. However, for leek and especially for garlic, the identification of beneficial traits in wild relatives, the exploitation of these traits via sexual hybridization and the use of marker-assisted breeding are still in their infancy. Therefore, the development of genomic-linkage maps of leek and garlic seems to be the obvious next step. The establishment of the relationships of these maps with the onion marker maps and with the maps of other monocots (synteny) will be very beneficial for leek and garlic breeding in general and will also assist in the isolation of as yet unidentified genes.

The next step in alien-introgression research will be to improve our understanding of the transmission of 'wild' chromatin into the cultivated species. What factors influence this process? Can the genes involved in species incongruency or in nucleocytoplasmic interactions be located and identified? The bridge cross between onion, *A. fistulosum* and *A. roylei* is very interesting for this type of research because advanced populations are available and the techniques to analyse the introgression process, e.g. molecular-marker and *in situ* hybridization technology, are currently in use in *Allium* studies.

From a fundamental point of view, the study of the genome organization of *Allium* and especially the evolution of repetitive DNA is very intriguing (see King *et al.*, 1998; van Heusden *et al.*, 2000b). *Allium* has one of the largest genomes in the plant kingdom and this makes these species uniquely suited for this type of research. How is the repetitive DNA distributed on the *Allium* chromosomes, and which repetitive-DNA families are present in *Allium*? Furthermore, where are the single-copy genes located? Are they distributed randomly on the chromosomes or in clusters, and where are these clusters located on the chromosomes? Moreover, what is the effect of randomly occurring chiasmata (*A. cepa* and *A. roylei* type) versus highly localized chiasmata (*A. fistulosum* and *A. ampeloprasum* leek group) on the genome organization? All in all, the future of *Allium* alien-introgression research looks very promising, both from a fundamental and from an applied point of view.

Acknowledgements

I would like to thank Drs A.G. Balkema-Boomstra, A.W. van Heusden, A.P.M. den Nijs, L.W.D. van Raamsdonk, R.E. Voorrips and Ing. W.A. Wietsma from Plant Research International and Prof. Dr L.I. Khrustaleva from the Timiryazev Agricultural Academy, Moscow, Russia, for critically reading this manuscript.

References

Albini, S.M. and Jones, G.H. (1990) Synaptonemal complex spreading in *Allium cepa* and *Allium fistulosum*. III. The F_1 hybrid. *Genome* 33, 854–866.

Anderson, R. (1949) *Introgressive Hybridization*. John Wiley & Sons, New York, 109 pp.

Arifin, N.S., Ozaki, Y. and Okubo, H. (2000) Genetic diversity in Indonesian shallot (*Allium cepa* var. *ascalonicum*) and *A.* × *wakegi* revealed by RAPD markers and the origin of *A.* × *wakegi* identified by RFLP analyses of amplified chloroplast genes. *Euphytica* 111, 23–31.

Bark, O.H., Havey, M.J. and Corgan, J.N. (1994) Restriction fragment length polymorphism (RFLP)

analysis of progeny from an *Allium fistulosum* × *A. cepa* hybrid. *Journal of the American Society for Horticultural Science* 119, 1046–1049.
Berninger, E. (1965) Contribution à l'étude de la stérilité-mâle de l'oignon (*Allium cepa* L.). *Annales de l'Amélioration des Plantes* 15, 183–199.
Bijl van Duyvenbode, J.R (1990) Breeding ornamental onions. *The Plantsman* 12, 152–156.
Block, E. (1992) The organosulfur chemistry of the genus *Allium* – implications for the organic chemistry of sulfur. *Angewandte Chemie Internationale, English Edition* 31, 1135–1178.
Brar, D.S. and Khush, G.S. (1997) Alien introgression in rice. *Plant Molecular Biology* 35, 35–47.
Buiteveld, J., Suo, Y., van Lookeren Campagne, M.M. and Creemers-Molenaar, J. (1998) Production and characterization of somatic hybrid plants between leek (*Allium ampeloprasum* L.) and onion (*Allium cepa* L.). *Theoretical and Applied Genetics* 96, 765–775.
Christou, P.(1993) Particle gun mediated transformation. *Current Opinion in Biotechnology* 4, 135–141.
Cryder, C.M., Corgan, J.N., Urquhart, N.S. and Clason, D. (1991) Isozyme analysis of a progeny derived from (*Allium fistulosum* × *A. cepa*) × *Allium cepa*. *Theoretical and Applied Genetics* 82, 337–345.
Currah, L. (1986) Leek breeding: a review. *Journal of Horticultural Science* 61, 407–415.
Currah, L. and Maude, R.B. (1984) Laboratory tests for leaf resistance to *Botrytis squamosa* in onions. *Annals of Applied Biology* 105, 277–283.
de Hertogh, A.A. and Zimmer, K. (1993) *Allium* – ornamental species. In: Halevy, A.H. (ed.) *Handbook of Flowering*. CRC Press, Boca Raton, Florida, pp. 22–33.
de Vries, J.N. and Jongerius, R. (1992) A pericentric inversion between a pair of homoeologous chromosomes from *Allium roylei* Stearn and *A. cepa* L. *Allium Improvement Newsletter* 2, 11–12.
de Vries, J.N. and Wietsma, W.A. (1992) *Allium roylei* Stearn restores cytoplasmic male sterility of Rijnsburger onion (*A. cepa* L.). *Journal of Genetics and Breeding* 46, 379–382.
de Vries, J.N., Wietsma, W.A. and de Vries, T. (1992a) Introgression of leaf blight resistance from *A. roylei* Stearn into onion (*A. cepa* L.). *Euphytica* 62, 127–133.
de Vries, J.N., Wietsma, W.A. and Jongerius, M.C. (1992b) Linkage of downy mildew resistance genes Pd_1 and Pd_2 from *Allium roylei* Stearn in progeny of its interspecific hybrid with onion (*A. cepa*). *Euphytica* 64, 131–137.
de Vries, J.N., Wietsma, W.A. and Jongerius, M.C. (1992c) Introgression of characters from *Allium roylei* Stearn into *A. cepa* L. In: Hanelt, P., Hammer, K. and Knüpffer, H. (eds) *The Genus* Allium – *Taxonomic Problems and Genetic Resources. Proceedings of an International Symposium Held at Gatersleben*. Institut für Pflanzengenetik und Kulturpflanzenforschung (IPK), Gatersleben, Germany, pp. 321–325.
de Vries, J.N., Jongerius, R., Sandbrink, H. and Lindhout, P. (1992d) RAPD markers assist in resistance breeding. *Prophyta* 2, 50–51.
de Vries, J.N., Wietsma, W.A. and Appels, M. (1992e) Direct and *Allium roylei* mediated transfer of *A. fistulosum* genes to onion. *Allium Improvement Newsletter* 2, 9–10.
Dowker, B.D. (1990) Onion breeding. In: Rabinowitch, H.D. and Brewster, J.L. (eds) *Onions and Allied Crops*, Vol. I. *Botany, Physiology and Genetics*. CRC Press, Boca Raton, Florida, pp. 215–232.
Dubouzet, J.G. and Shinoda, K. (1998) Phylogeny of *Allium* L. subg. *Melanocrommyum* (Webb et Berth.) Rouy based on DNA sequence analysis of the internal transcribed spacer region of nrDNA. *Theoretical and Applied Genetics* 97, 541–549.
Dubouzet, J.G., Arisumi, K., Yoshitake, T., Etoh, T., Sakata, Y. and Matsuo, E. (1993) Studies on the development of new ornamental *Allium* ssp. through interspecific hybridization II. Utilization of *Allium chinense* × *A. thunbergii* hybrids and *A. tuberosum* 'Tender-Pole' to develop heat tolerant ornamental *Allium* spp. *Journal of the Japanese Society for Horticultural Science* 61, 901–910.
Dubouzet, J.G., Arisumi, K., Etoh, T., Maeda, M. and Sakata, Y. (1994) Studies on the development of new ornamental *Allium* through interspecific hybridization III. Hybridization of autumn flowering species through pull-style pollination, cutflower culture and embryo rescue. *Memoirs of the Faculty of Agriculture, Kagoshima University* 30, 35–42.
Dubouzet, J.G., Etoh, T., Arisumi, K. and Yoshitake, T. (1996) A diagnostic test to confirm interspecific *Allium* hybrids using random amplified polymorphic DNA from crude leaf DNA extracts. *Journal of the Japanese Society for Horticultural Science* 65, 321–326.
Dubouzet, J.G., Shinoda, K. and Murata, R. (1998) Interspecific hybridization of *Allium giganteum* Regel: production and early verification of putative hybrids. *Theoretical and Applied Genetics* 96, 385–388.
Eady, C., Weld, R.J. and Lister, C.E. (2000) *Agrobacterium tumefaciens*-mediated transformation of onion (*Allium cepa* L.) and transgenic-plant regeneration. *Plant Cell Reports* 19, 376–381.

Emsweller, S.L. and Jones, H.A. (1935a) An interspecific hybrid in *Allium. Hilgardia* 9, 265–273.

Emsweller, S.L. and Jones, H.A. (1935b) Meiosis in *A. fistulosum*, *A. cepa* and their F_1 hybrid. *Hilgardia* 9, 277–294.

Etoh, T. (1983a) Accomplishment of microsporogenesis in a garlic clone. *Memoirs of the Faculty of Agriculture, Kagoshima University* 19, 55–63.

Etoh, T. (1983b) Germination of seeds obtained from a clone of garlic, *Allium sativum. Proceedings of the Japanese Academy* 59(B), 83–87.

Etoh, T. (1984) Hybrids between wild garlic (*Allium longicuspis* Regel) and garlic (*A. sativum* L.). In: van der Meer, Q.P., Wietsma, W.A. and Hollander, H.D.M. (eds) *Proceedings of the 3rd Eucarpia Allium Symposium*. IVT, Wageningen, The Netherlands, pp. 78–82.

Etoh, T. (1986) Fertility of the garlic clones collected in Soviet Central Asia. *Journal of the Japanese Society for Horticultural Science* 55, 312–319.

Etoh, T. (1997) True seeds in garlic. *Acta Horticulturae* 433, 247–255.

FAO (2001) Agrostat database, updated annually: http://apps.fao.org/

Fedak, G. (1999) Molecular aids for integration of alien chromatin through wide crosses. *Genome* 42, 584–591.

Feinbrun, N. (1943) *Allium* section *Porrum* of Palestine and neighbouring countries. *Palestine Journal of Botany Jerusalem* 3, 1–21.

Feinbrun, N. (1948) Further studies of *Allium* of Palestine and the neighbouring countries. *Palestine Journal of Botany Jerusalem* 4, 144–157.

Felix, E.L. (1933) Disease resistance in *Allium fistulosum* L. *Phytopathology* 23, 109–110.

Friesen, N. and Klaas, M. (1998) Origin of some minor vegetatively propagated *Allium* crops studied with RAPD and GISH. *Genetic Resources and Crop Evolution* 45, 511–523.

Friesen, N., Fritsch, R. and Bachmann, K. (1997) Hybrid origin of some ornamentals of *Allium* subgenus *Melanocrommyum* verified with GISH and RAPD. *Theoretical and Applied Genetics* 95, 1229–1238.

Galvan, G.A., Wietsma, W.A., Putrasemedja, S., Permadi, A.H. and Kik, C. (1997) Screening for resistance to anthracnose (*Colletotrichum gloeosporioides* Penz.) in *Allium cepa* and its wild relatives. *Euphytica* 95, 173–178.

Glimelius, K., Fahlesson, J., Landgren, M., Sjodin, G. and Sundberg, E. (1991) Gene transfer via somatic hybridization in plants. *Trends in Biotechnology* 9, 24–30.

Gohil, R.N. (1984) Extent of recombination possible in the cultivated leek. In: van der Meer, Q.P., Wietsma, W.A. and Hollander, H.D.M. (eds) *Proceedings of the 3rd Eucarpia Allium Symposium*. IVT, Wageningen, The Netherlands, pp. 99–105.

Gonzalez, L.G. and Ford-Lloyd, B.V. (1987) Facilitation of wide-crossing through embryo rescue and pollen storage in interspecific hybridization of cultivated *Allium* species. *Plant Breeding* 98, 318–322.

Grant, V. (1971) *Plant Speciation*. Columbia University Press, New York, 435 pp.

Hadley, H.H. and Openshaw, S.J. (1980) Interspecific and intergeneric hybridization. In: Fehr, W.R. and Hadley, H.H. (eds) *Hybridization of Crop Plants*. American Society of Agronomy, Madison, Wisconsin, pp. 133–159.

Hanelt, P. (1990) Taxonomy, evolution and history. In: Rabinowitch, H.D. and Brewster, J.L. (eds) *Onion and Allied Crops*, Vol. I. *Botany, Physiology, and Genetics*. CRC Press, Boca Raton, Florida, pp. 1–26.

Havey, M.J. (1991) Molecular characterization of the interspecific origin of viviparous onion. *Journal of Heredity* 82, 501–503.

Havey, M.J. (1999) Seed yield, floral morphology, and lack of male-fertility restoration of male-sterile onion (*Allium cepa*) populations possessing the cytoplasm of *Allium galanthum*. *Journal of the American Society for Horticultural Science* 124, 626–629.

Havey, M.J. and Lopes Leite, D. (1999) Toward the identification of cytoplasmic male sterility in leek: evaluation of organellar DNA diversity among cultivated accessions of *Allium ampeloprasum*. *Journal of the American Society for Horticultural Science* 124, 163–165.

Heiser, C.B. (1973) Introgression re-examined. *Botanical Review* 39, 347–366.

Hizume, M. (1994) Allodiploid nature of *Allium wakegi* Araki revealed by genomic *in situ* hybridization and localization of 5S and 18S rDNAs. *Japanese Journal of Genetics* 69, 407–415.

Holford, P., Croft, J.H. and Newbury, H.J. (1991) Structural studies of microsporogenesis in fertile and male-sterile onions (*Allium cepa* L.) containing the cms-S cytoplasm. *Theoretical and Applied Genetics* 82, 745–755.

Hong, C. and Etoh, T. (1996) Fertile clones of garlic (*Allium sativum* L.) abundant around the Tien-Shan mountains. *Breeding Science* 46, 349–353.

Hong, C., Etoh, T., Landry, B. and Matsuzoe, N. (1997) RAPD markers related to pollen fertility in garlic (*Allium sativum* L.). *Breeding Science* 47, 359–362.

Inden, H. and Asahira, T. (1990) Japanese bunching onion (*Allium fistulosum* L). In: Brewster, J.L. and Rabinowitch, H.D. (eds) *Onions and Allied Crops*, Vol. III. *Biochemistry, Food Science, and Minor Crops*. CRC Press, Boca Raton, Florida, pp. 159–178.

Jiang, J., Friebe, B. and Gill, B.S. (1994) Recent advances in alien gene transfer in wheat. *Euphytica* 73, 199–212.

Jones, G.H., Khazanehdari, K.A. and Ford-Lloyd, B.V. (1996) Meiosis in leek (*Allium porrum* L.) revisited. II. Metaphase I observations. *Heredity* 76, 186–191.

Jones, R.N. (1990) Cytogenetics. In: Rabinowitch, H.D. and Brewster, J.L. (eds) *Onions and Allied Crops*, Vol. I. *Botany, Physiology and Genetics*. CRC Press, Boca Raton, Florida, pp. 199–214.

Kalloo, G. and Chowdhury, J.B. (eds) (1992) *Distant Hybridization of Crop Plants*. Monographs on Theoretical and Applied Genetics, Vol. 16, Springer Verlag, Berlin, 271 pp.

Kamenetsky, R. and Rabinowitch, H.D. (2001) Floral development in bolting garlic. *Sexual Plant Reproduction* 13, 235–241.

Kaul, M.L.H. (1988) *Male Sterility in Higher Plants*. Monographs on Theoretical and Applied Genetics, Vol. 9, Springer Verlag, Berlin, 1005 pp.

Keller, E.R.J., Schubert, I., Fuchs, J. and Meister, A. (1996) Interspecific crosses of onion with distant *Allium* species and characterization of the presumed hybrids by means of flow cytometry, karyotype analysis and genomic *in situ* hybridization. *Theoretical and Applied Genetics* 92, 417–424.

Khazanehdari, K.A. and Jones, G.H. (1997) The causes and consequences of meiotic irregularity in the leek (*Allium ampeloprasum* ssp. *porrum*): implications for fertility, quality and uniformity. *Euphytica* 93, 313–319.

Khazanehdari, K.A., Jones, G.H. and Ford-Lloyd, B.V. (1995) Meiosis in the leek (*Allium porrum* L.) revisited. I. Prophase I pairing. *Chromosome Research* 3, 433–439.

Khrustaleva, L.I. and Kik, C. (1998) Cytogenetical studies in the bridge-cross *Allium cepa* × (*A. fistulosum* × *A. roylei*). *Theoretical and Applied Genetics* 96, 8–14.

Khrustaleva, L.I. and Kik, C. (2000) Introgression of *Allium fistulosum* into *A. cepa* mediated by *A. roylei*. *Theoretical and Applied Genetics* 100, 17–26.

Khush, G.S. and Brar, D.S. (1992) Overcoming the barriers in hybridization. In: Kalloo, G. and Chowdhury, J.B. (eds) *Distant Hybridization of Crop Plants*. Monographs on Theoretical and Applied Genetics, Vol. 16. Springer Verlag, Berlin, pp. 47–61.

Kik, C., Buiteveld, J. and Verbeek, W.H.J. (1997a) Biotechnological aspects of onion breeding. *Acta Horticulturae* 433, 291–297.

Kik, C., Samoylov, A.M., Verbeek, W.H.J. and van Raamsdonk, L.W.D. (1997b) Mitochondrial DNA variation and crossability of leek (*Allium porrum*) and its wild relatives from the *Allium ampeloprasum* complex. *Theoretical and Applied Genetics* 94, 465–471.

King, J.J., Bradeen, J.M., Bark, O., McCallum, J.A. and Havey, M.J. (1998) A low-density genetic map of onion reveals a role for tandem duplication in the evolution of an extremely large diploid genome. *Theoretical and Applied Genetics* 96, 52–62.

Klaas, M. (1998) Applications and impact of molecular markers on evolutionary and diversity studies in the genus *Allium*. *Plant Breeding* 117, 297–308.

Koch, H.P. and Lawson, L.D. (1996) *Garlic. The Science and Therapeutic Application of* Allium sativum *L. and related species*, 2nd edn. Williams and Wilkins, Baltimore, Maryland, 329 pp.

Kofoet, A., Kik, C., Wietsma, W.A. and de Vries, J.N. (1990) Inheritance of resistance to downy mildew (*Peronospora destructor* [Berk.] Casp.) from *Allium roylei* Stearn in the back-cross *Allium cepa* L. × (*A. roylei* × *A. cepa*). *Plant Breeding* 105, 144–149.

Kollmann, F. (1971) *Allium ampeloprasum* L., a polyploid complex. I. Ploidy levels. *Israel Journal of Botany* 20, 13–20.

Kollmann, F. (1972) *Allium ampeloprasum* L. – a polyploid complex. II. Meiosis and relationships between the ploidy types. *Caryologia* 25, 295–312.

Kononkov, P.F. (1953) The question of obtaining garlic seed. *Sad i Ogorod* 8, 38–40 (in Russian).

Labani, R.M. and Elkington, T.T. (1987) Nuclear DNA variation in the genus *Allium* L. (*Liliaceae*). *Heredity* 59, 119–128.

Levan, A. (1940) Meiosis of *Allium porrum*, a tetraploid species with chiasma localization. *Hereditas* 26, 454–462.

Levings, C.S. (1990) The Texas cytoplasm of maize: cytoplasmic male sterility and disease susceptibility. *Science* 250, 942–947.

Maaß, H.I. (1997) Genetic diversity in the top onion, *Allium × proliferum (Alliaceae)*, analysed by isozymes. *Plant Systematics and Evolution* 208, 35–44.

Maaß, H.I. and Klaas, M. (1995) Infraspecific differentiation of garlic (*Allium sativum* L.) by isozyme and RAPD markers. *Theoretical and Applied Genetics* 91, 89–97.

McCollum, G.D. (1971) Sterility of some interspecific *Allium* hybrids. *Journal of the American Society for Horticultural Science* 96, 359–362.

McCollum, G.D. (1974) Chromosome behavior and sterility of hybrids between common onion, *Allium cepa*, and the related wild *A. oschaninii*. *Euphytica* 23, 699–709.

McCollum, G.D. (1982) Experimental hybrids between *Allium fistulosum* and *A. roylei*. *Botanical Gazette* 143, 238–242.

Mathew, B. (1996) *A Review of* Allium *Section* Allium. Royal Botanic Gardens, Kew, UK, 176 pp.

Mes, T.H.M., Fritsch, R.M., Pollner, S. and Bachmann, K. (1999) Evolution of the chloroplast genome and polymorphic ITS regions in *Allium* subg. *Melanocrommyum*. *Genome* 42, 237–247.

Messiaen, C.M., Cohat, J., Pichon, M., Leroux, J.P. and Beyries, A. (1993) *Les Allium alimentaires reproduits par voie végétative*. INRA, Paris, 228 pp.

Michelmore, R.W., Paran, I. and Kesseli, R.V. (1991) Identification of markers linked to disease resistance genes by bulked segregant analysis: a rapid method to detect markers in specific genomic regions by using segregating populations. *Proceedings of the National Academy of Sciences USA* 88, 9828–9832.

Myers, J.M. and Simon, P.W. (1998) Microprojectile bombardment of garlic, *Allium sativum* L. *Proceedings of the 1998 National Onion (and other Allium) Research Conference*. University of California, Davis, California, pp. 121–126.

Netzer, D., Rabinowitch, H.D. and Weintal, C. (1985) Greenhouse technique to evaluate pink root disease caused by *Pyrenochaeta terrestris*. *Euphytica* 34, 385–391.

Nomura, Y. and Makura, K. (1996) Morphological and agronomical characteristics in interspecific hybrid plants between rakkyo (*Allium chinensis*) and other edible *Allium* species. *Breeding Science* 46, 17–23.

Ohsumi, C., Kojima, A., Hinata, K., Etoh, T. and Hayashi, T. (1993) Interspecific hybrid between *Allium cepa* and *Allium sativum*. *Theoretical and Applied Genetics* 85, 969–975.

Patil, B.S., Pike, L.M. and Yoo, K.S. (1995) Variation in the quercetin content in different colored onions (*Allium cepa* L.). *Journal of the American Society for Horticultural Science* 120, 909–913.

Patterson, A.H. (1996) *Genome Mapping in Plants*. Academic Press, San Diego, California, 330 pp.

Peffley, E.B. and Hou, A. (2000) Bulb-type onion introgressants possessing *Allium fistulosum* L. genes recovered from interspecific hybrid backcrosses between *A. cepa* L. and *A. fistulosum* L. *Theoretical and Applied Genetics* 100, 528–534.

Peffley, E.B. and Mangum, P.D. (1990) Introgression of *Allium fistulosum* L. into *Allium cepa* L.: cytogenetic evidence. *Theoretical and Applied Genetics* 79, 113–118.

Peterka, H. and Budahn, H. (1996) Crosses between cultivated *Allium* species. In: Pank, F. (ed.) *Proceedings of International Symposium on Breeding for Resistance in Medicinal and Aromatic Plants*. Quedlinburg, Germany, pp. 150–153.

Peterka, H., Budahn, H. and Schrader, O. (1997) Interspecific hybrids between onion (*Allium cepa* L.) with S-cytoplasm and leek (*Allium ampeloprasum* L.). *Theoretical and Applied Genetics* 94, 383–389.

Pike, L.M. (1986) Onion breeding. In: Bassett, M.J. (ed.) *Breeding Vegetable Crops*. AVI Publishing Company, Westport, Connecticut, pp. 357–394.

Pooler, M.R. and Simon, P.W. (1993) Characterization and classification of isozyme and morphological variation in a diverse collection of garlic clones. *Euphytica* 68, 121–130.

Pooler, M.R. and Simon, P.W. (1994) True seed production in garlic. *Sexual Plant Reproduction* 7, 282–286.

Rabinowitch, H.D. (1997) Breeding alliaceous crops for pest resistance. *Acta Horticulturae* 433, 223–246.

Randle, W.M., Lancaster, J.E., Shaw, M.L., Sutton, K.H., Hay, R.L. and Bussard, M.L. (1995) Quantifying onion flavor compounds responding to sulfur fertility – sulfur increases levels of alk(en)yl cysteine sulfoxides and biosynthetic intermediates. *Journal of the American Society for Horticultural Science* 120, 1075–1081.

Ricroch, A., Peffley, E.B. and Baker, R.J. (1992) Chromosomal location of rDNA in *Allium*: *in situ* hybridization using biotin- and fluorescein-labelled probe. *Theoretical and Applied Genetics* 83, 413–418.

Rieseberg, L.H. (1998) Molecular ecology of hybridization. In: Carvalho, C.R. (ed.) *Advances in Molecular Ecology*, IOS Press, Amsterdam, pp. 243–265.

Rieseberg, L.H. and Wendel, J.F. (1993) Introgression and its consequences in plants. In: Harrison, R.G. (ed.) *Hybrid Zones and the Evolutionary Process*. Oxford University Press, New York, pp. 70–109.

Rieseberg, L.H., Sinervo, B., Linder, R., Ungerer, M.C. and Arias, D.M. (1996) Role of gene interactions in hybrid speciation: evidence from ancient and experimental hybrids. *Science* 272, 741–745.

Roberts, H.F. (1929) *Plant Hybridization before Mendel*. Princeton University Press, Princeton, New Jersey, 374 pp.

Saini, S.S. and Davis, G.N. (1967) Compatibility in some *Allium* species. *Proceedings of the American Society for Horticultural Science* 91, 401–409.

Saini, S.S. and Davis, G.N. (1969) Male sterility in *Allium cepa* and some species hybrids. *Economic Botany* 23, 37–49.

Schweisguth, B. (1973) Étude d'un nouveau type de stérilité mâle chez l'oignon, *Allium cepa* L. *Annales de l'Amélioration des Plantes* 23, 221–233.

Sharma, H.C. (1995) How wide can a wide cross be? *Euphytica* 82, 43–64.

Shigyo, M., Imamura, K., Iiono, M., Yamashita, K. and Tashiro, Y. (1998) Identification of alien chromosomes in a series of *Allium fistulosum–A. cepa* monosomic addition lines by means of genomic *in situ* hybridization. *Genes and Genetic Systems* 73, 311–315.

Simon, P.W. (1995) Genetic analysis of pungency and soluble solids in long-storage onions. *Euphytica* 82, 1–8.

Smilde, W.D., van Heusden, W.A. and Kik, C. (1999) AFLPs in leek (*Allium porrum*) are not inherited in large linkage blocks. *Euphytica* 110, 127–132.

Smith, B.M. and Crowther, T.C. (1995) Inbreeding depression and F_1 hybrids in leeks (*Allium ampeloprasum* ssp. *porrum* L.). *Euphytica* 86, 87–94.

Stace, C. (1975) *Hybridization and the Flora of the British Isles*. Academic Press, San Diego, California, 626 pp.

Stack, S.M. (1993) Diploidization in autotetraploid *Allium porrum* by restricted crossing over. *American Journal of Botany* 80, 78–79.

Stack, S.M. and Roelofs, D. (1996) Localized chiasmata and recombination nodules in the tetraploid onion *Allium porrum*. *Genome* 39, 770–783.

Stearn, W.T. (1946) Notes on the genus *Allium* in the Old World. *Herbertia* 11, 11–34.

Stearn, W.T. (1978) European species of *Allium* and allied genera of *Alliaceae*: a synonymic enumeration. *Annales Musei Goulandris* 4, 83–198.

Stebbins, G.L. (1950) *Variation and Evolution in Plants*. Columbia University Press, New York, 643 pp.

Stevenson, M., Armstrong, S.J., Ford-Lloyd, B.V. and Jones, G.H. (1998) Comparative analysis of crossover exchanges and chiasmata in *Allium cepa* × *fistulosum* after genomic *in situ* hybridization (GISH). *Chromosome Research* 6, 567–574.

Ulloa, M., Corgan, J.N. and Dunford, M. (1994) Chromosome characteristics and behavior differences in *Allium fistulosum* L., *A. cepa* L., their F_1 hybrid and selected backcross progeny. *Theoretical and Applied Genetics* 89, 567–571.

Ulloa, M., Corgan, J.N. and Dunford, M. (1995) Evidence for nuclear–cytoplasmic incompatibility between *Allium fistulosum* and *A. cepa*. *Theoretical and Applied Genetics* 90, 746–754.

van der Meer, Q.P. (1984) Breeding for resistance to yellow stripe virus in leeks (*Allium porrum* L.) – a progress report. In: van der Meer, Q.P., Wietsma, W.A. and Hollander, H.D.M. (eds) *Proceedings of the 3rd Eucarpia Allium Symposium*. IVT, Wageningen, pp. 16–19.

van der Meer, Q.P. and de Vries, J.N. (1990) An interspecific cross between *Allium roylei* Stearn and *Allium cepa* L., and its backcross to *A. cepa*. *Euphytica* 47, 29–31.

van der Meer, Q.P. and Hanelt, P. (1990) Leek (*Allium ampeloprasum* var. *porrum*). In: Brewster, J.L. and Rabinowitch, H.D. (eds) *Onions and Allied Crops*, Vol. III. *Biochemistry, Food Science and Minor Crops*. CRC Press, Boca Raton, Florida, pp. 179–196.

van der Meer, Q.P. and van Bennekom, J.L. (1978) Improving the onion crop (*Allium cepa* L.) by transfer of characters from *Allium fistulosum*. *Biuletyn Warzywniczy* 22, 87–91.

van der Valk, P., de Vries, S.E., Everink, J.T., Verstappen, F. and de Vries, J.N. (1991a) Pre- and post fertilization barriers to backcrossing the interspecific hybrid between *Allium fistulosum* L. and *A. cepa* L. with *A. cepa*. *Euphytica* 53, 201–209.

van der Valk, P., Kik, C., Verstappen, F., Everink, J.T. and de Vries, J.N. (1991b) Independent segregation of two isozyme markers and inter-plant differences in nuclear DNA content in the interspecific backcross (*Allium fistulosum* L. × *A. cepa* L.) × *A. cepa* L. *Euphytica* 55, 151–156.

van Geyt, J.P.C., Lange, W., Oleo, M. and de Bock, T.S.M. (1990) Natural variation within the genus *Beta* and its possible use for breeding sugarbeet: a review. *Euphytica* 49, 57–76.

van Heusden, A.W., van Ooijen, J.W., Vrielink-van Ginkel, M., Verbeek, W.H.J., Wietsma, W.A. and Kik, C. (2000a) A genetic map of an interspecific cross in *Allium* based on amplified fragment length polymorphism (AFLP™) markers. *Theoretical and Applied Genetics* 100, 118–126.

van Heusden, A.W., Shigyo, M., Tashiro, Y., Vrielink-van Ginkel, R. and Kik, C. (2000b) AFLP linkage group assignment to the chromosomes of *Allium cepa* L. via monosomic addition lines. *Theoretical and Applied Genetics* 100, 480–486.

van Raamsdonk, L.W.D. and Kik, C. (1997) A European Union funded project on onion quality improvement. *Allium Improvement Newsletter* 6, 47–51.

van Raamsdonk, L.W.D., Wietsma, W.A. and de Vries, J.N. (1992) Crossing experiments in *Allium* L. section *Cepa*. *Botanical Journal of the Linnean Society* 109, 293–303.

van Raamsdonk, L.W.D., Smiech, M.P. and Sandbrink, J.M. (1997) Introgression explains incongruence between nuclear and chloroplast DNA-based phylogenies in *Allium* section *Cepa*. *Botanical Journal of the Linnean Society* 123, 91–108.

van Raamsdonk, L.W.D., Vrielink-van Ginkel, M. and Kik, C. (2000) Phylogeny reconstruction and hybrid analysis in *Allium* subgenus *Rhizirideum*. *Theoretical and Applied Genetics* 100, 1000–1009.

van Tuyl, J.M. (1997) Interspecific hybridization of flower bulbs: a review. *Acta Horticulturae* 430, 465–476.

Ved Brat, S. (1965) Genetic systems in *Allium* III. Meiosis and breeding systems. *Heredity* 20, 325–338.

Vijn, I., van Dijken, A., Sprenger, N., van Dun, K., Weisbeek, P., Wiemken, A. and Smeekens, S. (1997) Fructan of the inulin neoseries is synthesized in transgenic chicory plants (*Cichorium intybus* L.) harbouring onion (*Allium cepa* L.) fructan : fructan 6G-fructosyltransferase. *The Plant Journal* 11, 387–398.

Villanueva-Mosqueda, E. and Havey, M.J. (1998) FISH analyses of advanced backcross plants of an *Allium* interspecific hybrid (*Allium cepa* L. × *Allium fistulosum* L.) to *A. cepa*. *Proceedings of the 1998 National Onion (and other Allium) Research Conference*. University of California, Davis, California, pp. 84–86.

Virk, P.S., Ford-Lloyd, B.V. and Newbury, H.J. (1998) Mapping AFLP markers associated with subspecies differentiation of *Oryza sativa* (rice) and an investigation of segregation distortion. *Heredity* 81, 613–620.

von Bothmer, R. (1974) Studies in the Aegean Flora XXI. Biosystematic studies in the *Allium ampeloprasum* complex. *Opera Botanica* 34, 5–104.

Vos, P., Hogers, R., Bleker, M., Reijans, M., van der Lee, T., Hornes, M., Frijters, A., Pot, J., Peleman, J., Kuiper, M. and Zabeau, M. (1995) AFLP™: a new technique for DNA fingerprinting. *Nucleic Acids Research* 23, 4407–4414.

Vosa, C. (1976) Heterochromatic patterns in *Allium* I. The relationship between the species of the *Cepa* group and its allies. *Heredity* 36, 383–392.

Vvedensky, A.I. (1944) The genus *Allium* in the USSR. *Herbertia* 11, 65–218.

Wannamaker, M.J. and Pike, L.M. (1987) Onion responses to various salinity levels. *Journal of the American Society for Horticultural Science* 112, 49–52.

Wendel, J.F., Olsen, P.D. and McDonald Stewart, J. (1989) Genetic diversity, introgression, and independent domestication of old world cultivated cottons. *American Journal of Botany* 76, 1795–1806.

Wendelbo, P. (1971) *Alliaceae*. In: Rechinger, K.H. (ed.) *Flora Iranica*. Akademische Druck- und Verlaganstalt, No. 76, Graz, Austria, pp. 1–100.

Williams, J.G., Kubelik, A.R., Livak, K.J., Rafalski, J.A. and Tingey, V. (1990) DNA polymorphisms amplified by arbitrary primers are useful as genetic markers. *Nucleic Acids Research* 18, 6531- 6535.

Williams, C.G., Goodman, M.J. and Stuber, C.W. (1995) Comparative recombination distances among *Zea mays* L. inbreds, wide crosses and interspecific hybrids. *Genetics* 141, 1573–1581.

Xu, Y., Zhu, L., Xiao, J., Huang, N. and McCouch, S.R. (1997) Chromosomal regions associated with segregation distortion of molecular markers in F_2, backcross, doubled haploid, and recombinant inbred populations in rice (*Oryza sativa* L.). *Molecular and General Genetics* 253, 535–545.

Yamashita, K. and Tashiro, Y. (1999) Possibility of developing a male sterile line of shallot (*Allium cepa* L. Aggregatum Group) with cytoplasm from *A. galanthum* Kar. et Kir. *Journal of the Japanese Society for Horticultural Science* 68, 256–262.

Yamashita, K., Arita, H. and Tashiro, Y. (1999) Cytoplasm of a wild species, *Allium galanthum* Kar. et Kir., is useful for developing the male sterile line of *A. fistulosum* L. *Journal of the Japanese Society for Horticultural Science* 68, 788–797.

Zamir, D. and Tadmor, Y. (1986) Unequal segregation of nuclear genes in plants. *Botanical Gazette* 147, 355–358.

Zeven, A.C. and van Harten, A.M. (1978). *Broadening the Genetic Base of Crops*. Pudoc, Wageningen, The Netherlands, 347 pp.

Zheng, S.-J., Henken, B., Sofiari, E., Jacobsen, E., Kik, C. and Krens, F.A. (2001) *Agrobacterium tumefaciens* mediated stable transformation of *Allium cepa* L.: production of transgenic onions and shallots. *Molecular Breeding* 7, 101–115.

Zupan, J.R. and Zambryski, P. (1995) Transfer of T-DNA from *Agrobacterium* to the plant cell. *Plant Physiology* 107, 1041–1047.

5 Diversity, Fertility and Seed Production of Garlic

T. Etoh[1] and P.W. Simon[2]

[1]*Laboratory of Vegetable Crops, Faculty of Agriculture, Kagoshima University, 21–24 Korimoto 1, Kagoshima 890-0065, Japan;* [2]*USDA/ARS, Department of Horticulture, 1575 Linden Drive, University of Wisconsin, Madison, WI 53706, USA*

1. Origins of Garlic and the History of its Cultivation

1.1 Garlic in Central Asia and the Mediterranean basin

Garlic (*Allium sativum* L.) has been cultivated by humans since ancient times, but its progenitors and centre of origin were not known until recently. Early taxonomists considered garlic to be a Mediterranean species. Linnaeus (1753) believed that Sicily was the original habitat of garlic, while Don (1827) mentioned Sicily as the origin of *A. sativum*, and *A. ophioscorodon* (classed as *A. sativum* in modern taxonomy) as originating in Greece or Crete. However, Regel (1875) stated that wild *A. sativum* plants grew in southern Europe and also reported seeing specimens from Dzungaria, a large desert basin in Central Asia, north of the Tien-Shan Mountains. Later Regel (1887) mentioned

Dzungaria and the Pamirs of southern Tajikistan as the habitat of *A. sativum* L. *typicum*, and the mountainous areas near Tashkent as the habitat of *A. sativum* L. *subrotundum* Gr. et Godr. De Candolle (1886) agreed with Regel that garlic was not indigenous in the Mediterranean area, and also considered that south-western Siberia was its original habitat.

Most recent researchers consider Central Asia to be the original home of garlic. Sturtevant (1919) concluded from an article by Pickering (1879) that garlic was native to the plains of western Tartary – currently the region of eastern Europe and western Russia. Vavilov (1951) and Kazakova (1971) proposed that Central Asia is the primary centre of origin of *A. sativum*, with the Mediterranean basin or the Mediterranean and the Caucasus as secondary centres. Recently Etoh (1986) and Kotlinska *et al.* (1991) discovered a number of fertile clones of a primitive garlic type on the north-western side of the Tien-Shan Mountains in Central Asia, and Etoh concluded that this area was the centre of origin of garlic. This conclusion by Etoh (1986) and Kotlinska *et al.* (1991) was confirmed on the basis of the presence of both fertile plants and the most primitive cultivars (Pooler, 1991) and by studies with molecular and biochemical markers (Maaß and Klaas, 1995).

One of the reasons why the centre of origin of garlic was unclear is that the progenitor species was unknown. Moreover, the presumed primary centre of origin, Central Asia, was closed to foreign researchers for a long time.

1.2 *Allium longicuspis* in Central Asia and the Mediterranean basin

Allium longicuspis Regel, the closest relative of garlic, is morphologically and karyologically very similar to garlic (Vvedensky, 1935; Etoh and Ogura, 1984; Mathew, 1996), and this sterile species (McCollum, 1976) was considered by Vvedensky (1935) to be a wild race of garlic. *A. longicuspis* was originally distinguished from garlic by the long filaments or exserted anthers, compared with the filaments of garlic which are typically shorter than the perianths (Regel, 1875; Vvedensky, 1935). However, many examples of exserted anthers have now been observed in fertile garlic plants (Kononkov, 1953; Konvicka *et al.*, 1978; Etoh, 1983a; Kotlinska *et al.*, 1991). On the other hand, *A. longicuspis* does not always have open flowers, in which case the anthers are not exserted from the perianths (Vvedensky, 1935; Kazakova, 1978). So it is doubtful whether exserted anthers should be used as the key feature separating these two species. One recent taxonomy adopts the difference in leaf number as the key to distinguishing the two species (Mathew, 1996). However, garlic has a great variation in the number of leaves (Etoh, 1985), and the variation in leaf number of garlic and *A. longicuspis* certainly overlaps. Fritsch claims that there is no significant difference between the two species (R.M. Fritsch, Head of Taxonomy Group of the Institute of Plant Genetics and Crop Plant Research at Gatersleben, Germany, personal communication; see also Fritsch and Friesen, Chapter 1, this volume).

Karyotype, isozyme and randomly amplified polymorphic DNA (RAPD) analyses by Etoh (1984), Pooler and Simon (1993a), Maaß and Klaas (1995) and Hong (1999) clearly indicate that variation in these markers for *A. longicuspis* lies within the range found within *A. sativum* (garlic). Currently some researchers believe that *A. longicuspis* Regel is a subspecies or group of *A. sativum* L., while others still contend that *A. longicuspis* is a separate species. *A. longicuspis* should therefore be considered either the closest wild relative or the wild ancestor of garlic.

McCollum (1976) described *A. longicuspis* as a sterile plant, but several recent collection missions found fertile accessions of this species, occurring spontaneously in the western Tien-Shan (Kyrgyzstan), the Karatau mountain range (Kazakhstan) and the Chatkal Mountains (Uzbekistan), regions north-west of the Tien-Shan Mountains. This area may be the original habitat of garlic or the ancestor of garlic, as suggested by Etoh (1986). Another possibility is that *A. longicuspis* and garlic may have a common wild ancestor. In any case, the habitat of *A. longicuspis* includes areas where sterile clones are found.

When considering the origin of *A. longicuspis*, Regel (1875, 1876) drew his conclusion on the basis of *A. longicuspis* specimens from Kokania, probably the place known today as Kokand in the easternmost part of Uzbekistan. Regel (1875) also referred to Turkestan and Dzungaria. Later, it was accepted that the natural habitat of *A. longicuspis* was in Central Asia, stretching between the Kopet Dag Mountains (between Turkmenia and Iran) in the west and the Tien-Shan Mountains in the east, with the Pamir Alai Mountains in the middle (Vvedensky, 1935; Wendelbo, 1971; Kazakova, 1978). Engeland (1991) named this area 'the garlic crescent'. Mathew (1996) added the area between eastern Turkey and Central Asia as the main natural distribution area of garlic. Engeland (1991) called this broader area 'the extended garlic crescent'.

1.3 Conclusions on the origin of garlic and its immediate relatives

Mathew (1996) made the interesting suggestion that the fertile Turkish plant, *Allium tuncelianum* (also called *A. macrochaetum*), with non-bulbiferous inflorescences, might be the common ancestor of garlic and *A. longicuspis*. *A. tuncelianum* was originally identified as *A. macrochaetum* Boiss. & Hausskn. subsp. *tuncelianum* Kollmann. This plant has the typical smell of garlic and is used as such in Turkey. The three plants, *A. tuncelianum*, *A. sativum* and *A. longicuspis* apparently have certain features in common as well as the characteristic odour, notably the coiling of the flower stem before anthesis, pale-coloured, small, glabrous, rather narrow perianth segments and glabrous filaments with very long lateral cusps. There are many examples of bulbiferous plants derived from non-bulbiferous plants in *Allium* and this fact lends force to the argument of Mathew (1996).

Gvaladze (1961) in Georgia, the Caucasus, proposed a subclassification of *A. sativum* into three groups, as follows:

1. Flowering plants with no bulbils in the inflorescences.
2. Flowering plants with both flowers and bulbils in the inflorescences.
3. Plants that form no flower stalks.

This classification may support Mathew's theory, since eastern Turkey (with plants of group 1) borders Georgia in the Caucasus, where plants of groups 2 and 3 grow wild. Etoh *et al.* (1992) collected a few fertile garlic clones in the Caucasus (near the border with Turkey). Hence Etoh (1985) agreed with the above evolutionary route from fertile to sterile plants, and considers the Caucasus to be the secondary centre of origin of garlic. Maaß and Klaas (1995) tested a few hundred clones from areas close to the centre of origin in Central Asia, using isozymes and RAPDs. They concluded that the most primitive cultivars with fertile flowers were from Andizhan in the Fergana basin (in easternmost Uzbekistan) and from West and South Georgia in the Caucasus. They suggested that the Caucasian cultivar might have been brought to Georgia from Central Asia. Erenburg (cited by Kazakova, 1978) reports flowering and seed production of garlic in Kazakhstan (Central Asia) and Dagestan, part of the Caucasus. Therefore there is some evidence that the habitat of the ancestor of garlic was the large area stretching from the western Tien-Shan Mountains to the Caucasus, the area termed by Engeland (1991) 'the extended garlic crescent'. Mathew (1996) encouraged further exploration of his conjecture. It should be noted that a karyological survey revealed that both garlic and *A. longicuspis* have similar characteristic satellite (SAT)-chromosomes (Etoh and Ogura, 1984; for more details, see Klaas and Friesen, Chapter 8, this volume).

1.4 The spread and diversity of garlic around the world

1.4.1 South and west

Engeland (1991) studied the history of garlic and produced historical maps on the topic. He proposed that wild *A. longicuspis* might have been cultivated by semi-nomadic hunter-gatherers more than 10,000 years ago, in the well-travelled region of 'the garlic crescent', a major trading route between China and the Mediterranean. He also suggested that wild garlic might have been very widely dispersed in early times and that it could easily have been taken by nomadic tribes to southern villagers, and from there the spread of garlic might have continued to the Mediterranean basin and to India, within a few millennia after the last ice age. Having studied the wide diversity of garlic names, De Candolle (1886) proposed that garlic extended from its original home to other areas before the migrations of the Aryans (2000–1500 BC). Engeland (1991) stressed the importance of the Caucasus region as one of the primary centres of distribution for most of the western world.

From the mention of garlic in Sanskrit, Engeland (1991) estimated that garlic was introduced to India more than 5000 years ago. Burkill (1966) indicated that garlic had been consumed in India from distant times and that from there it spread to the east, probably to South-East Asia. Further to the west, unbaked clay garlic models painted white were found in predynastic Egyptian cemetery graves more than 5000 years old (Tackholm and Drar, 1954). A bundle of garlic with scapes and bulbs was discovered in a tomb of the 18th Egyptian dynasty (Tackholm and Drar, 1954). A bolting-type garlic was also grown at this time in Egypt.

1.4.2 North and west

The Caucasus is a natural bridge of dispersion northward into Russia, the Ukraine and eastern Europe, as well as south to the shores of the Mediterranean or south-west through Turkey to south-eastern Europe. Greek and Roman writings provide solid evidence of the long history of garlic use in these ancient countries (Sturtevant, 1919; McCollum, 1976). In the south of Europe, where the climate suits the crop, the strong odour of garlic is appreciated more than in the north: hence the modern distribution of production areas in Europe. Garlic was introduced from the Mediterranean region to sub-Saharan Africa and to the Americas with explorers and colonists. Most of the cultivars currently grown in these continents are of the Mediterranean type.

1.4.3 East

The eastern part of the Tien-Shan Mountains is within China. However, no description of wild *A. longicuspis* is known in Chinese scripts (Anon., 1976; Xu, 1980, 1990). Probably this species was not naturally part of the ancient Chinese flora. Chia (AD 530–550) reported that Chang Kien, a famous Chinese general, first introduced garlic to China in the second century BC, but some researchers doubt this legend (Laufer, 1919; Kitamura, 1950). Since the Chinese name for garlic indicates a western Chinese origin, it is most likely that garlic was introduced into China from Central Asia across the wide barrier of the western desert by wandering traders. There is also a legend that the native Chinese wild garlic crossed with the introduced garlic and that only the hybrid plants survived (Engeland, 1991). However, there is only a small chance that this legendary hybrid may be related to the existence of multivalent chromosomes in all the East Asian garlic clones (Etoh, 1979).

In South China and South Asia generally, garlic leaves are consumed as a green vegetable, and special clones have been selected for leaf production. Differentiation of axillary buds and their development into cloves requires low temperatures (Takagi, 1990). Therefore, selection for leaf-producing rather than bulb-producing plants may have taken place in warm or hot regions. Indeed, many subtropical garlic cultivars develop only small bulbs (Messiaen *et al.*, 1993).

Garlic was introduced to Japan through Korea, but it was little used in Japan for a long time, while it became very popular in Korea.

1.5 Ecology

The natural habitat of this species is in 'gullies shaded during the day' (Vvedensky, 1935) or 'rocky valleys and river flats, 1350–2100 m' (Mathew, 1996). Garlic and *A. longicuspis* develop their bulbs during the summer, which may indicate that hot and dry summers were typical of the centre of evolution of both species. Central Asia has this type of climate. Both plants grow well under fairly dry conditions with bright sunlight. However, a very dry desert climate cannot support their growth. The Dzungaria basin desert (Regel, 1875) as we know it today seems too dry to be the original habitat of garlic. Etoh (1986) reported that there were no fertile garlics among the accessions collected in a mission in 1983 to Ashkhabad, located in the desert, north of the Kopet Dag Mountains, between Turkmenistan and Iran. Hence, it seems likely that gullies, rocky valleys or river-beds, where some moisture is still available even in the arid or semi-arid areas of Central Asia, may be the original habitat of garlic. It is worth noting that garlic is more tolerant to cold than common onion, *A. cepa*, another plant species native to Central Asia. Perhaps, if this region received more rainfall in earlier times, wild garlic might have grown much more extensively in these mountains.

2. Sources of Genetic Variation

Garlic was probably highly variable in the primary centre of evolution, even before its dispersal from that region. Thereafter, intraspecific variation must have increased, and isolation must have accelerated diversification, presuming that sexual reproduction occurred outside the centre of origin. Today garlic has great variation for maturity date, bulb size, shape and colour, flavour and pungency, clove number and size, number of whorls of cloves, bolting capacity, scape height, number and size of topsets (inflorescence bulbils) and number of flowers and fertility (McCollum, 1976). From isozyme and RAPD analyses, Pooler and Simon (1993a) and Maaß and Klaas (1995) were able to show that great heterogeneity exists within the Central Asian cultivar group. Another RAPD analysis, with 72 accessions collected from around the world, also showed considerable genetic diversity in the Central Asian group (Hong, 1999). Perhaps cross-pollination within garlic types or to ancestral forms in the not-too-distant past generated some of the great variation we now observe in garlic plants, and which is also shown by several genetic marker systems.

3. Subclassification

Helm (1956) described three botanical varieties of *A. sativum* L.: var. *sativum*, var. *ophioscorodon* and var. *pekinense*. However, Jones and Mann (1963) noted that many garlic clones possessed combinations of the characteristics used by Helm to discriminate var. *sativum* from var. *ophioscorodon*, and therefore proposed that the latter two varieties should be designated as horticultural groups rather than botanical taxa. They did not challenge, however, the separation of var. *pekinensis* (East Asian group), because of its distinct characteristics. Helm (1956) studied another garlic-like plant, known as rocambole, and concluded that this name should only be applied to forms of garlic with coiled scapes, and not to *A. scorodoprasum*.

According to Kazakova (1978), the authors Zagorodskij (1935), Kuznetsov (1954) and Alekseeva (1960) independently divided garlic cultivars into two subspecies or groups: non-bolting and bolting. Among them, Kuznetsov (1954) further subdivided each group into three ecotypes: continental type, south type and south coast of Russia type for non-bolters, and Central Asian type, Caucasus type and east Caucasus type for bolters. Kazakova (1978) disagreed with these classifications, since in field trials she found bolting to be an unstable trait whose expression depends on environmental conditions. This was later supported by Pooler and Simon (1993a). Al-Zahim *et al.* (1997) tested garlic and *A. longicuspis* accessions for RAPD polymorphism in the UK, and showed significant differences (in all but one accession) between bolters and non-bolters, thus supporting the Zagorodskij (1935), Kuznetsov (1954) and Alekseeva (1960) classification into two distinct genetic groups. Etoh (1985) reported that complete bolters always develop scapes and that non-bolters never develop scapes, whereas incomplete-bolting cultivars exhibit an intermediate response, with some variation in bolting habit. Messiaen *et al.* (1993) also reported that bolting habit is a nearly stable trait in France.

Komissarov (1964, 1965) reclassified the bolting forms into three groups in accordance with their geographical distribution: Mediterranean, Central Asian and Sino-Mongolian. He suggested that selection in cultivation of *A. longicuspis* has resulted in larger bulbs, loss of fertility and finally the development of non-bolting forms. It is assumed that the latter were derived from corresponding bolting forms and that the non-bolting Mediterranean group had evolved in a broad region, including the Caucasus, the Balkans and the Crimea, after long domestication. This region was suggested to be a more recent evolutionary source of most Mediterranean and European garlic than Central Asia. Komissarov concluded that the Central Asian group has a promising potential for garlic improvement because of traits such as winter-hardiness, high yield, and resistance to diseases and pests.

Kazakova (1971, 1978) classified the garlic taxon into two geographical subspecies: ssp. *sativum* (*mediterraneum*) for the Mediterranean group with large bulbs and cloves, and ssp. *asiae-mediae* for the Central Asian group with small bulbs and cloves. Both groups include bolting and non-bolting types.

Hanelt (1990) agreed with Jones and Mann's (1963) classification of *A. sativum* into two groups, the common garlic group (including var. *sativum* and *A. pekinense* as synonyms) and the *ophioscorodon* group. For a better understanding of garlic classification, Engeland (1991) proposed that the garlic taxon consists of the two subspecies *ophioscorodon* and *sativum* and five of what he termed 'varieties', perhaps better regarded as subgroups. Subsp. *ophioscorodon* usually develops flower stalks, and includes three varieties: 'Rocambole', 'Continental' and 'Asiatic', though later Engeland (1995) put 'Asiatic' into subsp. *sativum*. 'Rocambole' has distinctively coiled flower stalks and 'Continental' has very tall flower stalks with numerous, very tiny topsets. Subsp. *sativum* develops partial or no flower stalks, and it includes two subgroups, 'Artichoke' and 'Silverskin'. 'Artichoke' frequently has sets (bulbils) in the false stems (incomplete-bolting type: Etoh, 1985) and early-maturing bulbs. 'Asiatic', which was classified as a group of this 'Artichoke' variety later by Engeland (1995), develops very thick and broad leaves, and flower stalks; it also has unique elongated bulbils. 'Silverskin' rarely develops topsets, and produces only late-maturing bulbs.

Burba (1993) classified Argentinian garlic (a typical South American garlic) as non- or incomplete-bolting types, like Mediterranean garlic.

Messiaen *et al.* (1993) and Lallemand *et al.* (1997) classified garlic clones by morphological and physiological characteristics and by isozyme polymorphism. Cultivars from the western world were classified into one eastern European group of the bolting type and five Mediterranean groups: one complete-bolting, two incomplete-bolting and two non-bolting types. Asian clones were not clearly classified, but Central Asian clones and the East Asian clones had isozyme types different from those of the western world. The Central Asian seed-producing clones had the greatest isozyme polymorphism.

Tsuneyoshi *et al.* (1992) had a different approach, and used chemotaxonomic methods for garlic classification. Comparisons of mitochondrial DNA (mtDNA) provided the basis for distinguishing five groups. Most of the fertile cultivars were classified into the Russian (Central Asia and Caucasus) group, and cultivars from Central Asia exhibited the greatest genetic variation.

Pooler and Simon (1993a) evaluated 110 garlic clones, including *A. longicuspis*, by morphological traits and isozyme polymorphism. Of the 17 different electrophoretic/phenotypic groups investigated, three of the *ophioscorodon* type developed fertile pollen.

These results led Engeland to modify his 1991 morphological classification (Engeland, 1995) and to incorporate one bolting variety, 'Asiatic', into subsp. *sativum*. Moreover, 'Continental' of subsp. *ophioscorodon* was divided into two varieties, 'Purple Stripe' and 'Porcelain'.

Maaß and Klaas (1995) analysed intra-specific differentiation of garlic by both isozymes and RAPD markers and by morphological features. They proposed grouping Old World cultivars into four taxa: the *sativum* group from the Mediterranean; the *ophioscorodon* group from middle and eastern Europe; the *longicuspis* group from Central Asia, including *A. longicuspis*; and the subtropical group from southern Asia. The *longicuspis* taxon includes the East Asian subgroup *pekinense*, with bolting plants and coiling scapes: some have more or less fertile flowers. The *longicuspis* group is considered comparatively primitive. The subtropical group possibly originated from this *longicuspis* group a long time ago in northern India, and the *ophioscorodon* group possibly also originated from this *longicuspis* group in Transcaucasia and in a region north of the Black Sea. All *ophioscorodon* plants tested by the German researchers bolted and coiled, but the flowers were often deformed and sterile (Maaß, 1994). The *sativum* group was probably also derived from the *longicuspis* group in West Asia, and it was morphologically divided into bolting and non- or incomplete-bolting cultivars, which were discriminated by isozyme and RAPD analyses. However, the tested accessions showed relatively high genetic homogeneity (Maaß, 1994). Similar results were obtained from Iberian cultivars by Etoh *et al.* (2001). It should be noted that, of these taxons, only the *longicuspis* group from Central Asia has any fertile accessions.

4. Flowering: Genetics and Environment

Garlic clones vary in scape length in many areas (Etoh, 1985, 1986; Engeland, 1991). *Longicuspis* or 'Continental' types develop very tall scapes (Engeland, 1991) while incomplete-bolting or 'Artichoke' types usually develop short scapes (Etoh, 1985; Engeland, 1991), which always bear topsets but few flower buds (Etoh, 1985).

Non-bolting clones never develop scapes in warmer areas, such as subtropical Kagoshima, Japan. However, with 2-month vernalization at 10°C before winter, Etoh (1985) successfully induced flower buds, which persisted to meiosis in one incomplete-bolting clone.

In contrast, scape length did not vary much among a large collection of bolting and non-bolting clones exposed to the very cold winter of Wisconsin (Pooler, 1991) and typically non-bolting clones often tend to start initiating inflorescences to some extent after several years of cultivation in this climate. Perhaps not the depth of cold, but rather the duration of inductive cold conditions, from soon after autumn planting until late spring, accounts for the induction in the non-bolters. High incidence of flowering (29%) was recorded when a non-bolting clone was exposed to a constant temperature of 10°C from October to May, as compared with 11% blooming in field-grown plants in Wisconsin, where average temperatures are 10°C or less only from November through to April (Pooler and Simon, 1993b).

5. Discovery and Description of Fertile Clones

5.1 Early studies suggesting fertility

All garlic clones were long thought to be completely sterile (Weber, 1929), and all early literature indicated that they hardly ever produced flowers. Hence, garlic has long been propagated asexually by cloves or by topsets. This sexual sterility poses some difficulties, the most serious one being concerned with garlic improvement, since only limited genetic variation can be introduced via mutations and it is very hard to make significant progress by mutation breeding alone.

Vegetative propagation has some significant disadvantages. The major ones are the low rate of propagation, thus resulting in costly planting material. Storage is expensive, and decay and sprouting lead to losses during storage. Vegetative propagation enables perpetuation of pests, such as viruses and nematodes, in the propagation material and prevents 'cleansing' of the vegetative tissues. Consequently, there is a gradual or rapid increase in virus contamination of cloves and topsets, with a subsequent decrease in yield (Walkey, 1990; Salomon, Chapter 13, this volume).

Some old publications reported seed-producing garlic (Stephenson and Churchill, 1835). More recently, a number of reports documented seed production in garlic. Cicina (1955) described production of seeds in two *A. longicuspis* types, 'Chimkent garlic' and 'Chokpar garlic', on plants growing outdoors at Alma Ata in Kazakhstan. Using a Russian bolting cultivar, Kononkov (1953) obtained several viable seeds as a result of cross-pollination. Katarzhin and Katarzhin (1978) obtained a few seeds from a single garlic plant in the field. The offspring produced 120 seeds, which germinated to produce a second generation. Later, Katarzhin and Katarzhin (1982) reported the results of their work in the Volga–Akhtuba flood-plain area, which indicated that garlic can set viable seed under natural conditions. They obtained seeds from a variety from Batumi, Georgia, in the Caucasus, and then from three local cultivars from Volonezh and Poltava provinces of Ukraine and from the town of Groznyi in north Caucasus. Since these plant materials were not available outside the then USSR, their fertility was not evaluated elsewhere and their current fate is unknown.

Gvaladze (1961) obtained garlic seeds from cv. 'Svanetskaya' when plant nutrients were supplemented with boron. In the absence of this element, the generative organs of this variety degenerated at various developmental stages.

In Germany, Konvicka (1973) and Konvicka *et al.* (1978) reported fertile pollen in garlic plants treated with the antibiotics tylosin and tetracycline. The treatment resulted in the formation of fertile flowers with regular meiosis. However, Novak and Havranek (1975) and Etoh (1980) were unable to reproduce these results. More recently a fertile Italian garlic clone was described by Bozzini (1991). Chromosome counts revealed that this bulbiferous plant is tetraploid, and its karyotype differs from that of garlic. It was therefore classified as a member of the *A. ampeloprasum* taxon. The preliminary and inconsistent nature of these reports led to the assumption that garlic was an obligate apomict. If this were to be proved true, then regained fertility could not contribute to garlic improvement through recombination and breeding (Koul *et al.*, 1979).

5.2 Discovery and confirmation of fertility

Garlic cultivars are categorized as bolting or non-bolting. However, even bolting cultivars do not necessarily develop mature flowers, as in most cases the flowers fail to develop beyond the young bud stage. Studies on meiosis in garlic are therefore rare. Takenaka (1931) was the first to observe garlic meiosis, which showed irregular chromosome pairing, and he attributed garlic sterility to this cause. Later, both regular (Katayama, 1936; Krivenko, 1938) and irregular (Katayama, 1936; Etoh, 1979) meioses, including multivalent chromosomes (Etoh, 1979), were observed in different genotypes. In subsequent studies Etoh (1983a) therefore expanded the gene pool studied, and eventually identified two clones with regular meiosis. One of the two clones, a Russian garlic from Moscow Central Botanical Garden, developed normal microspores that matured and developed into viable pollen grains in the violet anthers without any special treatment. These pollen grains germinated on artificial media, and the flowers produced viable seeds after self-pollination (Etoh, 1983b). The fertile clone also produced one particular unique peroxidase isozyme band (Etoh, 1985). With this work, the earlier reports by Russian researchers on pollen production in untreated garlic were confirmed outside the USSR for the first time.

Etoh (1985) listed a number of reasons for garlic sterility. The most fundamental reason is thought to be the existence of chromosomal deletions. This assumption is supported by the observed loss of numerous satellites from SAT chromosomes (Etoh, 1985) and by the frequently observed micronuclei in the tetrads or microspores of a sterile clone (Etoh, 1980). Regular meiosis was observed in the fertile clone, no. 130, but homologous chromosomes often differed in length (Etoh, 1985). Similarly, differences in the size of homologues were reported for garlic and *A. longicuspis* clones with regular meiosis (Etoh, 1984; Etoh and Ogura, 1984); the high frequency of heterogeneity between somatic homologues can be attributed to chromosomal deletions. Accumulated deletions could result in the loss of a number of genes involved in gametogenesis. A high incidence of multivalents was also observed by Takenaka (1931) in East Asian garlic clones.

Deletions, duplications, inversions and translocations are common in asexually propagated bulbous crops. During meiotic division, these processes yield duplicated or deficient chromosomes. The resulting genetic imbalance yields sterile gametes. Garlic seed production was significantly improved in recent trials by cross-pollination among fertile clones, and a dramatic increase in fertility was achieved after two to four sexual cycles (Inaba *et al.*, 1995; Jenderek, 1998), probably reflecting the elimination of deletions and duplications through sexual reproduction. The accumulation of multiple deletions over time may account for the reduced bolting ability in garlic, although non-bolting types are sometimes observed in sexual progeny (Pooler, 1991).

The onset of anther degeneration varied in different studies. Gvaladze (1961) reported that generative-tissue degeneration occurred at different times in different groups of garlic growing in ambient conditions. Novak (1972) reported hypertrophy of the tapetal layer of anthers at the post-meiotic stage in a sterile cultivar and in *A. longicuspis*, but this was not confirmed in other sterile cultivars (Etoh, 1985). Studying microgametogenesis in mostly European garlic, Pooler and Simon (1994) observed microspore degeneration at or before the tetrad stage in most of their clones. In many sterile clones, microspore degeneration was detected in developing pollen grains between the tetrad and the microspore stages, before the binucleate stage (Etoh, 1979, 1980; Gori and Ferri, 1982), and was accompanied by anther degeneration.

Koul and Gohil (1970) attributed garlic sterility to nutritional competition between the topsets and flowers. This competition also occurs in the fertile clones, and the removal of topsets is recommended to ensure seed production in some clones. Topset removal can improve seed productivity but the presence of topsets is not likely to be the primary cause of sterility. Konvicka (1973) and Konvicka *et al.* (1978) claimed that rickettsia-like microorganisms were the causes of sterility in garlic. However, Novak and Havranek (1975) and Etoh (1980) were not able to reproduce garlic-fertility restoration by using antibiotics. Interestingly, Konvicka also had a fertile garlic in his collection which produced seeds without antibiotic treatment. Other garlic researchers also succeeded in obtaining fertile plants without antibiotic treatment, indicating that microorganisms do not usually cause sexual sterility in garlic.

From his discovery of a fertile garlic clone and subsequent work, Etoh (1985) proposed a comprehensive hypothesis on the evolution of garlic, as follows. Ancestral garlic had normal meiosis, was fertile and developed numerous flowers and topsets in the long-scaped umbel. The long scape may have originated from *A. longicuspis*, a species resembling and having a common ancestor with garlic, or possibly even being the same species, as demonstrated by their similar zymograms of peroxidase isozymes (Pooler and Simon, 1993a; Al-Zahim *et al.*, 1997) and by the similarity of their basic karyotypes ($K(2n) = 10m + 2sm_1^{sc} + 2sm_2^{sc} + 2sm$), observed in fertile garlic and *A. longicuspis* clones (Etoh and Ogura, 1984; Etoh, 1985).

Primitive garlic probably had more cold-hardiness and heat tolerance, larger numbers of differentiated foliage leaves and later maturity than modern sterile cultivars. It is uncertain whether garlic was sterile before domestication, but chromosomal mutations that resulted in sterility have probably accumulated gradually during millennia of vegetative propagation. With the occurrence of sterility, garlic may have evolved towards shorter scapes and fewer flower buds and topsets through accumulated mutations and deletions.

Incomplete-bolting types, with their characteristic short scapes and fewer topsets, evolved when garlic lost the ability to form flower buds, and later garlic lost the ability to flower. During this evolutionary process, karyotypic changes, including deletions and reciprocal translocations, accumulated. Malformation of flower buds, often seen today, may have been an evolutionary consequence after the development of sterility. Domestication and subsequent cultivation of garlic would probably have accelerated selection for larger bulbs and promoted sterility, as production of scapes reduces bulb yield, and garlic producers have often eliminated flowering plants. In summary, garlic evolution began with sexually reproduced plants, continued with sterility and incomplete bolting and ended with non-bolting genotypes.

In humid areas, garlic bulbs are usually harvested long before flowering because the bulbs or the outer skins of the bulbs rot in wet soil. In many parts of the world, young garlic scapes are pulled out of the false stems to be used as a green vegetable and to improve bulb growth. In other words, farmers' intentional selection against sexual reproduction has probably accelerated the evolution of garlic towards the current situation.

From this evolutionary standpoint and from the discovery of a few fertile Russian garlic clones in the past, Etoh (1985) predicted that some more fertile garlic clones should still occur in the primary centre of origin, namely former Soviet Central Asia. Since then, Pooler (1991) found nearly all the Central Asian clones tested to be fertile.

5.3 Further developments in garlic fertility

Successful development of fertile garlic and seed production requires the use of the wide genetic diversity common in the centre of origin. Moreover, it is also important to gather information on primitive clones and on wild relatives for the exact identification of the primary centre of origin of garlic. For these reasons, Etoh (1986), Kotlinska *et al.* (1991) and P.W. Simon, T. Kotlinska, L.M. Pike and J.F. Swenson (1989, unpublished) made garlic collections in a wide area of former Soviet Central Asia. The *Allium* distribution map in the former USSR drawn by Stearn (1944) suggested that the regions most appropriate for collection would lie in the mountains of Turkmenia, Pamir-Alai and Tien-Shan. These Central Asian mountain areas are the home of a great number of *Allium* species, including *A. longicuspis*. Political difficulties made organizing such a search difficult. Therefore, Etoh (1986) initially collected garlic bulbs in the bazaars in Tashkent, Samarkand, Dushanbe, Alma Ata, Frunze, Ashkhabad and Moscow. The collection was grown in Kagoshima, Japan. At about the same time, multinational research teams collected garlic in cities, villages, rural areas and nature reserves (Kotlinska *et al.*, 1991; P.W. Simon, T. Kotlinska, L.M. Pike and J.F. Swenson, 1989, unpublished) and planted the collection in Skierniewice, Poland; Madison, Wisconsin; and Pullman, Washington.

The Central Asian garlic clones are striking. Of the 31 garlic clones Etoh collected, 27 had regular meiosis with eight bivalent chromosomes and 14 produced pollen, although some had low levels of fertility (Etoh, 1986). One clone was male-sterile. The fertile clones varied considerably in morphology, suggesting that they form a diverse genetic resource for garlic. In the winter, all fertile clones produced horizontal leaves with strong winter-hardiness, all were late-maturity and all had purple or violet anthers, like the Russian fertile clone described previously (Etoh, 1983a). Sterile Central Asian clones produced erect leaves in the winter and the anthers were yellow. Some of these sterile plants, however, had eight bivalents (Etoh, 1986).

Fertile clones produced thousands of seeds and, when pollinated with fertile pollen, the male-sterile clone also produced many seeds (Etoh *et al.*, 1988). Similarly,

Pooler (1991) found that the six wild garlic clones from the 1989 P.W. Simon *et al.* expedition were fertile, as were four out of five *A. longicuspis* from this region and four cultivated garlic samples collected in bazaars near Ashkhabad and Samarkand. In the Etoh collection, all the clones from Frunze and Alma Ata on the northern side of the Tien-Shan Mountains were fertile, while all the clones from Ashkhabad near the Kopet Dag Mountains were sterile. In a second expedition to Central Asia in search of fertile garlic, with a more focused collecting area, Hong and Etoh (1996) found fertile clones around Lake Issyk-Kul, Almaty (Alma Ata) and Bishkek (Frunze) on the northwestern side of the Tien-Shan Mountains. From the localization of the fertile garlic clones, Etoh (1986) suggested that garlic has a centre of origin around the Tien-Shan Mountains, because much of the material collected there was fertile. This region forms part of the area suggested by Regel (1887), Vvedensky (1935), Wendelbo (1971) and Kazakova (1978) as the original home of garlic.

Kazakova (1978) reported that garlic blooms and produces seeds in Kazakhstan and in Dagestan of the Caucasus at 700–825 m elevations. Gvaladze (1965) and Etoh *et al.* (1992) found a few fertile clones in the Caucasus, and Pooler (1991) found three out of four clones from this region to be fertile. After all these observations, there is little doubt that fertile garlic occurs in Central Asia and this supports the theory that perhaps the large mountainous region from the Tien-Shan to Caucasus, called 'the extended garlic crescent' by Engeland (1991), may be its primary centre of origin.

Most male-fertile garlic clones develop purple anthers at anthesis, and this is an important visible marker, although some very fertile clones lack purple pigmentation. In addition to this marker, Etoh (1985) found a particular isozyme band of peroxidase in his first fertile clone, and later all fertile clones in the Japanese collection had this band (Etoh and Nakamura, 1988). However, several sterile clones also had it and four of 12 fertile garlic clones in the US collection lack what seems to be the same band (Pooler, 1991). Hong *et al.* (1997) found two RAPD markers related to pollen fertility. The two markers, $OPJ12_{1300}$ and $OPJ12_{1700}$, were detected in all 31 pollen-fertile clones with the operon 10-mer primer, OPJ12 (5′-GTCCCGTGGT-3′). $OPJ12_{1700}$ was not detected in 28 of 29 sterile clones, while $OPJ12_{1300}$ was not detected in 26 of 29 sterile clones. These two RAPD markers were also absent in 30 Iberian sterile garlic clones (Hong *et al.*, 2000a, b). The two RAPD markers may be quite useful in evaluating fertile garlic clones for breeding. As we map more of the genetic factors influencing garlic fertility, it should not be surprising to find many regions of the genome involved in this trait, since deletions and duplications occurring in virtually all regions of the genetic map induced male sterility in well-studied crops, such as maize.

5.4 Seed production and breeding of garlic

5.4.1 First evidence for seed production

Speculation about and, later, proof of male fertility in garlic by researchers around the world in the last 45 years have raised hopes that seed production may be possible. In fact, in 1953 Kononkov had already reported some early success, but it was not until the 1980s that Etoh (1983b) and Konvicka (1984) presented convincing proof that garlic can produce true seed. Since then Etoh *et al.* (1988), Pooler and Simon (1994), Inaba *et al.* (1995), Etoh (1997) and Jenderek (1998) have all reported successful garlic-seed production, as have several other groups in Asia, Europe and North America.

The production of flowers and viable gametes is essential for successful seed production. Therefore, the genetic make-up and environmental conditions that interfere with the initiation of flowering, chromosome aberrations that cause pollen or ovule abortion, and perhaps the presence of pathogenic agents, led earlier researchers to conclude wrongly that male-fertile garlic, and consequently seed production, do not occur. Other factors, such as competition between topsets and reproduction processes, also limited the chances for success. Removal of topsets appears to improve the level of male fertility to some extent in clones with pollen-production potential (Cheng, 1982). Beyond this, the process of seed development and maturation is improved by removing topsets (Etoh, 1983b; Etoh *et al.*, 1988; Pooler and Simon, 1994; Jenderek, 1998). Topset removal usually has a more dramatic positive effect for clones with medium or large bulbils (> 5 mm) than for those with small bulbils, even though larger numbers of the smaller bulbils tend to develop. Another treatment sometimes used to improve garlic-seed production is decapitation of the scape above the leaf sheath and preservation of the detached scapes in water (Konvicka *et al.*, 1978; Pooler and Simon, 1994), but this procedure was less effective than topset removal. Both treatments reduce the competition with developing seed for photosynthate.

Garlic seeds are smaller and less viable than those of bulb onion, and germination can take several months. In a comprehensive study on seed treatments, Etoh (1983b) found that scarification, stratification and moist-chilling were quite effective in stimulating germination, but phytohormone treatments had little effect (Etoh *et al.*, 1988). Seed storage at 5°C for 3–6 months, especially in moist conditions, followed by germination at 5°C on filter-paper, yielded approximately 20% germination. Thus, garlic seeds appear to be dormant, like many wild alliums. As garlic experiences very cold winters in its native area, garlic seeds probably survive a long dormant period and grow rapidly in the spring. With rapid germination, garlic seeds can form a flower stalk and bulb within one season (Etoh *et al.*, 1988; Pooler and Simon, 1994).

Pooler and Simon (1994) stored garlic seed at 3°C for 1–12 months and germinated most of the seeds *in vitro* on tissue-culture medium. Only about 10% germinated. This technique is too labour-intensive for routine use but may be useful in situations where simpler methods are not successful.

In contrast to these earlier results, Inaba *et al.* (1995) succeeded in obtaining almost 80% seed germination by moist-chilling (0–3°C) for more than 2 weeks, followed by long-day treatment (16 h/3000 lux) at 22°C in the spring following the seed harvest. Jenderek (1998) obtained 67–93% seed germination (with unreported treatments). After several generations of seed production, seed germination is evidently one trait that responds to selection. Then seed storage at 5°C or room temperature for several weeks, followed by germination in soil in a greenhouse, may become routinely feasible.

5.4.2 Interspecific hybridization

The discovery of fertile garlic clones opened the way for producing interspecific hybrids between fertile garlic and other *Allium* species (see also Kik, Chapter 4, this volume). Ohsumi *et al.* (1993) obtained interspecific hybrids between common onion, *A. cepa*, and garlic, *A. sativum*, by conventional crossing, followed by embryo rescue. This was a very significant and interesting accomplishment, since the two species belong to sections *Cepa* and *Allium*. The hybrid plants had only 2% pollen viability and did not produce seeds. As this was a very wide cross, the high level of sterility is not surprising. However, there is a possibility of introducing or introgressing the garlic genome more broadly into the genus *Allium* by backcrossing, using common onion, followed by embryo rescue.

The interspecific cross between *A. longicuspis* and garlic was successfully accomplished, just after the discovery of the first fertile clone (Etoh, 1984), by pollination of sterile *A. longicuspis* with pollen from fertile garlic. The resulting hybrids, however, were also sterile.

As discussed earlier, since 1984 several studies using biochemical and molecular markers to analyse garlic and *A. longicuspis* have concluded that these probably belong to the same species, and several fertile clones of *A. longicuspis* have been identified. Offspring of crosses between *A. sativum* and *A. longicuspis* made by Pooler and Simon (1994) and by Jenderek (1998) were indistinguishable from those resulting from garlic × garlic crosses. *Allium longicuspis* plants tend to have higher flowering rates, smaller topsets, purple anthers and better seed production than garlic plants. Hence, whether this represents a bona fide interspecific cross or not, the ability to combine traits of these plants is beneficial for improvement of seed production in garlic.

Another successful interspecific hybridization was performed between leek, *A. ampeloprasum* leek group (female) and fertile *A. sativum* (Sugimoto *et al.*, 1991). The two species both belong to the section *Allium*, but they differ in ploidy level. Leek is a tetraploid plant with 32 chromosomes, while garlic is a diploid plant with 16 chromosomes. Hence, several interspecific triploids with 24 chromosomes and near-triploid aneuploids were recovered. Tetraploids and diploids were also obtained, but these may not be hybrids. Other interspecific hybrids between fertile garlic and more closely related species, such as *A. tuncelianum*, described by Mathew (1996), may be interesting for future studies.

5.4.3 Large-scale seed production and breeding of garlic

Evidence for male fertility and seed production in garlic is very important for the elucidation of the natural origins of this crop and its wild progenitors. Even rare production of true seeds permits genetic recombinations with 'wild *A. sativum*' (or *A. longicuspis*, as the case may be) and this may have a highly significant effect over the long path of evolution. To be useful in the short term 'evolution' that plant breeders rely upon, however, a system for large-scale seed production is necessary. This has now been achieved.

Thousands of true garlic seeds have now been generated by Etoh *et al.* (1988) and by Pooler and Simon (1994). Etoh *et al.* (1988) cultivated 17 clones, primarily from Central Asia, removed topsets to improve seed set and provided a plastic cover to protect the flowers and seed. Pooler and Simon (1994) grew 150–200 clones from many locations. They removed topsets and detached scapes to improve seed set, as well as supplying houseflies, to facilitate pollination, and protecting the plants with shade cloth or in an air-conditioned greenhouse, to protect flowers and seed from rain and excessive heat. In both studies, plants were open-pollinated without emasculation to maximize seed production and the diversity of the resulting progeny. Seed was stored and germinated as described earlier and the germination rate varied widely, from 3.3 to 39.2% among clones and from 2.7 to 82.5% with seed storage/germination treatments (Etoh *et al.*, 1988).

Inaba *et al.* (1995) obtained more than 50,000 garlic seeds with up to 80% germination after initial apical meristem culture (to free the stock from viruses), followed by four reproduction cycles of material obtained by Etoh (1986) in Central Asia. Plants were grown outdoors in natural conditions, and flowers were open-pollinated after the removal of topsets. Plants from the original clones produced fewer than 20 seeds per plant, but the seed productivity increased in the fourth seed generation up to 248 seeds per plant. In later generations, topset removal became unnecessary.

Jenderek (1998) in California (Basic Vegetable Products), USA, produced approximately 2 million garlic seeds in 7 years using 64 fertile clones from material she introduced, plus the US Department of Agriculture (USDA) collection and other material. The seeds were produced outdoors with topset removal. She found 27 clones that yielded over 400 seeds per umbel, with a maximum of 656 per umbel. Seed germination ranged from 67 to 93%. Seed weight varied from 339 to 384 seeds g^{-1}. Up to 43% visible defects, such as chlorosis, root death or unusual bulbing habit, were observed among the progenies. Fertility, seed production, seed germination rate and speed and the incidence of seedling defects all responded to selection. Topset removal was not necessary in later generations for successful seed production.

The remarkable improvement of garlic seed production after several cycles of selection indicates a highly significant genetic component underlying this important trait. This places garlic researchers in a situation in which garlic breeding is now a reality for the first time in modern history. These developments are too recent to provide a perspective of what impact this may have in future garlic improvement. The many possibilities include garlic inheritance studies, genome mapping, systematics and other basic genetic studies. Now that more new genetic combinations are available outside garlic's centre of diversity than in the whole of history, large-scale field testing of seedlings is under way. Extensive variation has already been observed in plant growth and bulb characteristics. Inbreeding depression, heterosis, variation for disease resistance and other traits can also now be evaluated. The first useful seedling selections will be asexually propagated, like conventional garlic, but seed propagation of the garlic crop may become feasible at some time in the future.

References

Alekseeva, M.V. (1960) *Cultivated Onion*. Kolos, Moscow, 303 pp (in Russian).

Al-Zahim, M., Newbury, H.J. and Ford-Lloyd, B.V. (1997) Classification of genetic variation in garlic (*Allium sativum* L.) revealed by RAPD. *HortScience* 32, 1102–1104.

Anon. (1976) *Iconographia Cormophytorum Sinicorum*, Vol. 5. Chinese Academy Beijing Botanical Institute Science Press, Beijing, pp. 465–491.

Bozzini, A. (1991) Discovery of an Italian fertile tetraploid line of garlic. *Economic Botany* 45, 436–438.

Burba, J.L. (1993) *Producción de 'Semilla' de Ajo*. Asociación Cooperadora EEA, La Consulta, Argentina, 163 pp.

Burkill, I.H. (1966) *A Dictionary of the Economic Products of the Malay Peninsula*, Vol. 1, reprint edn. Ministry of Agriculture and Cooperatives, Kuala Lumpur, pp. 102–103.

Cheng, S.S. (1982) Sexual process in garlic (*Allium sativum* L. cv. 'Chonan'). *Proceedings of the Tropical Region – American Society for Horticultural Science* 25, 69–72.

Chia, S.-H. (530–550) *Ch'imin Yaoshu (Essential Arts for the People – Chinese Book of Husbandry)*. Translated into Japanese with supplementary articles by Nishiyama, B. and Kumashiro, Y. (1969). Asian Economic Press, Tokyo, 346 pp. + supplements.

Cicina, S.I. (1955) Onion species from Kazakhstan and the prospects of their introduction. *Bjulluten Glavnogo Botanicheskogo Sada* 21, 30–35 (in Russian).

De Candolle, A. (1886) *Origin of Cultivated Plants*. Reprint from the English edn (1967). Hafner, New York, 468 pp.

Don, G. (1827) A monograph of the genus *Allium*. An advance reprint of *Memoirs of the Wernerian Natural History Society* 1, 1–102 (1832).

Engeland, R.L. (1991) *Growing Great Garlic*. Filaree Productions, Okanogan, Washington, 213 pp.

Engeland, R.L. (1995) *1995 Supplement to Growing Great Garlic*. Filaree Productions, Okanogan, Washington, 32 pp.

Etoh, T. (1979) Variation of chromosome pairings in various clones of garlic, *Allium sativum* L. *Memoirs of the Faculty of Agriculture of Kagoshima University* 15, 63–72.

Etoh, T. (1980) An attempt to obtain binucleate pollen of garlic, *Allium sativum* L. *Memoirs of the Faculty of Agriculture of Kagoshima University* 16, 65–73.

Etoh, T. (1983a) Accomplishment of microsporogenesis in a garlic clone. *Memoirs of the Faculty of Agriculture of Kagoshima University* 19, 55–63.

Etoh, T. (1983b) Germination of seeds obtained from a clone of garlic, *Allium sativum* L. *Proceedings of the Japan Academy* 59 (Series B), 83–87.

Etoh, T. (1984) Hybrids between wild garlic (*Allium longicuspis* Regel) and garlic (*A. sativum* L.). In: *Proceedings of EUCARPIA 3rd Allium Symposium*. IVT, Wageningen, The Netherlands, pp. 78–82.

Etoh, T. (1985) Studies on the sterility in garlic, *Allium sativum* L. *Memoirs of the Faculty of Agriculture of Kagoshima University* 21, 77–132.

Etoh, T. (1986) Fertility of the garlic clones collected in Soviet Central Asia. *Journal of the Japanese Society for Horticultural Science* 55, 312–319.

Etoh, T. (1997) True seeds in garlic. *Acta Horticulturae* 433, 247–255.

Etoh, T. and Nakamura, N. (1988) Comparison of the peroxidase isozymes between fertile and sterile clones of garlic. In: *Proceedings of EUCARPIA 4th Allium Symposium*. Institute of Horticultural Research, Wellesbourne, UK, pp. 115–119.

Etoh, T. and Ogura, H. (1984) Comparison between *Allium longicuspis* Regel and a fertile garlic clone and their hybrid seeds. In: *1984 Abstracts of the Japanese Society for Horticultural Science Spring Meeting*, Japanese Society for Horticultural Science, Tokyo, pp. 170–171 (in Japanese).

Etoh, T., Noma, Y., Nishitarumizu, Y. and Wakamoto, T. (1988) Seed productivity and germinability of various garlic clones collected in Soviet Central Asia. *Memoirs of the Faculty of Agriculture of Kagoshima University* 24, 129–139.

Etoh, T., Johjima, J. and Matsuzoe, N. (1992) Fertile garlic clones collected in Caucasia. In: Hanelt, P., Hammer, K. and Knüpffer, H. (eds) *The Genus Allium – Taxonomic Problems and Genetic Resources*. IPK, Gatersleben, Germany, pp. 49–54.

Etoh, T., Watanabe, H. and Iwai, S. (2001) RAPD variation of garlic clones in the center of origin and the westernmost area of distribution. *Memoirs of the Faculty of Agriculture of Kagoshima University* 37, 21–27.

Gori, O. and Ferri, S. (1982) Ultrastructural study of the microspore development in *Allium sativum* clone 'Piemonte'. *Journal of Ultrastructure Research* 79, 341–349.

Gvaladze, G.E. (1961) The embryology of the genus *Allium* L. *Bulletin of the Academy of Sciences of the Georgian SSR* 26, 193–200 (in Russian).

Gvaladze, G.E. (1965) Vivipary and the capability of generative reproduction in *Allium sativum* L. *Bulletin of the Academy of Sciences of the Georgian SSR* 40, 412–427 (in Russian).

Hanelt, P. (1990) Taxonomy, evolution, and history. In: Rabinowitch, H.D. and Brewster, J.L. (eds) *Onions and Allied Crops*, Vol. I. *Botany, Physiology, and Genetics*. CRC Press, Boca Raton, Florida, pp. 1–26.

Helm, J. (1956) Die zu Wurz- und Speisezwecken kultivierten Arten der Gattung *Allium* L. *Kulturpflanze* 4, 130–180.

Hong, C.-J. (1999) Fundamental studies on crossbreeding in garlic, *Allium sativum*, L. PhD thesis, Kagoshima University, Kagoshima, Japan.

Hong, C.-J. and Etoh, T. (1996) Fertile clones of garlic (*Allium sativum* L.) abundant around the Tien-Shan mountains. *Breeding Science* 46, 349–353.

Hong, C.-J., Etoh, T., Landry, B. and Matsuzoe, N. (1997) RAPD markers related to pollen fertility in garlic (*Allium sativum* L.). *Breeding Science* 47, 359–362.

Hong, C.-J., Watanabe, H., Etoh, T. and Iwai, S. (2000a) Morphological and karyological comparison of garlic clones between the center of origin and westernmost area of distribution. *Memoirs of the Faculty of Agriculture of Kagoshima University* 36, 1–10.

Hong, C.-J., Watanabe, H., Etoh, T. and Iwai, S. (2000b) A search of pollen fertile clones in Iberian garlic by RAPD markers. *Memoirs of the Faculty of Agriculture of Kagoshima University* 36, 11–16.

Inaba, A., Ujiie, T. and Etoh, T. (1995) Seed productivity and germinability of garlic. *Breeding Science* 45 (Suppl. 2), 310 (in Japanese).

Jenderek, M.M. (1998) Generative reproduction of garlic (*Allium sativum*). *Sesja Naukowa* 57, 141–145 (in Polish).

Jones, H.A. and Mann, L.K. (1963) *Onions and Their Allies*. Leonard Hill Books, London, 286 pp.

Katarzhin, M.S. and Katarzhin, I.M. (1978) Experiments on the sexual reproduction of garlic. *Byulleten' Vsesoyuznogo Ordena Lenina I Ordena Druzhby Narodov Instituta Rastenievodstva Imeni N.I. Vavilova* 80, 74–76 (in Russian).

Katarzhin, M.S. and Katarzhin, I.M. (1982) On generative reproduction of garlic. *Trudy po Prikladnoi Botanike, Genetike I Selektsii* 72(3), 135–136 (in Russian).

Katayama, Y. (1936) Chromosome studies in some Alliums. *Journal of the College of Agriculture, Imperial University of Tokyo* 13, 431–441.

Kazakova, A.A. (1971) Most common onion species, their origin and intraspecific classification. *Trudy po Prikladnoi Botanike, Genetike I Selektsii* 45(1), 19–41 (in Russian).

Kazakova, A.A. (1978) *Allium*. In: Brezhnev, D.D. (ed.) *Flora of Cultivated Plants*, Vol. 10. Kolos, Leningrad, USSR, 262 pp. (in Russian).

Kitamura, S. (1950) Origins of Chinese cultivated plants. *Toho-Gakuho Kyoto* 19, 76–101 (in Japanese).

Komissarov, V.A. (1964) Evolution of the cultivated garlic, *A. sativum* L. *Izvestiya Timirjazevskoi Sel'skokhozyaistvennoi Akademii* 4, 70–73 (in Russian).

Komissarov, V.A. (1965) Classification of garlic, *A. sativum* L. *Doklady Sel'skokhozyaistvennoi Akademii Timirjazev* 108, 351–357 (in Russian).

Kononkov, P.F. (1953) The question of obtaining garlic seed. *Sad i Ogorod* 8, 38–40 (in Russian).

Konvicka, O. (1973) Die Ursachen der Sterilität von *Allium sativum* L. *Biologia Plantarum (Praha)* 15(2), 144–149.

Konvicka, O. (1984) Generative Reproduktion von Knoblauch (*Allium sativum*). *Allium Newsletter* 1, 28–37.

Konvicka, O., Nienhaus, F. and Fischbeck, G. (1978) Untersuchungen über die Ursachen der Pollensterilität bei *Allium sativum* L. *Zeitschrift für Pflanzenzüchtung* 80, 265–276.

Kotlinska, T., Havranek, P., Navratill, M., Gerasimova, L., Pimakhov, A. and Neikov, S. (1991) Collecting onion, garlic and wild species of *Allium* in central Asia, USSR. *FAO/IBPGR Plant Genetic Resources Newsletter* 83/84, 31–32.

Koul, A.K. and Gohil, R.N. (1970) Causes averting sexual reproduction in *Allium sativum* Linn. *Cytologia* 35, 197–202.

Koul, A.K., Gohil, R.N. and Langer, A. (1979) Prospects of breeding improved garlic in the light of its genetic and breeding systems. *Euphytica* 28, 457–464.

Krivenko, A.A. (1938) A cytological study of garlic (*Allium sativum* L.) *Biologicheskij Zhurnal* 7, 47–68.

Kuznetsov, A.V. (1954) *Cultivated Garlic*. Selkhozgiz, Moscow, 119 pp. (in Russian).

Lallemand, J., Messiaen, C.M., Briand, F. and Etoh, T. (1997) Delimitation of varietal groups in garlic (*Allium sativum* L.) by morphological, physiological and biochemical characters. *Acta Horticulturae* 433, 123–132.

Laufer, B. (1919) *Sino-Iranica*. Reprint from the original edition (1973). Ch'eng Wen Publishing, Taipei, Taiwan, 630 pp.

Linnaeus, C. (1753) *Species Plantarum*, Vol. 1. (Reprinted 1957.) Ray Society, London, 560 pp.

Maaß, H.I. (1994) What is the true *Allium sativum* L. var. *ophioscordon* (Link) Döll? *Allium Improvement Newsletter* 4, 12–14.

Maaß, H.I. and Klaas, M. (1995) Infraspecific differentiation of garlic (*Allium sativum* L.) by isozyme and RAPD markers. *Theoretical and Applied Genetics* 91, 89–97.

McCollum, G.D. (1976) Onion and allies. In: Simmonds, N.W. (ed.) *Evolution of Crop Plants*. Longman, London, pp. 186–190.

Mathew, B. (1996) *A Review of* Allium *Section* Allium. Royal Botanic Gardens, Kew, Richmond, UK, 176 pp.

Messiaen, C.M., Cohat, J., Leroux, J.P., Pichon, M. and Beyries, A. (1993) *Les Allium Alimentaires Reproduits par Voie Végétative*. INRA, Paris, 228 pp + English summary 42 pp.

Novak, F.J. (1972) Tapetal development in the anthers of *Allium sativum* L. and *Allium longicuspis* Regel. *Experientia* 28, 363–364.

Novak, F.J. and Havranek, P. (1975) Attempts to overcome the sterility of common garlic (*Allium sativum*). *Biologia Plantarum (Praha)* 17, 376–379.

Ohsumi, C., Kojima, A., Hinata, K., Etoh, T. and Hayashi, T. (1993) Interspecific hybrid between *Allium cepa* and *Allium sativum*. *Theoretical and Applied Genetics* 85, 969–975.

Pickering, C. (1879) *Chronological History of Plants: Man's Record of his Own Existence Illustrated Through their Names, Uses and Companionship*. Little, Brown & Co., Boston, Massachusetts, 145 pp.

Pooler, M.P. (1991) Sexual reproduction in garlic (*Allium sativum* L.). PhD thesis, University of Wisconsin-Madison, USA.

Pooler, M.R. and Simon, P.W. (1993a) Characterization and classification of isozyme and morphological variation in a diverse collection of garlic clones. *Euphytica* 68, 121–130.

Pooler, M.R. and Simon, P.W. (1993b) Garlic flowering in response to clone, photoperiod, growth temperature, and cold storage. *HortScience* 28, 1085–1086.

Pooler, M.R. and Simon, P.W. (1994) True seed production in garlic. *Sexual Plant Reproduction* 7, 282–286.

Regel, E. (1875) Alliorum adhuc cognitorum monographia. *Acta Horti Petropolitani* 3, 1–266.

Regel, E. (1876) Flora Turkestans, E. Regel's Botanical Work on the Materials Collected by O. and A. Fedschenko. *Izvestiya Imperatorskago Obschchestva Lyubitelei Estestvoznaniya, Antropologii I Etnografii* 21, 38–40 (in Russian and Latin).

Regel, E. (1887) Allii species Asiae centralis. *Acta Horti Petropolitani* 10, 279–362.

Stearn, W.T. (1944) The floristic regions of the USSR with reference to the genus *Allium*. *Herbertia* 11, 45–63.

Stephenson, J. and Churchill, J.M. (1835) *Allium sativum*. In: Burnett, G.T. (ed.) *Medical Botany*, 3 vols. John Churchill, London.

Sturtevant, E.L. (1919) *Sturtevant's Notes on Edible Plants* (Hedrick, U.P., ed.). J.B. Lyon, Albany, New York, 686 pp.

Sugimoto, H., Tsuneyoshi, T., Tsukamoto, M., Uragami, Y. and Etoh, T. (1991) Embryo-cultured hybrids between garlic and leek. *Allium Improvement Newsletter* 1, 67–68.

Tackholm, V. and Drar, M. (1954) Flora of Egypt, Vol. 3. *Cairo University, Bulletin of the Faculty of Science* 30, 86–113.

Takagi, H. (1990) Garlic *Allium sativum* L. In: Brewster, J.L. and Rabinowitch, H.D. (eds) *Onions and Allied Crops*, Vol. III. *Biochemistry, Food Science, and Minor Crops*. CRC Press, Boca Raton, Florida, pp. 109–146.

Takenaka, Y. (1931) Further reports of the cytological investigations on the sterile plants. *Journal of the Chosen Natural Historical Society* 12, 25–41 (in Japanese.)

Tsuneyoshi, T., Nosov, A.V., Kajimura, Y., Sumi, S. and Etoh, T. (1992) RFLP analysis of the mtDNA in garlic cultivars. *Japanese Journal of Breeding* 42 (Suppl. 2), 164–165 (in Japanese).

Vavilov, N.I. (1951) The origin, variation, immunity and breeding of cultivated plants. *Chronica Botanica* 13, 1–364.

Vvedensky, A.I. (1935) *Allium* L. In: Komarov, V.L. (ed.) *Flora SSSR*, Vol. 4. Translated into English by Airy-Shaw, H.K., as 'The genus *Allium* in the USSR'. *Herbertia* 11, 65–218 (1944); also English edn, *Flora of the USSR*, Vol. 4, pp. 87–216 (1968), Israel Program for Scientific Translations, Jerusalem.

Walkey, D.G.A. (1990) Virus diseases. In: Rabinowitch, H.D. and Brewster, J.L. (eds) *Onions and Allied Crops*, Vol. II. CRC Press, Boca Raton, Florida, pp. 191–212.

Weber, E. (1929) Entwicklungsgeschichtliche Untersuchungen über die Gattung *Allium*. *Botanisches Archiv* 25, 1–44.

Wendelbo, P. (1971) *Alliaceae*. In: Rechinger, K.H. (ed.) *Flora Iranica*, Vol. 76. Academische Druck- und Verlagsanstalt, Graz, Austria, 100 pp.

Xu, J.M. (1980) *Allium* L. In: Wang, F.-T. and Tang, T. (eds) *Flora Reipublicae Popularis Sinicae*, Vol. 14, *Monocotyledoneae – Liliaceae (1)*. Science Press, Beijing, pp. 170–272 (in Chinese).

Xu, J.M. (1990) Key to the Alliums of China. *Herbertia* 46(2), 140–164 (translation by Hanelt, P. and Chun-Lin, L.).

Zagorodskij, P.F. (1935) Garlic. *Rukovodstvo po Aprobatsii Sejlskokhozyastvennikh Kultur* 5, 253–264 (in Russian).

6 Genetic Transformation of Onions

C.C. Eady
New Zealand Institute for Crop & Food Research Limited, Private Bag 4704, Christchurch, New Zealand

1. Introduction

This chapter reviews the progress made in the genetic transformation of *Allium* species, and mainly of onion (*A. cepa* L.) as this is the only system within the genus for which a reliable protocol has been published. This section covers developments since 1995. For an earlier review, see Eady (1995).

The term genetic transformation, in this chapter only, covers systems that transfer a particular set of characterized genes via *Agrobacterium*-mediated or biolistic gene-delivery techniques. Systems that transfer larger amounts of essentially non-characterized DNA, e.g. somatic hybridization and cybridization techniques (Kumar and Cocking, 1987; Buiteveld, 1998), are not covered here.

Genetic transformation of alliums is still in its infancy; hence a general outline of transformation techniques is presented, followed by a summary of *Allium* traits that are suitable for transformation and a discussion of the potential risks of transforming onions. The section on transgene analysis summarizes the types of transgenic *Allium* plants that have been produced and the ways in which they have been characterized.

1.1 Plant genetic transformation

Routine protocols for the transformation of model plant species, such as tobacco, have been available for over 15 years (Horsch *et al.*, 1985). Such plants have been transformed with a plethora of foreign and modified gene constructs (Conner *et al.*, 1997). Unfortunately, it has not been possible to simply transfer this technology to all crop species. For example, the initial protocols for *Agrobacterium tumefaciens*-mediated transformation did not work on monocotyledonous plants until the process had been extensively modified (Hiei *et al.*, 1997). As a result, many alternative techniques have been designed to transfer DNA from test-tube to plant. These can be split into two categories of DNA transfer: direct DNA delivery and vector-mediated DNA delivery.

Direct DNA delivery uses physical, chemical or electrical methods to deliver DNA directly into the plant cell (Songstad *et al.*, 1995). Once in the cell, only intracellular processes are available to facilitate DNA integration into the host genome. Of the many direct DNA delivery techniques available, the most commonly used is biolistic gene transfer, where a gene gun is used to shoot tiny DNA-laden gold bullets into the plant cell. Ironically, the system was developed by experimentation using onion epidermal cells (Klein *et al.*, 1987). By 1990, stable transformation of maize and soybean had been reported using this technique (McCabe *et al.*, 1988; Fromm *et al.*, 1990). Different types of gene guns have been developed (Vain *et al.*, 1993), but the PDS1000 helium biolistic gun (Dupont) is the most widely used. Since the 1990s, biolistic gene transfer has gained favour, particularly for the transformation of monocotyledonous crop species (Christou, 1995). However, it has not been without its problems and it has produced results that have been difficult to repeat. It has also produced transformants that contain large numbers of unwanted integration events, such as the insertion of multiple and/or faulty copies of the transgene into the host genome, which prevent the recovery of phenotypically normal plants (Spencer *et al.*, 1992).

Vector-mediated DNA delivery harnesses the natural ability of certain microorganisms and viruses to mediate the successful transfer and integration of foreign DNA into the host plant. By far the most frequently used of the vector-mediated techniques is *Agrobacterium*-mediated transformation. *Agrobacterium* strains, containing a tumour-inducing (Ti) plasmid, have the ability to transfer a specific region of that plasmid, the T-DNA, to plant genomes. Under natural conditions, the Ti plasmid contains virulence genes that, with the help of chromosomal-based bacterial genes, effect the transfer process. The T-DNA sequences transferred contain flanking DNA sequences that assist in the integration process and genes that enable the affected plant cell to proliferate and produce a carbon source for the *Agrobacteria*. By manipulating this process it has been possible to substitute the wild type T-DNA region with modified T-DNA containing genes or sequences of choice (Christou, 1996). Using particular strains of *Agrobacterium* in combination with specific virulence genes and susceptible host-cell tissue types, it has been possible to broaden the host range of the *Agrobacterium*-mediated gene-transfer process (Hooykaas *et al.*, 1984; Jarchow *et al.*, 1991; Regansbury-Twink and Hooykaas, 1993). In 1994, the first routine transformation system for monocotyledonous plants was developed (Hiei *et al.*, 1994) and this has led to the resurgence in popularity of this technique.

Nearly all transgenic crop species have been produced using versions of the biolistic or *Agrobacterium*-mediated transformation systems. While they are effective mechanisms for transferring genes, they are still difficult and costly for many important crop species, and endeavours are constantly being made to simplify the process (Hansen and Wright, 1999). Recent important breakthroughs include the development, in model systems, of *in vivo Agrobacterium*-mediated transformation of germ-line cells (Clough and Bent, 1998). This technique avoids the need for expensive and technically difficult *in vitro* culture systems. In addition, researchers are harnessing the natural ability of transposon sequences to 'jump' genes from extrachromosomal plasmid DNA and integrate into plant genomes (Houba-Herin *et al.*, 1994; Lebel *et al.*, 1995), a commonly used technique in insect transformation (Rubin and Spradling, 1985). Other methods of targeted integrations and site-directed recombinations are also being developed (Ow, 1996; Puchta, 1998). The development of *in vivo* techniques, 'transposomics' and targeted integrations may soon lead to transformation methods that do not require highly skilled technical input, which undoubtedly will lead to transformation becoming a simple addition to the plant breeder's tool kit, and in doing so will promote its use as a routine technique. At present, only *Agrobacterium* and biolistic methods of transforming alliums have been reported (see Section 2).

1.2 Traits suitable for genetic modification in onions

Applied genetic engineering is still very much in its infancy, and the types of modification that have been made to commercially available cultivars are still relatively modest and mainly limited to herbicide resistance, *Bacillus thuringiensis* (Bt) expression, male sterility, virus resistance and altered fruit ripening (Table 6.1).

As transformation techniques become more routine and as genes and their products become better understood, specific modification will be increasingly used to further improve crop cultivars. Three commercial examples of this are the introduction of the phytase gene into canola (BASF AG), the suppression of the genetically modified (GM) *Fad*2–1 gene in soybean (Dupont), and the introduction in canola of a 12 : 0 acyl carrier protein thioesterase gene (Calgene Inc.).

Modification of alliums by transformation with existing proven genes (such as those listed in Table 6.1) could produce many advantages for the allium industry. When herbicide, insect, virus and disease resistance, male sterility and other traits are successfully introduced into other crops, their potential for incorporation into the alliums is raised. Other traits that might be altered in alliums include their sulphur biochemistry, pigmentation, fructan metabolism, and susceptibility to environmental conditions and to specific pests and diseases. For additional information on traits that have already been altered in other crops, see reviews by Dunwell (1998, 1999) and Table 6.1.

1.2.1 Disease and pest resistance

HERBICIDE RESISTANCE. Weed competition in alliums can account for yield losses of up to 70% and control is usually achieved by pre- and post-emergence herbicides (Rubin, 1990). Some of these herbicides have high, non-specific toxicity and are not easily degraded by soil microorganisms. Once modified to be resistant to the new generation of biodegradable herbicides, such as glyphosate (e.g. Roundup®) or phosphinothricin (e.g. Basta®), it may be possible to achieve efficient weed control in allium crops by using a single application of low-dosage herbicide. The enzyme 5-enolpyruvylshikimate-3-phosphate synthase (EPSPS) is important in the production of aromatic amino acids in plants, and its activity is inhibited by glyphosate. The ability to tolerate glyphosate can be conferred to the plant by insertion and expression of the *CP-4* gene, which encodes overproduction of EPSPS (Hetherington *et al.*, 1999), and also by a glyphosate oxidoreductase (*GOX*) gene, which encodes an enzyme that degrades the herbicide (Saroha *et al.*, 1998). Versions of the *CP-4* gene have now been inserted into the genomes of many crop species (Table 6.1), although precise details of sequences inserted are not usually published due to

Table 6.1. New plant varieties developed by genetic modification and the new genes they contain (adapted from http://www.vm.cfsan.fda.cov, 14/1/00).

New variety	Trait gene and source
Canola/oilseed rape (*Brassica napus*)	
Phytaseed canola	The phytase gene from *Aspergillus niger* van Tieghem
Bromoxynil-tolerant canola	The nitrilase gene from *Klebsiella pneumoniae* subsp. *ozaenae*
Male-sterile or fertility-restorer and glufosinate-tolerant canola	The male-sterile canola contains the barnase gene and the fertility restorer canola contains the barstar gene from *Bacillus amyloliquefaciens*; both lines have the phosphinothricin acetyltransferase gene from *Streptomyces viridochromogenes*
Glufosinate-tolerant canola	Phosphinothricin acetyltransferase gene from *Streptomyces viridochromogenes*
Male-sterile and fertility-restorer oilseed rape	The male-sterile oilseed rape contains the barnase gene from *Bacillus amyloliquefaciens*; the fertility-restorer lines express the barstar gene from *Bacillus amyloliquefaciens*
Glyphosate-tolerant canola	Enolpyruvylshikimate-3-phosphate synthase gene from *Agrobacterium* sp. strain CP4
Laurate canola	The 12 : 0 acyl carrier protein thioesterase gene from California bay, *Umbellularia californica*
Cantaloupe (*Cucumis melo*)	
Modified fruit-ripening cantaloupe	*S*-adenosylmethionine hydrolase gene from *E. coli* bacteriophage T3
Maize (*Zea mays*)	
Insect-protected and glufosinate-tolerant maize	The *cry9C* gene from *Bacillus thuringiensis* subsp. *tolworthi* and the bar gene from *Streptomyces hygroscopicus*
Glyphosate-tolerant maize	A modified enolpyruvylshikimate-3-phosphate synthase gene from maize
Male-sterile maize	The DNA adenine methylase gene from *Escherichia coli*
Insect-protected maize	The *cryIA(c)* gene from *Bacillus thuringiensis*
Glufosinate-tolerant maize	Phosphinothricin acetyl transferase gene from *Streptomyces hygroscopicus*
Insect-protected maize	The *cryIA(b)* gene from *Bacillus thuringiensis* subsp. *kurstaki*
Glyphosate-tolerant/insect-protected maize	The enolpyruvylshikimate-3-phosphate synthase gene from *Agrobacterium* sp. strain CP4 and the glyphosate oxidoreductase gene from *Ochrobactrum anthropi* in the glyphosate-tolerant lines; the *CryIA(b)* gene from *Bacillus thuringiensis* subsp. *kurstaki* in lines that are also insect-protected
Male-sterile maize	The barnase gene from *Bacillus amyloliquefaciens*
Glufosinate-tolerant maize	Phosphinothricin acetyltransferase gene from *Streptomyces viridochromogenes*
Cotton (*Gossypium hirsutum*)	
Bromoxynil-tolerant/insect-protected cotton	Nitrilase gene from *Klebsiella pneumoniae* and the *cryIA(c)* gene from *Bacillus thuringiensis* subsp. *kurstaki*
Sulphonylurea-tolerant cotton	Acetolactate synthase gene from tobacco, *Nicotiana tabacum* cv. *Xanthi*
Glyphosate-tolerant cotton	Enolpyruvylshikimate-3-phosphate synthase gene from *Agrobacterium* sp. strain CP4
Bromoxynil-tolerant cotton	A nitrilase gene isolated from *Klebsiella ozaenae*
Insect-protected cotton	The *cryIA(c)* gene from *Bacillus thuringiensis* subsp. *kurstaki*
Flax (*Linum usitatissimum*)	
Sulphonylurea-tolerant flax	Acetolactate synthase gene from *Arabidopsis*
Papaya (*Carica papaya*)	
Virus-resistant papaya	Coat-protein gene of the papaya ringspot virus

Table 6.1. *Continued.*

New variety	Trait gene and source
Potato (*Solanum tuberosum*)	
Insect- and virus-protected potato	The *cryIIIA* gene from *Bacillus thuringiensis* sp. *tenebrionis* and the potato virus Y coat-protein gene
Insect-protected potato	The *cryIIIA* gene from *Bacillus thuringiensis* sp. *tenebrionis*
Insect-protected potato	The *cryIIIA* gene from *Bacillus thuringiensis*
Radicchio (*Cichorium intybus* var. *foliosum*)	
Male-sterile radicchio rosso	The barnase gene from *Bacillus amyloliquefaciens*
Soybean (*Glycine max*)	
Glufosinate-tolerant soybean	Phosphinothricin acetyltransferase gene from *Streptomyces viridochromogenes*
High-oleic-acid soybean	Sense suppression of the *GmFad2–1* gene, which encodes a delta-12 desaturase enzyme
Glyphosate-tolerant soybean	Enolpyruvylshikimate-3-phosphate synthase gene from *Agrobacterium* sp. strain CP4
Squash (*Cucurbita pepo*)	
Virus-resistant squash	Coat-protein genes of cucumber mosaic virus, zucchini yellow mosaic virus and watermelon mosaic virus 2
Sugarbeet (*Beta vulgaris*)	
Glufosinate-tolerant sugarbeet	Phosphinothricin acetyltransferase gene from *Streptomyces viridochromogenes*
Glyphosate-tolerant sugarbeet	The enolpyruvylshikimate-3-phosphate synthase gene from *Agrobacterium* sp. strain CP4, and a truncated glyphosphate oxidoreductase gene from *Ochrobactrum anthropi*
Tomato (*Lycopersicon esculentum*)	
Insect-protected tomato	The *cryIa(c)* gene from *Bacillus thuringiensis* subsp. *kurstaki*
Modified-ripening tomato	*S*-adenosylmethionine hydrolase gene from *E. coli* bacteriophage T3
Flavr Savr™ tomato	Antisense polygalacturonase gene from tomato
Improved-ripening tomato	A fragment of the aminocyclopropane carboxylic acid synthase gene from tomato
Delayed-softening tomato	A fragment of the polygalacturonase gene from tomato

commercial confidentiality. Some seed companies are pursuing the transformation of onion cultivars with this resistance trait.

Phosphinothricin (PPT) resistance is conferred by the *pat* or *bar* genes, isolated from *Streptomyces viridochromogenes* and *Streptomyces hygroscopicus*, respectively (Vinnemier *et al.*, 1995). Both of these genes code for phosphinothricin acetyltransferase, which detoxifies the herbicide. Many PPT-resistant crop plants have been produced (Table 6.1). We have produced a few transgenic onion plants (C.C. Eady, J. Farrant, Erasmuson and Reader, unpublished) containing the *bar* gene alongside the *gfp* reporter gene. These plants are currently being characterized for copy number, expression and resistance to PPT (see Section 3 on analyses of transformants).

For examples of other herbicides for which gene-based resistance has been developed, e.g. sulphonylureas, triazines and bromoxynil-based herbicides, the reader is referred to Dekker and Duke (1995) and Tsaftaris (1996).

VIRAL RESISTANCE. The development and insertion of engineered viral protein genes, e.g. *P1*, *P3* (Moreno *et al.*, 1998), or *CI* (Wittner *et al.*, 1998) or gene sequences, e.g.

Nib (Guo *et al.*, 1999) or *CP* (Hackland *et al.*, 1994), has proved to be effective for preventing viral overload in a number of plant species (Revers *et al.*, 1999).

These, or similar, techniques are now being used on potato, squash and papaya (Table 6.1) to confer resistance in commercially important crops. Under vegetative propagation, there is no 'cleansing' sexual round to eliminate non-seed-transmitted viruses. A consequence of this is the gradual build-up of virus and a significant decrease in yield in garlic and shallot (Walkey, 1990). Potyviruses (e.g. onion yellow-dwarf virus, leek yellow-stripe virus), carlaviruses (e.g. garlic latent virus, shallot latent virus) (Walkey, 1990) and garlic and shallot virus X (Song *et al.*, 1998) are the most devastating *Allium* viruses (see also Salomon, Chapter 13, this volume). While *in vitro* elimination is possible (Fletcher *et al.*, 1998; Robert *et al.*, 1998), inbuilt resistance would provide a simpler solution. Recently researchers have isolated and sequenced coat-protein gene sequences from *Allium* carla virus (Tsuneyoshi *et al.*, 1998b) and potyvirus types (Kobayashi *et al.*, 1996; Tsuneyoshi *et al.*, 1998a; van der Vlugt *et al.*, 1999). With this knowledge, it should be relatively straightforward to engineer and express these sequences in *Allium* species to induce resistance.

INSECT RESISTANCE. To confer resistance to insect pests in the *Lepidoptera*, *Diptera* and *Coleoptera* orders (and maybe the *Thysanoptera*), the introduction of specific insecticidal protein genes from *B. thuringiensis* (*Bt* genes) (Crickmore *et al.*, 1998) may provide a control strategy. Specific forms of *Bt* genes have been engineered and introduced into plants to confer resistance against specific insects (Table 6.1; Hilder *et al.*, 1987; Schnepf *et al.*, 1998; Hilder and Boulter, 1999). They are currently being used to improve the commercial production of cotton and maize, among other crops. The above orders and *Hemiptera* all contain pests of *Allium* species (Soni and Ellis, 1990), so *Bt* gene technology may be useful to confer insect resistance into onions.

Sap-sucking onion thrips (*Thrips tabaci*) are the major insect pest of *Allium* species. In addition to physical damage, they also spread viral disease (Soni and Ellis, 1990). Damage levels on untreated crops can reach up to 55%. Thrips are difficult to control by conventional means, although integrated pest management (IPM) strategies, including biological control, partial plant resistance and chemical control measures at defined threshold levels of the pest, can alleviate the problem. Recently, thrips resistant to the synthetic pyrethroids have been reported (A. Stewart, Lincoln University, New Zealand, 1999, personal communication), presenting a serious control problem. The transformation of *Allium* species with genes conferring resistance to thrips could reduce dependence on the limited existing control measures. While transgenic plants with thrips resistance have not been approved for general release in any crop species, much research has been undertaken to combat this pest. The insertion and expression of protease inhibitors can reduce insect feeding (Hilder *et al.*, 1987), as can the insertion of lectin genes, e.g. from snowdrop (Rao *et al.*, 1998). The insertion of the tryptophan decarboxylase gene from *Catharanthus roseus* into tobacco reduced whitefly (*Bemisia tabaci*) emergence by 98.5% compared with non-transgenic controls (Thomas *et al.*, 1994). By directing expression of these genes to the sap-using, phloem-specific promoters (Stoger *et al.*, 1999), it may be possible to target these insects without expressing the foreign genes in other parts of the plant.

FUNGAL RESISTANCE. Antifungal resistance genes are not yet being used commercially to combat fungal pathogens in transgenic plants, although some field trials are under way. This area of research is vast and complex, as there are so many types of fungal infections. Fungal diseases range from specific systems requiring gene-for-gene interactions between host *R* genes and corresponding *Avr* genes in the pathogen (Hammond-Kosack and Jones, 1997) to facultative pathogens that are preferentially combated by pathogenesis-related (PR) resistance genes (Yun *et al.*, 1997). Detailed knowledge of the particular host–pathogen

R/Avr interaction is needed to identify the genes involved so that genetic-engineering strategies can be devised (Evans and Greenland, 1998). This information is not yet available for onion fungal pathogens.

Many onion fungal pathogens cause damage via hyphal invasion (Entwistle, 1990; Maude, 1990a) and may be susceptible to control by PR resistance genes. Such PR genes have been identified in plants (Yun *et al.*, 1997) and include: PR genes *PR-2* and *PR-3* (chitinases and glucanases) (Jongedijk *et al.*, 1995; Masoud *et al.*, 1996; Schickler and Chet, 1997), which act synergistically to prevent hyphal growth (Marchant *et al.*, 1998); *PGIP* genes encoding polygalacturonidase-inhibiting proteins, which inhibit enzymes released by the fungus that break down the plant cell wall (Toubart *et al.*, 1992; Powell *et al.*, 1994); *PR-5* genes encoding ribosome-inactivating proteins, which specifically act on fungal ribosomes (Stripe *et al.*, 1992); and genes encoding plant defensins, a class of small polypeptides that interfere with fungal cell-wall extension (Conceição and Broekaert, 1999). In addition, oxalate oxidase and oxalic decarboxylase genes have been introduced into plants, where they inhibit fungal invasion by detoxifying oxalic acid, the toxin produced by the fungi (WO 99/04012) (Thompson *et al.*, 1995; Kesarwani *et al.*, 2000). A gene encoding a non-specific lipid transferase that has activity against 12 types of pathogenic fungi has been isolated from onion (Phillippe *et al.*, 1995) and PGIP protein with antifungal activity has also been identified (Favaron *et al.*, 1997). If not useful in onion research, they may prove useful in other crops. Gene-discovery programmes are advancing rapidly with the introduction of DNA-chip technology (Kurian *et al.*, 1999). The genetic codes for a plethora of new genes with potential applications, such as conferring fungal resistance, are accumulating in databases around the world. Testing their potential is difficult and is likely to slow down their introduction into crop species. Efficient transformation systems are essential if such genes are to be tested *in planta*. Such systems are necessary to ensure that the gene product not only targets the particular host–pathogen interaction, but also to check that the gene and gene products have no adverse effects, e.g. to ensure that they lack activity against beneficial vesicular-arbuscular (VA) mycorrhiza.

Entwistle (1990) and Maude (1990b) described the major fungal pathogens of onion roots and bulbs. At present, such pathogens are controlled by rotation and fungicides in combination with particular curing and storage regimes (Entwistle, 1990). For a major fungal disease, such as onion white rot (OWR) (*Sclerotium cepivorum*) and others, control by fungicides is becoming increasingly difficult. Hence, other disease-control options are being investigated (Crowe and McDonald, 1998; A. Stewart, Lincoln University, New Zealand, 1999, personal communication). The introduction of genes that could prevent OWR and other soil-borne diseases would be extremely beneficial and could substantially reduce the amount of fungicide used or the need to move to new land in order to avoid the problem.

Oxalic acid is the toxin produced when *S. cepivorum* infects onions, and oxalate oxidase converts substrate into carbon dioxide and hydrogen peroxide, thus preventing the reduction in pH caused by oxalic acid (Stone and Armentrout, 1985). Maintaining cellular pH prevents the fungal pathogenic enzymes from working effectively and there is evidence that the production of hydrogen peroxide also activates other defence-related gene-expression products. Hence, of the genes available to combat OWR, the use of oxalate oxidase or oxalate decarboxylase probably holds the most potential, as these enzymes have been used in field trials to reduce disease symptoms caused by *Sclerotinia* species on sunflowers and tomatoes (WO 99/04012) (Kesarwani *et al.*, 2000).

As an alternative approach for OWR control, our research team has recently isolated an *Allium* root alliinase that may be responsible for producing volatile sulphur compounds. When released into the soil, these products stimulate dormant fungal sclerotia to germinate. Experiments are under way to determine if antisense technology can be used to reduce the expression of this gene in the root and thus to reduce the release of sulphur volatiles into the soil and consequently reduce sclerotial germination. Unfortunately, many soil microbes seem capable of degrading the sulphur precursors (King and Coley-Smith, 1969) that may accumulate in onion roots, thus releasing sulphur volatiles. Should the antisense technology prove effective, this biotic activity may cause difficulties. *In vivo* experimentation will clarify whether or not this approach has potential, and further analysis of the pathway may determine whether it is possible to block the accumulation of precursors, e.g. alliin.

Fungal control will no doubt continue to be approached by a combination of the above techniques. Genetic modification provides an extra tool to help plant breeders keep one step ahead of evolutionary developments in fungal pathogenesis. If the technology proves to be effective, it should be possible to control fungal pathogens using a more sustainable approach than chemical methods currently allow.

BACTERIAL RESISTANCE. The opportunistic nature of *Allium* bacterial diseases, such as leaf blight (*Xanthomonas* spp.), soft rot (*Erwinia* spp.) and sour skin/slippery skin (*Burkholderia cepacia*), makes the development of spray-based control strategies difficult (Maude, 1990a, b; see also Mark *et al.*, Chapter 11, this volume). Controlled-atmosphere storage and heat treatments have little effect on these diseases (Maude, 1990a) and in wet seasons the disease may take hold while the crop is still in the field. Antimicrobial genes, such as those encoding small channel-forming peptides, e.g. the magainins and cecropins (Bechinger, 1997), have a potent antibacterial effect (Kristyanne *et al.*, 1997) when added to plant extracts. *T4* lysozyme is another antibacterial gene that has been demonstrated to confer tolerance to bacterial pathogens (de Vries *et al.*, 1999). An antimicrobial gene from onion has also been shown to be active against Gram-positive bacteria (Phillippe *et al.*, 1995). However, as with the fungal research described above, commercial crops containing modified antibacterial genes have not yet been produced. First, a greater understanding of temporal and spatial expression profiles of the gene in the new host is required to advance this technology. Antibacterial genes producing stable proteins with greater activity *in planta* are also needed. As with antifungal genes, it is likely that new genes will soon be discovered with potential antibacterial properties. Again, methods to rapidly assess these genes for activity *in vivo* will be required to check the efficiency of the technology in alliums.

NEMATODE RESISTANCE. Many types of nematode are capable of infecting *Allium*. They are not considered a major pest in temperate climates, but they are in hot-climate countries, e.g. Thailand; however, they are at present well controlled by soil fumigants (Green, 1990). Genetic engineering offers the opportunity to control parasitic nematodes of *Allium* species, such as root-knot nematodes, which feed from giant transfer cells that they induce in the plant. Two strategies have been developed, in model crop systems, to combat these types of nematode (Singh and Sansavini, 1998), although they are still a long way from commercial application. The first relies on expression of a gene product in the plant that is directed against the nematode or its secretions. The second relies on the expression of a specific phytotoxic product in the giant transfer cells that effectively destroys the cell so that the nematode has no structure upon which to feed.

1.2.2 Male sterility

Two types of engineered male sterility are being used to produce hybrid seed in crops other than *Allium*: engineered sterility, based on the *barnase/barstar* genes (Mariani *et al.*, 1990), is being used in canola, maize and radicchio rosso (Table 6.1); and the adenine methylase gene from *Escherichia coli* is also being used to produce male-sterile maize (Table 6.1). Either of these systems could be applied to onion-hybrid seed production.

The majority of onion cultivars now being released are hybrids, derived from crosses with male-sterile germplasm. The principal source of the male-sterile

S-cytoplasm in these hybrids can be traced back to a single plant identified in 1925 in Davis, California. While other sources have been reported (e.g. T-cytoplasm), they require complex fertility restoration (Havey, 1998b; see also Havey, Chapter 3, this volume). The ability to engineer sterility in *Allium* species would remove the limitation of so much reliance on a single source of cytoplasmic male sterility (CMS), thus greatly enhancing the potential for hybrid seed production. CMS has already been transferred from onion to leek by protoplast cybridization experiments. However, the functionality of such somatic fusions has not been clearly demonstrated (Buiteveld, 1998). Considerable effort is under way to find new sources of CMS in onion (Havey, 1998b) and leek (Buiteveld *et al.*, 1998a, b; Havey and Lopes Leite, 1999). Recently a new source of sterility has been backcrossed into onions from *A. galanthum* (Havey, 1998a). If this system proves effective, then there may be less of a need to engineer this trait.

1.2.3 Quality traits

PUNGENCY. A unique metabolic pathway in *Allium* plants converts cysteine into several forms of *S*-alk(en)yl-cysteine sulphoxides (ACSOs) (Lancaster and Boland, 1990; Block, 1992; see also Randle and Lancaster, Chapter 14, this volume). This secondary metabolic pathway leading to the production of onion pungency has been studied in great detail over the last 10 years (e.g. Lancaster and Boland, 1990; Lancaster *et al.*, 1998; Kopsell *et al.*, 1999). The cleavage of these ACSOs by the enzyme alliinase, upon disruption of the cell, produces volatile flavours, odours and lachrymatory compounds ('pungency'), as well as pyruvate and ammonia (Clark *et al.*, 1998). Some of the first compounds produced upon lysis of the cell are the thiosulphinates, which subsequently produce the cascade of additional organosulphur products that make up some of the above compounds. Work is currently being undertaken at the University of Wisconsin by Irwin Goldman's group to determine the thiosulphinates derived from particular ACSOs and to identify those responsible for particular health benefits, such as antiplatelet activity (Orvis *et al.*, 1998). This biochemical understanding, together with a knowledge of the genes responsible, may one day make it possible to increase the health benefits of a diet rich in alliums.

Specific oxidases that can oxidize *S*-allyl-L-cysteine, a precursor of *S*-allyl-cysteine sulphoxide, have been detected in garlic (Ohsumi *et al.*, 1993). Our research team and other groups around the world have isolated alliinase genes responsible for this process (van Damme *et al.*, 1992; Clark, 1993; Manabe *et al.*, 1998; Lancaster *et al.*, 2000). Manipulation of the levels of such compounds may in future allow onions with customized flavours and pungencies to be produced. We are currently regenerating transformed plants containing antisense versions of the alliinase gene in order to see whether this type of manipulation can result in gene silencing and be used to modify onion pungency (see Section 3.2.3). Other enzymes, such as gamma glutamyl cysteine synthetase, glutathione *S*-transferase and γ-glutamyl transpeptidase, involved in the production of the ACSOs, are also being investigated (Lancaster and Shaw, 1994). Other plant species, including the brassicas, are known to produce ACSOs (Maw, 1982) and are thought to have a similar sulphur pathway leading to the production of methyl cysteine sulphoxide. Research in our laboratory has demonstrated that ACSOs are also produced by *Arabidopsis*. The identification and manipulation of the genes in this model system will no doubt provide tools for the difficult task of understanding and manipulating the various components that regulate the pathway in onion.

CARBOHYDRATES. *Allium* species, like only 15% of flowering plants (Hendry and Wallace, 1993), store the majority of their carbohydrate reserves as fructose polymers, known as fructans (Darbyshire and Steer, 1990; Ernst *et al.*, 1998). This is probably due to the role of fructans as cryoprotectants, osmoregulants and a source of stored carbohydrates that can rapidly be mobilized during the breaking of dormancy (Shiomi *et al.*, 1997). Fructans are increasingly being recognized as a health-giving component of the human diet (Roberfroid and Delzenne, 1998). They are also used in products such as biodegradable plastics and wash softeners and, in the inulin form, as an artificial sweetener (Yun, 1996; Gupta and Kaur, 1997). It is possible to manipulate fructan products and levels in plants by introducing specific sucrose : sucrose-fructosyltransferases (SSTs) (Sévenier *et al.*, 1998) or fructan : fructan-fructosyltransferases (FFTs) isolated from onion (Vijn *et al.*, 1997). This work has led to the production of fructan-containing beets. SSTs or FFTs from onion (Vijn *et al.*, 1997, 1998) could be useful in the further customization of carbohydrate content in such beets. The manipulation, in allium crops, of fructan assimilation or degradation, via transformation, may give the potential to enhance the health benefits of these vegetables as well as their storage characteristics, solids content and sweetness. Again, a greater understanding of the biochemistry of the system (aided no doubt by illuminating transformation work) is required before concrete benefits can be achieved.

1.2.4 Other characteristics

Colour, size, shape, number, thickness and adhesion of skins, storage abilities, solids content, quercetin levels, pungency and sweetness are all traits that breeders would like to manipulate. Despite the large genetic variation within the *Allium* gene pool for many of these traits, the ability to precisely engineer any of them in alliums does not yet exist, although it may soon be possible to alter fructan composition and thus possibly affect storage (indirectly by affecting osmotic potential and thus water content), sweetness and solids content. The very existence of white, yellow and red onions indicates that the anthocyanin-based colour pathway is present and so it should be possible to introduce proven anthocyanin regulatory genes to specifically modify *Allium* colour, as has successfully been done in other plants (Tanaka *et al.*, 1998).

Other attributes sought by onion breeders include clonal seed production through apomixis and the ability to manipulate flowering. This latter characteristic could help in the production of hybrid seed, which is an often unreliable process due to asynchronous flowering. At present, these characteristics are beyond the scope of genetic engineers. However, it may one day be possible to manipulate them as our understanding of the physiology, biochemistry and genetics of apomixis and florogenesis improves.

1.3 Risks of producing GM onions

Concern exists among the public about the risks of genetic engineering, especially with regard to the production of food crops. However, in the last 20 years, microorganisms with potentially a far greater ability to escape and spread have been genetically modified to produce custom-made products, without any negative response from the community and very few, if any, proven side-effects. Genes shuffle naturally within and between species using various techniques, and have, through the millennia, created essentially every life form imaginable to fill all available ecological niches. It seems highly improbable that biotechnologists, trying to improve sustainable crop production, will develop a product with negative consequences that are greater than those of current agroindustry practices or greater than nature's own abilities to reduce agricultural production. On the other hand, genetic engineering does have the potential to help to feed the 3 billion people who will be born in the next three decades (Kendall *et al.*, 1997).

Deliberate introductions of highly specialized plants and animals have been made around the world, in fact the global agricultural system as we know it today has depended upon this (Diamond, 1998). The introduction of crop species has caused relatively little concern, as they are generally not adapted for existence outside the field environment. Without such introductions, our lifestyles would be very different. In contrast, the introduction of highly evolved wild plant and animal species into unmanaged ecosystems has caused severe modifications to the native flora and fauna. The manipulation of crop species, to alter only a few well-defined characteristics, is highly unlikely to convert them into organisms as invasive as the highly evolved wild species and therefore would not significantly improve their chances of surviving in a natural habitat.

There are fears that GM foods may be toxic, despite the requirement for rigorous testing regimes that are more comprehensive than any previously implemented for other crops. Yet, in the evolutionary struggle, life forms have become masters of biochemical warfare and even innocuous crops can contain toxic surprises (IFBC, 1990; Vetter, 2000) which need precise processing to ensure their elimination, e.g. in kidney beans, cassava and potatoes.

The 'developed' world is supported primarily by produce from intensive agriculture. This agroindustry relies heavily on the use of fossil fuels and the application of large quantities of toxic chemicals, each with its own risks (Carson, 1962; Colborn *et al.*, 1996). This scenario is unlikely to change rapidly, even with the adoption of more environmentally friendly methods of agricultural production. So the risks of developing or not developing GM crops should be compared with those presented by our current, less than perfect systems and with other emerging alternatives. What follows is a brief discussion of some of the key concerns expressed about GM crops as they relate to onion. For more general information on the subject, see Conner (1997).

1.3.1 The potential for GM onion crops to become weeds

Onions possess very few, if any, weedy characteristics, such as seed dormancy, broad adaptation, indeterminate growth, continuous flowering, seed production and dispersal. The occasional volunteers from leftover bulbs, which grow following onion production, rarely survive to produce seed. Seed viability declines quickly in open storage and onion seedlings will not thrive because they are not competitive with other plants. As it is not intended to introduce genes conferring weedy characteristics into onions, it is highly improbable that onions, GM or otherwise, will ever become a major weed problem. In the USA, only *A. vineale* from the *Allium* genus is considered a weed that is difficult to control and it multiplies by topsets rather than by seeds.

1.3.2 The possibility of horizontal gene transfer to other species

Although there are estimated to be about 600 members of the *Allium* genus (Davies, 1992), very few grow wild in the vicinity of temperate-grown crop alliums. In Central Asia, crop alliums do grow in close proximity to wild species and can be interfertile, although only *A. vavilovii* is readily interfertile. They may also cross with *A. fistulosum* and *A. roylei* in the foothills of the Himalayas. For specific detail on hybridization within the *Allium* species, see Kik (Chapter 4, this volume). The absence of reports of hybrid populations in these regions suggests that such events, if they do occur, are selectively disadvantaged.

Alliums are generally grown for their vegetative organs and harvested prior to flowering; again, this reduces the possible risk of spread via pollen. *Allium* crops, such as garlic and shallot, although capable of flowering, are usually vegetatively propagated, thus reducing the opportunity for introduced genes to spread via pollen transfer. For these reasons it is unlikely, though not impossible, that any genes from temperate-grown onions would spread via interspecific hybridization to wild *Allium* species. Even if such an event did occur, the genetic modification would still have to enhance the fitness of the new hybrid wild *Allium* in order for the gene to be successfully transferred to future generations. This is very unlikely, but should be considered when proposing any genetic modification. Horizontal gene transfer of DNA sequences via transposons, retroviruses or other means can occur in both GM and non-GM crops.

1.3.3 Interaction of the GM crop with other species and ecosystems

As GM onions would only be grown in a farm environment, the only species with which they would come into contact would be those coexisting in that already artificial environment, visitor species and the species with which they have subsequent contact. Coexisting species would include, among others, weed species, disease-causing pathogens and beneficial organisms.

Depending on the nature of the particular modification, it may be possible for the transformed plant to withstand the impact of either weed species or pathogens of onions (in fact, this would be a major aim of GM *Allium* production). Due to the specific nature of gene-product interactions, it is unlikely that there would be a detrimental effect on any beneficial organisms. For particular organisms, *in vitro* or *in vivo* experimentation can be used to determine possible interactions. However, it is impossible to determine interactions with all organisms and thus, as with any new technology, a calculated risk must be taken concerning its commercialization. Visiting species would include pests, such as thrips. Again, plants capable of resisting thrips attack would be the aim of some *Allium* GM work. Beneficial visitors could include insect pollinators. Any GM onion producing foreign proteins that could be harmful to such insects should be tested prior to a decision to release the crop for commercial use. Testing methods for the production of transgenes in pollen and techniques for the elimination of some potential problems have already been developed (Eady *et al.*, 1995; Wilkinson *et al.*, 1998). It may be argued that all GM problems cannot be foreseen. However, this argument holds true for all endeavours: problems that may emerge as a result of not pursuing the potential benefits of GM crops also cannot be foreseen.

1.3.4 Health risks from eating food derived from GM crops

The chemical composition of DNA is essentially the same in all living organisms. Since we consume millions of base pairs of DNA with each meal from a plethora of different organisms without harm, it is extremely unlikely that an altered sequence inserted into onions would be digested differently from an unaltered sequence. Contamination of food with insect- and other animal-cell origins is common. An extreme example of this is public concern over eating human genes; yet, every time we swallow, we inadvertently digest cells from our mouth lining, each containing roughly 6×10^9 base pairs of human DNA.

Genetic manipulation or conventional plant breeding could be used to develop foods that might accidentally contain new allergens or toxins. However, much research and characterization will be necessary before any GM onions will become available for commercial release. This research will have to show that the new onion poses no greater health risk than untransformed onions. It may not always be possible to identify the very small number of people who react differently to a particular novel food product. However, this risk is present whether the product is a GM entity or not.

1.3.5 Conclusion

This is a very brief overview of some of the possible and perceived risks of GM organisms as they relate to onions. It is unlikely that a crop species such as onion or its wild relatives, with their specialist niche adaptations, could become generalist weeds with environmentally detrimental consequences. Also, by careful characterization of GM *Allium* crops and knowledge of the genes inserted, it will be possible to produce alliums with desirable characteristics that are safe for both cultivation and human consumption. Genetic engineering has the capability to broaden the germplasm base for *Allium* crops and to speed up the introduction of new discrete genes into the top-performing cultivars. While careful risk management is required to monitor this technology, it is important that the perceived risks surrounding GM alliums are balanced accurately against current or alternative practices so that this promising technology is not unfairly disadvantaged.

2. Onion Transformation Protocols

2.1 Introduction

The field of transformation consists of many techniques, including artificial hy/cybridization, direct gene transfer and vector-mediated gene transfer. In this chapter, discussion is limited to the transfer of discrete genes. It does not include hy/cybridization. For a comprehensive review of this technique as it relates to alliums, see Buiteveld (1998).

In order to successfully transform plants with discrete genes, the DNA sequences have to be delivered to the cell, the gene then has to be regulated in the desired manner and the cell has to be both regenerated and selected. Finally, the regenerated putative transformant has to be hardened and grown outside the flask and characterized. These processes are outlined below in the context of the latest *Allium* transformation systems.

2.2 Gene delivery

Both vector-mediated and direct gene-transfer systems have been applied to alliums with some success (Eady, 2001). However, to date, only the vector-mediated *Agrobacterium* system has been reported to be repeatable and to work on more than one cultivar (Eady *et al.*, 1998a, 2000; Zheng *et al.*, 1999).

Myers and Simon (1998a) used the PDS 1000 helium particle gun (Dupont) as a direct gene-transfer system to produce a transgenic garlic plant. However, as with similar work in onions (Eady and Lister, 1997), this system is very inefficient and requires the transformation of a specific cell line. In the case of garlic, regeneration takes about 13 months, which increases the chance of producing undesirable somaclonal variation. There have also been claims that transgenic leek plants have been produced using particle bombardment. However, this research was undertaken for commercial clients and it is not clear how far it has progressed (B. Schrijver, Christchurch, New Zealand, 1999, personal communication).

Future developments in *Allium* gene delivery will probably use the methods developed in model plant systems for which transformation techniques are further advanced. Recently, in *Arabidopsis*, an *in vivo* technique has been developed whereby the floral tissues are simply dipped in a modified *Agrobacterium* solution and then allowed to develop. Up to 3% of the seed produced can be transgenic (Clough and Bent, 1998). In other developments, researchers are using transposon sequences to 'jump' genes into the desired genome (Houba-Herin *et al.*, 1994) or they may use homologous recombination systems to direct site-specific gene integration (Vergunst and Hooykaas, 1998, 1999). Ultimately one of these systems may prove to be more effective than using *Agrobacterium*.

2.3 Gene regulation

Numerous regulatory sequences are now available to direct foreign gene expression in plant cells. Many of these have been isolated and modified into a 'cassette' format so that the gene to be expressed can simply be slotted downstream of the promoter of choice, e.g. the *pBIN* series of binary vectors (Clontech Laboratories Inc., California, USA) and the *pCambia* series (Centre for Application of Molecular Biology to International Agriculture, Canberra, Australia). It is important to use sequences that produce a high level of selective gene product at a later stage when the transformed material is being selected. The ability of available plant regulatory sequences (promoters, introns, leader sequences) to direct gene expression in onion cells has not been studied in detail, although some information on commonly used promoter sequences has been obtained from bombardment studies in onion and garlic (Eady *et al.*, 1996; Myers and Simon, 1996; Barandiaran *et al.*, 1998). These reports concluded that the cauliflower mosaic virus (CaMV) 35S promotional sequence drives high levels of expression in *Allium* tissue. This sequence and the *nos* promotional sequence have both subsequently been used in successful *Allium* transformation studies (Eady *et al.*, 2000). Fusing gene promoters, enhancers and other regulatory sequences (either from *Allium* genes or from other origins) to reporter genes, such as the *gfp* gene (Haseloff *et al.*, 1997), and studying the expression of such introduced constructs, will make it possible to induce precise spatial and temporal specific transgene expression patterns in alliums. For example, work in our laboratory has recently identified a root-specific alliinase, which from Northern analysis appears to be primarily expressed in the root (Lancaster *et al.*, 2000). The promoter from this could be fused to a chosen structural gene to induce root-specific expression.

2.4 Culture systems

The *in vitro* culture of *Allium*, reviewed by Novak (1990) and more recently by Eady (1995), has been primarily concerned with clonal propagation from multicellular meristems. It is preferable to obtain transgenic plants by integrating DNA into a single totipotent cell and then regenerating a complete plant from that cell. The cell has to be competent both for accepting DNA and for regeneration. The alternative to this is when the cell is competent to accept DNA but can regenerate only as part of an existing multicellular structure. In this case, a chimeric tissue is produced as the primary transgenic material and independence to regenerate proceeds when the transgenic cell mass reaches a particular size or developmental stage. In reality, totipotency can only be truly observed in isolated protoplasts. In other systems, it is difficult to determine the precise role of adjacent cells, although it is obvious that some systems are more dependent on surrounding cells than others.

Callus (or dedifferentiated) cells provide useful sources of independent cells. However, regeneration from such starting material to a phenotypically normal plant can be difficult. The major *Agrobacterium*-based monocotyledonous transformation protocol claims to use embryo-derived callus material (WO 94/00977), which, by definition, is a dedifferentiated uniform cell line. Such a system is unlikely to work with *Allium* species, as dedifferentiated *Allium* cultures rarely, if ever, regenerate (Novak, 1990; Eady, 1995). Zheng *et al.* (1999) have based their transformation protocol on such a system. However, in reality the original protocol and Zheng's probably use cells from culture rather than callus. Evidence for the lack of regeneration from callus is found in regular reports of onion cultures losing their capacity to regenerate over time (Novak, 1990; Eady, 1995), i.e. eventually they lose the ability to differentiate.

Efforts to transform onion have focused on mature or immature embryo or embryo-derived cultures as a source of dual transformation/regeneration-competent cell types. These types of cultures have recently been reported for several *Allium* species (Silvertand *et al.*, 1996; Xue *et al.*, 1997; Eady *et al.*, 1998b; Saker, 1998; Zheng *et al.*, 1998). Our laboratory now uses a technique that delivers DNA as soon as possible after isolation of the immature embryo (see below). Initially, embryogenic cultures, similar to those produced from maize embryos (Welter *et al.*, 1995), were first derived from the immature embryos and then transformed. However, this process required longer *in vitro* culture (increasing the likelihood of somaclonal variation) and produced fewer stable transgenic tissues and no mature transgenic plants. The onion transformation system developed at Plant Research International, Wageningen, The Netherlands (Zheng *et al.*, 1999), also uses embryos that have been precultured for a few days in a similar fashion to the patented *Agrobacterium*-based monocotyledonous transformation system (WO 94/00977). Myers and Simon (1998b) developed a regeneration system from garlic root and shoot meristem tissue, presumably because embryo-derived tissue is not readily available, as most garlic is sterile.

Protoplast regeneration systems have been developed for leek cell-fusion systems (Buiteveld and Creemers-Molenaar, 1994) and it may be possible to adapt them for use in protoplast-based transformation protocols. One problem associated with this difficult culture process is that of somaclonal variation, which may arise from the long and complex culture regime.

Recent developments in model species have seen a shift away from *in vitro* culture-based transformation towards *in vivo* transformation (see Section 2.2). If this trend continues and the technologies can be readily transferred to alliums then it should be possible to circumvent difficulties currently encountered with the tissue-culture step of *Allium* transformation.

2.5 Selection of transgenic tissue

Transgenic plants are usually selected by using either antibiotic- or herbicide-resistant gene constructs. Initial investigations indicate that herbicides such as geneticin, hygromycin or phosphinothricin could all be useful selective agents for transgenic *Allium* selection (Eady and Lister, 1998). Since then, the *nptII* gene has been successfully used to confer resistance to the antibiotics paromycin (Myers and Simon, 1998a) or geneticin (Eady *et al.*, 2000). The *bar* gene has also been used to confer resistance to the herbicide phosphinothricin (see Section 3.2.2). Groups who are concerned about the use of antibiotic resistance to develop commercial crops favour the use of herbicide resistance as the selectable marker. This too has its limitations, especially if it becomes desirable to 'pyramid' genes (i.e. to insert additional genes into already transformed plants). Other selection systems have recently been developed in plants, including the use of specific nutritional requirements in the regeneration media, e.g. the phosphomannose isomerase (PMI) gene as the selectable gene and mannose as the selective agent (Joersbo *et al.*, 1998, 1999) and visual reporter genes (Vain *et al.*, 1998). These have not yet been tested on alliums. In addition, removable selection systems are being developed, e.g. by cotransformation, site-specific recombination and transposon-mediated systems (Daley *et al.*, 1998; Vergunst and Hooykaas, 1998; Weld, 2000). The selective gene can be removed at a later stage, leaving only the gene of choice. This process allows multiple alterations to be made to a particular cultivar. The speed with which these developments can be applied to alliums remains to be seen.

2.6 'Exflasking'

Transferring the primary transformant from *in vitro* culture to the glasshouse is often a technically difficult process. Fortunately, *Allium* plantlets in culture are quite robust and there are numerous reports of successful transfer to the glasshouse (Novak, 1990). Two techniques are used in our laboratory. They are based on either the transfer of vigorously growing plantlets or of *in vitro* bulbs produced by culturing the plantlets on $\frac{1}{2}$ Murishige and Skoog medium (MS) plus 120 g l^{-1} of sucrose (Seabrook, 1994). For these processes to be successful, it is essential that the glasshouse is warm (12–23°C day, 4–16°C night) and has at least 12 h of bright daylight.

2.7 An *Agrobacterium*-mediated onion transformation protocol

2.7.1 Bacterial strain and plasmids

Agrobacterium tumefaciens strain LBA4404 containing the plasmid binary vector *pBIN* or *pCambia* derivatives have been used in *Allium* transformation experiments. Overnight, *Agrobacterium* cultures grown in Luna broth (LB) media (Sambrook *et al.*, 1989) containing appropriate selective agents (e.g. Eady *et al.*, 2000) were replenished with an equal volume of LB containing antibiotic and 100 μM acetosyringone (virulence-gene-inducing factor) and grown until they reached an optical density of about 1.0 at 550 nm. *Agrobacteria* were isolated by centrifugation and resuspended in an equal volume of liquid embryogenic induction medium (P5) (Eady *et al.*, 1996) containing 200 μM acetosyringone.

2.7.2 Transformation procedure

Immature embryos from field-grown umbels of bulb onion cv. 'Canterbury Longkeeper' were isolated under a stereomicroscope (Eady *et al.*, 1996). The embryos used were from immature seeds at the stage of recently blackened seed-coat and with the endosperm still liquid. They were removed from the ovaries, cut into ~1 mm lengths and transferred in batches of 40 into 0.8 ml of *Agrobacterium* solution, vortexed for 30 s and placed under vacuum (~25 mmHg) for 30 min. These tissue pieces were then blotted dry on filter-paper before transfer to P5 (Eady *et al.*, 1998b) media. After 6 days of cocultivation with the bacteria at 28°C in the dark, embryo pieces were transferred to P5 containing appropriate selection agents in order to select for transgenic tissue and eliminate *Agrobacteria*. Embryo pieces were cultured in the dark under the same conditions described for the production of secondary embryos (Eady *et al.*, 1998b), with transfer to fresh medium every fortnight. After ~8–16 weeks, actively growing material (also identified using visual-marker gene expression, if appropriate) was transferred to regeneration medium (Eady *et al.*, 1998b) containing the selective agent 20 mg l^{-1} of geneticin when using the *nptII* gene. Shoot cultures were maintained for 12 weeks, and developing shoots were transferred to $\frac{1}{2}$ MS media (Murashige and Skoog, 1962) plus selective agent to induce rooting of transgenic shoots only. Rooted plants were either transferred to $\frac{1}{2}$ MS plus 120 g l^{-1} sucrose to induce bulbing or to soil in the glasshouse (12 h 12–23°C day, 12 h 4–16°C night). *In vitro* bulbs could be maintained for many months on the media and transferred to the glasshouse when appropriate.

Increasing day length induced bulbing in glasshouse-grown plants naturally. After 50% of the tops had fallen, bulbs were lifted and air-dried. Bulbs greater than 45 mm in diameter were cold-stored at 4°C for 3 months to induce floral meristems prior to planting. Plants from all transformants, produced using the above technique, have grown in a phenotypically normal fashion and produced scapes and umbels. Flowers were self-pollinated by enclosing individual umbels within microperforated plastic bread bags containing greenbottle flies. Seed was collected 2–3 months later from dried umbels.

3. Analyses of Transformants

3.1 Detection of the transgene

Initially, the presence of the transgene in putative transgenic onion tissue was screened using the polymerase chain reaction (PCR) in order to amplify specific fragments of a particular DNA sequence. PCR cannot be used to demonstrate conclusively the presence of transgene fragments, as the possibility of amplifying DNA contained in contaminating microorganisms cannot be absolutely eliminated. Adaptor-ligation PCR is a recent advance in PCR that may help solve this problem (Zheng *et al.*, 2000). However, at present, it is still routine to determine transgenic status conclusively by Southern blot analysis.

After about 2 months' growth in the glasshouse, transgenic leaf material (approximately 1 g) was collected and Southern blot analysis was performed on putative transgenic plants (Eady *et al.*, 2000). Plants have been screened for the presence of introduced *nptII*, *gfp* and *bar* genes (Fig. 6.1). In all cases, integrations have been observed in copy numbers (number of integrations per genome) similar to those observed in the transformation of other plant species.

3.2 Gene expression

Genes under the control of the *CaMV35s* and *nos* promoters have been introduced into onion. Transformants regenerated under selection indicated that both promoters are switched on in tissues produced during culture and regeneration. The profile of *CaMV35s* expression has been determined using the *pBINmgfpER* binary vector, which has, in its T-DNA, the visual reporter gene *mgfpER* under *CaMV35s* regulatory control.

3.2.1 Expression of gfp *gene*

Observation under a fluorescence microscope (excitation 475 nm, emission 510 nm) (Haseloff *et al.*, 1997) was used to identify

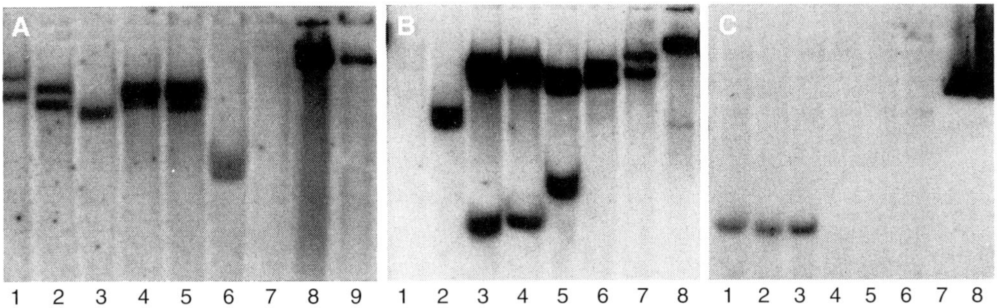

Fig. 6.1. Southern blot analysis of *Hin*dIII-digested onion DNA from plants transformed with: (A) the *mgfpER* reporter gene, (B) the *nptII* antibiotic-resistance gene and (C) the *bar* herbicide-resistance gene. (A) Lanes 1–6, transgenic plants; lane 7, non-transgenic control; lanes 8 and 9, five- and one-copy number equivalents of control plasmid DNA containing the *mgfpER* sequence. (B) Lane 1, non-transgenic control; lanes 2–7, transgenic plants; lane 8, five-copy control of equivalent plasmid DNA containing the *nptII* sequence. (C) Lanes 1–3, transgenic clonal plants; lanes 4–6, control non-*bar*-containing plants; lanes 7 and 8, five- and ten-copy equivalents of plasmid DNA containing the *bar* sequence.

tissues expressing green fluorescent protein (GFP). Larger tissues with high and low levels of expression were studied using fluorescence stereomicroscopes, and high-level expressing tissues were readily discernible by hand-held fluorescent lanterns. Studies with the *gfp* reporter gene demonstrated that transient gene expression is first observed after 3 days of cocultivation. Up to 16% of initial tissue pieces produced stable GFP. However, this frequency was affected by a number of factors, including genotype, condition of the embryo (i.e. whether isolated from healthy large umbels or weak diseased umbels), size of the embryo and cocultivation and selection conditions. Transgenic tissues were transferred to regeneration medium after 10–16 weeks. They responded in a manner similar to non-transgenic, embryo-derived cultures: up to 2.7% of transferred tissue produced shoots (Colour Plate 3). Shoot cultures placed on rooting medium containing the selective agent geneticin produced roots that fluoresced (Colour Plate 3). In all green differentiated structures, the presence of red fluorescing chlorophyll masked GFP observation. Hence, most fluorescence was observed in root tips, while in shoots fluorescence was obscured and was only apparent in young leaves or stomatal guard cells. In floral organs, GFP expression could be observed in petals, ovules, stamens and stigma tissues. GFP was not apparent in freshly dehisced pollen. In bulbs, GFP expression could be seen in epidermal skin cells (Colour Plate 3) and scale cells.

3.2.2 Expression of herbicide resistance

Onion plants containing a *CaMV35s–bar* gene construct have been regenerated by selection on media containing PPT. Expression of the *bar* gene in mature plants was confirmed by leaf paint assays and spraying. Initially, leaves were painted with a 0.5% solution of Basta® (a commercial herbicide containing PPT) (Fig. 6.2). Plants demonstrating resistance to Basta® were then sprayed with commercially recommended concentrations of the herbicide (Fig. 6.2) to confirm resistance. The level of resistance achieved indicates that the commercial production of transgenic onions containing a herbicide-resistance gene is feasible and that weeds could be controlled in such a crop using infrequent low-dosage applications of a single new-generation herbicide. This would effectively eliminate the need for the complex multi-herbicide pre- and post-emergence programmes (Rubin, 1990) that are currently used. Such a system could also considerably reduce the amount of fossil fuels required to achieve effective weed control by reducing the number of applications required.

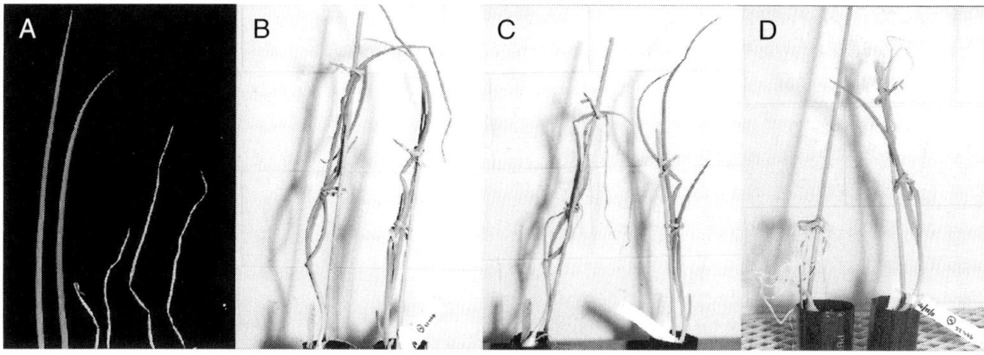

Fig. 6.2. The effect of the herbicide Basta® on transgenic onion plants containing the *bar* gene. (A) Transgenic (left two) and non-transgenic (right four) leaves 10 days after painting with a 0.5% solution of the herbicide. (B, C and D) Non-transgenic (left) and transgenic (right) onion plants 0, 3 and 10 days, respectively, after spraying with a 0.5% solution of the herbicide.

3.2.3 Antisense alliinase gene expression

Three antisense alliinase gene constructs have recently been introduced separately into onions. The presence of the individual constructs in transgenic plants has been determined by Southern blot detection of flanking T-DNA sequences (C.C. Eady, M. Pither-Joyce and J. Farrant, unpublished). These constructs – a *CaMV35s* anti-bulb alliinase, a bulb alliinase promoter, anti-bulb alliinase and a *CaMV35s* anti-root alliinase promoter – have been introduced in order to modify onion alliinase levels, allowing researchers to determine the impact of modified levels on the amount of volatile sulphur compounds subsequently produced. To date, results indicate that the antisense plants have reduced levels of alliinase activity (Fig. 6.3). Western blots probed with polyclonal antibodies specific to alliinase indicate that this reduction in activity is caused by a decrease in the levels of the enzyme (Fig. 6.3).

3.3 Stability of transgenes

Transformants produced in initial experiments have grown to maturity and appear phenotypically normal (Fig. 6.4). Twelve

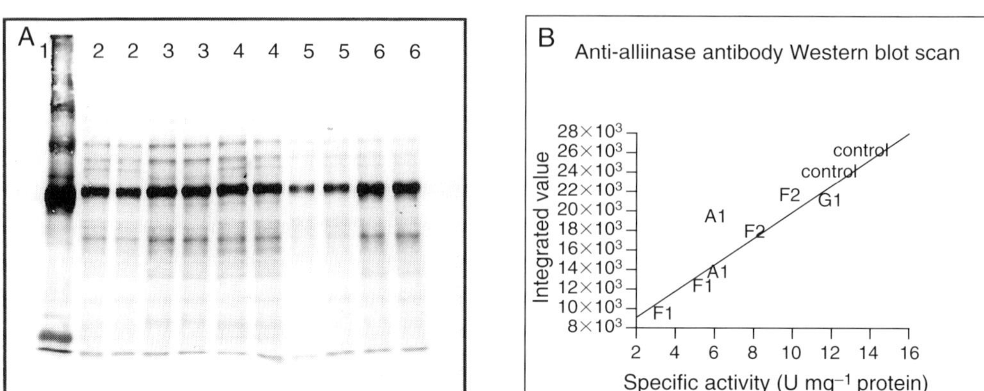

Fig. 6.3. (A) Western blot analysis of alliinase enzyme levels in the roots of transgenic plants containing an antisense alliinase gene. Lane 1, alliinase. Lanes 2–5, transgenic onion-root samples. Lane 6, non-transgenic control onion-root sample. (B) Correlation between specific alliinase enzyme activity per mg of root protein and the Western blot scan of transgenic and non-transgenic samples. (Specific activity: one unit = the conversion of 1 μmol of substrate to pyruvate.)

Fig. 6.4. Seed set in transgenic onions in containment. To be used for inheritance studies.

independent transformants from these plants have been selfed. F_1 seed has recently been collected and germinated. Initial results indicate that the transgene is usually inherited in a normal Mendelian fashion.

4. Concluding Remarks

The ability to genetically engineer alliums paves the way for biochemists to manipulate the key enzymes involved in the sulphur and carbohydrate pathways that are unique to this family. This will improve our understanding of the processes that distinguish alliums. It may also lead to the selection of alliums with improved characteristics through collaboration between biochemists, molecular biologists and plant breeders, or the development of modified cultivars via genetic modification. In the meantime, characteristics that have been modified in other crops, in order to improve the sustainability of production, will eventually be applied to *Allium* crops. There is an urgent need for introduced resistance to pests and diseases, because gene pools of most cultivated alliums lack sources of resistance, most cultivated alliums are biennials and the work with wild species is rather painstaking and complicated. This technology has the potential to reduce the levels of pesticides and fossil fuels currently used in the intensive production of this crop. Herbicide-resistant onions are still at least 8 years away from commercial production. This lag behind other crops may be an advantage because it will give the public more time to debate the issues surrounding the technology. Certainly genetic modification has a significant role to play in overcoming a number of technical problems encountered in *Allium* crop improvement.

Acknowledgements

Special thanks to Martin Shaw, Meeghan Pither-Joyce and Julie Farrant for their technical contribution to the results presented in this chapter; and to Tracy Williams, Nadene Winchester and Carla Appel for their patience and persistence in editing the text and preparing the illustrations.

References

Barandiaran, X., Dipietro, A. and Martin, J. (1998) Biolistic transfer and expression of a *uida* reporter gene in different tissues of *Allium sativum* L. *Plant Cell Reports* 17, 737–741.

Bechinger, B. (1997) Structure and functions of channel-forming peptides: magainins, cecropins, melittin and alamethicin. *Journal of Membrane Biology* 156, 197–211.

Block, E. (1992) The organosulfur chemistry of the genus *Allium* – implications for the organic chemistry of sulfur. *Angewandte Chemie, International Edition in English* 31, 1135–1178.

Buiteveld, J. (1998) Regeneration and interspecific somatic hybridization in *Allium* for transfer of cytoplasmic male sterility to leek. PhD thesis, Agricultural University of Wageningen, The Netherlands.

Buiteveld, J. and Creemers-Molenaar, J. (1994) Plant regeneration from protoplasts isolated from suspension cultures of leek (*Allium ampeloprasum* L.). *Plant Science* 100, 203–210.

Buiteveld, J., Kassies, W., Geels, R., Campagne, M.M.V., Jacobsen, E. and Creemers-Molenaar, J. (1998a) Biased chloroplast and mitochondrial transmission in somatic hybrids of *Allium ampeloprasum* L. and *Allium cepa* L. *Plant Science* 131, 219–228.

Buiteveld, J., Suo, Y., Campagne, M.M.V. and Creemers-Molenaar, J. (1998b) Production and characterization of somatic hybrid plants between leek (*Allium ampeloprasum* L.) and onion (*Allium cepa* L.). *Theoretical and Applied Genetics* 96, 765–775.
Carson, R. (1962) *Silent Spring*. Fawcett Publications, Greenwich, Connecticut, 304 pp.
Christou, P. (1995) Strategies for variety-independent genetic transformation of important cereals, legumes and woody species utilizing particle bombardment. *Euphytica* 85, 13–27.
Christou, P. (1996) Transformation technology. *Trends in Plant Science* 1, 423–431.
Clark, S.A. (1993) Molecular cloning of a cDNA encoding alliinase from onion (*Allium cepa* L.). PhD dissertation, University of Canterbury, Christchurch, New Zealand.
Clark, S.A., Shaw, M.L., Every, D. and Lancaster, J.E. (1998) Physical characterization of alliinase, the flavor generating enzyme in onions. *Journal of Food Biochemistry* 22, 91–103.
Clough, S. and Bent, A. (1998) Floral dip: a simplified method for *Agrobacterium*-mediated transformation of *Arabidopsis thaliana*. *The Plant Journal* 16, 735–743.
Colborn, T., Myers, J. and Dumanoski, D. (1996) *Our Stolen Future*. Dutton, New York, 306 pp.
Conceição, A.D. and Broekaert, W.F. (1999) Plant defensins. In: Datta, S.K. and Muthukrishnan, S. (eds) *Pathogenesis-related Proteins in Plants*. CRC Press, Boca Raton, Florida, pp. 247–260.
Conner, A. (1997) *Genetically Engineered Crops: Environmental and Food Safety Issues*. Miscellaneous Publications Series, 39, Royal Society of New Zealand, Wellington, 35 pp.
Conner, A., Jacobs, J. and Genet, R. (1997) Transgenic potatoes versus 'traditional' potatoes: what's the difference? In: McLean, G.D., Waterhouse, P.M., Evans, G. and Gibbs, M.J. (eds) *Commercialisation of Transgenic Crops: Risk, Benefit and Trade Considerations*. Australian Government Publishing Services, Canberra, pp. 23–36.
Crickmore, N., Zeigler, D.R., Feitelson, J., Schnepf, E., Van Rie, J., Lereclus, D., Baum, J. and Dean, D.H. (1998) Revision of the nomenclature for the *Bacillus thuringiensis* pesticidal crystal proteins. *Microbiology and Molecular Biology Reviews* 62, 807–813.
Crowe, F. and McDonald, M.R. (1998) White rot information in conjunction with the 2nd International Symposium on Edible Alliaceae, November 1997. In: Voss, R.E. (ed.) *Proceedings of the 1998 National Onion (and Other Allium) Research Conference, 10–12 December, 1998, Sacramento, California*. University of California, Davis, California, pp. 191–195.
Daley, M., Knauf, V.C., Summerfelt, K.R. and Turner, J.C. (1998) Co-transformation with one *Agrobacterium tumefaciens* strain containing two binary plasmids as a method for producing marker-free transgenic plants. *Plant Cell Reports* 17, 489–496.
Darbyshire, B. and Steer, B.T. (1990) Carbohydrate biochemistry. In: Brewster, J.L. and Rabinowitch, H.D. (eds) *Onions and Allied Crops*, Vol. III. *Biochemistry, Food Science, and Minor Crops*. CRC Press, Boca Raton, Florida, pp. 1–16.
Davies, D. (1992) *Alliums – the Ornamental Onions*. Timber Press, Portland, Oregon, 168 pp.
Dekker, J. and Duke, S. (1995) Herbicide-resistant field crops. *Advances in Agronomy* 54, 69–116.
de Vries, J., Harms, K., Broer, I., Kriete, G., Mahn, A., During, K. and Wackernagel, W. (1999) The bacteriolytic activity in transgenic potatoes expressing a chimeric T4 lysozyme gene and the effect of T4 lysozyme on soil- and phytopathogenic bacteria. *Systematic and Applied Microbiology* 22, 280–286.
Diamond, J. (1998) *Guns, Germs and Steel – a Short History of Everybody for the Last 13,000 Years*. Vintage, London, 480 pp.
Dunwell, J.M. (1998) Novel food products from genetically modified crop plants: methods and future prospects. *International Journal of Food Science and Technology* 33, 205–213.
Dunwell, J.M. (1999) Transgenic crops: the next generation, or an example of 2020 vision. *Annals of Botany* 84, 269–277.
Eady, C.C. (1995) Towards the transformation of onions (*Allium cepa*). *New Zealand Journal of Crop and Horticultural Science* 23, 239–250.
Eady, C.C. (2001) *Allium* transformation. In: Learmonth, R. and Khachatourians, G.G. (eds) *The Handbook of Transgenic Plants*. Marcel Dekker, New York, pp. 655–671.
Eady, C.C. and Lister, C.E. (1997) Transformation of onion (*Allium cepa* L.). In: *Papers of Alliums Australia 1997, 2nd International Symposium on Edible Alliaceae, Adelaide, Australia*, p. 58 (abstract).
Eady, C.C. and Lister, C.E. (1998) A comparison of four selective agents for use with *Allium cepa* L. immature embryos and immature embryo derived cultures. *Plant Cell Reports* 18, 117–121.
Eady, C.C., Twell, D. and Lindsey, K. (1995) Pollen viability and transgene expression following storage in honey. *Transgenic Research* 4, 226–231.

Eady, C.C., Lister, C.E., Suo, Y.Y. and Schaper, D. (1996) Transient expression of *uida* constructs in *in vitro* onion (*Allium cepa* L.) cultures following particle bombardment and *Agrobacterium*-mediated DNA delivery. *Plant Cell Reports* 15, 958–962.

Eady, C.C., Weld, R.J. and Lister, C.E. (1998a) Transformation of onion (*Allium cepa* L.). In: *Proceedings of the 1998 National Onion (and Other Allium) Research Conference, 10–12 December, Sacramento, California, USA*. University of California, Davis, California, pp. 91–94.

Eady, C.C., Butler, R.C. and Suo, Y. (1998b) Somatic embryogenesis and plant regeneration from immature embryo cultures of onion (*Allium cepa* L.). *Plant Cell Reports* 18, 111–116.

Eady, C.C., Weld, R.J. and Lister, C.E. (2000) *Agrobacterium tumefaciens*-mediated transformation and regeneration of onion (*Allium cepa* L.). *Plant Cell Reports* 19, 376–381.

Entwistle, A.R. (1990) Root diseases. In: Rabinowitch, H.D. and Brewster, J.L. (eds) *Onions and Allied Crops*, Vol. II. *Agronomy, Biotic Interactions, Pathology, and Crop Protection*. CRC Press, Boca Raton, Florida, pp. 103–154.

Ernst, M.K., Chatterton, N.J., Harrison, P.A. and Matitschka, G. (1998) Characterization of fructan oligomers from species of the genus *Allium* L. *Journal of Plant Physiology* 153, 53–60.

Evans, I.J. and Greenland, A.J. (1998) Transgenic approaches to disease protection: applications of antifungal proteins. *Pesticide Science* 54, 353–359.

Favaron, F., Castiglioni, C., Dovidio, R. and Alghisi, P. (1997) Polygalacturonase inhibiting proteins from *Allium porrum* L. and their role in plant tissue against fungal endo-polygalacturonases. *Physiological and Molecular Plant Pathology* 50, 403–417.

Fletcher, P.J., Fletcher, J.D. and Lewthwaite, S.L. (1998) *In vitro* elimination of onion yellow dwarf and shallot latent viruses in shallots (*Allium cepa* var. *ascalonicum* L.). *New Zealand Journal of Crop and Horticultural Science* 26, 23–26.

Fromm, M., Morrish, F., Armstrong, C., Williams, R., Thomas, J. and Klein, T. (1990) Inheritance and expression of chimeric genes in the progeny of transgenic maize plants. *Bio/Technology* 8, 833–839.

Green, C.D. (1990) Nematode pests of *Allium* species. In: Rabinowitch, H.D. and Brewster, J.L. (eds) *Onions and Allied Crops*, Vol. II. *Agronomy, Biotic Interactions, Pathology, and Crop Protection*. CRC Press, Boca Raton, Florida, pp. 155–171.

Guo, H.S., Lopez-Moya, J.J. and Garcia, J.A. (1999) Mitotic stability of infection-induced resistance to plum pox potyvirus associated with transgene silencing and DNA methylation. *Molecular Plant–Microbe Interactions* 12, 103–111.

Gupta, A.K. and Kaur, N. (1997) Fructan-storing plants – a potential source of high fructose syrups. *Journal of Scientific and Industrial Research* 56, 447–452.

Hackland, A.F., Rybicki, E.P. and Thomson, J.A. (1994) Coat protein mediated resistance in transgenic plants. *Archives of Virology* 139, 1–22.

Hammond-Kosack, K. and Jones, J. (1997) Plant disease resistance genes. *Annual Review of Plant Physiology and Plant Molecular Biology* 48, 575–607.

Hansen, G. and Wright, M.S. (1999) Recent advances in the transformation of plants. *Trends in Plant Science* 4, 226–231.

Haseloff, J., Siemering, K.R., Prasher, D.C. and Hodge, S. (1997) Removal of a cryptic intron and subcellular localization of green fluorescent protein are required to mark transgenic *Arabidopsis* plants brightly. *Proceedings of the National Academy of Sciences USA* 94, 2122–2127.

Havey, M.J. (1998a) Male sterile cytoplasms used commercially to produce hybrid-onion seed. In: *Proceedings of the 1998 National Onion (and Other Allium) Research Conference, 10–12 December, Sacramento, California, USA*. University of California, Davis, California, pp. 35–38.

Havey, M.J. (1998b) Seed yield, fertility restoration, and morphology of CMS onion populations possessing the cytoplasm of *A. galanthum*. In: *Proceedings of the 1998 National Onion (and Other Allium) Research Conference, 10–12 December, Sacramento, California, USA*. University of California, Davis, California, pp. 313–317.

Havey, M.J. and Lopes Leite, D. (1999) Toward the identification of cytoplasmic male sterility in leek: evaluation of organellar DNA diversity among cultivated accessions of *Allium ampeloprasum*. *Journal of the American Society for Horticultural Science* 124, 163–165.

Hendry, G.A.F. and Wallace, R.K. (1993) The origin, distribution, and evolutionary significance of fructans. In: Suzuki, M. and Chatterton, N.L. (eds) *Science and Technology of Fructans*. CRC Press, Boca Raton, Florida, pp. 119–139.

Hetherington, P.R., Reynolds, T.L., Marshall, G. and Kirkwood, R.C. (1999) The absorption, translocation and distribution of the herbicide glyphosate in maize expressing the *CP-4* transgene. *Journal of Experimental Botany* 50, 1567–1576.

Hiei, Y., Ohta, S., Komari, T. and Kumashiro, T. (1994) Efficient transformation of rice (*Oryza sativa* L.) mediated by *Agrobacterium* and sequence analysis of the boundaries of the T-DNA. *The Plant Journal* 6, 271–282.

Hiei, Y., Komari, T. and Kubo, T. (1997) Transformation of rice mediated by *Agrobacterium tumefaciens*. *Plant Molecular Biology* 35, 205–218.

Hilder, V.A. and Boulter, D. (1999) Genetic engineering of crop plants for insect resistance – a critical review. *Crop Protection* 18, 177–191.

Hilder, V.A., Gatehouse, A.M.R., Sheerman, S.E., Barker, R.F.A. and Boulter, D. (1987) A novel mechanism of insect resistance engineered into tobacco. *Nature* 330, 160–163.

Hooykaas, P., Hofker, M., den Dulk-Ras, H. and Schilperoot, R. (1984) A comparison of virulence determinants in an octopine Ti plasmid, a nopaline Ti plasmid, and an Ri plasmid by complementation analysis of *Agrobacterium tumefaciens* mutants. *Plasmid* 11, 195–205.

Horsch, R., Fry, J., Hoffman, N., Eichholtz, D., Rogers, S. and Fraley, R. (1985) A simple and general method for transferring genes into plants. *Science* 227, 1229–1231.

Houba-Herin, N., Domin, M. and Pedron, J. (1994) Transposition of a Ds element from a plasmid into the plant genome in *Nicotiana plumbaginifolia* protoplast derived cells. *The Plant Journal* 6, 55–66.

IFBC (International Food Biotechnology Council) (1990) Biotechnologies and food – assuring the safety of foods produced by genetic manipulation. *Regulatory Toxicology and Pharmacology* 12, s1–s196.

Jarchow, E., Grimsley, N. and Hohn, B. (1991) *virf*, the host-range-determining virulence gene of *Agrobacterium tumefaciens*, affects T-DNA transfer to *Zea mays*. *Proceedings of the National Academy of Sciences USA* 88, 10426–10430.

Joersbo, M., Donaldson, I., Kreiberg, J., Petersen, S.G., Brunstedt, J. and Okkels, F.T. (1998) Analysis of mannose selection used for transformation of sugar beet. *Molecular Breeding* 4, 111–117.

Joersbo, M., Petersen, S.G. and Okkels, F.T. (1999) Parameters interacting with mannose selection employed for the production of transgenic sugar beet. *Physiologia Plantarum* 105, 109–115.

Jongedijk, E., Tigelaar, H., van Roekel, J.S.C., Bres-Vloemans, S.A., Dekker, I., van den Elzen, P.J.M., Cornelissen, B.J.C. and Melchers, L.S. (1995) Synergistic activity of chitinases and β-1,3-glucanases enhances fungal resistance in transgenic tomato plants. *Euphytica* 85, 173–180.

Kendall, H., Beachy, R., Eisner, T., Gould, F., Herdt, R., Raven, P., Schell, J. and Swaminathan, M. (1997) *Bioengineering of Crops: Report of the World Bank Panel on Transgenic Crops*. Environmentally and Socially Sustainable Development Studies and Monographs Series No. 23, World Bank, Washington, DC, 40 pp.

Kesarwani, M., Natarajan, K., Mehta, A. and Datta, A. (2000) Overexpression of oxalate decarboxylase from *Collybia velutipes* confers resistance to fungal infection in transgenic tobacco and tomato. In: *6th International Congress of Plant Molecular Biology*. ISPMB, Québec, Abstract S09–42.

King, J. and Coley-Smith, J. (1969) Production of volatile alkyl sulphides by microbial degradation of synthetic alliin and alliin-like compounds, in relation to germination of sclerotia of *Sclerotium cepivorum* Berk. *Annals of Applied Biology* 64, 303–314.

Klein, T.M., Wolf, E.D., Wu, R. and Sanford, J.C. (1987) High-velocity microprojectiles for delivering nucleic acids into living cells. *Nature* 327, 70–73.

Kobayashi, K., Rabinowicz, P., Bravo Almonacid, F., Helguera, M., Conci, V., Lot, H. and Mentaberry, A. (1996) Coat protein gene sequences of garlic and onion isolates of the onion yellow dwarf potyvirus (OYDV). *Archives of Virology* 141, 2277–2287.

Kopsell, D.E., Randle, W.M. and Eiteman, M.A. (1999) Changes in the *S*-alk(en)yl cysteine sulfoxides and their biosynthetic intermediates during onion storage. *Journal of the American Society for Horticultural Science* 124, 177–183.

Kristyanne, E.S., Kim, K.S. and Stewart, J.M. (1997) Magainin 2 effects on the ultrastructure of five plant pathogens. *Mycologia* 89, 353–360.

Kumar, A. and Cocking, E. (1987) Protoplast fusion: a novel approach to organelle genetics in higher plants. *American Journal of Botany* 74, 1289–1303.

Kurian, K.M., Watson, C.J. and Wyllie, A.H. (1999) DNA chip technology. *Journal of Pathology* 187, 267–271.

Lancaster, J.E. and Boland, M.J. (1990) Flavor biochemistry. In: Brewster, J.L. and Rabinowitch, H.D. (eds) *Onions and Allied Crops*, Vol. III. *Biochemistry, Food Science, and Minor Crops*. CRC Press, Boca Raton, Florida, pp. 33–72.

Lancaster, J.E. and Shaw, M.L. (1994) Characterization of purified γ glutamyl transpeptidase in onions: evidence for *in vivo* role as a peptidase. *Phytochemistry* 36, 1351–1358.

Lancaster, J.E., Shaw, M.L., Pither, M.D., Farrant, J.P. and McCallum, J.A. (1998) A review of the regulation of sulfur metabolism and its effect on quality in onions and garlic. In: *Proceedings of the 1998 National Onion (and Other Allium) Research Conference, 10–12 December 1998, Sacramento, California, USA*. University of California, Davis, California, pp. 143–156.

Lancaster, J.E., Shaw, M., Pither-Joyce, M., McCallum, J. and McManus, M. (2000) A novel alliinase from onion (*Allium cepa* L.) roots: biochemical characterization and cDNA cloning. *Plant Physiology* 122, 1269–1279.

Lebel, E., Masson, J., Bogucki, A. and Paskowski, J. (1995) Transposable elements as plant transformation vectors for long stretches of foreign DNA. *Theoretical and Applied Genetics* 91, 899–906.

McCabe, D., Swain, W., Martinell, B. and Christou, P. (1988) Stable transformation of soybean (*Glycine max*) by particle acceleration. *Bio/Technology* 6, 923–926.

Manabe, T., Hasumi, A., Sugiyama, M., Yamazaki, M. and Saito, K. (1998) Alliinase [*S*-alk(en)yl-L-cysteine sulfoxide lyase] from *Allium tuberosum* (Chinese chive) – purification, localization, cDNA cloning and heterologous functional expression. *European Journal of Biochemistry* 257, 21–30.

Marchant, R., Davey, M.R., Lucas, J.A., Lamb, C.J., Dixon, R.A. and Power, J.B. (1998) Expression of a chitinase transgene in rose (*Rosa hybrida* L.) reduces development of blackspot disease (*Diplocarpon rosae* Wolf). *Molecular Breeding* 4, 187–194.

Mariani, C., Beuckeleer, M.D., Truettner, J., Leemans, J. and Goldberg, R.B. (1990) Induction of male sterility in plants by a chimaeric ribonuclease gene. *Nature* 347, 737–741.

Masoud, S.A., Zhu, Q., Lamb, C. and Dixon, R.A. (1996) Constitutive expression of an inducible beta-1,3-glucanase in alfalfa reduces disease severity caused by the oomycete pathogen *Phytophthora megasperma* F Sp *medicaginis*, but does not reduce disease severity of chitin-containing fungi. *Transgenic Research* 5, 313–323.

Maude, R.B. (1990a) Leaf diseases of onions. In: Rabinowitch, H.D. and Brewster, J.L. (eds) *Onions and Allied Crops*, Vol. II. *Agronomy, Biotic Interactions, Pathology, and Crop Protection*. CRC Press, Boca Raton, Florida, pp. 173–189.

Maude, R.B. (1990b) Storage diseases of onions. In: Rabinowitch, H.D. and Brewster, J.L. (eds) *Onions and Allied Crops*, Vol. II. *Agronomy, Biotic Interactions, Pathology, and Crop Protection*. CRC Press, Boca Raton, Florida, pp. 273–296.

Maw, G. (1982) Biochemistry of *S*-methyl-L-cysteine and its principal derivatives. *Sulfur Reports* 2(1), 1–32.

Moreno, M., Bernal, J.J., Jimenez, I. and Rodriguez Cerezo, E. (1998) Resistance in plants transformed with the *P1* or *P3* gene of tobacco vein mottling potyvirus. *Journal of General Virology* 79, 2819–2827.

Murashige, T. and Skoog, F. (1962) A revised medium for rapid growth and bio-assays with tobacco tissue cultures. *Physiologia Plantarum* 15, 473–497.

Myers, J.M. and Simon, P.W. (1996) Promoter activity and osmoticum effect in garlic and onion for transformation. *In Vitro* 32, 68A (abstract).

Myers, J.M. and Simon, P.W. (1998a) Microprojectile bombardment of garlic, *Allium sativum* L. In: *Proceedings of the 1998 National Onion (and Other Alliums) Research Conference, 10–12 December 1998, Sacramento, California, USA*. University of California, Davis, California, pp. 121–126.

Myers, J.M. and Simon, P.W. (1998b) Continuous callus production and regeneration of garlic (*Allium sativum* L.) using root segments and shoot tip derived plantlets. *Plant Cell Reports* 17, 726–730.

Novak, F.J. (1990) *Allium* tissue culture. In: Rabinowitch, H.D. and Brewster, J.L. (eds) *Onions and Allied Crops*, Vol. I. *Botany, Physiology, and Genetics*. CRC Press, Boca Raton, Florida, pp. 233–250.

Ohsumi, C., Hayashi, T. and Sano, K. (1993) Formation of alliin in the culture tissues of *Allium sativum*. Oxidation of *S*-allyl-L-cysteine. *Phytochemistry* 33, 107–111.

Orvis, K.S., Galmarini, C.R., Goldman, I.L. and Havey, M.J. (1998) A comparison of onion-induced platelet aggregation by plasma and whole blood aggregometry. In: *Proceedings of the 1998 National Onion (and Other Allium) Research Conference, 10–12 December, Sacramento, California, USA*. University of California, Davis, California, p.178.

Ow, D. (1996) Recombinase-directed chromosome engineering in plants. *Current Opinions in Biotechnology* 7, 181–186.

Phillippe, B., Cammue, B.P.A., Thevissen, K., Hendriks, M., Eggermont, K., Goderis, I.J., Proost, P., van Damme, J., Osborn, R.W., Guerbette, F., Kader, J.C. and Broekaert, W.F. (1995) A potent antimicrobial protein from onion seeds showing sequence homology to plant lipid transfer proteins. *Plant Physiology* 109, 445–455.

Powell, A., D'Hallewin, G., Hall, B., Stotz, H., Labivitch, J. and Bennet, A. (1994) Glycoprotein inhibitors of fungal polygalacturonases: expression of pear PGIP improves resistance in transgenic tomatoes. In: *Proceedings of the 4th International Congress of Plant Molecular Biology, Amsterdam, 19–24 June*. The International Society for Plant Molecular Biology, Abstract no. 1904.

Puchta, H. (1998) Towards targeted transformation in plants. *Trends in Plant Science* 3, 77–78.

Rao, K.V., Rathore, K.S., Hodges, T.K., Fu, X., Stoger, E., Sudhakar, D., Williams, S., Christou, P., Bharathi, M., Bown, D.P., Powell, K.S., Spence, J., Gatehouse, A.M.R. and Gatehouse, J.A. (1998) Expression of snowdrop lectin (Gna) in transgenic rice plants confers resistance to rice brown planthopper. *The Plant Journal* 15, 469–477.

Regansbury-Twink, A. and Hooykaas, P. (1993) Transgenic *Nicotiana glauca* plants expressing bacterial virulence gene *virF* are converted into hosts for nopaline strains of *A. tumefaciens*. *Nature* 363, 69–70.

Revers, F., Le Gall, O., Candresse, T. and Maule, A.J. (1999) New advances in understanding the molecular biology of plant/potyvirus interactions. *Molecular Plant–Microbe Interactions* 12, 367–376.

Roberfroid, M.B. and Delzenne, N.M. (1998) Dietary fructans. *Annual Review of Nutrition* 18, 117–143.

Rubin, B. (1990) Weed control. In: Rabinowitch, H.D. and Brewster, J.L. (eds) *Onions and Allied Crops*, Vol. II. *Agronomy, Biotic Interactions, Pathology, and Crop Protection*. CRC Press, Boca Raton, Florida, pp. 63–84.

Rubin, G. and Spradling, A. (1985) Genetic transformation of *Drosophila* with transposable element vectors. *Science* 218, 348–353.

Saker, M.M. (1998) *In vitro* regeneration of onion through repetitive somatic embryogenesis. *Biologia Plantarum* 40, 499–506.

Sambrook, J., Fritsch, E.F. and Maniatis, T. (1989) In: Ford, N., Nolan, C. and Ferguson, M. (eds) *Molecular Cloning, a Laboratory Manual*, Vol. 3, Appendix A, 2nd edn. Cold Spring Harbor Laboratory Press, Plainview, New York, Appendix A.1–E.39.

Saroha, M.K., Sridhar, P. and Malik, V.S. (1998) Glyphosate-tolerant crops – genes and enzymes. *Journal of Plant Biochemistry and Biotechnology* 7, 65–72.

Schickler, H. and Chet, I. (1997) Heterologous chitinase gene expression to improve plant defense against phytopathogenic fungi. *Journal of Industrial Microbiology and Biotechnology* 19, 196–201.

Schnepf, E., Crickmore, N., van Rie, J., Lereclus, D., Baum, J., Feitelson, J., Zeigler, D.R. and Dean, D.H. (1998) *Bacillus thuringiensis* and its pesticidal crystal proteins. *Microbiology and Molecular Biology Reviews* 62, 775–806.

Seabrook, J. (1994) *In vitro* propagation and bulb formation in garlic. *Canadian Journal of Plant Science* 74, 155–158.

Sévenier, R., Hall, R.D., van der Meer, I.M., Hakkert, H.J.C., van Tunen, A.J. and Koops, A.J. (1998) High level fructan accumulation in a transgenic sugar beet. *Nature Biotechnology* 16, 843–846.

Shiomi, N., Onodera, S. and Sakai, H. (1997) Fructo-oligosaccharide content and fructosyltransferase activity during growth of onion bulbs. *New Phytologist* 136, 105–113.

Silvertand, B., van Rooyen, A., Lavrijsen, P., van Harten, A.M. and Jacobsen, E. (1996) Plant regeneration via organogenesis and somatic embryogenesis in callus cultures derived from mature zygotic embryos of leek (*Allium ampeloprasum* L.). *Euphytica* 91, 261–270.

Singh, Z. and Sansavini, S. (1998) Genetic transformation and fruit crop improvement. *Plant Breeding Reviews* 16, 87–134.

Song, S.I., Song, J.T., Kim, C.H., Lee, J.S. and Choi, Y.D. (1998) Molecular characterization of the garlic virus X genome. *Journal of General Virology* 79, 155–159.

Songstad, D.D., Somers, D.A. and Griesbach, R.J. (1995) Advances in alternative DNA delivery techniques. *Plant Cell, Tissue and Organ Culture* 40, 1–15.

Soni, S.K. and Ellis, P.R. (1990) Insect pests. In: Rabinowitch, H.D. and Brewster, J.L. (eds) *Onions and Allied Crops*, Vol. II. *Agronomy, Biotic Interactions, Pathology, and Crop Protection*. CRC Press, Boca Raton, Florida, pp. 213–271.

Spencer, T., O'Brien, J., Start, W., Adams, T., Gordon-Kamm, W. and Lemaux, P. (1992) Segregation of transgenes in maize. *Plant Molecular Biology* 18, 201–210.

Stoger, E., Williams, S., Christou, P., Down, R.E. and Gatehouse, J.A. (1999) Expression of the insecticidal lectin from snowdrop (*Galanthus nivalis* agglutinin; GNA) in transgenic wheat plants: effects on predation by the grain aphid *Sitobion avenae*. *Molecular Breeding* 5, 65–73.

Stone, H. and Armentrout, V. (1985) Production of oxalic acid by *Sclerotium cepivorum* during infection of onion. *Mycologia* 77, 526–530.

Stripe, F., Barbieri, L., Battelli, L.G., Soria, M. and Lappi, D.A. (1992) Ribosome-inactivating proteins from plants – present status and future prospects. *Bio/Technology* 10, 405–412.

Tanaka, Y., Tsuda, S. and Kusumi, T. (1998) Metabolic engineering to modify flower color. *Plant and Cell Physiology* 39, 1119–1126.

Thomas, J.C., Adams, D., Nessler, C., Bohnert, H.J. and Brown, J.K. (1994) Reduced reproduction of whitefly (*Bemisia tabaci*) on transgenic tobacco expressing tryptophan decarboxylase. In: *Proceedings of the 4th International Congress of Plant Molecular Biology, Amsterdam, 19–24 June*. The International Society for Plant Molecular Biology, Abstract no. 1924.

Thompson, C., Dunwell, J.M., Johnstone, C.E., Lay, V., Ray, J., Schmitt, M., Watson, H. and Nisbet, G. (1995) Degradation of oxalic acid by transgenic oilseed rape plants expressing oxalate oxidase. *Euphytica* 85, 169–172.

Toubart, P., Desiderio, A., Salvi, G., Cervone, F., Daroda, L., De Lorenzo, G., Bergmann, C., Darvill, A.G. and Albersheim, P. (1992) Cloning and characterization of the gene encoding the endopolygalacturonase-inhibiting protein (PGIP) of *Phaseolus vulgaris* L. *The Plant Journal* 2, 367–373.

Tsaftaris, A. (1996) The development of herbicide-tolerant transgenic crops. *Field Crops Research* 45, 115–123.

Tsuneyoshi, T., Matsumi, T., Natsuaki, K.T. and Sumi, S. (1998a) Nucleotide sequence analysis of virus isolates indicates the presence of three potyvirus species in *Allium* plants. *Archives of Virology* 143, 97–113.

Tsuneyoshi, T., Matsumi, T., Deng, T.C., Sako, I. and Sumi, S. (1998b) Differentiation of *Allium* carlaviruses isolated from different parts of the world based on the viral coat protein sequence. *Archives of Virology* 143, 1093–1107.

Ucman, R., Zel, J. and Ravnikar, M. (1998) Thermotherapy in virus elimination from garlic – influences on shoot multiplication from meristems and bulb formation *in vitro*. *Scientia Horticulturae* 73, 193–202.

Vain, P., Keen, N., Murillo, J., Rathus, C., Nemes, C. and Finer, J. (1993) Development of the particle inflow gun. *Plant Cell, Tissue and Organ Culture* 33, 237–246.

Vain, P., Worland, B., Kohli, A., Snape, J.W. and Christou, P. (1998) The green fluorescent protein (GFP) as a vital screenable marker in rice transformation. *Theoretical and Applied Genetics* 96, 164–169.

van Damme, E.J.M., Smeets, K., Torrekens, S., van Leuven, F. and Peumans, W.J. (1992) Isolation and characterization of alliinase cDNA clones from garlic (*Allium sativum* L.) and related species. *European Journal of Biochemistry* 209, 751–757.

van der Vlugt, R.A.A., Steffens, P., Cuperus, C., Barg, E., Lesemann, D.-E., Bos, L. and Vetten, H.J. (1999) Further evidence that shallot yellow stripe virus (SYSV) is a distinct potyvirus and reidentification of Welsh onion yellow stripe virus as a SYSV strain. *Phytopathology* 89, 148–155.

Vergunst, A.C. and Hooykaas, P.J.J. (1998) *Cre/lox*-mediated site-specific integration of *Agrobacterium* T-DNA in *Arabidopsis thaliana* by transient expression of *cre*. *Plant Molecular Biology* 38, 393–406, 1269.

Vergunst, A.C. and Hooykaas, P.J.J. (1999) Recombination in the plant genome and its application in biotechnology. *Critical Reviews in Plant Sciences* 18, 1–31.

Vetter, J. (2000) Plant cyanogenic glycosides. *Toxicon* 38, 11–36.

Vijn, I., van Dijken, A., Sprenger, N., van Dun, K., Weisbeek, P., Wiemken, A. and Smeekens, S. (1997) Fructan of the inulin neoseries is synthesized in transgenic chicory plants (*Cichorium intybus* L.) harbouring onion (*Allium cepa* L.) fructan:fructan 6G-fructosyltransferase. *The Plant Journal* 11, 387–398.

Vijn, I., van Dijken, A., Lüscher, M., Bos, A., Smeets, E., Weisbeek, P., Wiemken, A. and Smeekens, S. (1998) Cloning of sucrose:sucrose 1-fructosyltransferase from onion and synthesis of structurally defined fructan molecules from sucrose. *Plant Physiology* 117, 1507–1513.

Vinnemier, J., Drogelaser, W., Pistorius, E.K. and Broer, I. (1995) Purification and partial characterization of the *Streptomyces viridochromogenes* Tu494 phosphinothricin-*N*-acetyltransferase mediating resistance to the herbicide phosphinothricin in transgenic plants. *Zeitschrift für Naturforschung C – A Journal of Biosciences* 50, 796–805.

Walkey, D. (1990) Virus diseases. In: Rabinowitch, H.D. and Brewster, J.L. (eds) *Onions and Allied Crops*, Vol. II. *Agronomy, Biotic Interactions, Pathology, and Crop Protection*. CRC Press, Boca Raton, Florida, pp. 191–212.

Weld, R. (2000) Transient expression following DNA transfer to plants: the phenomenon, its causes and some applications. PhD thesis, University of Canterbury, Christchurch, New Zealand.

Welter, M.E., Clayton, D.S., Miller, M.A. and Petolino, J.F. (1995) Morphotypes of friable embryogenic maize callus. *Plant Cell Reports* 14, 725–729.

Wilkinson, J.E., Lindsey, K. and Twell, D. (1998) Antisense-mediated suppression of transgene expression targeted specifically to pollen. *Journal of Experimental Botany* 49, 1481–1490.

Wittner, A., Palkovics, L. and Balazs, E. (1998) *Nicotiana benthamiana* plants transformed with the plum pox virus helicase gene are resistant to virus infection. *Virus Research* 53, 97–103.

Xue, H.M., Araki, H., Kanazawa, T., Harada, T. and Yakuwa, T. (1997) Callus formation and plantlet regeneration through *in vitro* culture of immature embryo and seedling in Chinese chive (*Allium tuberosum* Rottler). *Journal of the Japanese Society for Horticultural Science* 66, 353–358 (in Japanese).

Yun, D.-J., Bressan, R.A. and Hasegawa, P.M. (1997) Plant antifungal proteins. *Plant Breeding Reviews* 14, 39–87.

Yun, J.W. (1996) Fructo-oligosaccharides – occurrence, preparation, and application. *Enzyme and Microbial Technology* 19, 107–117.

Zheng, S.J., Henken, B., Sofiari, E., Jacobsen, E., Krens, F.A. and Kik, C. (1998) Factors influencing induction, propagation and regeneration of mature zygotic embryo-derived callus from *Allium cepa*. *Plant Cell, Tissue, and Organ Culture* 53, 99–105.

Zheng, S., Henken, B., Sofiari, E., Jacobsen, E., Kik, C. and Krens, F. (1999) Factors affecting *Agrobacterium tumefaciens* mediated transient expression of gusA in shallot (*Allium cepa* L.) and regeneration of transgenic plants. In: *Towards the Third Millennium – International Symposium on Plant Genetic Engineering, Centro de Ingeneria Genetica y Biotecnologia, 6–10 December 1999*. Elfos Scientae, Havana, p. 89.

Zheng, S.-J., Henken, B., Cato, S., Sofiari, E., Jacobsen, E., Krens, F. and Kik, C. (2000) Molecular characterization of transgenic shallots (*Allium cepa* L.) by amplification and sequencing of genomic DNA flanking T-DNA borders. In: *6th International Congress of Plant Molecular Biology*. ISPMB, Québec, Abstract S03–126.

7 Doubled-haploid Onions

B. Bohanec
Biotechnical Faculty, Centre for Plant Biotechnology and Breeding, University of Ljubljana, Jamnikarjeva 101, 1000 Ljubljana, Slovenia

1. Introduction

Hybrid cultivars of onion are considered to be superior to open-pollinated (OP) varieties, due to their higher uniformity and expressed heterosis. In contrast to some other cross-pollinated species, such as maize, where modern inbred lines express only minor inbreeding depression, onion populations still possess deleterious recessive genes and high inbreeding depression is obvious. Onion breeding lines are usually selfed only two or three times, rendering it difficult to obtain complete genetic and phenotypic uniformity in the resulting hybrid. Double haploids provide an alternative strategy that offers, for the first time in onion, complete homozygosity and phenotypic uniformity.

Haploid plants can be obtained from male or female gametic cells; however, species differ according to the ability to induce haploids via androgenesis or gynogenesis. As reviewed by Keller and Korzun (1996), in onion even large anther culture experiments failed to generate haploids, and R.C. Muren (California, USA, 1998, personal communication) reported that a major effort in his laboratory to generate haploids from anther tissue resulted in complete failure.

Successful haploid induction via gynogenesis was actually developed more or less simultaneously in three laboratories. Bruno Campion and his colleagues at the Research Institute for Vegetable Crops near Lodi, Italy, published their first results on cultured unpollinated onion ovules at the 4th EUCARPIA *Allium* symposium in 1988 and elsewhere (Campion and Alloni, 1990). Roger C. Muren, at the H.A. Jones Memorial Research Center of the Sunseeds Company in Oregon, USA, used unpollinated ovaries (Muren, 1989) and Joachim Keller tried unpollinated ovules, ovaries and whole flower buds at the Institute of Plant Breeding Research in Quedlinburg, Germany (Keller, 1990). Ten years after these first discoveries, media determined in 1989 by Muren, with only minor modifications, are still the most efficient in haploid generation and are used in most laboratories all around the world.

For practical use of doubled-haploid lines in plant breeding, procedures for haploid induction should be efficient, not too laborious, and genotype-insensitive. Regenerants should grow well in tissue culture and be easy to double from haploid to doubled-haploid level and the plants produced should be easily hardened off. The double haploids generated should maintain their genetic integrity and produce fertile seed. These demands are not easy to meet. I shall outline the limitations and acceptable solutions available at the present stage of knowledge.

2. Procedures for Gynogenic Embryo Induction

2.1 Choice of organ and culture procedure

Gynogenic haploid induction in onion can be achieved by culturing unpollinated ovules, ovaries or whole flower buds. Induction procedures consist of one or two steps with or without subculturing.

Ovule culture is the most laborious. Ovules can be extracted immediately after sterilization of the flowers (Keller, 1990) or after flower bud preculture (Campion and Alloni, 1990; Bohanec *et al.*, 1995). Ovary cultures have been prepared in two ways: (i) ovaries are isolated from immature flower buds (for information on flower physiological age, see Section 2.2 below) and cultured until embryo regeneration (Muren, 1989; Campion *et al.*, 1992); or (ii) immature buds are first cultured for 10–14 days following isolation of the ovaries, and then subcultured on a different medium until regeneration (Bohanec *et al.*, 1995; Jakše *et al.*, 1996). The second procedure has the advantage that the ovaries are already swollen and extraction is simple. Flower bud culture is the simplest way of inducing gynogenic haploids in onion and has been used in many recent studies (Cohat, 1994; Geoffriau *et al.*, 1997a; Javornik *et al.*, 1998; Bohanec and Jakše, 1999; Michalik *et al.*, 2000).

Unlike the situation in some other species, such as sugarbeet, the yield of haploid regenerants from ovule culture in onion is low compared with culture of whole ovaries or flower buds. Since the ovule-culture method is laborious and yields the lowest number of embryo regenerants, it is no longer used for haploid induction in onion. For the ovary-culture method, we estimate that extraction of ovaries from precultured flower buds, compared with whole-flower culture, requires three times more work, while the gynogenic response of ovary vs. flower culture is often similar. Self-pollination in cultured flowers is zero, since, as noted by Cohat (1994) for shallot, onion anthers do not dehisce within the culture vessel (presumably because of the high humidity). The only disadvantage of whole-flower culture over ovary culture is the growth of basal callus, which is often formed from the septal nectaries region when whole flowers are cultured. Flowers that form a basal callus may produce haploid embryos that are of lower quality. The negative aspect of flower culture is an increased possibility of somatic regeneration from the callus, which is genotype-dependent. The development of basal callus in 39 accessions of bulb onion was scored by Bohanec and Jakše (1999). They concluded that only in one accession could this negative associated phenomenon justify the laborious removal of the tepals in the ovary extraction method.

Subculturing of ovaries or flowers on hormone-free media after a certain period of culture and before regeneration of embryos (Campion *et al.*, 1992) has been reported, but it did not increase gynogenic efficiency.

2.2 Flower bud developmental stage

The first exact study of the appropriate flower bud developmental stage was published by Muren (1989). He concluded that flower buds 3–5 days before anthesis were superior to older or younger ones. Later, buds which at the time of culture were in the stage immediately prior to dehiscence or up to 3 days younger were preferred. Michalik *et al.* (2000) concluded that small (young) buds (2.8–3.0 mm long) produced significantly fewer embryos than older (3.5–4.5 mm long) ones, with a noticeable genotype specificity. For instance, medium-sized buds (3.5–3.8 mm) were optimal for cv. 'Kutnowska', while 'Wolska' and 'Fiesta' benefited from larger pre-anthesis buds (4.3–4.5 mm).

Klein and Korzonek (1999) studied correlations between flower bud size, mean anther length and the stage of pollen development with bud length, size of ovule and stage of embryo-sac development in cv. 'Kutnovska'. The smallest buds (2.8–3.0 mm long) had differentiating archaeosporal cells or megaspore mother cells in prophase I, while the largest unopened buds (up to 4.5 mm long) had mature embryo sacs.

Empirical studies of onion gynogenesis led to the assumption that haploid cells in the embryo sac are appropriately developed to give rise to haploid embryos in *in vitro* culture. However, no solid evidence is available to definitively support this conclusion. Until recently, there was no evidence as to which of the haploid cells within the ovule sac give rise to the haploid embryos. Musial *et al.* (1999) showed that ovules in fully developed flower buds, just prior to anthesis, consist of a mature megagametophyte with two polar nuclei. However, the secondary nucleus is also visible, with traces of degenerating antipodals at one pole and with the egg cell, accompanied by two unequal synergids, at the micropylar pole. After 7 days in culture, synergids were present in all the examined embryo sacs and, 7 days later (14th day), synergids and endosperm nuclei were detected in some ovules. Additional studies (Musial *et al.*, 2001) showed that, at the time of inoculation, ovules with ovaries sized 2.0–3.0 mm in diameter (from highly responsive donor plants) contained ovules with two- or four-nucleate embryo sacs (smallest ovaries) to mature embryo sacs in the largest ones. It seems likely that the embryos are actually induced from ovaries cultured at the immature stage. From the 2nd to the 7th week in culture, formation of haploid embryos (from globular to almost mature cylindrical stage) was detected in 5.7% of the ovules and their origin was, for several reasons, most probably the egg cell. In addition, ovules containing endosperm only (3.6%) and ovules containing the egg apparatus (0.5%) or both endosperm and embryo (0.4%) were detected. This last finding is probably unique and has not yet been reported in other species.

Two possible methods are used for flower bud collection: either the whole umbel is excised at the stage at which about 30% of the flower buds have reached the appropriate stage, or the buds are sheared off by scissors a few at a time, usually at 2-day intervals. This latter method, used in our laboratory for the last few years, has the advantage that larger numbers of appropriate-sized buds can be collected from single donor plants, with no negative effects.

2.3 Cultivation of donor plants

Use of unsoiled plants and properly sterilized explants is of paramount importance, since any contamination can result in a complete loss. Despite the fact that growth conditions have often been reported to have the most important effect on the success of androgenesis/gynogenesis of a number of donor-plant species, there are only few recommendations on the culture conditions for onion donor plants. Based on our experience, data in the literature and personal communication with several researchers in the field, we propose the following. Donor plants should preferably be maintained in a greenhouse, protected from pests and diseases and carefully watered at soil level, and not by sprinklers or rain, which may promote contamination. Field culture is an option only in a dry climate with little rain or dust. Muren (1989) reported no increase or a decrease in embryo yield when intact umbels were kept at 5°C for 5 or 15 days or 40°C for 4 h. However, more recently, in the UK, Puddephat *et al.* (1999) reported a tenfold increase in yield when flower buds were harvested from donor plants raised in growth chambers at 15°C, compared with 10°C or ambient conditions in a glasshouse. If such results are confirmed, the choice of culture conditions can be manipulated to significantly improve yields.

Thrips are the second major cause of both primary and secondary contamination, often noticed only after a few weeks in culture. The control of thrips is extremely difficult and our experience shows that regular watering of donor plants with a solution of Confidor® (imidacloprid) is the only effective measure.

2.4 Sterilization of explants and temperature treatments

A number of disinfectants can be used with similar efficiency. We prefer to immerse the sampled floral buds in a solution of dichloroisocyanuric acid (16.6 g l^{-1}) combined with a few drops of Tween 20, for 8–12 minutes. This disinfectant is superior to the solid organic chlorine and the unstable sodium hypochlorite, which may cause damage to the delicate tissue.

Incubation of cultured ovaries at an elevated temperature of 40°C for 1 or 3 days prior to culture did not improve embryo yield. Organs are usually cultured in Petri dishes 10 cm in diameter, usually with 30 flowers or ovaries per dish. Temperature and light regimes in the growth chambers are the standard conditions used for tissue culture, 25 ± 2°C and 16/8 h light/dark photoperiod. Light is provided by low intensity standard fluorescent lamps (30–100 $\mu mol\ s^{-1}\ m^{-2}$), or exceptionally (Campion *et al.*, 1995b) by photosynthetically optimized Grolux lamps. There are no reports of the effect of light conditions in culture affecting gynogenesis.

3. Media Composition

The effects of media components on haploid embryogenesis have been intensively studied in many plant species. Numerous potentially useful substances are proposed and available and it is impossible to test them all for optimal concentration, duration of treatment and combination effects. Additionally (especially in early experiments), yields of onion haploid embryos were low, reaching only six to seven induced embryos per 100 flowers. Hence, large experimental units (300–500 flowers per treatment) were needed, quantities that would require prohibitive volumes of large-scale tests for media combinations.

Keller and Korzun (1996) reviewed the media used in embryogenesis up to 1994. Some of the media used in the early experiments were later found to be suboptimal; hence a brief summary of media used in recent studies is given.

3.1 Basal mineral components

The three most often used combinations of macro- and micro-elements were B5 (Gamborg *et al.*, 1968), BDS (Dunstan and Short, 1977) and MS (Murashige and Skoog, 1962). There are no specific studies available, but it seems that the three basal media have similar effects on culture development and yield. In our laboratory, no differences in growth, development and yield were recognized between B5 and BDS (unpublished data).

3.2 Carbohydrates and gelling agents

Vinterhalter and Vinterhalter (1999) proposed that some sucrose effects are similar to hormone-like activities in some *in vitro* culture systems, so the role of sucrose in onion gynogenesis needs to be more carefully studied. In recent studies, optimal results were obtained with 10% sucrose in the culture medium. However, Geoffriau *et al.* (1997a) reported similar results with a lower concentration of 7.5%. There is no conclusive evidence on the effect of other sugars, such as maltose, glucose or fructose, on gynogenesis.

Following embryo emergence, carbohydrate content in the micropropagation medium is usually lowered. For instance, in our experiments, for cultivation of approximately 7000 embryos during 1998–2000, half-strength BDS medium supplemented with 30 g l^{-1} glucose was used, enabling adequate plantlet growth and eliminating hyperhydration.

In most studies, agar is the gelling agent. An increase in embryo yield was recorded by Jakše *et al.* (1996) when gellan-gum was used instead, but a higher proportion of vitreous regenerants resulted. Gellan-gum is a key substance in the culturing media, as it promotes the induction of somatic regenerants from onion buds or ovaries (Luthar and Bohanec, 1999). The adverse effect of this gelling agent should be considered when haploid induction is performed on gellan-gum-containing media.

3.3 Plant growth regulators

Muren (1989) applied 2,4-dichlorophenoxyacetic acid (2,4-D) at 2 mg l^{-1} and benzylaminopurine (BAP) at 2 mg l^{-1} in the culture medium. This combination has since been approved by several other researchers and has become the standard composition of growth regulators for embryogenesis. Other previously studied growth regulators, including naphthalene acetic acid (NAA), indolebutyric acid (IBA), glutamine, *N* 6-(2-isopentenyl)-adenine (2iP) and gibberellic acid (GA_3), were less effective. Jakše *et al.* (1996) demonstrated that 2,4-D can be replaced by phenylacetic acid in the induction medium and that thidiazuron can replace BAP in the second-stage medium, but these did not substantially improve Muren's combination.

Campion *et al.* (1995b) studied the effect of duration of ovary or bud cultures' exposure to plant growth regulators (15, 30 and 45 days) prior to transfer to growth-regulator-free media. The authors concluded that a 15-day treatment was sufficient for gynogenic stimulation of ovaries and flowers.

Martinez *et al.* (2000) tested the effects of polyamines as a substitute for auxins and cytokinins on onion gynogenesis. Media supplemented with 2 mM of putrescine were sufficient to induce haploid embryos, while addition of 0.1 mM spermidine promoted embryo maturation. Results are promising and offer an alternative to the standard hormone combination. Further studies are needed to test effects of polyamines on a broader range of genotypes.

An alternative approach to testing the influence of plant growth regulators was the application of 2,4-D to the onion scape (hollow inflorescence stalk) (Jakše *et al.*, 2001). Fifty or 100 mg l^{-1} of 2,4-D was injected at the time when the first flowers developed until the stage that can be inoculated for haploid induction. At 2-day intervals, the 60 most mature flowers were cut and inoculated following a previously published procedure (Bohanec and Jakše, 1999). Results indicate that the induction percentages of less responsive lines were improved when flower buds were cut 10–14 days after the injections. An unexpected result was achieved in a control treatment, where the optimal response was obtained when 2,4-D was also omitted in culture media, indicating that, for some 'low-responsive' lines, 2,4-D (present in most published induction procedures) is harmful. However, such preliminary results need to be confirmed in larger experiments.

4. Genotypic Effect

It was realized very early that the genetic make-up of the donor onion plants and the growth conditions play the most important roles in the success of gynogenesis. In onion haploid induction, frequencies are usually expressed as the number of embryos formed on 100 flowers; therefore, since ovaries contain six ovules, a theoretical maximum of 600% can be obtained. In the first experiments, yields were low, ranging from 0 to 3% for different genotypes. Improvement in culture conditions resulted in a significant increase, up to 7.6 % for Asgrow's experimental hybrid 'XPH 3371' (Bohanec *et al.*, 1995). The only higher yield (of 21.9%) was reported by Cohat (1994) for a shallot accession.

Later studies have focused on more variable genetic material from different regions of the world. In a 3-year study, Geoffriau *et al.* (1997a) analysed 18 onion cultivars and populations from eastern, northern and southern Europe and four from the USA. Two cultivars showed high gynogenic potential, but yields varied with years. The best yield for one cultivar was 17.4% in an optimal year. In less favourable years, four cultivars produced no embryos. In a similar study, Bohanec and Jakše (1999) analysed 39 accessions from Europe, North America and Japan. Two European and three Japanese accessions produced no embryos, and the highest gynogenic yield was obtained from North American cultivars and inbred lines. Two inbred lines and one F_1 hybrid produced mean numbers of 18.6, 19.3 and 22.6 embryos per 100 cultured flowers, respectively. Data were also recorded for individual donor plants. Very high variability was found within cultivars and even within inbred lines. Hence, minimum and maximum values for the five donor plants of the above-mentioned inbred line with the mean value of 19.3%, were 4.4 and 51.7%, respectively. When single plants were induced to flower in 2 consecutive years, variation in gynogenic yield within plants between seasons was much lower than that recorded between individuals of the same line. This and later results (B. Bohanec, unpublished) confirmed that genetic variability in gynogenesis is much higher than that brought about by culture conditions. Michalik *et al.* (2000) scored 11 Polish onion cultivars and 19 breeding lines for gynogenetic potential. The majority of the tested genotypes produced a very low or low embryo yield, except for the breeding line '601A', which had 10.0% embryo yield. Javornik *et al.* (1998) for the first time cultured flowers from selfed plants of three doubled-haploids, in order to generate a second cycle of haploid plants. Only one line produced a very high yield, with the mean number of 118.3 haploid embryos per 100 cultured flowers. Embryo yield within the responsive genotype ranged between 67 and 196 per 100 flowers, thus indicating that the variation was induced by growth conditions. The other two lines yielded only 2.3 and 0.3% haploid embryos. The results show that genes coding for low or high gynogenic potential are present in gynogenic regenerants.

Since the genotype effect is the key element for successful haploid induction, the genetic basis of this trait should be studied more carefully. Studies are in progress to determine key features, such as the number of genes involved and the mode of their expression. Bohanec *et al.* (1999) demonstrated that crossing of responsive and non-responsive onion lines resulted in increased gynogenic ability in the hybrid progeny.

5. Gynogenic Haploid Induction Processes in Onion

Sterilized flowers or ovaries are cultured as described above. At the time haploid embryos start to form on top of the ovaries, the latter usually change colour from green to pale yellow. On media supplemented with 2 mg l^{-1} 2,4-D and 2 mg l^{-1} BAP at high sucrose content and solidified with agar, it takes between 60 and 180 days for sprouting, with the majority of embryos emerging after *c.* 100 days of culture. The length of the embryogenic process is genotype-dependent. B. Bohanec, M. Jakše and Z. Luthar (unpublished) noticed that embryo emergence is much shorter (1 month) on media favourable for somatic regeneration. It is unclear whether media constituents can be optimized to shorten the regeneration process but without also promoting somatic regeneration.

In most cases, haploid embryos emerge from ovaries in a way similar to the germination of true seeds. The sprouting embryo forms a typical loop structure (Colour Plate 4A), which is clearly distinguishable from somatic regenerants. The latter occasionally proliferate at the flower base. When extracted from the ovary (Colour Plate 4B), a complete bipolar embryo is formed with already developed roots. However, some embryos develop into abnormal structures. Experience shows that up to 50% of the embryos fail to develop normally and the rate seems to be genotype-dependent. Colour Plate 4C shows all the embryos developed from a single harvest following their culture in the same Petri dish. Despite the fact that many embryos sprout from a single ovary, some are considerably smaller (half-size or less) than others (the normal ones). Some are also deformed and will probably not develop into normal plantlets.

Later, each sprout forms a single plantlet or, in some cases, multiple plantlets. Experience indicates that half-strength BDS supplemented with 30 g l^{-1} glucose provides the most favourable medium for this stage of development and also for minimizing the production of vitreous plantlets. Elongated plantlets with well-developed roots are then transferred to a greenhouse for conditioning and further growth (Fig. 7.1). All or almost all well-developed plants normally survive this step. Usually both haploid and doubled-haploid plants grow vigorously and finally form bulbs similar to those of normal heterozygous onion plants (Colour Plate 4D). Flowering occurs in the 1st or the 2nd year, and the inflorescence of haploids is clearly distinguished from that of doubled-haploids. The former produce only rudimentary floral structures (Fig. 7.2) as compared with the normal inflorescence of the latter plants (Colour Plate 4E).

Very little information exists on fertility restoration of doubled-haploid onion plants. Campion *et al.* (1995a) reported that the first seeds produced by doubled-haploid lines were obtained following spontaneous genome doubling of haploid plants. Our experience (B. Bohanec and M. Jakše, unpublished) with selfing of doubled-haploids (using oryzalin treatment (see Section 7 for details)) suggests that fewer than 50% of the regenerants set seeds. Some of these seeds are not viable. More information is needed to draw conclusions, but it seems

Fig. 7.1. Elongated plantlets with well-developed roots.

Fig. 7.2. Inflorescence of haploids with rudimentary floral structures.

likely that some deleterious recessive genes expressed in doubled-haploid plants code for low fertility.

Hybrid onion cultivars are usually produced with cytoplasmic male-sterile (CMS) genotypes as seed lines. These hybrid plants should not be used for gynogenic haploid induction, since the CMS cytoplasm is transmitted to the progeny with the consequent male sterility (Doré and Marie, 1993).

6. Determination of Ploidy and Homozygosity

Some earlier publications on onion gynogenesis reported that about 30% of the generated haploids underwent spontaneous chromosome doubling (Muren, 1989; Keller, 1990). Most recent studies report that about 90% of regenerants remain haploid.

Different methods have been used to analyse ploidy level. Chromosome counting was performed in root tips (Muren, 1989; Campion and Alloni, 1990; Keller, 1990; Bohanec *et al.*, 1995; Campion *et al.*, 1995b) or shoot-tip cells (Campion *et al.*, 1995b). The latter technique reflects a situation closer to reality as flowers are formed in shoot-tips. Doré and Marie (1993) and Keller and Korzun (1996) measured nucleus lengths in stomata and guard cells, and used scanning cytometry of epidermal cells. More recently, flow cytometry of leaf tissues has predominantly been used (Cohat, 1994; Bohanec *et al.*, 1995; Campion *et al.*, 1995a; Jakše *et al.*, 1996; Geoffriau *et al.*, 1997a, b; Javornik *et al.*, 1998; Bohanec and Jakše, 1999). This method has several advantages over other techniques for ploidy analysis: it is fast, non-destructive (unlike shoot-apex extraction), can be performed on different tissues and reflects the exact proportion of DNA quantities in the studied tissue. The only limitation is that, in cases where individual plants contain more than two nuclear stages, intermediate peaks represent a mixture of G2 phase of lower and G1 level of higher ploidy level, which cannot be determined separately.

The analysis of ploidy is somewhat complicated, since haploid, diploid and polyploid levels are often found in the same plant tissue. Different ploidy states can therefore be present within a single inflorescence. However, only the diploid flowers form seeds, the others remaining sterile. It is also unclear whether the ploidy of haploid plants can spontaneously double simply by prolonged vegetative growth over a few successive years. These questions are currently under investigation.

Homozygosity of regenerants can be determined in several ways. In view of the biennial growth habit of onion, analysis of progeny obtained after selfing of putative homozygous lines is a lengthy process. Isozyme analysis of polymorphic loci determined by one of the electrophoretic systems is an alternative method (see Klaas and Friesen, Chapter 8, this volume). Loaiza-Figueroa and Weeden (1991) studied polymorphism in onion by using 12 isozyme systems. Keller and Korzun (1996) reported the use of malate dehydrogenase (MDH), phosphoglucoisomerase (PGI), phosphoglucomutase (PGM) and galactosidase (GAL) systems on putative onion doubled-haploid regenerants, while our group has used an esterase (EST) system (Bohanec *et al.*, 1995; Campion *et al.*, 1995a; Jakše *et al.*, 1996; Bohanec and Jakše, 1999). In our esterase system, donor plants were selected according to the heterozygosity of the analysed locus, and the homozygous plants were excluded. Approximately 50% of the donor plants in the cultivars studied were heterozygous for EST. When such preselected donors are used, only one isozyme system is needed for accurate and reliable analysis for homozygosity of regenerants. It should be noted that esterase is active only in certain developmental stages. In normal plants, high esterase activity is common only in the first young sprouting leaves from bulbs and in flowers, while all tissues used from *in vitro*-grown regenerants express high esterase activity.

In bulb onion, the proportion of haploid regenerants is very high and haploid embryos are in most cases easily distinguished from somatic regenerants by different morphological characters. Therefore, determination of homozygosity is not essential or should be limited only to those regenerants that exhibit diploid chromosome numbers prior to induction of genome doubling. Alternatively, the progeny of fertile lines that underwent spontaneous chromosome doubling can be evaluated for uniformity either by morphological characteristics or by biochemical or molecular markers.

7. Genome-doubling Procedures and Fertility

Data on the percentage of haploids and of spontaneously doubled-haploids are not consistent, and in many cases diploid regenerants were not tested for their homozygous/heterozygous status. In our studies, at least 90% of the regenerants remain haploid. This figure is based on a large number of tested gynogenic plants of different genetic backgrounds. Therefore, it is safe to propose that spontaneous doubling of gynogenic plants is a rare event in the bulb onion. This is in agreement with data on other plant species, such as wheat or durum wheat, where gynogenesis (in contrast to androgenesis) of regenerants results in up to 100% haploids (M. Jäger-Gussen, Vienna, Austria 1996, personal communication; Mdarhri-Alaoui *et al.*, 1998).

The major problem in genome doubling in onion is inaccessibility of the apical meristem of adult field-grown plants. No information is available on successful chromosome doubling in such plants. Hence, chromosome doubling of haploid onion plantlets should be attempted during *in vitro* propagation. The use of colchicine for genome doubling was tested by Campion *et al.* (1995b). Rooted *in vitro*-cultured plantlets were longitudinally halved and the apices were exposed in colchicine-containing media (optimal treatment: 3 days at 10 mg l^{-1} colchicine). Genome doubling was measured by chromosome counting in root-tip and shoot-tip cells. Up to 46% of treated plants were diploid.

Geoffriau *et al.* (1997b) made two longitudinal cuts of micropropagated gynogenic plantlets to produce four slices. Colchicine and oryzalin were applied to the basal part of each quarter (optimal treatment: 24 h at 2.5 mM colchicine or 50 μM oryzalin). The two chemicals were equally effective, resulting in genome doubling at 65.7 and 57.1%, respectively. Following regeneration, however, oryzalin-treated plants produced higher-quality regenerants. Bohanec and Jakše (1997) tested the effect of the same chemicals on halved basal shoots. The treated tissues were placed in colchicine- or oryzalin-containing media for 3 days. The diploidization with oryzalin (10 μM) was more pronounced than that with colchicine (10 mg l^{-1}) resulting in 67% and 21% $2n$ plants, respectively. Higher concentrations of oryzalin had a negative effect on the proliferation of plantlets.

An alternative approach, based on treatment of embryos immediately after regeneration, was studied recently in our laboratory (Jakše and Bohanec, 2001). About 7000 embryos were induced in 1998 and 1999, and thereafter treated with amiprophos-methyl or oryzalin in liquid or on solid medium. Preliminary results indicate that amiprophos-methyl is efficient and less toxic than oryzalin.

8. Genetic Stability of Regenerants

For doubled-haploid lines that are intended for use in hybrid onion breeding, it is essential that only minimal novel genetic variation is induced during the process of haploid induction. It is well known that the use of tissue culture to produce plants from microspores or megaspores frequently results in genetic changes, termed gametoclonal variation (Evans *et al.*, 1984). Breeders try to minimize this gametoclonal variation, since in most cases breeders expect variation to be within the limits of the parent material.

In onion, the occurrence of gametoclonal variation has been tested primarily by the randomly amplified polymorphic DNA (RAPD) technique. In the large genome of onion, any molecular-marker technique covers only a small part of the genome; hence conclusions are based only on the analysis of a limited fragment of the genome studied. When DNA extract was amplified by primer OPA-04, an additional RAPD band was found in two out of 12 gynogenic regenerants (Bohanec *et al.*, 1995). In another study, Campion *et al.* (1995a) found no novel variation (scored by RAPD and α-esterase isozyme) among selfed progeny of a doubled-haploid line. In a more complex study (Javornik *et al.*, 1998), where the first and second cycle of gynogenic onions were compared, a low degree of induced genetic variation was detected in a less responsive line. No variation was detected, however, in a highly responsive line.

The data available so far have led us to conclude that the amount of undesired variation induced during haploid embryos' regeneration is generally low. However, existing studies have involved spontaneously doubled gynogenic lines. It is likely that the use of antimitotic substances, such as colchicine, oryzalin, amiprophos-methyl or others, may result in higher proportions of genetic changes; therefore procedures for genome doubling require further evaluation.

9. The Use of Doubled-haploids in Onion Breeding and Basic Research

The advantages of doubled-haploid breeding versus random-mating populations and other conventional methods were reviewed by Khush and Virmani (1996). The authors pointed out that doubled-haploid populations exhibit more additive variance and no dominance variance. In addition, comparison of different genetic models for selection for specific and general combining ability, shows that doubled-haploid selection is always more efficient than classical methods, even when population size is restricted.

Despite the limitations of the procedure, doubled-haploid lines have already been used in breeding hybrid onion varieties for some years in several seed companies; however, no results have been published from these sources. The use of doubled-haploid lines as parents for F_1 hybrids enabled for the first time the production of highly uniform onion varieties expressing maximal heterotic effect. At the moment, the major limiting factors are genotype-dependent induction frequency and severe inbreeding depression.

Breeding schemes using doubled-haploid lines for the creation of hybrid varieties need to be altered from the established procedures. First, hybrid varieties possessing CMS cytoplasm should not be used as donor plants, since the plants produced would be sterile and could not be selfed and multiplied. Doubled-haploids originating from fertile hybrid varieties can be used only as pollinators (C lines) possessing the restorer (*Ms*) gene. Therefore, lines designed to be used as seed parents (B lines) should be induced from plants possessing normal (N) cytoplasm and maintainer (ms) nuclear genes. Secondly, an appropriate CMS source should be used for the start of back-crossing using the B doubled-haploid line as the recurrent parent. Probably five back-crosses would be needed to obtain an isoline. Such a new breeding procedure would produce a much higher uniformity of hybrids, since present procedures starting with back-crossing to the non-inbred line can only reach a maximal inbreeding coefficient of 0.5 (Fehr, 1993).

One approach to overcoming low rates of gynogenic regeneration is based on genetic improvement. Genes from high-yielding lines can be introduced into low- or non-responsive genotypes. Lines with high induction potential have already been identified; hence this may be considered as a practical approach, even for the long cycle of breeding of the biennial onion. Using lines with high gynogenic potential as donor lines back-crossed to recurrent high-value breeding lines would require haploid induction in each back-cross cycle. However, if responsive lines were not to differ much from desired breeding lines, one or two cycles of back-crossing should be sufficient. The need to test progeny according to gynogenic ability in each cycle can only be overcome by identification of genetic markers indicating the presence of genes needed for gynogenic regeneration.

Another approach calls for alternative induction protocols, which would overcome the low response rates, but there are no indications of a major breakthrough so far.

Inbreeding depression has been a major problem in most cross-pollinated crops. Proposed solutions have mainly been associated with a number of cycles of selection to eliminate deleterious recessive genes. It seems that, in onion, deleterious genes do not affect the vegetative growth of gynogenic plants. It is likely that the haploid regeneration procedure *per se* provides a strong selection pressure against genotypes that possess such genes. Hence, both haploid and doubled-haploid plants are vigorous, and bulbs of such plants are often indistinguishable from those produced by heterozygous plants. However, the expression of inbreeding depression in the generative tissue results in low fertility of doubled-haploid lines. No reports are available as yet, but it seems likely that interpollination among selfed doubled-haploid lines will enable the creation of an improved population, which, following a few gynogenic cycles, will result in improved fertility of doubled-haploid lines formed from such populations, and so eliminate a serious bottleneck in non-selected populations.

Haploid or doubled-haploid plants can serve for purposes other than breeding hybrid cultivars. Genetic analysis of complex traits can be simplified when segregation among doubled-haploid homozygous plants is used instead of segregation in the standard F_2 generation. Such an analysis was performed by Pauls (1996) for seed colour in *Brassica napus*, which is based on a three-gene system involving epistasis and dominance effects. Segregation ratios in F_2s of doubled-haploids and 'normal' F_2 populations were 2 : 5 : 1 and 12 : 51 : 1, respectively. The author also proposed the use of doubled haploids in other genetic studies, such as the analysis of maternally controlled traits or determination of the effects of modifying genes.

During the last decade, doubled-haploid populations have frequently been used as a tool for studying and developing biochemical and molecular markers. The advantage of doubled-haploids in these studies is that, when two populations, one possessing and one not possessing specific characters, are studied, there are no phenotypic intermediates caused by heterozygosity. Young (1994) proposed that recombinant inbred lines derived from individual F_2 plants or in original doubled-haploid lines are better suited for analysis of quantitative traits compared with F_2 or back-cross populations. It should be noted that, starting from the F_2 generation, the selection of inbred lines by the single-seed-descent method takes five to six generations (10–12 years), whereas doubled-haploids are formed directly from gametic cells of F_1 plants.

Numerous studies using this approach have been published during the last 10 years. The majority of such studies have been made on crops for which haploid induction protocols were already well established. Mapping using doubled-haploid populations has thus been performed in barley, wheat, maize, rice, rapeseed and vegetable brassicas, asparagus, pepper and others. This method has been used for the identification of molecular markers for specific individual genes or quantitative-trait loci (QTL) characters and for the construction of genetic maps.

We can conclude that research on the use of doubled haploids in onion has undoubtedly shown substantial progress since the first publications in the late 1980s. Major advances were achieved in the optimization of gynogenic induction procedures, the identification of responsive genotypes and the establishment of genome-doubling protocols. As pointed out in previous sections, it is also clear that current protocols still need considerable improvement. An optimal protocol would be genotype-insensitive and produce large numbers of regenerants, which would efficiently double and exhibit fertility. So far, high induction frequencies can be achieved only in a limited number of genotypes, genome doubling can be problematic and fertility be restored only in a minor percentage of doubled lines. The first two points, induction percentage and genome doubling, are certainly a matter of improved tissue-culture practice. It can be predicted that, with more varied approaches being tested, substantial progress can be expected. For instance, recently improved microspore culture protocols have been applied very efficiently in several species, and this approach has not yet been tested in onion. The third point – low fertility of regenerants – is, however, more a problem of onion populations in general and not specific to the haploid-induction procedure. Inbreeding depression, expressed as low fertility, prevents conventional inbreeding procedures from being done by selfing to exclude deleterious genes. Actually, *in vitro* selection for vigour during emergence and growth of the haploid embryos is already the first step in the elimination of a proportion of the deleterious genes that affect vegetative plant growth. Deleterious traits associated with the generative stage need to be eliminated by recurrent selection for traits associated with fertility, as pointed out above.

Doubled-haploid protocols are therefore already available and it is up to breeders to decide whether they would like to incorporate this technique in their breeding scheme or wait for further improvements. Haploidy also offers several advantages for basic genetic and biotechnological studies, and therefore this technique actually has the potential to be used in a wide range of applications.

References

Bohanec, B. and Jakše, M. (1997) Characteristics of onion haploid induction procedure. In: *Proceedings of the 1st Congress of the Genetics Society of Slovenia, 2–5 September, Ljubljana, Slovenia*, Genetics Society of Slovenia, Ljubljana, Slovenia, pp. 47–49.

Bohanec, B. and Jakše, M. (1999) Variations in gynogenic response among long-day onion (*Allium cepa* L.) accessions. *Plant Cell Reports* 18, 737–742.

Bohanec, B., Jakše, M., Ihan, A. and Javornik, B. (1995) Studies of gynogenesis in onion (*Allium cepa* L.): induction procedures and genetic analysis of regenerants. *Plant Science* 104, 215–224.

Bohanec, B., Jakše, M. and Havey, M.J. (1999) Effects of genotype on onion gynogenesis and attempts of genome doubling at embryo stage – a progress report. In: *Gametic Embryogenesis in Monocots, COST-824 Workshop, 10–13 June 1999*, Jokioinen, Finland, p. 37–38.

Campion, B. and Alloni, C. (1990) Induction of haploid plants in onion (*Allium cepa* L.) by *in vitro* culture of unpollinated ovules. *Plant Cell, Tissue, and Organ Culture* 20, 1–6.

Campion, B., Azzimonti, M.T., Vicini, E., Schiavi, M. and Falavigna, A. (1992) Advances in haploid plant induction in onion (*Allium cepa* L.) through *in vitro* gynogenesis. *Plant Science* 86, 97–104.

Campion, B., Bohanec, B. and Javornik, B. (1995a) Gynogenic lines of onion (*Allium cepa* L.): evidence of their homozygosity. *Theoretical and Applied Genetics* 91, 598–602.

Campion, B., Perri, E., Azzimonti, M.T., Vicini, E. and Schiavi, M. (1995b) Spontaneous and induced chromosome doubling in gynogenic lines of onion (*Allium cepa* L.). *Plant Breeding* 114, 243–246.

Cohat, J. (1994) Obtention chez l'échalote (*Allium cepa* L. var. *aggregatum*) de plantes haploides gynogénétiques par culture *in vitro* de boutons floraux. *Agronomie* 14, 229–304.

Doré, C. and Marie, F. (1993) Production of gynogenic plants of onion (*Allium cepa* L.) after crossing with irradiated pollen. *Plant Breeding* 111, 142–147.

Dunstan, D.I. and Short, K.C. (1977) Improved growth of tissue cultures of onion, *Allium cepa*. *Physiologia Plantarum* 41, 70–72.

Evans, D.A., Sharp, W.R. and Medina-Filho, H.P. (1984) Somaclonal and gametoclonal variation. *American Journal of Botany* 71, 759–774.

Fehr, R.W. (1993) *Principles of Cultivar Development*, Vol. 1. Macmillan, New York, 536 pp.
Gamborg, O.L., Miller, R.A. and Ojima, K. (1968) Nutrient requirements of suspension cultures of soybean root cells. *Experimental Cell Research* 50, 151–158.
Geoffriau, E., Kahane, R. and Rancillac, M. (1997a) Variation of gynogenesis ability in onion (*Allium cepa* L.). *Euphytica* 94, 37–44.
Geoffriau, E., Kahane, R., Bellamy, C. and Rancillac, M. (1997b) Ploidy stability and *in vitro* chromosome doubling in gynogenic clones of onion (*Allium cepa* L.). *Plant Science* 122, 201–208.
Jakše, M. and Bohanec, B. (2001) Studies of alternative approaches for genome doubling in onion. In: Bohanec, B. (ed.) *COST Action 825 – Biotechnological Approaches for Utilization of Gametic Cells – Final Meeting, 1–5 July 2000, Bled, Slovenia*. Luxembourg, pp. 101–104.
Jakše, M., Bohanec, B. and Ihan, A. (1996) Effect of media components on the gynogenic regeneration of onion (*Allium cepa* L.) cultivars and analysis of regenerants. *Plant Cell Reports* 15, 934–938.
Jakše, M., Havey, M.J. and Bohanec, B. (2001) Advances in gynogenic haploid induction procedure in onion. In: Randle, W.M. (ed.) *Alliums 2000, Proceedings of 3rd International Symposium on Edible Alliaceae, University of Georgia, Athens, Georgia, 30 October–3 November 2000*. University of Georgia, Athens, Georgia, pp. 66–69.
Javornik, B., Bohanec, B. and Campion, B. (1998) Studies on the induction of a second cycle gynogenesis in onion (*Allium cepa* L.) and genetic analysis of the plants. *Plant Breeding* 117, 275–278.
Keller, E.R.J. and Korzun, L. (1996) Haploidy in onion (*Allium cepa* L.) and other *Allium* species. In: Jain, S.M., Sopory, S.K. and Veilleux, R.E. (eds) In vitro *Haploid Production in Higher Plants*, Vol. 3. Kluwer Academic Publishers, Dordrecht, The Netherlands, pp. 51–75.
Keller, J. (1990) Culture of unpollinated ovules, ovaries, and flower buds in some species of the genus *Allium* and haploid induction via gynogenesis in onion (*Allium cepa* L.). *Euphytica* 47, 241–247.
Khush, G.S. and Virmani, S.S. (1996) Haploids in plant breeding. In: Jain, S.M., Sopory, S.K. and Veilleux, R.E. (eds) In vitro *Haploid Production in Higher Plants*, Vol. 1. Kluwer Academic Publishers, Dordrecht, The Netherlands, pp. 11–33.
Klein, M. and Korzonek, D. (1999) Flower size and developmental stage of *Allium cepa* L. umbels. *Acta Biologica Cracoviensia Series Botanica* 41, 185–192.
Loaiza-Figueroa, F. and Weeden, N. (1991) Effects of seed increase procedures on isozyme polymorphism in *Allium*. *FAO/IBPGR Plant Genetic Resources Newsletter* 83/84, 1–3.
Luthar, Z. and Bohanec, B. (1999) Induction of direct somatic organogenesis in onion (*Allium cepa* L.) using a two-step flower or ovary culture. *Plant Cell Reports* 18, 797–802.
Martinez, L.E., Agüero, C.B., López, M.E. and Galmarini, C.R. (2000) Improvement of *in vitro* gynogenesis induction in onion (*Allium cepa* L.) using polyamines. *Plant Science* 156, 221–226.
Mdarhri-Alaoui, M., Saidi, N., Chlyah, A. and Chlyah, H. (1998) Obtention par gynogenèse *in vitro* de plantes haploides chlorophyliennes chez le blé dur. *Comptes Rendus de l'Académie des Sciences, III, Sciences de la Vie* 321, 25–30.
Michalik, B., Adamus, A. and Nowak, E. (2000) Gynogenesis in Polish onion cultivars. *Journal of Plant Physiology* 156, 211–216.
Murashige, T. and Skoog, F. (1962) A revised medium for rapid growth and bio assays with tobacco tissue cultures. *Physiologia Plantarum* 15, 473–497.
Muren, R. (1989) Haploid plant induction from unpollinated ovaries in onion. *HortScience* 24, 833–834.
Musial, K., Bohanec, B. and Przywara, L. (1999) Embryological analysis of *in vitro* cultured unpollinated ovules of *Allium cepa* L. In: *IX International Conference of Plant Embryologists, 20–22 September 1999, Crakow, Poland*, Polish Academy of Sciences, Crakow, poster abstract, p. 51.
Musial, K., Bohanec, B. and Przywara, L. (2001) Embryological study on gynogenesis in onion (*Allium cepa* L.). *Sexual Plant Reproduction* 13, 335–341.
Pauls, K.P. (1996) The utility of doubled haploid populations for studying the genetic control of traits determined by recessive alleles. In: Jain, S.M., Sopory, S.K. and Veilleux, R.E. (eds) In vitro *Haploid Production in Higher Plants*, Vol. 1. Kluwer Academic Publishers, Dordrecht, The Netherlands, pp. 125–144.
Puddephat, I.J., Robinson, H.T., Smith, B.M. and Lynn, J. (1999) Influence of stock plant pretreatment on gynogenic embryo induction from flower buds of onion. *Plant Cell, Tissue, and Organ Culture* 57, 145–148.
Vinterhalter, D.V. and Vinterhalter, B.S. (1999) Hormone-like effects of sucrose in plant *in vitro* cultures. *Phyton (Austria) Special Issue 'Plant Physiology'* 39, 57–60.
Young, N.D. (1994) Constructing a plant genetic linkage map with DNA markers. In: Phillips, R.L. and Vasil, I.K. (eds) *DNA-based Markers in Plants*. Kluwer Academic Publishers, Dordrecht, The Netherlands, pp. 39–57.

8 Molecular Markers in *Allium*

M. Klaas[1] and N. Friesen[2]
[1]*Gotthard Müller Straβe 57, D-70794 Filderstadt-Bernhausen, Germany;* [2]*Botanical Garden of the University of Osnabrück, Albrechtstrasse 29, D-49076 Osnabrück, Germany*

1. Introduction: Why Molecular Markers?

Evolution is a two-phase process, in which genetic variability accumulates in a random fashion, after which morphological, biochemical or physiological changes are induced and stabilized by environmental pressure or the plant breeder's efforts (Mayr, 1969). While the evolutionist's interest lies primarily with investigating the forces directing the second set of processes, molecular markers can be used to sample the underlying genetic variability when it is not directly being subjected to the action of evolutionary pressures. This pool of information gives the opportunity to reconstruct the course of the evolutionary process while avoiding the use of phenotypic markers, such as morphological or anatomical features, which may be influenced by the very mechanisms which they are required to elucidate. Consequently, there is an increased utilization of molecular markers in evolutionary and systematic studies. However, for efficiency reasons, the use of molecular markers in these studies depends on pre-existing data, such as taxonomic classification. Increased standardization of the techniques and availability of equipment and expertise have also promoted the application of molecular technologies for other purposes, such as for quick analysis of cytoplasm types, the verification of hybrid plants and extensive use in the construction of genetic maps. Concurrently, the recent years since Klaas (1998) reviewed the topic have seen continued and widespread application of molecular markers in *Allium* research, including the use of the most technically advanced procedures such as comparative sequencing of DNA markers and the development of amplified fragment length polymorphisms (AFLPs) and microsatellites. While the choice of a particular technique depends on both the biological question to be answered and the available laboratory equipment and expertise, the quality of data and their suitability for a particular study must be judged by the same rigorous standards. This review aims to give an overview of molecular-marker applications in *Allium*, and some judgements about the validity of the approaches taken. We hope that our reflections will be of some help in planning future studies, since there is usually no one single most useful technique. Choices have to be made based on the weighting of different markers' strengths and drawbacks and on the practical options available in a particular laboratory.

For more detailed information, several excellent monographs have been published, which include laboratory protocols of the techniques as well as broader topics, such as laboratory set-up and sampling strategies (Zimmer *et al.*, 1993) and the theory, scope and limits of applications (Hillis *et al.*, 1996). For concise descriptions of selected procedures, see Hoelzel (1992).

2. Markers

2.1 Isozymes

While isozyme analysis was historically the first application of molecular markers in *Allium*, it still holds some advantages today over the now more widely employed DNA markers in certain applications. For general introductions to the techniques, see May (1992) and Murphy *et al.* (1996). In particular, allozyme analysis, which detects polypeptide variants corresponding to different alleles at one locus, is a very cost-effective means of obtaining Mendelian molecular markers in a short time for large numbers of individuals. The isozyme investigation of large *Allium* collections (*A. sativum*: 300 accessions, Maaß and Klaas, 1995; 110 accessions, Pooler and Simon, 1993; *A. cepa* var. *ascalonicum*: 189 accessions, Arifin and Okubo, 1996; *A. cepa* and *A. fistulosum*: 188 accessions and 29 accessions, respectively, Peffley and Orozco-Castillo, 1987) and a larger study in *A. douglasii* (29 populations each with 30–60 analysed individuals, Rieseberg *et al.*, 1987), with numbers of accessions not yet paralleled in studies based on DNA markers, testify to the strength of the approach.

Briefly, plant tissue is squashed in a suitable buffer that preserves enzyme activity; this solution is applied to a starch gel and electrophoretically separated. Thereafter, protein bands with enzymatic activity are revealed by specific staining reactions. Changes of peptide amino acid sequence which result in altered electrophoretic mobility due to charge, size or conformation differences can be detected. Following separation, horizontal slicing of the gel allows for the scoring of up to three different enzyme systems, using separate staining reactions.

The genetic structures of the major enzyme systems are well characterized (Wendel and Weeden, 1989), so a thorough interpretation of the banding patterns yields Mendelian data that have been shown to correspond well with DNA-marker results (restriction fragment length polymorphism (RFLP): Chase *et al.*, 1991; randomly amplified polymorphic DNA (RAPD): Maaß and Klaas, 1995).

The major limitation of isozyme analysis is the small number (15 or less) of suitable enzyme systems, of which usually only a subset will exhibit sufficient variability. The suitability of isozyme markers for an intended plant study therefore has to be tested in advance. More isozyme alleles can be resolved by technological refinements, such as different gel-matrix pore sizes or different buffer systems, or by differential heat-stability tests (Murphy *et al.*, 1996), but then the advantage of simple and quick application is often lost. Since fresh plant material is preferable, isozyme analysis is still a good choice if a limited number of high-quality data points are needed for a large number of individuals in a population study or for the genetic characterization of larger living collections of crops. At taxonomic levels higher than species or close species complexes, the assignment of observed bands to homologous loci based on electrophoretic mobility seems dubious.

2.2 DNA markers

The use of DNA-based markers avoids detection problems due to uneven expression, which has been a major problem in developing additional isozyme systems. It also allows for the development of basically unlimited numbers of markers, and it enables prolonged storage of samples for later analysis, either as frozen tissue or even as dried material kept at room temperature.

2.2.1 RFLP

Apart from some earlier experiments on direct hybridization of DNAs from different taxa, yielding distance type of data (Werman *et al.*, 1996), the RFLP technique brought the first opportunity for DNA-based molecular markers. Purified DNA is cut by a restriction endonuclease at specific recognition sites, and then the digested DNA is electrophoretically separated according to size. RFLPs detect nucleotide substitution, which results in loss or gain of a recognition site, or insertions/deletions, which lengthen/shorten a specific fragment.

The direct visualization of separated restriction fragments is possible from digested purified chloroplast DNA (cpDNA) (Linne von Berg *et al.*, 1996, in a first DNA-based phylogeny of the genus *Allium*). The approach allows scoring of numerous bands from one gel, with virtually no possibility for contamination to influence the results, but it has been rarely applied, since it depends on the isolation of chloroplasts from fresh leaves prior to DNA extraction. More common is the transfer of the restricted and separated DNA fragments to a membrane, followed by hybridization with a specific labelled DNA probe. The detection of labelled bands on the membrane is much more sensitive than direct visualization, and can be extended by prolonged exposure. The hybridization with specific probes is a precondition for analysing nuclear DNA changes by RFLPs. The hybridization signal is also an indicator of overall sequence similarity, an important information aspect that is missing from polymerase chain reaction (PCR)-generated markers.

While a lot of high-quality data have been generated by RFLPs, their use was replaced to some extent by PCR after wider realization of the potential of this approach. RFLPs require larger quantities of relatively high-quality DNA, which has to be highly purified, since the restriction endonucleases are generally more sensitive to small impurities in the target DNA than the Taq DNA polymerase working at higher temperatures, and today the very same type of data can be generated faster by PCR.

2.2.2 PCR-based techniques

CAPS. Cleaved amplified polymorphic sequences (CAPS) simplify the gathering of RFLP data, avoid the complicated blotting/hybridization procedures of traditional Southern blots and require only small amounts of total genomic DNA. DNA regions known or suspected to contain polymorphisms are amplified from genomic DNA by specific PCR primers, followed by restriction analysis of the purified PCR products. Although the technique is expensive in evolutionary or genetic-diversity studies, because of primer costs, large data sets can be generated in a short time (Mes *et al.*, 1998, 1999; Friesen *et al.*, 1999, evolution within genus *Allium*), and it is an economical substitute for Southern blots in repeated tests for the presence of known polymorphisms. For example, it can be used as an indicator for the presence of certain cytoplasm types (Havey, 1995, identification of *A. cepa* cytoplasms; Dubouzet *et al.*, 1998, verification of *A. giganteum* hybrids).

RAPD. RAPD analysis requires no prior knowledge of the genome investigated and can thus be readily applied to different species (Williams *et al.*, 1990). It has also been applied in numerous *Allium* investigations. A PCR reaction is carried out using a single primer, usually of ten bases, and with purified total genomic DNA as the target. Sequences between primer binding sites within a suitable distance, generally less than 2 kilobases (kb), are amplified and scored for size differences after electrophoresis. Somewhat pointedly, the technique has been likened to practising PCR without a clue (see also Wolfe and Liston, 1998, for a general discussion of the technique), referring both to the lack of any pre-experiment sequence information about the target DNA and its use by many practitioners who are oblivious to the limitations of the approach. Nevertheless, RAPD analysis offers a quick and comparatively cheap approach for the detection of small genetic differences, since a larger proportion of the genome can be sampled than with other techniques. To avoid the shortfalls of RAPD analysis, such as low reproducibility of some bands and the uncertain homology of fragments comigrating in gel electrophoresis, rigorous laboratory standards are required. All reactions should be repeated, and all reactions should be analysed on the same gel for a reliable scoring of presence and absence of bands (Friesen and Klaas, 1998; Wolfe and Liston, 1998). Impurities in the genomic DNA may prevent the reproduction of some bands, and the banding pattern is reproducible only within a specific range of DNA concentrations. This therefore requires the determination of the DNA concentration either fluorometrically or by titration in several PCR reactions. However, with new DNA-isolation kits (such as the Qiagen DNeasy kit or the Macherey-Nagel DNA Plant Nucleospin kit), these problems are easily overcome, avoiding expensive procedures, such as CsCl density-gradient purification of DNA.

For the guaranteed reproduction of specific bands – for example, if linkage to genes of interest is assumed – a RAPD band can be transformed into a sequence-characterized amplified region (SCAR) band (Paran and Michelmore, 1993). The RAPD band is cloned and adjacent bases from genomic sequence are added to the RAPD sequence in order to obtain a PCR primer that should bind only at one locus in the genome.

For an investigation of genetic diversity, in our opinion at least three scored RAPD bands per taxon are required. More than three or four usually do not add more substantial information, due to the inherent noise in the data. Preferably these bands are scored from several primers, since, if more than about ten bands are scored per reaction, less reliable bands have to be included. In this case, due to differences of base composition within a genome, the genome may not be sampled homogeneously.

Mendelian inheritance of RAPD markers chosen for analysis of genetic distances has been demanded (Bradeen and Havey, 1995; Rieseberg, 1996) but might not be feasible for a project of limited size, especially if only initial information is being gathered about a little-known group. Similarly, homology of RAPD bands has been tested by hybridization (in interspecific applications: Inai *et al.*, 1993; Yamagishi, 1995; Lannér *et al.*, 1996), but this is feasible only for a small number of bands of special interest. Other approaches have been used to increase the data value from RAPD reactions – for example, by evaluating differences in band intensity (Demeke *et al.*, 1992) and/or by using more primers of different composition. If more accessions have to be analysed than can fit on to one gel (usually not more than 40–50), other methods should be used, such as hybridization of dot blots with RAPD probes (*Allium* subgenus *Rhizirideum*: Dubouzet *et al.*, 1997), AFLPs (in *Allium*: Smilde *et al.*, 1999) or microsatellites (in *A. cepa*: Fischer and Bachmann, 1998). At a higher taxonomic level, CAPS approaches (e.g. genus *Allium*: Mes *et al.*, 1998; Friesen *et al.*, 1999) are preferable to an extension of the RAPD approach, due to its limitations. While RAPDs are, with all the necessary repetitions and optimizations, not such a cheap procedure as they have sometimes been portrayed, they can still generate, in any laboratory with standard equipment, informative and reproducible data for a medium-sized study of closely related species or populations. RAPDs have been successfully used in investigations of infraspecific variation and the differentiation of close species, but the interpretation of the data depends on the assumption that amplification products of equal size are homologues. If the relationship between the taxa within a study is not well known, this assumption of band homology is generally hazardous without further tests.

MICROSATELLITES. For the development of microsatellite markers, genomic DNA is fragmented and cloned. From this material, clones can be detected by hybridization which carry highly repetitive sequences of two or three base-pair unit length. After sequencing, flanking non-repetitive sequences can be determined: these are used for the generation of PCR primers. They amplify the repetitive region. Changes in copy number of the repeat can be detected as length variations of the PCR product (Gupta and Varshney, 2000). The mutation of repeat copy numbers occurs at a rate several orders of magnitude higher than nucleotide substitution (Aquadro, 1997), and is therefore useful at the level of population studies. Even with enrichment procedures available that facilitate the detection of suitable repetitive DNA clones (Edwards *et al.*, 1996; for *A. cepa*: Fischer and Bachmann, 1998), the generation of microsatellites is still cumbersome. It is justified only for long-term projects with crops of economic importance – for example, for finding markers linked to genes of special interest – or to contribute towards the construction of a genetic map. The running costs are higher than those of other markers, since the alleles can only be separated on gels prepared from expensive high-resolution agaroses, such as Metaphor™, or, preferably, on polyacrylamide sequencing gels. A specific advantage is the detection of allelic variants at the same locus (as with isozyme analysis), but with virtually unlimited numbers of markers, limited only by material and manpower. In crops other than *Allium*, large data sets for diversity investigation have been generated by microsatellites (e.g. Chavarriaga-Aguirre *et al.*, 1999, also with comparisons with other techniques), but the usefulness of the technique is clearly limited by the availability of funds.

AFLPS. AFLPs basically transform RFLP-type data into PCR-generated markers (Vos *et al.*, 1995). Restriction-enzyme recognition sites are extended by adapter sequences, rendering a PCR reaction on genomic DNA templates less ubiquitous, so that a discrete number of amplification products are generated. These products are separated on sequencing-type gels, and the pattern is detected via labelled primers (radioactively or with a fluorescent label) or directly via silver staining of unlabelled PCR products. While the procedure is technically demanding, advance preparation of the markers is not required. As with other PCR markers, virtually unlimited numbers of markers can be generated in a short time by using different primer extensions flanking the restriction-site core. The technique has been successfully applied to the generation of molecular mapping data and for the generation of nuclear DNA markers for relationship and hybrid analysis. Once the technique is established in a laboratory, the generation of large data sets is straightforward, as the applications in *Allium* testify (Smilde *et al.*, 1999; van Raamsdonk *et al.*, 2000). Compared with microsatellites, AFLP's strength is in gathering large numbers of data points for smaller numbers of taxa – for example, in mapping experiments. Microsatellites should yield allelic markers with higher certainty across a larger number of investigated accessions, since the length and nucleotide sequence of both primer sequences of one microsatellite marker add to the specificity of the amplified locus.

DNA FINGERPRINTING. DNA fingerprinting has earlier been called the hybridization of a genomic DNA blot with labelled microsatellite sequences, which makes corresponding target sequences visible throughout the genome (see Bruford *et al.*, 1992, for an overview). The technique has been tried on *A. cepa* without success (Sharon *et al.*, 1995; M. Klaas, unpublished results), even though the presence of corresponding sequences in

A. cepa has been demonstrated by the successful generation of microsatellite markers (Fischer and Bachmann, 1998). Possibly the large genome size of *Allium* makes direct visualization of microsatellite sequences by hybridization under standard conditions, as tested in other plant groups, difficult. The term fingerprinting has subsequently been used in the literature for various PCR-based techniques, such as RAPD, microsatellite analysis (termed single-locus fingerprinting by Bruford *et al.*, 1992) and even isozyme application. Today, therefore, it stands more for the purpose, rather than for a specific technique, of characterizing a genome down to the level of a cultivar, since identification of individuals is usually not an issue in plant science. In *Allium*, cultivar or line identities have been checked by RAPD analysis (*A. cepa*: Campion *et al.*, 1995; Havey, 1995; *A. sativum*: Bradley *et al.*, 1996; Al-Zahim *et al.*, 1999, in tests for somaclonal variation; triploid onion: Puizina *et al.*, 1999). However, as discussed below, the genetic differences between even recognized botanical varieties might be too small to be detected by these general approaches. RFLPs with nuclear probes were also successfully used to distinguish *A. cepa* commercial inbreds (King *et al.*, 1998b). While the approach required considerable experimental effort, a high resolution was achieved by use of 69 anonymous complementary DNA (cDNA) probes and an alliinase clone.

COMPARATIVE DNA SEQUENCING. Potentially the most informative but also the most laborious marker technique is comparative DNA sequencing of specific loci, which has been greatly facilitated by use of PCR techniques. It has been applied in a number of studies on the molecular evolution of *Allium*. The technique is restricted to phylogenetic applications at the section level and above, and will be dealt with in Section 3.3. The comparison between nuclear DNA markers and chloroplast markers, in particular, allows insights into reticulate evolution and hybridogenic speciation common in *Allium*.

The third genome – mitochondrial DNA (mtDNA) – has not been used as a marker for molecular evolution. In plants, the nucleotide substitution is much slower than in animals, where mtDNA has often been applied to molecular evolution studies. In plants, mtDNA is prone to frequent rearrangements, which makes interpretation of data difficult. Since mtDNA is implicated in cytoplasmic male sterility (CMS) systems, RFLP-based detection systems have been developed to distinguish between different mtDNA types (*A. ampeloprasum*: Kik *et al.*, 1997; *A. cepa*/*A. ampeloprasum*: Buiteveld *et al.*, 1998; *A. schoenoprasum*: Engelke and Tatlioglu, 2000).

3. Applications in *Allium* Research

3.1 Phylogeny/taxonomy

During the late 1980s and the 1990s, molecular phylogenetics has dramatically reshaped our views of the relationships between organisms and of their evolution. Numerous DNA regions representing the nuclear and chloroplast genomes are now routinely used for phylogenetic inference for plants. Revised concepts of relationships based on phylogenetic analyses are resulting in revised classification in many groups of plants (Soltis and Soltis, 2000).

3.1.1 The genus Allium *and its subdivisions*

The position of the genus within the *Alliaceae* was investigated by Fay and Chase (1996), through a phylogenetic analysis of plastid DNA sequences coding for the large subunit of ribulose-1,5-biphosphate carboxylase (rbcL). This data set, comprising 52 species, also included sequences of *A. subhirsutum*, *A. altaicum* and *Nectaroscordum siculum*. According to Fay and Chase (1996), *N. siculum* should be included in the genus *Allium*, and *Milula spicata*, a rare Central Himalayan–south-eastern Tibetan endemic species, is the closest relative to the genus *Allium*. Its status has recently been revised, also by other molecular markers (Friesen *et al.*, 2000, discussed below).

A first approach to structuring the genus *Allium* itself by molecular markers was published by Linne von Berg *et al.* (1996). From

48 species representing the major subgenera, plastid DNA was isolated and digested with restriction enzymes and the fragment patterns were analysed phenetically, i.e. the presence/absence of each fragment is counted as an independent character of equal weight, contributing to an overall measure of genetic similarity based on the shared proportion of fragments. The major subgenera were identified as clusters in the UPGMA (unweighted pair-group method using arithmetic averages) dendrogram, with the notable exception that species of subgenera *Amerallium* and *Bromatorrhiza* were joined in a loosely associated cluster. RFLP experiments with heterologous plastid DNA probes were applied to investigate more closely the interrelationship of the *Amerallium–Bromatorrhiza* complex (Samoylov *et al.*, 1995, 1999). The subgenus *Bromatorrhiza*, originally circumscribed by Ekberg (1969) by the occurrence of fleshy roots as storage organs and the lack of true storage bulbs or rhizomes, again proved to be polyphyletic and is now partly integrated into the subgenus *Amerallium* (all species with x = 7) and partly included into subgenus *Rhizirideum* (species with x = 8). The distribution of *Amerallium* species between Old World and New World habitats was well reflected in the phylogenetic data, which was also supported by internal transcribed spacer (ITS) sequence analysis (Dubouzet and Shinoda, 1999).

Ohri *et al.* (1998) undertook a survey of the nuclear DNA content (2C values) in 86 species of all subgenera of the genus *Allium*. However, contrary to some earlier assumptions, little indication of phylogenetic information was found in these data; significant loss or gain of DNA amounts per genome was observed, and the 2C values seemed to be related more to ecological factors than to systematic affiliation. Some generalizations, such as a larger or smaller DNA content in certain subgenera, were possible, but there were no distinct discontinuities defining certain groups. In an earlier limited study of 25 *Allium* species, Jones and Rees (1968) had already found considerable differences between 2C values, but they did not attempt to investigate the possible correlation of DNA loss or gain with phylogeny, since at that time the taxonomy of *Allium* was little understood.

Mes *et al.* (1998) included 29 species of *Allium* and seven species of related genera in a phylogenetic study using RFLP data from PCR-amplified cpDNA. In a cladistic analysis, the large subgenera *Rhizirideum* and *Allium*, which had remained largely intact in the phenetic analysis of RFLP bands (Linne von Berg *et al.*, 1996), proved to be polyphyletic, and *N. siculum* was clearly placed in the genus *Allium*. Some deviating sections are affiliated to other groups: the subgenus *Rhizirideum* sect. *Anguinum* with *A. tricoccum* and *A. victorialis* and sect. *Butomissa* with *A. tuberosum* are now associated with subgenus *Melanocrommyum*, and the two subgenus *Allium* sections *Allium* and *Scorodon* are separated by several *Rhizirideum* sections. Earlier, unification or separation of taxa was based on morphological traits, thus leading to mistaken classifications. Hence, using molecular markers, Mes *et al.* (1999) confirmed the artificial nature of subgenera *Rhizirideum*, *Bromatorrhiza* and *Allium*. Some sections in the monophyletic subgenus *Melacrommyum* are also artificial. The subgenus *Bromatorrhiza* is subdivided between the x = 7 and the x = 8 species, in agreement with the earlier studies (Samoylov *et al.*, 1995, 1999; Linne von Berg *et al.*, 1996). In these studies the taxonomy at the level of sections remains more or less intact, but the affiliation of some deviating groups to larger-order structures is changed by the cladistic analysis of molecular markers. The phenetic analysis of RFLP data for a UPGMA clustering (Linne von Berg *et al.*, 1996) gave less reliable grouping at the level of subgenera. Their approach could also lead to the inclusion of misleading data, since bands of the same size were treated as homologues without verification by probe hybridization.

Dubouzet *et al.* (1997) proposed a first phylogeny of subgenus *Rhizirideum* based on nuclear DNA markers. Dot blots with genomic DNA of 44 species were successively hybridized with 55 RAPD fragments. These probes were isolated from separate PCR reactions from 11 *Rhizirideum* species and 11 RAPD primers. Most probes hybridized to sets of species from several sections, resulting in continuous rather than binomial signal distribution. The intensity of the hybridization signals was determined densitometrically and transformed into a distance matrix. The resulting UPGMA dendrogram largely confirmed the taxonomy as detailed in Hanelt *et al.* (1992). Unfortunately, no species from outside subgenus *Rhizirideum* were included, which might have been needed in order to reliably structure the subgenus itself, which is not a monophyletic group (Mes *et al.*, 1998, 1999). The approach of Dubouzet *et al.* (1997) avoids problems of band homology, as in standard RAPD experiments or RFLPs without hybridization, but the analysis is restricted to distance methods.

Other recent publications on molecular taxonomy (Dubouzet and Shinoda, 1998, subgenus *Melanocrommyum*; van Raamsdonk *et al.*, 2000, subgenus *Rhizirideum*) have the same shortcomings: no species from outside the studied subgenera were included. We regard this as crucial for the adequate positioning of taxa from polyphyletic groups. Another very important aspect in a molecular taxonomy study is the origin and quality of the studied plants. Often researchers collect seeds from botanical gardens, seed companies or other sources and use it without further checks. In the experience of the Gatersleben taxonomic group, about 50% of such material is incorrectly determined or has hybridogenic origins. Our experience indicates that species from genus *Allium* in particular are frequently hybridized in collections and are often wrongly named (N. Friesen, personal observation).

In an ongoing investigation of the phylogeny of *Allium* using molecular markers, we searched for a suitable outgroup taxon as close as possible to but outside the ingroup being studied, to be a part of the cladistic analysis. The results of Fay and Chase (1996) and the general morphological similarity indicated that *Milula* should be the appropriate candidate for this purpose. Phylogenetic relationships between *Allium* and the monotypic Himalayan genus *Milula* were analysed using sequences of the nuclear ribosomal DNA (rDNA) ITS region and of the intergenic spacers from the chloroplast *trn*D(GUC)–*trn*T(GGU) region (Friesen *et al.*, 2000). The comparison of ITS data with the independent cpDNA data set unambiguously placed *M. spicata* within *Allium* subgenus *Rhizirideum*, close to *A. cyathophorum*. Two major clades were found in *Allium* based on both data sets: subgenera *Nectaroscordum* (x = 9) and *Amerallium* (x = 7) on one side, with subgenera *Caloscordum*, *Rhizirideum* and *Milula* (all x = 8) on the other side. This result supports the division of *Allium* into two large groups, as suggested by earlier cpDNA analyses (Linne von Berg *et al.*, 1996; Mes *et al.*, 1999), and the breaking up of subgenus *Bromatorrhiza*, which appears to be an artificial taxon (Samoylov *et al.*, 1995, 1999). Only two small differences in the positions of *A. kingdonii* and *A. insubricum* between the analyses based on nuclear DNA and cpDNA data were found. Hybridization events or sample errors could explain the different positioning of these taxa. To resolve these conflicts a much larger sample of species would have to be analysed to avoid errors introduced by taxon selection.

3.1.2 Comparison of cpDNA and nuclear DNA

To study the relationships in the entire genus *Allium*, the ITS region of nuclear rDNA was sequenced from 216 samples that represented 195 *Allium* species, two species of *Nothoscordum*, and one species of each of the genera *Milula*, *Ipheion*, *Dichelostemma* and *Tulbaghia* (N. Friesen, M. Klaas and F.R. Blattner, unpublished data). The subgenera *Rhizirideum* and *Allium*, which are not monophyletic according to the cpDNA analysis, were represented by 162 accessions, and representatives of each section of the subgenera *Amerallium*, *Caloscordum*, *Nectaroscordum* and *Melanocrommyum* were also included. In all cases where species were placed in an unexpected position in the preliminary phylogenetic tree, we analysed more accessions from the particular species to avoid errors from taxon selection. Within the 195 *Allium* species, the lengths of the ITS regions are in a range from 612 bp in *A. cyathophorum* to 661 bp in *A. triquetrum*. Pairwise genetic distances (Kimura, 1980) are between 1% Kimura distance (between species from one section) to 53% (*A. haneltii* (subgen. *Allium*) to *A. bulgaricum* (subgen. *Nectaroscordum*)). These are unusually large intrageneric genetic distances for the ITS data within *Allium*: Kimura distances above 40% often characterize the most distant genera within subfamilies or even families. Intrageneric distances in other plant families are mostly less than 10%. These findings make *Allium* of very ancient origin, and molecular evolution has not been accompanied by the rise of comparable numbers of taxonomic categories.

A phylogenetic analysis of ITS sequences (Fig. 1.1 in Fritsch and Friesen, Chapter 1, this volume) supported a monophyletic origin of most circumscribed sections, with some exceptions (the morphologically variable sections *Reticulato-Bulbosa*, *Oreiprason* and *Scorodon* are polyphyletic) and a polyphyletic origin of subgenera *Rhizirideum* and *Allium*. Subgenera *Rhizirideum* and *Allium* are subdivided into six monophyletic groups which have different relationships: section *Anguinum* is a sister group of subgenus *Melanocrommyum*; sect. *Butomissa* (including some species from sect. *Reticulato-Bulbosa*) is a sister group to all other sections of subgenus *Rhizirideum* and *Allium*; sect. *Rhizirideum*, *Caespitosoprason*, *Tenuissima* and *A. eduardii* (sect. *Reticulato-Bulbosa*) are sister groups to all the other sections of subgenus *Rhizirideum* and *Allium*; most species from subgenus *Allium* form a monophyletic clade, excluding species from sect. *Scorodon sensu stricto* and *A. turkestanicum*.

In parallel to the beginning of our ITS project described above, a set of cpDNA sequences was gathered in order to detect possible differences between the two data sets, indicative of reticulate evolution events not detected by a single marker. The non-coding rbcL–atpB spacer from cpDNA was amplified and sequenced from 60 accessions belonging to 50 species of genus *Allium* and one species each of the genera *Nothoscordum*, *Tulbaghia* and *Bloomeria* (Fig. 8.1; see Table 8.1 for European Molecular Biology Laboratory (EMBL)* sequence accession numbers and information on the plant accessions, original data not yet published). The resolution of the *rbcL–atpB* marker overlaps to some extent with that of the ITS, but is generally more useful at a somewhat higher taxonomic level – below section level too few substitutions are found – on the other hand, inclusion into the alignment of related species outside *Allium* is still possible, due to conserved portions of the region (see Soltis and Soltis, 1998, for a discussion of different sequencing markers in phylogenetic studies). Three major clades were found in *Allium* based on these sequences: subgenera *Nectaroscordum* (x = 9) and *Amerallium* (x = 7, including species from former subgenus *Bromatorrhiza*); subgenera *Caloscordum*, *Melanocrommyum* and section *Anguinum* (all x = 8); and subgenera *Rhizirideum* (including species from former subgenus *Bromatorrhiza*) and *Allium* (all x = 8). The cpDNA data largely agree with the phylogeny based on the nuclear ITS sequences. Remarkably, in section *Cepa*, the species *oschaninii* and *pskemense* are not included and are more distant than in the ITS tree, indicating a different evolutionary origin of their nuclear and cytoplasmic DNA. The species of subgenus *Allium* are divided into three groups, compared with their monophyletic appearance in the ITS tree of these taxa (see Fig. 1.1, Fritsch and Friesen, Chapter 1, this volume). In theory, all phylogenies based on chloroplast markers should yield the same tree; however, the *rbcL/atpB* intergene sequence data provide a far better taxonomic resolution at the genus level compared with the CAPS-based analysis (Fig. 8.2; Mes *et al.*, 1999). At higher taxonomic levels of 'old genera' like *Allium*, restriction data such as those generated by CAPS apparently include increasingly homoeological characters, resulting in

*European Molecular Biology Laboratory (EMBL), an international network of research institutes funded by 15 countries, is dedicated to research in molecular biology. Apart from computational services such as analysing data sets and providing DNA-analysis software, one of its main goals is to establish a central computer database of DNA sequences as a resource for the scientific community.

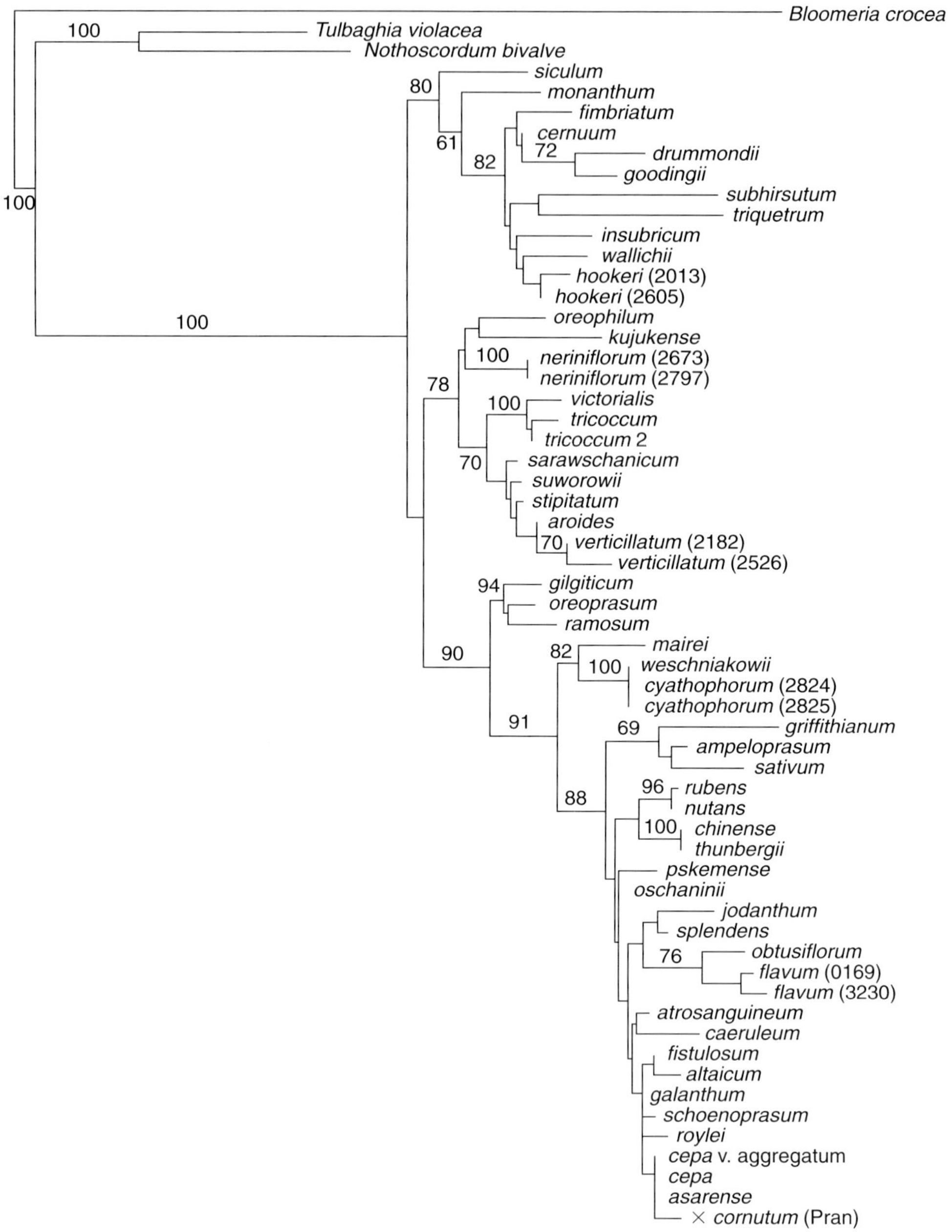

Fig. 8.1. Strict consensus tree of maximum parsimony analysis of the *rcbL–atpB* intergenic region, based on 388 trees. Consistency index (CI) 0.760; retention index (RI) 0.790. The non-coding region between the *rcbL* and *atpB* chloroplast genes was amplified from conserved sequences within the genes, as described by Savolainen *et al.* (1994), and manually sequenced in both directions. Intervening primers were synthesized based on *Allium*-specific sequences. From the sequence files, *c.* 880 bp per species, an alignment of *c.* 1100 bp was constructed with the ClustalW program (Thompson *et al.*, 1994), and this was used for a cladistic analysis of the data with PAUP 3.1 (Swofford, 1993). The figures above the branches indicate bootstrap values; only figures > 50 are given. bp, base pairs.

Table 8.1. The origin and taxonomy of the investigated accessions of the genus *Allium*.

Subgenus, Section	Species	2*n*	TAX	Origin	EMBL
Allium					
Allium	*ampeloprasum* L.	48	1025	Zugdidi, Caucasus, Georgia	AJ299086
	sativum L.	16	1319	North Tajikistan	AJ299088
Avulsea	*griffithianum* Boiss.	16	3660	Zaravshan Mts, Takhta-Karachi Pass, Uzbekistan	AJ299128
Caerulea	*caeruleum* Pall.	16	1525	BG Moscow, Russia	AJ299141
Codonoprasum	*flavum* L.	16	3230	BG Linz, Austria	AJ299120
	flavum L.	16	0169	Dizderica, Croatia	AJ299091
	obtusiflorum DC	16	3101	Piserra dello Zingaro-Scopollo, Italy	AJ299119
Amerallium					
Amerallium	*drummondii* Regel	14	0200	BG Uppsala, Sweden	AJ299144
Briseis	*triquetrum* L.	16	0933	BG Liege, Belgium	AJ299137
Bromatorrhiza	*hookeri* Thwaites	22	2013	Kunming, China	AJ299095
	hookeri Thwaites	22	2506	SW Lijiang, China	AJ299105
	wallichii Kunth	14	2441	BG Gatersleben, Germany	AJ299104
Caulorhizideum	*goodingii* Ownbey	14	3471	Arizona, USA	AJ299124
Lophioprason	*cernuum* Roth	14	0497	BG Strasbourg, France	AJ299133
Microscordum	*monanthum* Maxim.	32	5457	Vladivostok, Russia	AJ299135
Molium	*subhirsutum* L.	14	0023	Adiacenze di Petralia, Italy	AJ299103
Narcissoprason	*insubricum* Boiss. et Reut.	14	0230	BG Marburg, Germany	AJ299101
Rhopetoprason	*fimbriatum* Wats. v. *purdyi* Eastw.	14	3487	Lake County, California, USA	AJ299146
Caloscordum					
Caloscordum	*neriniflorum* Herbert	16	2379	Somon Chalchgol, Mongolia	AJ299102
	neriniflorum Herbert	16	2797	Dauria, Russia	AJ299115
Nectaroscordum					
Nectaroscordum	*siculum* Ucria	18	0093	Garden in Gatersleben, Germany	AJ299138
Melanocrommyum					
Acmopetala	*suworowii* Nabelek	16	3652	Alma-Ata–Dzhambul road, Kazakhstan	AJ299127
Aroidea	*aroides* M. Pop. et Vved.	16	2517	BG Tashkent, Uzbekistan	AJ299106
Megaloprason	*sarawschanicum* Regel	16	3673	Zaravshan Mts, Tajikistan	AJ299129
	stipitatum Regel	16	2257	Kholmon Valley, Tajikistan	AJ299100
Miniprason	*karataviense* Regel	18	2989	Chilchenboa Mts, Uzbekistan	AJ299118
Porphyroprason	*oreophilum* C.A. Mey.	16	0348	BG Graz, Austria	AJ299125

Continued.

Table 8.1. *Continued.*

Subgenus, Section	Species	2*n*	TAX	Origin	EMBL
Melanocrommyum (continued)					
Verticillata	*verticillatum* Regel	16	2182	Gazimajlik Mts, Tajikistan	AJ299099
	verticillatum Regel	16	2526	Dushanbe, Tajikistan	AJ299107
Vvedenskya	*kujukense* Vved.	20	3625	Karatau Mts, Kuyuk Pass, Kazakhstan	AJ299147
Rhizirideum					
Anguinum	*tricoccum* Sol.	16	2582	BG Glencoe, Minnesota, USA	AJ299109
	victorialis L.	16	2673	Caucasus, Georgia	AJ299112
Annuloprason	*atrosanguineum* Kar. et Kir.	16	2560	Kusawlisai Valley, Tajikistan	AJ299108
Butomissa	*ramosum* L.	32	2735	BG Alma-Ata, Kazakhstan	AJ299114
	gilgiticum Wang et Tang			Karakorum, Pakistan (Herbarium, Gatersleben)	AJ299140
Campanulata	*jodanthum* Vved.	16	1330	Kondara Valley, Tajikistan	AJ299089
Cepa	*altaicum* Pall.	16	0339	BG Kaunas, Lithuania	AJ299121
	cepa L.	16	A878	cv. Stuttgarter Riesen	AJ299139
	cepa Aggregatum group	16	1810	1986, No.4–1	AJ299093
	fistulosum L.	16	0266	Wisley Gardens, UK	AJ299111
	galanthum Kar. et Kir.	16	1729	BG Alma-Ata, Kazakhstan	AJ299092
	oschaninii B. Fedtsch.	16	2177	Varsob Valley, Tajikistan	AJ299098
	pskemense B. Fedtsch.	16	1994	BG Copenhagen, Denmark	AJ299094
	roylei Stearn	16	5152	Olomouc, Czech Republic	AJ299142
	asarense R.M.Fritsch et Matin	16	3900	Central Kopetdag, Turkmenistan	AJ299131
	× *cornutum* G.C. Clementi ex Vis.	24	5193	'Pran', Kashmir, India	AJ299134
Reticulato-Bulbosa	*oreoprasum* Schrenk	16	3643	Transili Mts, Turgen Valley, Kazakhstan	AJ299126
	splendens Schult. et Schult. f.	48	1288	BG Kyoto, Japan	AJ299087
Rhizirideum	*nutans* L.	32	2080	Gorno-Altaisk, Altai, Russia	AJ299096
	rubens Schrad.	16	1609	Temirtau, Kazakhstan	AJ299145
Sacculiferum	*chinense* G. Don	32	3407	Fukui, Japan	AJ299122
	thunbergii G. Don	16	3408	Kumamoto, Japan	AJ299123
Schoenoprasum	*schoenoprasum* L.	16	4214	Garden in Gatersleben, Germany	AJ299132
Coleoblastus	*mairei* Levl.	16	2104	BG Zurich, Switzerland	AJ299097
Cyathophora	*cyathophorum* Bur. et Franch.	16	2824	BG Oslo, Norway	AJ299116
	cyathophorum Bur. et Franch	16	2825	BG Jena, Germany	AJ299117
	weschniakowii			BG Moscow University; Tienshan (Lake Issuk-Kul, Kyrgyzstan)	AJ299143

Bloomeria	*crocea* (Torrey) Coville		2697	BG Santa Barbara, California, USA	AJ299113
Nothoscordum	*bivalve* (L.) Britton	18	594	BG Palermo, Italy	AJ299136
Tulbaghia	*violaceae* Harv.		1467	Chelsea Physic Garden, London, UK	AJ299090

TAX, accession numbers of the Department of Taxonomy of the Institute for Plant Genetic and Crop Plant Research, Gatersleben; EMBL, sequence accession numbers at the European Molecular Biology Laboratory; BG, botanical garden.

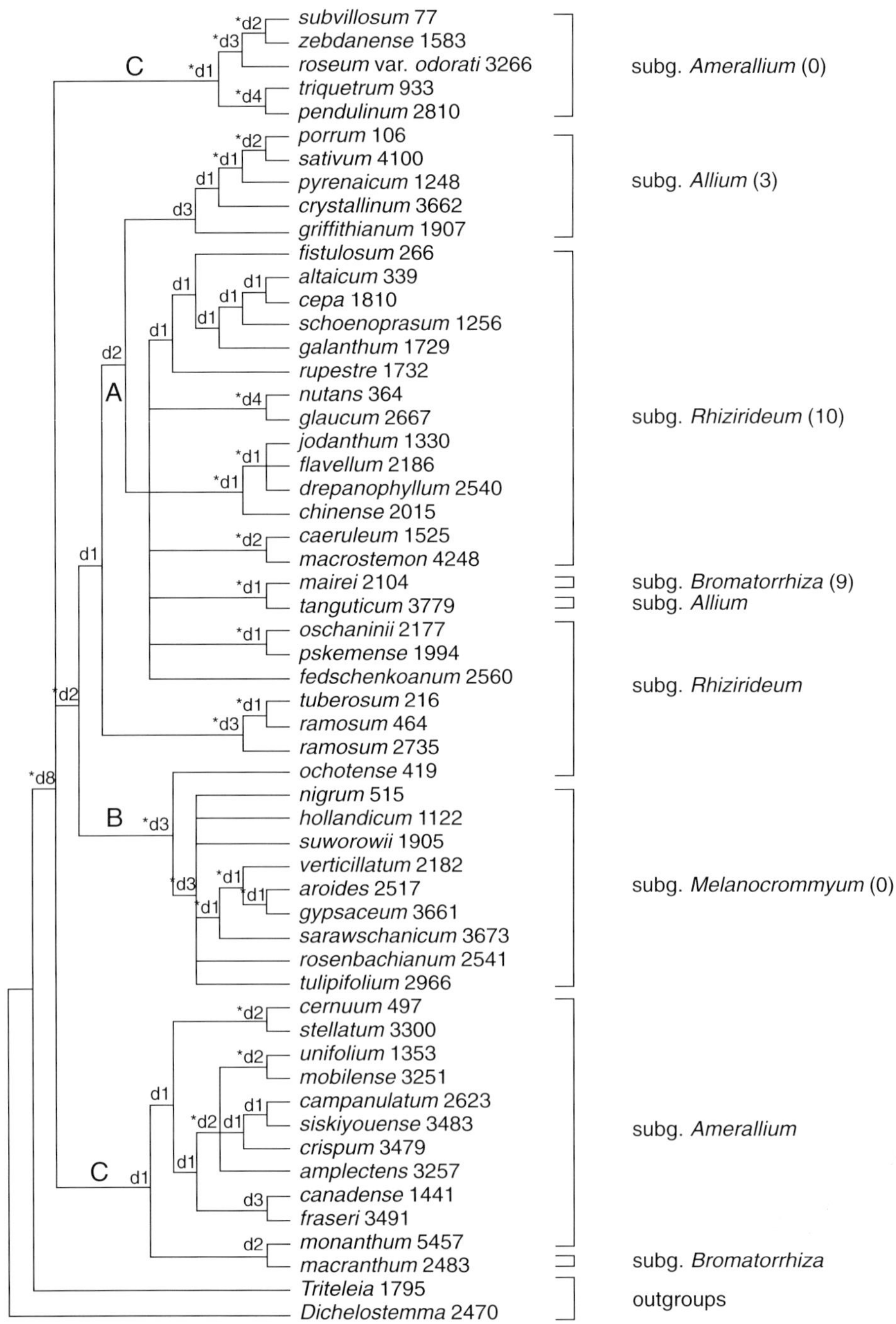

Fig. 8.2. Consensus cladogram based on restriction sites and length variants in the chloroplast DNA from *Allium* species (Mes *et al.*, 1999, with the permission of the publishers of *Genome*).

higher noise in the data and unresolved trees. However, CAPS data enable the reconstruction of phylogenies with excellent resolution at lower taxonomic levels (at least within subgenera), and large data sets can be generated in far less time compared with a sequence project with a comparable number of taxa. Comparative sequencing necessarily involves the generation of data from conserved (i.e. uninformative) sequences, which is required to ensure a correct alignment of sequences as the most important step in phylogenetic analysis.

3.1.3 GISH

Detailed information on the chromosomal composition of hybrid plants is possible by genomic *in situ* hybridization (GISH) analysis, a powerful method for the analysis of differentiation between genomes (Schwarzacher and Heslop-Harrison, 2000). Chromosome spreads from metaphase plates are hybridized with total labelled genomic DNA from one of the suspected parent species. By addition of different ratios of unlabelled blocking DNA from the other parent, even closely related genomes can be distinguished, so that single chromosomes or even parts thereof can be attributed to one or the other parent species.

3.2 Infraspecific applications

The molecular approach is often the only means of obtaining a sufficient number of unbiased markers for infraspecific investigations. In populations of wild species, an assessment of genetic diversity (usually only in part reflected in morphological differentiation) is essential for the investigation of the status of subspecies groups and problems of recent or ongoing speciation events. In an extensive isozyme study, Rieseberg *et al.* (1987) sampled populations from four varieties of *A. douglasii* (subgen. *Amerallium*) from a limited region in the north-western USA. With 12 enzyme systems, 22 loci were scored and allelic frequencies for a total of 26 populations determined. These molecular data clearly separated the two varieties *douglasii* and *nevii*, while vars *columbianum* and *constrictum* populations formed a third group in the clustering dendrogram. From two isozyme autapomorphies (derived character states that define a new evolutionary line) found in *constrictum*, this variety was concluded to be a recent derivative of *columbianum*. In a later report, Smith and Vuong Pham (1996) applied RAPD data in a similar investigation of the rare *A. aaseae*, endemic to Idaho, and its more common sister species *A. simillimum*. From 12 selected primers, 65 variable markers were scored in 14 populations from both species, but in this case the RAPD dendrogram did not confirm the species status of the populations as determined by morphology. Since the species are defined by ecological and morphological data, an explanation of the RAPD results could be a recent speciation event or might indicate multiple origins of *A. aaseae* from *A. simillimum*. Hybridization and introgression occur in common habitats but do not explain the lack of genetic differentiation in geographically distant populations of the two species.

Most studies of infraspecific differentiation in *Allium* have been aimed at crop plants of economic importance. Questions of crop evolution and the interrelationships of cultivars and varieties are addressed, and the relation of crops to close or ancestral wild species can be clarified. The determination of the genetic diversity of crop accessions is of direct use in a gene bank, both to assess the value of a collection and to direct future collecting missions.

3.2.1 Chives

Allium schoenoprasum is extremely widespread in Eurasia and North America. Different morphological types have been described, and section *Schoenoprasum* contains several closely related species, in part of polyploid nature, with partly unclear species status. In such a situation, molecular markers could bring some clarification difficult to obtain by other means. Using 233 RAPD markers derived from 11 primers, Friesen and Blattner (2000) investigated 38 accessions from section

Schoenoprasum, including 29 accessions from *A. schoenoprasum* (covering its geographical range in Eurasia), and representatives of seven other species. The molecular markers indicated genetic differentiation within *A. schoenoprasum* according to geographical distribution; however, the morphological types of chives described earlier (Stearn, 1978; Friesen, 1996) were not reflected in the dendrograms.

The raw data were analysed in several ways and their relative merits discussed. Pairwise genetic distances were calculated for the construction of UPGMA and neighbour-joining trees. In principal coordinate data analysis (PCA), new independent coordinates' axes were calculated in a process analogous to the construction of a regression line from a cloud of data points, so as to explain a maximum of the diversity of the underlying data. The taxa are graphically represented as points in a three-dimensional (3-D) space defined by the first three coordinates (see Fig. 8.3). This representation was well suited to demonstrating the reflection of geographical origin in the genetic grouping. Cladistic analysis is based on the reconstruction of a series of phylogenetic splitting events, each defined by gain or loss of characters common to at least two offspring taxa. The temporal order of these events is deduced from comparison with an outgroup species as close as possible to but outside the investigated group. The procedure is more commonly applied in analysing DNA sequence data, where it is generally

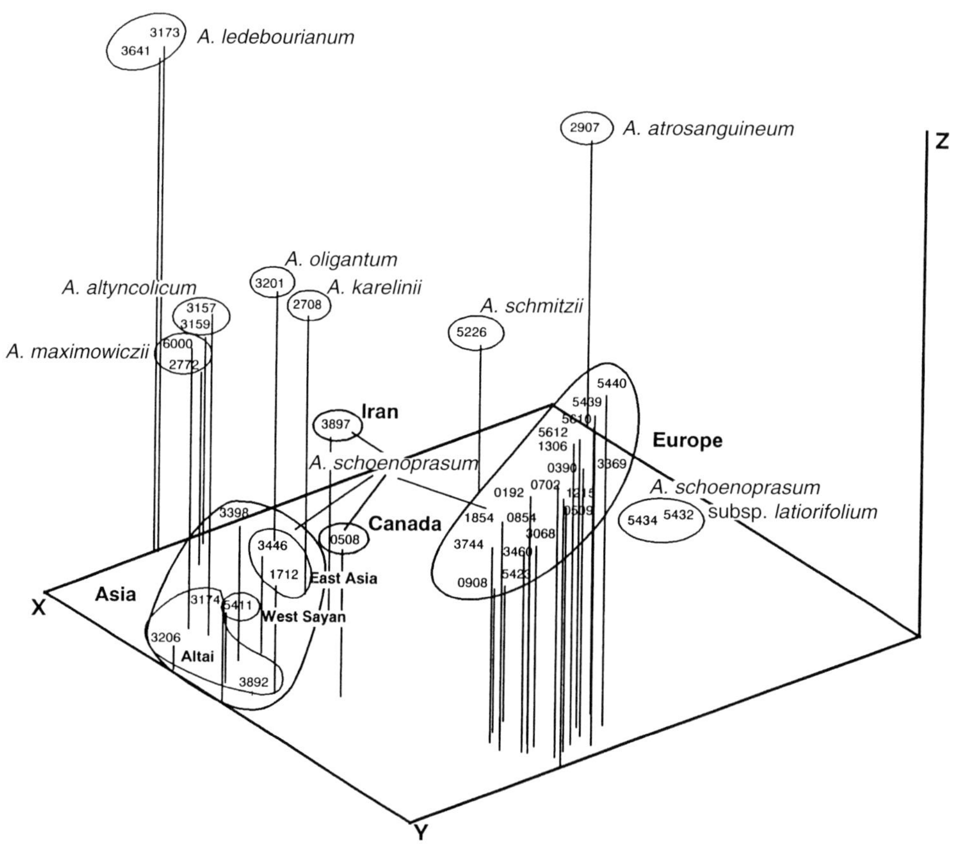

Fig. 8.3. Three-dimensional plot of the first three principal coordinates, calculated from Jaccard distances of 38 accessions of seven species of *Allium* sect. *Schoenoprasum* and *A. atrosanguineum*, based on 233 RAPD bands. The four-digit numbers here and in Fig. 8.4 are accession numbers of the living collection of the Department of Taxonomy, IPK, Gatersleben.

clearer what constitutes a change, and it is compared with the gain or loss of a RAPD band. Nevertheless, a cladistic tree of the *Schoenoprasum* RAPD data was constructed. The consensus tree yielded a grouping very similar to the neighbour-joining tree based on pairwise distances. The removal of known polyploid species based on theoretical objections in a phylogenetic analysis did not change the grouping of the remaining species.

3.2.2 Garlic

Allium sativum is a predominantly sterile species known only in cultivation (see Etoh and Simon, Chapter 5, this volume). Nevertheless, there is great variability in morphological and physiological features and varying degrees of bolting and flower formation, which led to the proposition of three botanical varieties. Presumably *A. longicuspis* is the wild progenitor, but whether its status should be as a separate species or just as feral plants derived from crops has been disputed. Pooler and Simon (1993) investigated a collection of 110 garlic clones with morphological and isozyme methods for an infraspecific classification. Thirteen isozyme systems were tested, although, because of inconsistent staining or lack of variability, only four were useful, and 17 different enzyme groups were detected. While flower characteristics correlated well with isozyme data, bulb-related traits or geographical origin had little predictive value for the genetic relationship of accessions. Maaß and Klaas (1995) tested 300 clones with isozymes, and 48 of these were tested with RAPDs as well, to compare the two marker systems. Their gene pool contained many accessions from areas close to the centre of origin in Central Asia and was suitable for investigating the genetic relationship between cultivated clones with primitive features, derived strains and a feral accession of *A. longicuspis*. Twelve isozyme systems were tested which identified 22 loci, ten of which were polymorphic and defined 16 isozyme groups. Predictably, the 125 RAPD markers allowed a more detailed distinction, but generally both markers gave a good delimitation of varieties *sativum* (bolting and non-bolting types could be separated) and *ophioscorodon*. The third variety, *pekinense* was not distinguishable by either marker from *longicuspis*-type plants; nor was an accession determined as *A. longicuspis* separated from more primitive (i.e. partially fertile) garlics, based on molecular markers.

A similar range of accessions was investigated by Al-Zahim *et al.* (1997). Their results differed in some important aspects. Twenty-seven named garlic cultivars were structured with 63 polymorphic RAPD bands generated from 26 primers. Eleven accessions were assigned to variety *ophioscorodon*, 11 to variety *sativum* and five to *A. longicuspis*. In agreement with Maaß and Klaas (1995), the accessions of var. *sativum* (only non-bolting accessions were included) grouped together: however, these workers found genetic differentiation within var. *ophioscorodon* and interspersal with *A. longicuspis* accessions. These findings were in contrast to the genetic homogeneity of the *ophioscorodon* group (80 accessions were investigated by isozymes, seven of these by RAPDs), being genetically clearly distinct from *longicuspis*-type accessions, as reported by Maaß and Klaas (1995). The different results can probably be explained by the different morphological classification of the material prior to the molecular study, rather than by a misapplication of the RAPD markers in either case, since a comparable number of primers and markers per taxon was used in both laboratories. In the well-characterized collection in Gatersleben, *ophioscorodon* was morphologically clearly distinguishable from *A. longicuspis* (Helm, 1956; Maaß and Klaas, 1995; Maaß, 1996b), while Al-Zahim *et al.* (1997) reported difficulties in distinguishing *ophioscorodon* from *A. longicuspis* based solely on exserted anthers. An interspersal of var. *ophioscorodon* accessions with plants from the *longicuspis* group would explain these data.

Bradley *et al.* (1996) investigated a collection of 20 Australian garlic accessions with five RAPD primers, resulting in 65 marker bands. The approach was well suited to grouping the major Australian cultivars according to bolting behaviour, early and late types, and places of origin.

3.2.3 Common onion and related crops

Allium cepa is the most important *Allium* crop in terms of economics and areas of production worldwide. Apart from the common bulb onion, shallots and some hybrid crops are derived from this species. Wilkie *et al.* (1993) demonstrated the applicability of RAPD markers in *Allium* in investigations of seven cultivars of *A. cepa* and one accession of each of four other species. Between all the species, 91 polymorphic band positions were scored in reactions with 20 random primers, but within *cepa* only seven bands were polymorphic, resulting in limited resolution at the infraspecific level. Roxas and Peffley (1992) also reported the successful application of RAPDs to onion-cultivar identification using six random primers, but no details were given.

RAPDs and one isozyme locus have been used to test the genetic integrity of a doubled-haploid (DH) line derived from the open-pollinated 'Dorata di Parma' onion cultivar (Campion *et al.*, 1995). While a high degree of RAPD polymorphism was observed among individuals of the parent cultivar population, no differences were found among individuals of the DH line. In a second gynogenic line, a haploid line derived from the Japanese cultivar 'Senshyu Yellow', no RAPD-detectable incidence of genetic instability was found during micropropagation (Campion *et al.*, 1995).

Hybrid-onion seeds are produced from inbred lines: these necessarily retain a relatively high heterozygosity level, since inbreeding depression leads to rapid loss of vigour in bulb onions. In a detailed study, Bradeen and Havey (1995) investigated the use of RAPDs for the testing of the integrity of inbred lines, which is essential to hybrid performance. From a cross between two distant cultivars differing in pungency, soluble solids and storage properties, 59 F_3 families were analysed for the segregation of RAPD markers. Of 580 tested random primers, only 53 detected polymorphisms, and 12 of these gave bands in the 3 : 1 segregation ratio expected for genetic markers inherited in a Mendelian way. In a test of four inbred lines, they were not clearly separated in a UPGMA clustering dendrogram based on data from these genetically characterized markers, and some incidence of contamination was found. Data generated in parallel from markers which were not genetically characterized (i.e. not segregating in a Mendelian fashion) agreed only poorly with these results and were discarded from the final analysis. The use of these and other markers in onion for the construction of a low-density genomic map was summarized (Havey *et al.*, 1996; King *et al.*, 1998a).

Considerably more polymorphisms were detected in genomic RFLP blots probed with random nuclear cDNAs, even though this approach met with some technical difficulties due to the high 2C value of onion (Bark and Havey, 1995). These workers investigated the genetic diversity in 17 open-pollinated populations of onions that bulbed under short-day (SD) and long-day (LD) conditions, and two inbred lines (SD and LD). Of 104 cDNA clones, 60 detected at least one polymorphism. In total, 146 fragments were scored for presence or absence. The raw data were analysed cladistically (parsimonious evolution) and by methods based on genetic distances (UPGMA, PCA). The populations were not clearly separated according to their day-length response, a trait conventionally used to classify onion groups, and yet populations known to be closely related were recognized from the DNA data. Generally, the SD populations were the more diverse, and it appeared that LD onions are a derived group.

A shallot was included in the analysis as an outgroup, but was possibly too close to the SD onions. One accession of *A. fistulosum* was also included, but only 14% of the detected fragments were identical to those of *A. cepa*. King *et al.* (1998b) applied the same technique to the investigation of 14 commercial *A. cepa* inbreds in an RFLP study with 69 anonymous cDNA clones. As few as ten polymorphic restriction enzyme/probe combinations were able to distinguish all the investigated inbreds, indicating a high resolving power of suitably chosen nuclear RFLP probes for the characterization of onion lines.

Shallots, while formerly considered a separate species (*A. ascalonicum*), are now considered part of *A. cepa* as the Aggregatum group. Maaß (1996a) compared the isozyme patterns of 30 individuals of a distinct type, the French grey shallot, with those of 466 bulb onions and other shallots, 15 *A. oschaninii* and 22 *A. vavilovii*. The allele distribution at four isozyme loci suggests that the grey shallot is more closely related to either *A. vavilovii* or *A. oschaninii* than to *A. cepa* and the other shallots, which appear as a closely related assemblage (Fig. 8.4; Messiaen *et al.*, 1993). Within *cepa*, only two loci were polymorphic, while all four loci were polymorphic within the wild species.

The relation of common onions to different types of shallots was investigated by RAPD markers from four primers and morphological traits (Le Thierry D'Ennequin *et al.*, 1997). The seed-propagated types of shallot proved to be closely related to the common onions, while the vegetative shallots grouped separately. In agreement with the isozyme data from Maaß (1996a), the grey shallot was clearly distinct from both types of shallots as well as from common onions.

Arifin and Okubo (1996) structured a large collection of 189 tropical shallots and the sterile wakegi accessions with five isozyme systems. They identified 25 enzyme patterns of wakegi and 18 patterns of shallots; the two groups were clearly distinct, even though the two crops are grown interchangeably in many areas where plants were collected.

While the number of easily scorable isozyme loci is clearly not sufficient for a detailed infraspecific analysis of *A. cepa* compared with the DNA-based approaches described earlier, isozyme analysis provides a powerful technique for investigating the relationships between close species, when large numbers of accessions have to be analysed in order to take into account the infraspecific variation within each species.

3.2.4 *Japanese bunching onion*

Three varietal groups of *A. fistulosum* were differentiated with seven loci from five enzyme systems (Haishima and Ikehashi, 1992; Haishima *et al.*, 1993) when 23 Japanese, one Chinese and one Korean accessions were analysed. Many loci were uniform among the Japanese accessions, probably due to a loss of diversity over the introduction and selection of this crop. The fixation of several otherwise rare alleles indicated random genetic drift due to 'founder' effects resulting from a population bottleneck.

3.3 Hybrids

Interspecific hybridization is known in many *Allium* groups as a mode of speciation through the evolution of the genus. Usually, the interspecific hybrid and offsprings were recognized from intermediate morphological features and increased chromosome numbers. In crops, deliberate hybridizations of close species have been carried out for the construction of maps (van Heusden *et al.*, 2000), to explore the possibility of introgression of desirable foreign traits such as disease resistance not available in the crop's germplasm (Khrustaleva and Kik, 2000), or to probe the relationships between species (van Raamsdonk *et al.*, 1992). In experimental crossings with known parents, it is usually sufficient to validate the hybrid plants by the detection of a few markers characteristic of each parent. These include morphological (e.g. leaf shape, pigmentation), physiological (resistance to pests, growth habit) and/or molecular markers, such as RAPD bands, ITS restriction-enzyme sites, or isozyme markers (van der Valk *et al.*, 1991; Buiteveld *et al.*, 1998; Dubouzet *et al.*, 1998). Similarly, the cytoplasm of the hybrid can be characterized by one or a few parent-specific PCR or RFLP markers from mtDNA or cpDNA (Holford *et al.*, 1991; Satoh *et al.*, 1993; Havey, 1995). More detailed information on the chromosomal composition of hybrid plants is possible by GISH analysis, a powerful method for the analysis of differentiation between genomes. Khrustaleva and Kik (2000) were able to identify by GISH the parental species to chromosomal regions in experimental hybrids with genomic contributions from three parents – *cepa*, *fistulosum* and *roylei*. (For a detailed discussion, see Kik, Chapter 4, this volume.)

For the reconstruction of species hybridization in wild populations or crops in the more distant past, this approach gives increasingly unreliable results. Processes of sequence elimination, intergenomic exchanges and independent molecular evolution in the lineages of putative parent species and the suspected hybrid tend in time to blur the clear distinction between all involved genomes, thus making clear identification by GISH more difficult (Friesen and Klaas, 1998). In these cases, discrete markers, such as RAPDs or AFLPs, may be the most useful, since these data are amenable to the detection of underlying information. For example, PCA can be used to identify segments of the genome that are the most closely related to those of another species (Fig. 8.4; see Fig. 8.3 for an intraspecific application; Friesen and Hermann, 1998; Friesen and Klaas, 1998; Friesen and Blattner, 2000).

3.3.1 Genetic structure of species complexes

INVESTIGATION OF WILD HYBRIDOGENIC SPECIES. Several cytological and molecular techniques have been tested to investigate the hybrid nature of *A. altyncolicum*, suspected to be a (4*n*) allopolyploid species derived from a spontaneous cross between the diploid species *schoenoprasum* × *ledebourianum* (Friesen *et al.*, 1997a). C-banding, ITS sequencing, PCR-RFLP of plastid DNA, GISH, RAPD analysis and rDNA RFLP were applied. GISH revealed the segmental allopolyploid nature of *A. altyncolicum* by specific hybridization of one parent's labelled genomic DNA only to the corresponding chromosomes in the hybrid. Due to the closeness of the parental genomes and their mutual adaptation (since this species originated many generations ago), the ratios

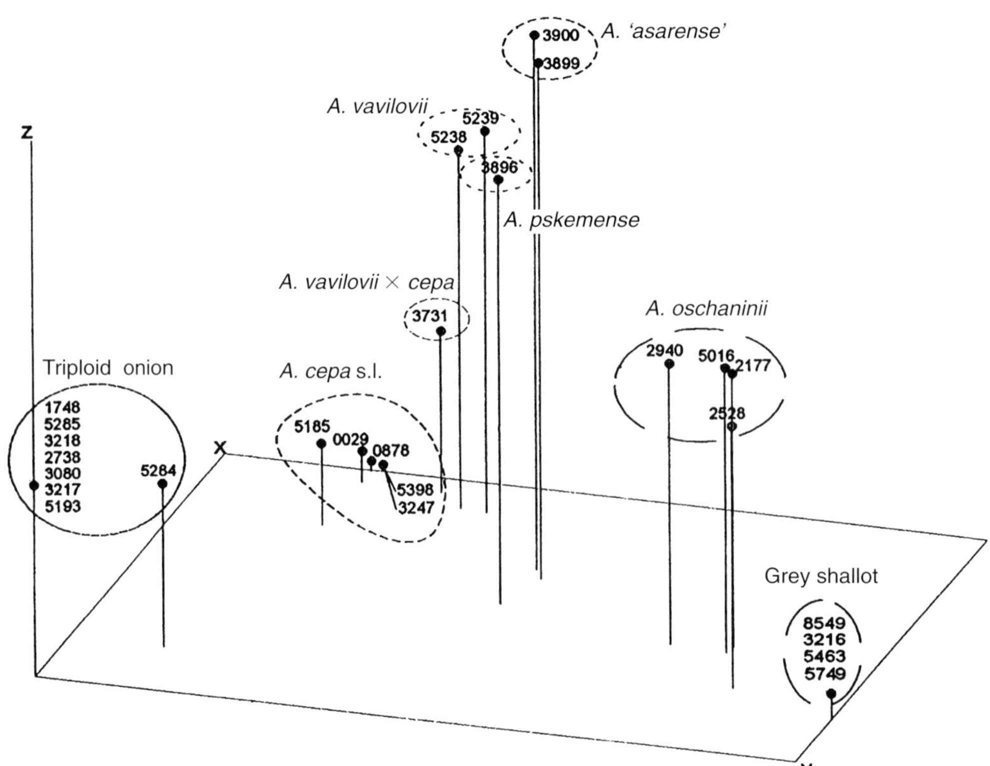

Fig. 8.4. Three-dimensional plot of the first three principal coordinates, calculated from a distance matrix based on RAPD data (Friesen and Klaas, 1998).

of labelled DNA and blocking DNA from the other parent required careful calibration to distinguish the parental chromosomes. The chromosomes were also identified by C-banding, the other approaches gave only little or no species-specific distinction. In a related study, ornamentals from subgenus *Melanocrommyum* which had been generated by uncontrolled pollination in breeder's fields were investigated (Friesen *et al.*, 1997b). Initial RAPD screens of the suspected hybrid plants and suspected parent species identified (or excluded) putative genomic contributors. In subsequent GISH experiments, the presence (or absence) of the parental genomes was unequivocally demonstrated (Friesen *et al.*, 1997b).

ANALYSIS OF HYBRID CROPS. Several vegetatively propagated crop species in *Allium* are of hybrid origin. They often arose spontaneously, to be subsequently selected and maintained by gardeners for their unusual properties. *Allium wakegi* is a sexually sterile ancient garden crop in Japan and China. Its hybrid nature (*A. fistulosum* × *A. cepa*) was long suspected because of the intermediate morphology of leaves, bulbs and flowers. The hybrid nature of this species has been proven by GISH (Hizume, 1994). Additional evidence for the hybrid character of *A. wakegi* was gathered by localization of 5S-RNA loci at chromosomal positions corresponding to *A. cepa* and *A. fistulosum* (Hizume, 1994). *A. fistulosum* was identified as the maternal parent of *A. wakegi* by RFLP experiments on purified plastid DNA that was hybridized to an *A. fistulosum* cpDNA probe (Tashiro *et al.*, 1995). Tested by this limited approach, all investigated *A. wakegi* accessions had an identical cytoplasm with a *fistulosum*-like RFLP pattern.

Top onions, or topsetting onions, and viviparous onions are other hybrid species of suspected *A. fistulosum* × *A. cepa* origin, long known from European botanical gardens and gardeners' books as locally cultivated minor garden crops. Havey (1991a) analysed two (2*n*) accessions with RFLP probes for the plastid and the nuclear genome. From the six restriction-enzyme sites in the cpDNA, *A. fistulosum* was determined as the seed parent, while seven out of 11 restriction-enzyme nuclear rDNA fragments were found both from *A. fistulosum* and from *A. cepa*. Maaß (1997a) used six isozyme assays to analyse a large collection of 164 top-onion accessions, six accessions of *A. wakegi*, the parental species *A. cepa* (59 accessions), *A. fistulosum* (27 accessions, including one population of *A. altaicum*) and some artificial hybrids. All allele combinations from the top onions were also found in the hybrids, in addition to some others. The recombination of the hybrids' genome (in crops as well as in artificial hybrids) from the parent species could also be verified by this approach. This required the prior analysis of the allelic diversity of representative collections of the parent species. Finally, both parental genomes of the topsetting onion were clearly identified in a GISH experiment (Friesen and Klaas, 1998).

Less clear is the parentage situation in the (3*n*) viviparous onion *Allium* × *cornutum* (known as 'Pran'), which has long been cultivated in Kashmir and proved to be a rather widespread garden crop, even cultivated sporadically in Europe. Isozyme analysis (Maaß, 1997b) and RAPD analysis (Friesen and Klaas, 1998; Puizina *et al.*, 1999) found no differences between 'Pran' and the Croatian cultivar 'Ljutica'. This crop had been suspected to be either an allotriploid (AAB) (Singh *et al.*, 1967) or a segmental allotriploid (AA′A″) (Koul and Gohil, 1971). The cpDNA pattern could not be attributed to any of the species analysed by Havey (1991b), but is identical with the pattern found in the S cytoplasm of CMS bulb onions (Havey, 1993, 1995), and the nuclear rDNA fragments indicate *A. cepa* as one parent. These data suggested that 'Pran' originated from a cross between a so far unknown seed parent and *A. cepa*. Isozyme analysis failed to identify the second parent of 'Pran' in a comparative study including accessions of sect. *Cepa* species *cepa*, *fistulosum*, *galanthum*, *pskemense*, *vavilovii* and *oschaninii* (Maaß, 1997b). However, *A. fistulosum* was excluded as the second parent (Havey, 1991b; Maaß, 1997b), as also were with some probability *A. schoenoprasum* and *A. roylei*. Friesen and Klaas (1998) investigated the relation of the (3*n*) onion to several sect. *Cepa* species with RAPD and GISH. They found GISH hybridization signals mainly from *A. cepa* (or the close *A. vavilovii*) probes in experiments with different blocking DNAs, rendering the crop a segmental allopolyploid; no GISH signal was obtained from *A. roylei*, which had given no common RAPD bands. The unknown parent contributing the non-*cepa* fraction of the (3*n*) onion remained unknown. Using GISH, Puizina *et al.* (1999) confirmed that a third of the 'Ljutica' genome belongs to *cepa*, another third to *roylei* and the remainder to an as yet unknown third parent. These workers observed signal (in eight chromosomes) from a *roylei* probe with *cepa* blocking DNA. These data are in complete disagreement with the above results of Friesen and Klaas (1998). The latter did not get a clear signal, even without blocking DNA, and found no common RAPD bands between *roylei* and eight studied triploid accessions. While the lack of any common RAPD bands argues against a close relation of *roylei* and triploid onion, the two groups' contradictory GISH results should be resolved by reanalysis of the enigmatic plant.

The diploid grey shallot is a distinct form of shallot long cultivated in France and Italy. Isozyme studies (Maaß, 1996a) proved insufficient to identify the genomic composition, but suggested a closer relationship to *A. vavilovii* and *A. oschaninii* than to *A. cepa*. While a first RAPD study with 24 markers (Le Thierry D'Ennequin *et al.*, 1996) indicated affiliation of grey shallots with other normal shallots belonging to *A. cepa*, both GISH and RAPD data (Friesen and Klaas, 1998) show that most of the chromosomes of grey shallot belong to *A. oschaninii*, with only one and a half chromosome arms derived from either *cepa* or *vavilovii* (see Fig. 8.4 for a 3-D representation of genetic distances between these species; see also Rabinowitch and Kamenetsky, Chapter 17, this volume).

4. Conclusions

During recent years, the application of molecular markers has become routine in *Allium* research, due to the increased ease of use and the standardization of the biochemical techniques and of the procedures for the evaluation of results. The power of established techniques, such as RFLP and repetitive DNA analysis, has been enhanced by combination with PCR approaches, enabling increased resolution in less experimental time. While undeniably substantial progress has been made, the extent of a diversity survey is still limited by the necessary effort involved in the generation of molecular markers and analysis for each sample. A breakthrough in this field, e.g. if a survey on large germplasm collections is attempted, will only be achieved by complete automatization of marker generation and analysis. The adaptation of microarray techniques as used at present for expression profiling might be suitable, or further development of genetic bit analysis as presented in *Allium* (Alcala *et al.*, 1997), which is able to detect single-site allelic polymorphisms colorimetrically.

The framework of the genus's phylogeny can be considered as validated, especially if the same groupings are resolved by nuclear as well as chloroplast markers. This also applies to the relationships of the subgenera within the genus and their circumscription. However, with the finer detail now available (see Figs 8.1 and 8.2 and the ITS-based tree in Fritsch and Friesen, Chapter 1, this volume), some arbitrariness has become apparent in the decisions made as to which groups are elevated to subgenus level. Final classification will depend not only on phylogenetic conclusions but on practical considerations.

More troubling are the contradictory groupings obtained in spite of thorough analysis with different markers: plastid DNA analysis should yield approximations of the very same tree, regardless of its origin from RFLP, CAPS or DNA sequences. Excluding experimental errors, these differences are most probably founded on use of the marker at levels of taxonomic resolution not suitable for its resolving power, i.e. when the phylogenetic signal from observed mutations is hidden by multiple changes affecting the same restriction site. The only alternative explanation would be the existence of irregular recombination of plastid DNA not yet documented.

Differences between nuclear DNA and cpDNA phylogenies are explained by reticulate evolution, leading to new recombinant types in the nuclear DNA but not the cpDNA. However, a slight unease remains at the interpretation of ITS sequences, which is at present the only nuclear marker established for investigations at the genus level on a larger scale. Its specific type of sequence evolution – concerted evolution in the repetitive rDNA cluster leading to homogenization if different types of ITS sequences are present – is not invariably representative of molecular evolution throughout the genome, even though ITS analysis yields reasonable groupings in agreement with other types of data, such as morphological and anatomical studies.

References

Alcala, J., Giovannoni, J.J., Pike, L.M. and Reddy, A.S. (1997) Application of Genetic Bit Analysis (GBA™) for allelic selection in plant breeding. *Molecular Breeding* 3, 495–502.

Al-Zahim, M.A., Newbury, H.J and Ford-Lloyd, B.V. (1997) Classification of genetic variation in garlic (*Allium sativum* L.) revealed by RAPD. *HortScience* 32, 1102–1104.

Al-Zahim, M.A., Ford-Lloyd, B.V. and Newbury, H.J. (1999) Detection of somaclonal variation in garlic (*Allium sativum* L.) using RAPD and cytological analysis. *Plant Cell Reports* 18, 473–477.

Aquadro, C.F. (1997) Insight into the evolutionary process from patterns of DNA sequence variability. *Current Opinions in Genetics and Development* 7, 835–840.

Arifin, N.S. and Okubo, H. (1996) Geographical distribution of allozyme patterns in shallot (*Allium cepa* var. *ascalonicum* Backer) and wakegi onion (*A.* × *wakegi* Araki). *Euphytica* 91, 305–313.

Bark, O.H. and Havey, M.J. (1995) Similarities and relationships among populations of the bulb onion as estimated by nuclear RFLPs. *Theoretical and Applied Genetics* 90, 407–414.

Bradeen, J.M. and Havey, M.J. (1995) Randomly amplified polymorphic DNA in bulb onion and its use to assess inbred integrity. *Journal of the American Society for Horticultural Science* 120, 752–758.

Bradley, K.F., Rieger, M.A. and Collins, G.G. (1996) Classification of Australian garlic cultivars by DNA fingerprinting. *Australian Journal of Experimental Agriculture* 36, 613–618.

Bruford, W.M., Hanotte, O., Brookfield, J.F.Y. and Burke, T. (1992) Single-locus and multilocus DNA fingerprinting. In: Hoelzel, A.R. (ed.) *Molecular Genetic Analysis of Populations. A Practical Approach.* IRL Press, Oxford, pp. 225–269.

Buiteveld, J., Kassies, W., Geels, R., van Lookeren Campagne, M.M., Jacobsen, E. and Creemers-Molenaar, J. (1998) Biased chloroplast and mitochondrial transmission in somatic hybrids of *Allium ampeloprasum* L. and *Allium cepa* L. *Plant Science* 131, 219–228.

Campion, B., Bohanec, B. and Javornik, B. (1995) Gynogenic lines of onion (*Allium cepa* L.): evidence of their homozygosity. *Theoretical and Applied Genetics* 91, 598–602.

Chase, C.D., Ortega, V.M. and Vallejos, C.E. (1991) DNA restriction fragment length polymorphisms correlate with isozyme diversity in *Phaseolus vulgaris* L. *Theoretical and Applied Genetics* 81, 806–811.

Chavarriaga-Aguirre, P., Maya, M.M., Tohme, J., Duque, M.C., Iglesias, C., Bonierbale, M.W., Kresovich, S. and Kochert, G. (1999) Using microsatellites, isozymes and AFLPs to evaluate genetic diversity and redundancy in the cassava core collection and to assess the usefulness of DNA based markers to maintain germplasm collections. *Molecular Breeding* 5, 263–273.

Demeke, T., Adams, R.P. and Chibbar, R. (1992) Potential taxonomic use of random amplified polymorphic DNA (RAPD): a case study in *Brassica*. *Theoretical and Applied Genetics* 84, 990–994.

Dubouzet, J.G. and Shinoda, K. (1998) Phylogeny of *Allium* L. subg. *Melanocrommyum* (Webb et Berth.) Rouy based on DNA sequence analysis of the internal transcribed spacer region of nrDNA. *Theoretical and Applied Genetics* 97, 541–549.

Dubouzet, J.G. and Shinoda, K. (1999) Relationships among Old and New World *Alliums* according to ITS DNA sequence analysis. *Theoretical and Applied Genetics* 98, 422–433.

Dubouzet, J.G., Shinoda, K. and Murata, N. (1997) Phylogeny of *Allium* L. subgenus *Rhizirideum* (G. Don ex Koch) Wendelbo according to dot blot hybridization with randomly amplified DNA probes. *Theoretical and Applied Genetics* 95, 1223–1228.

Dubouzet, J.G., Shinoda, K. and Murata, N. (1998) Interspecific hybridization of *Allium giganteum* Regel: production and early verification of putative hybrids. *Theoretical and Applied Genetics* 96, 385–388.

Edwards, K.J., Barker, J.H.A., Daly, A., Jones, C. and Karp, A. (1996) Microsatellite libraries enriched for several microsatellite sequences in plants. *BioTechniques* 20, 758–760.

Ekberg, L. (1969) Studies in the genus *Allium* II. A new subgenus and new sections from Asia. *Botaniska Notiser (Lund)* 122, 57–68.

Engelke, T. and Tatlioglu, T. (2000) Mitochondrial genome diversity in connection with male sterility in *Allium schoenoprasum* L. *Theoretical and Applied Genetics* 100, 942–948.

Fay, M.F. and Chase, M.W. (1996) Resurrection of *Themidaceae* for the *Brodiaceae* alliance, and circumscription of *Alliaceae*, *Amaryllidaceae* and *Agapanthoideae*. *Taxon* 45, 441–451.

Fischer, D. and Bachmann, K. (1998) Microsatellite enrichment in organisms with large genomes (*Allium cepa* L.). *BioTechniques* 24, 796–802.

Friesen, N. (1996) A taxonomic and chorological revision of the genus *Allium* L. sect. *Schoenoprasum* Dumort. *Candollea* 51, 461–473.

Friesen, N. and Blattner, F.R. (2000) Geographical isolation predominates over ecological differentiation in the phylogeny of *Allium schoenoprasum*. *Plant Biology* 2, 297–305.

Friesen, N. and Hermann, N. (1998) Taxonomy, chorology and evolution of *Allium lusitanicum* – the European '*A. senescens*'. *Linzer Biologische Beiträge* 30, 815–830.

Friesen, N. and Klaas, M. (1998) Origin of some minor vegetatively propagated *Allium* crops studied with RAPD and GISH. *Genetic Resources and Crop Evolution* 45, 511–523.

Friesen, N., Borisjuk, N., Mes, T.H.M., Klaas, M. and Hanelt, P. (1997a) Allotetraploid origin of *Allium altyncolium* (sect. *Schoenoprasum*) as investigated by karyological and molecular markers. *Plant Systematics and Evolution* 206, 317–335.

Friesen, N., Fritsch, R. and Bachmann, K. (1997b) Hybrid origin of some ornamentals of *Allium* subgenus *Melanocrommyum* verified with GISH and RAPD. *Theoretical and Applied Genetics* 95, 1229–1238.

Friesen, N., Pollner, S., Bachmann, K. and Blattner, F.R. (1999) RAPDs and non-coding chloroplast DNA reveal a single origin of the cultivated *Allium fistulosum* from *A. altaicum* (*Alliaceae*). *American Journal of Botany* 86, 554–562.

Friesen, N., Fritsch, R.M., Pollner, S. and Blattner, F.R. (2000) Molecular and morphological evidence for an origin of the aberrant genus *Milula* within Himalayan species of *Allium* (*Alliaceae*). *Molecular Phylogenetics and Evolution* 17, 209–218.

Gupta, P.K. and Varshney, R.K. (2000) The development and use of microsatellite markers for genetic analysis and plant breeding with emphasis on bread wheat. *Euphytica* 113, 163–185.

Haishima, M. and Ikehashi, H. (1992) Identification of isozyme genes in native varieties of Japanese bunching onion (*Allium fistulosum* L.). *Japanese Journal of Breeding* 42, 497–505.

Haishima, M., Kato, J. and Ikehashi, H. (1993) Isozyme polymorphism in native varieties of Japanese bunching onion (*Allium fistulosum* L.). *Japanese Journal of Breeding* 43, 537–547.

Hanelt, P., Schultze-Motel, J., Fritsch, R., Kruse, J., Maaß, H.I., Ohle, H. and Pistrick, K. (1992). Infrageneric grouping of *Allium* – the Gatersleben approach. In: Hanelt, P., Hammer, K. and Knüpffer, H. (eds) *The Genus Allium – Taxonomic Problems and Genetic Resources. Proceedings of an International Symposium, Gatersleben, 11–13 June 1991*. IPK, Gatersleben, Germany, pp. 107–123.

Havey, M.J. (1991a) Molecular characterization of the interspecific origin of viviparous onion. *Journal of Heredity* 82, 501–508.

Havey, M.J. (1991b) Phylogenetic relationships among cultivated *Allium* species from restriction enzyme analysis of the chloroplast genome. *Theoretical and Applied Genetics* 81, 752–757.

Havey, M.J. (1993) A putative donor of S-cytoplasm and its distribution among open-pollinated populations of onion. *Theoretical and Applied Genetics* 86, 128–134.

Havey, M.J. (1995) Identification of cytoplasms using the polymerase chain reaction to aid in the extraction of maintainer lines from open-pollinated populations of onion. *Theoretical and Applied Genetics* 90, 263–268.

Havey, M.J., King, J.J., Bradeen, J.M. and Bark, O. (1996) Molecular markers and mapping in bulb onion, a forgotten monocot. *HortScience* 31, 1116–1118.

Helm, J. (1956) Die zu Würz- und Speisezwecken kultivierten Arten der Gattung *Allium* L. *Kulturpflanze* 4, 130–180.

Hillis, D.M., Moritz, C. and Mable, B.K. (eds) (1996) *Molecular Systematics*, 2nd edn. Sinauer Associates, Sunderland, Massachusetts, 655 pp.

Hizume, M. (1994) Allodiploid nature of *Allium wakegi* Araki revealed by genomic *in situ* hybridization and localization of 5S and 18S rDNAs. *Japanese Journal of Genetics* 69, 407–415.

Hoelzel, A.R. (1992) *Molecular Genetic Analysis of Populations, A Practical Approach*, 1st edn. IRL Press, Oxford, 315 pp.

Holford, P., Croft, J.H. and Newbury, H.J. (1991) Differences between, and possible origins of, the cytoplasms found in fertile and male-sterile onions (*Allium cepa* L.). *Theoretical and Applied Genetics* 82, 737–744.

Inai, S., Ishikawa, K., Nunomura, O. and Ikehashi, H. (1993) Genetic analysis of stunted growth by nuclear–cytoplasmic interaction in interspecific hybrids of *Capsicum* by using RAPD markers. *Theoretical and Applied Genetics* 87, 416–422.

Jones, R.N. and Rees, H. (1968) Nuclear DNA variation in *Allium*. *Heredity* 23, 591–605.

Khrustaleva, L.I. and Kik, C. (2000) Introgression of *Allium fistulosum* into *A. cepa* mediated by *A. roylei*. *Theoretical and Applied Genetics* 100, 17–26.

Kik, C., Samoylov, A.M., Verbeek, W.H.J. and van Raamsdonk, L.W.D. (1997) Mitochondrial DNA variation and crossability of leek (*Allium porrum*) and its wild relatives from the *Allium ampeloprasum* complex. *Theoretical and Applied Genetics* 94, 465–471.

Kimura, M. (1980) A simple method for estimating evolutionary rates of base substitutions through comparative studies of nucleotide sequences. *Journal of Molecular Evolution* 16, 111–120.

King, J.J., Bradeen, J.M., Bark, O., McCallum, J.A. and Havey, M.J. (1998a) A low-density map of onion reveals a role for tandem duplication in the evolution of an extremely large diploid genome. *Theoretical and Applied Genetics* 96, 52–62.

King, J.J., Bradeen, J.M. and Havey, M.J. (1998b) Variability for restriction fragment-length polymorphisms (RFLPs) and relationships among elite commercial inbred and virtual hybrid onion populations. *Journal of the American Society for Horticultural Science* 123, 1034–1037.

Klaas, M. (1998) Applications and impact of molecular markers on evolutionary and diversity studies in the genus *Allium*. *Plant Breeding* 117, 297–308.

Koul, A.K. and Gohil, R.N. (1971) Further studies on natural triploidy in viviparous onion. *Cytologia* 36, 253–261.

Lannér, C., Bryngelsson, T. and Gustafsson, M. (1996) Genetic validity of RAPD markers at the intra- and inter-specific level in wild *Brassica* species with $n = 9$. *Theoretical and Applied Genetics* 93, 9–14.

Le Thierry D'Ennequin, M., Panaud, O., Robert, T. and Ricroch, A. (1997) Assessment of genetic relationships among sexual and asexual forms of *Allium cepa* using morphological traits and RAPD markers. *Heredity* 78, 403–409.

Linne von Berg, G., Samoylov, A., Klaas, M. and Hanelt, P. (1996) Chloroplast DNA restriction analysis and the infrageneric grouping of *Allium* (*Alliaceae*). *Plant Systematics and Evolution* 200, 253–261.

Maaß, H.I. (1996a) About the origin of the French grey shallot. *Genetic Resources and Crop Evolution* 43, 291–292.

Maaß, H.I. (1996b) Morphologische Beobachtungen an Knoblauch. *Der Palmengarten* 60, 65–69.

Maaß, H.I. (1997a) Genetic diversity in the top onion, *Allium* × *proliferum*, analysed by isozymes. *Plant Systematics and Evolution* 208, 35–44.

Maaß, H.I. (1997b) Studies on triploid viviparous onions and their origin. *Genetic Resources and Crop Evolution* 44, 95–99.

Maaß, H.I. and Klaas, M. (1995) Infraspecific differentiation of garlic (*Allium sativum* L.) by isozyme and RAPD markers. *Theoretical and Applied Genetics* 91, 89–97.

May, B. (1992) Starch gel electrophoresis of allozymes. In: Hoelzel, A.R. (ed.) *Molecular Genetic Analysis of Populations. A Practical Approach*. IRL Press, Oxford, pp. 1–27.

Mayr, E. (1969) Grundgedanken der Evolutionsbiologie. *Naturwissenschaften* 56, 14–25. Reprinted in: Mayr, E. (1976) *Evolution and the Diversity of Life*, 2nd printing. The Belknap Press of Harvard University Press, Cambridge, Massachusetts, pp. 9–29.

Mes, T.H.M., Friesen, N., Fritsch, R.M., Klaas, M. and Bachmann, K. (1998) Criteria for sampling in *Allium* (*Alliaceae*) based on chloroplast DNA PCR-RFLPs. *Systematic Botany* 22, 701–712.

Mes, T.H.M., Fritsch, R.M., Pollner, S. and Bachmann, K. (1999) Evolution of the chloroplast genome and polymorphic ITS regions in *Allium* subg. *Melanocrommyum*. *Genome* 42, 237–247.

Messiaen, C.M., Cohat, J., Leroux, J.P., Pichon, M. and Beyries, A. (1993) *Les Allium Alimentaires Reproduits par Voie Végétative*. INRA, Paris, 230 pp.

Murphy, R.W., Sites, J.W., Buth, D.G. and Haufler, C.H. (1996) Proteins: isozyme electrophoresis. In: Hillis, D.M., Moritz, C. and Mable, B.K. (eds) *Molecular Systematics*, 2nd edn. Sinauer Associates, Sunderland, Massachusetts, pp. 51–120.

Ohri, D., Fritsch, R. and Hanelt, P. (1998) Evolution of genome size in *Allium* L. (*Alliaceae*). *Plant Systematics and Evolution* 210, 57–86.

Paran, I. and Michelmore, W. (1993) Development of reliable PCR-based markers linked to downy mildew resistance genes in lettuce. *Theoretical and Applied Genetics* 85, 985–993.

Peffley, E.B. and Orozco-Castillo, C. (1987) Polymorphism of isozymes within plant introductions of *Allium cepa* L. and *A. fistulosum* L. *HortScience* 22, 956–957.

Pooler, M.R. and Simon, P.W. (1993) Characterization and classification of isozyme and morphological variation in a diverse collection of garlic clones. *Euphytica* 68, 121–130.

Puizina, J., Javornik, B., Bohanec, B., Schweizer, D.M., Maluszynska, J. and Papeš, D. (1999) Random amplified polymorphic DNA analysis, genome size, and genomic *in situ* hybridization of triploid viviparous onions. *Genome* 42, 1208–1216.

Rieseberg, L.H. (1996) Homology among RAPD fragments in interspecific comparisons. *Molecular Ecology* 5, 99–105.

Rieseberg, L.H., Peterson, P.M., Soltis, D.E. and Annable, C.R. (1987) Genetic divergence and isozyme number variation among four varieties of *Allium douglasii* (*Alliaceae*). *American Journal of Botany* 74, 1614–1624.

Roxas, V.P. and Peffley, E.B. (1992) Short-day onion varietal identification using molecular (RAPD) markers. *Allium Improvement Newsletter* 2, 15–17.

Samoylov, A., Klaas, M. and Hanelt, P. (1995) Use of chloroplast DNA polymorphisms for the phylogenetic study of the subgenera *Amerallium* and *Bromatorrhiza* (genus *Allium*). *Feddes Repertorium* 106, 161–167.

Samoylov, A., Friesen, N., Pollner, S. and Hanelt, P. (1999) Use of chloroplast DNA polymorphisms for the phylogenetic study of *Allium* subgenus *Amerallium* and subgenus *Bromatorrhiza* (*Alliaceae*) II. *Feddes Repertorium* 110, 103–109.

Satoh, Y., Nagai, M., Mikami, T. and Kinoshita, T. (1993) The use of mitochondrial DNA polymorphism in the classification of individual onion plants by cytoplasmic genotypes. *Theoretical and Applied Genetics* 86, 345–348.

Savolainen, V., Manen, J.F., Douzery, E. and Spichiger, R. (1994) Molecular phylogeny of families relating to *Celastrales* based on rbcL 5′ flanking sequences. *Molecular Phylogeny and Evolution* 3, 27–37.

Schwarzacher, T. and Heslop-Harrison, P. (2000) *Practical* in situ *Hybridization*. BIOS Scientific Publishers, Oxford, 203 pp.

Sharon, D., Adato, A., Mhameed, S., Lavi, U., Hillel, J., Gomolka, M., Epplen, C. and Epplen, J.T. (1995) DNA fingerprinting in plants using simple-sequence repeat and minisatellite probes. *HortScience* 30, 109–112.

Singh, F., Ved Brat, S. and Khoshoo, T.N. (1967) Natural triploidy in viviparous onions. *Cytologia* 32, 403–407.

Smilde, W.D., van Heusden, A.W. and Kik, C. (1999) AFLPs in leek (*Allium porrum*) are not inherited in large linkage blocks. *Euphytica* 110, 127–132.

Smith, J.M. and Vuong Pham, T. (1996) Genetic diversity of the narrow endemic *Allium aaseae* (*Alliaceae*). *American Journal of Botany* 83, 717–726.

Soltis, D.E. and Soltis, P.S. (1998) Choosing an approach and an appropriate gene for phylogenetic analysis. In: Soltis, D.E., Soltis, P.S. and Doyle, J.J. (eds) *Molecular Systematics of Plants. II. DNA Sequencing*. Kluwer Academic Publishing, Boston, Massachusetts, pp. 1–42.

Soltis, D.E. and Soltis, P.S. (2000) Contributions of plant molecular systematics to studies of molecular evolution. *Plant Molecular Biology* 42, 45–75.

Stearn, W.T. (1978) European species of *Allium* and allied genera of *Alliaceae*: a synonymic enumeration. *Annales Musei Goulandris* 4, 83–198.

Swofford, D.L. (1993) *PAUP: Phylogenetic Analysis using Parsimony, Version 3.1.1*. Illinois Natural History Survey, Champaign, Illinois.

Tashiro, Y., Oyama, T., Iwamoto, Y., Noda, R. and Miyazaki, S. (1995) Identification of maternal and paternal plants of *Allium wakegi* Araki by RFLP analysis of chloroplast DNA. *Journal of the Japanese Society for Horticultural Science* 63, 819–824.

Thompson, J.D., Higgins, D.G. and Gibson, T.J. (1994) ClustalW: improving the sensitivity of progressive multiple sequence alignment through sequence weighting, position-specific gap penalties and weight matrix choices. *Nucleic Acids Research* 22, 4673–4680.

van der Valk, P., Kik, C., Verstappen, F., Everink, J.T. and de Vries, J.N. (1991) Independent segregation of two isozyme markers and inter-plant differences in nuclear DNA content in the interspecific backcross (*Allium fistulosum* L. × *A. cepa* L.) × *A. cepa* L. *Euphytica* 55, 151–156.

van Heusden, A.W., van Ooijen, J.W., Vrielink-van Ginkel, R., Verbeek, W.H.J., Wietsma, W.A. and Kik, C. (2000) A genetic map of an interspecific cross in *Allium* based on amplified fragment length polymorphism (AFLP™) markers. *Theoretical and Applied Genetics* 100, 118–126.

van Raamsdonk, L.W.D., Wietsma, W.A. and de Vries, J.N. (1992) Crossing experiments in *Allium* L. section *Cepa*. *Botanical Journal of the Linnean Society* 109, 293–303.

van Raamsdonk, L.W.D., Vrielink-van Ginkel, M. and Kik, C. (2000) Phylogeny reconstruction and hybrid analysis in *Allium* subgenus *Rhizirideum*. *Theoretical and Applied Genetics* 100, 1000–1009.

Vos, P., Hogers, R., Bleeker, M., Reijans, M., van de Lee, T., Hornes, M., Frijters, A., Pot, J., Peleman, J., Kuiper, M. and Zabeau, M. (1995) AFLP: a new technique for DNA fingerprinting. *Nucleic Acids Research* 23, 4407–4414.

Wendel, J.F. and Weeden, N.F. (1989) Visualization and interpretation of plant isozymes. In: Soltis, D.E. and Soltis, P.S. (eds) *Isozymes in Plant Biology*. Dioscorides Press, Portland, Oregon, pp. 5–45.

Werman, S.D., Springer, M.S. and Britten, R.J. (1996) Nucleic acids I: DNA–DNA hybridization. In: Hillis, D.M., Moritz, C. and Mable, B.K. (eds) *Molecular Systematics*, 2nd edn. Sinauer Associates, Sunderland, Massachusetts, pp. 169–203.

Wilkie, S.E., Isaac, P.G. and Slater, R.J. (1993) Random amplified polymorphic DNA (RAPD) markers for genetic analysis in *Allium*. *Theoretical and Applied Genetics* 86, 497–504.

Williams, J.G.K., Kubelik, A.R., Livak, K.J., Rafalski, J.A. and Tingey, S.V. (1990) DNA polymorphisms amplified by arbitrary primers are useful as genetic markers. *Nucleic Acids Research* 18, 6531–6535.

Wolfe, A.D. and Liston, A. (1998) Contributions of PCR-based methods to plant systematics and evolutionary biology. In: Soltis, D.E., Soltis, P.S. and Doyle, J.J. (eds) *Molecular Systematics of Plants, II. DNA Sequencing*. Kluwer Academic Publishers, Boston, Massachusetts, pp. 43–86.

Yamagishi, M. (1995) Detection of section-specific random amplified polymorphic DNA (RAPD) markers in *Lilium*. *Theoretical and Applied Genetics* 91, 830–835.

Zimmer, E.A., White, T.J., Cann, R.L. and Wilson, A.C. (eds) (1993) *Molecular Evolution: Producing the Biochemical Data*. *Methods in Enzymology* 224, Academic Press, San Diego, California.

9 Agronomy of Onions

A.-D. Bosch Serra[1] and L. Currah[2]

[1]*Departament de Medi Ambient i Ciències del Sòl, Universitat de Lleida, Av. Alcalde Rovira Roure 177, E-25198 Lleida, Spain;* [2]*Currah Consultancy, 14 Eton Road, Stratford-upon-Avon CV37 7EJ, UK*

PART 1. INITIAL CONSIDERATIONS AND CROP ESTABLISHMENT

1. Introduction

Onion agronomy was reviewed comprehensively during the early 1990s (Brewster, 1990, 1994; Corgan and Kedar, 1990; Currah and Proctor, 1990; Uzo and Currah, 1990). Here, we will update the topic, with emphasis on recent changes.

Two important and interrelated trends can be distinguished. One is the quantification of management aspects, including irrigation scheduling, weed- and pest-damage forecasting and growth modelling. A concrete example of the improved productivity made possible by applying this new knowledge will be described from Spain.

The second trend is the movement towards greater environmental awareness. The major challenge today is how to produce onion crops in ways that are sustainable and environmentally responsible and still provide an economic return to the grower.

Today's high yields have been achieved by the use of crop-protection chemicals as a substitute for costly hand labour, particularly for weed control. Total inputs of all pesticides have been estimated at 23 kg ha^{-1} per crop in The Netherlands (E. Steenge, The Netherlands, 2000, personal communication) and 10.5 kg ha^{-1} of pesticide active ingredients during the post-transplant cropping season in Norway (Saethre *et al.*, 1999). Under these systems, onions also need substantial quantities of mineral nutrients. Emissions of pesticides and of nitrogen (N) in various forms from onion fields can be higher than from other crops, as can the impact on soil and aquatic organisms (Wijnands and van Asperen, 1999).

Integrated crop management (ICM) systems are refined and less wasteful versions of traditional production methods, with more rational use of resources in response only to defined needs. ICM permits the careful usage of pesticides, but also demands increased efficiency of use of all external crop-production resources, including fuel, water and chemical inputs.

Even while this refining process is taking place in conventional production, increased consumer demand for 'organic' vegetables means that growers who convert to organic production, in search of a better market, need to find ways of supplying adequate crop nutrition and controlling weeds, pests and diseases without using synthetic chemicals. Reports on the methods being developed are included in this chapter.

2. Establishing an Onion Crop: What, Where and How

2.1 Diversity and uses

Consumers and processors use onions as green or salad onions with or without bulbs, as fresh bulbs soon after harvest or as dry bulbs stored for later use when fresh onions are not available; others are destined for processing, either as fresh-cut products, for pickling, for freezing, for dehydration as flakes and powder, or as onion oil after distillation. The distance of the farm from the factory is of economic importance.

Some distinctive local onion products include blanched onion sprouts ('calçots'), produced in Tarragona, in north-eastern Spain, from the long-day (LD) and short-dormancy cv. 'Blanca Grande Tardía de Lérida'. Large bulbs from the mid-July harvest are stored briefly in the open and then replanted in August to September at a 40 cm × 30 cm spacing. When sprouts are 30 cm long, they are earthed up repeatedly two or three times between October and December. White, sweet, smooth-tasting shoots (about four to seven per bulb) are harvested from November to March, before inflorescence elongation. The quality of the product, sold as 'Calçot de Valls. Denominació de Qualitat', is regulated and certified (DARP, 1995) and controlled by an authorized company, which performs an EN45011 standard assessment. Groups of 25–50 shoots, with a 15–25 cm white part, 1.7–2.5 cm in diameter at 5 cm from the roots, are taped together with a numbered certified label, and are sold for cooking directly in the flames from burning vine shoots, and consumed as a main dish with a special sauce. 'Calçots' are a restaurant speciality in Spain (http://www.altcamp.info/calçotada.htm; http://www.gencat.es/darp/c/departam/revista/cgabir40.htm) and just one example of the varied Mediterranean allium cuisine.

2.1.1 *Choice of cultivar*

For each latitude and altitude zone, onion cultivars can be chosen to suit particular markets and growing seasons. Consumers have rather conservative preferences and often reject unfamiliar-looking bulbs unless onions are in very short supply. For example, imports to northern Europe, where yellow-brown onions are common, are mostly of similar-looking onions, e.g. from New Zealand and Tasmania.

Where winters are mild, such as in Israel or on the Spanish coast, distinct seasonal groups of cultivars are grown successively to provide supplies for most of the year. In other areas, for example Scandinavia or Russia, there may be just one summer production season.

The most common way of classifying onion cultivars is by day-length sensitivity: short-day (SD), intermediate-day (ID), LD (as characterized by US 'long-day' onions) and very long-day (VLD) (northern Europe). The related crop, shallot (*A. cepa* Aggregatum group), is described by Rabinowitch and Kamenetsky (Chapter 17, this volume). Within the principal day-length groupings, we can distinguish early, main-crop and late variants for each major season. Varieties within defined day-length types vary in shape, size, firmness, bolting, skin and scale colours – white, light yellow, dark yellow/brown, bronze, pink, red and dark purplish red – pungency, sweetness and juiciness and in their potential for storage.

There has been a move recently from open-pollinated (OP) to hybrid onion cultivars in many countries and wide ranges of hybrids are now available (Havey, 1999). The advantages for seed companies are obvious: they keep control of the parental lines and minimize the risk of pirating of varieties. In Israel, the change to hybrids was followed by a doubling of yields from 20–50 to 50–100 t ha^{-1} in fresh market onions and from 8 up to 15 t ha^{-1} of dry matter in onions for processing. These yield increases can be attributed both to improved genotypes and hybrid vigour and to superior management (H.D. Rabinowitch, Israel, 2000, personal communication). Hybrids have taken over much of the north-western European market and are being bred in countries such as Holland and Poland (where until recently only improved OPs from landrace stocks were normally grown). In the USA, some OPs are still common, especially in areas where transplants are grown (South Texas, southern California), but elsewhere there has been a general swing towards hybrids. Yet some scientists have disputed the advantages of hybrids to growers (Dowker and Gordon, 1983; van der Meer, 1993). Table 9.1 lists the main onion types and some modern cultivars derived from them, and Table 9.2 provides details of onions grown in Russia and the Central Asian republics. We would welcome more information from our readers on the onions currently grown in temperate countries, as Table 9.1 is far from comprehensive.

For processing, the choice is more limited than for fresh onions. The Basic Vegetable Products Company (California, USA) supplies contract growers with a range of white, high-dry-matter cultivars for dehydration purposes to suit different environments. In Mendoza, Argentina, 'Southport White Globe' ('SWG') has been adopted (Belettieri, 1997); it is also grown in New Zealand (Rogers, 1989) and Spain (Bosch Serra, 1999). In New Zealand one of the advantages of 'SWG' is its prolific root system, which gives it relatively good tolerance to pink root rot. In the north-east of Spain, dehydration cultivars such as 'Staro', 'Albeno' and 'Albion', are also recommended (Bosch Serra, 1999).

SD onions for fresh consumption are discussed and listed by Currah (Chapter 16, this volume): they include West African, Indian and Creole onions, as well as US 'Grano' and 'Granex' types.

2.1.2 Onion markets and their preferences

The great diversity of onion cultivars offers a wide range of size, shape, flavour and quality. In north-western Europe, most 'cooking'-quality onions sold are of medium size, globe-shaped and yellow-brown, with a preference in the UK for darker-coloured skins. There are smaller market segments for mild, large Spanish onions, shallots, red and white onions, and an ever-expanding market for salad onions, which are now imported into Europe during the winter. Not all salad onions are *A. cepa*, as *A. fistulosum* offers the possibility for summer production under long days without the risk of bulbing. Cv. 'White Lisbon' and selections from it are the main type of *A. cepa* salad onion. A developing market exists for organically produced onions, so far mainly satisfied by a few countries such as Argentina. In Europe, onion quality standards are high and regulated (Commission of the European Communities, 1983, 1997).

Table 9.1. Onions of temperate and Mediterranean climates: principal commercial types and some recent cultivars developed from them. For short-day onions of the USA, see Table 16.8, Chapter 16.

Country	Cultivar type and examples	Description
Europe		
UK, Belgium, Netherlands and Germany	Rijnsburger (VLD) Hyton, Hysam, Hygro, Hyfast, Hyskin, Augusta, Jumbo, Wijbo, Heldis, Balstora, Caribo, Promo, Mercato, Robusta, Renate, Durco, Produrijn, Luctor	Brown, globular bulbs with well-developed skins, suitable for storage
	Stuttgarter Riesen (VLD) Sturon, Diskos, Plano	Brown, rather flat, for early summer production from sets
	Zittauer (LD) Zirius, Beno, Luna	
	Brunswick Blood Red (VLD) Red Baron, Rodo, Mimo	Somewhat flat, good for storage
	White Lisbon, Winter Hardy Lisbon	Salad onions
	Turbo, Centurion	Used for sets
	Calypso, Super Bear, Spirit, Summit, Armstrong, Albion, Fantasy, Trafford, Hyfield, Taletom Takmark, Takstar	Recent cvs from trial reports
	Imai Early Yellow, Buffalo, Ocean, Siberia, Felix, Radar, Amigo, Juno, Swift, Arctic, Wapiti, Glacier, Baltic, Nordic, Dynasty, Continental, Yellow Stone	Overwintering onions from Japanese cvs
	Shallots: Hollandse Gele, Topper, Atlantic, Santé	Shallots grown in NW Europe excluding Brittany
France	Auxonne, de Mulhouse (LD)	Yellow/brown, suitable for production from sets
	Jaune Paille des Vertus (LD) Verton	Yellow/brown, slightly less long-day than Rijnsburger types
	Rouge de Niort (LD)	Pink regional variety
	Rouge de Toulouse (ID)	Pink regional variety
	Jaune hâtif de Valence (SD)	Similar to Spanish Babosa type
	Jaune d'Espagne (ID/LD)	Similar to Spanish summer onions
	Shallots: Mikor, Jermor, Longor, Arvro, Trégor, Delvad	Selections from Jersey-type reddish shallots, available as virus-tested stocks
Spain[1]	Reina de Abril, Babosa (SD)	Overwintering early fresh onion, low dry matter
	Liria (ID)	Second early yellow/green with brown tones, summer onion
	Valenciana de Grano, Recas (LD)	Summer storage onion, brown skins, large-size bulbs
	De la Reina, Blanca de Paris, Amonquelina, Amarilla Paja Virtudes (SD)	Regional SD cvs Overwintering cultivar maturing in June, copper skin
	Dulce de Fuentes, Blanca del País, Amarilla Achatada (ID)	Regional ID cvs
	Colorada de Figueras, Colorada de Vich, Morada de Amposta, Morada de Zalla (LD)	Regional LD cvs (summer onions for Spanish conditions)
	Blanca Grande Tardía de Lérida (LD)	

Europe—*Continued*		
Poland	Wolska, Rawska, Blonska, Sochaczewska (LD), Czerniakowska, Zytawska	Summer brown storage cvs with good skins and long dormancy
Denmark	Owa	Elongated brown summer onion
Switzerland	Wädenswil	
Hungary and Austria	Makói (LD) Aroma, Alsógödi	Long-storing light-brown globular bulbs
	Makói fehér (white), Makói bronz	Colour variants of Makói
	Makomi	Recent selection from Makói
Czech Rep.[3]	Vsetana	
	Obrovska Zluta	
	Sumperska	
	Zazriva 2	
	Moravska Polhora	
	Alice	
Bulgaria[2]	Lyaskovski, numbered selections	Flat–round, 15–18% DM; suitable for sets and storage
	Slivenski, numbered selections	More globe-shaped bulbs than Lyaskovski
	Starozagorski	Flatter bulbs
	Asenovgradski zhult, sel. Plovdivski 10	Best for long transport, shaped like Slivenski
	Melnishki	Strongly flattened bulbs, yellow skin, good storability
	Samovodska kaba (syn. Bulgarska kaba, Kantar Topuz), numbered selections	Bulbs with conical shape (wider at top?), sweet taste, average winter storability
	Shumenska burzitska	Earlier than Samovodska, sweet
	Gyumyurdinska kaba	Flat bulbs, firm, with good storage, resistant to downy mildew
	Asenovgradska, numbered selections	Red salad onion for S. Bulgaria
	Trimontsium	Bulbs flat/round, good yields, suitable for storage; from cross between 'Plodivski 10' and a Spanish type ex USA
	Pioner	Bulbs nearly round, pungent, good storability; from cross of Bulgarian and US cvs
	Konkurent	Bulbs near round, high dry matter; from cross of 'Makoi' and 'Plovdivski 10'
	Jubilei 50	Nearly round bulbs, good storage quality; from cross of 'Makoi' and Austrian cv. 'Bernsteinfarbige'
	Plovdivski cherven	Red, from cross of 'Red Wethersfield' (US) and 'Red Flavour' (Dutch cv.)
	Ispanski 482	Developed from US Sweet Spanish type

Continued.

Table 9.1. *Continued.*

Country	Cultivar type and examples	Description
Europe—*continued*		
Slovenia	Belokrankja	Summer onions from sets
	Holandska Rumena	Summer onions from sets
Yugoslavia (former)	Kupuzinski Jabucar	
	Prizrenski Pogacar	Good storage onion, high dry matter
Romania[4]	Diamant, Aurie de Buau, Rosie de Aries, Rosie de Gagaras	
	Rosie de Turda[3]	
	Gurghiu[3]	
Italy	Dorata di Parma (LD)	Summer storage onion
	Tropea, Tropeana (ID)	Long-bulbed red onion, low dry matter, sweet
	Italian Flat Red	Sweet red summer onion, low dry matter
	Barletta, Pompei	Small white pickling onions, grown in North Europe
	Bianca di Giugno	Early summer onion
	Giarratana, Density, Blanco Duro	Onions cultivated in Italy recently
	Rossa di Firenze, Fuego, Sterling, Borettana, Himera, Ruby	
	Rossa di Lucca[3]	
Norway	Lafort, Lava, Laskala	Very-long-day selections from Rijnsburger types
Finland	Potato onion, e.g. Helsingin Yliopisto	Yellow-skinned multiplier onion type, vegetatively propagated
	Lemi[3]	
Greece	Vatikiotiko	Red-skinned overwintering onion
Croatia	Fuseor[3]	
Republic of Macedonia	Jesenachki, Moldavski, Skopski, Srebrenjak, Sidra	
Asia		
Turkey	Yalova numbered varieties	Yellow/brown storage and export onions
	Çorum, Akgün 12	
	Kantartopu	

Asia—*Continued*		
Iran	Hamadan, Arak, Zanjan	
	Sefide Kashan, Ghermez Azarshahr, Toupaz	
	Azimi, Massiha, Moghaddam, Valizadeh	
	Dorcheh	Long-storing cv.
Georgia	Ravalsviliani[3]	
Pakistan (NW)	Swat-1 (ID)	Red, globe-shaped, short storage life
Ukraine	Chebotarskii, Luganskii, Moldavskii	
Russia[5]	Krasnodar	Used to force for green tops in Siberia
	Ispanskij, Karatalskij, Chernomorskij	Mediterranean ecological group
	Andizhanj, Masallinskiij, Samarkand	Central Asian ecological group
	Vertiuzhanskij, Dnestrovskij, Kaba, Skwirskij	Central European ecological group
	Arzamasskij, Bessonovskij, Danilovskij, Kilichinskij, Mstersky, Rannij Zheltyi, Rostovskij repchatyi mestnyi, Strigunovskij	Central Russian ecological group
	Pogarskij[3]	
	Sibirskii Skoropelyi[3]	
	Tschernuschka[3]	
Japan	Senshyu (ID)	Yellow globe type (via USA)
	Sapporoki (LD)	Yellow globe from northern Japan, summer onion
	Extra Early Imai, Kaizuka-wase (ID)	Yellow flattened globe for overwintered production
	Tsukisapu, Tsukihikari, Toyohira (LD)	Cvs for Hokkaido, North Japan
	Kitamiku-25	
	Takanishiki	Bred for South Japan
	Kairyo-Unzenmaru	
	Kamui, Rantaro	
	Highgold Nigou, Eskimo	
P R China[6]	(NW and NE) Topaz, Xiongyue	For overwinter or summer production
	(Yellow R) Tianjin, Dashuitao, Beijing purple and yellow	

Continued.

Table 9.1. *Continued.*

Country	Cultivar type and examples	Description
Asia—*Continued*		
Republic of Korea	Cheongdangnang	
	Paechong-Joseng	
	Chenju-Whang	
	Kinkyu	Early-maturing overwintering cv.
	Bonganghwang	Late-maturing overwintering cv.
	Yongangwhang	
	Samda	
North America		
USA and Canada	Yellow Globe types (LD), e.g. YG Danvers, Early Yellow Globe, Wolverine, Ebenezer, Copra, Corona, Prince, Spartan Banner	Similar to North European brown onions with slightly shorter DL response
USA	Sweet Spanish types (ID/LD), e.g. Olé, Riverside, Ringmaker, Big Mac, Valiant, Gringo, Daytona, Orogrande, Robin, Celebrity, Sweet Sandwich, Colorado No. 6, Sweet Spanish, Buffalo, Walla Walla	Similar to Spanish summer storage onions, many selections including hybrids
	Southport White Globe (LD), Dehydrators	White dehydrator onion, selections and hybrids
	Red Wethersfield	Red cv. from New England
	Norstar 210B, Rocket	Grown in Canada and midwest of USA
	Shallot: Red California; Frog's Legs[3]	
South America		
Argentina	Sintetica 14, Valcatorce INTA (LD)	Similar to Spanish summer-storage onions
	Antartica INTA, Valenciana	
Chile	Valenciana (LD) Valenciana-INIA, Valcatorce	Similar to Spanish summer-storage onions
	Perla	White US cv.
South Brazil (RS, SC, PA)	EMPASC-351 Seleçao Crioula	Cvs bred in Santa Catarina State from local traditional varieties
	EMPASC-352 Bola Precoce	
	EMPASC-355 Juporanga	
	EMPASC-356 Rosada	
	Baia Periforme, southern selections	

South America—*Continued*		
South Brazil (RS, SC, PA)	Aurora, Petrolini	Mentioned in recent reports
	EPAGRI-362 Crioula Alto Vale	
	EPAGRI-363 Superprecoce	
	Norte 14	
Australasia		
Australia	Creamgold, Early Creamgold (ID/LD) and selections, hybrids from it	Thick-skinned brown onions for export; smaller than Spanish storage but thought to have been derived from them
	White Spanish (ID), SA White Globe	White globe storage onion of South Australia
	Southport White Globe (LD)	Dehydration
New Zealand	Pukekohe Long Keeper (LD) and selections from it	Very-long-storing onion, related to Creamgolds but slightly longer-day response
	Southport White Globe (LD)	For dehydration
Africa		
Morocco	Rouge d'Amposta	Reddish/purple Spanish ID cv.
Algeria	Rouge d'Amposta	
Egypt	Giza 6, Giza 20, Shandaweel, Beheri (= local)	ID cvs resembling Spanish storage onions
South Africa	Caledon Brown, Australian Brown (ID)	Local selections of Spanish storage types
	Radium	Locally bred hybrid from Creole type
	Cape Yellow Flat	Related to Dutch Straw Yellow flat onion

[1]A.-D. Bosch Serra.
[2]Todorov (1999a, b).
[3]Fischer and Bachmann (2000).
[4]Scurtu (1999)
[5]I.G. Tarakanov (2001); see Table 9.2 for details.
[6]Xu *et al.* (1994).
The remainder were collected from articles noted in *Horticultural Abstracts* 1994–2000 and from L. Currah's literature collection.

Table 9.2. Main Russian onion cultivars grown in the states on the territory of the former USSR (from Tarakanov, 2001).

Cultivar	Type	Maturity	Bulb shape	Bulb index	Skin colour	Flesh colour	Storage ability
Mediterranean ecological group							
Ispanskij cv. group							
Ispanskij 313*†	ID	Late	Globe	0.9–1.0	Light yellow, sometimes pinkish	White, greenish	Poor
Krasnodarskij G-35*	LD	Medium	Globe and high globe	0.9–1.3	Yellow, sometimes pinkish	White	Poor
Oranhzhevyi	LD	Medium	Globe and flattened globe	1.0–1.1	Deep yellow	White	Medium
Karatalskij cv. group							
Donetskij zolotistyi	LD	Medium	Flattened globe	0.8–1.1	Yellow and straw-yellow	White	Medium
Karatalskij*†	LD	Early	Globe and flattened globe	0.9–1.1	Yellow and golden-yellow	White	Medium
Oktiabrskij*	LD	Medium	Globe	0.9–1.1	Yellow-brown	White	Good
Chernomorskij cv. group							
Chebotarskij mestnyi	LD	Medium late	Flattened globe	0.8–0.9	Dark purple, red	White, purple	Good
Central Asian ecological group							
Andizhan cv. group†(only)							
Andizhanskij†	ID	Late	Globe and high globe	0.8	White	White	Medium
Margelanskij kruglyi†	ID	Late	Flattened globe	0.8–1.0	White	White, sometimes greenish	Medium
Masallinskij cv. group							
Masallinskij mestnyi†	ID	Late	Flattened globe	0.6	Purple	Light purple	Medium
Samarkand cv. group							
Kartlis *(only)	ID	Medium	Flattened globe	0.6–0.7	Purple	Light purple	Medium
Samarkandskij krasnyi†(only)	ID	Late	Flattened globe	0.6–0.7	Purple, reddish	Light purple	Medium
Central European ecological group							
Vertiuzhanskij cv. group							
Vertiuzhanskij tiraspolskij	LD	Early	Thick flat	0.5–0.75	Light brown	White	Good
Dnestrovskij cv. group							
Dnestrovskij	LD	Medium late	Globe	0.9–1.0	Brown-yellow, light yellow	White	Good
Luganskij*	LD	Late	Globe and flattened globe	0.9–1.1	Yellow-brown	White	Good
Kaba cv. group							
Kaba (Kaba zheltyi)†	LD	Late	Flattened globe and globe	0.6–1.1	Yellow, sometimes brownish	White, greenish	Poor
Kaba 132†(only)	ID	Late	Globe	0.9–1.1	Yellow, sometimes brownish	White, cream-coloured	Medium

Skwirskij cv. group							
Odnoletnij sibirskij	LD	Early	Flattened	0.8	Yellow, sometimes light yellow	White	Good
Odnoletnij khavskij 74	LD	Early	Globe and flattened globe	0.8–1.0	Yellow and yellow-brown	White	Good
Skwirskij	LD	Medium	Flattened globe	0.7–0.9	Yellow, brownish	White	Good
Central Russian ecological group							
Arzamasskij cv. group							
Arzamasskij mestnyi	LD	Medium	Globe and high globe	0.9–1.1	Yellow, brownish	White	Good
Ufimskij mestnyi	LD	Medium	Globe	0.95–1.1	Yellow, brownish	White	Good
Bessonovskij cv. group							
Bessonovskij mestnyi	LD	Early	Thick flat, flattened globe	0.6–0.7	Yellow	White	Good
Pogarskij mestnyi uluchshenny	LD	Early	Thick flat, flattened globe	0.7–0.8	Yellow	White	Good
Danilovskij cv. group							
Danilovskij 301	LD	Medium	Thick flat, flattened globe	0.6–0.7	Deep red, purple	Light purple	Medium
Kilinchinskij cv. group							
Kilinchinskij mestnyi	LD	Medium	Thick flat, flattened globe	0.65–0.85	Purple, deep purple	White, purple	Good
Msterskij cv. group							
Msterskij mestnyi	LD	Early	Thick flat, flattened globe	0.6–0.7	Yellow, sometimes pinkish	White	Good
Miachkovskij	LD	Early	Thick flat, flattened globe	0.6–0.7	Yellow, sometimes pinkish	White	Medium
Rannij zheltyi cv. group							
Rannij zheltyi	LD	Early	Thick flat, flattened globe	0.6–0.7	Yellow	White	Medium
Rostovskij repchatyi mestnyi cv. group							
Rostovskij repchatyi mestnyi	LD	Early	Flat and flattened globe	0.5–0.9	Yellow	White	Good
Spasskij mestnyi uluchshennyi	LD	Medium	Thick flat, flattened globe	0.6–0.8	Yellow, brownish	White	Medium
Timiryazevskij	LD	Early	Flattened globe	0.75–0.8	Light brown, straw-yellow	White	Good
Strigunovskij cv. group							
Strigunovskij mestnyi	LD	Early	Globe	1.0–1.2	Yellow	White	Good
Chernigovskij	LD	Medium late	Globe and flattened globe	0.95	Deep yellow	White	Good

Continued.

Table 9.2. *Continued.*

Cultivar	Type	Maturity	Bulb shape	Bulb index	Skin colour	Flesh colour	Storage ability
Recently selected varieties							
Peshpazak†(only, autumn-sown)	SD	Early	Flattened globe	0.6–0.8	Yellow, pinkish or light grey	White	Medium
Dusti†	ID	Medium	Globe	0.9–1.1	Brown	White	Medium
Odintsovets	LD	Medium	Globe and flattened globe	0.7–0.8	Yellow	White	Good
Zolotnichek	LD	Medium	Globe	0.9–1.1	Yellow	White	Good
Rannij rozovyi	LD	Medium	Globe	0.9–1.0	Pink, with yellowish spots	White	Good

*Besides the temperate zone of Russia, these varieties are grown in the Caucasian republics.
†Besides the temperate zone of Russia, these varieties are grown in the Central Asian republics (Andizhan cv. group – only in these republics).

The most important registered (in 2000) varieties of Russian selection are given in the table. Besides these, a number of cultivars of foreign selection are registered in the Russian Federation (not all of them are grown on a wide scale):
F_1 hybrids: Banco, Brandy, Brenda, Hyton, Jungo, Durko, Copra, Corona, Prince, Summit, Spirit, Stardust, Tamara, Bonilla.
OP varieties: Bulcato, Grandina, Della Rocca Bruna, Kutnowska, Musona, Olina, Red Baron, Sochaczewska, White Sweet Globe, Hiberna, Shetana, Stuttgarter Riesen, Exhibition.

The cultivars in the table are classified according to Professor A.A. Kazakova's classification system, which is accepted in Russia. Thus, it is used in the last edition of *The Guide for Approbation of Vegetable Crops* (Moscow, 1982). However, the application of this classification may be a subject for discussion (e.g. see P. Hanelt, in *Onions and Allied Crops* (Boca Raton, 1990), vol. 1, pp. 19–20). Nevertheless, headings in the table provide the reader with useful information about the geographical distribution and ecological adaptedness of the cultivars.

Vegetatively propagated shallots are also traditionally popular in Russia; they are grown mostly in the western and north-western parts of the country, in the Urals, Siberia and the Far East. A great number of locally adapted clones, producing up to 20 bulbs in a cluster, with good storage ability, can be found there. Often, they are also used for forcing. There are several registered varieties as well: Belozerets 94, Kubanskij zheltyi D-332, Kunak, Sibirskij zheltyi, Sir 7, Sprint.

In Mexico, white bulb onions are preferred. In the USA, large, very mild, yellow onions are popular, though in the north-eastern and central states, large volumes of yellow pungent onions, similar to their European counterparts, but with slightly shorter day-length response, are grown and stored for winter supply. There is a major production area in the Rocky Mountain states of the USA, from Colorado up to Washington, where large, sweet Spanish-type soft onions are grown in the summer and stored in bins rather than in bulk for the winter.

Sweet onions of the 'Grano'/'Granex' type are produced in Georgia (Vidalia area) as well as in southern Texas and southern California (see Currah, Chapter 16 this volume). In Washington State, 'Walla Walla Sweet' (ID) is produced.

In Japan, large, sweet, yellow onions are also in demand, and producers in South-East Asian countries compete to supply them. Japanese consumers appreciate the sweet, juicy, overwintered onions with resistance to bolting, which have been selected in Japan from North American sources. Longer-day summer Japanese onions include cv. 'Sapporoki' and hybrids with superior disease resistance, which have been developed from it (Table 9.1).

Better communication about the areas planted in major production zones, the effects of weather conditions in real time and the state of markets worldwide are beginning to evolve. Models can be used, in producer countries, in order to predict some of the impacts on national onion economies as a consequence of more open markets. The effect of the liberalized US–Mexico onion trade has been analysed by Fuller *et al.* (1996), using an intertemporal equilibrium model of the North American dry-onion economy.

Even the best models will not completely solve the problems of unforeseen weather conditions and other factors that affect world onion prices. Swings in supply and demand from year to year and unforeseen gluts on international markets are likely to remain part of the onion supply picture for some time to come. Commercial web sites that supply price information on an international basis are starting to appear.

2.2 Onion-crop establishment

2.2.1 Rotations

The onion crop is susceptible to many root diseases. Problems associated with pathogenic fungi become serious when onions are grown as a monoculture or when growers use the same fields for seed-beds year after year (Sumner *et al.*, 1997). Rotations are a key aspect for sustainable agricultural production systems: they can help to control nitrate leaching and minimize herbicide and pesticide applications.

The order of the rotation is important: crops with high-volume residues, e.g. maize, may affect onion emergence, and some crops, such as lucerne, should be avoided as a preceding crop, since they may reappear as weeds. Maroto Borrego (1995) considered that vegetables such as tomato, pepper, aubergine, melon, cucumber, cauliflower, lettuce, beans or peas are good preceding crops: he suggested a rotation sequence, covering 3 years, of early cauliflower, onion cv. 'Babosa', chufa (*Cyperus esculentus*), early potatoes, lettuce, artichoke and melon, for the central-eastern coast of Spain.

In the north-east of Spain (Pla d'Urgell, Catalunya), a rotation of onions–maize–wheat/lucerne–wheat is a common practice. In Valais (Switzerland), onions are grown in a rotation that includes carrots, cabbages and, in a less common rotation, tomato and celery (Rossier *et al.*, 1994). In Palampur (India), a three-crop sequence, aubergine–Chinese cabbage–onion or okra–radish–onion, gave higher gross returns than other onion sequences evaluated (Arya and Bakashi, 1999). In New York State, USA, in muck (organic) soils, improvements in onion yield and quality were found after a rotation to Sudan grass (*Sorghum sudanensis*) in run-down soil (http://www.nysipm.cornell.edu/reports/ann_rpt/AR97/com_veg.html).

Environmental criteria can also be considered in planning onion crop rotations. Shock *et al.* (2000a) in eastern Oregon found that sugarbeet is a good scavenger of residual soil nitrate to follow shallow-rooted onions: the beet benefits from N residues and diminishes the potential for nitrate pollution of groundwater. In Israel, damage by *Sclerotium rolfsii* to crops of groundnuts, tomatoes and beans could be reduced by including onions in the rotation (Zeidan *et al.*, 1986).

Onions grown as an intercrop can reduce pest damage. In Bulgaria, intercropping beans with onions significantly reduces the density of the damaging bean weevil (*Acantoscelides obtectus*), leaf aphids (*Aphis* spp.) and red spider mite (*Tetranychus urticae*) (Mateeva *et al.*, 1998). Xu *et al.* (1994) listed several intercrops grown with onions in different parts of China.

2.2.2 Land preparation and soil management

Onions are cultivated in all kinds of soils: sandy (Georgia, USA; Norfolk, UK), heavy clay (Venezuela), peat organic soils (USA and Canada) or volcanic soils (Chile). Heavy and stony soils hamper mechanical harvesting. The onion crop requires a homogeneous, fine soil structure in the surface layer. Hence, the most suitable soils are sandy loams to loams with a fair content of organic matter and good soil structure. Various implements can be used for field preparation (mould-board plough, subsoiling plough, cultivator, rotavator) but soil compaction must be prevented. In sandy soils, a roller may be used after preparatory work on the soil. Elevated beds (15 cm or more high) allow good drainage and can be constructed to combat soil erosion. Sometimes serpentine furrows are set up inside larger-sized beds, which allows water to be supplied to a limited area at one time.

On acid soils with pH < 6.0, lime is applied 2–3 months before land preparation to bring the soil pH into the 6.2–6.5 range (Kelley and Granberry, 2000). These pre-plant applications affect onion bulb quality and storage by increasing bulb firmness or decreasing the percentage of bolting (Randle, 1995). On organic soils, a pH of 5.5 is considered sufficient.

Crust formation may impede onion seedling emergence, occasionally making a second sowing necessary. In parts of the northern USA and in Canada, onions are often grown on muck (i.e. organic or peat) soils, which can maintain a tilth that prevents crusting. Surface protection from the impact of raindrops by mulching, the incorporation of organic matter or high-frequency but low-intensity irrigation during early growth stages helps to prevent crust formation or mitigate its effects. Spraying the soil surface with 500×10^{-6} g ml^{-1} of polyacrylamide at a rate of 4 l m^{-2} and phosphogypsum (PG) (a calcium sulphate by-product that results from the process of wet acid phosphorus (P) production) spread 24 h later at a rate to 5 t ha^{-1}, prior to a simulated rainstorm, were useful for preventing crust formation and increasing onion emergence in non-sodic, loamy-textured soils; PG alone was ineffective (Rapp *et al.*, 2000). Nevertheless, in other soil conditions, with water having an electrical conductivity of 0.65 decisiemens (dS) m^{-1} on average, similar PG applications have been reported to be useful (Ramírez *et al.*, 1997) in onion nurseries for improving emergence.

Vermiculite can be also used as an anti-crustant. It can be applied over the seed row (1.3 m^3 ha^{-1}) and then covered with a thin soil layer (5 mm), which also prevents wind and water erosion; H_3PO_4 banded over the row (57 kg P ha^{-1}) is also effective (Hemphill, 1982).

In Korea, Jo *et al.* (1997) found that the application of an emulsified plastic gel on a drying, but still moist, soil promoted the formation of water-stable soil aggregates. This decreased soil cracking and increased emergence rates of Japanese bunching onions. In Thailand, rice straw mulch is used.

Solarization is useful for efficiently sterilizing soil prior to sowing in climates where it can be applied effectively (Katan and DeVay, 1991). Moist soil is heated (> 40°C) over 4 weeks by mulching with transparent polyethylene, killing pathogens and weed seeds (Rabinowitch *et al.*, 1981). In Egypt, this method has been used to treat onion seedbeds and production fields successfully (Satour *et al.*, 1989; Abdallah, 1998).

In Florida (USA), Vavrina and Roka (2000) showed that plastic mulches improved the net return from SD onions by increasing bulb weight and marketable yield.

Using white-on-black plastic mulch resulted in the highest yields of jumbo (> 10 cm) onions.

Onions are very sensitive to salinity: a significant decline of growth was recorded even at 1.4 dS m^{-1} treatment solution (Wannamaker and Pike, 1987). Therefore saline soils should be avoided if possible. Cultivars vary in salinity tolerance during germination or early growth (Palaniappan *et al.*, 1999). Nevertheless, in some irrigated areas, from arid to subhumid, onions are cropped in (sub)saline soils. This usually depresses yields (Ramírez and Rodríguez, 1997), even when techniques to combat it are used – for example, planting seedlings on the side of the ridges. Salinity affects yields more when onions are grown at higher temperatures and at low air humidity (Maas, 1990). Shannon and Grieve (2000) regard onion as one of the crops most sensitive to poor water quality. An adequate drainage system to remove saline runoff water is essential.

Soil characteristics may affect onion storability. Rossier *et al.* (1994) evaluated 10-year data from four soil classes in Switzerland. Onions from soils with more than 5.4 mg sodium (Na) 100 g^{-1} of soil showed the highest storage aptitude, developing fewer sprouts, but were smaller than those from soils with lower Na content; and calcium (Ca) level in the cell sap was related to sprouting. However, soil salinity during crop growth must be maintained lower than 370 mg of salts 100 g^{-1} of soil in order to avoid any toxicity to onion plants.

Soil properties are also the determining factor for the mobility and persistence of pesticides in soils. In Florida, Buttler *et al.* (1998) published an electronic extension guide to help onion growers to select pesticides according to soil properties, such as leaching or runoff ratings, so as to safeguard water supplies.

2.2.3 Direct sowing and factors affecting seedling emergence

Direct sowing is the dominant method for salad-onion production, for bulb onions under mechanized cultivation systems in humid temperate climates, for processing onions and where subsequent irrigation can be well managed. Seeds of high germination capacity, often pelleted, are sown to the desired stand, using a seed rate that takes into account a 'field factor'. This allows local soil conditions to be used to modify the tested germination percentage (ISTA, 1985). In loamy sand and silty soils, Kretschmer (1996) found a good correlation between a vigour test and field onion emergence.

For salad onions, a bed system with several rows at thick spacing is usual. For bulb production, beds can also be used or machinery can be adapted to other systems – for example, ridges and furrows. On the flat, precision or pneumatic seeders can be used. The Stanhay belt-type seeder had a better seeding uniformity overall, but a higher percentage of misses (15%), than the Carraro vacuum-type model (5%) (Bracy and Parish, 1998).

Ideally, soil should be kept damp until the seedlings emerge. Drying out at this stage can lead to uneven emergence, which reduces uniformity of development throughout the life of the crop. A well-worked but fairly dense layer of soil is needed below the seeds so that water can reach them by capillarity. The seed is firmed in by a press wheel or a roller, as close contact of the seed with the ground improves water uptake, gives more even germination and allows good primary root development. Under the risk of crust formation, flat profiles must be avoided; hence, concave bed profiles above the seed are preferred. If stagnating water is a problem, a convex profile is recommended.

Field temperature and soil moisture content affect mean emergence time and final emergence (Kretschmer, 1994). High emergence of 90% and more was achieved at water capacities from 40 to 80% and at temperatures from 10 to 20°C. Kretschmer and Strohm (1996) found in Germany that, though 1–2 cm drilling depth can give optimal emergence, under dry conditions deeper sowing depths (down to 4–5 cm) might be considered.

Rowse *et al.* (1999) developed a model of the effect of water stress on germination rate of carrot and onion. The model assumes that radicle growth can initiate once the difference between the ambient water potential and the virtual osmotic potential exceeds a certain threshold ($\Psi_{\pi\nu} \geq \Psi - Y$, where Y is a constant for the population). The time between the initiation of onion radicle growth and its actual emergence is assumed to be inversely related to water potential ($-0.45\Psi_{\pi\nu}$) once radicle growth is initiated. The model can be used to predict some of the events that occur during onion-seed priming, as well as those in the soil.

Whalley *et al.* (1999) showed that mechanical impedance (caused by soil compaction) from 0.19 to 0.75 MPa of mean penetrometer pressure reduced the rate and extent of onion shoot development, but affected roots less than shoots. Water stress induced by polyethylene glycol solution (–0.4 MPa osmotic potential) in sand mixtures reduced the ability of seedlings to penetrate impeded soils. Onion shoots recovered better than carrot shoots from soil impedance damage and, on recovery, gave rates of elongation that were faster than those from non-impeded seedlings. The data obtained allowed differential equations to be developed, describing the mean elongation rate of onion shoots as a function of mechanical impedance, water stress, shoot length, temperature and time.

In order to avoid these physical problems, all the crop management practices that can promote high rates of emergence and rapid early growth of the onion seedling are of interest. Seed priming and the use of starter fertilizers are discussed in Section 2.2.7.

2.2.4 Nursery beds

This method is used where hand-labour is relatively cheap, and is useful in hot dry climates, where greater care can be given to the seedlings by amendments to the soil (e.g. farmyard manure or sand) or sterilization. It is also easier to apply mulches or mobile shade covers to protect the seeded areas from heavy rains or direct sun irradiation. When the seedlings reach the three- to four-leaf stage or pencil thickness, they can be transplanted. Seed-beds can be either raised in wet seasons or sunk to allow flood irrigation in dry weather. Transplant use allows the fields to be devoted to other crops for a longer period: this is important in systems where a staple crop must be harvested before onions can be planted out.

Mettananda and Fordham (1999) showed that time of bulbing, bulb quality and yield of onions can be correlated to plant size at transplanting, with different relationships depending on environmental conditions. They suggest that plant size manipulation through different sowing rates or sowing dates (earlier sowing leads to larger transplants) in the nursery may be a useful management practice in order to optimize final bulb size.

Transplants are often trimmed for ease of handling after lifting, though experiments have repeatedly shown that this has an adverse effect on growth and causes more of a check than planting untrimmed seedlings. However, trimming facilitates hand-planting. Bare-root transplants normally suffer a delay of 2–3 weeks in growth compared with direct-sown seedlings. Transplants give an evenly planted field without random gaps, while (in theory) weak seedlings can be rejected (an advantage where open-pollinated cultivars are grown).

2.2.5 Transplants in plugs or blocks

Transplants can also be raised as cell modules or in soil blocks under rain shelters or tunnels, where the additional heat allows rapid early growth of the seedlings. Root disturbance on planting is minimal, compared with traditional bare-root transplants. Onions can produce up to seven bulbs from a single module, as the bulbs arrange themselves into a cluster. Multiseeded modules are useful because they need smaller amounts of protected space to raise and can be planted out mechanically. Although the method is about 3.5 times more expensive than direct drilling, better returns are obtained from the earlier crop. Early weed problems are also avoided.

In Australia, Chung (1989a) obtained the highest total bulb yields (83–88 t ha^{-1}) from transplanting 25–50 modules m^{-2} with two to five plants per module (100–125 plants m^{-2}) at a 20 cm row spacing. Because four plants per module (25 modules m^{-2}) produced an average of 56% of bulbs between 50 and 70 mm and was the least costly system, it was the most appropriate for export onions. Compared with direct drilling, transplanting allowed land preparation to be delayed by about 10 weeks (in early September) and minimized possible soil erosion during the most vulnerable period of the year in Tasmania, but without changing the maturity period (mid-January).

In Michigan, two plants per cell (12 weeks; 4.6 cm^3) optimized the yield of bulbs ≥ 76 mm in diameter, while not adversely affecting bulb shape (Herison *et al.*, 1993). Leskovar and Vavrina (1999) in Texas found that returns were better from 7.1 cm^3 cells, in which plants could be raised for 10–12 weeks, rather than smaller ones (4 cm^3), which needed to be planted out at 8–10 weeks in December, using cv. 'Texas Grano 1015Y'. In a similar experiment in Florida, bulbs harvested from seedlings transplanted in October and grown in 20 cm^3 cells had a significantly higher percentage of bolters than seedlings from 6.5 cm^3 cells. Evidently, the choice of cell size and number of plants per cell must be decided according to the plant size required for frost survival, the market preference for bulb size and bolting control. In Korea, the optimum transplant-growing period for overwintered 'Changnyongdang' onions was 45–55 days. Plastic mulch and early sowings promoted doubling and bolting, as did the use of large transplants with neck diameter over 7 mm (Ha *et al.*, 1998).

2.2.6 Sets

Sets are small, dry, onion bulbs (less than 25 mm diameter) raised specially for planting in the following growing season, when they mature considerably earlier than onions from drilled seed. The post-dormant sets can start into growth very fast, and the plants produced are more 'ready to bulb', i.e. they can be induced to start bulbing by a shorter day length, compared with seedlings. However, large sets may be prone to bolting, since they are close to the critical threshold for flowering. Details of the operations involved in set production in the main European producer country, Holland, were summarized in Anon. (1998). A specialized market exists in north-western Europe for onion sets of cvs 'Stuttgart Giant', 'Sturon' and others. In Israel, the autumn production of cv. 'Beit Alpha' depends on sets produced earlier in the spring (Kedar *et al.*, 1975; Corgan and Kedar, 1990). In Zimbabwe, cv. 'Pyramid' is favoured for sets: they are stored at high temperature (about 27°C) and dipped in fungicide against white rot before planting. An advancement of 1–2 months in harvest date can be achieved compared with the earliest seeded crop. Cultivars which store reasonably well and are rather flat in shape are used for set production worldwide (L. Currah, personal observations). The soil-tillage method is also important for onions grown from sets. In southern Norway, on loam soil, Dragland (1989) showed that soil compaction by repeated tractor wheeling reduced yields by 6%, while autumn ploughing followed by spring harrowing gave the highest yields.

2.2.7 Choice of field strategies and additional practices

The field strategies selected for producing onions depend on season, water availability, soil type and the price premium available for early onions in the country or for jumbo (77–101 mm) and colossal (≥ 102 mm) diameter bulbs. The conformation of the ground must also be taken into account: level fields are needed for certain irrigation layouts – sloping ones may not allow direct drilling with irrigation.

TIMING. Timing the crop to avoid defects such as bolting and doubling is of great importance where onions are produced during a long cool season and there is substantial year-to-year variation in the monthly mean temperatures during the early part of the growing season. Classic experiments by Robinson (1971, 1973) in the lowveld of Zimbabwe demonstrated this, using a range of South African varieties. Later, in Botswana

(Madisa, 1994) and Zimbabwe (R.L. Msika, Zimbabwe, 1992, unpublished data), experiments at higher altitudes confirmed the difficulties of avoiding inducing high percentages of bolting while still making the best use of a potentially long growing season. Sowing in February in Marondera, Zimbabwe, in 1992, for example, resulted in high bolting in almost all cultivars except for the very early cvs 'Early Lockyer Brown' and 'Early Lockyer White' (Yates, Australia) and 'Ori' (Hazera, Israel) (R.L. Msika, Zimbabwe, 1992, unpublished data). In theory, it should be possible to control the factors that affect inflorescence induction by adjusting the sowing date (Castell, 1974; Abd El-Rehim *et al.*, 1996). However, in practice it is not possible to forecast the coming night temperatures accurately: historical records may allow them to be estimated as probabilities. Corgan and Kedar (1990) suggested that a low rate of bolting (< 10%) in a subtropical onion crop showed that it had been sown at about the right time to optimize bulb yields. The risk of increasing bolting and doubling bulbs in early sowings in Korea is increased under transparent polyethylene-film mulch cultivation (Ha *et al.*, 1998), probably because of the more rapid initial growth. Japanese breeders have successfully tackled bolting problems by developing a range of overwintered onions with reliable bolting resistance (Brewster *et al.*, 1977).

PLANT DENSITY. Selection of optimum plant populations is a critical decision, as onion yields increase and bulb size decreases with higher plant populations (McGeary, 1985; Galmarini and Della Gaspera, 1995; Stoffella, 1996) and disease severity becomes higher (Boff *et al.*, 1998). Market preferences for bulb shape and size (Grant and Carter, 1997) may be the final criteria in recommending densities.

An unexpected finding was that with increased productivity through planting density came an increase in the percentage of bolting (premature flowering) in the Spanish cv. 'Valenciana de Grano' (Bosch Serra and Domingo Olivé, 1999). The threshold mean diurnal temperature for vernalization was over 16°C (night minima close to 10°C were registered). One factor associated with increased bolting was higher leaf area index (LAI). It appears that direct competition for nutrients, in particular N, under high densities with adequate water, was not the main factor, but that a light factor might be partially responsible, possibly involving changes in light quality at high LAI. This response may be one that is peculiar to 'Valenciana de Grano', when grown under conditions of high density.

Sowing or planting distances can be established according to mathematical calculations of the size-grade distribution of onions, based on the total yield and plant density with a dynamic onion growth model (de Visser and van den Berg, 1998), allowing the optimum plant density to be calculated, depending on the size grades desired. For instance, in The Netherlands, for onions larger than 40 mm, the simulated optimum plant density was 101 plants m^{-2}. The method offers a tool for calculating the financially optimum plant density, taking into account the balance between yields and prices for different onion sizes.

SEED PRIMING. The uniformity of emergence at the desired density depends on the quality of the seed, its performance in the soil and its ability to escape physical or chemical constraints. The use of primed seeds reduces the time spread and increases the median rate of emergence. Seed priming is the process whereby seed is allowed to imbibe water before sowing, in a controlled manner. Sometimes the seed is redried after priming; in this case, the priming process serves to bring all the seeds within a seed lot to the same stage of readiness to germinate, hence improving the evenness of germination. But after 6 or 12 months of storage the number of abnormal seedlings increases (Drew *et al.*, 1997), so priming should be done only in quantities sufficient for use in a single season.

Various priming methods can be used. Haigh *et al.* (1986) used an aerated salt solution, but it gave decreased percentages of emergence in onions, probably because a large proportion of onion seeds reached a stage of development that was no longer tolerant of desiccation before they were field-sown. Solid matrix priming, i.e. mixing seeds with a solid material and water in known proportions, is another possibility (Taylor *et al.*, 1988). A successful method developed within the past 10 years is drum priming (Rowse, 1996). This is being used commercially – for example, by Elsoms, UK, for leek seed (R. Dobbs, UK, 2000, personal communication). Priming seeds by the drum method and by aerated polyethylene glycol (PEG) solutions in bubble-columns gave virtually identical improvements in seed and seedling leek performance compared with untreated seeds (Gray *et al.*, 1990). Drum priming has the advantage that controlled amounts of water are added to seed as it revolves in a drum, such that the seed can take in moisture without ever becoming really wet. The imbibed seed is then dried with warm air and can be sown with a normal seed drill.

SEED COATINGS. It is common to coat onion seed, prior to sowing, with fungicides and sometimes also with insecticides, thus reducing the amount of chemical used per unit area of land. The coating applied is usually coloured. Pelletization is a further stage, in which the seed is coated to alter size and shape so that it is easier to sow using a precision drill. Future developments may include treatments against insects (such as thrips), which will offer long-term protection to the crop by using new-generation pesticides, such as fipronil, which interfere with the growth processes of the insects. However, the widespread practice of treating seed is unacceptable for 'organic' growers and methods for the production of 'organic' seed are currently being investigated in Europe (B.M. Smith, Wellesbourne, UK, 2000, personal communication).

The coating of onion seeds with insecticides was recently reviewed by Taylor *et al.* (2001), who described the development of effective coatings containing cyromazine (*N*-cyclopropyl-1,3,5-triazine-2,4,6-triamine), sold as Trigard®, plus fungicides, to control onion maggot, *Delia* (*Hylemya*) *antiqua*. The new coatings or pellets incorporate finely ground sphagnum peat moss to protect the seedlings from the phytotoxic effects of the fungicide. This enables a relatively high loading rate for cyromazine, necessary to protect onions throughout the season in the north-eastern production regions of the USA.

STARTER FERTILIZERS. The onion seedling is relatively slow-growing. Its roots branch very little and rarely develop root hairs. It is therefore difficult to supply onion seedlings with adequate amounts of soil nutrients to sustain an optimum growth rate in the early growth stages (unless fertigation is used). Historically, quite large quantities of granular fertilizers have been applied as base dressings before onions are sown. This is wasteful, particularly for N, since much of the nitrate content of the fertilizer, as well as the existing mineral N in the soil, may be washed away by leaching before the onion roots can use it. A more effective method is to place fertilizers (usually liquid rather than granular) in a zone below or close to the initial root zone: this is the starter-fertilizer concept. Implements for the application of starter fertilizer need to be designed to cause minimum disruption to the firmness of the seed bed: a slim applicator that drips the starter solution into the ground just ahead of the seed coulter is used.

In northern Europe (Henriksen, 1987; Stone and Rowse, 1992; Sørensen, 1996; Salo, 1999), the use of starters accelerates onion shoot growth and improves yields. In the UK, Brewster *et al.* (1991) showed that the maturity date of an onion crop was advanced a few days by the use of injected phosphates-of-ammonium starter fertilizer (N: 27.4 and P: 35.9 kg ha^{-1}), even when a basal N–P–K (potassium) dressing had been applied and the crops were kept supplied with adequate N and water throughout the season. Both early (8 March)- and late (19 May)-sown crops of cv. 'Hyton', a Rijnsburger-type summer onion, were cultivated. The later crop showed a lower percentage of late-maturing 'thick-necked' bulbs following the starter-fertilizer treatment, though in neither case was the total yield significantly increased. 'Thick necks' are a quality defect, often a hazard in the tropics probably due to borderline day length, and in northerly production zones, either due to late sowing or to cool or overcast weather late in the season. Starter fertilizers may help to avoid this condition.

Stone (1998) tested the effect of a number of complex fertilization designs and concluded that much fertilizer could be saved on soils of low residual nutrient status by using starter fertilizers applied directly under the crop, rather than by broadcasting them in the quantities normally recommended (P: 65 and K: 230 kg ha^{-1}, in the UK at the time of the experiments). For the two nutrients, P and K, thresholds of starter-solution benefit were determined at P > 5 and K > 3 in terms of Ministry of Agriculture, Fisheries and Food (MAFF) indices (MAFF, 1994). With soils of lower initial status for the two nutrients, there were considerable benefits from using the starter solutions.

In the Pla d'Urgell, Catalunya, Bosch Serra (1996) studied the use of starter fertilizers, in soils with low initial P and K levels, on cv. 'Valenciana de Grano' in an early sowing (January) under border irrigation (a system in which long, gently sloping, nearly level beds constructed across the slope of the land are flood-irrigated). Basal dressings were of 74 kg ha^{-1} N, 138 kg ha^{-1} P_2O_5 and 138 kg ha^{-1} K_2O in all plots, applied 3 weeks prior to sowing. All treatments were top-dressed in June. Initial shoot growth was accelerated by using phosphates-of-ammonium fertilizer solution (N: 22 and P: 50.4 kg ha^{-1}) and shoots were 58% heavier at the start of bulbing (July) than those without starter. The crop maturity date was advanced by 2.5 days. Bulb yields with starter fertilizer were significantly increased (35%) compared with the control without starters. Under such an irrigation system, in which water stress often occurs during the bulbing period, initial enhancements of seedling growth by starter fertilizers translate into differences in final yield, despite the slightly shorter growing period. The benefit of starter solutions in relation to water-supply was also studied with cv. 'Hysam' in the UK. The response to irrigation was greater where the starter solutions (N: 20 and P: 27 kg ha^{-1}) had been applied (Rahn *et al.*, 1996).

Starters ensure that adequate P for early growth is accessible for the young roots, but the chemical composition is important. Ammonium phosphate (AP) improved early growth and final yield compared with broadcast ammonium nitrate, but urea ammonium nitrate (UAN) showed no additional benefits (Stone, 2000), indicating that the response to starters is mainly attributable to better access to P supply for the seedlings. These experiments were on spring-sown onions in an N-depleted soil at Wellesbourne, UK. Nevertheless, from fitted response curves, AP in combination with broadcast N or injected UAN enabled comparable yields to be achieved with approximately half the application rate of broadcast N. N was used more efficiently in the presence of starter fertilizer, thus helping to prevent nitrate leaching and reducing water contamination.

In order to maximize starter benefits, additional N applications later during the growth season are required, so that growth is not limited by shortages of N at times when the crop needs it (see Section 4.3.1 in Part 2).

Thompson *et al.* (1990) demonstrated the compatibility of carbosulphan and chlorfenvinphos formulations with AP starter-fertilizer solution applied under radish seed at sowing. Their experiment suggests that an opportunity exists to optimize insecticide application for onion-fly (*D.* (*H.*) *antiqua*) control.

PART 2. FIELD AGRONOMY

3. Plant Growth and Development

3.1. Whole-plant growth models

The 1990s saw advances in the techniques of modelling the growth of onions by several research groups. Developments in computational technology and access to the Internet are allowing modelling to enter a new era, in which it is indispensable for addressing and integrating the complexity of onion growth and development.

The first model for onion growth at potential production level in the field was by de Visser (1992, 1994a) in The Netherlands. His model ALCEPAS makes use of concepts first explored in the Dutch general crop-growth simulation model SUCROS87 (Spitters *et al.*, 1989). The influence on bulbing of the specific factors day length, temperature and LAI (the ratio of leaf area to ground area) were quantified by a temperature sum which was adjusted for day length and LAI, the latter via the red/far-red ratio of the light. The model is directly applicable to VLD cultivars comparable to 'Rijnsburger Robusta' and must be adapted for other cultivars. ALCEPAS was further tested (de Visser, 1994b) by comparing predicted and actual yields. Bulb dry-matter production was correctly simulated under non-stressful conditions, but LAI was overestimated and the time of 50% fall-over was underestimated at low plant densities.

In ALCEPAS, the light extinction coefficient (KDF) was calculated under overcast skies and it was given the value 0.54, which is higher than that found in the UK (0.47 ± 0.04: Tei *et al.*, 1996a). In Spain, under clear skies, the best fitting for the transmitted photosynthetically active radiation (I_{trans}) was a second-degree polynomial regression equation ($\ln(I_{trans}) = -0.68\ LAI + 0.10\ (LAI)^2$); thus the reduction of I_{trans} per unit of LAI increment decreases with onion crop growth (Bosch Serra and Casanova, 2000).

In the field, bulb respiration is very low. This is coupled with the uniform distribution of the radiation inside the canopy and with earlier termination of blade growth, and consequently allows higher conversion efficiency of absorbed photosynthetically active radiation (PAR) than in lettuce and red beet crops (Tei *et al.*, 1996a). At later onion growth stages, 5.08 ± 0.25 g MJ^{-1} was recorded. Tei *et al.* (1996b) also found that increases in onion dry weight were best described by an expolinear function and no substantial differences were found using time, day-degrees (DD) (T_{base} = 5.9°C) or effective day-degrees (0.136 DD $m^2\ MJ^{-1}$) as the independent variable.

In New Zealand, Lancaster *et al.* (1996) tried to simplify the prediction of bulb size and maturity under non-limiting agronomic conditions. Data from cvs 'Pukekohe Longkeeper' and 'Early Longkeeper' (base temperature of 5°C), at 40 plants m^{-2}, were used. Onset of bulbing (defined as a bulb : neck ratio > 1.2) occurred when degree-days were more than 600 and photoperiod > 13.75 h. This dual-threshold relationship, combined with the number of leaves produced after the start of bulbing and measurements of plant size at bulbing, was a good predictor of final bulb size. The number of leaves to appear after the start of bulbing was correlated with time to maturity, but this was not validated with an independent data set. These simple relationships can be useful under defined growth conditions but they would be difficult to apply widely – for instance, by changing the density – in a more intensive system.

Brewster (1997a, b) pointed out how in theory several existing growth models might be amalgamated as modules in order to explain the response of the crop to the key environmental factors affecting onions right through their life cycle, including flowering and seed production. The models include those covering vernalization, the rate of development of the seed stalk and time to seed maturation. This approach has so far mainly been applied to VLD European onions, and it still needs to be developed and tested on ID and SD onions (Fig. 9.1).

Models can predict onion growth, but non-destructive measurements in real time allow new possibilities of quantifying actual growth and yields. Field reflectance measurements at 660.9 nm (ρ_r) and 813.2 nm (ρ_{ir}) spectral bands were used for monitoring the above-ground biomass and LAI during a period of onion rapid leaf growth: from six leaves to the onset of bulbing (Bosch Serra and Casanova, 2000). The method is suitable for a range of commercial Spanish field densities (30, 60, 90 plants m^{-2}) under non-limiting growth conditions, as a tool for monitoring onion real growth at a regional level.

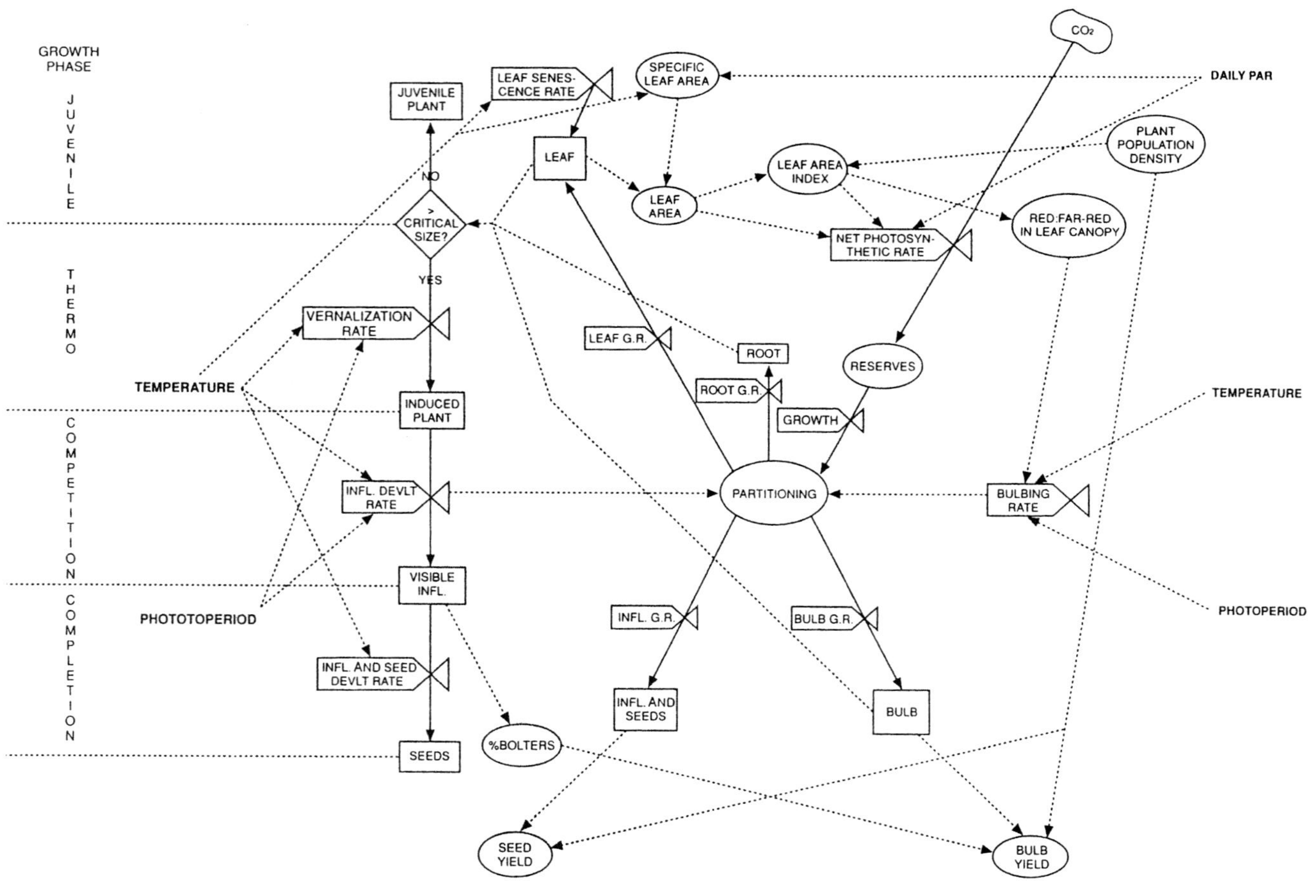

Fig. 9.1. The main environmental controls of growth and development in onions, showing how growth, bulbing and flowering interrelate to determine bulb or seed yield. Devlt, development; infl., inflorescence; G.R., growth rate (based on dry matter); PAR, photosynthetically active radiation. (From Brewster, 1997, with permission.)

3.2 Measuring the effects of leaf loss

Leaf loss in onions, whether due to biotic (pests, diseases) or abiotic (e.g. sand blast, hail) causes has a direct effect in decreasing yield. The greatest impact on total marketable yields and yield of individual market classes occurs near the onset of bulbing (Bartolo *et al.*, 1994). Leaf loss can also delay crop maturity and predispose plants to disease infection, which further reduces functional leaf area.

In Navarra, Spain, the effect of different degrees of leaf loss by manual defoliation at various stages of onion development was studied to evaluate the effects on final crop yield, and regression equations were developed (Muro *et al.*, 1998). The results help with insurance claims for hail damage or other types of leaf damage.

3.3 Studies on roots

Bosch Serra *et al.* (1997) compared the root growth of the two dehydration cvs 'Staro' and 'SWG' and the large fresh-market cv. 'Valenciana de Grano' in laboratory and field trials. Frequent irrigation promoted substantially more root growth than was previously reported in the UK for 'Rijnsburger'-type onions (Greenwood *et al.*, 1982). Onion root length (RL) was related to shoot dry weight (SDW) ($\ln RL = a + b \times \ln SDW$). The value of the intercept (a) was higher in 'Valenciana de Grano' than in the other cultivars.

Onion cultivars can have different strategies, investing more or less energy in root vs. shoot growth. Maximum average root density in the top 20 cm of soil, in a crop grown at a density of 80 plants m^{-2}, was between 8 and 9 cm cm^{-3}, but 90% of the root system was still concentrated in the top 40 cm of soil depth, and only 2–3 % of total root length was recorded below 60 cm depth, as found by Greenwood *et al.* (1982). During the 15 days prior to the start of bulbing, root length in the first 20 cm depth approximately doubled in the case of 'Staro' and 'SWG' and tripled for cv. 'Valenciana de Grano': therefore, this period is considered critical for water-supply in order to permit elongation of the roots. These findings contrast with received wisdom on onion root growth, which holds that root formation stops at the start of bulbing. Root elongation, however, may continue at a rapid rate.

3.4 Onions and climate change

Studies of the effect of increasing temperatures and higher CO_2 levels in the atmosphere on the growth of cvs 'Hysam' and 'Sito' were performed in the UK, in polyethylene-covered tunnels (Daymond *et al.*, 1997). Mean temperatures warmer than ambient by 2.5°C reduced yields (by 3.4–4.4% $°C^{-1}$ and 8.7–11.8% $°C^{-1}$ in cvs 'Hysam' and 'Sito', respectively), presumably because they shortened the duration of growth. Enrichment with CO_2 at 532 μmol (vs. ambient concentration of 374 μmol) increased bulb dry weight (by 29.0–37.4% and 35.3–51.0% in cvs 'Hysam' and 'Sito', respectively) because it increased the rate of leaf expansion and the rate of photosynthesis until bulbing and extended the duration of bulb growth. From comparison of the temperature rise needed to offset entirely the yield increases of each cultivar due to elevated CO_2 (8.5–10.9°C and 4.0–5.8°C for cvs 'Hysam' and 'Sito', respectively), it was concluded that a future concentration of 560 μmol mol^{-1} CO_2 associated with a 2.1°C rise in global temperatures should be beneficial for bulb onion production in the UK, particularly for long-season cultivars.

The advantages of climate change for the commercial production of bulb onions (cv. 'Hysam') in Britain were corroborated by Wurr *et al.* (1998) in a more comprehensive experiment. However, Wheeler *et al.* (1998) concluded that warmer crop-production temperatures would be detrimental to postharvest bulb quality in the UK because sprouting in storage might increase.

In New Zealand, in the context of agricultural onion practices and greenhouse gas fluxes, van der Weerden *et al.* (2000) started a new onion research line focusing on N_2O emissions and showed that onion yield averages 10 t kg^{-1} N_2O-N.

4. Crop Management

4.1. Conventional and integrated versus organic methods

In the 20th century, modern agriculture achieved great success in the reduction of human starvation and famine through the genetic improvement of crops and the introduction of more energy (in the form of fuel, machinery, manufactured pesticides and fertilizers and pumped water) into the agricultural systems. Energy was widely available and heavily consumed, often with low efficiency. The lack of sustainability was highly criticized, and nowadays producers in many countries are adopting new management systems for crop production. These systems involve much lower or no applications of pesticides and a more rational use of fertilizers, water and other agricultural resources. Two approaches are currently being developed for the bulb onion to address this problem.

4.1.1 Conventional and integrated systems

The ICM system avoids wasteful use of resources by tailoring applications of inputs to the actual requirements of the crop at different stages, while minimizing dispersion of polluting chemicals. Total or partial replacement of these materials reduces pollution and lowers production costs and the risks to human health. The objectives of ICM have been well defined (El Titi *et al.*, 1993), but no standard rules exist and produce may be tagged with 'integrated' or 'controlled' labels that are far from self-explanatory.

ICM methods can be accepted for a single crop or may be practised at the whole-farm level (Gysi, 1996). Integrated onion weed and pest management (IPM) is the aspect of controlled production that has been most developed (e.g. Anon., 1996; Delahaut and Marcell, 1999; Reiners *et al.*, 2000). Adopting an IPM programme can significantly reduce synthetic chemical application with minimum effect on quality or yields (Hoffmann *et al.*, 1995; see also Lorbeer *et al.*, Chapter 12, this volume).

4.1.2. Organic onion production – a real possibility?

This approach prohibits the use of synthesized chemical pesticides and readily soluble mineral fertilizers. Only certain approved pesticides and 'organic' fertilizers can be applied to the crop. Genetically modified organisms are forbidden and sewage sludge or 'grey water' from domestic use may not be used as fertilizer. Nevertheless, organic production rules can allow the use of certain 'pure' chemicals, such as sulphur, for disease control.

These days organic agriculture can incorporate the advances in understanding of the interactions between agricultural systems components, and technical innovations that do not compromise 'organic' status. Many farmers in the tropics use low-input production methods from necessity. In developed countries, although the importance of the sustainable use of resources and of organic matter in the soil was recognized from the 1950s onwards, only in recent years have low-input growing systems been readopted, under pressure from consumer groups, environmental protection agencies and supermarkets. Commercial buyers see a new way to attract customers and to satisfy people's awareness of health (see Keusgen, Chapter 15, this volume).

The Council of the European Communities (1991) produces regulatory guidelines in the wide sense for organic certification schemes. Some of its regulations were implemented later or amended, particularly those covering imports. There are many European Union (EU) certification bodies, such as Ecocert in France, the Soil Association and the Organic Food Federation in the UK and Naturland in Germany.

In the USA, the Federal Organic Foods Production Act of 1990 (http://www.ams.usda.gov/nop/orgact.htm), known as OFPA, and its amendments are the basis of standards for organically produced products, and several certification bodies exist. Klonsky *et al.* (1994) published a guide for growers of organic vegetables in the central coast region of California, and Greer and Kuepper (1999) published a guide for organic allium production. The International Federation of Organic Agriculture Movements (IFOAM) has published IFOAM basic standards for organic agriculture and food processing (http://www.ifoam.org/letter.html), which are considered to be the trend-setter for the organic movement.

We include information in the UK from a popular publication on the topic, since little has yet been published about it in scientific journals.

AN EXAMPLE OF ORGANIC ONION PRODUCTION METHODS IN WALES, UK. Roberts (1998) produces onions for the organic market. He has difficulty in obtaining pesticide-free seeds without a special order. The seedlings are raised as modules, five seeds per cell, sown in early to mid-February to raise the summer crop. Nitrogen shortage in the early stages is difficult to solve with purely organic methods: a worm compost or top dressings with (organic) dried blood mixed with seaweed meal may be used.

Crop rotation has been helpful in keeping downy mildew (*Peronospora destructor*) at bay. A rotation of at least 5 years is recommended, since resting spores of the disease can survive that long in the soil. Onions follow brassicas in the rotation, with a light application of farmyard manure. Sometimes early-germinating weeds can be skimmed off the tops of the beds with a rotavator before the modules are planted. Weeds are controlled between the rows with a steerage hoe/brush weeder and by hand-hoeing.

The bulbs are harvested at 50% tops down, and dried either in the field or in a home-made heated tunnel until cured.

ORGANIC ONION PRODUCTION IN NORTH-EASTERN SPAIN. In Pla d'Urgell, Catalunya, a family company (Cal Valls, Vilanova de Bellpuig, 2000, personal communication), local pioneers of organic farming, produce onions after tomatoes grown for processing as an essential part of a field rotation. Onions are followed by hairy vetch (*Vicia villosa*), which is buried and used as a green fertilizer for a new tomato crop. Chopped tomato residues are surface-spread prior to chisel-ploughing. At the end of January, 1-year-old, mostly manure-based, compost is applied at a rate of 18 t ha^{-1} and the land is subsoiled. Seedlings are produced in modules (4.6 cm^3), with three plants per cell. Cultivars are mainly the LD 'Colorada de Figueras' and 'Morada de Amposta', popular in Catalunya; other traditional cultivars can be included, according to consumer demand. Mechanical transplanting is done at about mid-March, when seedlings have two to three leaves. Weeding with a mechanical cultivator starts about 15 days later, when the young weeds are visible, and is repeated once every 2–3 weeks according to need, until the time when it may damage the onion leaves. After that, hoes are used between rows if necessary. Thrips, the main pest problem, are controlled by avoiding any plant water stress, using sprinkler irrigation. Sometimes onion-fly attacks occur but they are not serious and no treatment is applied.

At maturity, but when leaf blades are still upright, bulbs are undercut with a non-vibrating knife to allow foliage to protect bulbs while they dry and thus reduce the risk of sun-scald. After field-curing, the plants are topped with a cutter. Yields, harvested by hand in late July or early August, are around 45 t ha^{-1}.

CAN ORGANIC ONION PRODUCTION BE ECONOMICAL? An important point on which data are needed is the economical feasibility of onion production under the different farming methods: conventional, integrated and organic. In Finland, Stenberg (1999) was unable to identify clear differences in onion production costs between them, probably because the dearest item in the Finnish onion production system was the onion sets, which made up about half the costs. Pesticide costs ranged from 16% of the total costs in conventional production to 2% in organic farming. Under organic production, cost depends mostly on the plant establishment method, weed management and labour, and it can vary a lot.

At a study day in Italy, several speakers discussed the compromises needed between profit and environmental impact, and

D'Ercole and Cembalo (1999) concluded that, even without EU subsidies, a price 20% above that of conventionally produced vegetables should be sufficient to make organic production financially viable. More detailed data from well-recorded comparisons are needed to enable growers to make informed decisions on whether to convert to 'organic' growing methods.

4.2 Water management

Irrigation scheduling has become a more precise technique in recent years. Several research groups have reported on experiments to determine onion water needs under different environmental conditions from Norway (Riley, 1989), Jordan (Abu-Awwad, 1994), India (Hegde, 1988) and Brazil (Coelho *et al.*, 1996). We will examine a selection of irrigation scheduling recommendations in greater detail.

4.2.1 Irrigation criteria

In Botswana, on sandy soil, Imtiyaz *et al.* (2000) found under sprinkler irrigation that a fixed amount of 18 mm of irrigation application at cumulative Class A pan evaporation (CPE) of 11–22 mm resulted in the highest average onion marketable yield, between 49 and 57 t ha^{-1}, with mean irrigation production efficiencies (fresh yield/water applied) of between 4 and 6 kg m^{-3}.

De Santa Olalla *et al.* (1994) studied water balance in a selection of 'Valenciana de Grano', using transplants at a density of 25 plants m^{-2}. Crop evapotranspiration (ETc) was obtained from Penman's formula, as modified by the Food and Agriculture Organization (FAO) (Doorenbos and Pruitt, 1977), and crop coefficients (kc) were chosen and adjusted from Doorenbos and Kassam (1979). Applying water at 100% ETc (T1) and 120% ETc (T2) gave the highest yields – 6.4 and 7.7 kg m^{-2}, respectively. At least 355–415 mm of water was needed in total, and this was provided over 17–20 applications. In T1 and T2, 11.2 or 13.8 kg of fresh bulbs (91% moisture content) m^{-3} of irrigation water plus rainwater were attained, respectively. These regimes gave onion bulbs of 6–9 cm in diameter, the target size for this type of onion. This irrigation scheduling can be improved when changes in seasonal kc values can be anticipated. In this context Al-Jamal *et al.* (1999) related onion kc to growing degree-days.

Bosch Serra (1999) analysed the drip-irrigation practice in Pla d'Urgell, Spain, using onion cv. 'Valenciana de Grano' in late sowings (March, when the soil is dry) at a density of 80 plants m^{-2}. The high-frequency irrigation practised gave complete wetting of the seed-bed surface. Irrigation maintained the water matric potential higher than −18 kPa until shortly before undercutting at harvest, when tops had fallen but were still green. Overall agronomic efficiency of water use (WUE_{ag}) was 27.23 kg of fresh yield (91.6% moisture content) m^{-3} of irrigation water applied (616 mm), thus 19.15 kg m^{-3} of total water received (including 260 mm of seasonal rainfall). These figures imply, at least, double or threefold the WUE previously shown or estimated by other authors (Renault and Wallender, 2000), because density is higher and onions grow better than in previous experiments, with larger diameters than before at higher density. Mean bulb diameter (7.7 cm) and mean bulb weight (209.7 g) were still acceptable. The benefits from irrigation at the end of the growing season are in agreement with the results of Shock *et al.* (2000b) in eastern Oregon (see Section 3.3 above). They found, using a subsurface drip-irrigation system, that soil water potential (SWP) should not be reduced on LD onions below −20 kPa, at 20 cm depth, during the later part of the onion growing period, because it reduced bulb size and had no beneficial effect on reducing bulb decomposition in storage. Gaviola *et al.* (1998) compared the time for terminating irrigations on an onion crop for dehydration, from 5 weeks to 1 week before harvest, and the highest total yields were obtained when irrigation was maintained until 7–8 days before harvest.

Three-year furrow-irrigation studies (Shock *et al.*, 1998a) in the Treasure Valley of Oregon and south-western Idaho, USA, on cv. 'Great Scott' (also an LD Spanish storage type) showed that total yields, marketable yields and profits increased with increasing irrigation threshold: under warm, dry growing conditions, an irrigation threshold higher than −12.5 kPa was recommended, as the optimum yield was still compatible with the best storage performance. Under cooler, wetter conditions, marketable yields and profits were maximized by a calculated threshold of −27 kPa. At higher irrigation thresholds under these conditions, more onions deteriorated in storage. To maintain an irrigation threshold higher than −12.5 kPa requires a highly efficient irrigation system, such as drip irrigation.

Buried drip irrigation is a possibility, although it may not be suitable for 1.5–2 m wide onion beds, in a vegetable crop rotation with lettuce and processing tomatoes, unless sprinkler irrigation can be used to supply extra water for emergence and during bulbing (May *et al.*, 1994). Studies in Oregon by Shock *et al.* (2000c) on cv. 'Vision' showed that onion profits were greatest with a calculated SWP of −17 kPa in 1997 and at the highest application level of −10 kPa in 1998, a season in which severe hail damage in June was followed by a hot dry spell. The premium prices for colossal (> 102 mm diameter) bulbs in the USA naturally tend to affect the conclusions of these studies by Shock and co-workers.

High yields are associated with high irrigation frequency, which avoids any water stress: onions are particularly sensitive at the time of bulbing (de Lis *et al.*, 1967; Chung, 1989b; Guerber-Cahuzac, 1992). Irrigation scheduling can usefully be based on onion ETc values, from measures of crop reference evapotranspiration (ETo) which can also be derived from pan evaporation, adjusted by typical onion coefficients, as established by Doorenbos and Pruitt (1977) in the initial stages and by Doorenbos and Kassam (1979) and Allen *et al.* (1998) as crop canopy increases. During the maturity/harvest period, kc may be increased according to time of harvesting. Soil matric water potential* (i.e. the water held in the soil pores by capillarity) (SWP) seems to be a good tool for improving schedule irrigation for high yields. However, for each set of local circumstances, an adjustment of the amount of water applied will be needed in order to take local factors into account – for example, salinity; SWP was measured at 0.2 m depth. Nevertheless, direct measurements of SWP and the use of high SWP thresholds as irrigation criteria can be recommended on onion fields.

Assuming high irrigation frequency, better scheduling may be expected to increase applied fertilizer use efficiency, to reduce leaching and to improve onion yields by increasing bulb size. It is also a way to increase the nutritional productivity of water, a concept in which the onion crop has been identified as being highly productive (Renault and Wallender, 2000).

4.2.2 Water-economy measures in arid climates

In arid and semi-arid regions, water availability is generally the most important factor for onion production and it is very important to optimize water use efficiency. In semi-arid north-eastern Nigeria, Adetunji (1994) compared soil mulches of transparent polyethylene film and organic mulches (groundnut shell and millet stover): they maintained a higher SWP regime (between −10 kPa and −30 kPa) than bare plots (−75 kPa) during irrigation cycles, and significantly increased bulb-onion yields, by 80, 44 and 50%, respectively, compared with bare-plot yields of 4.5 t ha^{-1}. In a glasshouse pot experiment, Abu-Awwad (1999) found that covering the soil surface reduced the amount of irrigation water needed by 70% compared with the open soil surface; however, onion yields from both surface treatments became comparable as the total water applied increased.

*Soil matric potential is defined as 'the energy per unit of volume of water required to transfer an infinitesimal quantity of water from a reference pool of soil water at the elevation of the soil to the point of interest in the soil at reference air pressure'.

In Lanzarote, Canary Islands, dry-farming of onions is the main agricultural activity (Niebla Tomé and Viera Paramio, 1990). Growers use natural mulches consisting of 1–2 cm manure and 8–12 cm of 'lapilli' – volcanic-ash particles 1–2 mm in diameter. This technique was developed in the 18th century, when the Timanfaya volcanic eruption covered the most important agricultural areas, and is used to protect the soil from evaporation and to harvest water from 'horizontal precipitation' (condensation of water from low clouds) (Acosta Baladón, 1973). Seedlings are watered only at transplanting time. Yields (13–15 t ha^{-1}) are much higher than can be expected with an annual rainfall of 140 mm, a temperature range of 16–24°C and strong winds (23 km h^{-1}). The April harvest time allows onions to be exported. Nevertheless, this production system is in crisis, due to low yields and competition with other economic sectors, though it is considered an example of natural environment preservation by generations of farmers, who have built a peculiar and beautiful agricultural landscape.

4.2.3 Measures to combat salinity and other irrigation problems

Low water quality, mainly saline water, is another potential problem: relative yields decline as salt accumulation from the salinity of the water increases, as described by Abu-Awwad (1996) in Jordan. In order to prevent salt accumulation in the root zone, deliberate leaching can be a useful tool, although in dry climates, with water of high salt content (> 3 dS m^{-1}), efficient leaching requires soils with high permeability, good drainage and salinity-tolerant crops (Ayers and Westcot, 1985). Low water infiltration may be a problem under traditional furrow-irrigation systems, as it increases water runoff and erosion. In order to avoid these problems in furrows compacted by wheels, and also to increase lateral water movement and soil moisture, straw applied in irrigation furrows (630–900 kg ha^{-1}) after planting can significantly improve bulb size and onion yield by up to 74% (Shock *et al.*, 1999). The final benefits must be evaluated, depending on soil type and slope, rates of straw applied and the duration of irrigation. Sometimes, low water penetration and ponding are due to the deterioration of soil hydraulic properties caused by irrigation with water of low electrical conductivity (EC) (< 0.5 dS m^{-1}) combined with the formation of a depositional crust (made of dispersed clay, dominated by illite, and silt particles from a slightly calcareous soil), as a consequence of furrow-irrigation erosion. In Venezuela, Ramírez *et al.* (1999) found that PG applied at 2 t ha^{-1} spread over the surface before irrigation began, increased the electrolyte concentration of the water and the rate of infiltration in the furrows. Onion yields rose by 25% compared with the untreated control (18.1 t ha^{-1}).

4.3 Fertilizer requirements of onions

Onion nutrient contents and bulb mineral exportation were examined comprehensively in the 1960s by Zink (1966), using the dehydration cv. 'SWG' in California. These classic studies form the baseline for later work. Fink *et al.* (1999) published figures from recent studies summarizing the N, P, K and magnesium (Mg) contents of field vegetables, including onions, for use in calculating fertilizer requirements and nutrient balances. A second approach to determining nutritional status for optimum yields is the diagnosis and recommendation integrated system (DRIS), which focuses on nutrient balances. Foliar DRIS norms for onions were developed for N, P, K, Mg and copper (Cu) by Caldwell *et al.* (1994), with data obtained from 'Granex 33' onion fields on sandy Ultisols in the USA.

Bosch Serra (1999) reported that changing the irrigation system frequency from border to drip gave improved P, iron (Fe) and manganese (Mn) contents in cv. 'Valenciana de Grano'. P exports from the field in dry bulbs increased from 1.8 to 3.3 kg t^{-1} as irrigation frequency increased, and yields doubled. Fe and Mn contents trebled and zinc (Zn) contents had a tendency to diminish.

Drip-irrigation experiments were done with a plant density of 80 plants m^{-2}, bulb dry-matter yields around 11–13 t ha^{-1} and mean bulb diameters of 7, 5.6 and 5.7 cm in 'Valenciana de Grano', 'Staro' and 'SWG', respectively. In these conditions, the equilibrium N : P : K : Ca : Mg for bulb nutrient exports was 8 : 1 : 9 : 2 : 0.3. For K, luxury consumption may exist. P exports were 35–38 kg ha^{-1}. On the other hand, plant P contents were maintained constant from the onset of bulbing at around 0.3% (± 0.03). Decline in P content was only noticeable under border irrigation if P supply was limited. The P removal per tonne of edible drip-irrigated onions agrees with the results of Alt *et al.* (1999) from 19 harvests in Germany (assuming 10% of onion-bulb dry matter), although some differences exist in K and Mg removals, perhaps due to the different onion types in the trials or to some K/Mg antagonism in Spanish fields. The equilibrium P : K : Mg was 1 : 7.75 : 0.44 in German harvests.

In the Spanish experiments (Bosch Serra, 1999), maximum N extractions were close to 300 kg ha^{-1}. It also appeared that the N dilution curve proposed by Greenwood *et al.* (1992) must be revalidated under fertigation when light interception is lower than 60%. Greenwood *et al.* (1992) obtained a relationship between the percentage of organic N in the plant (Nc) and plant mass (W) when growth was N-sufficient (Nc = $1.35 + 4.05^{-0.26W}$). When the curve was compared with the Spanish results obtained in two experimental years, under fertigation, and with different densities (Bosch Serra, 1999), although some points demonstrated the dilution concept, N contents were always lower than the critical N content at biomass between 1 and 4 t ha^{-1}. This fact was not related to a lack of N, because there was ample soil N availability. Increasing the biomass by increasing the density, at a similar developmental plant stage and size, changed the data point above the N critical dilution curve. Maximum uptake rates per day occurred at the start of bulbing.

In order to improve fertilization practices, dynamic approaches can be considered: these aim to supply the roots with nutrients to promote an optimum uptake rate, depending on the actual growth stage throughout crop development. Bosch Serra (1999), in the drip-irrigation system described above, combined assessments of onion root characteristics and growth with nutrient uptake rates during the growing period. Nutrient concentrations in the soil solution necessary to sustain the nutrient fluxes into the root system by diffusive supply were calculated according to the classical model of Baldwin *et al.* (1973). On a silty clay loam soil, with a volumetric soil content of 0.32, maximum extractions per root-length unit were found at the early stages. This fact highlights the usefulness of starter fertilizers (discussed earlier) in the early stages; later, root length/unit growth increment increases and critical nutrient content decreases as the crop grows. The peak value of predicted concentration differences to sustain P inflows as orthophosphate ($H_2PO_4^-$), in the top 20 cm of soil, was 33×10^{-6} mol l^{-1}.

4.3.1 Nitrogen

N fertilization is of great importance for onion production but also needs considering in relation to environmental protection. Simulation models can be useful tools for achieving both goals. SOIL and SOILN models developed at the Swedish University of Agricultural Sciences have been adapted to simulate the growth and N dynamics of an onion crop (Salo, 1996). An N simulation model developed by soil scientists at Horticulture Research International (Wellesbourne, UK) is available. The N-ABLE model calculates crop responses for a range of crops, including onions, to the incorporation of fertilizer N (ammonium- and nitrate-based fertilizers) and plant debris. The advantage of N-ABLE is that the model calculates for each day the distribution of water and mineral N down the soil profile and the amounts of nitrate leached below different depths, in addition to the increments in crop growth and N uptake. Most of the algorithms in the model are given in Greenwood and Draycott (1989) and Greenwood *et al.* (1996). The Internet address for the model (with links to others for K and P) is http://www.qpais.co.uk/nable/nitrogen.htm.

In the Arkansas valley of Colorado, Ells *et al.* (1993), on a soil with 2.5% organic matter, concluded that yields higher than 50 t ha^{-1} could be obtained without N fertilizer when more than 42 ppm NO_3^--N was present in the top 33 cm of soil and up to 1120 mm of irrigation water was applied.

Environmental concerns about N management are important. For example, in Hokkaido, Japan, Hayashi and Hatano (1999) calculated that the N leached annually from an onion field can correspond to 58% of applied N. In-depth practical studies on the use of N were undertaken in The Netherlands by de Visser *et al.* (1995) and de Visser (1998). From 26 multilevel N-fertilizer trials, they looked for a relationship between the amount of soil mineral nitrogen before sowing and the optimum amount of N fertilizer (de Visser *et al.*, 1995). The work did not change the current advice in The Netherlands for a fixed rate of 100–120 kg N ha^{-1}. However, in view of an apparent recovery rate of 32% for the fertilizer at 100 kg ha^{-1}, the risk of N leaching is still serious. The risk can be higher if the uptake of N by the crop ceases early due to disease.

Later, de Visser (1998) concluded, from field experiments in 1991–1994, taking into account an N-uptake curve during crop growth and the rate of mineralization of N in the soil, that split applications of N could save on the quantities applied and that possibilities existed for reducing the threshold application rates for N in use previously: rates ranging from 72 to 110 kg ha^{-1} could be applied as two or three split dressings. Information on N residues in the soil, a very sensitive point now in Holland, was also obtained during these studies. It was estimated that about half of the quantity of N applied was leached out of the soil (0–60 cm) before it could be used by the crop (de Visser, 1998), compared with about 36% (0–90 cm) as estimated by Greenwood *et al.* (1992) in the UK. De Visser's results showed that N dressings could be split without serious risks of yield loss. Using a two-way split based on about 140 kg N ha^{-1} at the four-leaf stage gave maximum yields. Basal applications of 30–50 kg ha^{-1} were proposed, with later applications timed to meet the current needs of the crop.

Bosch Serra (1999) took this concept further. Under border irrigation, N basal fertilization is of limited use, because more than half of the N available in the first 20 cm depth can be lost in the first irrigation, when the soil is initially dry and the profile is then wetted. Thus, under this system, recommendations based on soil mineral N before sowing do not ensure optimum fertilization. The basal fertilization proposed by de Visser (1998) can be recommended. The best period to obtain soil samples in order to adjust application quantities is at the four-leaf stage, before the period of high N demand. The amount applied between the four- and six-leaf stages depends on both plant N content and crop biomass. N measurements at different depths throughout the growing period suggested that the later recommendations of de Visser (1998) could reduce N environmental pollution during the crop cycle and after crop harvest as well.

In experiments over two seasons, split N applications (as ammonium nitrate) were also shown by Wiedenfeld (1986) to be advantageous for onion yields in one season compared with preplanting broadcast slow-release fertilizer applications (as methylene urea or sulphur-coated urea), but not in the other. Weather conditions seemed to influence the responses, and split applications have an advantage under climatic conditions that increase N losses, such as in rainy years. Under dry climates, slow-release fertilizers can usefully reduce the application costs of fertilizer top-dressings. Another possibility is the use of ammonium-based fertilizers applied as solid pellets in the soil or as discrete fluid bands: this has been successful in other horticultural crops (Bacher and Lenz, 1996).

Related to the efficient use of N for onion-crop production, Fenn *et al.* (1991) found, in a calcareous silty clay loam, that the use of urea combined with $CaCl_2$ at a 0.25 molar proportion of Ca^{2+} : urea, applied before planting 10 cm below the seeding depth, and side-dressed twice later on, increased yields of cv. 'Yellow Sweet' by 64% compared with urea alone. Fenn *et al.* (1993) continued their studies on the effect of Ca^{2+} on NH_4^+ absorption on onion plants

34, 60 and 134 days old. Pots filled with a calcareous soil were grown with a nitrate nutrient solution. Irrigating with $Ca^{2+} : NH_4^+$ at molar ratios from 0 to 2 for 30 h was carried out at a desired growth stage.

Young seedlings responded best to added N in the presence of increasing Ca^{2+} concentrations, significantly increasing top and bulb dry weight. By increasing $Ca^{2+} : NH_4^+$ molar ratios or by the addition of Ca^{2+} with urea, plant N use efficiency was increased and more dry matter was produced with the same amount of N. It should be noted that Ca^{2+} stimulated NH_4^+ uptake, a result that is also observed when they are applied together in bands (Fenn *et al.*, 1991).

In contrast, Gamiely *et al.* (1991) reported that 70-day-old onion transplants had decreased leaf area, root, leaf and bulb dry weight and pungency, in an NH_4^+-N solution culture as the sole N source, compared with growth with nitrate alone or in combination with NH_4^+-N. At the N-form ratio 3 : 1 (NH_4-N : NO_3-N), water uptake decreased without decreasing yield.

The importance of N mineral soil ionic forms, in relation to onion-plant requirements at different growth stages and to root growth, was studied by Abbès *et al.* (1995a). In a growth-chamber experiment, they found that maximum N influx (I_{max}) for ammonium was considerably higher than I_{max} for nitrate from day 28 to day 42, they were similar during the 42–56-day period and thereafter I_{max} for nitrate exceeded considerably the one for ammonium. In the early stages, ammonium N gave maximum onion-plant growth. Later, maximum growth was obtained as the proportion of nitrate increased in the nutrient solution. Ammonium reduced the uptake of cations and increased the uptake of P, although no apparent nutrient deficiency was observed whatever $NH_4^+ : NO_3^-$ ratio was applied.

In order to improve N uptake in the early onion growth stages, Abbès *et al.* (1995b) studied the effect of different ammonium sources on 84-day-old onions in acid and chemically poor soils with different textures, in growth chambers. Ammoniated peat increased onion N uptake and root growth over ammonium sulphate or peat treated with ammonium sulphate. At rates higher than 266 mg NH_4^+-N kg^{-1}, or earlier in sandy soils, ammonium toxicity appeared. In a similar experiment, Abbès *et al.* (1995c) found that the nitrification inhibitor nitrapyrin did not entirely decrease ammonia toxicity by peat–ammonia–mineral fertilizers, but the highest granule size (10 mm) gave better results than the smaller ones (2–4 mm). The results obtained on N uptake in Abbès *et al.*'s (1995a, b) experiments were close to the results of an onion model on N uptake developed by Abbès *et al.* (1996). The mechanistic model describes simultaneous NO_3^- and NH_4^+ uptake as NH_4^+ is released into the soil–plant system and is transformed into NO_3^-. It can also assess toxicity risks. Onions take up nitrate in much greater amounts than NH_4^+, especially in soils where volumetric water content (m^3 of water m^{-3} of soil) is less than 0.21. Nevertheless, the model supports the finding that the early growth stage, when the onion root system has the highest affinity for NH_4^+ and soil moisture content is high enough to promote NH_4^+ diffusion in the soil solution, is the period when ammonium-bearing fertilizers are most effective.

Improvement of onion agricultural practices in order to reduce N pollution does not finish at harvest. In the Treasure Valley region of eastern Oregon, USA, the potential nitrate contamination by unlined landfills (5000–6000 m^3), where cull onions were historically disposed of, have also been analysed (Hutchings *et al.*, 1998).

4.3.2 Potassium

Greenwood and Stone (1998) argued that in a wide range of C3 crops grown in the UK, the critical K percentage (K%) declines with the increase in plant mass in much the same way as critical N%, and that both are linked to photosynthesis. Also, a linear relationship exists between the decline of critical N% and the concentration of total cations. Site-specific factors may influence the maximum and critical K concentrations. In onions, their results showed some discrepancies with the predictions: these may be attributable to the way the measurements were done.

4.3.3 Copper and other microelements

Cu deficiency can occur on peat or acid mineral soils. During a 5-year period in New York State, USA, Ellerbrock (1997) correlated onion yield on organic (i.e. peat or muck) soils with Cu soil-test levels, in order to develop guidelines for Cu use. When available Cu levels are measured using HCl extraction, it was concluded that, when levels are higher than 0.3 ppm, it is not necessary to apply copper sulphate as a fertilizer, because quality and yield will not be affected. Furthermore, Cu applications, when not necessary, result in a substantial increase in extractable Cu levels, which can lead to adverse effects on plants or soil. Areas where Cu sprays are used regularly against bacterial diseases of onions may risk a build-up of Cu to damaging levels.

In Egypt, Sliman *et al.* (1999) found that foliar sprays of the nutrients Zn, boron (B) and Mn resulted in improved onion yields over two seasons, while Cu sprays gave no improvement. Zn and Mn were applied as sulphates and B as borax in solution. Zn and B gave the best yield improvements.

4.3.4 Biofertilizers

The use of readily soluble 'artificial' fertilizers is now being questioned in some parts of the world and in some agricultural systems. Biofertilizers, a term used to describe living organisms that can be applied to the soil to promote increases in nutrient uptake, provide an alternative method of supplying nutrients to an onion crop without the direct use of artificial chemical fertilizers, or can be used to supplement them. They are of particular interest for organic production systems. They are receiving serious attention in India (Tandon, 1999), for example, where there are worries about the continued use of 'artificials' as the sole source of nutrients, as well as problems of cost and supply. The first biofertilizers to be widely used were the *Rhizobium* bacteria, cultures used with specific legume crops. Since their introduction, many others, such as *Azotobacter* and *Azospirillum* have become available commercially.

Microbial biofertilizers were studied on main-crop onions in Nasik, India (Bhonde *et al.*, 1997). The 3-year study in 1993–1996 involved comparisons of artificial manures and/or farmyard manure (FYM) combined with various proportions of *Azotobacter* biofertilizer (A). The *Azotobacter* treatment was applied by dipping the transplant roots into a solution made up of 1500 g *Azotobacter* in 50 l of water ha^{-1} for 5 min prior to planting. The soil was low in P, and various rates of N up to the maximum recommended level were added as well as the *Azotobacter* treatment. The results showed that both A + 100% and 50% recommended N gave significantly higher marketable yields than the other treatments, and that A + 50% N was superior in economic return while giving a marketable yield of 23.06 t ha^{-1} of cv. 'Agrifound Light Red', compared with 19.4 t ha^{-1} from the FYM + normal N treatment used as control.

Another type of organic soil amendment, vermicompost (compost made by earthworms from organic garbage), was also trialled at Nasik and Karnal (Bhonde, 1997). Compost at rates of 2, 3 or 4 t ha^{-1} gave significantly lower yields than 100 : 50 : 50 NPK, but the former gave bulbs with higher dry-matter content and superior storage ability. Trials in Rahuri, Maharashtra State, India, by Shete *et al.* (1993) on the white onion cv. 'Phule Safed', found that 5 t ha^{-1} vermicompost resulted in lower yields compared with using 20 t ha^{-1} FYM (29.3 and 36.7 t ha^{-1}, respectively) and was equivalent to the no-manure treatment (28.7 t ha^{-1}). Vermicompost is regarded as a useful possible fertilizer for organic onion production in India. More trials are needed to establish the optimum dosage rate for organic fertilizer, according to its nutrient composition, to achieve higher yields.

A recent report from Connecticut, USA, on the use of leaf compost as an organic manure for onions, applied annually as a 2 cm mulch, together with applications of artificial fertilizers, gave some interesting results. In a sandy soil, the compost had a buffering action against dry or wet years and tended to stabilize onion yields. Percentages of bulbs in the bigger size classes were also increased in two out of three years with leaf compost. Onions produced with leaf compost also suffered less from soft rot (cause not identified) than those produced under a conventional system (Maynard and Hill, 2000).

The *Azospirillum* and 'phosphobacteria' (not defined) effect was tested on the production of multiplier onions, cv. 'Co-4', at Madurai, southern India (Thilakavathy and Ramaswamy, 1998). A basal dressing of K at 30 kg ha^{-1} was applied to all treatments. The highest yields were produced with the two biofertilizers at similar rates, together with 45 kg each of N and P fertilizer, which resulted in yields of 18.37 t ha^{-1}. There were improvements in the appearance and pungency of the bulbs following the biofertilizer treatments.

Stribley (1990) comprehensively reviewed the use of vesicular-arbuscular mycorrhizae (VAM) in onion production and made a convincing case for the advantages to be obtained from their use. The benefits in terms of P uptake and drought resistance seem to be attractive. Onions' VAM dependence was stressed by Krikun *et al.* (1990) in a P-sorbing soil in the northern Negev of Israel. They studied the influence of soil fumigation (with 60 g m^{-2} of 98% methyl bromide and 2% chloropicrin mixture) and P fertilization on onion yield and P tissue contents. At all P application rates (0 to 330 kg P ha^{-1}), fumigation significantly reduced yields and P content in plant tissues (Bendavid-Val *et al.*, 1997). However, VAM use does not seem to have been taken up widely in bulb-onion production at present, possibly because problems of mass VAM inoculum production (Sullia, 1991) are still not solved, or for commercial reasons.

In Prague (Czech Republic), Vosátka (1995) obtained positive onion growth responses after inoculation with different arbuscular mycorrhizal fungi at field level in sterilized, non-sterilized or fumigated soils. She also found that this positive effect can be obtained by inoculating the substrate of cell trays for seedling production before sowing. In her study, 20–30 g of inoculum l^{-1} of module substrate was a sufficient initial dose for preinoculation of plants, a dose that can be even lower when the fungal strain *Glomus etunicatum* S 329 is applied. This method may be more practical than applying field inoculum.

In Japan, three commercial VAM preparations are used by growers of Japanese bunching onions (K. Tawaraya, Japan, 2000, personal communication). These preparations are also produced in India (Tandon, 1999).

4.4 Weed control

4.4.1 *Using herbicides*

Since onions compete poorly with weeds, the use of herbicides is widespread and the economic advantages of their use have been demonstrated (Menges, 1987). Rubin (1990) reviewed the topic of weed control and the herbicides then in use. From different experiments in weed control in France and Spain, recommendations on herbicide selectivity and when to apply in order to improve their efficiency have been built up (Vergniaud *et al.*, 1989). The programme for herbicide weed control requires a combination of different herbicides, taking into account the season and the predominant local weed flora. In pre-emergence, propaclor, chlorpropham and chlorthal and, in post-emergence, propaclor (broad-leaf and grass herbicide), ioxynil octanoate (broad-leaf herbicide), oxyfluorfen (broad-leaf and grass herbicide), diclofop-methyl and fluazifop-butyl (grass herbicides) are widely used, although some of them can be mixed with other herbicides. In Spain, extension advice is based on 10 years of onion herbicide experiments (MAPA, 1993).

The persistence of herbicide residues in the soil and in onion bulbs has recently been evaluated for pendimethalin under Mediterranean conditions (Tsiropoulos and Miliadis, 1998). Some soil-applied onion herbicides can be transported into sediment through runoff or can reach aquifers through movement in the soil profile. Herbicides and other pesticides from various crops, including onions, have been found in groundwater in different US states and in Europe (Goodrich *et al.*, 1991; RIVM/RIZA, 1991; CAST, 1992). Management practices for furrow-irrigated onions, in order to reduce and prevent the aquifers from contamination by metabolites of the herbicide dimethyl 2,3,5,6-tetra-chloro-1,4-benzenedicarboxylate (DCPA), commonly used in onion fields in Oregon, were studied by Shock *et al.* (1998b). The combination of straw mulching and DCPA banding was an effective measure. Straw used with DCPA reduced the herbicide losses in both sediment and runoff solution and also herbicide movement into the soil when DCPA was banded.

Some common onion herbicides give poor weed control at low soil temperatures and with high soil organic-matter content, as in the Fraser Valley (Canada). Under these soil conditions, the potential use of ammonium nitrate as a contact herbicide on onions (cv. 'White Lisbon') was studied (Bitterlich *et al.*, 1996). Ammonium nitrate solutions (7.5, 10, 15 and 20% N) were sprayed (800 l ha^{-1}) on sunny summer days approximately 3–4 weeks after sowing. The weed species *Capsella bursa-pastoris*, *Gnaphalium uliginosum* and *Amaranthus retroflexus* were very susceptible to ammonium nitrate, while *Chenopodium album*, *Portulaca oleracea* and *Poa annua* were tolerant. Onion dry weight was not affected or increased slightly with the application of the ammonium nitrate solution.

Mathematical models can be useful tools in optimizing herbicide management at the right time and dose and in optimizing the net growers' margin. Dunan *et al.* (1999) in Colorado, USA, published a study on the development of a plant-process economic model for weed-management decisions in irrigated onion. The model simulates the dynamics of the competition for light between the Spanish-type onion crop and five annual weed species (*A. retroflexus*, *C. album*, *Echinochloa crus-galli*, *Helianthus annuus* and *Panicum miliaceum*), assuming that there are no water or nutrient limitations to plant growth. By comparing the cost of weeding with the economic results of not controlling weeds at a particular time, the model enables decision-making on onions on a rational basis throughout the growing season, based on calculations of the effect of differing degrees of competition applied by weeds on the different growth stages of the onion crop. The model is available from P. Westra at pwestra@lamar.colostate.edu.

4.4.2 Using non-chemical methods

As environmental problems take on greater importance, scientific studies are starting to appear on the effects of organic production methods. Bond *et al.* (1998a, b, c) in the UK described the effects of various methods of weed control on onions, and changes in the weed seed-bank, in organic and conventional systems. In salad-onion production of cv. 'White Lisbon', removing the weeds at 4 weeks after 50% crop emergence avoided weed interference with the crop consistently in both systems, although the optimum period was relatively narrow. In transplanted onion crops of cv. 'Promo', a single weeding at 5, 6 or 7 weeks after planting prevented reductions in yield. Thus transplanting gave greater flexibility in the timing of weed removal. Yield loss was mainly attributed to competition by weeds before hand-weeding or to mechanical damage to the crop when weed removal was carried out later. A single weeding that removed inter-row weeds but left within-row weeds in place did little to reduce yield loss (Bond *et al.*, 1998a). This is a major limiting factor in organic horticulture, because the single hoeing treatment must be done together with within-row hand-weeding. The development of an effective method for removing weeds selectively from within the crop row is needed to make non-chemical methods of weed control viable. In an integrated cropping system, inter-row cultivation can be combined with a selective herbicide band application.

In studies of changes in the soil weed seed-bank after various production methods, Bond *et al.* (1998b, c) found that, after unweeded transplanted bulb onions and unweeded drilled salad onions, there was a 15-fold and between two- and 70-fold increase in weed seed numbers in the soil, respectively. Weed seed numbers were lower following single or multiple weedings of the crop. Neglecting to control weeds results in a rapid increase in the weed seed-bank. With one single weeding, late-germinating ephemeral weeds or those having wind-blown propagules may become a problem.

Solarization controls most annual weeds, such as *Amaranthus* spp., *P. oleracea* (Horowitz *et al.*, 1983; Katan and DeVay, 1991) and also *Sorghum halepense* (Rubin and Benjamin, 1984) in Israel. In Portugal, solarization also helped to reduce weeds in onion seedling production and benefited final yields of cv. 'Valenciana' (Silveira *et al.*, 1990/91).

Flaming and infrared radiation techniques for thermal weed control are further possibilities. The methods rely on heating plants (0.1 s, at 70–80°C) until the cells burst. They can be used for pre-emergence weeding and during crop growth. Thermal methods are selective, because onion plants are more resistant to heat than many weeds. Douzals *et al.* (1994) developed a prototype thermal weeder using propane flame burners for onions, which was tested with success on all stages of organically grown onions. The best attack angle was 30–40°. Reflectors were added in order to protect fallen onion leaves. As with herbicide applications, the success of this technique is linked to the knowledge of weed stages susceptible to heat. However, some weeds, such as *Cirsium arvense*, are heat-tolerant and others can reappear after thermal weeding, e.g. *Agropyrum repens*. Flame weeding can be useful, particularly for organic farmers, because it is relatively less labour-intensive than hand-weeding and therefore more profitable (Rifai *et al.*, 1996).

A new method for mechanical intrarow weeding was developed by Melander (1997) in Denmark. He studied the efficiency of a vertical rotating-axis brush weeder at different settings for yield of onions and weed control in the field. The adjustment of the direction of brush rotation allowed the grower to select the type of work done, either primarily uprooting the weeds or primarily covering them with soil.

In a 2-year field experiment, Melander (1998a) studied the interactions between different non-chemical control methods in onions (cv. 'Hyton') grown on a sandy loam soil. By combining different strategies (flaming and brush weeding, or harrowing and sowing in darkness plus brush weeding), 80–90% weed control was achieved. Future optimization of these strategies or combining them with other weed-control techniques, such as hoeing close to the row (Melander and Hartvig, 1997), will allow better weed control without herbicides. The economics of non-chemical methods and in which circumstances it pays to use them were discussed by Melander (1998b).

4.5 Harvest

Harvesting by hand is still used (see Gubb and MacTavish, Chapter 10, this volume), but, because of its expense, the process has gradually been mechanized. First, undercutters with a flat blade were used; later, oscillating blades and rotating bar undercutters were introduced (Maw and Smittle, 1986). Sometimes, tops are cut off before the undercutting, but this practice can reduce storability (Füstös *et al.*, 1994) if the bulbs cannot be dried immediately.

The timing of undercutting in relation to harvest time is another factor to consider for maximum yields and also for preserving skin quality. Dehydrator onions in New Mexico should be undercut just prior to harvest and harvest should not be delayed more than 15 days after 80% fall-down (Wall and Corgan, 1999) as, in the dry climate, bulb diseases, such as fusarium basal rot (*Fusarium oxysporum* Schl. f.sp. *cepae* (Hanz.) Snyder and Hansen), can be responsible for yield reductions after undercutting if harvesting is delayed.

Mechanical harvesters for onions were first adapted from potato harvesters but, in many cases, they were not successful because of soil characteristics or bulb damage. Any mechanical onion-harvesting operation must keep onion damage to a minimum. This is important if the crop is for fresh consumption, and even more if it is intended for storage. Herold *et al.* (1996, 1998) showed that multiple mechanical loads had a significant effect on onion respiration rate as well as resulting in additional mass losses in storage. Under practical conditions (measurements from lifting to delivery to storage), too many loads during handling and excessively high peak forces due to rough transfers were the critical load sources.

For sweet onions ('Granex'–'Grano' type), Maw *et al.* (1996) measured different physical and mechanical properties, such as crushing load and puncture resistance, in order to understand the factors behind mechanical damage. Later, Maw *et al.* (1998) developed the principles of operation of a mechanical harvester for sweet onions that included grasping the onion plants by their tops, lifting them from the ground, shaking soil from the roots, severing the leaves above each bulb and then delivering bulbs to a container and tops to the ground. Further information on curing and storage is in Gubb and MacTavish (Chapter 10, this volume).

5. A Practical Example of Onion Agronomy Improvement: Pla D'Urgell, Spain

The Pla d'Urgell district is located in the east of the Ebro Valley in north-eastern Spain. The climate is semi-arid, with low rainfall (400 mm year^{-1}) and high between-year variability (230–660 mm). Summer temperatures are high: the average July temperature is 24°C but 41°C can be recorded. Heavy frosts occur in winter, from October to April, and a minimum temperature of −19°C has been registered. Soils are calcareous, low in organic matter (< 2%) and loamy or finer (silty loam and silty clay loam), with significant areas salt-affected to different degrees (Herrero *et al.*, 1993). Water from Urgell channels, built 100 years ago, is of high quality, although sometimes underground water with higher salinity content (1.5 dS m^{-1}) may be pumped to supply the district.

Onions (mainly cv. 'Recas', a selection of cv. 'Valenciana de Grano') are cultivated for storage. The most important diseases are basal rot and the saprophytic *Alternaria alternata* (Fr.) Keissler (J.P. Marín, Lleida, Spain, 1991, unpublished results). *Peronospora destructor* (Berk.) Casp., *Sclerotinia squamosa* (Vien.-Bourg.) Dennis and *Botrytis allii* Munn rarely occur. *Thrips tabaci* Lind. is the most important pest. *A. retroflexus* is the most important weed; others are *Convolvulus arvensis*, *Cynodon dactylon* and *S. halepense*, which are serious in some fields (Consola and Recasens, 1989). The use of herbicides can be combined with hand-weeding.

Onion cultivation used to be done with border irrigation, in rotation with field crops, the best preceding crop being wheat. Under the traditional system, onions were sown at the end of January or in February, without irrigation. In many years, emergence was much delayed and uneven. Irrigation started in April and continued until August. Due to limitations connected with water rights (farmers schedule irrigation by turns) and the border-irrigation method, onions very often suffered water stress.

Under the traditional system, irrigation after sowing was the limiting factor. The introduction of starter fertilizers and the priming of seeds were essential improvements in order to assure emergence at an acceptable plant density and rapid initial growth. Substantial nitrate residues can remain in the soil postharvest, even when recommended fertilizer practices are followed. Under such conditions, yields of about 50 t ha^{-1} were considered satisfactory, though very often yield was only about 30 t ha^{-1}. Onion harvesting was done by hand after passing a horizontal blade below the bulbs. The crop was a profitable one for many years.

During the 1980s, temporary mobile-sprinkler irrigation was introduced in order to ensure emergence. Lack of know-how and soils prone to crust formation (especially when some salinity occurs) made sprinkler irrigation less successful than expected, but eventually, in suitable soils it allowed yields to be stabilized closer to 50 t ha^{-1}.

A major improvement was the introduction in suitable areas in 1990 of drip irrigation, where the whole of the seed-bed surface is watered. The new irrigation system allowed farmers to delay sowings until the end of March because it guaranteed quick and uniform emergence. The increase of irrigation frequency also allowed an advancement of maturity and harvest time, compared with the traditional system. Sometimes harvests were as much as a month earlier, probably due to the higher LAI achieved under the new system. Faster growth and development led to increased bolting problems, which are now controlled by adjusting densities to between 60 and 70 plants m^{-2}. It may prove necessary to use selection against bolting in the production area, to improve the genetic adaptation of the traditional cultivar to the new methods of cultivation. Under this system, yields were raised to 90 t ha^{-1} with great stability between years. Onions are stored under ambient conditions, although some farmers use cold storage for late selling.

Under the drip-irrigation system, correct rotations are even more important than before. Farmers, in order to maximize the return on their drip-line investments, may try to maintain the high-value crop in the same field for 2, 3 or 4 years: this can lead to increased soil fungal problems.

In recent years, although farmers have made great efforts to utilize the principles of integrated crop production, profitability has been hard to maintain because of low onion prices in the market. A major constraint, which still has not been overcome in the area, is the manual lifting of onions. Farmers cannot find a suitable machine to harvest onions on these loamy to silty loam soils, which are rather sticky when wet and very hard when dry. So the cost of labour is still very high and currently prevents the costs of production of this late, long-storage onion crop from being competitive.

6. Conclusions

Onion agronomy continues to pose problems for scientists, but increasingly these problems concern finding methods of economizing on costs and making the best use of scarce resources, including water. The rapid take-up of fertigation delivered by drip pipes in regions where the terrain permits their use shows that growers are alert to improvements that give economies of labour and inputs and are ready to put them into practice. Professional advisers are increasingly being used by growers to keep technically up to date and to make use of predictions for timing the spraying of crops (see Lorbeer *et al.*, Chapter 12, this volume) and also for the scheduling of irrigation through direct field measurements of plant demand. At the same time, older technologies are being revived to tackle problems such as thrips on onions through cultural methods rather than by chemical sprays.

We would like to emphasize here the need to regard onion growing as part of the farming system, to include the aspect of rotations in particular, and to suggest that the underground parts of the onion and its coworkers, the mycorrhizae and other soil-living organisms, which may have a protective action against root disease, are still comparatively neglected at the present time and merit greater scientific attention. There is no doubt that breeding for pest and disease resistance and tolerance will take a higher profile during the next few years, but, equally, developing a better understanding of the soil inhabitants and their mediating role between the crop and the soil deserves to receive more resources.

Acknowledgements

We thank all of our colleagues who have generously helped us with information and comments, including Dr I.G. Tarakanov, Dr J.L. Brewster, Dr S.R. Bhonde, Professor H.D. Rabinowitch, Dr Z. Füstös, Professor D.J. Greenwood, Dr W. Bond, Dr K. Tawaraya and Dr B. Melander in particular.

References

Abbès, C., Parent, L.E., Karam, A. and Isfan, D. (1995a) Effect of NH_4^+ : NO_3^- ratios on growth and nitrogen uptake by onions. *Plant and Soil* 171, 289–296.

Abbès, C., Parent, L.E., Karam, A. and Isfan, D. (1995b) Onion response to ammoniated peat and ammonium sulfate in relation to ammonium toxicity. *Canadian Journal of Soil Science* 75, 261–272.

Abbès, C., Parent, L.E., Karam, A. and Isfan, D. (1995c) N uptake and recovery by onions from peat-mineral fertilizers. *Canadian Journal of Soil Science* 75, 273–280.

Abbès, C., Robert, J.L. and Parent, L.E. (1996) Mechanistic modeling of coupled ammonium and nitrate uptake by onions using the finite element method. *Soil Science Society of America Journal* 60, 1160–1167.

Abdallah, M.M.F. (1998) Improving vegetable transplants using soil solarization. II. Onion '*Allium cepa*'. *Annals of Agricultural Science Cairo* 3, 831–843.

Abd El-Rehim, G.H., Ahmed, F.A., Shalaby, G.I. and Waly, A.A. (1996) Effect of transplanting date and planting density on bulb yield and quality of Giza 20 onion in upper Egypt. *Egyptian Journal of Agricultural Research* 75(39), 681–695.

Abu-Awwad, A.M. (1994) Irrigation water management of trickle-irrigated onion. *Dirasat Series B, Pure and Applied Sciences* 21(6), 187–199.

Abu-Awwad, A.M. (1996) Irrigation water management for onion trickle irrigated with saline drainage water. *Dirasat Series B, Pure and Applied Sciences* 23(1), 46–54.

Abu-Awwad, A.M. (1999) Irrigation water management for efficient water use in mulched onion. *Journal of Agronomy and Crop Science* 183, 1–7.

Acosta Baladón, A.N. (1973) *Cultivos enarenados*. Instituto Nacional de Meteorología, Madrid, 224 pp.

Adetunji, I.A. (1994) Response of onion to soil solarization and organic mulching in semi-arid tropics. *Scientia Horticulturae* 60, 161–166.

Al-Jamal, M.S., Sammis, T.W., Ball, S. and Smeal, D. (1999) Yield-based, irrigated onion crop coefficients. *Applied Engineering in Agriculture* 15, 659–668.

Allen, R.G., Pereira, L.S., Raes, D. and Smith, M. (1998) *Crop Evapotranspiration. Guidelines for Computing Crop Water Requirements*. Irrigation and Drainage Paper No. 56, FAO, Rome, 300 pp.

Alt, D., Ladebusch, H. and Melzer, O. (1999) Long-term trial with increasing amounts of phosphorus, potassium and magnesium applied to vegetable crops. *Acta Horticulturae* 506, 29–36.

Anon. (1996) *IPM Manual for Carrots, Onions, Celery and Lettuce in Ontario*. Ontario Ministry of Agriculture, Food and Rural Affairs, Guelph, Ontario, 67 pp.

Anon. (1998) *Teelt van plantuien*. Teelthandleiding Praktijkonderzoek voor de Akkerbouw en de Vollegrondsgroenteteelt 81, Lelystad, The Netherlands, 64 pp.

Arya, P.S. and Bakashi, B.R. (1999) Onion based cropping system studies under mid-hill conditions of Himachal Pradesh. *Advances in Horticulture and Forestry* 6, 79–85.

Ayers, R.S. and Westcot, D.W. (1985) *Water Quality for Agriculture*. Irrigation and Drainage Paper No. 29, FAO, Rome, 174 pp.

Bacher, W. and Lenz, F. (1996) Ammonium deposit fertilization to reduce nitrate leaching. *Acta Horticulturae* 428, 123–130.

Baldwin, J.P., Nye, P.H. and Tinker, P.B. (1973) Uptake of solutes by multiple root systems from soil. III. A model for calculating the solute uptake by randomly dispersed root system developing in a finite volume of soil. *Plant and Soil* 38, 621–635.

Bartolo, M.E., Schwartz, H.F. and Schweissing, F.C. (1994) Yield and growth response of onion to simulated storm damage. *HortScience* 29, 1465–1467.

Belettieri, C. (1997) The *Allium* dehydrated industry in Argentina. *Acta Horticulturae* 433, 483–489.

Bendavid-Val, R., Rabinowitch, H.D., Katan, J. and Kapulnik, Y. (1997) Viability of VA-mycorrhizal fungi following soil solarization and fumigation. *Plant and Soil* 195, 185–193.

Bhonde, S.R. (1997) To study the effect of vermi-compost manure on yield and storage quality of rabi onion. *NHRDF News Letter* 17(4), 1–5.

Bhonde, S.R., Sharma, S.B. and Chougule, A.B. (1997) Effect of bio-fertilizer in combination with nitrogen through organic and inorganic sources on yield and quality of onion. *NHRDF News Letter* 17(2), 1–3.

Bitterlich, I., Upadhyaya, M.K. and Shibairo, S.I. (1996) Weed control in cole crops and onion (*Allium cepa*) using ammonium nitrate. *Weed Science* 44, 952–958.

Boff, P., Stuker, H. and Gonçalves, P.A.S. (1998) Influencia da densidade de plantas na ocorrência de doenças foliares e produção de bulbos de cebola. *Fitopatologia Brasileira* 23, 448–452.
Bond, W., Burston, S., Bevan, J.R. and Lennartsson, M.E.K. (1998a) Optimum weed removal timing in drilled salad onions and transplanted bulb onions grown in organic and conventional systems. *Biological Agriculture and Horticulture* 16, 191–201.
Bond, W., Burston, S., Moore, H.C., Bevan, J.R. and Lennartsson, M.E.K. (1998b) Changes in the weed seedbank following different weed control treatments in transplanted bulb onions grown organically and conventionally. In: Champion, G.T., Grundy, A.C., Jones, N.E., Marshall, E.J.P. and Froud-Williams, R.J. (eds) *Weed Seedbanks: Determination, Dynamics and Manipulation, Oxford, UK, 23–24 March 1998. Aspects of Applied Biology* 51, 273–278.
Bond, W., Moore, H.C., Atkinson, R.J., Bevan, J.R. and Lennartsson, M. (1998c) Changes in the weed seedbank following different weeding treatments in drilled salad onion and carrot crops grown in organic and conventional systems. *Biological Agriculture and Horticulture* 16, 203–215.
Bosch Serra, A.D. (1996) Inyección de fertilizante nitrógeno-fosforado bajo semilla en cebolla (*Allium cepa* L.) In: GMACS (ed.) *IW Congreso de la Sociedad Española de la Ciencia del Suelo, Información de suelos para el siglo XXI*. Edicions Universitat de Lleida, Lleida, Spain, pp. 163–168.
Bosch Serra, A.D. (1999) Bases ecofisiológicas de la producción de cebolla (*Allium cepa* L.): Aportaciones para la mejora de las técnicas de cultivo en el Pla d'Urgell (Lleida). Doctoral thesis, Escola Tècnica Superior d'Enginyeria Agrària, Departament de Medi Ambient i Ciències del Sòl, The University of Lleida, Spain, 461 pp.
Bosch Serra, A.D. and Casanova, D. (2000) Estimation of onion (*Allium cepa* L.) biomass and light interception from reflectance measurements at field level. *Acta Horticulturae* 519, 53–59.
Bosch Serra, A.D. and Domingo Olivé, F. (1999) Ecophysiological aspects of nitrogen management in drip-irrigated onion (*Allium cepa* L.). *Acta Horticulturae* 506, 135–140.
Bosch Serra, A.D., Bonet Torrens, M., Domingo Olivé, F. and Melines Pagès, M.A. (1997) Root growth of three onion cultivars. In: van Ittersum, M.K. and van der Geijn, S.C. (eds) *Perspectives for Agronomy. Adopting Ecological Principles and Managing Resource Use*. Elsevier, Amsterdam, pp. 123–133.
Bracy, R.P. and Parish, R.L. (1998) Seeding uniformity of precision seeders. *HortTechnology* 8, 182–185.
Brewster, J.L. (1990) Cultural systems and agronomic practices in temperate climates. In: Rabinowitch, H.D. and Brewster, J.L. (eds) *Onions and Allied Crops*, Vol. II. *Agronomy, Biotic Interactions, Pathology, and Plant Protection*. CRC Press, Boca Raton, Florida, pp. 1–30.
Brewster, J.L. (1994) *Onions and Other Vegetable Alliums*. CAB International, Wallingford, UK, 236 pp.
Brewster, J.L. (1997a) Environmental physiology of the onion: towards quantitative models for the effects of photoperiod, temperature and irradiance on bulbing, flowering and growth. *Acta Horticulturae* 433, 347–373.
Brewster, J.L. (1997b) Onions and garlic. In: Wein, H.C. (ed.) *The Physiology of Vegetable Crops*. CAB International, Wallingford, UK, pp. 581–619.
Brewster, J.L., Salter, P.J. and Darby, R.J. (1977) Analysis of growth and yield of overwintered onions. *Journal of Horticultural Science* 52, 335–346.
Brewster, J.L., Rowse, H.R. and Bosch, A.D. (1991) The effects of sub-seed placement of liquid N and P fertilizer on the growth and development of bulb onions over a range of plant densities using primed and non-primed seed. *Journal of Horticultural Science* 66, 551–557.
Buttler, T.M., Hornsby, A.G., Stall, W.M., Johnson, F.A., Noling, J.W. and Kucharek, T.A. (1998) *Managing Pesticides for Onion Production and Water Quality Protection*, 2nd edn. Circular 990, The University of Florida, Gainesville, Florida. Electronic publication at: http://edis.ifas.ufl.edu/BODY_SS030
Caldwell, J.O'N., Summer, M.E. and Vavrina, C.S. (1994) Development and testing of preliminary foliar DRIS norms for onions. *HortScience* 29, 1501–1504.
CAST (1992) *Water Quality: Agriculture's Role*. Council for Agricultural Science and Technology, Ames, Iowa, 103 pp.
Castell, V. (1974) La floración prematura de la cebolla valenciana temprana (variedad Babosa). I. Influencia de las fechas de siembra y transplante. *Anales INIA. Serie: Producción Vegetal* 4, 71–89.
Chung, B. (1989a) Multi-plant module transplants of bulb onions. *Acta Horticulturae* 247, 187–191.
Chung, B. (1989b) Irrigation and bulb onion quality. *Acta Horticulturae* 247, 233–237.
Coelho, E.F., de Souza, V.A.B. and Conceição, M.A.F. (1996) Comportamento da cultura da cebola em três regimes de irrigação e cinco espaçamentos. *Pesquisa Agropecuaria Brasileira* 31, 585–591.

Commission of the European Communities (1983) Commission Regulation (EEC) No. 2213/83 of 28 July 1983 laying down quality standards for onions and witloof chicory. *Official Journal of the European Communities* L213, 04/08/1983, 0013–0021.

Commission of the European Communities (1997) Commission Regulation (EC) No 2390/97 of 1 December 1997 amending Regulation (EEC) No. 2213/83 laying down quality standards for onions and witloof chicory. *Official Journal of the European Communities* L330, 02/12/1997, 0012–0012.

Consola, J. and Recasens, J. (1989) Valoración florística y biología de las malas hierbas de los cultivos de cebolla (*Allium cepa* L.) de la comarca del Urgell (Cataluña Occidental). In: EWRS (ed.) *Proceedings of 4th Symposium on Weed Problems in Mediterranean Climates*, Vol. II. EWRS, Valencia, Spain, pp. 62–68 (in Spanish).

Corgan, J.N. and Kedar, N. (1990) Onion cultivation in subtropical climates. In: Rabinowitch, H.D. and Brewster, J.L. (eds) *Onions and Allied Crops*, Vol. II. *Agronomy, Biotic Interactions, Pathology, and Plant Protection*. CRC Press, Boca Raton, Florida, pp. 31–47.

Council of the European Communities (1991) Council regulation (EEC) No. 2092/91 of 24 June 1991 on organic production of agricultural products and indications referring thereto on agricultural products and foodstuffs. *Official Journal of the European Communities* L 198, 22/07/1991, 0001–0015.

Currah, L. and Proctor, F.J. (1990) *Onions in Tropical Regions*. Bulletin 35, Natural Resources Institute, Chatham, UK, 245 pp.

DARP (1995) Ordre de 31 de març de 1995, per la qual s'aprova el Reglament de la Denominació de Qualitat Calçot de Valls i del seu Consell. *Diari Oficial de la Generalitat de Catalunya* 2038, 12/04/1995, 2990–2993.

Daymond, A.J., Wheeler, T.R., Hadley, P., Ellis, R.H. and Morison, J.I.L. (1997) The growth, development and yield of onion (*Allium cepa* L.) in response to temperature and CO_2. *Journal of Horticultural Science* 72, 135–145.

Delahaut, K.A. and Marcell, S.L. (1999) Onions in Wisconsin. Electronic publication at: http://ipcm.wisc.edu/piap/onions.htm

de Lis, B.R., Ponce, I., Cavagnaro, J.B. and Tizio, R.M. (1967) Studies of water requirements of horticultural crops: II. Influence of drought at different growth stages of onion. *Agronomy Journal* 59, 573–576.

D'Ercole, E. and Cembalo, L. (1999) Orticultura biologica: un compromesso tra redditività e impatto ambientale. *Italus Hortus* 6(1/2), 22–27.

de Santa Olalla, F.M., de Juan Valero, J.A. and Fabeiro Cortés, C. (1994) Growth and production of onion crop (*Allium cepa* L.) under different irrigation schedulings. *European Journal of Agronomy* 3, 85–92.

de Visser, C.L.M. (1992) *Bestudering van het groeiverloop van zaaiuien en bouw van een groeimodel*. Proefstation voor de Akkerbouw en de Groenteteelt in de Vollegrond, Verslag No. 142, Lelystad, The Netherlands, 264 pp.

de Visser, C.L.M. (1994a) ALCEPAS, an onion growth model based on SUCROS87. I. Development of the model. *Journal of Horticultural Science* 69, 501–518.

de Visser, C.L.M. (1994b) ALCEPAS, an onion growth model based on SUCROS87. II. Validation of the model. *Journal of Horticultural Science* 69, 519–525.

de Visser, C.L.M. (1998) Effects of split application of nitrogen on yield and nitrogen recovery of spring-sown onions and on residual nitrogen. *Journal of Horticultural Science and Biotechnology* 73, 403–411.

de Visser, C.L.M. and van den Berg, W. (1998) A method to calculate the size distribution of onions and its use in an onion growth model. *Scientia Horticulturae* 77, 129–143.

de Visser, C.L.M., van den Berg, W. and Niers, H. (1995) Relation between soil mineral nitrogen before sowing and optimum nitrogen fertilization in onions. *Netherlands Journal of Agricultural Science* 43, 333–345.

Doorenbos, J. and Kassam, A.H. (1979) *Yield Response to Water*. Irrigation and Drainage Paper No. 33, FAO, Rome, 193 pp.

Doorenbos, J. and Pruitt, W.O. (1977) *Guidelines for Predicting Crop Water Requirements*. Irrigation and Drainage Paper No. 24, FAO, Rome, 144 pp.

Douzals, J.P., Lemière, J.P., Juncker, E., Coquille, J.C. and Thomas, J.M. (1994) Development of a thermal weeding machine. In: IFOAM (ed) *Communications de la quatrième conference internationale I.F.O.A.M.* Association Colloque IFOAM, Quetigny Cedex, France, pp. 157–161.

Dowker, B.D. and Gordon, G.H. (1983) Heterosis and hybrid cultivars in onion. In: Frankel, R. (ed.) *Heterosis*. Monographs on Theoretical and Appplied Genetics, Vol. 6, Springer-Verlag, Berlin, pp. 220–233.

Dragland, S. (1989) The effects of tillage and soil compaction on cabbage, onion and swede. *Norsklandbruksforsking* 3, 145–152.

Drew, R.L.K., Hands, L.J. and Gray, D. (1997) Relating the effects of priming to germination of unprimed seeds. *Seed Science and Technology* 25, 537–548.

Dunan, C.M., Westra, P. and Moore, F.D. (1999) A plant process economic model for weed management decisions in irrigated onion. *Journal of the American Society for Horticultural Science* 124, 99–105.

Ellerbrock, L.A. (1997) Copper requirements for onions grown on organic soils in New York. *Acta Horticulturae* 433, 567–571.

Ells, J.E., McSay, A.E., Soltanpour, P.N., Schweissing, F.C., Bartolo, M.E. and Kruse, E.G. (1993) Onion irrigation and nitrogen leaching in the Arkansas valley of Colorado, 1990–1991. *HortTechnology* 3, 185–187.

El Titi, A., Boller, E.F. and Gendrier, J.P. (1993) Integrated production. Principles and technical guidelines. *IOBC/WPRS Bulletin* 16(1), 1–39.

Fenn, L.B., Taylor, R.M., Binzel, M.L. and Burks, C.M. (1991) Calcium stimulation of ammonium absorption in onion. *Agronomy Journal* 83, 840–843.

Fenn, L.B., Taylor, R.M. and Burks, C.M. (1993) Influence of plant age on calcium stimulated ammonium absorption by radish and onion. *Journal of Plant Nutrition* 16, 1161–1177.

Fink, M., Feller, C., Scharpf, H.-C., Weier, U., Maync, A., Ziegler, J., Paschold, P.-J. and Strohmeyer, K. (1999) Nitrogen, phosphorus, potassium and magnesium contents of field vegetables – recent data for fertilizer recommendations and nutrient balances. *Journal of Plant Nutrition and Soil Science* 162, 71–73.

Fischer, D. and Bachmann, K. (2000) Onion microsatellites for germplasm analysis and their use in assessing intra- and interspecific relatedness within the subgenus *Rhizirideum*. *Theoretical and Applied Genetics* 101, 153–164.

Fuller, S., Gillis, M. and Ziari, H.A. (1996) Effect of liberalized U.S.–Mexico dry onion trade: a spatial and intertemporal equilibrium analysis. *Journal of Agricultural and Applied Economics* 28, 135–147.

Füstös, Z., Gilinger, M.P. and Ombodi, A. (1994) Effects of postharvest handling and cultivars on keeping quality of onions (*Allium cepa* L.) in storage. *Acta Horticulturae* 368, 212–219.

Galmarini, C.R. and Della Gaspera, P.G. (1995) Efecto de la época de transplante y la densidad de plantación en el cultivo de cebolla tipo Valenciana. *Horticultura Argentina* 14(37), 23–29.

Gamiely, S., Randle, W.M., Mills, H.A., Smittle, D.A. and Banna, G.I. (1991) Onion plant growth, bulb quality, and water uptake following ammonium and nitrate nutrition. *HortScience* 26, 1061–1063.

Gaviola, S., Lipinski, V. and Galmarini, C.L. (1998) Efecto de diferentes regimenes de riego sobre cebolla para deshidratar. *Ciencia del Suelo* 16, 115–118.

Goodrich, J.A., Lykins, B.W. and Clark, R.M. (1991) Drinking water from agriculturally contaminated groundwater. *Journal of Environmental Quality* 20, 707–717.

Grant, D.G. and Carter, B.V. (1997) The influence of cultural factors on the bulb shape of the onion (*Allium cepa* L.) cultivar 'Pukekohe Longkeeper'. *Acta Horticulturae* 433, 527–532.

Gray, D., Rowse, H.R. and Drew, R.L.K. (1990) A comparison of two large-scale seed priming techniques. *Annals of Applied Biology* 116, 611–616.

Greenwood, D.J. and Draycott, A. (1989) Experimental validation of an N-response model for widely different crops. *Fertilizer Research* 18, 153–174.

Greenwood, D.J. and Stone, D.A. (1998) Prediction and measurement of the decline in the critical-K, the maximum-K and total plant cation concentrations during the growth of field vegetable crops. *Annals of Botany* 82, 871–881.

Greenwood, D.J., Gerwitz, A., Stone, D.A. and Barnes, A. (1982) Root development of vegetable crops. *Plant and Soil* 68, 75–96.

Greenwood, D.J., Neeteson, J.J., Draycott, A., Wijnen, G. and Stone, D.A. (1992) Measurement and simulation of the effects of N-fertilizer on growth, plant composition and distribution of soil mineral-N in nationwide onion experiments. *Fertilizer Research* 31, 305–318.

Greenwood, D.J., Rahn, C.R., Draycott, A., Vaidyanathan, L.V. and Paterson, C.D. (1996) Modelling and measurement of the effects of fertilizer-N and crop residue incorporation on N-dynamics in vegetable cropping. *Soil Use and Management* 12, 13–24.

Greer, L. and Kuepper, G. (1999) *Organic Allium Production*. Appropriate Technology Tranfer for Rural Areas, Fayetteville, Arkansas. Electronic publication at: http://www.attra. org/attra-pub/allium.html

Guerber-Cahuzac, B. (1992) *Réponse variétale de l'oignon à des stress hydriques*. Document Institut Supérieur Agricole de Beauvais, Beauvais, France, 14 pp.

Gysi, C. (1996) Integrated crop production of field-grown vegetables. A comparison of guidelines from some European countries. *Acta Horticulturae* 428, 15–20.

Ha, I.-J., Suh, J.-K., Hwang, H.-J., Kim, W.-I. and Kim, B.-S. (1998) Effect of sowing date and seedling age at planting on growth and yield for growing plug seedling in onion (*Allium cepa* L.). *RDA Journal of Horticultural Science (I)* 40, 90–97.

Haigh, A.M., Barlow, E.W.R. and Milthorpe, F.L. (1986) Field emergence of tomato, carrot, and onion seeds primed in an aerated salt solution. *Journal of the American Society for Horticultural Science* 111, 660–665.

Hanelt, P. (1990) Taxonomy, evolution and history. In: Rabinowitch, H.D. and Brewster, J.L. (eds) *Onions and Allied Crops,* Vol. I. *Botany, Physiology, and Genetics*. CRC Press, Boca Raton, Florida, pp. 19–20.

Havey, M.J. (1999) Onion. In: Wehner, T.C. (ed.) *Vegetable Cultivar Descriptions for North America*. List 25, 1999. *HortScience* 34, 961–968.

Hayashi, Y. and Hatano, R. (1999) Annual nitrogen leaching to subsurface water from clayey aquic soil cultivated with onions in Hokkaido, Japan. *Soil Science and Plant Nutrition* 45, 451–459.

Hegde, D.M. (1988) Effect of irrigation and nitrogen fertilization on yield, quality, nutrient uptake and water use of onion (*Allium cepa* L.). *Singapore Journal of Primary Industries* 16, 111–123.

Hemphill, D.D. (1982) Anticrustant effects on soil mechanical resistance and seedling emergence. *HortScience* 17, 391–393.

Henriksen, K. (1987) Effect of N- and P-fertilization on yield and harvest time in bulb onions (*Allium cepa* L.). *Acta Horticulturae* 208, 207–215.

Herison, C., Masabni, J.G. and Zandstra, B.H. (1993) Increasing seedling density, age, and nitrogen fertilization increases onion yield. *HortScience* 28, 23–25.

Herold, B., Truppel, I., Siering, G. and Geyer, M.A. (1996) Pressure measuring sphere for monitoring handling of fruit and vegetables. *Computers and Electronics in Agriculture* 15, 73–88.

Herold, B., Oberbarnscheidt, B. and Geyer, M. (1998) Mechanical load and its effect on bulb onions due to harvest and post-harvest handling. *Journal of Agricultural Engineering Research* 71, 373–383.

Herrero, C., Boixadera, J., Danés, R. and Villar, J.M. (1993) *Mapa de sòls de Catalunya 1:25000*. Sheet no. 360–1-2 (65–28) Bellvís, DGPIA and ICC, Barcelona, 198 pp.

Hoffmann, M.P., Petzoldt, C.H., MacNeil, C.R., Mishanec, J.J., Orfanedes, M.S. and Young, D.H. (1995) Evaluation of an onion thrips pest management program for onions in New York. *Agriculture, Ecosystems and Environment* 55, 51–60.

Horowitz, M., Regev, Y. and Herzlinger, G. (1983) Solarization for weed control. *Weed Science* 31, 170–179.

Hutchings, J.J., Hammel, J.E. and Osiensky, J.L. (1998) Nitrogen leaching from unlined cull-onion landfills. *Journal of Environmental Quality* 27, 1254–1260.

Imtiyaz, M., Mgadla, N.P., Chepete, B. and Manase, S.K. (2000) Response of six vegetable crops to irrigation schedules. *Agricultural Water Management* 45, 331–342.

ISTA (1985) International rules for seed testing. *Seed Science and Technology* 13, 299–355.

Jo, I.-S., Hyun, B.-K., Jang, Y.-S., Cho, H.-J., Shin, J.-S. and DeBodt, M.F. (1997) Development of plastic emulsion as a soil conditioner and its effects on soil improvement. *RDA Journal of Agro Environmental Science* 39(2), 8–13.

Katan, J. and DeVay, J.E. (eds) (1991) *Soil Solarization*. CRC Press, Boca Raton, Florida, 267 pp.

Kedar, N., Levy, D. and Goldschmidt, E.E. (1975) Photoperiodic regulation of bulbing and maturation of Bet Alpha onions (*Allium cepa* L.) under decreasing daylength conditions. *Journal of Horticultural Science* 50, 373–380.

Kelley, W.T. and Granberry, D.M. (2000) *Dry Bulb Onions. Commercial Vegetable Production*. Circular 801, The University of Georgia Agricultural and Environmental Sciences, Cooperative Extension Service, Georgia, USA. Electronic publication at: http://www.ces.uga.edu/pubcd/c801-w.html

Klonsky, K., Tourte, L., Chaney, D., Livingston, P. and Smith, R. (1994) *Cultural Practices and Sample Costs for Organic Vegetable Production on the Central Coast of California*. Giannini Foundation Information Series No. 94–2, Department of Agricultural Economics, University of California, Davis, California, 84 pp.

Kretschmer, M. (1994) Influence of temperature and soil water capacity on the emergence of onion seeds. *Acta Horticulturae* 362, 181–188.

Kretschmer, M. (1996) Ein Triebkrafttest für Zwiebelsaatgut. *Gartenbauwissenschaft* 61(4), 184–187.

Kretschmer, M. and Strohm, E. (1996) Aussaattiefe bei Zwiebeln.Teil I. *Gemüse* 32(8), 488–491.

Krikun, J., Haas, J.H., Dodd, J. and Kinsbursky, R. (1990) Mycorrhizal dependence of four crops in a P-sorbing soil. *Plant and Soil* 122, 213–217.

Lancaster, J.E., Triggs, C.M., de Ruiter, J.M. and Gandar, P.W. (1996) Bulbing in onions: photoperiod and temperature requirements and prediction of bulb size and maturity. *Annals of Botany* 78, 423–430.

Leskovar, D.I. and Vavrina, C.S. (1999) Onion growth and yield are influenced by transplant tray cell size and age. *Scientia Horticulturae* 80, 133–143.

Maas, E.V. (1990) Crop salt tolerance. In: Tanji, K.K. (ed.) *Agricultural Salinity Assessment and Management*. American Society of Civil Engineers, New York, pp. 262–304.

McGeary, D.J. (1985) The effects of plant density on the shape, size, uniformity, soluble solids content and yield of onions suitable for pickling. *Journal of Horticultural Science* 60, 83–87.

Madisa, M.E. (1994) Onion cultivar trials for yield and storage in Botswana, 1992–93. *Onion Newsletter for the Tropics* 6, 38–44.

MAFF (1994) *Fertilizer Recommendations for Agricultural and Horticultural Crops*, 6th edn. Reference book No. 209, HMSO, The Stationery Office, Norwich, UK, 112 pp.

MAPA (1993) *Eficacia y selectividad de herbicidas en cebolla. Diez años de ensayos en España*. MAPA, Madrid, 36 pp.

Maroto Borrego, J.V. (1995) *Horticultura herbácea especial*, 4th edn. Mundi-Prensa, Madrid, 611 pp.

Mateeva, A., Svetleva, D., Stratieva, S. and Andonov, D. (1998) Influence of intercropping of maize, onion, garlic and bean on population density of some bean pests. *Mededelingen Faculteit Landbouwkundige, Universiteit Gent* 63(2b), 507–510.

Maw, B.W. and Smittle, D.A. (1986) Undercutting onions. *HortScience* 21, 432–434.

Maw, B.W., Hung, Y.C., Tollner, E.W., Smittle, D.A. and Mullinix, B.G. (1996) Physical and mechanical properties of fresh and stored sweet onions. *Transactions of the ASAE* 39, 633–637.

Maw, B.W., Smittle, D.A., Mullinix, B.G. and Cundiff, J.S. (1998) Design and evaluation of principles for mechanically harvesting sweet onions. *Transactions of the ASAE* 41, 517–524.

May, D.M., Hanson, B.R., Gonzales, J., Egbert, J. and Jones, R.T. (1994) Problems and considerations necessary in crop rotation using buried drip irrigation in vegetable crops grown in the Central Valley of California. In: *Proceedings of the 25th National Agricultural Plastics Congress*. American Society for Plasticulture, St Augustine, Florida, pp. 90–95.

Maynard, A.A. and Hill, D.E. (2000) Cumulative effect of leaf compost on yield and size distribution in onions. *Compost Science and Utilization* 8(1), 12–18.

Melander, B. (1997) Optimization of the adjustment of a vertical axis rotary brush weeder for intra-row weed control in row crops. *Journal of Agricultural Engineering Research* 68, 39–50.

Melander, B. (1998a) Interactions between soil cultivation in darkness, flaming and brush weeding when used for in-row weed control in vegetables. *Biological Agriculture and Horticulture* 16, 1–14.

Melander, B. (1998b) Economic aspects of of physical intra-row weed control in seeded onions. In: Foguelman, D. and Lockeretz, W. (eds) *Organic Agriculture – the Credible Solution for the 21st Century. Proceedings of the 12th IFOAM Scientific Conference, 15–19 November, Mar del Plata, Argentina*. IFOAM, Tholey Theley, Germany, pp. 180–185.

Melander, B. and Hartvig, P. (1997) Yield responses of weed-free seeded onions (*Allium cepa* L.) to hoeing close to the row. *Crop Protection* 16, 687–691.

Menges, R.M. (1987) Weed seed population dynamics during six years of weed management systems in crop rotations on irrigated soil. *Weed Science* 35, 328–332.

Mettananda, K.A. and Fordham, R. (1999) The effects of plant size and leaf number on the bulbing of tropical short-day onion cultivars (*Allium cepa* L.) under controlled environments in the United Kingdom and tropical field conditions in Sri Lanka. *Journal of Horticultural Science and Biotechnology* 74, 622–631.

Muro, J., Irigoyen, I. and Lamsfus, C. (1998) Effect of defoliation on onion crop yield. *Scientia Horticulturae* 77, 1–10.

Niebla Tomé, J.E. and Viera Paramio, J.J. (1990) El cultivo de la cebolla en la isla de Lanzarote. Enarenado y jable: un ejemplo de creatividad técnica. *Agricultura* 701, 1048–1052.

Palaniappan, R., Yerriswamy, R.M. and Varalakshmi, L.R. (1999) Germination of seeds early and seedling growth in onion in a saline environment. *Agricultural Science Digest, Karnal* 19(1), 31–34.

Rabinowitch, H.D., Katan, J. and Rotem, I. (1981) The response of onion to solar heating, agricultural practices and pink root disease. *Scientia Horticulturae* 15, 331–340.

Rahn, C.R., Shepherd, M.A. and Hiron, R.W.P. (1996) The effect of water supply on the response of onions and calabrese to starter solutions. *Acta Horticulturae* 428, 141–150.

Ramírez, H. and Rodríguez, O. (1997) Evaluation of some physical and chemical parameters of a soil under an onion crop. *Acta Horticulturae* 433, 537–542.
Ramírez, H., Pérez, M., Ferrer, R. and Suarez, N. (1997) Effect of phosphogypsum on the crusting of seedbed of onion (*Allium cepa* L.). *Acta Horticulturae* 433, 533–536.
Ramírez, H., Rodriguez, O. and Shainberg, I. (1999) Effect of gypsum on furrow erosion and intake rate. *Soil Science* 164, 351–357.
Randle, W.M. (1995) Preplant calcium affects onion bulb quality and shelf-life. *HortScience* 30, 768 (abstract).
Rapp, I., Shainberg, I. and Banin, A. (2000) Evaporation and crust impedance role in seedling emergence. *Soil Science* 165, 354–364.
Reiners, S., Petzoldt, C.H. and Hoffmann, M.P. (2000) *Integrated Crop and Pest Management Guidelines for Commercial Vegetable Production*. Cornell Cooperative Extension Publication, Ithaca, New York. Electronic publication at: http://www.nysaes.cornell.edu/recommends
Renault, D. and Wallender, W.W. (2000) Nutritional water productivity and diets. *Agricultural Water Management* 45, 275–296.
Rifai, M.N., Lacko-Bartosová, M. and Puskárová, V. (1996) Weed control for organic vegetable farming. *Rostlinná Výroba* 42(10), 463–466.
Riley, H. (1989) Irrigation of cereals, potato, carrot and onion on a loam soil at various levels of moisture deficit. *Norwegian Journal of Agricultural Sciences* 3, 117–145.
RIVM/RIZA (1991) *Sustainable Use of Groundwater. Problems and Threats in the European Communities*. Report no. 600025001, RIVM/RIZA, Bilthoven/Lelystad, The Netherlands, 80 pp.
Roberts, J. (1998) Growing onions without tears. *New Farmer and Grower* 57, 18–19.
Robinson, J.C. (1971) Studies on the performance and growth of various short-day onion varieties (*Allium cepa* L.) in the Rhodesian lowveld in relation to date of sowing. 1. Yield and quality analysis. *Rhodesian Journal of Agricultural Research* 9, 31–38.
Robinson, J.C. (1973) Studies on the performance and growth of various short-day onion varieties (*Allium cepa* L.) in the Rhodesian lowveld in relation to date of sowing. 2. Growth analysis. *Rhodesian Journal of Agricultural Research* 11, 51–68.
Rogers, B.T. (1989) Testing suitability of onion cultivars for dehydration. *Acta Horticulturae* 247, 157–162.
Rossier, N., Palasthy, A. and Schwarz, A. (1994) Influence des types de sols valaisans sur l'aptitude à la conservation de l'oignon de garde. *Revue Suisse de Viticulture, Arboriculture et Horticulture* 26, 199–205.
Rowse, H.R. (1996) Drum priming. A non-osmotic method of priming seeds. *Seed Science and Technology* 24, 281–294.
Rowse, H.R., McKee, J.M.T. and Higgs, E.C. (1999) A model of the effects of water stress on seed advancement and germination. *New Phytologist* 143, 273–279.
Rubin, B. (1990) Weed competition and weed control in *Allium* crops. In: Rabinowitch, H.D. and Brewster, J.L. (eds) *Onions and Allied Crops*, Vol. II. *Agronomy, Biotic Interactions, Pathology, and Crop Protection*. CRC Press, Boca Raton, Florida, pp. 63–84.
Rubin, B. and Benjamin, A. (1984) Solar heating of the soil: involvement of environmental factors in the weed control process. *Weed Science* 32, 138–142.
Saethre, M.G., Orpen, H.M. and Hofsvang, T. (1999) Action programmes for pesticide risk reduction and pesticide use in different crops in Norway. *Crop Protection* 18, 207–215.
Salo, T. (1996) Simulated and measured nitrogen status in soil and in onion crops. *Acta Horticulturae* 428, 193–204.
Salo, T. (1999) Effects of band placement and nitrogen rate on dry matter accumulation, yield and nitrogen uptake of cabbage, carrot and onion. *Agricultural and Food Science in Finland* 8, 157–232.
Satour, M.M., Grinstein, A., Rabinowitch, H.D., Abdel-Rahim, M.F., Katan, J., El-Yamani, T. and Radwan, A. (1989) Soil solarization in onion fields in Egypt and Israel: short- and long-term effects. *Acta Horticulturae* 255, 151–159.
Scurtu, I. (1999) The *Allium* collection at the Research Institute for Vegetable and Flower Growing, Romania. In: Maggioni, L., Astley, D., Rabinowitch, H., Keller, J. and Lipman, E. (compilers) *Report of a Working Group on* Allium, *6th meeting, 23–25 October 1997, Plovdiv, Bulgaria*. International Plant Genetic Resources Institute, Rome, pp. 59–60.
Shannon, M.C. and Grieve, C.M. (2000) Options for using low-quality water for vegetable crops. *HortScience* 35, 1058–1062.

Shete, M.B., Chaudhary, S.M. and Warade, S.D. (1993) A note on use of FYM and vermicompost on yield of white onion cv. Phule Safed. *Allium Improvement Newsletter* 3, 36–38.
Shock, C.C., Feibert, E.B.G. and Saunders, L.D. (1998a) Onion yield and quality affected by soil water potential as irrigation threshold. *HortScience* 33, 1188–1191.
Shock, C.C., Seddigh, M., Hobson, J.H., Tinsley, I.J., Shock, B.M. and Durand, L.R. (1998b) Reducing DCPA losses in furrow irrigation by herbicide banding and straw mulching. *Agronomy Journal* 90, 399–404.
Shock, C.C., Jensen, L.B., Hobson, J.H., Seddigh, M., Shock, B.M., Saunders, L.D. and Stieber, T.D. (1999) Improving onion yield and market grade by mechanical straw application to irrigation furrows. *HortTechnology* 9, 251–253.
Shock, C.C., Seddigh, M., Saunders, L.D., Stieber, T.D. and Miller, J.G. (2000a) Sugarbeet nitrogen uptake and performance following heavily fertilized onion. *Agronomy Journal* 92, 10–15.
Shock, C.C., Feibert, E.B.G. and Saunders, L.D. (2000b) Onion storage decomposition unaffected by late-season irrigation reduction. *HortTechnology* 10, 176–178.
Shock, C.C., Feibert, E.B.G. and Saunders, L.D. (2000c) Irrigation criteria for drip-irrigated onions. *HortScience* 35, 63–66.
Silveira, H.L., Gomes, R., Aguiar, L., Caixinhas, M.L., Bica, J. and Bica, M. (1990/91) Soil solarization under polyethylene film: cultivation of lettuce and onions. *Plasticulture* 85, 35–42.
Sliman, Z.T., Abdelhakim, M.A. and Omran, A.A. (1999) Response of onion to foliar application of some micronutrients. *Egyptian Journal of Agricultural Research* 77, 983–993.
Sørensen, J.N. (1996) Improved N efficiency in vegetable production by fertilizer placement and irrigation. *Acta Horticulturae* 428, 131–140.
Spitters, C.J.T., van Keulen, H. and van Kraalingen, D.W.G. (1989) A simple universal crop growth simulator: SUCROS87. In: Rabbinge, R., Ward, S.A. and van Laar, H.H. (eds) *Simulation and Systems Management in Crop Protection*. PUDOC, Wageningen, The Netherlands, pp. 147–181.
Stenberg, M. (1999) The economy of good vegetable production in open cultivation. In: Hagg, M., Ahvenainen, R. and Evers, A.M. (eds) *Agri-Food Quality II: Quality Management of Fruits and Vegetables – from Field to Table*. Royal Society of Chemistry, Cambridge, pp. 25–27.
Stoffella, P.J. (1996) Planting arrangement and density of transplants influence sweet Spanish onion yields and bulb size. *HortScience* 31, 1129–1130.
Stone, D.A. (1998) The effects of 'starter' fertilizer injection on the growth and yield of drilled vegetable crops in relation to soil nutrient status. *Journal of Horticultural Science and Biotechnology* 73, 441–451.
Stone, D.A. (2000) The effects of starter fertilizers on the growth and nitrogen use efficiency of onion and lettuce. *Soil Use and Management* 16, 42–48.
Stone, D.A. and Rowse, H.R. (1992) Reducing the nitrogen requirement of vegetable crops by precision fertilizer injection. *Aspects of Applied Biology* 30, 399–402.
Stribley, D.P. (1990) Mycorrhizal associations and their significance. In: Rabinowitch, H.D. and Brewster, J.L. (eds) *Onions and Allied Crops*, Vol. II. *Agronomy, Biotic Interactions, Pathology, and Crop Protection*. CRC Press, Boca Raton, Florida, pp. 85–101.
Sullia, S.B. (1991) Use of vesicular-arbuscular mycorrhiza (VAM) as biofertilizer for horticultural plants in developing countries. In: Prakash, J. and Pierik, R.L.M. (eds) *Horticulture – New Technologies and Applications*. Kluwer Academic Publishers, Dordrecht, The Netherlands, pp. 49–53.
Sumner, D.R., Gitaitis, R.D., Gay, J.D., Smittle, D.A., Maw, B.W., Tollner, E.W. and Hung, Y.C. (1997) Control of soilborne pathogenic fungi in fields of sweet onion. *Plant Disease* 81, 885–891.
Tandon, H.L.S. (1999) *Organic Fertilizers and Biofertilizers – A Techno-commercial Sourcebook*. Fertilizer Development and Consultation Organisation, New Delhi, 191 pp.
Tarakanov, I. (2001) A brief survey of the onion cultivars growth in Russian Federation and other republics of the former USSR. *Allium Information Newsletter* 10, 10–13.
Taylor, A.G., Klein, D.E. and Whitlow, T.H. (1988) SMP: solid matrix priming of seeds. *Scientia Horticulturae* 37, 1–11.
Taylor, A.G., Eckenrode, C.J. and Straub, R.W. (2001) Seed coating technologies and treatments for onion: challenges and progress. *HortScience* 36, 199–205.
Tei, F., Scaife, A. and Aikman, D.P. (1996a) Growth of lettuce, onion and red beet. 1. Growth analysis, light interception, and radiation use efficiency. *Annals of Botany* 78, 633–643.
Tei, F., Aikman, D.P. and Scaife, A. (1996b) Growth of lettuce, onion and red beet. 2. Growth modelling. *Annals of Botany* 78, 645–652.

Thilakavathy, S. and Ramaswamy, N. (1998) Effect of inorganic and biofertilizer treatments on yield and quality parameters of multiplier onion (*Allium cepa* L. var. *aggregatum*). *NHRDF News Letter* 18(2), 18–20.

Thompson, A.R., Rowse, H.R., Springer, P.H. and Edmonds, G.H. (1990) Compatibility of liquid insecticide treatments and starter fertilizer solution applied under radish at sowing. *Mededelingen Faculteit Landbouwkundige, Universiteit Gent* 55(2b), 647–655.

Todorov, Y. (1999a) Results of onion breeding and introduction in Bulgaria. In: Maggioni, L., Astley, D., Rabinowitch, H., Keller, J. and Lipman, E. (compilers) *Report of a Working Group on* Allium, *6th meeting, 23–25 October 1997, Plovdiv, Bulgaria*. International Plant Genetic Resources Institute, Rome, pp. 85–88.

Todorov, Y. (1999b) Quality of Bulgarian onion varieties. In: Maggioni, L., Astley, D., Rabinowitch, H., Keller, J. and Lipman, E. (compilers) *Report of a Working Group on* Allium, *6th meeting, 23–25 October 1997, Plovdiv, Bulgaria*. International Plant Genetic Resources Institute, Rome, pp. 89–90.

Tsiropoulos, N.G. and Miliadis, G.E. (1998) Field persistence studies on pendimethalin residues in onions and soil after herbicide postemergence application in onion cultivation. *Journal of Agricultural and Food Chemistry* 46, 291–295.

Uzo, J.O. and Currah, L. (1990) Cultural systems and agronomic practices in tropical climates. In: Rabinowitch, H.D. and Brewster, J.L. (eds) *Onions and Allied Crops*, Vol. II. *Agronomy, Biotic Interactions, Pathology, and Plant Protection*. CRC Press, Boca Raton, Florida, pp. 49–62.

van der Meer, Q.P. (1993) Onion hybrids: evaluation, prospects, limitations, and methods. *Acta Horticulturae* 358, 243–248.

van der Weerden, T.J., Sherlock, R.R., Williams, P.H. and Cameron, K.C. (2000) Effect of three contrasting onion (*Allium cepa* L.) production systems on nitrous oxide emissions from soil. *Biology and Fertility of Soils* 31, 334–342.

Vavrina, C.S. and Roka, F.M. (2000) Comparison of plastic mulch and bare-ground production and economics for short-day onions in a semitropical environment. *HortTechnology* 10, 326–330.

Vergniaud, P., Acosta, T., Le Quillec, S., Montegano, B., Pardo, A., Pelletier, J., Suso, M.L., Taberner, A. and Zaragoza, C. (1989) Methods and techniques of direct sown onion weed control. In: EWRS (ed.) *Proceedings of 4th Symposium on Weed Problems in Mediterranean Climates*, Vol. II. EWRS, Valencia, Spain, pp. 69–82.

Vosátka, M. (1995) Influence of inoculation with arbuscular mycorrhizal fungi on the growth and mycorrhizal infection of transplanted onion. *Agriculture, Ecosystems and Environment* 53, 151–159.

Wall, A.D. and Corgan, J.N. (1999) Yield and dry weight of dehydrator onions after uprooting at maturity and delaying harvest. *HortScience* 34, 1068–1070.

Wannamaker, M.J. and Pike, L.M. (1987) Onion responses to various salinity levels. *Journal of the American Society for Horticultural Science* 112, 49–52.

Whalley, W.R., Finch-Savage, W.E., Cope, R.E., Rowse, H.R. and Bird, N.R.A. (1999) The response of carrot (*Daucus carota* L.) and onion (*Allium cepa* L.) seedlings to mechanical impedance and water stress at sub-optimal temperatures. *Plant, Cell and Environment* 22, 229–242.

Wheeler, T.R., Daymond, A.J., Ellis, R.H., Morison, J.I.L. and Hadley, P. (1998) Postharvest sprouting of onion bulbs grown in different temperature and CO_2 environments in the UK. *Journal of Horticultural Science and Biotechnology* 73, 750–754.

Wiedenfeld, R.P. (1986) Rate, timing, and slow-release nitrogen fertilizers on cabbage and onions. *HortScience* 21, 236–238.

Wijnands, F.G. and van Asperen, P. (1999) Milieubelasting verminderen door gerichte middelenkeuze. *PAV-Bulletin-Akkerbouw* June, pp. 28–37.

Wurr, D.C.E., Hand, D.W., Edmondson, R.N., Fellows, J.R., Hannah, M.A. and Cribb, D.M. (1998). Climate change: a response surface study of the effects of CO_2 and temperature on the growth of beetroot, carrot and onions. *Journal of Agricultural Science (Cambridge)* 131, 125–133.

Xu, P., Sun, H., Sun, R. and Yang, Y. (1994) *Allium* production and research in China. *Acta Horticulturae* 358, 127–132.

Zeidan, O., Elad, Y., Hadar, Y. and Chet, I. (1986) Integrating onion in crop rotation to control *Sclerotium rolfsii*. *Plant Disease* 70, 426–428.

Zink, F.W. (1966) Studies on the growth rate and nutrient absorption of onion. *Hilgardia* 37, 203–218.

10 Onion Pre- and Postharvest Considerations

I.R. Gubb[1] and H.S. MacTavish[2]
[1]*Fresh Produce Consultancy, Mulberry Lodge, Culmstock, Cullompton, Devon EX15 3JB, UK;* [2]*ADAS Arthur Rickwood, Mepal, Ely CB6 2AB, UK*

1. Introduction

In recent times, much research on the storage of onions has focused on developing alternatives to the use of maleic hydrazide (MH) treatment to maintain onion dormancy, since consumers are becoming intolerant of chemical residues in food. Reviews on onion storage since the late 1980s include Komochi (1990), Currah and Proctor (1990) (tropics), Maude (1990) (diseases) and Brice *et al.* (1997). In this chapter we selectively review research published since the preparation of *Onions and Allied Crops* (Rabinowitch and Brewster, 1990), presenting significant advances in onion pre- and postharvest science.

2. Onion Quality

The principal aims of bulb-onion storage are to maintain the 'quality capital' present at harvest (Guerber-Cahuzac, 1996) and to satisfy consumer demand for extended availability of onions of satisfactory quality. The bulbs of edible alliums are naturally dormant organs adapted to maintaining plant viability during periods unfavourable to growth (Brewster, 1994). Following dormancy breaking, they normally resume growth and progress towards flowering and seed production. Appropriate pre- and postharvest treatments can slow down or arrest this process.

A wide choice of cultivars is available on the world market: their storage potential varies from short to very long. Onions grown specifically for storage are cured, dried and held in long-term stores before being cleaned, trimmed, graded and bagged for marketing (Timm *et al.*, 1991). Sweeter and softer onions, historically grown for the fresh bulb market, need special treatment to keep them dormant if they are to be sold later. Recent advances in the science and technology of onion storage have extended the potential life of onion bulbs of both types.

Criteria for onion quality differ between countries. In the UK (Love, 1995) and Australia (Jackson *et al.*, 1989), size and skin finish are paramount. Skin colour is important: a range from pale straw through to a deep copper colour is acceptable for most European markets (Gorini and Testoni, 1990) and other temperate countries. For the UK market, bulb shape should be the globe, with only moderate variations: completely oval or very flat bulbs are not acceptable. Thick and badly trimmed necks are also rejected. In France, lack of internal bulb defects, homogeneity of size, acceptable trimming and firmness are the main marketing criteria (Guerber-Cahuzac, 1996). A neck length up to 4 cm is allowed under European Community (EC) quality standards. Firmness and at least one complete skin are required, and skin cracking should not be evident. Sprouting is not allowed in Class I, but early signs of sprouting are allowed in Class II, provided that the shoots would not become visible for at least 10 days after purchase. Bacterial rots, watery scale and fungal storage rots make the bulbs unsaleable.

In the USA, No. 1 grade onions ('Bermuda', 'Granex' and 'Grano' types) should have typical cultivar characteristics, be mature, fairly firm and well shaped and free from decay, stains or sunscald damage, doubles (more than one distinct bulb joined only at the base) and bottlenecks (elongated bulbs with abnormally thick necks) (USDA, 1997). The onions should be free from seedstems, splits (bulbs with more than one obvious neck), dry sunken areas, sunburn, sprouting, staining, dirt or foreign material, tops and roots, translucent or watery scales, moisture, disease and insects.

Quality factors can be affected by mineral nutrition, timing of irrigation or rainfall (Chung, 1989), cultivar differences and the use of MH (Love, 1995). Onion flavour, defined by pungency and sweetness, varies with cultivar and growing conditions: there is an increasing demand in the USA and the UK for sweeter onions with low pungency.

Postharvest chemical application is best avoided, as it is too close to the consumer; controlled-atmosphere (CA) storage is therefore of increasing interest, since it can extend storage life beyond that achievable with cold storage alone. It also influences sweetness and pungency (H.S. MacTavish, in preparation). The shelf-life of onions after consumer purchase can be affected by the conditions of warming to ambient temperature after cold storage, conditions throughout the marketing chain and the packaging used.

Maintenance of skin integrity and the firmness, colour and flavour of onions is of paramount importance during curing and in the choice of storage regime. Respiration, resumption of growth and pathological breakdown are the biological factors involved in the deterioration of onions. Bulbs also lose water by evaporation or may be physically damaged. Careful handling and the choice of a suitable storage method for the cultivar type in question are vital to ensure that the product retains its quality until it reaches the consumer. 'Cosmetic quality', i.e. retaining an attractive appearance, is of increasing importance in competitive retail markets.

3. Preharvest Factors that Affect Storage

Brewster (1994) described a web of complex interactions among factors contributing to quality of bulbs in postharvest storage, including cultivar, stage of bulb development, premature defoliation, skin integrity and conditions during maturation, harvesting and curing. Preharvest use of MH will be dealt with in Section 8.1.

3.1 Cultivars

Onion cultivars vary greatly in their inherited storage ability (Abdalla and Mann, 1963; Currah and Proctor, 1990; Peters *et al.*, 1994; Havey and Randle, 1996; Galmarini *et al.*, 2000), so correct cultivar choice is essential for successful storage. Factors influencing storage life are bulb composition, the dry-matter (DM) content within a genotype, the number and toughness of the outer skins after curing and the depth of dormancy of the mature bulbs. Most of these factors are controlled genetically, but they are also significantly affected by the environment, so year-to-year variation is common.

In temperate regions, long-storing onions have been developed over hundreds of years: their use spread from Europe to North and South America and Australasia and eventually to Japan (Bosch Serra and Currah, Chapter 9, this volume). In the tropics, locally adapted cultivars tend to store better than short-day (SD) cultivars brought in from the USA, such as 'Grano' and 'Granex' types (Nabos, 1976; Brice *et al.*, 1997; Rouamba *et al.*, 2001). The soft, sweet bulbs of 'Grano'/'Granex' sprout quickly and are susceptible to pathogen attack and to mechanical injury. However, in recent years, high-yielding longer-storing SD onions from Israel are gaining popularity in Africa, South-East Asia and Central and South America (Peters *et al.*, 1994; Kariuki and Kimani, 1997; Msika and Jackson, 1997; Currah, Chapter 16, this volume).

In India, Patil *et al.* (1987) tested 45 cultivars for their yield, growth traits and storability. The thin-neck trait was correlated well with storage potential. In cultivars producing large bulbs, the occurrence of bolting and twin bulbs was directly related to losses during storage (Patil *et al.*, 1987).

3.2 Mineral nutrition

Many researchers found that high levels of nitrogenous fertilizer resulted in reduced onion storage life (Kato *et al.*, 1987; Singh and Dhankar, 1991; El-Gizawy *et al.*, 1993; Wright, 1993; Batal *et al.*, 1994), though others have produced differing results, perhaps dependent on the requirements of specific cultivars. Zafrir (1992) in Israel demonstrated that biweekly applications of nitrogen throughout the growing season up to final amounts of 300 and 500 kg ha^{-1} had no adverse effect on quality and keeping ability of long-storing cvs 'Ben Shemen' (intermediate-day (ID)) and 'RAM 710' (SD), grown from autumn to spring and from early spring to summer, respectively.

In India, Singh and Dhankar (1991) and Pandey and Pandey (1994) found that increasing the rate of applied nitrogen (N) from 50 to 150 kg ha^{-1} led to significant increases in storage loss of onion during 4–5 months under ambient conditions. The timing of N applications is important: bacterial storage rots in New Zealand were more severe after late N applications (Wright, 1993).

In Georgia, USA, bulb rots of cv. 'Granex 33' were reduced by splitting the N application between early and late growth periods (Batal *et al.*, 1994): bulb decay was highest after ammonium nitrate and lowest after calcium and sodium nitrate use (Batal *et al.*, 1994).

Increasing applications of phosphorus fertilizer from 25 to 100 kg ha^{-1} resulted in decreased weight loss, sprouting and rotting in onions stored for up to 160 days in India (Singh *et al.*, 1998a). Rossier *et al.* (1994) in Switzerland analysed the results from 10 years of trials on various soil types in the Valais region: they found that phosphate, present at > 1.3 mg soluble P_2O_5 100 g^{-1} soil, and N as NH_4, tended to promote fungal diseases in store. Onions from slightly saline or sodic soils stored better than those from soils with a more balanced nutrient regime. (Yields and bulb size, however, suffer under sodic and saline conditions.) Good storage quality was negatively correlated with CaO concentration in the cell sap (Rossier *et al.*, 1994).

Adequate sulphur (S) fertility is needed for the development of pungent onion flavours (Randle, 1997; Randle and Lancaster, Chapter 14, this volume) and for healthy growth. Lancaster *et al.* (2001) showed that onions grown with very low S produced softer bulbs than those grown with adequate S supplies. However, in India, excessive S adversely affected storage quality, resulting in increased bulb-neck thickness and moisture content when applied at 30 or 60 kg ha^{-1}. Zinc (Zn) fertilizer at 10 kg ha^{-1} Zn-ethylenediamine tetra-acetic acid (EDTA) reduced sprouting, rotting and weight loss after 90 days in storage (Kumar *et al.*, 1998) but, in another Indian study, the addition of 25 kg ha^{-1} $ZnSO_4$ to potassium fertilizer (100 kg ha^{-1}), resulted in poorer storage performance than the use of potassium fertilizer alone; the latter treatment was further enhanced, in terms of improved storage quality, by addition of 80 kg ha^{-1} N (Singh and Dhankar, 1991). In northern Egypt, foliage treatment of 6–10-week-old onion transplants or of bulbs after harvest with boric acid at 250 or 500 ppm reduced weight loss and decay throughout 6 months of storage at room temperature (28 ± 3°C, 55–65% relative humidity (RH) (Alphonse, 1997). In India, copper (3 ppm, split application at 60 and 70 days after transplanting) reduced storage losses in onion, while zinc, boron, iron and manganese had no or deleterious effects, although boron treatment reduced sprouting (Singh and Tiwari, 1992). Copper treatment reduced yields through decreased bulb size.

Fertilizer and microelement treatments should aim to provide adequate nutrition for the onion crop – ideally tailored to field conditions throughout cropping (Bosch Serra and Currah, Chapter 9, this volume). Further studies are needed to elucidate the interesting question of mineral content and its influence on storage quality, suggested by Rossier *et al.* (1994).

3.3 Soil texture and irrigation

In Poland, irrigation increased the firmness of bulbs grown on a mineral soil compared with an organic soil, and subsequently those from the mineral soil stored better (Perlowska and Kaniszewski, 1988). Smart (1986, quoted by Brice, 1994) compared onions from three contrasted fenland soils in the UK: a sandy soil gave earlier-maturing onions with superior skin quality to those from both peaty and silty soils.

In Oregon, USA, a calculated irrigation threshold of 27 kilopascals (kPa) was recommended: decay due to neck rot (*Botrytis allii*) increased during storage when field water use was higher (Shock *et al.*, 1998). In Maharashtra, India, the effect of withholding irrigation for 12 days prior to harvest, followed by 3 days' curing, resulted in lower storage losses compared with later irrigation and longer curing times (Bhonde *et al.*, 1996). Current recommendations are to apply the final irrigation 10–15 days before onion harvest (Bhonde, 1998).

3.4 Temperature and humidity in the field

There are few reports on the effect of the field temperature on postharvest quality. In the UK, an increase of 1°C over ambient temperatures during production of cvs 'Sito' and 'Hysam' reduced bulb yields by 3–12%, due to a shortened period of crop growth (Daymond *et al.*, 1997). The rate of postharvest sprouting rose following increases in ambient field temperature (Wheeler *et al.*, 1998).

In general, in temperate climates, hot dry weather at the end of the onion cropping season speeds up leaf drying and allows harvesting of already partly dry bulbs with the skins unmarked by rain. Under wet conditions, *Botrytis cinerea* can cause 'brown stain' of storage onions (Sherf and MacNab, 1986). Controlled temperature curing has been widely adopted to prevent this condition and to reduce neck rot in stored bulbs (Maude *et al.*, 1984). In Israel, longer dormancy and better storage result when bulbs are ripened under mild to warm than under low to mild temperatures.

3.5 Carbon dioxide

Bulb yield increases resulted from a rise in CO_2 from 374 to 532 ppm (Daymond *et al.*, 1997), but the time to dormancy break at an average temperature of 11.6°C was unchanged (Wheeler *et al.*, 1998).

3.6 Harvest time in relation to bulb maturity

The timing of harvesting strongly influences both yield and storability. Highest yields are achieved when plants remain intact until the leaves are completely dry. However, for long storage life, it is now common in developed countries to harvest the crop when 50–90% of the tops have fallen: some yield is sacrificed in order to produce an adequate number of attractive intact skins, which will be retained until the time the onions are sold.

In New Zealand, bulbs of cv. 'Pukekohe Longkeeper' lifted at 10% fallen tops retained significantly more skins (95% of bulbs with three or more intact outer skins) than those lifted at 70% tops down, but the earlier crop had the highest incidence of bacterial soft rots caused by *Pseudomonas marginalis* and *P. viridiflava* (Wright and Grant, 1997, 1998). Bulbs harvested at 90%

fallen tops and field-cured had the worst skin retention. Recommendations were to harvest at 70–90% tops down and allow the foliage to dry in the field before topping.

Late lifting (at 90% or more tops down) of European storage-type onions results in increased sprouting and rooting, storage rots and weight loss and higher incidence of watery scale (Böttcher, 1999). In some Hungarian cultivars, the loss of yield resulting from 'early' harvest (i.e. before 100% tops down) was more than compensated for by the increase in storability (Füstös *et al.*, 1994).

Bulbs harvested too soon (i.e. when still developing) have low levels of growth inhibitors (Isenberg *et al.*, 1987), high moisture content in the foliage leaves and bulb necks (providing an environment which favours pathogen infestation) and thicker necks, are not yet fully dormant and thus are simply unsuitable for storage. For example, in Thailand, bulbs of cvs 'Granex 33' and 'RAM-710' are normally harvested at 10% fallen tops for economic reasons, whereas when left in the field until 100% maturity they had significantly longer shelf-life in ambient storage (Peters *et al.*, 1994).

The effect of maturity at harvest on quality during the following month was studied in Florida on fresh-market cvs 'Granex 33' and 'TG1015Y': they were harvested at 10-day intervals, beginning 94 and 115 days, respectively, after transplanting (Sargent *et al.*, 1991, 2001). Harvest maturity stage significantly affected initial bulb weight, respiration and incidence of sprouting, decay and cumulative weight loss. Initial respiration rates in store declined markedly between the first and third harvests, subsequently stabilizing until the fifth harvest (Fig. 10.1). Sprouting was rapid and significant from harvests 1 and 2, but dormancy was established by harvest 3. There was little storage decay over 1 month in cv. 'Granex 33', but significant decay caused by bacterial soft rot after harvests 4 and 5 in 'TG1015Y', attributed to its thicker necks. Cumulative weight loss was negatively correlated with harvest maturity (Sargent *et al.*, 1991). The recommendations were that in Florida, cv. 'Granex 33' should be harvested with at

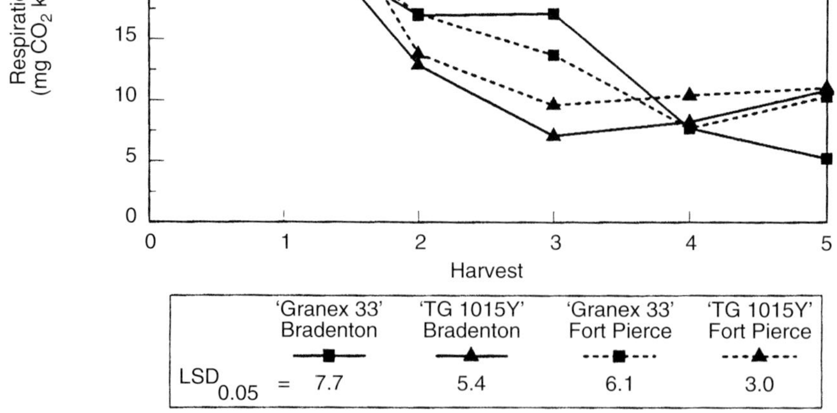

Fig. 10.1. Effect of harvest time on initial respiration rate of onion cvs 'Granex 33' and 'TG1015Y' grown at two sites in Florida (from Sargent *et al.*, 1991, with permission).

least 20% tops down (115 days after transplanting (DAT)) and TG1015Y at 5–25% tops down, at about 132 DAT; non-dried bulbs from these maturity stages could be stored for 2 weeks at 1°C and 80% RH and then withstood 2 weeks at 20°C, allowing time for transport and retailing (Sargent *et al.*, 2001). The bulbs should be handled carefully, to reduce injury, and be cleaned prior to trimming, to reduce contamination (Sargent *et al.*, 1991).

Smittle and Maw (1988) in Georgia found similar decreases in the percentage of bulbs marketable with percentage fallen tops in cvs 'Granex 33' and 'Sweet Georgia', after 1 month at 22–25°C (see also Section 8.3 on CA storage). In short-day 'Grano'-type onions in New Mexico, USA, average bulb weight increased and firmness decreased with delayed harvest, beyond 20% of bulbs with mature (collapsed) necks (Wall and Corgan, 1994) and harvesting at 80% tops down was recommended.

In both Europe and the USA, the consensus is that the optimum harvest time for storage onions is at 80–90% tops down (Büttcher, 1999). Timing of harvesting in temperate growing regions is usually described in current advisory literature and is also discussed in Bosch Serra and Currah (Chapter 9, this volume).

4. The Harvesting Process

Mechanized lifting is the common practice in temperate climates where long-storing, long-day (LD) and ID onions are grown. Modified potato-lifters are commonly used, and the distances that the onions fall during harvesting and in later operations should be kept to a minimum to reduce damage (Geyer *et al.*, 1994; Oberbarnscheidt *et al.*, 1997; Herold *et al.*, 1998).

Large volumes of onions are still harvested by hand in many parts of the world, including sweet onions in the southern USA and elsewhere if soils are difficult to manage. However, with care, mechanization can be used successfully on softer onions (Maw *et al.*, 1999). Studies were made of the physical properties of sweet onions (Maw *et al.*, 1996) before formulating recommendations to improve their treatment at harvest and beyond in Georgia, USA (Maw *et al.*, 1997a, b, 1999).

Undercutting onions (i.e. running a blade below bulb level to separate the bulbs from most of the roots) is commonly done prior to lifting. In New Mexico, USA, dehydration onions should be undercut immediately before harvest to minimize yield losses (Wall and Corgan, 1999).

Onions and shallots are sometimes hung up by their leaves or plaited into strings for storage, particularly in the tropics (Currah and Proctor, 1990; Peters *et al.*, 1994; Rabinowitch and Kamenetsky, Chapter 17, this volume). However, when stored in bins or in bulk, tops are usually removed, in order to improve air flow through the bulbs and reduce trash in the store. Topping can be carried out before, during or after harvest and before store loading. In temperate countries, it is often done before harvest, provided that a heated, forced-air-ventilated store can be used for immediate curing. In the tropics, where controlled drying facilities are rare, it is safer to top after most of the leaves have dried (Brice *et al.*, 1997). In India, sun-curing of bulbs with tops or storing bulbs with dried foliage minimized storage losses in Kharif (summer wet season) onions compared with other treatments (Pandey *et al.*, 1992). Bhonde (1998) recommended shade-curing for several days before onions are stored. Other studies have also examined the benefits for postharvest storage for leaving tops on the onions until after curing (Füstös *et al.*, 1994; Bhonde and Bhadauria, 1995; Chauhan *et al.*, 1995).

Clearly, it is important to distinguish between types of onions and their subsequent use for long- or short-term storage when interpreting the results of maturity-date trials. Local practice may vary according to whether optimum yield or a good appearance out of storage is the principal requirement. The timing of harvesting should be decided according to the importance of these considerations: relatively early harvesting favours better skin retention while later harvesting maximizes yields.

5. Curing and Drying

5.1 The curing process

During curing, the thin outer layers of the bulb are dried to form one or more complete dry skins, which act as a barrier to water loss and microbial infection. Ideally, the dirty outer skins can be removed after storage to show a clean, intact, inner dry skin before retail sale. Even for fresh-market onions, at least one complete skin should be present.

Initial curing (surface and neck drying) may take between a few hours and several days under forced, heated air ventilation, depending on the temperature and RH of the ventilating air and the stage of maturation of the bulbs. Curing is complete when the necks have dried out and are tightly closed and the skins rustle and have developed an attractive colour. An index of cure was developed for sweet onions in Georgia, USA, graded 1 (neck not dry) to 5 (neck will easily bend and flatten to bulb) (Maw *et al.*, 1997b).

Traditional field-curing was done by 'windrowing': detached bulbs shaded by their tops were laid on their sides to dry for 1 or 2 weeks (Thompson, 1996). In hot climates, the bulbs are sometimes covered with straw or leaves while curing. In wet weather, drying takes longer, the bulbs risk becoming watermarked and are prone to rot in subsequent storage; roots may also regrow. Good practice in temperate regions is to move onions into storage straight after lifting and dry them by passing air at 30–32°C through bulk stores or bins – the 'direct harvest' method (Tatham, 1982; Maude *et al.*, 1984). After 3–5 days at higher temperature, the heat is gradually lowered to the safer level of 27°C and 70–75% RH for the completion of drying over about 20 days; this continues gradually until the outer skins reach the 'rustling dry' stage on top of the stack. The temperature can then be lowered to near 0°C for long-term storage.

Biochemical changes during skin curing are being investigated in Japan (Hirota *et al.*, 1999; Takahama and Hirota, 2000). The flavonoid constituents of the drying outer skins of onion bulbs are oxidized and antifungal compounds are formed (Takahama and Hirota, 2000).

5.2 Temperature and humidity during curing: effects on quality and on pathogens

In India, bulbs of cv. 'Nasik Red' were cured at 47, 50 or 53°C for 3 h or at 47 or 50°C for 2, 3, 4 or 5 h and then stored for 4–5 months under ambient conditions (20–30°C, 50–80% RH) or at 21 ± 1°C and 60–65% RH (Thamizharasi and Narasimham, 1993). Treatments for 2–4 h at 47 or 50°C were optimal for increased marketability, resulting in only 2.8% of decay. However, there are hazards to hot-air drying and the timing of the treatment is crucial. In Norwegian trials with European storage onions, late harvesting (at 100% tops down), followed by long drying at 30°C, produced the highest incidence of translucent scales, highest internal CO_2 levels and lowest O_2 levels (Solberg and Dragland, 1998). Late harvesting and prolonged field-curing increased the incidence of leathery scales, a severe quality defect.

Storability of cv. 'Granex 33' onions in Georgia, USA, was improved by curing for 48 h at 35–38°C with low humidity (Maw *et al.*, 1997a). In early-harvested onions, curing for at least 72 h at 35–38°C was required; for late-harvested onions, 24 h was sufficient (Maw *et al.*, 1997b). The depth of bulbs in the stack (20–120 cm) influenced curing, because the drying front moved through the stack in the direction of the air flow and curing was not complete until it had moved through the bulbs completely (Maw *et al.*, 1997b).

In New Zealand, bulbs of cv. 'Pukekohe Longkeeper' were either field-cured, with water applied by sprinkler for 1 h day^{-1} for 10 days, or cured in store at 25°C for 5 days, and then stored for 4 months (12–27°C, 70–85% RH) (Wright and Grant, 1997). The wet field cure resulted in 79% bulbs with skin staining and 70% suffered from soft rot, compared with only 6% stains and 6% soft rot in bulbs cured at 25°C. This emphasizes the value of curing under controlled conditions to retain skins and to reduce the potential for rots to develop. It is important to keep the onions dry throughout curing and to avoid storing bulbs that were exposed to excess moisture during the field-curing process.

In the UK, the severity of neck rot

(*B. allii*) was substantially reduced when bulbs were dried after topping at 30°C with an air flow of 425 m^3 h^{-1} t^{-1}, compared with field-curing followed by storage at ambient temperature (18°C) (Maude *et al.*, 1984). Curing at high temperature was most effective within 48 h of topping, ensuring that the neck area was dried rapidly to prevent the fungus from spreading from the leaves into the neck of the bulb. The same argument applies when there is danger from the spread of bacterial disease into the bulbs (Mark *et al.*, Chapter 11, this volume).

6. Composition and Changes in Bulbs during Curing and Storage

6.1 Fresh weight and moisture loss

Freshly harvested onions contain 80–93% water (according to cultivar), and water removal from the outer skins during curing causes a rapid loss of up to 5% of total weight. Kopsell and Randle (1997) found that cv. 'Dehydrator No. 3' lost 2.1% and cv. 'Granex 33' lost 4.2% of their prestorage mass during the first month of storage. Weight loss continues in healthy dormant bulbs at a low rate, due to respiration and evaporation. Moisture loss from stored onions is lowered by a reduced water-pressure deficit between the bulb and the storage environment (Thamizharasi and Narasimham, 1991), but it is important to maintain the RH of the air below the threshold that encourages pathogens to develop (roughly < 80% RH). Water-vapour loss from onion bulbs was greater during 45 days' storage at 21–35°C, 20% RH, than at ambient RH (50–80%). Moisture losses occurred via the neck and the basal region and also through the sides, which accounted for almost half of all moisture losses (Thamizharasi and Narasimham, 1988).

In Polish onions, the DM content of the true scales increased towards the centre of the bulb during storage at 5°C (Ostrzycka and Perlowska, 1992), consistent with moisture loss from the sides. Cultivar-specific weight losses of between 2 and 5% $month^{-1}$ were recorded in warm ambient storage in Zimbabwe (overall average 3.3%) (Msika and Jackson, 1997). The relatively low initial rate represents loss of water through the skin and by low-level respiration of dormant bulbs; this was followed by a change to a steeper slope, indicating more rapid weight loss, associated with the resumption of sprout growth and senescence of older fleshy scales (N. Hyde and J. Reeves, from unpublished data of R.L. Msika, 1991; Fig. 10.2). Such records can help to identify cultivars with superior storage potential.

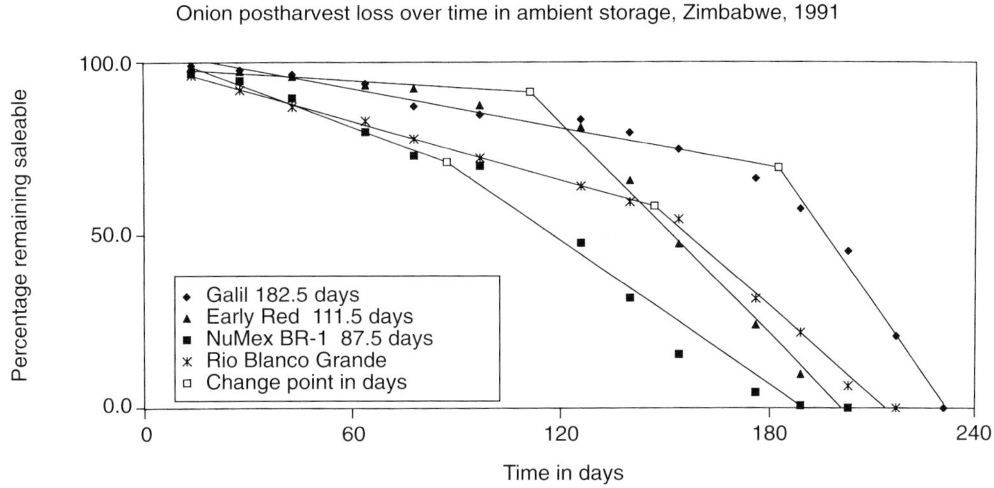

Fig. 10.2. Percentage weight loss over time for onions stored under ambient conditions in Marondera, Zimbabwe, 1991. Unpublished data of R.L. Msika, diagrams by N. Hyde and J. Reeves.

One of the aims of postharvest management is to keep the skins on and intact and dry enough to act as an effective barrier to water loss. Skins tend to crack and fall off at air RH < 55%, and good control of the RH at 55–80% in the air circulating in the store or post-storage conditioning room is essential to retain them (Hole *et al.*, 2000).

6.2 Respiration

Respiration rate is related exponentially to increased storage temperature between 0 and 20°C and is generally a good indication of postharvest quality degradation (Peiris *et al.*, 1997). Respiration of damaged bulbs is more rapid than that of intact ones and this can result in higher water-vapour production in the storage environment; if not controlled by ventilation, this can lead to rooting and then to sprouting.

Benkeblia *et al.* (2000) measured the respiration rates of untreated onions cv. 'Rouge d'Amposta' in France stored at 4°C with 85% RH, 10°C with 80% RH and 20°C with 65% RH. Though the respiratory quotient increased with temperature, the Q_{10} of the untreated onions was only 1.67 at the start of storage. After 2 months at 4°C, it had reached a value of 2.4. This is similar to values quoted by van den Berg and Lentz (1972) of Q_{10} = 2.5 after 1–2 months and 3.5 after 4–6 months of storage in the USA. The respiration rates found after 2 months of storage were 0.21 and 0.32 mmol kg^{-1} h^{-1} at 20°C for unsprouted and sprouted onions, respectively (Benkeblia *et al.*, 2000).

6.3 Carbohydrates

The water-soluble carbohydrates (WSC) in onion bulbs consist of fructose, glucose and sucrose and a series of oligofructans, the maximum degree of polymerization (DP) reached being between 10 and 15 (Suzuki and Cutcliffe, 1989; Ernst *et al.*, 1998). The simple carbohydrates and the lower-DP fructans are present in the largest proportions (Darbyshire and Steer, 1990) and sweet, low-DM onions may contain few or no fructans (Suzuki and Cutcliffe, 1989). Throughout storage, fructans are gradually hydrolysed to fructose, and at the time of sprouting sucrose is synthesized and transported to the sprout and basal plate for growth (Pak *et al.*, 1995). Various studies have examined the relationship between fructose content and storability (Rutherford and Whittle, 1982, 1984; Suzuki and Cutcliffe, 1989; Salama *et al.*, 1990; Horbowicz and Grzegorzewska, 1995). Throughout storage of US cv. 'Sentinel' storage-type onions, fructose increased for up to 15 weeks at 0°C, indicating low-temperature hydrolysis of fructans; fructose increased slightly or hardly at all at 15 and 30°C (Salama *et al.*, 1990). In Russia, cultivars with a high ratio of di- to monosaccharides stored better, as did those that metabolized high-polymer carbohydrates slowly (Anan'ina, 1986). Subtle differences in ratios of the various soluble carbohydrates influence osmotic potential in onion bulbs (Sinclair *et al.*, 1995a) and storage performance across a wide range of cultivars (Sinclair *et al.*, 1995b). The relationships were not straightforward but, when better understood, should provide a basis for the selection of onions for specific uses.

Rapid sprouting in storage in the UK was associated with lower levels of total WSC in the centre of bulbs at the time of harvest (Wheeler *et al.*, 1998). In Germany, losses in disaccharides and the increase in monosaccharides were greater in refrigerated onions than in ambient-ventilated stores (Böttcher, 1992).

In the USA, Kopsell and Randle (1997) found significant differences in soluble-solids content (SSC) during storage, dependent on the cultivar. Prestorage SSC ranged from 13.7% for cv. 'Dehydrator No. 3' to 7.6% for cv. 'Granex 33'. SSC increased and subsequently decreased quadratically during maturation and storage of SD cvs 'Dehydrator No.3', 'Rio Unico' and 'Granex 33', while for ID cvs 'Walla Walla Sweet' and 'Candy' and LD cvs 'Pukekohe', 'Zenith' and 'Sweet Sandwich', SSC decreased linearly over time (Kopsell and Randle, 1997; Table 10.1).

Table 10.1. Effects of storage time at 5 ± 3°C, and 80 ± 5% relative humidity on bulb enzymatically-formed pyruvic acid (EPY) (μmol ml^{-1} fresh mass) of several short-day (SD), intermediate-day (ID) and long-day (LD) onion cultivars.

Months	Dehydrator No. 3 (SD)	Granex 33 (SD)	Sweet Sandwich (LD)	Pukekohe (LD)	Walla Walla Sweet (ID)
0	12.8 ± 1.8	6.3 ± 0.7	6.6 ± 0.5	11.0 ± 2.1	9.5 ± 1.3
1	11.1 ± 0.6	6.3 ± 0.7	7.8 ± 1.7	12.8 ± 2.2	10.8 ± 0.7
2	11.1 ± 0.7	7.1 ± 0.7	7.6 ± 1.0	11.7 ± 0.9	9.0 ± 0.9
3	10.6 ± 0.2	7.8 ± 1.0	6.6 ± 0.9	10.3 ± 1.6	7.3 ± 2.2
5	–	–	7.0 ± 2.4	10.9 ± 1.8	7.1 ± 1.8
6	–	–	5.0 ± 0.4	10.0 ± 0.9	7.9 ± 2.5
Regression significance					
Linear	$P = 0.05$	$P = 0.03$	$P = 0.11$	$P = 0.10$	$P = 0.1$
Quadratic	–	–	$P = 0.13$	–	–

Reprinted in abbreviated form with permission from Kopsell and Randle (1997).

6.4 Organic acids

The major organic acids in onions are malic and citric with small amounts of fumaric and succinic acids. They increased throughout storage of cv. 'Sentinel' for 15 weeks at 30°C (Salama *et al.*, 1990). The ratio of citrate to malate varied from 1 : 5 at 0°C to 2 : 1 at 30°C; MH and RH had no effect on organic acids throughout storage (Table 10.2).

6.5 Pungent flavours

Development of pungent flavours in onions is well understood (Randle and Lancaster, Chapter 14, this volume). Upon cellular disruption, the enzyme alliinase, present in the vacuole, is released to hydrolyse the flavour precursors collectively termed *S*-alk(en)yl-L-cysteine sulphoxides (ACSOs), present in the cytoplasm (Kopsell *et al.*, 1999). The products of hydrolysis are unstable alk(en)yl sulphenic acids, which rearrange non-enzymatically to form thiosulphinates, contributing to perceived 'flavour'. Pyruvic acid and ammonia are non-flavour products of the enzymatic reaction (Kopsell *et al.*, 1999); there is a good correlation between enzymatically determined pyruvic acid (EPY) and overall taste perception (Wall and Corgan, 1992). Expression of onion flavour is dominated by organic S compounds and modified by simple and complex sugars (Randle, 1997).

Table 10.2. Organic acid levels (mg g^{-1} dry weight) in the inner and outer leaves of onion bulbs.

Leaf location	Organic acids (mg g^{-1})		
	Malate	Citrate	Total
Inner	26	23	49
Outer	33	13	46
Significance	***	***	*

Means of 144 values (4 storage times × 3 temperatures × 2 RH × 2 MH treatments × 3 replicates).
*, *** *F* tests were significant at $P = 0.05$ and 0.01 respectively.
From Salama *et al.* (1990) with permission.

Onion pungency changes during storage (Kopsell *et al.*, 1999), increasing in some cultivars (Shekib *et al.*, 1986) and declining in others, especially in pungent storage cultivars (Kopsell and Randle, 1997). Bulbs of Polish cvs 'Sochaczewska' and 'Blonska' harvested when 90% of leaves were still green were more pungent than those harvested when more mature, and during 2–3 months of drying/storage in ambient conditions, followed by 0–1°C, pungency increased (Horbowicz, 1998).

During storage, EPY decreased linearly in cv. 'Dehydrator No. 3' but linearly increased in the sweet cv. 'Granex 33', while 'Rio Unico' (all SD cultivars) showed no distinct trend. In ID and LD cultivars, EPY decreased linearly or quadratically during storage (Kopsell and Randle, 1997). Changes in pungency during storage were found, due to dissimilarities in ACSOs associated with individual onion phenotypes: however, the relationship between EPY and ACSO content was not stoichiometric (Lancaster *et al.*, 1998). During storage, content of individual ACSOs altered: (+)-*S*-methyl-L-cysteine sulphoxide (MCSO) generally decreased while *trans*-(+)-*S*-(1-propenyl)-L-cysteine sulphoxide (PrenSCO) increased, corresponding with a linear decrease in γ-L-glutamyl-*S*-(1-propenyl)-L-cysteine sulphoxide (γGPECSO) (Kopsell *et al.*, 1999). The rates of change of ACSOs varied between cultivars. The trends were indicative of activity of γ-glutamyl transpeptidase throughout bulb storage, although other research suggested this enzyme was minimally active during storage (Lancaster and Shaw, 1991). Two major γ-glutamyl peptides, γ-glutamyl *trans*-(+)-*S*-(1-propenyl)-L-cysteine sulphoxide and *S*-2-carboxypropyl glutathione, may function as storage compounds for the dormant onion bulbs and are biosynthetic intermediates in the production of ACSOs (Lancaster and Shaw, 1991). After 6 months of storage at 0°C, concentrations of ACSOs in the inner scales and at the top and bottom of each bulb were increased, compared with no changes in the dead brown skin and senescent outer tissues of cvs 'Hysam' and 'Grano de Oro' in the UK (Bacon *et al.*, 1999).

6.6 Flavonol glucosides

The major flavonol glucosides in the edible portion of the onion include quercetin 3,4′-*O*-diglucoside (QDG) and quercetin 4′-*O*-monoglucoside (QMG) (Price and Rhodes, 1996). These compounds were reasonably resistant to degradation during drying, storage and processing of onion bulbs, with a loss of 25% due to boiling or frying, and 50% of QMG was lost during the initial drying process at 28°C for 10 days (cvs 'Red Baron' and 'Cross Bow') (Price *et al.*, 1997). The loss during drying may be due in part to mobilization of QMG towards the drying skins, which in this case were removed for the extraction process. Even after sprouting, little change in the flavonol glucosides occurred in the edible portion of the onion (Price *et al.*, 1997). Hirota *et al.* (1999) and Takahama and Hirota (2000) described the formation of compounds found in brown onion skins, in particular 3,4-dihydrobenzoic acid and 2,4,6-trihydroxyphenylglyoxylic acid. These compounds are thought to be created by the enhanced peroxidase-dependent oxidation of quercetin during the drying-down processes, which produce the protective skins of onion bulbs. 3,4-Dihydrobenzoic acid is an antifungal agent.

6.7 Colours

White onions tend to develop chlorophyll when exposed to light, whether in the field or in store. The pigments of red onion skins are simple and consist mainly of malonated anthocyanins (Donner *et al.*, 1997). Fossen *et al.* (1996, 1998) described red pigments which included novel peonidins.

In red and brown onions, the colour intensifies with curing and this can be manipulated for particular markets. For example, Dutch consumers prefer lighter-coloured onions than those in the UK and this can be achieved by varying the curing time, temperature and humidity (Bleasdale and Thompson, 1966). When stored in well-lit conditions or out of doors, coloured onions tend to become rather bleached after some months (L. Currah, UK, 2001, personal communication).

6.8 Vitamins

Ascorbic acid (vitamin C) content in stored onions in Germany increased linearly by 0.5 mg 100 g^{-1} fresh weight $month^{-1}$, irrespective of the storage temperature (Böttcher, 1992). In Cuba, ascorbic acid content of several SD cultivars decreased with increasing storage temperatures (0, 10, 20–25°C, 22–32°C) (Iglesias *et al.*, 1987).

6.9 Physical and chemical properties of onions and onion-skins

Bulb firmness may be partly related to the adhesion of cell-wall fibrils to one another within the fleshy scales, due to the presence of non-uronide carbohydrates and the strength of the middle lamella (Mann *et al.*, 1986). Changes in carbohydrate metabolism or damage during storage may have bad effects on firmness and onion quality.

Ha *et al.* (1997) used nuclear magnetic resonance techniques to establish that, in dry onion cell walls, cellulose/xyloglucan microfibrils acted as solid rods while dry pectins were in a glassy state. On hydration, the pectins became gel-like but the microfibrils continued to provide rigidity. Lancaster *et al.* (2001) studied onion firmness in relation to S nutrition and found that lack of adequate S gave smaller and softer onions with a smaller proportion of DM in the cell-wall material. The authors deduced that the S composition of the cellular components (including ACSOs) is maintained at the expense of bulb growth.

The mechanical properties of onion skins in relation to humidity were studied by Hole *et al.* (2000). They discovered that, following exposure to air at 95% RH, the damp skins were much more elastic in several directions and consequently could resist stretching better than very dry skins, which were brittle. Therefore, controlled RH of the air during post-storage conditioning can be manipulated to slightly dampen the outer skins, with a target of 75% RH to enhance skin retention (Brice, 1994).

Considerable research on onion physicochemical properties is being done at Norwich, UK (Ng *et al.*, 1998, 2000), in connection with possible uses for skins from brown onions. Cell-wall materials from skins of cv. ‘Sturon’ included several phenolic compounds, such as protocatechuic acid (the most abundant), vanillic acid and *p*-hydroxybenzoic acid. In the outer epidermis of adjacent fleshy scales, the most abundant phenolics were *trans*-ferulic, *trans*-coumaric and vanillic acids. Flavonoids were also present and it is postulated that these may be involved in peroxidative cross-linking in the cell walls of dry onion skins.

6.10 Mechanical injury

One of the main issues in the postharvest handling of onion is how to limit the number and intensity of mechanical impacts on the bulbs (Herold *et al.*, 1998). Impact and pressure bruising damages both surface and internal tissues, thus providing an entry for pathogens and stimulating respiration (Yoo and Pike, 1995a), which can rise up to 250% compared with that of undamaged bulbs, with higher rates maintained for 30–35 days after impact. This also reduces DM content (Geyer *et al.*, 1994; Yoo and Pike, 1995a). A damaged basal plate and missing scales were associated with rapid breaking of dormancy (Füstös, 1997). Cutting off the tops of bulbs to encourage sprouting for seed production is a well-known practice (Currah and Proctor, 1990). In Texas trials, onion bulbs with the top halves removed sprouted immediately after harvest at 15 or 24°C, but not at 30°C (Yoo and Pike, 1995b).

Mechanical damage during harvest and handling often becomes evident once bulbs are brought out of storage. Bruising occurs to a greater extent after curing, as a result of handling firmer onions (Hak and Ludwig, 1988; Timm *et al.*, 1991). Impacts on the bulbs can be transmitted through the scales to the bulb interior (Maw *et al.*, 1995). Methods of assessing internal bulb quality by using X-rays are now being developed (Tollner and Shahin, 2000).

Within stores, damage can occur at the base of stacks if the depth of onions is too high for the bulbs in the lower layers to sustain. This problem can be alleviated by using bins for storage that keep the layers of onions at a safe height.

6.11 Growth substances

Some classic accounts of growth substance in onions are those of Thomas and Isenberg (1972) and Isenberg *et al.* (1987). They detailed the movement of inhibitors from the leaves into the bulb tissues at maturity, the gradual change to lower inhibitor- and higher growth-substance levels during overwinter storage at 5–8°C in the UK, in the order: gibberellins (with a first peak in December), cytokinins and auxins. A second gibberellin peak accompanied sprouting in March. Isenberg *et al.* (1987) concluded that UK storage onions have a rest period from harvest until midwinter, with several weeks' cessation of sprout growth even under favourable conditions. The peaks of activity of growth substances in sequence, roughly 30 days apart, were thought to correspond to internal development of the growth apex, representing floral initiation under cold conditions (initial gibberellic acid (GA) peak), cell multiplication (cytokinins), the initiation of sprout growth (auxins) and the appearance of a visible floral initial; the second gibberellin peak accompanied actual sprouting at the end of the dormant period. Thus, the apparently inert dormant onion actually undergoes important internal changes, leading towards flower production (Kamenetsky and Rabinowitch, Chapter 2, this volume).

Growth inhibitors increased and gibberellins decreased throughout drying in cv. 'Sochaczewska' (Kielak and Bielinska-Czarnecka, 1987). In Japan, abscisic acid (ABA) levels were high at the onset of dormancy, reaching a maximum 1 month after storage, gradually decreased during storage and increased again during sprouting (Matsubara and Kimura, 1991; see also Section 7 below).

7. Dormancy and Dormancy Breaking

7.1 The nature of onion dormancy and changes over time

Within the onion bulb, a succession of internal changes take place, preparing it for regrowth (see Section 6.11). In agreement with Brewster (1987), Miedema (1994a, b) and Miedema and Kamminga (1994) showed with Japanese, Dutch and US cultivars that dormancy exists in bulbs soon after maturation, followed by rest, during which slow internal preparation for rooting and sprouting takes place, unless temperatures are very low (near 0°C) or above 25°C. Onion dormancy can be rapidly broken under favourable conditions for regrowth (e.g. Abdalla and Mann, 1963; Pak *et al.*, 1995); for example, resumption of root growth is promoted when onions get wet in the field during curing.

Sprout meristems were mitotically active from lifting throughout storage at 4 or 10°C for up to 25 weeks, with greatest activity 5 weeks after harvest (Matejko and Dahlhelm, 1991). Earlier, Abdalla and Mann (1963), in the USA, found that, in cv. 'Excel', the average number of mitoses per apex detected in the weeks before harvesting was 10–13. The number declined to <1–4 by 3 weeks later, with more divisions continuing at 15 than at 0 and 30°C, where division practically stopped; but Abdalla and Mann (1963) found that at no time was the shoot apex morphologically inactive. In northern European cultivars in The Netherlands, mitotic activity of the apex decreased before harvest, was low for the 3 weeks after harvest and increased after that at 4–8 weeks as sprouts were initiated, when onions were stored at 16°C. The pause in mitosis was comparatively short and was not regarded as a true dormant period but rather as a transition between storage-scale and foliar-leaf formation in the bulb (Pak *et al.*, 1995; Fig. 10.3). Mitotic activity was connected with leaf initiation and elongation in the inner bud of bulbs; the extent of sprout growth was dependent on temperature (Abdalla and Mann, 1963; Matejko and Dahlhelm, 1991). Although carbohydrates and enzymes were available for fast sprouting, sprout growth remained linear rather than exponential during dry storage at 16°C, and was considered to be limited by lack of external water (Pak *et al.*, 1995).

Starch has been found in *A. cepa* in the primary thickening meristem (PTM) during sprouting, but not during dormancy; absence of starch may therefore be useful as a marker for dormancy (Ernst and Bufler, 1994). Ernst *et al.* (1999) studied four cultivars stored at 0, 15 and 30°C. Low starch in the PTM indicated primarily root dormancy, and only indirectly sprout dormancy. Starch in the PTM increased before sprouting at the low and intermediate temperatures but

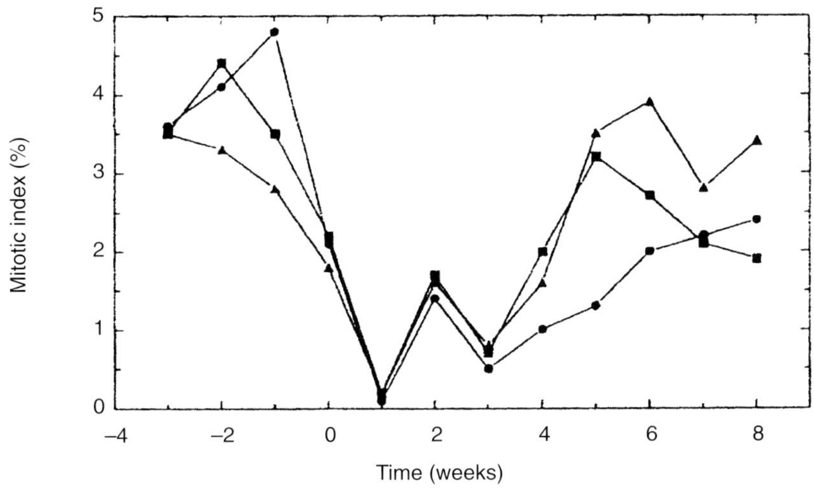

Fig. 10.3. Mitotic activity in the meristems of onion cultivars 'Hysam' (circles), 'Hystar' (squares) and 'Centurion' (triangles). Percentage of dividing cells (mitotic index) starting 3 weeks before harvest (harvested at time 0) and during storage at 16°C. Mean values of five apices. (From Pak *et al.*, 1995, with permission.)

was not detectable at the highest temperature. This provides an interesting clue to the mechanism of high-temperature dormancy. Miedema (1994b) considered that lack of cytokinin, due to root dormancy, was its immediate cause.

7.2 Cultivars

Onion cultivars can be characterized by the toughness of the dry scales, the colour, thickness and number of which are mainly genetically determined. Skin quality is an important factor in determining storability and has a significant role in maintaining dormancy (Füstös, 1997).

In Poland, Adamicki (1998) considers that late-maturing cultivars generally store better than early-maturing cultivars. In Zimbabwe, midseason-maturing SD cultivars stored better than most early- and all late-maturing ones (R.L. Msika, unpublished data). In general, 'Grano'/ 'Granex' have thinner and fewer skins than traditional LD storage cultivars and, normally, a shorter storage life. In the Republic of Macedonia, bulbs for seed production were stored in non-controlled conditions for 5 months; cv. 'Texas Grano' had the greatest losses in terms of number of affected bulbs (91%) and bulb mass (93.9%), with cv. 'Moldavski' having the fewest losses (59% sprouted, 60.1% mass reduction) (Agic *et al.*, 1997). In Iran, the local cv. 'Dorcheh' stores longer at both low and high temperature than cv. 'Texas Early Grano' (Ramin, 1999). In Holland, the range in time to 50% rooting at 10°C in ten cultivars was from 8 to 63 days, and to 50% sprouting, 40 to 156 days, with considerable bulb-to-bulb variation (Miedema, 1994a). In the tropical countries, storage lives of different types of short-day onions vary considerably (Peters *et al.*, 1994). They averaged 1–2 months for 'Grano/ Granex', 4–5 months for 'Red Creoles' and up to 10 months for local cultivars in Egypt (means calculated from a survey: Currah and Proctor, 1990).

7.3 Temperature and dormancy breaking

Temperature plays a critical role in the spoilage of onions in stores (Abdalla and Mann, 1963; Komochi, 1990; Mondal and Pramanik, 1992; Tanaka *et al.*, 1996). While both low and high temperatures maintain onion dormancy, intermediate temperatures between about 5 and 20°C are effective in breaking dormancy, with some variation due to cultivars; in many studies, 15°C has been found the optimum temperature for promoting sprouting. At room temperature in Georgia, USA, the quantity of marketable bulbs of 'Granex'-type onions decreased by 12–25% month^{-1}, due to water loss and black mould damage (Smittle, 1988).

Many wild alliums show high-temperature dormancy in hot seasons, and the reaction of bulb onions is probably related to this behaviour. In the tropics, in the absence of refrigerated stores, the storage of onions at 25°C within the range of 50–70% RH produces the least spoilage (Mondal and Pramanik, 1992). Miedema and Kamminga (1994) found that low cytokinin concentrations occurred under high temperature (30°C) conditions (Tables 10.3, 10.4). After 6 or 12 weeks of storage at 30°C, rooting and subsequent sprouting of cv. 'Augusta' (Rijnsburger type) were more rapid than after storage at 5 or 15°C; however, at the latter temperatures, cytokinin activity was six and almost nine times greater, respectively, after 18 weeks than in 30°C storage. Increased levels of cytokinins, probably generated during root initiation, were associated with onion sprouting (Miedema and Kamminga, 1994).

Sprouting in onion bulbs was thought to be inhibited by ABA (Yamazaki *et al.*, 1995, 1999a) and promoted by cytokinins (Kuraishi *et al.*, 1989), with dormant cultivars having increased sensitivity to ABA compared with non-dormant cultivars (Yamazaki *et al.*, 1999b). The importance of *cis-* rather than *trans*-ABA in the breaking of bulblet dormancy has been suggested (Kuraishi *et al.*, 1989).

Roots start to grow within the base plate and do not emerge until sufficient outside moisture is available to support them. Tanaka *et al.* (1985) described and distinguished 'external' and 'internal' (new) roots within the basal plate and showed that external moisture was the cue to start the internal roots growing at temperatures of 5°C up to 15°C, but that temperatures of 2 or 30°C strongly suppressed their development. Miedema (1994b) found that substituting benzyl adenine (BA) for roots was effective in stimulating sprouting if the roots themselves were trimmed off.

The time lapse between the appearance of visible roots and visible sprouts varies between cultivars. For example, in the Japanese cv. 'Radar', rooting was followed about a month later by visible sprouting, whereas in cv. 'Hyduro' (Rijnsburger storage type) there was a lapse of about 3 months before sprouts became visible, after rooting had started (Miedema, 1994a; Fig. 10.4)

In temperate countries, storage at low temperatures near or even below 0°C is commonly used to keep both onions and pathogens inactive. Ambient air can be used to keep onions dormant during the winter but refrigeration must be used in the spring to further delay sprouting. Onions with relatively high DM content can tolerate temperatures just below 0°C, but those with low DM may be damaged by freezing.

In Algeria, 9°C treatment of cv. 'Rouge d'Amposta' promoted sprouting faster than

Table 10.3. The effects of storage temperature and duration on dormancy characteristics in bulb samples of onion cv. 'Augusta'. Time to rooting and sprouting were estimated on bulbs planted in moist vermiculite at 15°C; three replicates of 20 bulbs were used per temperature and sampling date. Values followed by the same letter are not significantly different at $P \leq 0.05$.

Duration of storage (weeks)	Days to 50% rooting			Days to 50% sprouting		
	5°	15°	30°	5°	15°	30°
0	21.7a	21.7a	21.7a	52.3a	52.3a	52.3a
6	11.3bc	14.0b	7.8c	40.7b	46.2ab	29.1c
12	5.0d	4.9d	3.5de	37.6b	23.9cde	22.1de
18	2.7e	2.5e	3.0e	27.3cd	16.5f	19.6ef

From Miedema and Kamminga (1994), with permission.

Table 10.4. The effects of storage temperature and duration on cytokinin activity in bulb samples of onion cv. 'Augusta', estimated with the *Amaranthus* bioassay. Values are means ± standard error of three bioassays.

Duration of storage (weeks)	Cytokinin activity (nmol zeatin eq. g^{-1} fresh weight)		
	5°C	15°C	30°C
0	0.10 ± 0.00	0.10 ± 0.00	0.10 ± 0.00
6	0.17 ± 0.03	0.23 ± 0.03	0.27 ± 0.03
12	1.80 ± 0.12	3.93 ± 0.15	0.33 ± 0.03
18	2.90 ± 0.10	4.23 ± 0.19	0.50 ± 0.06

From Miedema and Kamminga (1994), with permission.

storage at 0°C. When the concentrations of phenolics and peroxidase activity were relatively high, inner bud development was inhibited; sprouting was accompanied by high concentrations of oligosaccharides and glucose (Benkeblia and Selselet-Attou, 1999a). In a further study of cv. 'Rouge d'Amposta' onions during storage at 4 and 20°C, an inverse relationship between phenolic content and the amount of sprouting development of bulbs was observed. Low temperature had a stimulatory effect on phenylalanine ammonia-lyase (which is involved in phenolic metabolism) and peroxidase activity, both of which are highly involved in onion-bulb sprouting (Benkeblia, 2000a).

Cold treatment of *A. victoralis* L. ssp. *platyphyllum* Hult. (a wild species used as a food plant in East Asia) at 0°C was more effective in breaking dormancy than 5°C (Kanazawa *et al.*, 1997).

The expression of histone 2A has been inversely correlated with dormancy in cv. 'Robusta' (Carter *et al.*, 1999). High levels were found in basal tissues and in the inner, meristematically active parts of bulbs, and expression levels increased throughout storage as onions began to emerge from dormancy. A comparison of early- and late-sprouting onion UK breeding lines showed that histone 2A levels peaked at around the same time of year, irrespective of sprouting time, suggesting that differences in storage longevity are not related to different times of dormancy breakage. Factors controlling the rate of sprout emergence post-dormancy (primarily temperature) are likely to be major determinants of storage capability (Carter *et al.*, 1999).

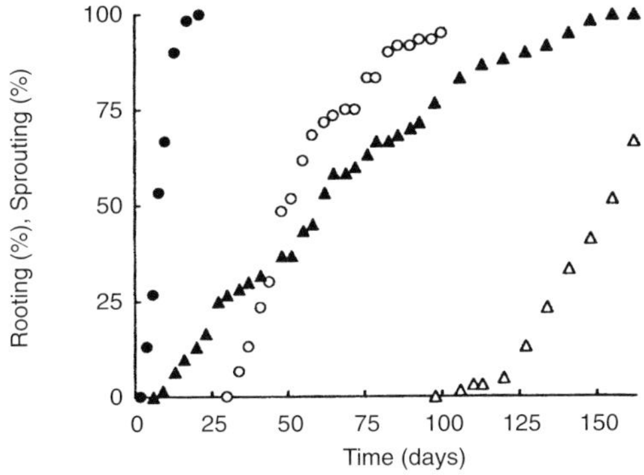

Fig. 10.4. Time course of rooting (solid symbols) and subsequent sprouting (open symbols) of onion cvs 'Radar' (circles) and 'Hyton' (triangles) at 10°C in moist vermiculite (from Miedema, 1994a, with permission).

7.4 Relative humidity

Control of humidity during storage is important for three main reasons. One is concerned with discouraging disease development. Pathogens can attack onion skins when the moisture content rises above a percentage that can be reached when the skin is in equilibrium with air at RH > 80% (see Section 10). The second reason is the prevention of rooting, also encouraged by high air humidity or free water in store (the start of rooting also involves shape changes at the base of the bulb, which can lead to skin cracking). The third, related reason is the need to retain sufficient skins on onion bulbs out of store. The moisture content of the skin is mainly controlled by the RH of the surrounding air, in equilibrium with moisture from the interior of the bulb. When dry skins are lost, a new equilibrium is reached after higher initial water loss and, for this, manipulation of the air RH may be needed throughout storage (see Section 6.9).

Ideally, the air RH in the store should be between 65 and 70% (Mondal and Pramanik, 1992), though in practice wider limits than these are used. In Brazilian experiments, rates of fresh-weight loss were higher when bulbs were stored at < 55% RH, because very dry onion-skins crack easily, so exposing wetter interior skins until a new equilibrium is reached. The optimum storage conditions in a 30-day trial were 20°C (from a range from 20 to 35°C) with RH between 55 and 70%. The water content of the skins increased dramatically when the air RH rose above 75% and, at high skin moisture content, both skin permeability and rates of fresh-weight loss increased (de Matos *et al.*, 1997).

High RH inside stores encourages root development and therefore may tend to break dormancy in onions that would keep well if dry. Methods of storage that keep onion basal plates dry, e.g. hanging in strings, avoid this difficulty.

7.5 Internal atmosphere

Since the meristem of an onion bulb is surrounded by many layers of bulb scales, it may be subjected to an environment of high CO_2 concentrations. After 3 months of storage at 20°C, Ladeinde and Hicks (1988) found that the internal atmosphere in bulbs was 3.1% CO_2 and 16.2% O_2. In Georgia, USA, shoot growth was inhibited by keeping sweet-onion bulbs under low O_2 and high CO_2 concentrations in CA similar to that used in apple storage (Smittle, 1988; see also Section 8.4).

In Texas, the effects of internal CO_2 atmospheres on shoot growth and respiration rates in cv. 'TG 1015Y' stored at 1, 7, 13, 20, 27 or 34°C for 12 weeks were measured (Yoo *et al.*, 1997). Maximum shoot growth occurred at 13 and 20°C, coinciding with maximum respiration rates during the first 8 weeks of storage. Internal CO_2 concentration ranged from 2 to 5%, with the centre scale tissues at 11–17% CO_2, a figure that increased with higher storage temperatures, while the internal gas volume decreased (Yoo *et al.*, 1997; Fig. 10.5). Sealing the neck area at 1 or 27°C increased the CO_2 concentrations, but had no effect on sprouting, indicating that elevated internal

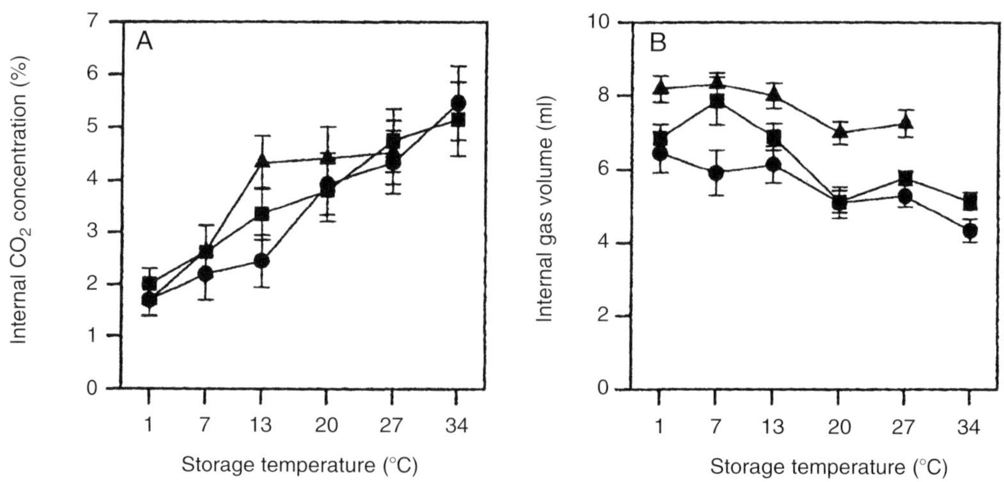

Fig. 10.5. Changes in internal CO_2 concentration (A) and internal gas volume (B) in onion bulbs stored at different temperatures for 4 (circles), 8 (squares) and 12 (triangles) weeks. Vertical bars indicate estimates of the standard deviation of the population ($n = 10$). Data point for 34°C at 12 weeks is missing due to decay of the bulbs. (From Yoo *et al.*, 1997, with permission.)

CO_2 concentrations at higher temperatures were not the sole cause of inhibited shoot growth at high storage temperatures. Respiration was unaffected by concentrations of 10–30% CO_2, although 30% CO_2 accelerated ethylene evolution, perhaps due to injury (Pal and Buescher, 1993).

8. Effect of Chemical and Radiation Treatments on Storage and Quality

8.1 Maleic hydrazide (MH)

The benefit of preharvest application of the mitotic inhibitor MH on onion storage was demonstrated during the 1950s (Isenberg, 1956). MH acts by inhibiting mitosis in the meristematic regions, thus preventing sprout development (Masters *et al.*, 1984) and inhibiting bulb respiration (Salama and Hicks, 1987), leading to lower losses of water from the bulbs. After several months, blackening of the growing point and desiccation of the bulbs makes them unsaleable. MH does not directly prevent pathogen development but it does retard sprouting, so there is less senescent material in the bulb (i.e. dying outer scales) for pathogens to attack.

Responses to MH vary with cultivars. In a study in Mauritius 'Red Creole' (bulbs with 10–12% DM) had the best and cv. 'Yellow Dessex' (a 'Granex' type) the worst storage potential, after being treated with up to 2000 ppm MH (Goburdhun, 1995). Optimum MH rates for control of sprouting, rotting and total weight loss are of the order of 1600 ppm for cv. 'Local Red' (Goburdhun, 1995) and, in India, up to 2000 ppm for cv. 'MDU.1' (Shanthi and Balakrishnan, 1989) and cv. 'Co. 4' (Vijayakumar *et al.*, 1987) and 4000 ppm for cv. 'Pusa Red' (Singh *et al.*, 1998b). In India, MH was effective only when applied 2 and 3 weeks before harvest (Singh and Dhankhar, 1995) and, in Poland, application of MH is recommended when 50–60% of tops are fallen (Kepka *et al.*, 1989).

The timing of MH application has a direct effect on the amount of residues. MH has low toxicity to humans: the acceptable daily intake is 5 mg kg^{-1} body weight (Codex Alimentarius Commission, 1991, quoted in Miedema, 1994a). The minimum detection limit for MH is 0.5 ppm (Kubilius and Bushway, 1999). Levels of MH may increase in processed food products because of moisture loss (Lewis *et al.*, 1998).

The major metabolite of MH in plants is its β-D-glucoside, which is hydrolysed by the acid nature of stomach conditions, so the effective dose ingested by humans may be higher than that calculated from tissue-residue analysis (Komossa and Sandermann, 1995).

8.2 Ethylene and cytokinins

There are conflicting reports on the effect of ethylene during storage. Sprouting in onion bulbs (cv. 'Hyton') in The Netherlands was slightly stimulated by ethephon (1 μmol) and strongly promoted by benzyl adenine at 250 nmol injected into the cavity in the centre of the bulb, followed by storage at 25°C (Miedema and Kamminga, 1994). In Algeria, ethylene production from stored bulbs averaged 4.49 ± 0.3 nmol kg^{-1} h^{-1} (Benkeblia and Selselet-Attou, 1999b), perhaps not enough to stimulate sprouting. If endogenous ethylene actually stimulates sprouting, research into the use of ethylene blockers, such as methyl cyclopropene (MCP), may be worthwhile.

8.3 Other chemicals

Several chemicals, including fungicides, have been tried out for improving onion storage life. In general, these are not in wide use because they would appear unattractive to consumers by leaving visible residues. However, they may be very useful for preserving mother bulbs for seed production. Among substances with beneficial effects are S as SO_2 (Thamizharazi and Narasimham, 1992), lime (Tanaka and Nonaka, 1981), carbendazim (Srivastava *et al.*, 1997), iprodione and streptocycline (Srivastava and Tiwari, 1997). Boron, applied postharvest as borax, extended the storage life of onions in Egypt (Alphonse, 1997).

8.4 Research into controlled atmosphere storage

The use of low temperature alone for long-term storage of onions is limited by the lack of inhibition of inner rooting and swelling of the base plate after dormancy has been broken. CA storage, using relatively high levels of CO_2 combined with low levels of O_2 at low temperatures, can increase onion storage life.

In the USA, CA storage was first developed for extending the marketing life of valuable sweet onions from Vidalia, Georgia, which have a naturally short life under ambient conditions (Smittle, 1988). Now, such stores are increasingly being constructed to extend the life of long-storing cultivars (Adamicki and Saltveit, 1997).

In Georgia, USA, sugar concentrations in sweet onions decreased and pungency increased during storage, reducing the quality of the bulbs; quality was best preserved by an atmosphere of 3% O_2 + 5% CO_2 at either 1 or 5°C (Smittle, 1988, 1989). Following 7 months' storage of 'Granex' cultivars at 1°C in the CA conditions detailed above, more than 92% of bulbs remained in a marketable condition after a further 3 weeks under ambient conditions (Smittle, 1988, 1989). Substantial storage capacity has now been constructed using these recommendations, in order to extend the marketing season for sweet onions.

Very low storage O_2 concentrations (0.7%) may subsequently result in increased rates of sprouting under ambient conditions (Sitton *et al.*, 1997). Neck rot was significantly reduced at low O_2 and with CO_2 above 8.9%, but CO_2 injury was significant when the gas concentration was above 4.1% for cv. 'Walla Walla' in the USA (Sitton *et al.*, 1997). In the UK, CO_2 above 10% for short-term storage and above 1% for long-term storage caused injury, accelerated softening and led to rots and a putrid odour, while O_2 at < 1% caused off odours and breakdown (Gadalla, 1997).

CA considerably increased storage life of onions cv. 'Momiji No. 3' in Saga, south-western Japan: the onions kept for 4 months under ambient storage, for 8 months under ambient atmosphere at 1°C or for 1 year with 1% O_2 + 1% CO_2 at 1°C (Tanaka *et al.*, 1996). In CA storage, weight loss was low and rotting caused by grey mould, neck rot and black mould was negligible; rooting was inhibited for 6 months, to the extent that after 12 months there was no swelling of the basal plate and sprouting was reduced. Rooting and sprouting incidence rose with O_2 concentration. CA storage therefore effectively maintains innate dormancy (Tanaka *et al.*, 1996).

In Russia, the optimum storage conditions for cv. 'Strigunovski Nosovskii' were at 1–5°C, 80–92% RH and 5% O_2 + 1–2% CO_2, when the lowest losses in DM, sugar and vitamins occurred (Polishchuk *et al.*, 1988). Bishop (1996) stated that typical regimes within commercial CA stores are now 0°C with 65–75% RH and 3% O_2 + 5% CO_2.

Rooting susceptibility of sweet onions after CA storage (3% O_2 + 5% CO_2, 1°C, 70% RH) increased in later-harvested onions (cv. 'Granex 33') as the duration of CA storage lengthened (Smittle *et al.*, 1994). The RH of the circulating air for root inhibition required lowering as the duration of CA storage increased. Under CA, onions need a low RH (about 70%), which can be achieved with a large differential between the refrigerant and air temperature of about 11–12°C with natural air circulation or between 9 and 10°C where air is circulated by a fan (Thompson, 1998).

8.5 Irradiation

Gamma irradiation is not popular for treating foods in most countries but can be an effective way to prolong onion storage life. The length of food exposure to ionizing energy and the strength of the source determine the irradiation dose, measured in kilograys (kGy). The World Health Organization (WHO) has approved the use of 0.15 kGy gamma irradiation to prevent onions from sprouting during storage (Kobayashi *et al.*, 1994). Irradiation at 0.03 kGy had the potential to reduce losses from 80% to 5.8% in cv. 'Valenciana Sintetica', grown in Argentina (Piccini *et al.*, 1987).

In Egypt, irradiation with 0.04, 0.06 or 0.08 kGy completely inhibited sprouting in stored bulbs (El-Gizawy *et al.*, 1993). In Poland, ionizing radiation at 0.05–0.06 kGy prevented onion sprouting and also inhibited reproduction and development of the bulb mite *Rhizoglyphus echinopus* (Ignatowicz, 1998). When Polish onions were treated with 0.08 kGy gamma rays, a slight darkening of the apex occurred: this did not affect the commercial value of the bulbs (Smierzchalska *et al.*, 1988). The best results after irradiation and storage of cv. 'Sochaczewska' (89.5% marketable bulbs vs. 20.5% in non-irradiated bulbs) were found with 0.06 kGy treatment 1 month after harvest, followed by storage at 1°C (Gajewski, 1994).

Irradiation at 5 kGy increased the content of disulphides and trisulphides; however, irradiation at permitted levels was not observed to change the most unstable aroma compounds in onions (Xi *et al.*, 1994).

Bulbs of cv. 'Valenciana Sintetica 14' grown in Argentina were irradiated within 30–40 days of harvest with 0.06 kGy and subsequently stored in Brazil for 180 days at 20–25°C, 50–100% RH (Walder *et al.*, 1997). Weight loss of irradiated and control bulbs were 13 and 32%, respectively, and gave 92.3% (treated) and 52.3% (untreated) marketable bulbs. In France, treatments with 0.15 and 0.30 kGy resulted in a decline in the respiration rates of cv. 'Rouge d'Amposta' onions at 4, 10 and 20°C, in contrast to untreated bulbs, where respiration rates rose over time (Benkeblia *et al.*, 2000). A new facility in Algeria will be used commercially to irradiate fruit and vegetables (Benkeblia, 2000b).

9. Methods of Curing and Storage

For reviews, see Currah and Proctor (1990), Komochi (1990) and Brice *et al.* (1997).

9.1 Field storage

In many countries with dry climates, onions are stored in the field (e.g. in dry areas of Argentina), as a cheap and effective method of storage. The heaps are often covered with a plastic sheet, straw or earth to protect the bulbs from condensation and occasional rains. In northern Europe, such methods were used until the second half of the 20th century, when the practice of keeping onions dry from harvest time onwards and controlling the curing and drying temperatures were shown to be beneficial for quality, an increasingly important factor in the marketing of onions (Currah and Proctor, 1990). These practices are now followed in many parts of the developed world where onions are harvested at times of year that are likely to be wet (Tatham, 1982).

In Egypt, Bahnasawy *et al.* (1998) recorded considerable diurnal fluctuations and temperature differences inside the field-stored onion heaps, due apparently to evaporative cooling during the day and to the insulating effects of the rice straw covering the heaps by night. Extremes ranged between 20 and 31°C in the centre of the heap and 23 and 35°C in the exterior of the heap, by night and day, respectively. Weight loss stabilized at about 0.5% week^{-1} after the first 3 weeks. The authors recommended ventilating the centres of heaps to reduce air RH and discourage pathogen development. In the hot, dry conditions of Pakistan, onions of cv. 'Phulkara' for seed production are often stored and keep well on the open ground in thin layers (L. Currah, UK, 2000, personal communication).

In the tropics, many traditional types of onion stores have been developed by farmers; examples are given in Currah and Proctor (1990) and illustrated in Brice *et al.* (1997). Traditional Indian onion stores made from locally available materials were described by Warade *et al.* (1997). There is no ventilation at the base or top of the structure and, throughout storage, losses are often high. Losses are reduced by insulating with straw, applying Dithane to the bulbs, providing perforated pipes for ventilation and curing bulbs prior to storage (Warade *et al.*, 1997). In Sudan, where onions are kept in traditional straw huts without ventilation, raising platforms from 0.15 to 0.5 m above ground level resulted in less sprouting and reduced weight loss (Musa *et al.*, 1994).

Traditional bamboo structures (called 'chawls' in some areas) are used for bulk storing of onions in India and Thailand. In an Indian study, onions in a one- or two-tiered wire-mesh structure lost 26 and 38% in weight, respectively, and in the conventional chawl system 47%, indicating that the wire-mesh structures were superior (Maini *et al.*, 1997). Commercial stores constructed by the National Horticultural Research and Development Foundation (NHRDF) in India now accommodate thinner layers of bulbs and allow good through ventilation.

9.2 Ventilation with forced ambient air

Forced-air ventilation improves the removal of excessive humidity and heat. It helps to keep the outside layers of the onions dry and the bulbs dormant.

In Yemen experiments, stacking of onion sacks was compared with bulk storage in 5 t wooden bins with positive forced-air ventilation. By timing the periods for ventilation to coincide with favourable outside ambient temperatures and RH, the onions in bins were kept at a temperature of 28–34°C and at 50–70% RH. Significant reductions were achieved in storage losses of local red cultivars over 33 weeks (Brice *et al.*, 1995).

Forced ventilation in storage bins in Honduras increased marketable bulbs in four 'Grano'/'Granex'-type cultivars after 3 months' storage, from an average of 23 to 62%, due to maintenance of temperature and RH near optimum levels (Medlicott *et al.*, 1995). However, during very wet weather, this method could not provide dry enough conditions to extend storage life further. In Sudan, marketable onions after 4 months in insulated stores ventilated with dry air totalled 90%, compared with 74% of those ventilated with humid air for 4 h each day (Musa *et al.*, 1994). The slowest ventilation speed (114 m^3 h^{-1} t^{-1}) was the most effective at reducing weight loss and sprouting through 12 h night-time ventilation (Musa *et al.*, 1994). In a Brazilian study, air flows of 60 and 75 m^3 h^{-1} m^{-3} were the most effective, and 30 m^3 h^{-1} m^{-3} was insufficient for onions stored in 0.56 m diameter × 2.15 m high silos with a depth of 2 m (de Matos *et al.*, 1998). Where onions are stored under ambient conditions, without refrigeration, it is therefore recommended that conditions be improved by natural or, better, forced ventilation.

9.3 Ventilation with heated air

Forced heated air as a curing method reduces weight loss and enhances colour compared with field-curing (Sanguansri and Gould, 1990).

In southern Africa, farmers have devised stores using heated air (27°C) to maintain high enough temperatures to keep the onions dormant while avoiding excessive temperatures, which can encourage black mould and bacterial diseases (L. Currah, UK, 2000, personal communication). In southern Brazil, wood-burning stoves are used during cool humid weather to keep onions in stores warm and dry (de Matos, 1987).

Solar energy is sometimes used for drying onions. Ting *et al.* (1987) devised a solar dryer capable of supplying a 7.1°C temperature rise for 9 h day^{-1} and with the ability to cure a 2300 kg batch in 4–5 days. An inexpensive solar dryer for onions is illustrated in an extension publication from Panama (Sánchez and Serrano, 1994).

9.4 Stores with controlled-temperature facilities

Stores in which onions are stored in bulk or in bins are now commonly used in developed countries in the temperate zone in connection with the 'direct harvest' system. Bulk stores are supplied with under-floor duct ventilation, used initially for high-temperature (30–32°C) dry curing. The airflow rate and temperature are then reduced in stages to allow the temperature of the bulbs to be reduced to near 0°C for long-term storage (Matson *et al.*, 1985; Brice *et al.*, 1997). Stores with the CA facility are similar but with greater airtightness and the possibility of removing excess CO_2 by scrubbers;

certain firms specialize in CA technology. Computer controls are now commonly installed in these stores but, even so, they need regular attention to make sure that all the equipment is functioning correctly.

10. Diseases of Storage

Postharvest development of pathogens already present from the field is mainly determined by the temperature and RH in onion stores (Hayden and Maude, 1997). Some infections, however, vary from year to year according to whether they were introduced on seeds and also according to weather conditions during growth which favour their build-up (Maude, 1990).

Recent work on some of the important storage pathogens of onions include a comprehensive illustrated account of onion-storage defects by Snowdon (1991) and a general review of onion pathogens and modern methods to combat them in Lorbeer (1997). Hayden and Maude (1997) summarized recent findings on the control of storage pathogens.

10.1 Black mould

Black mould (*Aspergillus niger*) commonly occurs on onions stored at temperatures above about 25°C, with an optimum at about 30°C, especially at > 80% RH (Hayden *et al.*, 1994). This disease is the main cause of the rejection of onion bulbs (van Konijnenburg and Ardizzi, 1997) from hot production areas, such as Texas, Egypt, India and Sudan (Hayden and Maude, 1997). Regulation of the storage environment effectively controls black mould in the UK. However, in Sudan, for example, where suitable conditions for the growth of the pathogen occur in the field and in uncontrolled storage environments, it is difficult to limit the inoculum entering the store. The fungus is apparent on bulbs within a few days of storage. The symptoms are the abundant black conidia produced on and sometimes under the outer skins of onion bulbs in store (Hayden and Maude, 1997). The source of the inoculum is infected seed and soil. Seed treatments are partially effective. Reducing mechanical damage and wounding and using short-term high-temperature drying, followed by storage at < 80% RH, are the most effective treatments.

Preharvest treatment of bulbs with sulphur dioxide at 1% (v/v) for 72 h or heat treatment at 50°C for 3 h can reduce losses due to *A. niger* in store (Thamizharasi and Narasimham, 1992); sulphur dust treatment is an effective long-term storage control method (Chavan *et al.*, 1992; Padule *et al.*, 1996).

10.2 Neck rot

Botrytis allii, the causal organism of neck rot, grows optimally at 21°C and is therefore a problem in temperate climates, such as northern Europe and Canada. Symptoms occur after 8–10 weeks in store, with a softening and rotting of neck tissues. Numerous small (1–5 mm) black sclerotia develop beneath the outer dry skins (Hayden and Maude, 1997). The source of inoculum is infected seed, with the fungus present within rather than on the surface of the seed (Stewart and Franicevic, 1994). In the UK, effective seed treatment (benomyl plus thiram) virtually eliminates the disease from the stored crop, at least in dry years. Neck rot is more severe when infection can occur early in the growing season (often because of wet field conditions) and, after this, artificial curing may not be effective (de Visser *et al.*, 1994). There is some evidence that *B. allii* conidia produced at low temperatures cause more rapid and destructive rots than conidia produced at higher temperatures (Bertolini and Tian, 1997).

Film-coat applications of *Enterobacter agglomerans* to naturally infected onion seed resulted in control levels similar to those on benomyl-treated seed (Peach *et al.*, 1994). Other means of discouraging pathogen development include postharvest treatment by heated-air drying at 30–32°C during the early stage of storage, before the temperature is reduced to about 27°C for second-stage drying and then to a low level for long-term storage in the UK and in Holland (Hayden and Maude, 1997).

10.3 Other pathogens

Aspergillus fumigatus and *Penicillium* spp. frequently occur in the microflora of stored temperate onions, but the former flourish only at > 40°C and the latter at 1–5°C or 20–25°C (Hayden and Maude, 1997). In-store losses may also be caused by bacteria of the genera *Erwinia*, *Lactobacillus* and *Pseudomonas* (see Mark *et al.*, Chapter 11, this volume). Application of high hydrostatic pressure (200–400 MPa) at 5°C for 30 min inactivated isolated microorganisms, including Gram-positive and Gram-negative bacteria, moulds and yeasts, and reduced the viable populations of the organisms on onion tissue (Arroyo *et al.*, 1999).

From stored onions in Yemen, Maude *et al.* (1991) isolated 11 distinct bacterial or yeast organisms, several of which were also human pathogens or which live in the gut (e.g. *Pseudomonas aeruginosa*, *Enterococcus faecalis*); many of them were found in combination in the rotting bulbs. They concluded that, in the prevailing high temperatures, senescent onion tissues were likely to be invaded by a wide range of opportunistic organisms, which speed up the break-down of the dying bulb scales. Better husbandry practices, including cutting the tops off further from the flesh of the necks, were advised. This method is also recommended in India (Bhonde, 1998).

11. Conclusions and Future Directions

Space restrictions do not allow us to deal with onion storage and transport technology issues here: a review that will include these topics is being prepared by Currah (2002). Onion storage and transport present several unanswered problems. Breeding for better storage qualities is already solving some of these, such as the short dormancy of some desirable types of onions. Others, such as identifying the correct conditions for sea transport of onions of different types, require cooperation between exporters, shippers and importers to pinpoint the optimum conditions attainable under shipboard conditions. The treatment of onions out of store is also the province of commerce rather than research at present, but research-based recommendations on this topic are badly needed. The onion trade continues to expand internationally and the advent of a distinct 'organic onion' as a commodity may well need research inputs to deal with the attendant diseases and disorders in an environmentally friendly manner in the near future.

Most SD onions derived from 'Grano' types suffer from short dormancy and thus tend to sprout within a few weeks after harvest, unless expensive control measures (CA and cold storage) are used. Breeding SD cultivars with inherited long-storage capacity will provide the optimal solution to reducing storage losses and costs. In recent years, attempts have been made in Brazil, India and Israel to develop high-quality SD long-keeping cultivars. Indeed, some of these perform better than traditional cultivars, as well as outyielding the popular 'Grano', 'Granex' and 'Creole' types (Peters *et al.*, 1994), thus cutting losses, increasing growers' incomes and providing a continuous supply at reasonable price at times when bulb onions cannot be grown due to climatic conditions (e.g. the monsoon period in Thailand).

References

Abdalla, A.A. and Mann, L.K. (1963) Bulb development in the onion (*Allium cepa* L.) and the effect of storage temperature on bulb rest. *Hilgardia* 35, 85–112.

Adamicki, F. (1998) Comparison of quality and storage ability in some Polish cultivars of onions. *Biuletyn Warzywniczy* 48, 89–100 (in Polish).

Adamicki, F. and Saltveit, M.E. (1997) Effect of ultra low oxygen on the storage and quality of some vegetables. In: *Postharvest Horticulture Series*, No. 18. Department of Pomology, University of California, Davis, California, pp. 26–33.

Agic, R., Gjorgjievska, M.C., Martinovski, G., Jevtic, S. and Lazic, B. (1997) Dynamics of losses during bulb storage from semi-acrid onion cultivars. *Acta Horticulturae* 462, 565–570.

Alphonse, M. (1997) Response of stored onions to different boron treatments. *Alexandria Journal of Agricultural Research* 42, 171–183.

Anan'ina, M.N. (1986) Variation in the chemical composition of onion varieties during ripening and storage. *Nauchno Tekhnicheskii Byulleten' Vsesoyuznogo Ordena Lenina i Ordena Druzhby Narodov Nauchno Issledovatel'skogo Instituta Rastenievodstva Imeni N.I. Vavilova* 166, 67–70 (in Russian).

Arroyo, G., Sanz, P.D. and Prestamo, G. (1999) Response to high pressure, low temperature treatment in vegetables: determination of survival rates of microbial populations using flow cytometry and detection of peroxidase activity using confocal microscopy. *Journal of Applied Microbiology* 86, 544–556.

Bacon, J.R., Moates, G.K., Ng, A., Rhodes, M.J.C., Smith, A.C. and Waldron, K.W. (1999) Evaluation of flavour potential of different tissues from onion (*Allium cepa* L.). In: *Agri-Food Quality II: Quality Management of Fruits and Vegetables – from Field to Table, Turku, Finland, 22–25 April 1998*, pp. 271–273.

Bahnasawy, A.H., Ghaly, A.E., El-Haddad, Z.A. and El-Ansawy, M.Y. (1998) Evaluating the current system of onions storage in Egypt. In: *Northeast Agricultural and Biological Engineering Conference, Halifax, Nova Scotia, Canada*. American Society of Agricultural Engineers, St Joseph, Minnesota, Paper 9815, 30 pp.

Batal, K.M., Bondari, K., Granberry, D.M. and Mullinix, B.G. (1994) Effects of source, rate, and frequency of N application on yield, marketable grades and rot incidence of sweet onion (*Allium cepa* L. cv. Granex-33). *Journal of Horticultural Science* 69, 1043–1051.

Benkeblia, N. (2000a) Phenylalanine ammonia-lyase, peroxidase, pyruvic acid and total phenolics variations in onion bulbs during long-term storage. *Lebensmittel-Wissenschaft und -Technologie* 33, 112–116.

Benkeblia, N. (2000b) Food irradiation of agricultural products in Algeria. Present situation and future developments. *International Agrophysics* 14, 259–261.

Benkeblia, N. and Selselet-Attou, G. (1999a) Effects of low temperatures on changes in oligosaccharides, phenolics and peroxidase in inner bud of onion *Allium cepa* L. during break of dormancy. *Acta Agriculturae Scandinavica Section B – Soil and Plant Science* 49, 98–102.

Benkeblia, N. and Selselet-Attou, G. (1999b) Role of ethylene on sprouting of onion bulbs (*Allium cepa* L.). *Acta Agriculturae Scandinavica Section B – Soil and Plant Science* 49, 122–124.

Benkeblia, N., Varoquaux, P., Gouble, B. and Selselet-Attou, G. (2000) Respiratory parameters of onion bulbs (*Allium cepa*) during storage. Effects of ionising radiation and temperature. *Journal of the Science of Food and Agriculture* 80, 1772–1778.

Bertolini, P. and Tian, S.P. (1997) Effect of temperature of production of *Botrytis allii* conidia on their pathogenicity to harvested white onion bulbs. *Plant Pathology* 46, 432–438.

Bhonde, S.R. (1998) Storage of onion and salient features of post-harvest technology. *News Letter, National Horticultural Research and Development Foundation* 18(1), 10–15.

Bhonde, S.R. and Bhadauria, J.S. (1995) Effect of curing on keeping quality of small onions. *News Letter, National Horticultural Research and Development Foundation* 15(4), 1–4.

Bhonde, S.R., Srivastava, K.J., Sharma, H.K. and Chougule, A.B. (1996) Effect of withholding irrigation before harvesting on storage life of onion. *News Letter, National Horticulture Research and Development Foundation* 16(4), 1–4.

Bishop, D. (1996) Controlled atmosphere storage. In: Dellino, C.J.V. (ed.) *Cold and Chilled Storage Technology*. Blackie, London.

Bleasdale, J.K.A. and Thompson, R. (1966) Onion skin colour and keeping quality. In: *Annual Report for 1965*. National Vegetable Research Station, Wellesbourne, UK, pp. 47–49.

Böttcher, H. (1992) Quality changes of onions (*Allium cepa* L.) during storage. 1. Nutritional quality. *Die Nahrung* 36, 346–356 (in German).

Böttcher, H. (1999) Influence of harvest date on the postharvest responses of *Allium* vegetable species. *Gartenbauwissenschaft* 64, 220–226 (in German).

Brewster, J.L. (1987) The effect of temperature on the rate of sprout growth and development within stored onion bulbs. *Annals of Applied Biology* 111, 463–467.

Brewster, J.L. (1994) *Onions and Other Vegetable Alliums*. CAB International, Wallingford, UK, 236 pp.

Brice, J.R. (1994) Investigations into onion skin quality, 2 vols. PhD thesis, Postharvest Technology Department, Silsoe College, Cranfield University of Science and Technology, UK.

Brice, J.R., Bisbrown, A.J.K. and Curd, L. (1995) Onion storage trials at high ambient temperatures in the Republic of Yemen. *Journal of Agricultural Engineering Research* 62, 185–192.

Brice, J., Currah, L., Malins, A. and Bancroft, R. (1997) *Onion Storage in the Tropics*. NRI Publications, University of Greenwich, UK, 120 pp.
Carter, C.E., Partis, M.D. and Thomas, B. (1999) The expression of histone 2A in onion (*Allium cepa*) during the onset of dormancy, storage and emergence from dormancy. *New Phytologist* 143, 461–470.
Chauhan, K.P.S., Singh, S.P. and Chougule, A.B. (1995) Studies on the effect of windrow curing, neck cut and shade curing on export quality of onion bulbs during storage. *News Letter, National Horticultural Research and Development Foundation* 15(4), 5–7.
Chavan, V.B., D'Souza, T.F., Kokate, S.B. and Sawant, D.M. (1992) Efficacy of chemicals in restricting onion bulb rots in storage. *Maharashtra Journal of Horticulture* 6, 92–94.
Chung, B. (1989) Irrigation and bulb onion quality. *Acta Horticulturae* 247, 233–237.
Currah, L. (2002) Onions. In: *Postharvest*. Vol. 3. Blackwell, Oxford, (in press).
Currah, L. and Proctor, F.J. (1990) *Onions in Tropical Regions*. Bulletin 35, Natural Resources Institute, Chatham, UK, 232 pp.
Darbyshire, B. and Steer, B. (1990) Carbohydrate biochemistry. In: Brewster, J.L. and Rabinowitch, H.D. (eds) *Onions and Allied Crops*, Vol. III. *Biochemistry, Food Science, and Minor Crops*. CRC Press, Boca Raton, Florida, pp. 1–16.
Daymond, A.J., Wheeler, T.R., Hadley, P., Ellis, R.H. and Morison, J.I.L. (1997) The growth, development and yield of onion (*Allium cepa* L.) in response to temperature and CO_2. *Journal of Horticultural Science* 72, 135–145.
de Matos, A.T. (1987) *Cura e armazenamento de cebola com utilizaçao de ventilaçao forçada – Armazem Modelo EMPASC*. Comunicado Técnico No. 113, EMPASC, SC, Brazil, 12 pp.
de Matos, A.T., Finger, F.L. and Dalpasquale, V.A. (1997) Perda de materia fresca e isotermas de sorção em bulbos de cebola. *Pesquisa Agropecuária Brasileira* 32, 235–238.
de Matos, A.T., Dalpasquale, V.A. and Finger, F.L. (1998) Armazenamento de bulbos de cebola sob diferentes taxas de aeração intermitente. *Pesquisa Agropecuāria Brasileira* 33, 599–603.
de Visser, C.L.M., Hoekstra, L. and Hoek, D. (1994) *Research into Effective Chemical Control of Leaf Spot and Neck Rot and into Methods to Predict Neck Rot in Onions*. Verslag Proefstation voor de Akkerbouw en de Groenteteelt in de Vollegrond No. 178, Proefstation voor de Akkerbouwende Groenteteelt in de Vollegrond, Lelystad, The Netherlands, 85 pp. (in Dutch).
Donner, H., Gao, L. and Mazza, G. (1997) Separation and characterization of simple and malonated anthocyanins in red onions, *Allium cepa* L. *Food Research International* 30, 637–643.
El-Gizawy, A.M., Abdallah, M.M.F., El-Oksh, I.I., Mohamed, A.R.A.G. and Abdalla, A.A.G. (1993) Effect of soil moisture and nitrogen levels on chemical composition of onion bulbs and on onion storability after treatment with gamma radiation. *Bulletin of Faculty of Agriculture, University of Cairo* 44, 169–182.
Ernst, M. and Bufler, G. (1994) Stems of *Allium cepa* L. contain starch. *New Phytologist* 128, 403–406.
Ernst, M.K., Chatterton, N.J., Harrison, P.A. and Matitschka, G. (1998) Characterization of fructan oligomers from species of the genus *Allium* L. *Journal of Plant Physiology* 153, 53–60.
Ernst, M.K., Bufler, G., Röske, M., Ritzkowski, M., Metzger, C. and Lilbig, H.-P. (1999) Cultivar and temperature effects on starch in the primary thickening meristem of bulb onions. *Journal of Horticultural Science and Biotechnology* 74, 588–593.
Fossen, T., Andersen, Ø.M., Øvstedal, D.O., Pedersen, A.T. and Raknes, Å. (1996) Characteristic anthocyanin pattern from onions and other *Allium* spp. *Journal of Food Science* 61, 703–706.
Fossen, T., Pedersen, A.T. and Andersen, Ø.M. (1998) Flavonoids from red onion (*Allium cepa* L.). *Phytochemistry* 47, 281–285.
Füstös, Z.S. (1997) The role of the dry scale in the dormancy of onions (*Allium cepa* L.). *Acta Horticulturae* 433, 445–446.
Füstös, Z., Pankotai-Gilinger, M. and Ombodi, A. (1994) Effects of postharvest handling and cultivars on keeping quality of onions (*Allium cepa* L.) in storage. *Acta Horticulturae* 368, 212–219.
Gadalla, S.O. (1997) Inhibition of sprouting of onions during storage and marketing. PhD thesis, Cranfield University of Science and Technology, Silsoe, UK.
Gajewski, M. (1994) Effect of irradiation on storage results with onions (*Allium cepa* L.) in different conditions. *Biuletyn Warzywniczy* 41, 177–189 (in Polish).
Galmarini, C.R., Havey, M.J. and Goldman, I.L. (2000) Genetic analyses of correlated carbohydrate, flavor, and health enhancing traits in onion (*Allium cepa* L.). In: *Proceedings of Alliums 2000, 3rd International Symposium on Edible* Alliaceae, *University of Georgia, USA, 30 October–3 November 2000*. University of Georgia, Athens, Georgia, pp. 127–132.

Geyer, M., Oberbarnscheidt, B. and Herold, B. (1994) Methods to determine mechanical load of onions during harvest and postharvest. In: COST 94. *The Postharvest Treatment of Fruit and Vegetables: Quality Criteria. Proceedings of a COST workshop, 19–21 April, 1994, Bled, Slovenia*. Directorate-General for Information, Communication and Audiovisual, Commission of the European Communities, Brussels, pp. 179–186.

Goburdhun, S. (1995) Effect of maleic hydrazide on onion storage potential. *Revue Agricole et Sucrière de l'Ile Maurice* 74, 54–60.

Gorini, F.L. and Testoni, A. (1990) The relation between colour and quality of vegetables. *Acta Horticulturae* 259, 31–60.

Guerber-Cahuzac, B. (1996) La qualité de l'oignon. *Fruits* 51, 341–346.

Ha, M.-A., Apperley, D.C. and Jarvis, M.C. (1997) Molecular rigidity in dry and hydrated onion cell walls. *Plant Physiology* 115, 593–598.

Hak, P.S. and Ludwig, J.W. (1988) *Development of a Hardness Meter for Onions*. Publikatie No. 379, Instituut voor Bewaring en Verwerking van Landbouwprodukten, Wageningen, The Netherlands, 3 pp. (in Dutch).

Havey, M.J. and Randle, W.M. (1996) Combining abilities for yield and bulb quality among long- and intermediate-day open-pollinated onion populations. *Journal of the American Society for Horticultural Science* 121, 604–608.

Hayden, N.J. and Maude, R.B. (1997) The use of integrated pre- and post-harvest strategies for the control of fungal pathogens of stored temperate onions. *Acta Horticulturae* 433, 475–479.

Hayden, N.J., Maude, R.B. and Proctor, F.J. (1994) Strategies for the control of black mould (*Aspergillus niger*) on stored tropical onions. *Acta Horticulturae* 358, 271–274.

Herold, B., Oberbarnscheidt, B. and Geyer, M. (1998) Mechanical load and its effect on bulb onions due to harvest and postharvest handling. *Journal of Agricultural Engineering Research* 71, 373–383.

Hirota, S., Shimoda, T. and Takahama, U. (1999) Distribution of flavonols and enzymes that participate in the metabolism in tissues of onion bulbs. Mechanism of accumulation of quercetin and its glucosides in the abaxial epidermis. *Journal of Food Science and Technology* 5, 384–387.

Hole, C.C., Drew, R.L.K. and Gray, D. (2000) Humidity and mechanical properties of onion skins. *Postharvest Biology and Technology* 19, 229–237.

Horbowicz, M. (1998) Effect of stage of onion maturity on pungency changes during storage. *Biuletyn Warzywniczy* 48, 121–129 (in Polish).

Horbowicz, M. and Grzegorzewska, M. (1995) Effect of cooling and storage conditions on the contents of soluble carbohydrate and dry matter in onions. *Biuletyn Warzywniczy* 43, 45–58 (in Polish).

Iglesias, I., Salcines, R.M. and Garriga, E. (1987) Influencia de las condiciones de almacenamiento sobre el comportamento de la cebolla, cultivares Red Creole C-5, Texas Early Grano Strain 502 y White Majestic. *Agrotecnia de Cuba* 19, 65–74.

Ignatowicz, S. (1998) Control of pests of stored onion with irradiation used for inhibition of onion sprouting. *Biuletyn Warzywniczy* 48, 65–76 (in Polish).

Isenberg, F.M. (1956) The use of maleic hydrazide on onions. *Proceedings of the American Society for Horticultural Science* 66, 331–333.

Isenberg, F.M.R., Ludford, P.M. and Thomas, T.H. (1987) Hormonal alterations during the post-harvest period. In: Weichmann, J. (ed.) *Post-Harvest Physiology of Vegetables*. Marcel Dekker, New York, and Basle, Switzerland, pp. 45–94.

Jackson, K.J., Harper, T.W., Schrodter, G.N. and Duff, A.A. (1989) Marketing aspects of heavy vegetable research in Queensland. *Acta Horticulturae* 247, 137–142.

Kanazawa, T., Araki, H., Harada, T. and Yakuwa, T. (1997) Effect of low temperature storage on breaking the rest of *Allium victorialis* L. ssp. *platyphyllum* Hult. bulbs. *Journal of the Japanese Society for Horticultural Science* 66, 527–533.

Kariuki and Kimani, P.M. (1997) Yield and storage potential of onion cultivars in Kenya. In: Rabinowitch, H.D., Kimani, P.M. and Peters, R. (eds) *Proceedings of the East Africa Regional Alliums Workshop, 21–22 September 1994, Nairobi, Kenya*, MASHAV/CINADCO, Tel Aviv, Israel, pp. 50–56.

Kato, T., Yamagata, M. and Tsukahara, S. (1987) Nitrogen nutrition, its diagnosis and postharvest bulb rot in onion plant. *Bulletin of the Shikoku National Agricultural Experiment Station* 48, 26–49 (in Japanese).

Kepka, A., Adamicki, F. and Perlowska, M. (1989) The application of a new technique for onion storage. *Biuletyn Warzywniczy* Supplement 1, 97–105 (in Polish).

Kielak, E. and Bielinska-Czarnecka, M. (1987) Effects of length of drying on hormone activity in onion (*Allium cepa* L.) during storage. *Proceedings of the IV International Symposium of Plant Growth Regulators* 1, 155–159.

Kobayashi, A., Itagaki, R., Tokitomo, Y. and Kubota, K. (1994) Changes in aroma character of irradiated onion during storage. *Journal of the Japanese Society for Food Science and Technology* 41, 682–686 (in Japanese).

Komochi, S. (1990) Bulb dormancy and storage. In: Rabinowitch, H.D. and Brewster, J.L. (eds) *Onions and Allied Crops, II. Botany, Physiology, and Genetics*. CRC Press, Boca Raton, Florida, pp. 89–111.

Komossa, D and Sandermann, H. (1995) Plant metabolic studies of the growth regulator maleic hydrazide. *Journal of Agricultural and Food Chemistry* 43, 2713–2715.

Kopsell, D.E. and Randle, W.R. (1997) Onion cultivars differ in pungency and bulb quality changes during storage. *HortScience* 32, 1260–1263.

Kopsell, D.E., Randle, W.M. and Eiteman, M.A. (1999) Changes in the *S*-alk(en)yl cysteine sulfoxides and their biosynthetic intermediates during onion storage. *Journal of the American Society for Horticultural Science* 124, 177–183.

Kubilius, D.T. and Bushway, R.J. (1999) Determination of maleic hydrazide in potatoes and onions by fluorescence high performance liquid chromatography. *Journal of Liquid Chromatography and Related Technologies* 22, 593–601.

Kumar, M., Munsi, P.S., Das, D.K. and Chattopadhyay, T.K. (1998) Effect of zinc and sulphur application on the yield and postharvest quality of onion (*Allium cepa* L.) under different methods of storage. *Journal of Interacademicia* 2, 158–163.

Kuraishi, S., Yamashita, D., Sakurai, N. and Hasegawa, S. (1989) Changes of abscisic acid and auxin as related to dormancy breaking of *Allium wakegi* bulblets by vacuum infiltration and BA treatment. *Journal of Plant Growth Regulation* 8, 3–9.

Ladeinde, F. and Hicks, J.R. (1988) Internal atmosphere of onion bulbs stored at various oxygen concentrations and temperatures. *HortScience* 23, 1035–1037.

Lancaster, J.E. and Shaw, M.L. (1991) Metabolism of γ-glutamyl peptides during development, storage and sprouting of onion bulbs. *Phytochemistry* 30, 2857–2859.

Lancaster, J.E., Shaw, M.L. and Randle, W.M. (1998) Differential hydrolysis of alk(en)yl cysteine sulphoxides by alliinase in onion macerates: flavour implications. *Journal of the Science of Food and Agriculture* 78, 367–372.

Lancaster, J.E., Farrant, J. and Shaw, M.L. (2001) Sulfur nutrition affects cellular sulfur, dry weight distribution, and bulb quality in onion. *Journal of the American Society for Horticultural Science* 126, 164–168.

Lewis, D.J., Thorpe, S.A., Wilkinson, K. and Reynolds, S.L. (1998) The carry-through of residues of maleic hydrazide from treated potatoes, following manufacture into potato crisps and jacket potatoes. *Food Additives and Contaminants* 15(95), 506–509.

Lorbeer, J.W. (1997) Management of diseases in alliums. *Acta Horticulturae* 433, 585–591.

Love, J. (1995) *Quality Vegetables: a Review of the Factors Affecting the Quality and Shelf-Life of UK Field Vegetables for the Fresh Market*. Horticultural Development Council, Petersfield, UK, 47 pp.

Maini, S.B., Sagar, V.R., Chandan, S.S. and Rajesh, K. (1997) Evaluation of different structures for storage of onions. *Vegetable Science* 24, 73–74.

Mann, J.D., Monro, J.H. and Grant, D.R. (1986) Onion bulb composition and onion bulb firmness. *Proceedings of the Annual Conference, Agronomy Society of New Zealand* 16, 107–110.

Masters, L.R., Hicks, J.R. and Isenberg, F.M.R. (1984) Effect of maleic hydrazide on the cellular structure of the shoot apex of onion (*Allium cepa* L., cv. Northern Oak). *Acta Horticulturae* 157, 251–255.

Matejko, C. and Dahlhelm, H. (1991) Polyamine synthesis and its relation to dormancy in *Allium cepa* L. *Biochemie und Physiologie der Pflanzen* 187, 217–226 (in German).

Matson, W.E., Mansour, N.S. and Richardson, D.G. (1985) *Onion Storage – Guidelines for Commercial Growers*. Pacific Northwest Extension Publication No. 277, Pacific Northwest Cooperative Extension, Oregon State University, Cornwallis, Oregon, 15 pp.

Matsubara, S. and Kimura, I. (1991) Changes in ABA content during bulbing and dormancy and *in vitro* bulbing in onion plant. *Journal of the Japanese Society for Horticultural Science* 59, 757–762.

Maude, R.B. (1990) Storage diseases of onions. In: Rabinowitch, H.D. and Brewster, J.L. (eds) *Onions and Allied Crops, II. Agronomy, Biotic Interactions, Pathology, and Crop Protection*. CRC Press, Boca Raton, Florida, pp. 273–296.

Maude, R.B., Shipway, M.R., Presly, A.H. and O'Connor, D. (1984) The effect of direct harvesting and drying systems on the incidence and control of neck rot (*Botrytis allii*) in onions. *Plant Pathology* 33, 263–268.

Maude, R.B., Lyons, N.F., Curd, L., Abu Baker El Muallem and Bamakrama, H. (1991) Disease problems of onions in the Republic of Yemen. *Onion Newsletter for the Tropics* 3, 34–38.

Maw, B.W., Hung, Y.-C., Tollner, E.W., Smittle, D.A. and Mullinix, B.G. (1995) Detecting impact damage of sweet onions using muriatic acid and X-rays. *Applied Engineering in Agriculture* 11, 823–826.

Maw, B.W., Hung, Y.-C., Tollner, E.W., Smittle, D.A. and Mullinix, B.G. (1996) Physical and mechanical properties of fresh and stored sweet onions. *Transactions of the ASAE* 39, 633–637.

Maw, B.W., Smittle, D.A. and Mullinix, B.G. (1997a) The influence of harvest maturity, curing and storage conditions upon the storability of sweet onions. *Applied Engineering in Agriculture* 13, 511–515.

Maw, B.W., Smittle, D.A. and Mullinix, B.G. (1997b) Artificially curing sweet onions. *Applied Engineering in Agriculture* 13, 517–520.

Maw, B.W., Sumner, P.E. and Torrance, R.L. (1999) Commercial mechanization for harvesting sweet onions. In: *ASAE Annual International Meeting, Toronto, Canada, 18–21 July 1999*, Paper No. 99-1076. American Society of Agricultural Engineers, St Joseph, Michigan, 12 pp.

Medlicott, A., Brice, J., Salgado, T. and Ramírez, D. (1995) Forced ambient air storage of different onion cultivars. *HortTechnology* 5, 52–57.

Miedema, P. (1994a) Bulb dormancy in onion. I. The effects of temperature and cultivar on sprouting and rooting. *Journal of Horticultural Science* 69, 29–39.

Miedema, P. (1994b) Bulb dormancy in onion. III. The influence of the root system, cytokinin and wounding on sprout emergence. *Journal of Horticultural Science* 69, 47–52.

Miedema, P. and Kamminga, G.C. (1994) Bulb dormancy in onion. II. The role of cytokinins in high-temperature imposed sprout inhibition. *Journal of Horticultural Science* 69, 41–45.

Mondal, M.F. and Pramanik, M.H.R. (1992) Major factors affecting the storage life of onion – a review. *International Journal of Tropical Agriculture* 10, 140–146.

Msika, R.L. and Jackson, J.E. (1997) Onion production and research in Zimbabwe: country report. In: Rabinowitch, H.D., Kimani, P.M. and Peters, R. (eds) *Proceedings of the East Africa Regional Alliums Workshop, 21–22 September 1994, Nairobi, Kenya*. MASHAV/CINADCO, Tel Aviv, Israel, pp. 42–48.

Musa, S.K., Abdalla, Y.M., Haimoura, E. and Suleiman, Y. (1994) Improvement of onion storage in the Sudan. *Tropical Science* 34, 185–190.

Nabos, J. (1976) L'amélioration de l'oignon (*Allium cepa* L.) au Niger. *Agronomie Tropicale* 31, 387–397.

Ng, A., Smith, A.C. and Waldron, K.W. (1998) Cell wall composition of different onion (*Allium cepa* L.) varieties. *Food Chemistry* 63, 17–24.

Ng, A., Parker, M.L., Parr, A.J., Saunders, P.K., Smith, A.C. and Waldron, K.W. (2000) Physicochemical characteristics of onion (*Allium cepa* L.) tissues. *Journal of Agricultural and Food Chemistry* 48, 5612–5617.

Oberbarnscheidt, B., Herold, B. and Geyer, M. (1997) Wirkung mechanischer Belastungen auf Speisezwiebeln. *Landtechnik* 52(3), 134–135.

Ostrzycka, J. and Perlowska, M. (1992) Changes in the contents of dry matter, soluble solids and sugars in outer, middle and inner leaf-bases of onions during storage. *Biuletyn Warzywniczy* 39, 13–23 (in Polish).

Padule, D.N., Lohate, S.R. and Kotecha, P.M. (1996) Control of spoilage of onion bulbs by postharvest fungicidal treatments during storage. *Onion Newsletter for the Tropics* 7, 44–48.

Pak, C., van der Plas, L.H.W. and de Boer, A.D. (1995) Importance of dormancy and sink strength in sprouting of onions (*Allium cepa*) during storage. *Physiologia Plantarum* 94, 277–283.

Pal, R.K. and Buescher, R.W. (1993) Respiration and ethylene evolution of certain fruits and vegetables in response to carbon dioxide in controlled atmosphere storage. *Journal of Food Science and Technology* 30, 29–32.

Pandey, R.P. and Pandey, A. (1994) Storage losses of onion bulbs. *Madras Agricultural Journal* 81, 603–605.

Pandey, U.B., Singh, L., Singh, S.P. and Mishra, P.K. (1992) Studies on the effect of curing on storage life of kharif onion (*Allium cepa* L.). *News Letter, Associated Agricultural Development Foundation* 12(3), 14–16.

Patil, J.D., Desale, G.Y. and Kale, P.N. (1987) Correlation studies on morphological and storage characters of some onion varieties. *Journal of Maharashtra Agricultural Universities* 12, 114–115.

Peach, L., Maude, R.B. and Petch, G.B. (1994) Biocontrol of seed-borne *Botrytis allii* using an antagonistic bacterium. In: Martin, T. (ed.) *Seed Treatment: Progress and Prospects*. Monograph No. 57, British Crop Protection Council, Farnham, UK, pp. 345–350.

Peiris, K.H.S, Mallon, J.L. and Kays, S.J. (1997) Respiratory rate and vital heat of some specialty vegetables at various storage temperatures. *HortTechnology* 7, 46–49.

Perlowska, M. and Kaniszewski, S. (1988) The effect of different soil moisture levels on onion yield and storage. Part II. The effect of soil moisture level on quality and storage. *Biuletyn Warzywniczy* 32, 63–67 (in Polish).

Peters, R.J., Kowithayakorn, T., Chalard, T. and Rabinowitch, H.D. (1994) The effect of date of harvest on shelf-life of onions stored by hanging from leaves. *Acta Horticulturae* 358, 365–368.
Piccini, J.L., Evans, D.R. and Quaranta, H.O. (1987) L-Malate content in irradiated onions (*Allium cepa* L.) cv. Valenciana Sintetica 14. *Journal of Food Science and Technology (India)* 24, 91–93.
Polishchuk, S.F., Mel'nik, V.N., Tretyak, S.V., Kirichenko, V.I. and Rozhanchuk, V.N. (1988) Onion storage in a modified gas atmosphere. *Kartofel' i Ovoshchi* 6, 33–34 (in Russian).
Price, K.R. and Rhodes, M.J.C. (1996) Analysis of the major flavonol glycosides present in four varieties of onion (*Allium cepa*) and changes in composition resulting from autolysis. *Journal of the Science of Food and Agriculture* 74, 331–339.
Price, K.R., Bacon, J.R. and Rhodes, M.J. (1997) Effect of storage and domestic processing on the content and composition of flavonol glucosides in onion (*Allium cepa*). *Journal of Agricultural and Food Chemistry* 45, 938–942.
Rabinowitch, H.D. and Brewster, J.L. (eds) (1990) *Onions and Allied Crops*, 3 Vols. CRC Press, Boca Raton, Florida, 273, 320 and 265 pp.
Ramin, A.A. (1999) Storage potential of bulb onions (*Allium cepa* L.) under high temperatures. *Journal of Horticultural Science and Biotechnology* 74, 181–186.
Randle, W.M. (1997) Genetic and environmental effects influencing flavor in onion. *Acta Horticulturae* 433, 299–311.
Rossier, N., Palasthy, A. and Schwarz, A. (1994) Influence des types de sols valaisans sur l'aptitude à la conservation de l'oignon de garde. *Revue Suisse de Viticulture, Arboriculture et Horticulture* 26, 199–205.
Rouamba, A., Gbene, R.H., Bâ, D., Dembele, D., Ricroch, A. and Currah, L. (2001) Agronomic and physiological evaluation of some regional populations of onion in field and storage trials in West Africa. *Tropical Science* 41, 78–84.
Rutherford, P.P. and Whittle, R. (1982) The carbohydrate composition of onions during long term cold storage. *Journal of Horticultural Science* 57, 349–356.
Rutherford, P.P. and Whittle, R. (1984) Methods of predicting the long-term storage of onions. *Journal of Horticultural Science* 59, 537–543.
Salama, A.M. and Hicks, J.R. (1987) Respiration and fresh weight of onion bulbs as affected by storage temperature, humidity and maleic hydrazide. *Tropical Science* 27, 233–238.
Salama, A.M., Hicks, J.R. and Nock, J.F. (1990) Sugar and organic acid changes in stored onion bulbs treated with maleic hydrazide. *HortScience* 25, 1625–1628.
Sánchez, E. and Serrano, C.E. (1994) *Manual del Cultivo de la Cebolla para las Tierras Altas de Chiriquí*. IDIAP, Panamá, 42 pp.
Sanguansri, P. and Gould, I.V. (1990) Onions: artificial curing system and objective quality evaluation. In: *Agricultural Engineering Conference, Toowoomba, Queensland, Australia, 11–14 November 1990*, pp. 333–337.
Sargent, S.A., Zoellner, J.J., Stoffella, P.J. and Maynard, D.N. (1991) Harvest maturity affects storage quality of fresh, short-day onions. *Proceedings of the Annual Meeting of the Florida State Horticultural Society* 104, 64–68.
Sargent, S.A., Stoffella, P.J. and Maynard, D.N. (2001) Harvest date affects yield and postharvest quality of nondried shortday onion. *HortScience* 36, 112–115.
Shanthi, K. and Balakrishnan, R. (1989) Effect of nitrogen, spacing and maleic hydrazide on yield, nutrient uptake, quality and storage of MDU.1 onion. *Indian Journal of Horticulture* 46, 490–495.
Shekib, L.A, Shehata, A.A.Y. and El-Tabey, A. (1986) The effect of storage of fresh Egyptian onions on some of its quality aspects. *Alexandria Journal of Agricultural Research* 31, 167–174.
Sherf, A.F. and MacNab, A.A. (1986) *Vegetable Diseases and Their Control*, 2nd edn. Wiley, New York, 728 pp.
Shock, C.C., Feibert, E.B.G. and Saunders, L.D. (1998) Onion yield and quality affected by soil water potential as irrigation threshold. *HortScience* 33, 1188–1191.
Sinclair, P.J., Blakeney, A.B. and Barlow, E.W.R. (1995a) Relationship between bulb dry matter content, soluble solids concentration and non-structural carbohydrate composition in the onion (*Allium cepa*). *Journal of the Science of Food and Agriculture* 69, 203–209.
Sinclair, P.J., Neeson, R.J. and Barlow, E.W.R. (1995b) Osmotic potential and soluble solids concentration in onion (*Allium cepa*) bulbs. *Journal of the Science of Food and Agriculture* 69, 211–214.
Singh, J. and Dhankhar, B.S. (1991) Effect of nitrogen, potash and zinc on storage loss of onion bulbs (*Allium cepa* L.). *Vegetable Science* 18, 16–23.

Singh, J. and Dhankhar, B.S. (1995) Effect of pre-harvest chemical treatment on storage loss of onion. *Advances in Horticulture and Forestry* 4, 119–126.

Singh, J.V., Kumar, A. and Singh, C. (1998a) Studies on the storage of onion (*Allium cepa* L.) as affected by different levels of phosphorus. *Indian Journal of Agricultural Research* 32, 51–56.

Singh, J.V., Chetan, S. and Singh, C. (1998b) Studies on the storage of onion (*Allium cepa* L.) as affected by different concentrations of maleic hydrazide. *Indian Journal of Agricultural Research* 32, 81–87.

Singh, S. and Tiwari, R.S. (1992) Effect of micronutrients on storage of onion bulbs (*Allium cepa* L.) cv. Pusa Red. *Progress in Horticulture* 24, 135–140.

Sitton, J.W., Fellman, J.K. and Patterson, M.E. (1997) Effects of low-oxygen and high-carbon dioxide atmospheres on postharvest quality, storage and decay of 'Walla Walla' sweet onions. In: Saltveit, M.E. (ed.) *Postharvest Horticulture Series*, No. 18. Department of Pomology, University of California, Davis, California, pp. 20–25.

Smart, R. (1986) Studies on factors influencing skin quality of spring sown long term storage onions (*Allium cepa* L.). MPhil thesis, Silsoe College, Cranfield Institute of Technology, Silsoe, UK.

Smierzchalska, K., Perlowska, N., Wojniakiewicz, E. and Habdas, H. (1988) Application of ionizing radiation for prolonging the shelf-life of certain vegetables. *International Agrophysics* 4, 339–347.

Smittle, D.A. (1988) Evaluation of storage methods for 'Granex' onions. *Journal of the American Society for Horticultural Science* 113, 877–880.

Smittle, D.A. (1989) Controlled atmosphere storage of Vidalia onions. In: *Proceedings of the 5th International Controlled Atmosphere Research Conference, Wenatchee, Washington, USA, 14–16 June 1989*, Vol. 2. Coastal Plain Experiment Station, Tifton, Georgia, pp. 171–177.

Smittle, D.A and Maw, B.W. (1988) Effects of maturity and harvest methods on storage and quality of onions. *HortScience* 23, 141–143.

Smittle, D.A., Hayes, M.J. and Mercer, M.D. (1994) Susceptibility of Vidalia onions to rooting before and after controlled atmosphere storage. In: *Research Report No. 612*. Georgia Agricultural Experiment Stations, pp. 1–6.

Snowdon, A.L. (1991) *A Colour Atlas of Post-Harvest Diseases and Disorders of Fruits and Vegetables*, Vol. 2. *Vegetables*. Wolfe Publishing, London, 416 pp.

Solberg, S.O. and Dragland, S. (1998) Effects of harvesting and drying methods on internal atmosphere, outer scale appearance and storage of bulb onions (*Allium cepa* L.). *Journal of Vegetable Crop Production* 4(2), 23–25.

Srivastava, P.K. and Tiwari, B.K. (1997) Effect of pre-harvest fungicidal spray on the control of storage diseases of onion. *News Letter, National Horticultural Research and Development Foundation* 17(2), 4–6.

Srivastava, P.K., Gupta, R.P., Tiwari, B.K. and Sharma, R.C. (1997) Effect of systemic fungicides/bactericides on the control of diseases of onions in storage. *News Letter, National Horticultural Research and Development Foundation* 17(3), 4–6.

Stewart, A. and Franicevic, S.C. (1994) Infected seed as a source of inoculum for *Botrytis* infection of onion bulbs in store. *Australasian Plant Pathology* 23, 36–40.

Suzuki, M. and Cutcliffe, J.A. (1989) Fructans in onion bulbs in relation to storage life. *Canadian Journal of Plant Science* 69, 1327–1333.

Takahama, U. and Hirota, S. (2000) Deglucosidation of quercetin glucosides to the aglycone and formation of antifungal agents by peroxidase-dependent oxidation of quercetin on browning of onion scales. *Plant Cell Physiology* 41, 1021–1029.

Tanaka, K. and Nonaka, F. (1981) Studies on the rot of onion bulbs caused by *Aspergillus niger* and its control by lime application. *Bulletin of the Faculty of Agriculture of Saga University* 51, 47–51 (in Japanese).

Tanaka, K., Matsuo, Y. and Egashira, J. (1996) Controlled atmosphere storage for onions. *Acta Horticulturae* 440, 669–674.

Tanaka, M., Villamil, J. and Komochi, S. (1985) Studies on the storage of autumn harvested onion bulbs. III. Influence of storage temperature and humidity on the rooting and swelling on the stem plate of onion bulbs. *Research Bulletin of the Hokkaido National Agricultural Experiment Station* 144, 9–30 (in Japanese).

Tatham, P.B. (1982) *Bulb Onions*. Reference Book 348, ADAS/MAFF, London, 84 pp.

Thamizharasi, V. and Narasimham, P. (1988) Water vapour losses from different regions of onion (*Allium cepa* L.) bulb during storage. *Journal of Food Science and Technology* 25, 49–50.

Thamizharasi, V. and Narasimham, P. (1991) Water vapour sorption and transmission by onion (*Allium cepa* L.) scale under different temperature and humidity conditions. *Scientia Horticulturae* 46, 185–196.

Thamizharasi, V. and Narasimham, P. (1992) Growth of *Aspergillus niger* on onion bulbs and its control by heat and sulphur dioxide treatments. *Tropical Science* 33, 45–55.

Thamizharasi, V. and Narasimham, P. (1993) Effect of heat treatment on the quality of onions during long term tropical storage. *International Journal of Food Science and Technology* 28, 397–406.

Thomas, T.H. and Isenberg, F.M. (1972) Hormone physiology of onion bulbs during dormancy. *Experimental Horticulture* 23, 48–51.

Thompson, A.K. (1996) *Postharvest Technology of Fruit and Vegetables*. Blackwell Science, Oxford, 410 pp.

Thompson, A.K. (1998) *Controlled Atmosphere Storage of Fruits and Vegetables*. CAB International, Wallingford, UK, 278 pp.

Timm, E.J., Brown, G.K., Brook, R.C., Schulte, N.L. and Burton, C.L. (1991) Impact bruise estimates for onion packing lines. *Applied Engineering in Agriculture* 7, 571–576.

Ting, K.C., Flory, R.G., Singley, M.E. and Gaskell, M. (1987) Combined photovoltaic/air-heating flat plate collectors for onion curing. *American Society of Agricultural Engineers* 87, 4534–4550.

Tollner, E.W. and Shahin, M.A. (2000) X-ray imaging for classifying onions based on internal defects. *Book of Abstracts, Alliums 2000, Third International Symposium on Edible* Alliaceae, *30 October–3 November, 2000, University of Georgia, USA*. University of Georgia, Athens, Georgia, p. 29 (abstract).

USDA (1997) *US Standards for Grades of Bermuda–Granex–Grano Type Onions*. Washington, DC.

van den Berg, L. and Lentz, C.P. (1972) Respiratory heat production of vegetables during refrigerated storage. *Journal of the American Society for Horticultural Science* 97, 431–432.

van Konijnenburg, A. and Ardizzi, M.C.P. (1997) Variables which influence the occurrence of black mould (*Aspergillus niger*) on onion bulb in the Rio Negro Valley. *Acta Horticulturae* 433, 635–638.

Vijayakumar, R.M., Sundararajan, S. and Kumar, N. (1987) Effect of growth regulators on the postharvest physiology of small onion (*Allium cepa* L. var. *aggregatum* Don). *South Indian Horticulture* 35, 299–303.

Walder, J.M.M., Curzio, O.A., Croci, C.A., Domarco, R.E., Spoto, M.H.F. and Blumer, L. (1997) Avaliação da qualidade da cebola irradiada na Argentina e armazenada no Brasil. *Pesquisa Agropecuária Brasileira* 32, 565–569.

Wall, M.M. and Corgan, H.R. (1992) Relationship between pyruvate analysis and flavor perception for onion pungency determination. *HortScience* 27, 1029–1030.

Wall, M.M. and Corgan, H.R. (1994) Postharvest losses from delayed harvest and during common storage of short-day onions. *HortScience* 29, 802–804.

Wall, M.M. and Corgan, H.R. (1999) Yield and dry weight of dehydrator onions after uprooting at maturity and delaying harvest. *HortScience* 34, 1068–1070.

Warade, S.D., Desale, S.B. and Shinde, K.G. (1997) Effects of different storage recommendations on storability of onion bulbs. *Journal of Maharashtra Agricultural Universities* 22, 283–285.

Wheeler, T.R., Daymond, A.J., Ellis, R.H., Morison, J.I.L. and Hadley, P. (1998) Postharvest sprouting of onion bulbs grown in different temperature and CO_2 environments in the UK. *Journal of Horticultural Science and Biotechnology* 73, 750–754.

Wright, P.J. (1993) Effects of nitrogen fertilizer, plant maturity at lifting, and water during field-curing on the incidence of bacterial soft rot of onions in store. *New Zealand Journal of Crop and Horticultural Science* 21, 377–381.

Wright, P. and Grant, D. (1997) Effects of cultural practices at harvest on onion bulb quality and incidence of rots in storage. *New Zealand Journal of Crop and Horticultural Science* 25, 353–358.

Wright, P.J. and Grant, D.G. (1998) Evaluation of *Allium* germplasm for susceptibility to foliage bacterial soft rot caused by *Pseudomonas marginalis* and *Pseudomonas viridiflava*. *New Zealand Journal of Crop and Horticultural Science* 26, 17–21.

Xi, Y.F., Qian, D.M., Bian, Q.J. and Ying, T.J. (1994) Effects of gamma ray irradiation on cellular and subcellular structures of apical meristems in *Allium sativum* and *Allium cepa*. *Acta Agriculturae Nucleatae Sinica* 7, 134–138.

Yamazaki, H., Nishijima, T. and Koshioka, M. (1995) Changes in abscisic acid content and water status in bulbs of *Allium wakegi* Araki throughout the year. *Journal of the Japanese Society for Horticultural Science* 64, 589–598.

Yamazaki, H., Nishijima, T., Yamato, Y., Koshioka, M. and Miura, H. (1999a) Involvement of abscisic acid (ABA) in bulb dormancy of *Allium wakegi* Araki I. Endogenous levels of ABA in relation to bulb dormancy and effects of exogenous ABA and fluridone. *Plant Growth Regulation* 29, 189–194.

Yamazaki, H., Nishijima, T., Yamato, Y., Hamano, M., Koshioka, M. and Miura, H. (1999b) Involvement of abscisic acid in bulb dormancy of *Allium wakegi* Araki II. A comparison between dormant and nondormant cultivars. *Plant Growth Regulation* 29, 195–200.

Yoo, K.S. and Pike, L.M. (1995a) Postharvest losses of mechanically injured onions after curing. *HortScience* 30, 143.

Yoo, K.S. and Pike, L.M. (1995b) Effect of cross-cutting and temperature on shoot and root growth of onion bulbs. *HortScience* 30, 144.

Yoo, K.S., Andersen, C.R. and Pike, L.M. (1997) Internal CO_2 concentrations in onion bulbs at different storage temperatures and in response to sealing of the neck and base. *Postharvest Biology and Technology* 12, 157–163.

Zafrir, G. (1992) Effect of amount and distribution of nitrogen fertilization on the yield, quality and keeping ability of the bulb onion (*Allium cepa* L.). PhD thesis, The Hebrew University of Jerusalem, Israel.

11 Bacterial Diseases of Onion

G.L. Mark,[1*] R.D. Gitaitis[2] and J.W. Lorbeer[1]
[1]Department of Plant Pathology, Cornell University, Ithaca, NY 14853, USA;
[2]Department of Plant Pathology, University of Georgia, Coastal Plain Experiment Station, Tifton, GA 31793-0748, USA

*Current address: BIOMERIT Research Centre, Microbiology Department, N.U.I., Cork, Republic of Ireland.

1. Introduction

Bacterial pathogens of onion infect aerial parts of onion plants as well as onion bulbs. Infection of leaves can lead to bulb infection and decay while the plants are at different growth stages. Unless the inside neck tissues of onion plants are completely dry prior to the topping procedure at harvest, infection of the moist neck-wound tissue of healthy onion bulbs by bacterial pathogens can occur, resulting in decay of the bulbs under either field or indoor storage. When temperatures are optimum (30–35°C), bacterial decays of onion bulbs occur rapidly. Even when temperatures are somewhat lower, decays caused by bacterial pathogens render the bulbs unmarketable in a short period of time. Infected bulbs can decay rapidly when in transit to the market and thus are unacceptable to the buyer.

It appears that, at present, all onion cultivars are susceptible to bacterial infection and bulb decay. The most promising control procedures for bacterial diseases of onions involve the use of pathogen-free seed; undercutting and windrowing onions prior to harvest to effectively dry the onion tops and necks; crop rotation; and effective sanitation programmes for harvesting and grading equipment, as well as for storage containers and storage facilities. Although some bacterial diseases are reportedly managed by copper biocides, others are not controlled by foliar application of copper compounds (Schwartz and Mohan, 1995; Mark *et al.*, 1999a).

2. Sour Skin and Bacterial Canker

The name derives from the sour smell of onion bulbs infected with *Burkholderia cepacia* (formerly known as *Pseudomonas cepacia*). Extensive tissue maceration with water release is evident. Recent research in New York (Lorbeer *et al.*, 1998; Mark *et al.*, 1999b) has indicated that *B. cepacia* can infect onion plants growing under field conditions through the leaf axil of the plant, causing a disease named bacterial canker. This form of infection results in the death of the leaf infected and then in subsequent infection of the neck and bulb tissue of the plant. Bacterial canker of onion plants and sour skin of onion bulbs are progressive disease stages that can occur when onion plants growing in the field are infected by *B. cepacia*. Infection of succulent neck-wound tissue of onion plants topped at harvest ultimately results in the sour-skin stage of the disease in onion bulbs in storage.

2.1 History and distribution

B. cepacia was first isolated in the USA by Burkholder in 1950 from decayed onions in

New York and it is considered the primary bacterial onion pathogen in the Philippines (Daengsubha and Quimio, 1980). It was first isolated in Australia in 1985 (Cother and Dowling, 1985), and sporadic occurrences have been reported in Hungary (Füstös and Szarka, 1985) and in Mexico (Manrique *et al.*, 1991). Bacterial bulb rot of onion by *Pseudomonas* spp. was also reported in Korea (Choi and Han, 1990).

2.2 Mechanisms of infection

A wound is probably required for infection by *B. cepacia* to take place (Kawamoto and Lorbeer, 1972b; Gonzalez *et al.*, 1997), and the symptoms may indicate a hypersensitive response (Kawamoto and Lorbeer, 1972a). Water-congested tissue at the junction between the leaf blade and the sheath (the blade axil) was very susceptible to infection when stab-inoculated by *B. cepacia* (Kawamoto and Lorbeer, 1974). Young onion leaves appear to be the primary site of ingress for this bacterium, prior to infection of the bulb. This is in agreement with the 'pathogen–congenial host combination' (Klement and Lovrekovich, 1962) and the 'eusymbiotic relationship' (Klement, 1963). The interaction is more complex in mature leaves, where it is difficult to differentiate between *B. cepacia* behaving as a pathogen and as a saprophyte.

Bacteria require a mechanism to induce and maintain water congestion in the intercellular spaces (Rudolph *et al.*, 1994). Extracellular polysaccharides embed bacteria in intercellular spaces and these polysaccharides may be involved in the induction and maintenance of water congestion within these spaces (Rudolph *et al.*, 1994). Intercellular fluid from resistant but not from susceptible bean leaves degrades or inactivates *Pseudomonas phaseolicola* extracellular proteins (El-Banoby *et al.*, 1981). Sasser *et al.* (1970) suggested that a high osmotic potential of the intercellular fluid of pepper leaves prevented multiplication of *Xanthomonas vesicatoria* and that water congestion reduced this potential so that bacteria could multiply and cause disease. *B. cepacia* has been shown to spread more rapidly in water-soaked tissue than in uncongested onion tissue (Kawamoto and Lorbeer, 1974), and histological studies suggested that it moves via the intercellular spaces as free-swimming cells or as small clumps (Kawamoto and Lorbeer, 1972b). *B. cepacia* has been observed frequently in the substomatal cavities, and invasion occurs from within the leaf through the intercellular spaces. Bacteria in substomatal cavities were connected by strands of bacteria to other masses of bacteria and were usually traced to larger areas of bacterial colonization nearer to the site of inoculation (Kawamoto and Lorbeer, 1972b). Hence, onion bulbs with several infected scales may be a result of a single primary infection rather than due to multiple infections (Kawamoto and Lorbeer, 1972a).

Pectolytic enzymes may be responsible for the soft rot produced by this pathogen in onion. Phytopathogenic strains of *B. cepacia* produce endopolygalacturonase (PG) (Gross and Cody, 1985), whereas non-pathogenic strains do not (Gonzalez *et al.*, 1997). Hence, PGs are responsible for maceration of both scale and leaf tissue and are implicated in disease development (Ulrich, 1975). As acidic molecules do not readily penetrate tissues or cause membrane damage, it is likely that the acidic *B. cepacia* PGs penetrate the tissue via pectolysis of the cell walls (Ulrich, 1975), which in turn lowers the tissue's pH from 5.5 to 4.0, thus facilitating the enzyme activity, which is optimum at pH 4.4–4.6. The optimum temperature for pathogenesis by *B. cepacia* in onion is approximately 32°C, and at elevated temperatures the phytopathogenic *B. cepacia* strain ATCC 25416 produced non-pectolytic derivatives (Gonzalez *et al.*, 1997). Gonzalez *et al.* (1997) reported that the gene encoding the production of polygalacturonase activity by phytopathogenic *B. cepacia* 25416 is plasmid-determined.

2.3 Symptoms

Burkholder (1950) suggested that *B. cepacia* entered onion bulbs via wounds in the neck, when tops were removed at harvest. However, infection also occurs before harvest, suggesting that bacterial ingress also occurs in the upper plant parts, due to management practices or certain environmental conditions (Kawamoto, 1966; Kawamoto and Lorbeer, 1972b). Infected bulbs exhibit bacterial soft rot in the outer onion scales, with colours ranging from a pale yellow to brown. This decay remains localized to the infected scales, with no between-scale movement. In advanced infections, the outer infected scales can slip off during handling to expose yellow ooze on the underside of the scale, with a grainy texture.

B. cepacia can cause leaf blights in onions. Kawamoto and Lorbeer (1974) reported that artificial inoculation of leaf parts with *B. cepacia* via a wound resulted in lesions, which rapidly expanded only when the leaf tissue was water-congested. Young onion leaves are more susceptible to *B. cepacia*, whereas the majority of mature leaves fail to produce symptoms (Kawamoto and Lorbeer, 1972b). When *B. cepacia* infects wounded water-soaked onion tissue, the symptoms are typical of soft rot, whereas, in uncongested tissue, infection is observed as a dry leaf blight (Kawamoto and Lorbeer, 1974).

Disease symptoms are correlated with *B. cepacia* population levels within the onion leaf and are independent of time (Kawamoto and Lorbeer, 1972a). Population levels of *B. cepacia* decreased after the development of symptoms and this became more pronounced when the leaves dried out following the soft-rot stage in the bulb tissue (Kawamoto and Lorbeer, 1972a). In 1998, bacterial canker-like lesions were observed in the field on the leaf-blade axil of the outermost leaf, and *B. cepacia* was implicated as the causal agent (Lorbeer *et al.*, 1998). When the inner blade axil, as opposed to the outer one, was inoculated with *B. cepacia*, disease progression was seen to increase. It is believed that when an infected inner leaf becomes an outer leaf, as the onion matures, the canker then becomes evident in the field (Mark *et al.*, 1999b).

2.4 Epidemiology

Under environmental conditions providing extended periods of water congestion of the infected leaf-axil tissue, *B. cepacia* invades the bulb tissue of onion plants. If conditions are conducive to disease development, the infection will progress and result in bacterial soft-rot in the scales. However, the lesion will desiccate in dry weather and the infection progress will be halted. In controlled experiments when the conditions were not conducive, i.e. low humidity, the canker lesion dried up and the infected leaf sloughed off (Lorbeer *et al.*, 1998; Mark *et al.*, 1999b).

B. cepacia can be associated with organic soil particles and contaminated irrigation water. The bacterium usually enters freshly cut bulb necks that are still green and succulent. If proper undercutting and windrowing is followed, the necks will become dry, *B. cepacia* will not survive and infection will not occur. Infection can occur due to contaminated water striking the young leaves and moving into the leaf lacuna to the leaf axil and then into the outer scales. Infection appears to progress more rapidly when the inner leaves, rather than the outer leaves, are inoculated with *B. cepacia*, and advances into outer bulb scales via infected leaves and the corresponding bulb scales. The bacteria tended to spread more rapidly in water-soaked tissues when temperatures exceeded 30°C.

2.5 Causal organism – *Burkholderia cepacia*

2.5.1 Taxonomic and biochemical characteristics

B. cepacia belongs to the *Proteobacteria* group in the beta subclass, synonyms *Pseudomonas cepacia* and *Pseudomonas kingii* (Palleroni, 1992; Yabuuchi *et al.*, 1992; Lessie *et al.*, 1996). It is an obligate aerobe forming Gram-negative rods measuring 1.6–3.2 × 0.8–1.0 μm in size (Davis, 1995), producing non-fluorescent yellow, cream or white pigments. Optimum growth temperature *in vitro* is 30–35°C but the majority of strains are capable of growth at 40°C. They occur singly or paired in culture and are motile via one or more polar flagella. *B. cepacia* is oxidase positive, catalase positive and can grow in sterile deionized water (Gelbart *et al.*, 1976). Various strains make up a heterogeneous group and have been previously cited as some of the most nutritionally versatile of all the pseudomonads (Stanier *et al.*, 1966; Ballard *et al.*, 1970; Palleroni, 1984), as they can utilize more than 100 different carbon sources (Davis, 1995). *B. cepacia* can produce acid from D-fructose, D-arabinose and cellobiose but not from L-rhamnose in oxidative/fermentative medium, and has the ability to use sebacate 2,3-butanediol, mucate, saccharate, meso-tartrate and L-tartrate as sole carbon sources. *B. cepacia* can grow on D-tartrate and mesaconate but not on decarboxylated arginine. The bacterium has been shown to accumulate poly-β-hydroxybutyrate as an extracellular carbon source (Palleroni and Holmes, 1981; Palleroni, 1984) and can reduce nitrate to nitrite but does not denitrify and liquefy gelatin (Davis, 1995). Some strains of *B. cepacia* have exhibited multiple resistance to antibiotics while others produce antibiotic substances, such as bacteriocin (Gonzalez and Vidaver, 1979), xylocandins (Meyers *et al.*, 1987), A and B cepacins (Parker *et al.*, 1984) and pyrrolnitrin (Janisrewicz and Roitman, 1988).

2.5.2 *Genetic characteristics*

B. cepacia was assigned to the beta subclass of the *Proteobacteria* group, which differentiates into at least five distinct genomovars, referred to collectively as the *Burkholderia cepacia* complex, based on ribosomal (rRNA) (*rrn*) gene sequence analysis (Palleroni, 1992; Yabuuchi *et al.*, 1992; Lessie *et al.*, 1996). Phenotypic diagnostics have resulted in *Burkholderia multivorans* being proposed for Genomovar II, while Genomovar V was identified as the recently described *Burkholderia vietnamiensis*. The remaining Genomovars I, III and IV are dependent on differential phenotypic tests (Vandamme *et al.*, 1997). *B. cepacia* exhibits a high genomic plasticity, having a large complex genome approximately 4–9 Mb in size (McArthur *et al.*, 1988) containing many insertion sequences (IS) (Gaffney and Lessie, 1987). The high adaptability and catabolic function potential of *B. cepacia* could be due to the prevalence of IS elements in the genome and associated plasmids (Lessie *et al.*, 1990; Wood *et al.*, 1990). These IS elements have been implicated in the evolution of metabolic pathways and in plasmid and chromosomal rearrangement of various strains (Hendrickson *et al.*, 1996). A number of IS elements have been identified in *B. cepacia* based on their ability to promote genetic rearrangement (Gaffney and Lessie, 1987) and to employ foreign genes by the fusion of replicons (Barsomian and Lessie, 1986). This may explain the fast development of its adaptive capacity and its endurance. It has also been proposed that genes are transferred laterally among *Burkholderia* and other genera and that new metabolite capabilities are produced by genetic variation, as well as modification of existing pathways for degradation of toxic compounds (Hendrickson *et al.*, 1996).

The chromosomes of the species in the *Burkholderia* genera contain multiple replicons (Michaux *et al.*, 1993; Zuerner *et al.*, 1993; Cheng and Lessie, 1994) and this leads to variation in different genomovars. The number of replicons and overall genome size have been shown to vary between the genomovars of the *Burkholderia* complex (Lessie and Manning, 1995; Yao and Lessie, 1998; Table 11.1).

Macrorestriction fragment mapping of the *B. cepacia* genome showed that *B. cepacia* ATCC 17616 comprised three replicons, 3.4, 2.5 and 0.9 Mb in size (Cheng and Lessie, 1994). Different biosynthetic and degradable functions were associated with the 3.4 and 2.5 Mb replicons and all three contained rRNA genes (Cheng and Lessie, 1994). Three chromosomes were found in *B. cepacia* ATCC 25416 (Rodley *et al.*, 1995). Southern hybridization experiments indicated that all three replicons in *B. cepacia* ATCC 17616 contained genes for the 16S and 23S RNA regions of the rRNA operon (Cheng and Lessie, 1994). The 3.4 Mb replicon appeared to contain three sets of *rrn* genes, while the 2.5 and 0.9 Mb replicons each appeared to contain six sets of *rrn* genes (Lessie and Manning, 1995).

Table 11.1. Size (Mb) and number of repliccns in the five genomovars of the *Burkholderia* complex.

Genomovar	Number of replicons	Replicon size	Genome size
I	2–3	1.1–3.7	5.7–7.9
II (*B. multivorans*)	2–3	2.0–3.6	5.1–7.0
III	3	0.8–3.9	7.0–8.1
IV	3	1.3–3.9	8.2–8.6
V (*B. vietnamensis*)	3	1.1–3.9	6.7–7.5

2.6 Biochemical and physiological diagnostic techniques for identification

2.6.1 Conventional biochemical tests, Biolog, analytical profile index and multilocus enzyme electrophoresis

The majority of *B. cepacia* isolates exhibit a number of biochemical characteristics, and conventional tests can be used to rapidly identify this phytopathogen. These include the ability to utilize penicillin G, L-threonine and disaccharides, such as trehalose and cellobiose, as sole carbon sources (Lessie *et al.*, 1996); *B. cepacia* can also metabolize orthophthalate and D-serine (Lessie and Gaffney, 1986). Motility can be tested rapidly, using stab inoculation of motility medium (bioMérieux Inc., Hazelwood, Missouri). As *B. cepacia* is strictly aerobic, is oxidase-positive and has the ability to grow at 41°C, it can be distinguished from the *Enterobacteriaceae*. Many strains produce non-fluorescent pigments which distinguish them from the fluorescent pseudomonads on King's medium B (KMB). Their failure to produce xanthomonadins can separate them from the xanthomonads. Pseudomonads fail to grow in acidic conditions; however, *B. cepacia* can grow at pH 5.5.

Two diagnostic methods for identification of *B. cepacia* are the Biolog GN microplate system (Biolog Inc., Hayward, California) and the analytical profile index (API) 20NE diagnostic strip (bioMérieux Inc., Hazelwood, Missouri). The former provides a standardized micromethod, using 95 carbohydrate utilization tests to identify a range of enteric, non-fermenting and fastidious Gram-negative bacteria. If oxidation of the predried dehydrated carbohydrate source by the bacteria occurs, a burst of respiration reduces the tetrazolium violet (redox dye) indicator, resulting in a purple coloration. The colour intensity is measured relative to a reference well that has no carbohydrate source. A 'metabolic fingerprint' (Bochner, 1989a, b) results and the pattern is read by Microlog software via a microplate reader at 590 nm after incubation at 30°C of 4, 6 and 16–24 h (Biolog Inc., 1990; Gadzinski, 1990). Quality control should be conducted with a known *B. cepacia* isolate. One disadvantage is that *B. cepacia* strains tend to produce 'false positives', which, however, mostly show as a lighter colour change than that normally achieved for a false positive, and so the pattern can still be read. However, because bacteria from the soil can readily store extracellular carbohydrates, 'false positives' may be observed as a darker colour change, making the pattern unreadable. 'False positives' can occur after incubation periods of as little as 4 h. Manufacturers recommend the use of the protocol adopted for *Klebsiella*, *Enterobacter* and *Serratia* (Adams and Martin, 1964; Bryan *et al.*, 1986), for which the bacterial inoculum is diluted 20-fold prior to use in the test. The growth of the bacteria on minimal medium before the test can reduce the occurrence of false positives.

The API 20NE is a standardized micromethod consisting of eight conventional and 12 assimilation tests for the identification of non-fastidious non-enteric Gram-negative rods. The eight conventional tests are reduction of nitrates, indole production, acidification of glucose, arginine dihydrolase, urease, β-glucosidase hydrolysis, protease hydrolysis and β-galactosidase production. A profile index is calculated and analysed by the profile-index database. This diagnostic system has no false positives and it is efficient for rapid biochemical testing (Gilardi, 1983). However, some strains may exhibit atypical biochemical reactions due to unusual nutritional requirements or mutations.

Neither of the diagnostic tests can be used for epidemiological purposes to establish identity between isolates of the same bacterial species. Multilocus enzyme electrophoresis can be used for the separation of genetically defined units of population structure, and can discriminate between closely related bacterial strains of the same species (McArthur *et al.*, 1988; Whittam, 1989; Carson *et al.*, 1991; Yohalem and Lorbeer, 1994).

2.6.2 Molecular diagnostics – B. cepacia

Biochemical diagnostics of *B. cepacia* suffer from certain disadvantages, as certain strains can exhibit atypical phenotypic reactions, particularly clinical strains (Baxter *et al.*, 1997). Recently, there has been a move towards the use of molecular diagnostics in the identification of *B. cepacia* and this has concentrated on clinical isolates.

Polymerase chain reaction (PCR) diagnostics can give rapid detection and identification of bacterial pathogens, including bioassays that target the species-specific rRNA genes in the *Burkholderia* complex (Karparti and Jonasson, 1996; Clode *et al.*, 1999; LiPuma *et al.*, 1999). Several methods for the preparation of the *B. cepacia* DNA template used in PCR include emulsification (Clode *et al.*, 1999), boiling, and sonication (Karpati and Jonasson, 1996) of the bacterial colonies. In diagnostic PCR, primer sequences that encode the 16S and the 23S rRNA ribosomal chromosomal genes are often used for obtaining species-specific bacterial DNA targets (Karpati and Jonasson, 1996). Several oligonucleotide primers specific to the two rRNA regions in the *B. cepacia* genome have been constructed (Table 11.2).

Clode *et al.* (1999) reported that, of 78 bacterial cultures biochemically diagnosed as *B. cepacia*, 75 reacted with specific *B. cepacia* primers. However, three of the bacterial cultures produced an amplicon with specific *Burkholderia gladioli* primers. Fifteen asaccharolytic isolates were confirmed as *B. cepacia* using diagnostic PCR, but with other non-fermenting Gram-negative species no amplification was found with the primer sets used. No false positives resulted in PCR in the diagnostics of *B. cepacia*. However, Karpati and Jonasson (1996) reported lower sensitivity in detecting *B. cepacia* in the sputum of cystic fibrosis patients in relation to laboratory strains and this may be due to genetic heterogeneity. LiPuma *et al.* (1999) reported that an assay based on 16S and 23S rRNA gene analysis of *B. cepacia* ATCC 25416 (Genomovar I) proved useful in identifying Genomovars I, III and IV as a group with

Table 11.2. Oligonucleotide primer sequences used in diagnostic PCR for the identification of *B. cepacia*.

Primer name	RNA sequence 5' ➡ 3' direction	Nucleotide position	Genomic DNA	Amplicon size	Reference
PSR1	TTTCGAGCACTCCCGCCTCTCAG		16S rRNA	209 bp	Clode *et al.*,
PSL1	AACTAGTTGTTGGGGATTCATTTC		*B. cepacia*		1999
FK1	GTGCCTGCAGCCGCGGTAAT	515–534	Universal	FK1–FK5	
FK5	TCCCGCCTCTCAGCAAGGATTCC	1000–1022 bp	16S rDNA *B. cepacia*	535 bp	Karpati and Jonasson, 1996
PC-SSR	GCCATGGATACTCCAAAAGGA	Not applicable	23S rRNA *B. cepacia* (Genomovar I)		LiPuma *et al.*, 1999
PC-SSF	TCGGAATCCTGCTGAGAGGC	994–1013	16S rRNA *B. cepacia* (Genomovar I)		
PC1	GCTGCGGATGCGTGCTTTGC		23S rRNA	323 bp	*Clode et al.*,
PC2	GCCTTCTCCAATGCAGCGAC		*B. cepacia*		1999

100% sensitivity and 99% specificity and that development of PCR assays to distinguish *B. cepacia* genomovars is under way.

By subtyping *B. cepacia* PCR products, using a range of endonucleases in restriction fragment length polymorphism (RFLP) analysis (Singleton, 1999), it may also be possible to detect differences between the strains. In PCR ribotyping, the length of the spacer region, which is located between the 16S and 23S regions of the rRNA operon, can vary. This variation can occur in different copies of the rRNA operon within the same chromosome and therefore, when using electrophoresis, more than one band may result. The variation in the length of the spacer DNA in different bacterial isolates has the potential to be used for typing purposes (Kostman *et al.*, 1992; Ryley *et al.*, 1995; Shreve *et al.*, 1997; Singleton, 1999). PCR ribotyping is a rapid and accurate method for typing *B. cepacia* and is less time-consuming to carry out than standard ribotyping (Daser *et al.*, 1994).

2.6.3 Use of semi-selective media

B. cepacia colonies exhibit an intense green metallic sheen on Eosin methylene blue (EMB) glucose indicator medium, due to high levels of a constitutively formed glucose dehydrogenase (Sage *et al.*, 1990). There are several other semi-selective media for *B. cepacia* biotypes from soil. PCAT medium (Lumsden *et al.*, 1986) permits the growth of organisms capable of utilizing specific compounds and could limit the diversity of *B. cepacia* biotypes recovered. TB-T (Hagedorn *et al.*, 1987) is based on a combination of trypan blue (TB) and tetracycline (T), uses a basal medium of glucose and L-asparagine, and includes crystal violet and nystatin. Twenty-eight per cent of facultative organisms can also grow on this medium and can be separated from *B. cepacia* by anaerobic glucose fermentation and by their inability to grow at 41°C. Efficiency of recovery on TB-T is 78–86%, and recovery of *B. cepacia* biotypes can occur from low soil concentrations (i.e. 10^1–10^3 ml^{-1}). Clinical identification relies on a semi-selective medium containing colistin for *B. cepacia* (Henry *et al.*, 1997). However, it has been reported that other Gram-negative isolates resistant to colistin can grow on this medium (Hutchinson *et al.*, 1996). Clode *et al.* (1999) have reported the use of another semi-selective medium, MAST (MAST Diagnostics, MAST Group, Merseyside, UK), for identification of *B. cepacia*.

2.7 Host range of pathogen

As a phytopathogen, *B. cepacia* seemed for years to be specific to onion. However, in 1980, Dittapongpitch and Daengsubha reported that it was also pathogenic to Chinese cabbage. Recently it has been shown to infect shallot and wild leek (*A. tricoccum*) (Mark *et al.*, 1999b). Yohalem and Lorbeer (1997) found that pathogenicity of *B. cepacia* to onions is a variable characteristic, that pectolytic activity and pathogenicity by a range of *B. cepacia* isolates were highly correlated and that strains isolated from hospital environments were non-pathogenic to onion.

B. cepacia has been isolated from the rhizospheres of a wide range of plants, such as maize (DiCello *et al.*, 1997; Nacamulli *et al.*, 1997), pea (King and Parke, 1996), cucumber (Bevivino *et al.*, 1997), soybean, lettuce, tobacco (Tsuchiya *et al.*, 1995), barley, rye (Mark *et al.*, 1999b) and cotton (Heydari *et al.*, 1997). Plant development significantly affected the biodiversity of a *B. cepacia* population on maize roots, where higher polymorphism of *B. cepacia* was observed at early stages of growth (DiCello *et al.*, 1997).

2.8 Survival and behaviour in the soil

B. cepacia can persist in the soil for long periods of time (Sangodkar *et al.*, 1988), independent of the amount of water percolation (Hekman *et al.*, 1994). Its genetic diversity increased with environmental variability (McArthur *et al.*, 1988). Yohalem and Lorbeer (1997) reported that *B. cepacia* was isolated from all the agricultural soils they sampled, but that the pathogen was less frequent and more difficult to isolate in fields not cropped to onion.

3. Bacterial Streak and Bulb Rot

Bacterial streak and bulb rot of onion are caused by *Pseudomonas viridiflava* (Burkholder) Dowson.

3.1 History and distribution

P. viridiflava was first observed on onions in Georgia, USA, in 1990 (Gitaitis *et al.*, 1991). Since the initial report, the disease has been found in Florida, Colorado (Schwartz and Otto, 1998) and Venezuela (Hidalgo, Barquisimeto, 1999, personal communication).

3.2 Disease description and symptoms

The disease can be highly destructive and cause a total yield loss due to foliage damage and bulb decay in the field and during storage. In Georgia, Florida and Venezuela the disease affected fresh-market, sweet onions. However, in Colorado, the disease was associated with dry-bulb, pungent onions (Schwartz and Otto, 1998).

Generally, lesions develop as dark, water-soaked streaks that traverse most of the length of the leaf. In most cases, the first leaf to display a symptom is the third from the outside; however, under favourable conditions for the bacterium, the entire plant can be blighted. Early lesions, associated with small wounds or from infection through stomata, are small ($\leq$ 1.0 cm), oval and olive-green to tan in colour, and can be surrounded by a chlorotic area. Once infection has occurred, lesions develop rapidly and streak downward to the neck of the bulb. Infected leaves may appear wet or soft, are very dark green or black in colour, and collapse with the veins being prominent. Symptoms of this type are often associated with a soft-rot of the leaf-base, and when gently pulled, such leaves easily break off from the plant. However, under some conditions, blighted leaves are dry, are tan to a light brown in colour and tend to curl backwards from the leaf tip. It is not clear if leaf-blade dieback symptoms are associated with this disease.

When weather conditions are favourable for disease development, it is common for bulbs to rot in the field. Bulbs of severely infected plants are almost impossible to harvest, as the leaves pull away from the bulb during the lifting process and the rotting bulb remains in the ground. In plants with milder infections, bulbs can appear normal at harvest time, but the rot may have progressed into the neck of the bulb by means of only one infected leaf and may enter the inner scales before the bulbs are harvested or cured. If the bulbs are not yet infected at harvest, but the plants are immature or improperly cured, the bacterium can work its way through the neck and into the inner scales of the bulb, resulting in postharvest decay.

Typically, during the earliest stages of bulb rot, inner scales develop a distinct, pale, lemon-yellow colour. The discoloration rapidly becomes reddish-brown to brown and difficult to distinguish from rots caused by other pathogens. During the early stages of bulb rot, colonized tissues will fluoresce when viewed under ultraviolet light. In many cases, this will be followed by the production of a brilliant, shiny, blue-green, metallic-looking material, which can stain the inner onion scales. However, other soft-rot pathogens or secondary micro-organisms can quickly colonize damaged tissues. Consequently, the bulbs are softer and wetter and exhibit numerous colour variations.

3.3 Mechanisms of infection

Although the bacteria may enter natural openings such as stomata, new lesions are frequently associated with some type of physical damage, as in wounds caused by insect feeding, wind-blown sand or mechanical scraping by farm equipment, or with the crease that forms in the upper portion of the leaf as the foliage collapses (flagging symptoms). The type and degree of physical damage may have a direct bearing on the severity of infection, as well as on symptom appearance.

Although *P. viridiflava* was occasionally isolated from necrotic leaf tips, it was suspected that its presence in overfertilized plants was as a secondary colonizer rather than as a primary pathogen. The bacterium could not be detected in adjacent plants that had similar-appearing tip dieback symptoms and had received lower rates of nitrogen.

3.4 Epidemiology

In general, phytopathogenic bacteria within the '*syringae*' group do not survive well in soil and, when host tissues decompose, the bacterium can no longer be detected. This appears to hold true for *P. viridiflava* in infected onion tissues as well (Gitaitis *et al.*, 1997). In Georgia, USA, *P. viridiflava* was found as a resident epiphyte on weeds within and adjacent to onion fields and also in remote non-agricultural sites. It is known to survive between seasons in association with several weed species (Gitaitis *et al.*, 1998a), especially on cutleaf evening primrose (*Oenothera laciniata*), which is also the most significant weed problem in onion fields in Georgia.

Airborne dissemination is a possibility, but the exact distance the bacterium can be disseminated is not known. A significant disease reduction has been observed with improved weed control within onion fields. This suggests that the bacteria are normally disseminated for rather limited distances. The bacterium can also be mechanically transmitted by farm equipment that makes contact with weeds along field perimeters and by field workers during harvest. When onion clipping shears contaminated with *P. viridiflava* were used to remove onion tops at harvest, immature onions or onions that were not allowed to 'field-cure' for a minimum of 48 h prior to clipping developed significantly more rot during storage. It is very likely that properly cured mature bulbs have sufficiently dried tissues in the neck to act as a barrier to infection through contaminated onion shears.

In Georgia, USA, outbreaks of bacterial streak and bulb rot most frequently occur between January and April, when there are extended periods of rain and temperatures are mild. In contrast, this disease occurs in Venezuela (Hidalgo, Barquisimeto, 1999, personal communication) and in Colorado (Schwartz and Otto, 1998) under much warmer conditions. It is not known what factors are responsible for this disparity, but it is tempting to speculate that different biotypes may occur in the different regions. However, further research is required to answer this and other questions regarding the epidemiology of this pathogen. Like any pathogen, disease progress slows considerably or is halted altogether when weather conditions become less favourable for disease development.

3.5 Causal organism – *Pseudomonas viridiflava*

3.5.1 Taxonomic and biochemical characteristics

P. viridiflava is a Gram-negative, aerobic rod with one to three polar flagella. When grown on an iron-deficient medium, such as KMB (King *et al.*, 1954), it produces a water-soluble, yellow-green, fluorescent pigment (Lelliott *et al.*, 1966). On KMB and semi-selective medium T-5, colonies are initially pale cream but turn yellow with age (Gitaitis *et al.*, 1997). The bacterium is negative for the enzymes oxidase and arginine dihydrolase but is positive for gelatin hydrolysis (Lelliott *et al.*, 1966). Most strains are active for ice nucleation and utilize DL-lactate and erythritol (Jones *et al.*, 1984, 1986). These are useful characteristics for identifying the bacterium as a member of the fluorescent pseudomonads similar to *Pseudomonas syringae*.

P. syringae is a species represented by a diverse number of pathovars, some of which have distinct biochemical and physiological traits, as well as differences in host range. Therefore, it is difficult to separate *P. syringae* from *P. viridiflava*. The following are useful tests that help distinguish these pathogens: the production of pectinolytic enzymes at pH 8.5 but not at pH 5.0 (Hildebrand, 1971); the rot of carrot (Gitaitis *et al.*, 1991) and potato slices (Lelliott *et al.*, 1966); absence of levan production and the utilization of D-tartrate (Hildebrand and Schroth, 1972); and the production of rust-coloured lesions on bean pods of the cultivar 'Bush Blue Lake 274' (Cheng *et al.*, 1989).

Sucrose utilization is a relatively simple and rapid test that distinguishes *P. syringae* from *P. viridiflava* (Lelliott *et al.*, 1966; Billing, 1970; Hildebrand and Schroth, 1972; Jones *et al.*, 1984). However, the onion-pathogenic strains of *P. viridiflava* from the south-eastern USA produce acid (a presumptive test for utilization of a sugar) in sucrose after extended incubation (10–17 days) (Gitaitis *et al.*, 1991). This is very slow compared with many bacterial species that produce acid in sucrose in a matter of 24–72 h. In many instances, substrate utilization results may not be recorded after 7 days: the onion strains would be characterized as negative for sucrose utilization if the tests had been terminated within that time frame. Additional reports (Clara, 1934; Wilkie and Dye, 1973; Suslow and McCain, 1981) indicate that strains of *P. viridiflava* isolated from certain other hosts also utilize sucrose very slowly.

Another method for characterizing and identifying bacteria used extensively for the past 20 years is fatty acid analysis by gas–liquid chromatography (Moss *et al.*, 1980; Sasser *et al.*, 1984; Miller and Berger, 1985; Gitaitis and Beaver, 1990; Sasser, 1990). *P. viridiflava* and *P. syringae* have very similar fatty acid profiles and could easily be confused with one another. However, *P. viridiflava* strains isolated from onions typically contain delta-*cis*-9,10-methylene hexadecanoic acid (C-9,10 17:0), and the ratio of alpha-hydroxylauric acid (2-OH 12:0) to lauric acid (12:0) was greater than 1. In contrast, the ratio of 2-OH 12:0 to 12:0 acids was less than 1 in the strains of *P. syringae* that were tested (Gitaitis *et al.*, 1991). Variation between strains was also observed through the plotting of principal components derived from cluster analysis of the fatty acid profiles. Strains of *P. viridiflava* recovered from weeds and diseased onions in Georgia clustered more closely to each other than to strains of *P. syringae* or *P. viridiflava* from other hosts and geographical origins (Gitaitis *et al.*, 1998a).

3.5.2 Genetic characteristics

If available, some of the most rapid and reliable methods of identifying the pathogen are with enzyme-linked immunosorbent assay (ELISA) or PCR. Antisera or even prepared ELISA plates can be obtained from a variety of commercial vendors, including Agdia Inc. (Elkhart, Indiana). Primers from the pectate lyase gene (5′-TATTGCTGGTGTTACCC-3′ and 5′-GGTATCCAGAAACGACAC-3′) were capable of amplifying an amplicon of 606 base pairs (bp) in length from approximately 95% of the strains recovered from onions or from weeds in onion-growing regions of Georgia (Gitaitis *et al.*, 1998b). The same set of primers failed to amplify strains of *P. viridiflava* from bean, bell pepper, parsnip, tomato and watermelon either from Georgia or from elsewhere in the USA. These results support the groupings from previous fatty acid characterizations, in that weed and onion strains from Georgia were more similar to each other than to strains from other hosts or from different geographical origins.

3.6 Host range of pathogen

Burkholder first described *P. viridiflava* as a pathogen of bean (*Phaseolus vulgaris*) in New York (Burkholder, 1930). Since then, it has been reported from around the world as a pathogen of many different plant species. In addition to bean and onion, the bacterium has a natural host range of lucerne (*Medicago sativa*), bird's-foot trefoil (*Lotus corniculatus*), cabbage, cauliflower, cherry (*Prunus* spp.), Chinese gooseberry (*Actinidia chinensis*), dill (*Anethum graveolens*), chrysanthemum, grape, lettuce, lupin (*Lupinus angustifolius*), parsley (*Petroselinum crispum*), parsnip (*Pastinaca sativa*), passion fruit (*Passiflora edulis*), pea, pear, sweet pepper, poinsettia (*Euphorbia *pulcherrima*), poppy (*Papaver somniferum*), pumpkin (*Cucurbita maxima*), tansy (*Tanacetum coccineum*), rape (*Brassica napus* var. *napus*), tomato and watermelon. In addition to the above list, the bacterium has also been reported to infect the following upon artificial inoculation: buckwheat (*Fagopyrum esculentum*), clover (*Trifolium pratense*), cowpea (*Vigna unguiculata*), safflower (*Carthamus tinctorius*), sorghum (*Sorghum vulgare*), soybean (*Glycine max*) and zinnia (*Zinnia elegans*) (Billing, 1970; Wilkie and Dye, 1973; Suslow and McCain, 1981; Lukezic *et al.*, 1983; Jones *et al.*, 1984; Bradbury, 1986b). Finally, the bacterium has been recovered from several weed species as a resident epiphyte, i.e. the bacterium lives freely on the plant's surface in either a commensal or a protocooperative relationship. These plants include cutleaf evening primrose, dandelion (*Taraxacum officinale*), common fumitory (*Fumaria officinalis*), purple cudweed (*Gnaphalium purpureum*), Virginia pepperweed (*Lepidium virginicum*) and wild radish (*Raphanus raphanistrum*) (Gitaitis *et al.*, 1998a).

P. viridiflava is a weak pathogen or a secondary invader that colonizes behind other pathogens (Billing, 1970), or an opportunistic pathogen that invades wounded plants or those under extreme stress (Hunter and Cigna, 1981; Suslow and McCain, 1981; Lukezic *et al.*, 1983; Jones *et al.*, 1984). However, in onions, *P. viridiflava* is an aggressive primary pathogen that is particularly destructive on succulent plants receiving abundant levels of nitrogen. The bacterium has been responsible for losses of entire fields and has also been quite destructive as a postharvest pathogen.

4. Centre Rot

Centre rot of onion is caused by *Pantoea ananatis* (Serrano) Mergaert *et al.* (1993).

4.1 History and distribution

Centre rot was first observed on sweet onions in Georgia, USA, in 1997 (Gitaitis and Gay, 1997). A year later it was observed in dry-bulb, pungent onions in Colorado (Schwartz and Otto, 1998, 2000a). An almost identical disease in South Africa was attributed to the closely related bacterium *Erwinia herbicola*, which is synonymous with *Pantoea agglomerans* (Hattingh and Walters, 1981). Although *P. agglomerans* and *P. ananatis* are closely related (at one time they were considered to be the same species) and the seed of the variety that centre rot was first observed on in Georgia was produced in South Africa, there is no evidence to date that the disease is seed-borne. Rather, it appears that the bacterium is endemic to the south-eastern USA.

A bacteriophage specific to *P. ananatis* – presumptive evidence of the presence of the bacterium – has been recovered from several lakes in Florida and Texas (Eayre *et al.*, 1995). Using PCR, it was determined that *P. ananatis* was in Georgia several years prior to the 1997 epidemic in Vidalia onions. The evidence for its earlier presence came from screening the University of Georgia's Coastal Plain Experiment Station Culture Collection, where two strains from peach leaves and onion bulbs from 1986 and 1992, respectively, were identified as *P. ananatis*.

4.2 Disease description and symptoms

As the name suggests, the disease quite often affects the centre leaves of the plant. Affected leaves become water-soaked, soft and bleached white as the rot progresses. Surrounding tissues may appear tan to a darker brown. Advanced stages of the disease result in complete wilting and bleaching of all leaves. Bulb interiors may become soft and watery and produce a foul odour. Attempts to lift the plant by grabbing the leaves may result in liquefied tissues oozing from the neck and leaves breaking away from the plant. Unlike most other bacterial diseases of onion, 'centre rot' also infects seed stalks in a similar manner to the leaves, which results in scape lodging and loss of seed heads.

4.3 Causal organism – *Pantoea ananatis*

4.3.1 Taxonomic and biochemical characteristics

The name of this organism evolved from *Bacillus ananas* Serrano (1928), *Bacterium ananas* (Serrano) Burgvits (1935), *Chromobacterium ananas* (Serrano) Krasil'nikov (1949), *Pectobacterium ananas* (Serrano) Patel & Kulkarni (1951), *Erwinia herbicola* var. *ananas* (Serrano) Dye (1969) to *Erwinia ananas* Serrano (1928), where it remained for some time. In the 1984 edition of *Bergey's Manual of Systematic Bacteriology*, the designation of *E. ananas* was the only species within the '*herbicola*' group that was phytopathogenic (Lelliott and Dickey, 1984). Then the genus name was changed to *Pantoea* (Mergaert *et al.*, 1993) and the spelling was corrected to *Pantoea ananatis* (Trüper and De'Clari, 1997).

P. ananatis is a Gram-negative rod with yellow pigmentation when grown on nutrient agar. It utilizes glucose in both an oxidative and a fermentative manner and is positive for catalase and negative for oxidase, typical of the *Enterobacteriaceae* (facultative anaerobes). Typically, strains utilize cellobiose, melibiose, inositol, glycerol and sucrose but do not hydrolyse pectin, starch or gelatin. Key characteristics that separate *P. ananatis* from *P. agglomerans* are its ability to produce indole, and the lack of phenylalanine deaminase and nitrate reductase (Bradbury, 1986a). The bacterium is also ice-nucleation-active (Abe *et al.*, 1989).

4.3.2 Genetic characteristics

A sense primer (Pan ITS1) has been developed in Georgia, USA, for the intergenic transcribed spacer (ITS) region between the 16S and 23S rRNA genes to be used in conjunction with universal antisense primers EC5 or EC7 from *Escherichia coli* (Gurtler and Stanisich, 1996) in the 23S rRNA gene. The Pan ITS1 sequence is 5′-GTCTGATAGAAAGATAAAGAC-3′, the sequence of EC5 is 5′-TGCCAGGGCATCCACCG-3′ and the sequence of EC7 is 5′-GGTACTTAGATGTTTCAGTTC-3′. These primers have been used successfully to conduct PCR of leaf washes from numerous weeds and from crushed thrips. Although no specific insect relationship is known for *P. ananatis* so far, other members of this genus, most notably *Pantoea stewartii* and *Pantoea tracheiphila*, survive in association with and are vectored by corn flea beetles and cucumber leaf beetles, respectively (Leach, 1964; Pepper, 1967). Watanabe and Sato (1999) found that *P. ananatis* will inhabit the gut of mulberry pyralid larvae and suggested the use of the bacterium as a biocontrol agent for that insect.

4.4 Host range of pathogen

P. ananatis was originally reported as a pathogen of pineapple but the host range includes cantaloupe, honeydew melon, onion and sugarcane (Bradbury, 1986a; Wells *et al.*, 1987, 1993; Bruton *et al.*, 1991; Gitaitis and Gay, 1997). The geographical distribution includes Brazil, Guyana, Guatemala, Haiti, Malaysia, Mexico, Nigeria, the Philippines, Puerto Rico, Queensland (Australia), Taiwan and the USA (Bradbury, 1986a).

Using PCR, *P. ananatis* was detected as an epiphytic resident on 23 weed species, Bermuda grass (*Cynodon dactylon*) and soybean (*G. max*). The latter two are significant because they are used in a rotation between onion crops. Some of the weeds *P. ananatis* has been found on in Georgia include bristly starbur (*Acanthospermum hispidum*), broadleaf signalgrass (*Brachiaria platyphylla*), carpetweed (*Mollugo verticillata*), crabgrass (*Digitaria sanguinalis*), common cocklebur (*Xanthium pensylvanicum*), common ragweed (*Ambrosia artemisiifolia*), curly dock (*Rumex crispus*), Florida beggarweed (*Desmodium tortuosum*), Florida pusley (*Richardia scabra*), sicklepod (*Cassia obtusifolia*), spiny amaranth (*Amaranthus spinosus*), smallflower morning glory (*Jaquemontia tamnifolia*), Texas panicum (*Panicum texanum*), vaseygrass (*Paspalum urvillei*), verbena (*Verbena* spp.) and yellow nutsedge (*Cyperus esculentus*). Not only was the bacterium found on these weeds within and adjacent to onion production sites, but it was also detected on weeds as far away as 242 km from the nearest commercial onion production.

5. Bacterial Soft-rot

Bacterial soft-rot diseases are caused by *Erwinia* pathogens belonging to the soft-rot '*carotovora*' group, which is widespread (Voronkevitch, 1960; Graham, 1962).

5.1 History and distribution

In 1995, *Erwinia chrysanthemi* caused severe economic losses to onions in New York (Lorbeer, 1996; Lorbeer *et al.*, 1996). *E. chrysanthemi* has previously been isolated from crops in tropical and subtropical regions or observed as a pathogen of ornamental greenhouse crops in cooler climates (Pérombelon and Kelman, 1980), and in 1991 it was isolated from onion in Mexico (Manrique *et al.*, 1991). *Erwinia herbicola* was first isolated from infected onion seed in 1994 in Cuba, its first record for the Americas (Morales *et al.*, 1994). Another species, *Erwinia carotovora* subsp. *carotovora* is found in temperate and tropical regions on a range of host plants (Salmond, 1994).

5.2 Disease description and symptoms

The soft-rot erwinias produce typical soft-rot symptoms, mainly in the inner scales of the bulb onion (Mohan, 1995). The infected tissues become water-soaked and exhibit rot ranging from a pale yellow to light brown in colour. The soft-rot can progress from the inner scales until the whole onion bulb disintegrates, accompanied by the release of a watery, viscous fluid with a fetid odour. In the case of *E. chrysanthemi*, the inner bulb tissues may completely dissolve so that when the bulb is pulled it separates from the basal plate, which remains in the soil. Infection of the bulb by *E. carotovora* subsp. *carotovora* also results in wilting and whitening of the onion foliage. *E. herbicola* produces lesions on the onion flower stalks, leaves and pedicels and this will progress to soft-rot in the onion bulb (Morales *et al.*, 1994).

5.3 Mechanisms of infection

The pathogen attaches to the plant cells via fimbriae or pili (Romantschuk *et al.*, 1994) and, like many other plant pathogens (Garibaldi and Bateman, 1971), produces a complex mixture of pectic enzymes. Hydrolase and lyases degrade, in a random manner, the alpha-(1–4) linkages in the uronic acid polymers of pectic substances and appear to be the primary agents responsible for the maceration of tissue due to infection by *Erwinia* (McClendon, 1964; Zaitlin and Coltrin, 1964; Sato, 1968). Garibaldi and Bateman (1971) reported that *E. chrysanthemi* produces a number of polygalacturonic acid transeliminases in culture and that different isolates of this pathogen can vary in the number of isozymes of a given enzyme that they produce. The enzymes produced by *E. chrysanthemi* macerate plant tissue, induce electrolytic leakage and release soluble unsaturated uronides from intact plant tissue thus causing cell death (Garibaldi and Bateman, 1971). The enzymes involved in cell death that are produced by *E. chrysanthemi* have isoelectric points at or greater than pH 4.6 (Garibaldi and Bateman, 1971). The optimum temperature for pathogenesis by *E. chrysanthemi* is 32°C (Jovanovic, 1998).

5.4 Epidemiology

The erwinias are present in soil, crop residues and contaminated water (Mohan, 1995) and can spread through overhead water irrigation and via insects, such as the onion maggot *Delia antiqua* (Mergen). *E. carotovora* subsp. *carotovora* can survive in the intestinal tract of the onion-maggot larvae and the adult flies. Temperatures of 20–30°C and high humidity are optimum for infection of onion by *Erwinia* soft-rot pathogens, which can also continue when the onion bulbs are stored at temperatures greater than 3°C.

5.5 Causal organism – *Erwinia chrysanthemi*

5.5.1 *Taxonomic and biochemical characteristics*

The Gram-negative, facultatively aerobic *E. chrysanthemi* belongs to the *Enterobacteriaceae* family and it forms straight rods, which are 0.5–1.0 × 1–3 μm in size. The cells are motile by peritrichous flagella. They occur singly, in pairs or occasionally in short chains. *E. chrysanthemi* has a fermentative metabolism and optimally grows *in vitro* at 32°C. It also grows relatively well at 39°C, is oxidase-negative and catalase-positive and produces indole and hydrogen sulphide (Pérombelon and Kelman, 1980). The bacterial cells hydrolyse gelatin at 22°C, do not hydrolyse urea and are phenylalanine-deaminase-negative. They catabolize D-glucose to produce both acid and gas, and utilize a range of carbon sources, such as D-adonitol, cellobiose, glycerol, D-mannitol, D-mannose, melibiose, raffinose, L-rhamnose, salicin, D-sorbitol and D-xylose to produce acids. *E. chrysanthemi* can be distinguished from other erwinias by its ability to reduce nitrates and the majority of strains produce extracellular polysaccharides, even on sugar-rich media (Pérombelon and Kelman, 1980).

5.6 Biochemical and physiological diagnostic techniques for identification

5.6.1 *Biochemical and physiological tests (conventional tests, Biolog, analytical profile index, serological methods, FAME analysis)*

E. chrysanthemi can be distinguished from the pseudomonads by its inability to produce oxidase and from *E. carotovora* by its ability to grow at 39°C (Pérombelon and Kelman, 1980). Pectate medium has been used to isolate *E. carotovora* subsp. *carotovora* from onions exhibiting bacterial soft-rot (Taraka and Tsuboki, 1982). Two commercial diagnostic methods can be used for the identification of *Erwinia* pathogens in onion. These are the Biolog GN microplate system (Biolog Inc., Hayward, California) and the API 20E diagnostic strip (bioMérieux Inc., Hazelwood, Missouri). The latter is used for the identification of *Enterobacteriaceae* and other Gram-negative rods. It consists of 11 biochemical tests and nine carbohydrate assimilation tests. Serological methods have also been used in the identification of *Erwinia*. Monoclonal antibodies have been used in the detection of *E. carotovora* subsp. *carotovora* in potato (DeBoer and McNaughton, 1987; DeBoer *et al.*, 1988). Pectolytic enzymes produced by *E. carotovora* subsp. *carotovora* have been characterized by thin-layer isoelectric-focusing activity, gel overlays and qualitative enzyme assays (Willis *et al.*, 1987). Four *pel* genes and the *pel* 1 gene were recovered from 71 *E. carotovora* subsp. *carotovora* gene libraries constructed in *E. coli* HB101: these genes are clustered within the genome. Kori *et al.* (1992) carried out fatty acid analysis of the bacterial cellular membrane via gas–liquid chromatography and found that the ratio of amounts of lauric acid and myristic acid in *E. chrysanthemi* and *E. carotovora* subsp. *carotovora* were reversed relative to each other. They also showed that the fatty acid profiles were different for *E. chrysanthemi* depending on the host from which it had been isolated.

5.6.2 *Molecular diagnostics*

PCR-based methods have been used for the detection of *E. chrysanthemi* and *E. carotovora* subsp. *carotovora* (Nassar *et al.*, 1991). Nassar *et al.* (1996) characterized *E. chrysanthemi* by pectinolytic isozyme polymorphism and RFLP analysis of PCR-amplified fragments of *pel* genes. Randomly amplified polymorphic DNA (RAPD)-PCR has been carried out with *E. carotovora* subsp. *carotovora* (Maki-Valkama and Karjalainen, 1994). Host specificity of *E. chrysanthemi* has been investigated using PCR-RFLP methods (Nassar *et al.*, 1994). Darrasse and Bertheau (1994) have also detected *Erwinia* by using PCR and RFLP.

5.7 Host range of pathogen

E. carotovora subsp. *carotovora* lacks specificity in the host–pathogen interaction and the majority of the erwinias are considered opportunistic phytopathogens. *Erwinia* species can act as primary pathogens of a range of growing crops, harvested crops and plant residues. *E. chrysanthemi* has relatively high host specificity for maize or ornamentals (Salmond, 1994). *E. chrysanthemi* and *E. carotovora* subsp. *carotovora* are relatively major onion pathogens, whereas *Erwinia rhapontica* is a minor onion pathogen with a restricted host range (Pérombelon and Kelman, 1980). *E. carotovora* subsp. *carotovora* infects a range of host plants, such as onion, potato, carrot, radish, cucumber, Chinese cabbage, pepper, cabbage and lettuce (Dittapongpitch and Daengsubha, 1980). *E. herbicola* causes a leaf and stalk necrosis of onion and is mainly isolated from infected seed (Morales *et al.*, 1994). Sweet-onion F_1 hybrids, such as 'Granex' and 'Golden', are susceptible to *E. carotovora* subsp. *carotovora* (Jones, 1981).

5.8 Survival and behaviour in the soil

Persistence of *Erwinia* in the soil during summers in temperate regions is probably short. However, small numbers of the bacteria may overwinter in the colder soil. Survival the following summer is unlikely unless a susceptible crop is planted (Collmer and Keen, 1986). Soft-rot erwinias are not endemic in the soil and their widespread distribution may be due to the recurrent introduction of infected plant material (Logan, 1968; DeBoer *et al.*, 1979).

Unlike *B. cepacia*, *Erwinia* species do not accumulate energy-rich compounds such as glycogen and poly-β-hydroxybutyrate. Therefore, their ability to survive periods of low nutrient availability may be limited (Pérombelon, 1973), and yet they can survive indefinitely in the plant rhizosphere, particularly in tropical regions, where plant growth is often continuous and diverse. *Erwinia* can overwinter in infected plant residues that remain in the soil after harvest, as long as the plant material is not completely decomposed (Pérombelon, 1973). They are usually absent from seed, except for *E. herbicola* and occasionally *E. chrysanthemi* (Yáñez-Morales and Lorbeer, 1993; Yáñez-Morales *et al.*, 1994), and move readily in soil water. They are superficially attached to soil particles (Kikumoto and Sakamoto, 1970) and are readily dislodged by percolating soil water.

6. Onion Leaf Blights

6.1 History and distribution

Xanthomonas campestris was first observed as a leaf-blight pathogen of onion in Hawaii in 1978 (Alvarez *et al.*, 1978) and similar symptoms were evident on onion in Barbados 15 years later (Paulraj and O'Garro, 1993). Recently the pathogen has been described as occurring in the continental USA (Isakeit *et al.*, 2000; Schwartz and Otto, 2000c).

6.2 Disease description and symptoms

Symptoms of infection by *X. campestris* include a range of lesions, usually on mature onion leaves. These can be white flecks, pale spots or lenticular-shaped lesions, which develop into visible chlorotic streaks on the lower part of the onion leaf. As the disease progresses, tip dieback can occur and then extensive blighting of the outer mature onion leaves, which results in stunted plants with unmarketable-sized bulbs (Mohan, 1995).

6.3 Mechanisms of infection

X. campestris can spread via rain or sprinkler-irrigation water and infection is enhanced by the presence of dew. Wounding via wind or sandblasting on the leaves increases the possibility of infection (Mohan, 1995). Unlike *X. campestris*, *P. syringae* produces brown lesions on onion (Mohan, 1995).

6.4 Causal organisms

6.4.1 Taxonomic and biochemical characteristics

X. campestris is a yellow-pigmented, Gram-negative, aerobic and motile rod-shaped bacterium (Mohan, 1995). It can be distinguished from the pseudomonads by the fact that it is oxidase-negative and from *E. chrysanthemi* due to its inability to reduce nitrates to nitrites. It can be distinguished from the soft-rot '*carotovora*' group of *Erwinia* by its inability to hydrolyse pectate.

7. Soft-rot Pathogens of Onion

Burkholderia gladioli pv. *alliicola* (formerly known as *Pseudomonas alliicola*) causes what is commonly known as slippery skin (Roberts, 1973).

7.1 History and distribution

A soft-rot of onion bulbs believed to be caused by a bacterium was described in New York by Stewart (1899). *B. gladioli* pv. *alliicola* was isolated from onions in New York by Burkholder (1942) and has been reported in many regions of the world since the original description.

7.2 Disease description and symptoms

The bacterium infects the inner bulb scales, producing a water-soaked appearance, which eventually progresses to soft-rot of the entire internal bulb tissue. It derives its name from the fact that the infected core may slip out of the top of the onion when the base of the bulb is squeezed (Mohan, 1995). *Burkholderia gladioli* pv. *alliicola* is a Gram-negative, rod-shaped bacterium. It infects leaves and maturing bulbs in the field or postharvest through a wound. Wet and rainy conditions are conducive for this pathogen. Mature bulbs are very susceptible to the bacterium and can decay completely at room temperature in 10 days (Mohan, 1995). *Pseudomonas aeruginosa* is present in the soil and causes a brown internal soft rot in onion. *Pseudomonas marginalis* infects the onion foliage and appears as small water-soaked lesions. These lesions expand rapidly, resulting in a slimy, grey-brown rot, which may progress down to the leaf-base and decay the entire plant, which then exudes a characteristic vinegar-like odour (Wright and Hale, 1992). *Lactobacillus* is an opportunistic pathogen and enters, like the *Erwinia*, via a wound in the neck of the onion bulb or with insects, such as the onion maggot (Brewster, 1994). *Lactobacillus* produces a soft-rot that rapidly progresses throughout the onion bulb at higher temperatures (Brewster, 1994). *Enterobacter cloacae* was reported to cause bulb decay in onions in Colorado recently (Schwartz and Otto, 2000b).

8. Control Strategies and the Future

Copper bactericides have been used in the control of both bacterial soft-rot and leaf-blight pathogens of onion with varying degrees of success. Pyle *et al.* (1992) reported that *B. cepacia* was inactivated by low copper and silver ion concentrations when in combination with iodine. Kidambi *et al.* (1995) provided evidence that *B. cepacia* was resistant to copper. Mark *et al.* (1999a) also found that *B. cepacia* exhibited resistance *in vitro* to the majority of copper-based bactericides. They found, however, that ReZist (Stoller Enterprise Inc., Texas), when tested *in vitro*, inhibited the growth of *B. cepacia* isolates 97–36(A) and 97–38(A). When combined with Kocide 2000 (Griffin Corporation Inc., Georgia), isolate 97–36(A), the more pathogenic of the two *B. cepacia* isolates, was inhibited to an even greater extent. ReZist contains 2% chelated copper, 2% chelated manganese and 2% chelated zinc derived from copper hydroxide, manganese oxide, and zinc oxide chelated with ethanol 2-amino-2-hydroxy-1,2,3-propanetricarboxylate. Kocide 2000 contains 53.8% copper hydroxide with a metallic copper equivalent of 35%. In Colorado, Schwartz and Otto (1998) reported that high rates of Kocide 2000 provided excellent control of the leaf blight *Xanthomonas* on onion. Scheck and Pscheidt (1998) reported a 50% reduction in the population size of *P. syringae* pv. *syringae* with a cupric hydroxide and mancozeb treatment. However, a similar formulation – Mankocide (Griffin Corporation Inc., Georgia) – did not inhibit *B. cepacia*, even at double the manufacturer's recommended rate (Mark *et al.*, 1999a). *B. cepacia* has exhibited multiple resistance to antibiotics. Bactericidal sprays with fixed coppers will reduce epiphytic survival on onion leaves and control secondary dissemination. However, copper-tolerant strains can develop quickly in response to standard copper bactericides. So far, the mixture of fixed coppers with carbamate-based fungicides, such as maneb, has been effective against all strains of *P. viridiflava* (Burkholder) Dowson.

Mark *et al.* (1999a) reported that levels of *B. cepacia* in organic soils in New York remained low throughout the growing season in onion fields that previously had a rotation crop, such as lettuce or Sudan grass (*S. vulgare* Pere. var. *sudanense* Hitchc.), marketed by DeKalb-Pfizer Genetica (Illinois) as Sudex sudangrass hybrid. In fields continuously cropped to onion, amounts of *B. cepacia* increased significantly to high levels from the end of June to the end of July.

Managing fertility levels, particularly nitrogen, has been extremely important in reducing bacterial-streak and bulb-rot levels in Georgia. However, there is no formula that can be extrapolated to all soil types or onion cultivars. Growers afflicted with onion bacterial diseases will have to rely on their local extension agents to determine empirically the correct fertility-management strategy for their soil type and cultivars.

Reduction of leaf wounding by reliable insect control, particularly against thrips, is essential to disease reduction. Avoidance of mechanical wounding by increasing the clearance between plant leaves and farm machinery during the growing season would also be beneficial.

A programme that combines all of these measures, i.e. inspection and use of clean transplants, control of insects and weeds, use of the proper levels of fertilizer and sprays with a fixed copper plus maneb, is the most effective control strategy to use in the field at present. In addition, postharvest rots can be reduced by harvesting onions at the proper stage of maturity, allowing them to cure prior to topping, avoiding rough handling that could cause wounds or bruises and drying the bulbs with forced hot air.

Little is known about the control of centre rot, as it is a relatively new problem in onion. Control strategies used for bacterial diseases in general are currently recommended. Bactericidal sprays with fixed coppers should reduce epiphytic survival on onion leaves and control secondary dissemination. However, as is the case with most bacterial pathogens, copper-tolerant strains can develop. Thus, the inclusion of an ethylene bisdithiocarbamate (EBDC) fungicide, such as maneb, is recommended. The bacterium has been found on numerous weeds, so weed control may be beneficial. Likewise, good insect control and the avoidance of physical damage to plants with machinery should also be helpful.

Development of cultivars resistant to a range of bacterial pathogens as part of an integrated management strategy may ultimately be the answer to controlling or reducing these pathogens in onion. However, to date there are only limited reports of resistance to onion bacterial pathogens. Several cultivars have displayed a high level of tolerance to centre rot. O'Garro and Paulraj (1997) reported resistance to *X. campestris* in two onion cultivars (H-942 and H-508).

An integrated management strategy is necessary to control most bacterial diseases, and that is true for bacterial streak and bulb rot of onion. Efficient weed control is necessary both in seed-beds and production fields. Reduction of weed populations with post-emergence-type herbicides has had a beneficial effect in disease control by reducing the levels of initial inoculum.

Editors' Note

Within Europe, diagnostic kits for the identification of some onion bacterial pathogens are available from Adgen Ltd, UK. The catalogue is available on request from info@adgen.co.uk.

References

Abe, K., Watabe, S., Emori, Y., Watanabe, M. and Arai, S. (1989) An ice nucleation active gene of *Erwinia ananas*. Sequence similarity to those of *Pseudomonas* species and regions required for ice nucleation activity. *FEBS Letters* 258, 297–300.

Adams, G.A. and Martin, S.M. (1964) Extracellular polysaccharides of *Serratia marcescens*. *Canadian Journal of Biochemistry* 42, 1403–1413.

Alvarez, A.M., Buddenhagen, I.W., Buddenhagen, E.S. and Domen, H.Y. (1978) Bacterial blight of onion. A new disease caused by *Xanthomonas* sp. *Phytopathology* 68, 1132–1136.

Ballard, R.W., Palleroni, N.J., Doudoroff, M., Stanier, R.Y. and Mandel, M. (1970) Taxonomy of the aerobic *pseudomonads*: *Pseudomonas cepacia*, *P. marginata*, *P. alliicola*, *P. caryophylli*. *Journal of Genetic Microbiology* 60, 199–124.

Barsomian, G. and Lessie, T.G. (1986) Replicon fusions promoted by insertion sequences on *Pseudomonas cepacia* plasmid pTGL 6. *Molecular and General Genetics* 204, 273–280.

Baxter, I.A., Lambert, P.A. and Simpson, I.N. (1997) Isolation from clinical sources of *Burkholderia cepacia* possessing characteristics of *Burkholderia gladioli*. *Journal of Antimicrobial Chemotherapy* 39, 169–175.

Bevivino, A., Tabacchioni, S., Chiarini, L., Carusi, M.V., Del Gallo, M. and Visca, P. (1997) Phenotypic comparison between rhizosphere and clinical isolates of *Burkholderia cepacia*. *Microbiology* 140, 1069–1077.

Billing, E. (1970) *Pseudomonas viridiflava* (Burkholder, 1930; Clara, 1934). *Journal of Applied Bacteriology* 33, 492–500.

Biolog Inc. (1990) *Use of the Biolog System – Helpful Tips*. Biolog Inc., Hayward, California.

Bochner, B.R. (1989a) Sleuthing out bacterial identities. *Nature* 339, 157–158.

Bochner, B.R. (1989b) Breathprints at the microbiological level. *ASM News* 55, 536–539.

Bradbury, J.F. (1986a) Description of *Erwinia ananas*. In: *Guide to Plant Pathogenic Bacteria*. CAB International Mycological Institute, Kew, UK, p. 63.

Bradbury, J.F. (1986b) Description of *Pseudomonas viridiflava*. In: *Guide to Plant Pathogenic Bacteria*. CAB International Mycological Institute, Kew, UK, pp. 183–184.

Brewster, J.L. (1994) *Onions and Other Vegetable Alliums*. CAB International, Wallingford, UK, 236 pp.

Bruton, B.D., Wells, J.M., Lester, G.E. and Patterson, C.L. (1991) Pathogenicity and characterization of *Erwinia ananas* causing a postharvest disease of cantaloupe fruit. *Plant Disease* 75, 180–183.

Bryan, B.A., Linhardt, R.J. and Daniels, L. (1986) Variation in composition and yield of exopolysaccharide produced by *Klebsiella* serotype K32 and *Acinetobacter calcoaceticus* BD4. *Applied and Environmental Microbiology* 51, 1304–1308.

Burkholder, W.H. (1930) *The Bacterial Diseases of Bean: A Comparative Study*. Memoir No. 127, Cornell University Agricultural Experimental Station, Ithaca, New York, 88 pp.

Burkholder, W.H. (1942) Three bacterial plant pathogens: *Phytomonas caryophylli* sp. n., *Phytomonas alliicola* sp. n., and *Phytomonas manihotis* (Arthaud-Berthet et Bondar) Viegas. *Phytopathology* 32, 141–149.

Burkholder, W.H. (1950) Sour skin, a bacterial rot of onion bulbs. *Phytopathology* 40, 115–117.

Carson, L.A., Anderson, R.L., Panlilio, A.L., Beck-Saque, C.M. and Miller, J.M. (1991) Isoenzyme analysis of *Pseudomonas cepacia* as an epidemiological tool. *American Journal of Medicine* Suppl. 3B, 252–255.

Cheng, G.Y., Legard, D.E., Hunter, J.E. and Burr, T.J. (1989) Modified bean pod assay to detect strains of *Pseudomonas syringae* pv. *syringae* that cause bacterial brown spot of snap bean. *Plant Disease* 73, 419–423.

Cheng, H.P. and Lessie, T.G. (1994) Multiple replicons constituting the genome of *Pseudomonas cepacia* 17616. *Journal of Bacteriology* 176, 4034–4042.

Choi, J.E. and Han, K.S. (1990) Studies on the bacterial soft rot disease of *Liliaceae* crops in Korea. 4: Bacterial bulb rot of onion caused by *Pseudomonas* spp. *Korean Journal of Plant Pathology* 6, 358–362.

Clara, F.M. (1934) *A Comparative Study of the Green-fluorescent Bacterial Plant Pathogens*. Memoir No. 159, Cornell University Agricultural Experimental Station, Ithaca, New York, 34 pp.

Clode, F.E., Kaufman, M.E., Malnick, H. and Pitt, T.L. (1999). Evaluation of three oligonucleotide primer sets in PCR for the identification of *Burkholderia cepacia* and their differentiation from *Burkholderia gladioli*. *Journal of Clinical Pathology* 52, 173–176.

Collmer, A. and Keen, N.T. (1986) The role of pectic enzymes in plant pathogenesis. *Annual Review of Phytopathology* 24, 383–409.

Cother, E.J. and Dowling, V. (1985) Association of *Pseudomonas cepacia* with internal breakdown of onion – a new record for Australia. *Australian Journal of Plant Pathology* 14, 10–12.

Daengsubha, W. and Quimio, A.J. (1980) Vegetable soft rot bacteria in the Philippines. In: *Proceedings of the Second Southeast Asian Symposium on Plant Diseases in the Tropics, 20–26 October 1980, Bangkok, Thailand*. Kasetsart University, Bangkok, p. 109.

Darrasse, A. and Bertheau, Y. (1994) Detection and identification of pectolytic *Erwinias* using PCR and RFLP. In: *Plant Pathogenic Bacteria, Proceedings of the 8th International Congress, 9–12 June 1992, Versailles, France*. INRA, Paris, p. 427.

Daser, S.E., LiPuma, J.J., Kostman, J.R. and Stull, T.L. (1994) Characterization of PCR-ribotyping for *Burkholderia (Pseudomonas) cepacia*. *Journal of Clinical Microbiology* 32, 2422–2424.

Davis, R.M. (1995) Sour skin. In: Schwartz, H.F. and Mohan, S.K. (eds) *Compendium of Onion and Garlic Diseases*. American Phytopathological Society Press, St Paul, Minnesota, pp. 32–33.

DeBoer, S.H. and McNaughton, M.E. (1987) Monoclonal antibodies to the liposaccharide of *Erwinia carotovora* subsp. *atroseptica* serogroup I. *Phytopathology* 77, 828–832.

DeBoer, S.H., Allan, E. and Kelman, A. (1979) Survival of *Erwinia carotovora* in Wisconsin soils. *American Potato Journal* 56, 243–252.

DeBoer, S.H., Wieczorek, A. and Kummer, A. (1988) An ELISA test for bacterial ring rot of potato with a new monoclonal antibody. *Plant Disease* 72, 874–878.

DiCello, F., Bevivino, A., Chiarini, L., Fani, R., Paffetti, D., Tabacchioni, S. and Dalmastri, C. (1997) Biodiversity of a *Burkholderia cepacia* population isolated from the maize rhizosphere at different plant growth stages. *Applied Environmental Microbiology* 63, 4485–4493.

Dittapongpitch, V. and Daengsubha, W. (1980) Onion rot bacteria in Thailand. In: *Proceedings of the Second Southeast Asian Symposium on Plant Diseases in the Tropics, 20–26 October 1980, Bangkok, Thailand*. Kasetsart University, Bangkok, p. 116.

Eayre, C.G., Bartz, J.A. and Concelmo, D.E. (1995) Bacteriophages of *Erwinia carotovora* and *Erwinia ananas* isolated from freshwater lakes. *Plant Disease* 79, 801–804.

El-Banoby, F.E., Rudolph, K. and Mendgen, K. (1981) The fate of extracellular polysaccharides from *Pseudomonas phaseolicola* in leaves and leaf extracts from halo-blight susceptible and resistant bean plants (*Phaseolus vulgaris* L.). *Physiological Plant Pathology* 18, 91–98.

Füstös, Z. and Szarka, J. (1985) Reactions of onion varieties to bacterial infection and the inheritance of resistance. In: Darvas, B., Szentesi, A. and Viranyi, F. (eds) *Proceedings of the 1983 International Conference on Integrated Plant Protection*. Budapest, Hungary, pp. 106–110.

Gadzinski, P. (1990) *Microlog 2N™ Instruction Manual. Release 2.00*. ©Biolog Inc., Hayward, California.

Gaffney, T.D. and Lessie, T.G. (1987) Insertion-sequence-dependent rearrangement of *Pseudomonas cepacia* plasmid pTGL1. *Journal of Bacteriology* 169, 224–230.

Garibaldi, A. and Bateman, D.F. (1971) Pectic enzymes produced by *Erwinia chrysanthemi* and their effects on plant tissue. *Physiological Plant Pathology* 1, 25–40.

Gelbart, S.M., Reinhardt, G.F. and Greenlee, H.B. (1976) *Pseudomonas cepacia* strains isolated from water reservoirs of untreated nebulizers. *Journal of Clinical Microbiology* 3, 62–66.

Gilardi, G.L. (1983) *Pseudomonas cepacia*: culture and laboratory identification. *Laboratory Management* 21, 29–32.

Gitaitis, R.D. and Beaver, R.W. (1990) Characterization of fatty acid methyl ester content of *Clavibacter michiganensis* subsp. *michiganensis*. *Phytopathology* 80, 318–321.

Gitaitis, R.D. and Gay, J.D. (1997) First report of a leaf blight, seed stalk rot, and bulb decay of onion by *Pantoea ananas* in Georgia. *Plant Disease* 81, 1096.

Gitaitis, R.D., Baird, R.E., Beaver, R.W., Sumner, D.R., Gay, J.D. and Smittle, D.A. (1991) Bacterial blight of sweet onion caused by *Pseudomonas viridiflava* in Vidalia, Georgia. *Plant Disease* 75, 1180–1182.

Gitaitis, R., Sumner, D., Gay, D., Smittle, D., McDonald, G., Maw, B., Johnson, W.C., III, Tollner, B. and Hung, Y. (1997) Bacterial streak and bulb rot of onion: I. A diagnostic medium for the semiselective isolation and enumeration of *Pseudomonas viridiflava*. *Plant Disease* 81, 897–900.

Gitaitis, R., McDonald, G., Torrance, R., Hartley, R., Sumner, D.R., Gay, J.D. and Johnson, W.C. (1998a) Bacterial streak and bulb rot of onion: II. Epiphytic survival of *Pseudomonas viridiflava* in association with multiple weed hosts. *Plant Disease* 82, 935–938.

Gitaitis, R.D., Sanders, F.H. and Walcott, R.R. (1998b) Polymerase chain reaction and ELISA for the identification of *Burkholderia cepacia* and *Pseudomonas viridiflava*, bacterial pathogens of onion. In: *Papers and Abstracts*, Vol. 3, *7th International Congress of Plant Pathology, Edinburgh, UK, 9–16 August 1998*. Paper number 3.3.33. Electronic publication at: http://www.bspp.org.uk/icpp98/abstracts/3.3/33.html

Gonzalez, C.F. and Vidaver, A.K. (1979) Bacteriocin, plasmid and pectolytic diversity in *Pseudomonas cepacia* of clinical and plant origin. *Journal of Genetic Microbiology* 110, 161–170.
Gonzalez, C.F., Pettit, E.A., Valadez, V.A. and Provin, E.M. (1997) Mobilization, cloning and sequence determination of a plasmid-encoded polygalacturonase from a phytopathogenic *Burkholderia (Pseudomonas) cepacia*. *Molecular Plant–Microbe Interactions* 10, 840–851.
Graham, D.C. (1962) Blackleg disease of potatoes. *Scottish Agriculture* 41, 211–215.
Gross, D.C. and Cody, V.S. (1985) Mechanisms of plant pathogenicity by *Pseudomonas* species. *Canadian Journal of Microbiology* 31, 403–410.
Gurtler, V. and Stanisich, V.A. (1996) New approaches to typing and identification of bacteria using the 16S–23S rDNA spacer region. *Microbiology* 142, 3–16.
Hagedorn, C., Gould, W.D., Bardinelli, D.R. and Gustavson, D.R. (1987) A selective medium for enumeration and recovery of *Pseudomonas cepacia* biotypes from soil. *Applied and Environmental Microbiology* 53, 2265–2268.
Hattingh, M.J. and Walters, D.F. (1981) Stalk and leaf necrosis of onion caused by *Erwinia herbicola*. *Plant Disease* 65, 615–618.
Hekman, W.E., Heijnen, C.E., Burgen, S.L.G.E., van Veen, J.A. and van Elsas, J.D. (1994) Transport of bacterial inoculants through intact cores of two different soils as affected by water percolation and presence of wheat plants. *FEMS Microbiology Ecology* 16, 143–158.
Hendrickson, W., Hübner, A. and Kavanaugh-Black, A. (1996) Chromosome multiplicity in *Burkholderia cepacia*. In: Nakazawa, T., Furukawa, K., Haas, D. and Silver, S. (eds) *Molecular Biology of Pseudomonads*. American Society of Microbiology, Washington, DC, pp. 259–269.
Henry, D.A., Campbell, M.E., LiPuma, J.J. and Speert, D.P. (1997) Identification of *Burkholderia cepacia* isolates from patients with cystic fibrosis and the use of a simple new selective medium. *Journal of Clinical Microbiology* 35, 614–619.
Heydari, A., Misaghi, I.J. and McCloskey, W.B. (1997) Effects of three soil-applied herbicides on population of plant disease suppressing bacteria in the cotton rhizosphere. *Plant and Soil* 195, 75–81.
Hildebrand, D.C. (1971) Pectate and pectin gels for differentiation of *Pseudomonas* spp. and other bacterial plant pathogens. *Phytopathology* 61,1430–1436.
Hildebrand, D.C. and Schroth, M.N. (1972) Identification of the fluorescent pseudomonads. In: Geesteranus, H.P.M. (ed.) *Proceedings of the 3rd International Conference on Plant Pathogenic Bacteria, Wageningen, 14–21 April 1971*. Centre for Agricultural Publishing and Documentation, Wageningen, The Netherlands, pp. 281–287.
Hunter, J.E. and Cigna, J.A. (1981) Bacterial blight incited in parsnip by *Pseudomonas marginalis* and *Pseudomonas viridiflava*. *Phytopathology* 71, 1238–1241.
Hutchinson, R.G., Parker, S., Pryor, J.A., Duncan-Skingle, F., Hoffman, P.N., Hodson, M.E., Kaufman, M.E. and Pitt, T.L. (1996) Home use nebulizers: a potential primary source of *Burkholderia cepacia* and other colistin resistant, Gram-negative bacteria in patients with cystic fibrosis. *Journal of Clinical Microbiology* 34, 584–587.
Isakeit, T., Miller, M.E., Barnes, L.W., Dickstein, E.R. and Jones, J.B. (2000) First report of leaf blight of onion caused by *Xanthomonas campestris* in the continental United States. *Plant Disease* 84, 201.
Janisrewicz, W.J. and Roitman, J. (1988) Biological control of blue and gray mold on apple and pear with *Pseudomonas cepacia*. *Phytopathology* 78, 1697–1700.
Jones, J.B., Jones, J.P., McCarter, S.M. and Stall, R.E. (1984) *Pseudomonas viridiflava*: causal agent of bacterial leaf blight of tomato. *Plant Disease* 68, 341–342.
Jones, J.B., Gitaitis, R.D. and McCarter, S.M. (1986) Fluorescence on single-carbon sources for separation of *Pseudomonas syringae* pv. *syringae*, *P. syringae* pv. *tomato* and *P. viridiflava* on tomato transplants. *Plant Disease* 70, 151–153.
Jones, J.E. (1981) Postharvest losses in Barbados. In: *Report of a Consultative Meeting on Post Harvest Losses in the Caribbean, 19–24 July 1981, St Augustine, Trinidad and Tobago*, Vol. 2. Commonwealth Secretariat, London, pp. 157–164.
Jovanovic, O. (1998) Pathogenic and biochemical–physiological characteristics of the bacteria group '*Erwinia carotovora*' of different origin. Short version of a PhD thesis, Faculty of Agriculture, University of Belgrade, Yugoslavia. *Review of Research Work at the Faculty of Agriculture* 43(1), 7–24.
Karpati, F. and Jonasson, J. (1996). Polymerase chain reaction for the detection of *Pseudomonas aeruginosa*, *Stenotrophomonas maltophilia* and *Burkholderia cepacia* in the sputum of patients with cystic fibrosis. *Molecular and Cellular Probes* 10, 397–403.
Kawamoto, S.O. (1966) Studies of bacteria associated with decayed onions. MS thesis, Department of Plant Pathology, Cornell University, Ithaca, New York, USA.

Kawamoto, S.O. and Lorbeer, J.W. (1972a) Multiplication of *Pseudomonas cepacia* in onion leaves. *Phytopathology* 62, 1263–1265.

Kawamoto, S.O. and Lorbeer, J.W. (1972b) Histology of onion leaves infected with *Pseudomonas cepacia*. *Phytopathology* 62, 1266–1271.

Kawamoto, S.O. and Lorbeer, J.W. (1974) Infection of onion leaves by *Pseudomonas cepacia*. *Phytopathology* 64, 1440–1445.

Kidambi, S.P., Sundin, G.W., Palmer, D.A., Chakrabarty, A.M. and Bender, C.L. (1995) Copper as a signal for alginate synthesis in *Pseudomonas syringae* pv. *syringae*. *Applied and Environmental Microbiology* 61, 2122–2179.

Kikumoto, T. and Sakamoto, M. (1970) Ecological studies on the soft rot bacteria of vegetables X. The distribution of soft rot bacteria within the soil aggregates. *Annals of the Phytopathological Society of Japan* 36, 207–213.

King, E.B. and Parke, J.L. (1996) Population density of the biocontrol agent *Burkholderia cepacia* AMMDR1 on four pea cultivars. *Soil Biology and Biochemistry* 28, 307–312.

King, E.O., Ward, M.K. and Raney, D.E. (1954) Two simple media for the demonstration of pyocyanin and fluorescin. *Journal of Laboratory and Clinical Medicine* 44, 301–307.

Klement, Z. (1963) Methods for the rapid detection of the pathogenicity of phytopathogenic *Pseudomonads*. *Nature* 199, 299–300.

Klement, Z. and Lovrekovich, L. (1962) Studies on host–parasite relations in bean pods infected with bacteria. *Phytopathologische Zeitschrift* 45, 81–88.

Kori, Y., Furuya, N., Tsuno, K. and Matsuyama, N. (1992) Differentiation of *Erwinia chrysanthemi* and *E. carotovora* subsp. *carotovora* by cellular fatty acid analysis. *Journal of the Faculty of Agriculture, Kyushu University* 37, 173–178.

Kostman, J.R., Edlind, T.D., LiPuma, J.J. and Stull, T.Z. (1992) Molecular epidemiology of *Pseudomonas cepacia* determined by polymerase chain reaction ribotyping. *Journal of Clinical Microbiology* 30, 2084–2087.

Leach, J.G. (1964) Observations on cucumber beetles as vectors of cucurbit wilt. *Phytopathology* 54, 606–607.

Lelliott, R.A. and Dickey, R.S. (1984) Genus VII. *Erwinia* Winslow, Broadhurst, Buchanan, Krumwiede, Rogers, Smith 1920, 209AL. In: Krieg, N.R. and Holt, J.G. (eds) *Bergey's Manual of Systematic Bacteriology*, Vol. 1. Williams and Wilkins, Baltimore, Maryland, pp. 469–476.

Lelliott, R.A., Billing, E. and Hayward, A.C. (1966) A determinative scheme for the fluorescent plant pathogenic pseudomonads. *Journal of Applied Bacteriology* 29, 470–489.

Lessie, T.G. and Gaffney, T.D. (1986) Catabolic potential of *Pseudomonas cepacia*. In: Sokatch, J.R. and Ornston, L.N. (eds) *The Bacteria, A Treatise on Structure and Function*, Vol. 10. *The Biology of* Pseudomonas. Academic Press, New York, pp. 439–476.

Lessie, T.G. and Manning, B.D. (1995) Chromosomal multiplicity in *Burkholderia cepacia*. Electronic publication at: http://inbiap.biochem.vt.edu/brarg/brasym95/lessie 95.htm

Lessie, T.G., Wood, M.S., Byrne, A. and Ferrante, A. (1990) Transposable gene-activating elements in *Pseudomonas cepacia*. In: Silver, S., Chakrabatry, A.M., Iglewski, B. and Kaplan, S. (eds) Pseudomonas *Biotransformation, Pathogenesis and Evolving Biotechnology*. American Society for Microbiology, Washington, DC, pp. 279–291.

Lessie, T.G., Hendrickson, W., Manning, B.D. and Devereux, R. (1996) Genomic complexity and plasticity of *Burkholderia cepacia*. *FEMS Microbiology Letters* 144, 117–128.

LiPuma, J.J., Dulaney, B.J., McMenamin, J.D., Whitby, P.W., Stull, T.L., Coenye, T. and Vandamme, P. (1999) Development of rRNA-based PCR assays for the identification of *Burkholderia cepacia* complex isolates recovered from cystic fibrosis patients. *Journal of Clinical Microbiology* 327, 3167–3170.

Logan, C. (1968) The survival of the potato blackleg pathogen overwinter. *Records of Agricultural Research, Ministry for Agriculture in Northern Ireland, UK* 17, 115–121.

Lorbeer, J.W. (1996) Recent occurrences of bacterial bulb decay caused by *Erwinia chrysanthemi*. In: *Proceedings of the 1995 National Onion Research Conference. Wisconsin Center, Madison, Wisconsin, USA, 6–9 December*. University of Wisconsin, Madison, Wisconsin, pp. 135–139.

Lorbeer, J.W., LoParco, D.P. and Zumoff, C.H. (1996) Occurrence of a bacterial bulb decay of onion caused by *Erwinia chrysanthemi* in New York. *Phytopathology* 86(11) (Suppl.), S3 (abstract).

Lorbeer, J.W., Gundersheim, N.A. and Mark, G.L. (1998) Present status of bacterial diseases of onion and their control in New York. In: *Proceedings of the 1998 National Onion (and other Allium) Research Conference. 10–12 December, Sacramento, California, USA*. University of California, Davis, California, pp. 209–213.

Lukezic, F.L., Leath, K.T. and Levine, R.G. (1983) *Pseudomonas viridiflava* associated with root and crown rot of alfalfa and wilt of birdsfoot trefoil. *Plant Disease* 67, 808–811.

Lumsden, R.D., Bowie, M.D., Sasser, M. and Newark, D.E. (1986) Medium for isolation of *Pseudomonas cepacia* biotype from soil and the isolated biotype. US Patent number 4,588,584. Issued 13 May 1986, US Patent Office, Washington, DC.

McArthur, J.V., Kovacic, D.A. and Smith, M.H. (1988) Genetic diversity in natural populations of a soil bacterium across a landscape gradient. *Proceedings of the National Academy of Sciences of the USA* 85, 9621–9624.

McClendon, J.H. (1964) Evidence for the pectic nature of the middle lamella of potato tuber cell walls based on the chromatography of macerating enzymes. *American Journal of Botany* 51, 628–633.

Maki-Valkama, T. and Karjalainen, R. (1994) Differentiation of *Erwinia carotovora* subsp. *atroseptica* and *carotovora* by RAPD-PCR. *Annals of Applied Biology* 125, 301–309.

Manrique, G., Jose, M. and Fuciovsky-Zak, L. (1991) Soft-rot of onion (*Allium cepa*) caused by bacteria at Monticello, Mexico State. *Agrociencia (Mexico), Serie Proteccion Vegetal (Technical Note)* 2, 129–135.

Mark, G.L., Gundersheim, N.A. and Lorbeer, J.W. (1999a) Sensitivity of a range of phenotypically variable *Burkholderia cepacia* isolates to copper and non copper-based bactericides. *Phytopathology* 89 (Suppl.), S48 (abstract).

Mark, G.L., Lorbeer, J.W. and Gundersheim, N.A. (1999b) Characterization and quantification of *Burkholderia cepacia* isolated from onions and organic soil previously cropped to onions. *Phytopathology* 89 (Suppl.), S48 (abstract).

Mergaert, J., Verdonck, L. and Kersters, K. (1993) Transfer of *Erwinia ananas* (synonym, *Erwinia uredovora*) and *Erwinia stewartii* to the genus *Pantoea* emend. as *Pantoea ananas* (Serrano 1928) comb. nov. and *Pantoea stewartii* (Smith 1898) comb. nov., respectively, and description of *Pantoea stewartii* subsp. *indologenes* subsp. nov. *International Journal of Systematic Bacteriology* 43, 162–173.

Meyers, E., Bisacchi, G.S., Dean, L., Liu, W.G., Mirassian, B., Shisarchyk, D.S., Sykes, R.B., Taraka, S.K. and Trejo, W. (1987) Xylocardin: a new complex antifungal peptide. I. Taxonomy, isolation and biological activity. *Journal of Antibiotics* 40, 1515–1519.

Michaux, S., Pallison, J., Carles-Nuit, M.J., Borg, G., Allandet-Servent, A. and Ramuzz, M. (1993) Presence of two independent chromosomes in *Brucella melitensis* 16M genome. *Journal of Bacteriology* 175, 701–705.

Miller, L. and Berger, T. (1985) *Bacteria Identification by Gas Chromatography of Whole Cell Fatty Acids*. Hewlett-Packard Gas Chromatography Application Note 228–238, Hewlett-Packard Co., Palo Alto, California.

Mohan, S.K. (1995) Soft-rot, slippery skin and other bacterial diseases. In: Schwartz, H.F. and Mohan, S.K. (eds) *Compendium of Onion and Garlic Diseases*. American Phytopathological Society Press, St Paul, Minnesota, p. 32.

Morales, N., Vietinghoff, J., de los Angeles Péres, M., de la Rosa, N.R., Moreno, A., Cuello, I. and Seidel, D. (1994) *Erwinia herbicola* – a new pathogen of the onion seed production (*Allium cepa* L.) in Cuba. *Archives of Phytopathology and Plant Protection (Germany)* 29, 29–40.

Moss, C.W., Dees, S.B. and Guerrant, G.O. (1980) Gas–liquid chromatography of bacterial fatty acids with a fused-silica capillary column. *Journal of Clinical Microbiology* 12, 127–130.

Nacamulli, C., Bevivino, A., Dalmastri, C., Tabacchioni, S. and Chiarini, L. (1997) Perturbation of maize rhizosphere microflora following seed bacterization with *Burkholderia cepacia* MCI7. *FEMS Microbiology Ecology* 23, 183–193.

Nassar, A., Celli, J., Darrasse, A., Lemattre, M. and Bertheau, Y. (1991) Detection of *Erwinia chrysanthemi* by PCR (polymerase chain reaction). *Phytoma – La Défense des Végétaux, France* 430, 32.

Nassar, A., Lemattre, M. and Bertheau, Y. (1994) Studies of the host specificity of *Erwinia chrysanthemi* using PCR and RFLP methods. In: *Plant Pathogenic Bacteria, 8th International Conference, 9–12 June 1994, Versailles, France*. INRA, Paris, p. 428.

Nassar, A., Darrasse, A., Lemattre, M., Kotoujansky, A., Dervin, C., Vedel, R. and Bertheau, Y. (1996) Characterization of *Erwinia chrysanthemi* by pectinolytic isozyme polymorphism and restriction fragment length polymorphism analysis of PCR-amplified fragments of *pel* genes. *Applied and Environmental Microbiology* 62, 2228–2235.

O'Garro, L.W. and Paulraj, L.P. (1997) Onion leaf blight caused by *Xanthosomas campestris*: alternative hosts and resistant onion genotypes. *Plant Disease* 81, 978–982.

Palleroni, N.J. (1984) Gram-negative aerobic rods and cocci. Family I *Pseudomonadaceae* Winslow, Broadhurst, Buchanan, Krumwrede, Rogers, and Smith 1917555AL. In: King, N.R. and Holts, J.E. (eds) *Bergey's Manual of Systematic Bacteriology*, Vol. 1. Williams and Williams, Baltimore, Maryland, pp. 141–219.

Palleroni, N.J. (1992) Present situation in the taxonomy of aerobic *Pseudomonads*. In: Galli, E., Silver, S. and Witholt, B. (eds) Pseudomonas: *Molecular Biology and Biotechnology*. American Society for Microbiology, Washington, DC, pp. 105–115.

Palleroni, N.J. and Holmes, B. (1981) *Pseudomonas cepacia* sp. nov. nom. rev. *International Journal of Systematic Bacteriology* 31, 479–481.

Parker, W.L., Rathnum, M.L., Seiner, V., Trejo, W.H., Principe, P.A. and Sykes, R.B. (1984) Cepacin A and cepacin B, two new antibiotics produced by *Pseudomonas cepacia*. *Journal of Antibiotics* 37, 431–440.

Paulraj, L. and O'Garro, L.W. (1993) Leaf blight of onions in Barbados caused by *Xanthomonas campestris*. *Plant Disease* 77, 198–201.

Pepper, E.H. (1967) *Stewart's Bacterial Wilt of Corn*. Monograph No. 4, American Phytopathological Society Press, St Paul, Minnesota, 36 pp.

Pérombelon, M.C.M. (1973). Studies on the epidemiology and etiology of black leg (*Erwinia carotovora* var. *atroseptica* (van Hall) Dye) of potato. PhD thesis, Dundee University, Dundee, UK.

Pérombelon, M.C.M. and Kelman, A. (1980) Ecology of the soft-rot *Erwinias*. *Annual Review of Phytopathology* 18, 361–387.

Pyle, B.H., Broadaway, S.C. and McFeters, G.A. (1992) Efficacy of copper and silver ions with iodine in the inactivation of *Pseudomonas syringae* pv. *syringae*. *Applied and Environmental Microbiology* 61, 2122–2179.

Roberts, P. (1973) A soft-rot of imported onions caused by *Pseudomonas alliicola* (Burkh.) Starr and Burkh. *Plant Pathology* 22, 98.

Rodley, P.D., Romling, V. and Tummler, B. (1995). A physical genome map of the *Burkholderia cepacia* type strain. *Molecular Microbiology* 17, 57–67.

Romantschuk, M., Roine, E., Ojanen, T., van Doorn, J., Louhelainen, J., Nurmiaho-Lassila, E.-L. and Haahtela, K. (1994) Fimbria (pilus) mediated attachment of *Pseudomonas syringae*, *Erwinia rhapontica* and *Xanthomonas campestris* to plant surfaces. In: Kado, C.I. and Crosa, J.H. (eds) *Molecular Mechanisms of Bacterial Virulence*. Kluwer Academic Press, Dordrecht, The Netherlands, pp. 67–77.

Rudolph, K.W.E., Gross, M., Ebrahim-Nesbat, F., Nöllenburg, M., Zomorodian, A., Wydra, K., Neingebauer, M., Hettwer, U., El-Showny, W., Sonnerbry, B. and Klement, Z. (1994). The role of extracellular polysaccharides as virulence factors for phytopathogenic *Pseudomonads* and *Xanthomonads*. In: Kado, C.I. and Crosa, J.H. (eds) *Molecular Mechanisms of Bacterial Virulence*. Kluwer Academic Publishers, Dordrecht, The Netherlands, pp. 357–378.

Ryley, H.C., Millar Jones, L., Paull, A. and Weeks, J. (1995) Characterization of *Burkholderia cepacia* from cystic fibrosis patients living in Wales by PCR ribotyping. *Journal of Medical Microbiology* 43, 4336–4341.

Sage, A., Linker, A., Evans, L.R. and Lessie, T.G. (1990) Hexose phosphate metabolism and exopolysaccharide formulation in *Pseudomonas cepacia*. *Current Microbiology* 20, 191–198.

Salmond, G.P.C. (1994) Factors affecting the virulence of soft-rot *Erwinia* species: the molecular biology of an opportunistic phytopathogen. In: Kado, C.I. and Crosa, J.H. (eds) *Molecular Mechanisms of Bacterial Virulence*. Kluwer Academic Press, Dordrecht, The Netherlands, pp. 193–206.

Sangodkar, V., Chapman, P. and Chakrabarty, A. (1988) Cloning, physical mapping and expression of chromosomal genes specifying degradation of the herbicide 2,4,5-T by *Pseudomonas cepacia* AC1100. *Gene* 71, 267–277.

Sasser, J.M., Miller, R.W. and Fieldhouse, D.J. (1970) Osmotic potential, a controlling factor in the development of bacterial spot disease. *Phytopathology* 60, 1311–1312 (abstract).

Sasser, J.M., Fieldhouse, D.J. and Carter, C.N. (1984) Computer assisted identification of bacteria based on fatty acid analysis. *Phytopathology* 74, 882 (abstract).

Sasser, M. (1990) Identification of bacteria through fatty acid analysis. In: Klement, Z., Rudolph, K. and Sands, D. (eds) *Methods in Phytobacteriology*. Akadémia Kiadó, Budapest, Hungary, pp. 199–204.

Sato, S. (1968) Enzymatic maceration of plant tissue. *Physiologia Plantarum* 21, 1067–1075.

Scheck, H.J. and Pscheidt, J.W. (1998) Effect of copper bactericides on copper resistant and sensitive strains of *Pseudomonas syringae* pv. *syringae*. *Plant Disease* 82, 397–407.

Schwartz, H.F. and Mohan, S.K. (1995) *Compendium of Onion and Garlic Diseases*. APS Press, St Paul, Minnesota, 54pp.

Schwartz, H.F. and Otto, K.J. (1998) Onion bacterial disease management in Colorado. In: *Proceedings of the 1998 National Onion (and Other Allium) Research Conference, 10–12 December, Sacramento, California, USA*. University of California, Davis, California, pp. 214–218.

Schwartz, H.F. and Otto, K.J. (2000a) First report of a leaf blight and bulb decay of onion by *Pantoea ananitas* in Colorado. *Plant Disease* 84, 808.

Schwartz, H.F. and Otto, K.J. (2000b) First report of a bulb decay of onion by *Enterobacter cloacae* in Colorado. *Plant Disease* 84, 808.

Schwartz, H.F. and Otto, K.J. (2000c) First report of a leaf blight of onion caused by *Xanthomonas campestris* in Colorado. *Plant Disease* 84, 922.

Shreve, M.R., Johnson, S.J., Milla, C.E., Wielinski, C.L. and Regelmann, W.E. (1997) PCR ribotyping and endonuclease subtyping in the epidemiology of *Burkholderia cepacia* infection. *American Journal of Respiratory Critical Care Medicine* 155, 984–989.

Singleton, P. (1999) *Bacteria in Biology, Biotechnology and Medicine*, 5th edn. John Wiley & Sons, New York, 489 pp.

Stanier, R.Y., Palleroni, N.J. and Doudoroff, M. (1966) The aerobic *Pseudomonads*, a taxonomic study. *Journal of Genetic Microbiology* 43, 159–271.

Stewart, F.C. (1899) A bacterial rot of onion. *New York (Geneva) Agricultural Experiment Station Bulletin* 164, 209–212.

Suslow, T.V. and McCain, A.H. (1981) Greasy canker of poinsettia caused by *Pseudomonas viridiflava*. *Plant Disease* 65, 513–514.

Taraka, T. and Tsuboki, K. (1982) A simple diagnosis of bacterial soft-rot (caused by *Erwinia carotovora*) of onion by pectate medium. *Bulletin of Hokkaido Prefectural Agricultural Experiment Stations* 48, 32–39.

Trüper, H.G. and De'Clari, L. (1997) Taxonomic note: necessary correction of specific epithets formed as substantives (nouns) 'in apposition'. *International Journal of Systematic Bacteriology* 47, 908–909.

Tsuchiya, K., Homma, Y., Komoto, Y. and Suzui, T. (1995) Practical detection of *Pseudomonas cepacia* from rhizosphere antagonistic to plant pathogens with a combination of selective medium and ELISA. *Annals of the Phytopathological Society of Japan* 62, 318–324.

Ulrich, J.M. (1975) Pectic enzymes of *Pseudomonas cepacia* and penetration of polygalacturonase into cells. *Physiological Plant Pathology* 5, 37–44.

Vandamme, P., Holmes, B., Vancanneyt, M., Coenye, T., Hoste, B., Coopman, R., Revets, H., Lauwers, S., Gillis, M., Kersters, K. and Govan, J.R.W. (1997) Occurrence of multiple genomovars of *Burkholderia cepacia* in cystic fibrosis patients and proposal of *Burkholderia multivorans* sp. nov. *International Journal of Systematic Bacteriology* 47, 1188–1200.

Voronkevitch, I.V. (1960) On the survival of the *Erwinia* genus bacteria, the inciters of soft-rots in plants. *Bulletin of the Moscow Society of Naturalists Biological Series* 65, 95–105.

Watanabe, K. and Sato, M. (1999) Gut colonization by an ice nucleation active bacterium, *Erwinia (Pantoea) ananas* reduces the cold hardiness of mulberry pyralid larvae. *Cryobiology* 38, 281–289.

Wells, J.M., Sheng, W.-S., Ceponis, M.J. and Chen, T.A. (1987) Isolation and characterization of strains of *Erwinia ananas* from honeydew melons. *Phytopathology* 77, 511–514.

Wells, J.M., Butterfield, J.E. and Revear, L.G. (1993) Identification of bacteria associated with post-harvest diseases of fruits and vegetables by cellular fatty acid composition: an expert system for personal computers. *Phytopathology* 83, 445–455.

Whittam, T.S. (1989) Clonal dynamics of *Escherichia coli* in its natural habitat. *Antonie van Leeuwenhoek* 55, 23–32.

Wilkie, J.P. and Dye, D.R.W. (1973) Further hosts of *Pseudomonas viridiflava*. *New Zealand Journal of Agricultural Research* 16, 315–323.

Willis, J.W., Engwall, J.K. and Chatterjee, A.K. (1987) Cloning of genes for *Erwinia carotovora* subsp. *carotovora* pectolytic enzymes and further characterization of the polygalacturonases. *Molecular Plant Pathology* 77, 1199–1205.

Wood, M.S., Lory, C. and Lessie, T.G. (1990) Activation of the *lac* genes of Tn *951* by insertion sequences from *Pseudomonas cepacia*. *Journal of Bacteriology* 172, 1719–1724.

Wright, P.J. and Hale, C.N. (1992) A field and storage rot of onion caused by *Pseudomonas marginalis*. *New Zealand Journal of Horticultural Science* 20, 435–438.

Yabuuchi, E., Kosako, Y., Oyaizu, H., Yano, I., Hotta, H., Hasimoto, Y., Ezaki, T. and Arakawa, M. (1992) Proposal of *Burkholderia* gen. nov. and transfer of seven species of the genus *Pseudomonas* homology group II to the new genus with type species *Burkholderia cepacia* (Palleroni and Holmes, 1981) Comb. Nov. *Microbiological Immunology* 36, 1251–1275.

Yáñez-Morales, M.J. and Lorbeer, J.W. (1993) Identification of *Erwinia chrysanthemi* causing a soft rot of onion bulb in the Huasteca region of Mexico. In: *Proceedings of the 20th National Conference of Phytopathology*. Mexican Society of Plant Pathology, Zacatecas, Mexico, p. 11 (abstract).

Yáñez-Morales, M.J., Fucikovsky, Z.L., Lorbeer, J.W. and Gonsález, J.A. (1994) Identification of bacteria causing soft rot of onion bulbs in commercial onion seeds in the east of the state of San Luis Potosi, Mexico. In: *Proceedings of the 21st National Congress of Phytopathology*. Mexican Society of Plant Pathology, Cuernavaca, Mexico, p. 52 (abstract).

Yao, F. and Lessie, T.G. (1998) Large replicons in different *B. cepacia* isolates. In: *Burkholderia*, friend or foe. Electronic publication at: APSnethttp://www.scisoc.org/feature/BurkholderiaCepacia/replicon.htm

Yohalem, D.S. and Lorbeer, J.W. (1994) Multilocus isoenzyme diversity among strains of *Pseudomonas cepacia* isolated from decayed onions, soils and clinical sources. *Systematic Applied Microbiology* 17, 116–124.

Yohalem, D.S. and Lorbeer, J.W. (1997) Distribution of *Burkholderia cepacia* phenotype by niche, method of isolation and pathogenicity to onion. *Annals of Applied Biology* 130, 467–479.

Zaitlin, M. and Coltrin, D. (1964) Use of pectic enzymes in a study of the nature of intercellular cement of tobacco leaf cells. *Plant Physiology* 39, 91–95.

Zuerner, R.L., Herrman, J.L. and Sainte Girons, I. (1993) Comparison of genetic maps for the two *Leptospira interrogans* serovars provides evidence for two chromosomes and intraspecies heterogeneity. *Journal of Bacteriology* 175, 5445–5451.

12 Monitoring and Forecasting for Disease and Insect Attack in Onions and *Allium* Crops within IPM Strategies

J.W. Lorbeer,[1] T.P. Kuhar[2] and M.P. Hoffmann[2]
[1]*Department of Plant Pathology;* [2]*Department of Entomology, Cornell University, Ithaca, NY 14853, USA*

1. Introduction

Onions and related *Allium* crops are subject to a variety of diseases and attack by arthropod pests that can reduce crop yield and quality. Growers in North America and Europe typically rely heavily on prophylactic applications of fungicides and insecticides to prevent damage that could lead to yield loss or crop rejection. This inevitably results in unnecessary use of pesticides. In New York, for instance, onions rank highest in pesticide use per unit area among vegetables and second among all crops (Anon., 1999). In Central America, onion growers apply foliar insecticides nine to 12 times each cropping season (Rueda, 2000).

Integrated pest management (IPM) is a sustainable approach to managing diseases and arthropod pests; it promotes the use of a variety of strategies and tactics, including pest-resistant varieties and biological, cultural and chemical controls, in a way that reduces costs and minimizes health and environmental risks. Decision-making is a key component of IPM programmes (Binns and Nyrop, 1992). Pest-management decision-making typically involves a procedure for assessing the pathogen and insect population levels, an economic threshold (pathogen and/or insect population levels at which control measures should be taken) and/or a phenological forecast to determine when to sample. Accurate knowledge of pathogen and pest biology, life history and interactions with factors within the agro-ecosystem is critical to disease and insect pest-management decision-making.

A number of tools and procedures have been developed for IPM decision-making in *Allium* crops (Chaput, 1993; Hoffmann *et al.*, 1996). These include forecasting systems for disease outbreaks, which incorporate various climatic and agronomic data (Vincelli and Lorbeer, 1988a, b, 1989), simple plant inspections (Shelton *et al.*, 1987; Nyrop *et al.*, 1989; Theunissen and Legutowska, 1992; Petzoldt, 1994) and sampling devices for pests (Coudriet *et al.*, 1979; Vincent and Stewart, 1981; Gerson *et al.*, 1985).

The ability to predict disease outbreaks or to estimate insect pest-population levels can lead to a more judicious use of pesticides. In New York, the use of the BLIGHT-ALERT forecasting scheme (Vincelli and Lorbeer, 1989) reduced fungicide use by up to 44% (Hoffmann and Petzoldt, 1993). Also in New York, insecticide applications were reduced by 52 and 38% in 1986 and 1993, respectively, in IPM demonstration fields where insect-scouting programmes were used (Hoffmann *et al.*, 1995). In Michigan, insecticide use on onions was decreased by more than 50%, primarily because most growers stopped applying foliar insecticides to control onion maggots (Hoffmann *et al.*, 1996). Even greater reductions in insecticide use occurred in Canada after a forecasting method for onion maggot was implemented (Andaloro *et al.*, 1984).

The benefits of using IPM in onion production include cost savings, accurate and early detection of diseases and insect pests before damage can occur, and reduced pesticide use. The latter is of particular importance because of the development of resistance problems in certain pests (Harris and Svec, 1976; Carroll *et al.*, 1983; Gangloff, 1999), the introduction of non-target effects of the chemicals (Carruthers *et al.*, 1984), fewer new chemicals being registered for use on onions, loss of existing products and increasing socio-environmental pressure against the use of pesticides.

2. Forecasting for Onion Diseases

In the states of New York, Michigan and Ohio in the USA and the provinces of Ontario and Quebec in Canada, fungicides have historically been applied on 7–10-day spray schedules to control *Botrytis* leaf blight (*Botrytis squamosa*), downy mildew (*Peronospora destructor*) and purple blotch (*Alternaria porri*) during mid-June to early September (Lorbeer, 1992, 1997a), the time frame most favourable for the occurrence of these foliage diseases (Lorbeer, 1992; Lacy and Lorbeer, 1995). If weather conditions favourable for an outbreak of the disease occur only late in the growing season, fungicide sprays are not needed until those climatic conditions prevail (Shoemaker and Lorbeer, 1977a). If dry weather conditions occur intermittently for extended periods after the critical disease level (CDL) of the disease has been reached (Shoemaker and Lorbeer, 1977a), fungicide sprays for control of *Botrytis* leaf blight are not needed during those periods.

Effective control of the above three foliage diseases allows otherwise healthy onion plants to produce maximum-sized bulbs of the cultivar grown under the prevailing environmental conditions. Although many onion growers continue to control these three diseases with fungicide spray schedules based on the calendar, during the past 15 years the development, testing and adoption of forecast systems for the occurrence of the diseases have allowed growers who embrace these systems to apply fungicide sprays only when needed to effectively manage the diseases. Depending on the weather patterns for each growing season, this generally results in a reduction of the number of sprays utilized compared with calendar-based and grower-determined spray schedules.

In addition to the three foliar diseases of onion mentioned above, a fourth leaf disease, *Stemphylium* leaf blight, caused by *Stemphylium vesicarium*, has become a serious disease at times in many onion fields throughout the world (Rao and Paugi, 1975; Miller *et al.*, 1978; Shishkoff and Lorbeer, 1989; Miller, 1995; Basallote-Ureba *et al.*, 1999). The fungus also attacks asparagus and pears and the considerable studies conducted to date (Falloon *et al.*, 1987; Montesinos *et al.*, 1995) have suggested the possibility of adapting forecasting systems for the disease. However, the nature of the disease and the biology of its pathogen in onions are not yet well understood, so forecast systems for the occurrence of *Stemphylium* leaf blight of onion have not yet been developed.

Cladosporium leaf blotch (*Cladosporium allii-cepae*) is another important leaf disease of onion and other *Allium* species (Hill, 1995), which has been reported to occur regularly in the British Isles. Although a series of studies have developed considerable information concerning the nature of the disease and the biology of the pathogen, a forecast system is not utilized for predicting the disease. Rather, control is achieved by utilizing cultural procedures and fungicide applications late in the growing season at intervals of 14 days or less.

2.1 *Botrytis* leaf-blight forecasting systems

BLIGHT-ALERT, developed in New York (Vincelli and Lorbeer, 1989), BOTCAST, in Ontario (Sutton *et al.*, 1984; Sutton, 1986), and the conidial release predictor system in Michigan (Lacy, 1991), are effective forecast systems for the occurrence of *Botrytis* leaf blight. When adopted, these systems allow growers to apply fungicides only when there is the possibility of outbreaks of the disease.

2.1.1 BLIGHT-ALERT

BLIGHT-ALERT is based on the monitoring of weather conditions regulating the formation and release of conidia of *B. squamosa* as well as their deposition and germination on onion leaves and subsequent penetration of the host tissues (Shoemaker and Lorbeer, 1977b; Vincelli and Lorbeer, 1988a). BLIGHT-ALERT also incorporates the prediction of incoming weather in the form of precipitation-probability forecasts for a 30% or more chance of rain during the next 36 h (Vincelli and Lorbeer, 1988b). The improvement in weather forecasting with state-of-the-art satellite systems may allow the arbitrary extension of this time frame to several days if desired. Precipitation probability has been effectively utilized in BLIGHT-ALERT in New York, since storms tracking across the USA into New York usually follow the path predicted rather than veering off course. BLIGHT-ALERT is also based on continued field scouting throughout the onion-growing season to first detect the CDL, which is an average of one lesion per leaf, for the disease (Shoemaker and Lorbeer, 1977a; Vincelli and Lorbeer, 1987). The CDL calls for the initiation of fungicide sprays and then a continued evaluation by field-scouting on a weekly or biweekly basis to determine the effectiveness of subsequent fungicide sprays mandated by the weather-based portion of BLIGHT-ALERT. When disease levels determined by field-scouting appreciably increase above the CDL at any time, a grower-determined fungicide spray can be applied to reduce the incidence of the disease to approach the CDL and afterwards the grower can continue to base the application of future fungicide sprays on BLIGHT-ALERT forecasts. After considerable testing (DeMilia, 1993; Lorbeer, 1997b; Anon., 1998; Lorbeer *et al.*, 2001), commercial implementation of BLIGHT-ALERT has now been achieved in New York. IPM personnel and/or a grower membership organization called the Northeast Weather Association operate the system. Weather data are automatically downloaded daily from a network of electronic weather monitors via phone lines. Software automatically runs the monitored weather data through a BLIGHT-ALERT analysis scheme and the prediction for the occurrence of *Botrytis* leaf blight for each individual farm enrolled in the programme is available on a daily basis through the Internet. A flow chart (DeMilia, 1993) for operating the BLIGHT-ALERT decision process is depicted in Fig. 12.1.

2.1.2 BOTCAST

The BOTCAST system also uses monitored weather to predict the occurrence of outbreaks of *Botrytis* leaf blight (Sutton *et al.*, 1978, 1983, 1986). This forecast system incorporates, as in BLIGHT-ALERT, the monitoring of temperature, leaf wetness, relative humidity and rainfall, but commences environmental monitoring at the time of crop emergence. Daily weather data are utilized to predict whether the fungus has sporulated, whether it has infected leaves and, if that is the case, the severity of the infection. The data gathered are utilized to compute a daily disease-severity index. These indices are then evaluated on a cumulative basis until a disease-warning level is reached and the need for a fungicide spray is determined (Sutton *et al.*, 1985; Sutton, 1986). Since the occurrence of *Botrytis* leaf blight in both Ontario and New York is characterized by a slow initial increase during the early part of the growing season, followed by an explosive stage (Sutton *et al.*, 1985; Sutton, 1986), which in BLIGHT-ALERT is determined as the CDL, fungicide sprays are not necessary until that level is reached. BOTCAST determines that occurrence on the basis of prior weather, as recorded from the time of crop germination, while BLIGHT-ALERT determines that occurrence by field-scouting. Once the CDL of the disease is reached, both BOTCAST and BLIGHT-ALERT are weather-based predictive systems. However, field-scouting for *Botrytis* leaf-blight monitoring always continues with BLIGHT-ALERT and can be implemented into the BOTCAST system as desired.

2.1.3 NEOGEN ENVIROCASTER

This weather-monitoring instrument incorporates a software package for predicting the airborne presence of inoculum of *B. squamosa* and thus predicts the occurrence of *Botrytis* leaf blight of onion when weather conditions favour the occurrence of the disease. The disease model and software package for *Botrytis* leaf blight incorporated in the NEOGEN ENVIROCASTER utilizes a *B. squamosa* conidial-release predictor (Lacy, 1991) developed in Michigan (Alderman and Lacy, 1983, 1984; Lacy and Pontius, 1983). The system consists of a portable stand-alone weather-monitoring instrument, PESTCASTER, manufactured by the Neogen Corporation in Lansing, Michigan, which can be placed anywhere within an onion field. The instrument assimilates weather data on a daily basis, automatically processes the data and then displays a forecast for the possibility of the occurrence of the disease.

Although the NEOGEN ENVIROCASTER is no longer manufactured, many units are still utilized throughout the USA, with disease models and software packages for predicting the occurrence of specific diseases on a number of different agricultural and horticultural crops, as well as for *Botrytis* leaf blight of onion. Disease models and software packages for predicting the possible occurrence of purple blotch and downy mildew of onion have also been incorporated into the NEOGEN ENVIROCASTER.

2.2 Downy-mildew forecasting systems

Onion plants infected by *Peronospora destructor* suffer heavy leaf damage and frequently also develop spongy necks, which usually cause the resulting onion bulbs to be discarded at harvest or to lack keeping quality when placed in storage (Schwartz, 1995).

2.2.1 DOWNCAST

DOWNCAST, developed in Canada, predicts sporulation–infection periods for the downy mildew fungus *P. destructor*, but does not predict the duration of infectivity of the pathogen (Sutton, 1986; Jesperson and Sutton, 1987). It is used to time scouting to detect the first appearance of downy mildew in onion crops. Successful timing of fungicide application in relation to sporulation–

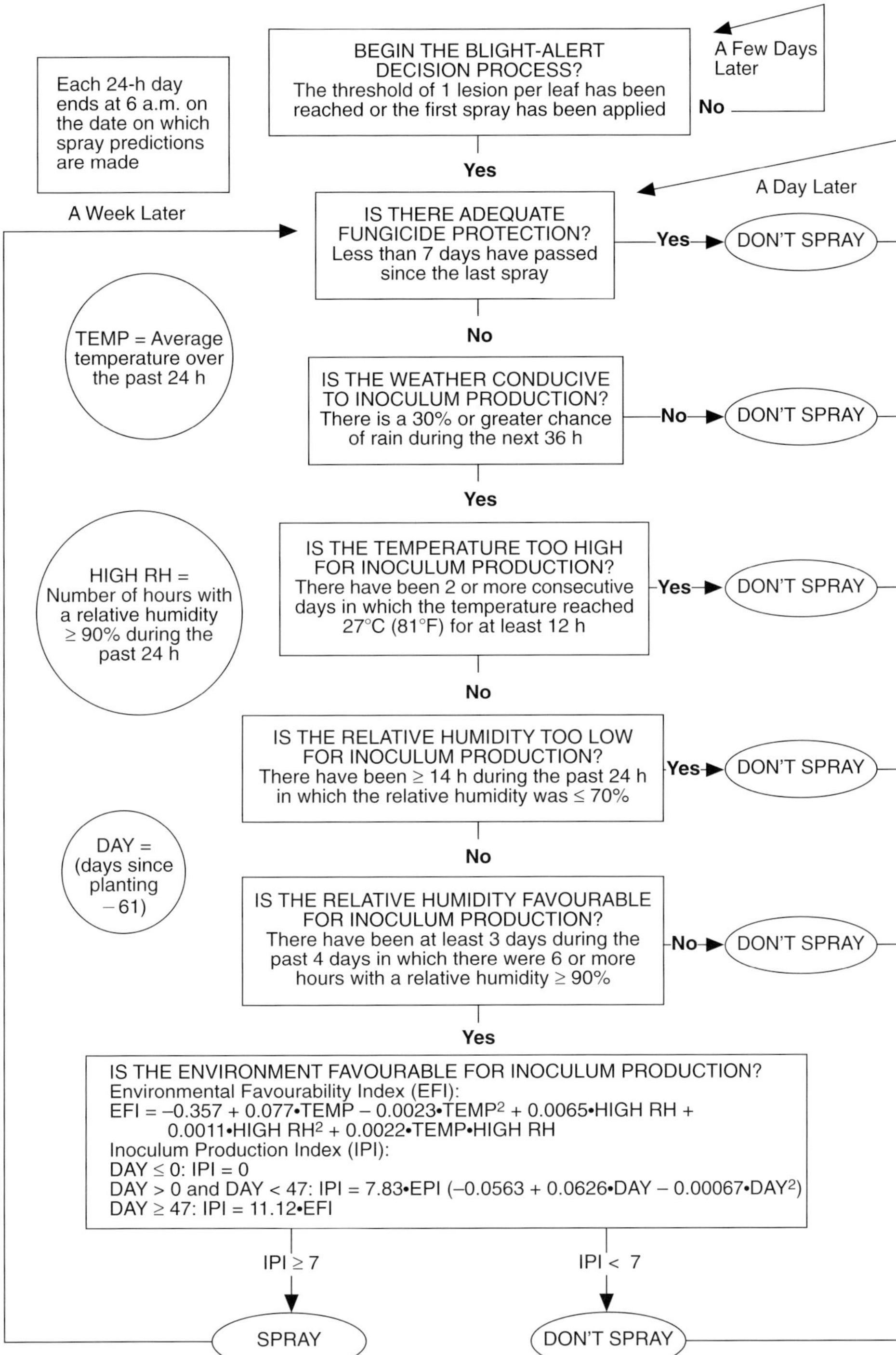

Fig. 12.1. Flow chart for the BRIGHT-ALERT decision process. Figure prepared by Dr Michael S. DeMilia. RH, relative humidity.

infection periods in the DOWNCAST system is critical to successful control of downy mildew. In Canada, it was determined that *P. destructor* sporulates at night under favourable environmental conditions and that infrared light from the rising sun and decreasing humidity during the morning hours trigger the release of the spores and their subsequent deposition on onion leaves (Leach *et al.*, 1982; Hildebrand and Sutton, 1984c). However, in tests on the DOWNCAST system in field trials in Holland (de Visser, 1998) and in the UK (T. Gilles, UK, 2001, personal communication), the model often failed to predict nights when sporulation occurred. Furthermore, DOWNCAST only predicts whether sporulation will occur or not, and does not predict it quantitatively. Currently, the effects of environmental factors on sporulation are being studied in more detail, with the aim of developing a quantitative model for sporulation (T. Gilles, UK, 2001, personal communication).

Leaf wetness must prevail for extended periods during the morning hours for downy-mildew infection to occur (Hildebrand and Sutton, 1984a). In Canada, brief 1–2 day periods of the sporulation–infection cycle are followed by 9–16 days of fungal growth in the leaves before another cycle of sporulation commences (Hildebrand and Sutton, 1984b). Thus, mildew outbreaks develop in stepwise increments of increasing levels of sporulation and disease presence. DOWNCAST allows for the detection of the first appearance of downy mildew on crops and, after that appearance, signals the need for effective fungicide spray schedules. Forecasts for the occurrence of downy mildew utilize the NEOGEN ENVIROCASTER. Another system for predicting the occurrence of downy mildew was developed by Palti (1989).

2.3 Purple-blotch forecasting systems

Research conducted in Michigan has provided the basis for a forecast system utilized in the NEOGEN ENVIROCASTER for predicting the potential occurrence of purple blotch of onion caused by *Alternaria porri* (Everts and Lacy, 1990a, b, 1996). Research in Texas and Nebraska on the biology of *A. porri* has also provided information for the development of procedures to predict the production and release of fungal conidia (Meredith, 1966) and the susceptibility of onion leaves of different ages to infection after deposition of the conidia on the leaves (Miller, 1983).

Onion leaves become increasingly susceptible to *A. porri* as the plants age; hence purple blotch becomes much more difficult to control as the bulbs mature (Miller, 1983). The total number of uninterrupted leaf-wetness hours (LWH) each day has been used as a measurement for the application of fungicides in a disease-control programme. The intervals between fungicide sprays are adjusted in relation to the potential for purple-blotch development (Miller *et al.*, 1986). When the number of LWH is consistently fewer than 12 h day^{-1}, the intervals between fungicide applications can be increased. Conversely, when the number of LWH is consistently greater than 12 h day^{-1}, the intervals between fungicide sprays should be shortened (Miller and Lacy, 1995).

If the resistance to *B. squamosa* in *Allium roylei* and in specific onion germplasm (de Vries *et al.*, 1992; Walters and Lorbeer, 1995; Walters *et al.*, 1996; Mutschler *et al.*, 1998) can eventually be transferred to and utilized in commercial onion cultivars, fungicide sprays previously mandated by any one of the forecast systems for *Botrytis* leaf blight could be eliminated. This development would allow for the use of a predictive system(s) for purple blotch that would not be superseded at times by the need for additional fungicide sprays mandated by the *Botrytis* leaf-blight predictors. Until the latter disease is controlled by the use of resistant varieties, the frequent problem of a disease prediction for the occurrence of *Botrytis* leaf blight and the need for a fungicide spray when one is not needed to control purple blotch will continue.

3. Monitoring and Decision-making for Arthropod Pests of *Allium* Crops

Onions and related crops are attacked by a number of arthropod species, but only a few cause serious crop damage on such a consistent basis as to warrant IPM programmes. Pest-management decision-making for the primary arthropod pests attacking *Allium* crops worldwide are discussed.

3.1 Onion maggot

The onion maggot (*Delia* (= *Hylemya*) *antiqua*) (Meigen) (*Diptera*: *Anthomyiidae*), or onion fly, as it is called in Europe, is one of the most important insect pests of onions in temperate and subtropical zones around the world (Finch *et al.*, 1986b; Narkiewicz-Jodko, 1988; Straub and Emmett, 1992; Gupta *et al.*, 1994). *Delia platura* (Meigen) and *Delia florilega* (Zetterstedt) also attack *Allium* crops, but to a lesser degree than *D. antiqua*. Some authors may refer to 'onion maggots' meaning a mixture of any of the three *Delia* species (Finch, 1989).

Finch (1989) and Straub and Emmett (1992) have reviewed the biology and management of *Delia* spp. Onion-maggot larvae damage plants by feeding on the root systems and burrowing into onion bulbs. In the northern USA and Canada, three generations of onion maggot occur each year (Eckenrode *et al.*, 1975). The pupae of the third generation overwinter. In late May, adults emerge and begin oviposition in onion fields. Prophylactic applications of soil insecticides at planting or as a seed treatment are standard practices for control of the first-generation onion-maggot larvae.

Foliar sprays against adults are sometimes used in an attempt to control subsequent generations of onion maggot (Finch *et al.*, 1986b). However, researchers in the USA concluded that flies spend very little time in onion fields and, because of high levels of resistance, foliar insecticides have relatively little effect on onion-maggot populations (Whitfield *et al.*, 1985; Finch *et al.*, 1986a). Reiners *et al.* (2000) recommend spraying only if > 5% of onion seedlings have been damaged by onion maggot. Without the aid of monitoring and forecasting, up to 12 sprays per year were applied to control this pest in Canada (Andaloro *et al.*, 1984).

3.1.1 Forecasting systems

Practical systems for forecasting the times of infestation by *Delia* spp. have been devised so that insecticide sprays may be applied to coincide with pest attacks. Trapping adults on sticky cards of various designs and colours forms the basis of the pest-monitoring system (Finch, 1989). IPM guidelines for Ontario growers recommend placing a set of sticky traps along each field border just above the crop canopy (Chaput, 1993). Vernon and Bartel (1985) and Vernon (1986) concluded that blue and purple traps increased selectivity towards *Delia* spp. adults. Trap catch also can be improved with the use of onion baits or surrogate (artificial) onions laced with sulphur-containing compounds (Harris and Miller, 1983). However, it is not helpful to catch more flies than are needed to obtain a reasonable estimate of fluctuations in population size.

In the north-eastern USA, cone traps (Vincent and Stewart, 1981; Throne *et al.*, 1984) baited with onions are used for monitoring fly abundance and peak flight times. Two cone traps per field should be placed strategically around field borders starting in early May and be checked throughout the summer.

Eckenrode *et al.* (1975) attempted to improve the timing of foliar sprays by computing degree-day accumulations from air temperatures exceeding 4°C and comparing them with peak adult flights of each of the onion-maggot generations. In Ontario, Liu *et al.* (1982) refined the degree-day predictions by correlating them with peak flights and ovarian development of captured females. These models can be used to determine the start of monitoring programmes and to avoid unnecessary expenditure of time and resources (Table 12.1). In Ontario, growers reduced the number of foliar sprays from approximately ten to two per season with the use of degree-day predictions and monitoring (Andaloro *et al.*, 1984). Degree-day calculations can also be used to adjust planting dates to reduce the time available for egg-laying by flies.

Table 12.1. Prediction and timing of onion-maggot adult flight based on degree-day (°D) accumulations (adapted from Eckenrode *et al.*, 1975; Liu *et al.*, 1982).

	New York		Ontario	
Event	Accumulated °D above 4.4°C	Date	Accumulated °D above 4.0°C	Date
1st flight	190	Late May to early June	210	Late May to late June
2nd flight	801	Mid- to late July	1025	Mid-July to mid-August
3rd flight	1468	Late August through October	1772	Late August to mid-September

3.2 Onion thrips

Onion thrips, *Thrips tabaci* (Lindeman) (*Thysanoptera*: *Thripidae*), is a major pest of *Allium* crops. These polyphagous insects occur worldwide and attack virtually all *Allium* crops (Lall and Singh, 1968; Soni and Ellis, 1990; Straub and Emmett, 1992; Baudoin *et al.*, 1994; Gupta *et al.*, 1994). Onion thrips injure plants directly by feeding on leaf tissue and occasionally by vectoring disease-causing organisms, such as onion yellow dwarf virus and purple blotch (*A. porri*) (Straub and Emmett, 1992). In some regions of the world, onion thrips can reduce onion yield and bulb size by more than 55% when they are not controlled (Rueda, 2000).

Western flower thrips, *Frankliniella occidentalis* Pergande (*Thysanoptera*: *Thripidae*), can also cause damage, but occurs less frequently on *Allium* crops than *T. tabaci*. Pest problems with *F. occidentalis* appear to be more localized in certain regions, such as the southern USA (Sites *et al.*, 1992). Biology and management of the two species is similar, although *F. occidentalis* has shown greater resistance to insecticides.

3.2.1 Forecasting systems

Onion thrips have a broad host range and populations move from one crop to another when conditions change – for example, when neighbouring crops are harvested (Shelton and North, 1986). Thus, the temporal and spatial arrival of onion-thrips populations into onion fields is variable and relatively unpredictable (Gangloff, 1999). Sampling procedures rather than degree-day forecasts are therefore used to monitor onion thrips.

Coudriet *et al.* (1979) used white sticky cards to sample onion thrips in the USA, but concluded that insect counting on plants was a more reliable technique for estimating populations. IPM guidelines in the USA and Canada suggest the use of yellow sticky traps around field borders to monitor thrips movement into fields. Frequent plant sampling is necessary from mid-June throughout the summer (in the northern USA) to estimate population levels (Hoffmann *et al.*, 1996). An action threshold of five to ten nymphs per plant has been suggested for the plant-count method (Straub and Emmett, 1992). Because the timing of thrips infestations is variable, some plants are attacked at a younger developmental stage than others. Consequently, economic thresholds for treatment should be dynamic, based on the developmental stage of the plant. Shelton *et al.* (1987) studied the spatial dispersion of onion thrips and determined that the insect was randomly distributed within an onion field. Experience has shown that field edges often have higher populations of thrips, because of immigration from surrounding vegetation. Field edges should therefore be included, but not form the basis of sampling sites.

A sequential-sampling plan was linked to a dynamic economic threshold of three thrips per leaf (Table 12.2). For Spanish and green bunching onions, the threshold is one thrips per leaf (Hoffmann *et al.*, 1996). For sweet onions in Honduras, Rueda (2000) calculated an action threshold of 0.5–1.6 thrips

Table 12.2. Onion thrips sequential-sampling chart (adapted from Shelton *et al.*, 1987).

	2-leaf stage		6-leaf stage		12-leaf stage	
Plant sample	Lower limit	Upper limit	Lower limit	Upper limit	Lower limit	Upper limit
15	35	145	120	420	250	830
20	60	180	185	535	385	1055
25	80	220	255	645	525	1275
30	105	255	325	755	670	1490
35	130	290	400	860	815	1705
40	155	325	475	965	965	1915
45	175	365	550	1070	1115	2125
50	200	400	625	1175	1270	2330

per leaf. Alternatively, in Canada, a binomial sequential-sampling plan has been developed, based on the presence of five insects per plant (Fournier *et al.*, 1994). Binomial sequential-sampling plans were found to be as reliable as the Iwao type sequential-sampling plans developed by Shelton *et al.* (1987). With either plan, a decision to treat or not to treat can usually be made after sampling only ten or 15 plants. However, if after 50 plants the cumulative number of thrips found still falls between the lower and upper limits, a treatment decision should be based on other factors, such as the developmental stage of the crop and weather conditions (Edelson *et al.*, 1989).

In the temperate zone, there are from three to five overlapping generations of thrips each season. Fields should be monitored weekly throughout the summer, and more frequently during hot, dry conditions. When used, onion-thrips IPM can significantly reduce insecticide inputs without adversely affecting onion yield or quality (Hoffmann *et al.*, 1995). Monitoring methods for thrips on leeks are discussed by De Clercq and Van Bockstaele (Chapter 18, this volume).

3.3 Leek moth

The leek moth, *Acrolepiopsis assectella* (Zeller) (*Lepidoptera*: *Yponomeutidae*), occurs throughout most of Europe and has also been reported in Hawaii (Carter, 1984). The life history of the insect is summarized by Straub and Emmett (1992). The leek moth is multivoltine, with four to six generations per year. Larvae feed primarily on leeks (*A. ampeloprasum*, leek group), but also attack other plants in the genus *Allium* (Lecomte *et al.*, 1998). Young larvae mine inside the leaf tissues, leaving the epidermis of the leaf intact. As larvae mature, they bore through the folded leaves in the pseudostem to feed near the centre of the plant. Severely attacked leaves may rot, causing plants to wilt. In onions, leek-moth larvae feed inside the hollow leaves, where they cause little damage, but they may bore into the bulb, causing direct damage to the crop.

3.3.1 Forecasting systems

Studies in France and Spain indicated that a synthetic pheromone, (Z)-11-hexadecenal, attracted adult males and provided early warning of leek-moth attack (Rahn, 1982). However, no relationship was established between pheromone-trap catch and egg or larval occurrence in the field. Later studies showed that pheromone monitoring did not provide reliable estimates of leek-moth density (Gill, 1985). Sampling for eggs or larvae is difficult and impractical for an IPM scouting programme. Thus, Nyrop *et al.* (1989) developed a sequential-sampling plan based on field counts of injured plants. Their results showed that a higher percentage of plants was damaged in border areas of fields, particularly by first-generation leek moth. Thus, as with the onion thrips, sampling sites should include, but not be restricted to, field edges. Nyrop *et al.* (1989) also analysed the economic value of the sampling plan for leek moth. Due to the high value of the crop and the relatively low cost of control, there was little difference in cost between a sampling-based IPM programme and prophylactic treatments for leek moth. However, the former could reduce insecticide use on leeks. As with many IPM programmes, observers' bias and scouting errors can be a problem with the plant-injury sampling method (Theunissen and Legutowska, 1992). Proper training of field scouts is critical to the effectiveness of the sampling programme.

3.4 Cutworms

Allium crops are occasionally attacked by cutworms, most notably *Agrotis* Ochsenheimer spp., including *Agrotis ipsilon* (Hufnagel), and *Euxoa* Huebner spp. (*Lepidoptera*: *Noctuidae*) (Soni and Ellis, 1990; Straub and Emmett, 1992). Outbreaks are sporadic and typically occur in wet seasons or regions, on poorly drained soils or in weedy patches of fields. Cutworms are generally an early-season pest. Moths are active in the spring and lay eggs on leaves or on the soil surface. Larvae of most species sever the seedling just above or below the soil line and may pull the plant down as they feed. Larvae feed at night and hide in the soil near the base of plants during the day.

3.4.1 Forecasting systems

Current monitoring programmes for cutworms involve simply inspecting fields for severed plants and searching for larvae in the soil and litter to confirm their presence in the field (Hoffmann *et al.*, 1996). Specific action thresholds have not been formally worked out for cutworms in onions. Archer and Musick (1977) and Story and Keaster (1983) have developed more accurate sampling techniques for cutworms in maize, using various baits and pitfall traps. However, given the sporadic nature of cutworm damage in onions and the relatively high labour input of these alternative sampling techniques, they are generally not feasible for onion IPM programmes.

Pheromone trapping for adults is also being investigated. The sex pheromone of *A. ipsilon* has been identified as a blend of (Z)-7-dodecenyl acetate (Z7–12:Ac) and (Z)-9-tetradecenyl acetate (Z9–14:Ac). Economically feasible synthetic blends are being tested for future use in monitoring traps (Gemeno and Haynes, 1998).

3.5 Beet armyworm

Beet armyworm, *Spodoptera exigua* (Hübner) (*Lepidoptera: Noctuidae*), can be a serious pest of *Allium* crops in the tropics (Grubben, 1994; Lin, 1994). Monitoring programmes for this pest generally involve making counts of larvae on plants. In Korea, Goh (1994) developed a sequential sampling plan for *S. exigua* on Chinese cabbage and Welsh (Japanese bunching) onion (*A. fistulosum*). An economic threshold of two larvae per plant in spring and five to six larvae per plant in autumn was suggested. Alternative sampling methods for *Spodoptera frugiperda* (J.E. Smith), a similar species, have been well studied in maize and other crops. These include presence/absence sampling (O'Neil *et al.*, 1989), beat-sheets and sweep-netting (Linker *et al.*, 1984), and adult monitoring using sugar-line trapping and synthetic pheromones (Linduska and Harrison, 1986; Chowdhury *et al.*, 1987).

3.6 Aster leafhopper

The aster leafhopper, *Macrosteles quadrilineatus* (*fascifrons*) Stål (*Homoptera*: *Cicadellidae*), is a concern in the USA because it transmits aster yellows, a mycoplasma-like disease of many crops (Madden *et al.*, 1995). Onions are not a preferred host for this pest and disease incidence is sporadic. In the USA, aster leafhoppers overwinter in the southern states and are carried northward on summer storm fronts (Hoy *et al.*, 1992). The insects arrive in the northern states in late June to July. Aster leafhoppers pass through several generations per season and pick up the disease-causing organisms by feeding on various weeds. In onions, disease incidence is usually low (< 1%) even when it may be high (10–20%) in other crops, such as lettuce and celery (Hoffmann *et al.*, 1996).

3.6.1 Forecasting systems

Monitoring programmes for the aster leafhopper have been developed on other crops, but not onions. O'Rourke *et al.* (1998) and Burkness *et al.* (1999) recently devised sequential-sampling plans for this pest in carrots, using a sweep-netting technique.

3.7 Aphids

Allium crops are also attacked by several aphids, including *Myzus persicae* (Sulzer), *Rhopalosiphum maidis* (Fitch) and *Schizaphis graminum* (Rondani) (*Homoptera*: *Aphididae*). Onions are not a preferred host crop and feeding injury in itself is not a concern. However, many aphid species can vector onion yellow dwarf virus (Fischer and Lockhart, 1974; Strobbs and van Diel, 1999). This disease is of most concern to producers of onion seed. Aphids have numerous natural enemies, which keep populations below damaging levels most of the time. Sampling schemes and thresholds for aphids are not generally used in onion IPM programmes.

3.8 Mites

Tetranychid mites, such as the two-spotted spider mite, *Tetranychus urticae* Koch (*Acari: Tetranychidae*), are generally of minor concern on onions, but populations can build up to damaging levels on occasion, particularly if natural predators are killed off by insecticides. In parts of the north-eastern USA, the bulb mite, *Rhyzoglyphus robini* Claparède (*Acari*: *Acaridae*), can be a damaging pest to onion bulbs (Diaz, 1998). Bulb mites are a cosmopolitan pest with a broad host range and are difficult to sample.

3.8.1 Monitoring systems

A hand-lens can be used to confirm the presence of tetranychid mites in fields, but no formal sampling protocol or economic threshold has been developed for *Allium* crops. Traditional monitoring programmes for bulb mites have relied on manual inspection of plants (Latta, 1939; Rawlins, 1955). More recently, Gerson *et al.* (1985) utilized garlic-baited traps to sample and monitor *R. robini* populations in *Allium* crops in Israel. Baited traps can provide relative estimates of bulb-mite populations in fields prior to planting.

4. Conclusions and Future Directions

IPM will continue to be the preferred strategy for management of many of the diseases and insect pests of onions. A variety of cost-effective IPM tactics are available, including cultural, biological and chemical controls. The integration of these tactics is essential to minimize risks of pathogen and insect-pest resistance developing to any single tactic. Further integration, particularly across disciplinary approaches, is still needed. For example, studies should be undertaken to determine the impact of fungicides on naturally occurring entomopathogenic fungi, such as *Entomophthora muscae*, which is an important mortality factor of onion maggot flies (Carruthers *et al.*, 1984, 1985). The impact of weeds and surrounding vegetation on pathogens, insect pests and beneficial insect populations should also be assessed. More research is needed to understand better the aetiology and ecology of diseases and the biology of insect pests of onions. The recent discovery of apparently unique populations of onion thrips, one within and the other at the periphery of onion fields (Gangloff, 1999), and the implications of this for onion thrips management need to be better understood. This discovery is one example that reinforces the need to take a whole-systems approach to insect pest management (Lewis *et al.*, 1997). Multitrophic interactions between pathogens and insect pests and their natural enemies/antagonists within the crop, as well as how these organisms have an impact at the landscape level, are all part of the whole-systems approach.

Development of cost-effective single-component pathogen and insect-pest management tactics should continue and be incorporated appropriately into a whole-systems approach. New pesticides that are applied at lower rates and with reduced environmental hazards are becoming available and will probably displace many compounds used today. Many of the traditional pesticides may be lost due to the development of pathogen and insect-pest resistance or to new government regulations. In the USA, full implementation of the Food Quality Protection Act of 1996 could result in the loss of several major categories of pesticides, resulting in fewer chemical options for disease and insect-pest control on onions in the future. Thus, it is imperative that alternative strategies for pest control, including novel approaches, be investigated.

The utilization of a new class of non-toxic, naturally occurring proteins called harpins, which when applied topically to plant surfaces activate the plant's own defence and growth systems, offers great promise for incorporation into future IPM programmes (Wei *et al.*, 1992; Kim and Beer, 2000). The first commercial harpin product, Messenger®, is currently being produced and marketed by EDEN Bioscience Corporation, Bothell, Washington. EDEN is now conducting efficacy studies on other proprietary harpin proteins that are many times more potent than Messenger®. Harpin proteins trigger a plant's natural defence system, which then protects the plant against pathogens and insect pests, also activating growth systems within the plant. Harpin proteins provide the benefits of modern technology without modifying the plant's DNA. The timing for application of harpin proteins could probably be based on forecast and monitoring systems for the occurrence of diseases and insect pests.

Another novel alternative to chemical pesticides is the use of non-woven fibre barriers for insect control (Hoffmann *et al.*, 2000). These barriers consist of arrangements of minute strands loosely intertwined in 'web' form, which act as a physical or behavioural barrier to insect oviposition or feeding. Tests using hot-melt extrusion methods to generate fibres in the field have shown promise against onion maggot. Non-woven fibres may also be used as a novel delivery system for compounds such as attractants or repellents.

Breeding disease- and insect-pest-resistant varieties is important for the protection of many crops. Onions and other *Allium* species are no exception. The recent identification of genes for resistance to *Botrytis* leaf blight within *A. roylei* and specific *A. cepa* cultivars is expected to lead to partial or full control of the disease by utilizing traditional plant-breeding techniques for the development of resistant varieties to the disease (Walters and Lorbeer, 1995; Walters *et al.*, 1996; Kik, Chapter 4, this volume). Recent advances in biotechnology and genetic engineering should provide a platform for the development of novel types of resistance mechanisms against diseases as well as insects.

Lastly, it will be critical that new knowledge generated through research is effectively delivered to the onion producer. Traditional means of extending information to producers should continue, but the Internet and other forms of electronic communication will undoubtedly grow in importance as a means of delivering information. In the not too distant future, field personnel will be able to provide instantaneous reports on crop status as well as disease levels and insect infestations to producers, via wireless e-mail systems. The World Wide Web will be a major source of information for onion producers, with hyperlinked disease, insect and crop-management sites functioning as 'one-stop-shopping' resources for information and guidelines.

Many disease and insect-pest management challenges face onion production worldwide. A concerted public–private effort will be required to protect onions effectively, in a sustainable manner and with a minimal impact on the environment. Key to the success of this endeavour will be continued support of research and extension from public sources, as well as from industry.

References

Alderman, S.C. and Lacy, M.L. (1983) Influence of dew period and temperature on infection of onion leaves by dry conidia of *Botrytis squamosa*. *Phytopathology* 73, 1020–1023.

Alderman, S.C. and Lacy, M.L. (1984) Influence of leaf position and maturity on development of *Botrytis squamosa* in onion leaves. *Phytopathology* 74, 1461–1463.

Andaloro, J.T., Rose, K.B. and Eckenrode, C.J. (1984) *Suppressing Onion Maggot in Commercial Fields and Research Plots, and Monitoring with Air Thermal Unit Accumulations*. Search: Agriculture No. 29, New York State Agricultural Experiment Station, Geneva, New York, 5 pp.

Anon. (1998) *Wallkill – Rondout Watershed Demonstration Project – NY-BLIGHT-ALERT*. USDA Water Quality Initiative Report – MJP 980901, Cornell Cooperative Extension, Middletown, New York, 4 pp.

Anon. (1999) *Agricultural Chemical Usage: 1998. Vegetable Summary*. New York Agricultural Statistics Service, Albany, New York, 48 pp.

Archer, T.L. and Musick, G.J. (1977) Evaluation of sampling methods for black cutworm larvae in field corn. *Journal of Economic Entomology* 70, 447–449.

Basallote-Ureba, M.J., Prados-Ligero, A.M. and Melero-Vara, J.M. (1999) Aetiology of leaf spot of garlic and onion caused by *Stemphylium vesicarium* in Spain. *Plant Pathology* 48, 139–145.

Baudoin, W., Bâ, M.L. and Jeangille, P. (1994) Onion production and constraints in the Sahelian countries of Africa. *Acta Horticulturae* 358, 37–42.

Binns, M.R. and Nyrop, J.P. (1992) Sampling insect populations for the purpose of IPM decision making. *Annual Review of Entomology* 37, 427–453.

Burkness, E.C., Venette, R.C., O'Rourke, P.K. and Hutchison, W.D. (1999) Binomial sequential sampling for management of aster leafhopper (*Homoptera: Cicadellidae*) and aster yellows phytoplasma in carrot: impact of tally threshold on the accuracy of treatment decisions. *Environmental Entomology* 28, 851–857.

Carroll, K.A., Harris, C.R. and Morrison, P.E. (1983) Resistance shown by a parathion-resistant onion maggot (*Diptera: Anthomyiidae*) strain to some other insecticides. *Canadian Entomologist* 115, 1519–1522.

Carruthers, R.I., Whitfield, G.H. and Haynes, D.L. (1984) Pesticide-induced mortality of natural enemies of the onion maggot, *Delia antiqua* (*Dip. Anthomyiidae*). *Entomophaga* 30, 151–161.

Carruthers, R.I., Haynes, D.L. and Macleod, D.M. (1985) *Entomophthora muscae* (*Entomophthorales: Entomophthoraceae*) mycosis in the onion fly, *Delia antiqua* (*Diptera: Anthomyiidae*). *Journal of Invertebrate Pathology* 45, 81–93.

Carter, D.J. (1984) *Pest* Lepidoptera *of Europe with Special Reference to the British Isles*. Junk, Dordrecht, The Netherlands, 431 pp.

Chaput, J. (1993) *Integrated Pest Management for Onions, Carrots, Celery and Lettuce in Ontario: A Handbook for Growers, Scouts and Consultants*. Ontario Ministry of Agriculture and Food Field Services, Central and North Region Pest Management Section, Toronto, 67 pp.

Chowdhury, M.A., Chalfant, R.B. and Young, J.R. (1987) Comparison of sugarline sampling and pheromone trapping for monitoring adult populations of corn earworm and fall armyworm (*Lepidoptera: Noctuidae*) in sweet corn. *Environmental Entomology* 16, 1241–1243.

Coudriet, D.L., Kishaba, A.N., McCreight, J.D. and Bohn, G.W. (1979) Varietal resistance in onions to thrips. *Journal of Economic Entomology* 72, 614–615.

DeMilia, M.S. (1993) The implementation of the BLIGHT-ALERT predictive system for the timing of fungicide applications for control of Botrytis leaf blight of onion. MSc thesis, Department of Plant Pathology, Cornell University, New York, USA.

de Visser, C.L.M. (1998) Development of a downy mildew advisory model based on DOWNCAST. *European Journal of Plant Pathology* 104, 933–943.

de Vries, J.N., Wietsma, W.A. and de Vries, T. (1992) Introgression of leaf blight resistance from *Allium roylei* Stearn into onion (*A. cepa* L.). *Euphytica* 62, 127–133.

Diaz, A. (1998) Aspects of the biology, ecology, and behavior of the bulb mite, *Rhizoglyphus robini* Claparède (*Acari: Acaridae*): a pest of onions in New York. PhD dissertation, Cornell University, Ithaca, New York, USA.

Eckenrode, C.J., Vea, E.V. and Stone, K.W. (1975) Population trends of onion maggots correlated with air thermal unit accumulations. *Environmental Entomology* 4, 785–789.

Edelson, J.V., Cartwright, B. and Royer, T.A. (1989) Economics of controlling onion thrips (*Thysanoptera: Thripidae*) on onions with insecticides in south Texas. *Journal of Economic Entomology* 82, 561–564.

Everts, K.L. and Lacy, M.L. (1990a) The influence of dew duration, relative humidity, and leaf senescence on conidial formation and infection of onion by *Alternaria porri*. *Phytopathology* 80,1203–1207.

Everts, K.L. and Lacy, M.L. (1990b) Influence of environment on conidial concentration of *Alternaria porri* in air and on purple blotch incidence of onion. *Phytopathology* 80, 1387–1391.

Everts, K.L. and Lacy, M.L. (1996) Factors influencing infection of onion leaves by *Alternaria porri* and subsequent lesion expansion. *Plant Disease* 80, 276–280.

Falloon, P.G., Falloon, L.M. and Grogan, R.G. (1987) Etiology and epidemiology of *Stemphylium* leaf spot and purple spot of asparagus in California. *Phytopathology* 77, 407–413.

Finch, S. (1989) Ecological considerations in the management of *Delia* pest species in vegetable crops. *Annual Review of Entomology* 34, 117–137.

Finch, S., Eckenrode, C.J. and Cadoux, M.E. (1986a) Behavior of onion maggot (*Diptera: Anthomyiidae*) in commercial fields treated regularly with parathion sprays. *Journal of Economic Entomology* 79, 107–113.

Finch, S., Cadoux, M.E., Eckenrode, C.J. and Spittler, T.D. (1986b) Appraisal of current strategies for controlling onion maggot (*Diptera: Anthomyiidae*) in New York State. *Journal of Economic Entomology* 79, 736–740.

Fischer, H.U. and Lockhart, B.E.L. (1974) High incidence of onion yellow dwarf in areas of commercial onion production in Morocco. *Plant Disease Reporter* 58, 252–253.

Fournier, F., Boivin, G. and Stewart, R.K. (1994) Comparison of binomial and Iwao type sequential sampling plans for monitoring onion thrips (*Thrips tabaci*) (*Thysanoptera: Thripidae*) in onions. *Phytoprotection* 75, 69–78.

Gangloff, J.L. (1999) Population dynamics and insecticide resistance of onion thrips, *Thrips tabaci* Lindeman (*Thysanoptera: Thripidae*) in onions. PhD dissertation, Cornell University, Ithaca, New York, USA.

Gemeno, C. and Haynes, K.F. (1998) Chemical and behavioral evidence for a third pheromone component in a North American population of the black cutworm moth, *Agrotis ipsilon*. *Journal of Chemical Ecology* 24, 999–1011.

Gerson, U.S., Yathom, S., Capua, S. and Thorens, D. (1985) *Rhizoglyphus robini* Claparède (*Acari: Astigmata: Acaridae*) as a soil mite. *Acarologia* 26, 371–380.

Gill, D. (1985) Les maladies et ravageurs du poireau. *Phytoma – Défense des Cultures* September–October, 29–41.

Goh, H.G. (1994) Sequential sampling method on the beet armyworm, *Spodoptera exigua* (Hubner) (*Lepidoptera: Noctuidae*) in Chinese cabbage and Welsh onion. *RDA Journal of Agricultural Science, Crop Protection* 36, 337–340.

Grubben, G.J.H. (1994) Constraints for shallot, garlic, and Welsh onion in Indonesia: a case study on the evolution of *Allium* crops in the equatorial tropics. *Acta Horticulturae* 358, 333–339.

Gupta, R.P., Srivastava, K.J. and Pandey, U.B. (1994) Diseases and insect pests of onion in India. *Acta Horticulturae* 358, 265–269.

Harris, C.R. and Svec, H.J. (1976) Onion maggot resistance to insecticides. *Journal of Economic Entomology* 69, 617–620.

Harris, M.O. and Miller, J.R. (1983) Color stimuli and oviposition behavior of the onion fly, *Delia antiqua* (Meigen) (*Diptera: Anthomyidae*). *Annals of the Entomological Society of America* 76, 766–771.

Hildebrand, P.D. and Sutton, J.C. (1984a) Effects of weather variables on spore survival and infection of onion leaves by *Peronospora destructor*. *Canadian Journal of Plant Pathology* 6, 119–126.

Hildebrand, P.D. and Sutton, J.C. (1984b) Relationships of temperature, moisture, and inoculum density to the infection cycle of *Peronospora destructor*. *Canadian Journal of Plant Pathology* 6, 127–134.

Hildebrand, P.D. and Sutton, J.C. (1984c) Interactive effects of the dark period, humid period, temperature, and light on sporulation of *Peronospora destructor*. *Phytopathology* 74, 1444–1449.

Hill, J.P. (1995) Cladosporium leaf blotch. In: Schwartz, H.F. and Mohan, S.K. (eds) *Compendium of Onion and Garlic Diseases*. American Phytopathological Society Press, St Paul, Minnesota, pp. 21–22.

Hoffmann, M.P. and Petzoldt, C.H. (1993) Demonstration of integrated pest management programs for onions in New York. In: *Proceedings of the 1993 National Onion Research Conference, 9–11 December, Ithaca, New York, USA*. pp. 38–43.

Hoffmann, M.P., Petzoldt, C.H., MacNeil, C.R., Mishanec, J.J., Orfanedes, M.S. and Young, D.H. (1995) Evaluation of an onion thrips pest management program for onions in New York. *Agriculture, Ecosystems and Environment* 55, 51–60.

Hoffmann, M.P., Petzoldt, C.H. and Frodsham, A.C. (1996) *Integrated Pest Management for Onions*. New York State IPM Program Publication No. 119, Cornell University, Ithaca, New York, 78 pp.

Hoffmann, M.P., Schwartz, P. and Baird, J.M. (2000) *Fiber Barriers for Control of Agricultural Pests*. US Patent No. 6,052,943, US Patent Office, Washington, DC.

Hoy, C.W., Heady, S.E. and Koch, T.A. (1992) Species composition, phenology, and possible origins of leafhoppers (*Cicadellidae*) in Ohio vegetable crops. *Journal of Economic Entomology* 85, 2336–2343.

Jesperson, G.D. and Sutton, J.C. (1987) Evaluation of a forecaster for downy mildew of onion (*Allium cepa* L.). *Crop Protection* 6, 95–103.

Kim, J.F. and Beer, S.V. (2000) *hrp* Genes and harpins of *Erwinia amylovora*: a decade of discovery. In: Vanneste, J.L. (ed.) *Fire Blight and its Causative Agent,* Erwinia amylovora. CAB International, Wallingford, UK, pp. 141–161.

Lacy, M.L. (1991) *Timing Fungicide Sprays for Control of* Botrytis *Leaf Blight of Onion with a Conidial Release Predictor*. Michigan Agricultural Experiment Station Research Report No. 513, East Lansing, Michigan, 6 pp.

Lacy, M.L. and Lorbeer, J.W. (1995) *Botrytis* leaf blight. In: Schwartz, H.F. and Mohan, S.K. (eds) *Compendium of Onion and Garlic Diseases*. American Phytopathological Society Press, St Paul, Minnesota, pp. 16–18.

Lacy, M.L. and Pontius, G.A. (1983) Prediction of weather mediated release of conidia of *Botrytis squamosa* from onion leaves in the field. *Phytopathology* 73, 670–676.

Lall, B.S. and Singh, L.M. (1968) Biology and control of the onion thrips in India. *Journal of Economic Entomology* 61, 676–679.

Latta, R. (1939) Observations on the nature of bulb mite attack on easter lilies. *Journal of Economic Entomology* 32, 125–128.

Leach, C.M., Hildebrand, P.D. and Sutton, J.C. (1982) Sporangium discharge by *Peronospora destructor*: influence of humidity, red-infrared radiation, and vibration. *Phytopathology* 72, 1052–1056.

Lecomte, C., Pierre, D., Pouzat, J. and Thibout, E. (1998) Behavioral and olfactory variations in the leek moth, *Acrolepsis assectella*, after several generations of rearing under diverse conditions. *Entomologia Experimentalis et Applicata* 86, 305–311.

Lewis, W.J., van Lenteren, J.C., Phatak, S.C. and Tumlinson, J.H. (1997) A total system approach to sustainable pest management. *Proceedings of the National Academy of Sciences of the USA* 94, 12243–12248.

Lin, C.H. (1994) The progress and problems of *Allium* crops in Taiwan. *Acta Horticulturae* 358, 53–59.

Linduska, J.J. and Harrison, F.P. (1986) Adult sampling as a means of predicting damage levels of fall armyworm (*Lepidoptera: Noctuidae*) in grain corn. *Florida Entomologist* 69, 487–491.

Linker, H.M., Johnson, F.A., Stimac, J.L. and Poe, S.L. (1984) Analysis of sampling procedures for corn earworm and fall armyworm (*Lepidoptera: Noctuidae*) in peanuts. *Environmental Entomology* 13, 75–78.

Liu, H.J., McEwen, F.L. and Ritcey, G. (1982) Forecasting events in the life cycle of the onion maggot, *Hylemya antiqua* (*Diptera: Anthomyiidae*): application to control schemes. *Environmental Entomology* 11, 751–755.

Lorbeer, J.W. (1992) *Botrytis* leaf blight of onion. In: Chaube, H.S., Singh, U.S., Mukhopadhyay, A.N. and Kumar, J. (eds) *Plant Diseases of International Importance*, Vol. II. *Diseases of Vegetables and Oil Seed Crops*. Prentice Hall, Englewood Cliffs, New Jersey, pp. 186–211.

Lorbeer, J.W. (1997a) Management of diseases of alliums. *Acta Horticulturae* 433, 585–591.

Lorbeer, J.W. (1997b) Disease monitoring and forecasting in vegetable crops emphasizing *Botrytris* leaf blight of onion. In: Bourgeois, G. and Guibord, M.O. (eds) *Proceedings of International Symposium on Agricultural Pest Forecasting and Monitoring, 10–13 October 1995, Québec City, Canada*. Ministère de l'Agriculture, des Pêcheries et de l'Alimentation du Québec, Québec, pp. 185–189.

Lorbeer, J.W., Petzoldt, C.H. and Walters, T.W. (2001) Integrated pest management of *Botrytis* leaf blight of onion. *Acta Horticulturae* 555, 129–132.

Madden, L.V., Nault, L.R., Murral, D.J. and Apelt, M.R. (1995) Spatial pattern analysis of the incidence of aster yellows disease in lettuce. *Researches on Population Ecology* 37, 279–289.

Meredith, D.S. (1966) Spore dispersal in *Alternaria porri* (Ellis) Neerg. on onions in Nebraska. *Annals of Applied Biology* 57, 67–73.

Miller, M.E. (1983) Relationship between onion leaf age and susceptibility to *Alternaria porri*. *Plant Disease* 67, 284–286.

Miller, M.E. (1995) *Stemphylium* leaf blight and stalk rot. In: Schwartz, H.F. and Mohan, S.K. (eds) *Compendium of Onion and Garlic Diseases*. American Phytopathological Society Press, St Paul, Minnesota, pp. 25–26.

Miller, M.E. and Lacy, M.L. (1995) Purple blotch. In: Schwartz, H.F. and Mohan, S.K. (eds) *Compendium of Onion and Garlic Diseases*. American Phytopathological Society Press, St Paul, Minnesota, pp. 23–24.

Miller, M.E., Taber, R.A. and Amador, J.M. (1978) *Stemphylium* blight of onion in South Texas. *Plant Disease Reporter* 62, 851–853.

Miller, M.E., Bruton, B.D. and Amador, J.M. (1986) Effects of number and timing of chlorothalonil applications on onion yield. *Plant Disease* 70, 875–876.

Montesinos, E., Moragrega, C., Llorente, I., Vilardel, P., Bonaterra, A., Ponti, I., Bugiani, R., Cavanni, P. and Brunelli, A. (1995) Development and evaluation of an infection model for *Stemphylium vesicarium* on pear based on temperature and wetness duration. *Phytopathology* 85, 586–592.

Mutschler, M.A., Cobb, E.D. and Lorbeer, J.W. (1998) Towards improved control of *Botrytis* leaf blight of onion. In: Voss, R.E. (ed.) *Proceedings of the 1998 Onion (and other* Allium*) Research Conference*. Vegetable Research and Information Center, University of California, Davis, California, pp. 227–230.

Narkiewicz-Jodko, J. (1988) Chemical control of the onion fly (*Delia antiqua* Meig.) in Poland. *Acta Horticulturae* 219, 59–65.

Nyrop, J.P., Shelton, A.M. and Theunissen, J. (1989) Value of a control decision rule for leek moth infestations in leek. *Entomologia Experimentalis et Applicata* 53, 167–176.

O'Neil, R.J., Andrews, K.L., Barfield, C.S. and Sobrado, C.E. (1989) Sampling program for fall armyworm in maize. *Journal of Economic Entomology* 82, 134–138.

O'Rourke, P.K., Burkness, E.C. and Hutchison, W.D. (1998) Development and validation of a fixed-precision sequential sampling plan for aster leafhopper (*Homoptera: Cicadellidae*) in carrot. *Environmental Entomology* 27, 1463–1468.

Palti, J. (1989) Epidemiology, prediction and control of onion downy mildew caused by *Peronospora destructor*. *Phytoparasitica* 17, 31–48.

Petzoldt, C. (1994) *Onion IPM Scouting Procedures for New York*. IPM Publication No. 106, New York State Agricultural Experiment Station, Geneva, New York, 28 pp.

Rahn, R. (1982) Sex-pheromone trapping of the leek moth, *Acrolepiopsis assectella* Z. (*Lepid. Yponomeutidae-Acrolepiinae*), using Z 11 HDAL. Results of the 1981 programme. *Agronomie* 2, 957–962.

Rao, N.N.R. and Paugi, M.S. (1975) *Stemphylium* leaf blight of onion. *Mycopathologia* 56, 113–118.

Rawlins, W.A. (1955) *Rhizoglyphus solani*, a pest of onions. *Journal of Economic Entomology* 48, 334.

Reiners, S., Petzoldt, C.H. and Hoffmann, M.P. (eds) (2000) *Integrated Crop and Pest Management Guidelines for Commercial Vegetable Production 2000*. Cornell Cooperative Extension Publication, Ithaca, New York, 309 pp.

Rueda, A.A. (2000) Developing the research and education components for an integrated pest management program for sweet onions in Honduras. PhD dissertation, Cornell University, Ithaca, New York, USA.

Schwartz, H.F. (1995) Downy mildew. In: Schwartz, H.F. and Mohan, S.K. (eds) *Compendium of Onion and Garlic Diseases*. American Phytopathological Society Press, St Paul, Minnesota, pp. 20–21.

Shelton, A.M. and North, R.C. (1986) Species composition and phenology of *Thysanoptera* within field crops adjacent to cabbage fields. *Environmental Entomology* 15, 513–519.

Shelton, A.M., Nyrop, J.P., North, R.C., Petzoldt, C. and Foster, R. (1987) Development and use of a dynamic sequential sampling program for onion thrips, *Thrips tabaci* (*Thysanoptera: Thripidae*), on onions. *Journal of Economic Entomology* 80, 1051–1056.

Shishkoff, N. and Lorbeer, J.W. (1989) Etiology of *Stemphylium* leaf blight of onion. *Phytopathology* 79, 301–304.

Shoemaker, P.B. and Lorbeer, J.W. (1977a) Timing initial fungicide application to control *Botrytis* leaf blight epidemics on onions. *Phytopathology* 67, 409–414.

Shoemaker, P.B. and Lorbeer, J.W. (1977b) The role of dew and temperature in the epidemiology of *Botrytis* leaf blight of onion. *Phytopathology* 67, 1267–1272.

Sites, R.W., Chambers, W.S. and Nichols, B.J. (1992) Diel periodicity of thrips (*Thysanoptera: Thripidae*) dispersion and the occurrence of *Frankliniella williamsi* on onions. *Journal of Economic Entomology* 85, 100–105.

Soni, S.K. and Ellis, P.R. (1990). Insect pests. In: Rabinowitch, H.D. and Brewster, J.L. (eds) *Onions and Allied Crops*, Vol. II. *Agronomy, Biotic Interactions, Pathology, and Crop Protection*. CRC Press, Boca Raton, Florida, pp. 213–271.

Story, R.N. and Keaster, A.J. (1983) Modified larval bait trap for sampling black cutworm (*Lepidoptera: Noctuidae*) populations in field corn. *Journal of Economic Entomology* 76, 662–666.

Straub, R.W. and Emmett, B. (1992) Pests of monocotyledon crops. In: McKinlay, R.G. (ed.) *Vegetable Crop Pests*. Macmillan, Basingstoke, UK, pp. 213–262.

Strobbs, L.W. and van Diel, L. (1999) First report of onion yellow dwarf virus in Ontario. *Plant Disease* 83, 782.

Sutton, J.C. (1986) Forecasting onion diseases: a key to efficient fungicide use. *Highlights* 9, 13–16.

Sutton, J.C., Swanton, C.J. and Gillespie, T.J. (1978) Relation of weather variables and host factors to incidence of airborne spores of *Botrytis squamosa*. *Canadian Journal of Botany* 56, 2460–2469.

Sutton, J.C., James, T.D.W. and Rowell, P.M. (1983) Relation of weather and host factors to an epidemic of *Botrytis* leaf blight in onions. *Canadian Journal of Plant Pathology* 5, 256–265.

Sutton, J.C., Gillespie, T.J. and Hildebrand, P.D. (1984) Monitoring crop microclimate in relation to plant disease. *Plant Disease* 68, 78–84.

Sutton, J.C., James, T.D.W. and Rowell, P.M. (1985) *BOTCAST: A Forecaster for Timing Fungicides to Control* Botrytis *Leaf Blight of Onions*. Circular EB 07 85, Department of Environmental Biology, University of Guelph, Ontario, 15 pp.

Sutton, J.C., James, T.D.W. and Rowell, P.M. (1986) BOTCAST: a forecasting system to time the initial fungicide spray for managing *Botrytis* leaf blight of onions. *Agriculture, Ecosystems and Environment* 18, 123–143.

Theunissen, J. and Legutowska, H. (1992) Observers' bias in the assessment of pest and disease symptoms in leek. *Entomologia Experimentalis et Applicata* 64, 101–109.

Throne, J.E., Robbins, P.S. and Eckenrode, C.J. (1984) *An Improved Screen Cone Trap for Monitoring Activity of Flying Insects*. New York's Food and Life Sciences Bulletin No. 106, New York Agricultural Experiment Station, Cornell University, Geneva, New York, 3 pp.

Vernon, R.S. (1986) A spectral zone of color preference for the onion fly, *Delia antiqua* (*Diptera: Anthomyiidae*) with reference to the reflective intensity of traps. *Canadian Entomologist* 118, 849–856.

Vernon, R.S. and Bartel, D.L. (1985) Effect of hue, saturation, and intensity on color selection by the onion fly, *Delia antiqua* (Meigen) (*Diptera: Anthomyiidae*) in the field. *Environmental Entomology* 14, 210–216.

Vincelli, P.C. and Lorbeer, J.W. (1987) Sequential sampling plan for timing initial fungicide application to control *Botrytis* leaf blight of onion. *Phytopathology* 77, 1301–1303.

Vincelli, P.C. and Lorbeer, J.W. (1988a) Forecasting spore episodes of *Botrytis squamosa* in commercial onion fields in New York. *Phytopathology* 78, 966–970.

Vincelli, P.C. and Lorbeer, J.W. (1988b) Relationship of precipitation probability to infection potential of *Botrytis squamosa* on onion. *Phytopathology* 78, 1078–1082.

Vincelli, P.C. and Lorbeer, J.W. (1989) BLIGHT-ALERT: a weather-based predictive system for timing fungicide applications on onion before infection periods of *Botrytis squamosa* in commercial onion fields in New York. *Phytopathology* 79, 493–498.

Vincent, C. and Stewart, R.K. (1981) Evaluation of two types of traps used in monitoring adults of the cabbage maggot, *Hylemya brassicae* (*Diptera: Anthomyiidae*). *Annales de la Société Entomologique du Québec* 26, 41–50.

Walters, T.W. and Lorbeer, J.W. (1995) Sources of resistance to *Botrytis* leaf blight of onion. In: *Proceedings of the 1995 National Onion Research Conference*. Wisconsin Center, Madison, Wisconsin, pp. 118–122.

Walters, T.W., Ellerbrock, L.A., van der Heide, J.J., Lorbeer, J.W. and LoParco, D.P. (1996) Field and greenhouse procedures to evaluate onions for *Botrytis* leaf blight resistance. *HortScience* 31, 436–438.

Wei, Z.-M., Lady, R.J., Zumoff, C.H., Bauer, D.W., He, S.Y., Collmer, A. and Beer, S.W. (1992) Harpin, elicitor of the hypersensitive response produced by the plant pathogen *Erwinia amylovora*. *Science* 257, 85–88.

Whitfield, G.H., Carruthers, R.I. and Haynes, D.L. (1985) Phenology and control of the onion maggot (*Diptera: Anthomyiidae*) in Michigan onion production. *Agriculture, Ecosystems and Environment* 12, 189–200.

13 Virus Diseases in Garlic and the Propagation of Virus-free Plants

R. Salomon
Agricultural Research Organization, The Volcani Center, Department of Virology, POB 6, Bet Dagan 50250, Israel

1. Introduction

Garlic (*Allium sativum* L.) is one of the most ancient cultivated herbs, and it has been vegetatively propagated since before the historical period. This mode of clonal propagation allows the production of a uniform crop that preserves quality traits, such as flavour and the nutraceutical properties of the plant (see Fritsch and Friesen, Chapter 1, Etoh and Simon, Chapter 5, and Randle and Lancaster, Chapter 14, this volume). However, there are a number of disadvantages of clonal propagation. The most important ones are as follows:

1. Low propagation rate: the number of cloves per bulb ranges in modern clones between seven and ten, resulting in the propagation material being very expensive.
2. A large volume of storage is required for the bulky bulbs, and losses in storage due to pests, rotting and premature sprouting are common.
3. The most severe disadvantage is the transmission by the propagules of pests and diseases from one field to another and the accumulation of intracellular parasites, notably viruses. Walkey (1990) reviewed the economic importance of viruses in reducing garlic yields.

Most of the plant viruses known today, including the *Potyviridae* group, are not seed-transmitted, or are seed-transmitted only to a very limited extent. Many viruses, however, survive in other living tissues even when dormant. Thus, plants propagated from seeds, such as onion, leek and chives, start their life cycle free of viruses. In contrast, the vegetatively propagated garlic and shallots (Rabinowitch and Kamenetsky, Chapter 17, this volume) accumulate viruses and perpetuate them from one generation to the next. Several pathogenic viruses are common in all garlic-growing areas, while others may be localized in one or a few geographical regions.

Until recently, virus identification was based on symptoms, host range and serology. These methods are inaccurate, and may consequently result in mistaken identifications. Similar or even identical viruses from different countries were thus characterized as different species (Yamashita *et al.*, 1995; Salomon *et al.*, 1996). For example, the sequence of the gene coding for the coat protein (CP) of garlic virus 2 (GV2) from Japan is essentially identical to that of leek yellow-stripe virus (LYSV) from Israel (Nagakubo *et al.*, 1994; Salomon *et al.*, 1996). Similarly, sequence analysis of garlic mosaic virus (GMV) from Japan was identical to that of LYSV (Barg *et al.*, 1995; Yamashita *et al.*, 1995; Takaichi *et al.*, 1998; Tsuneyoshi *et al.*, 1998b). Therefore, it is safe to conclude that GV2, GMV and LYSV represent the same virus isolated in different countries. The incorrect identification of onion yellow dwarf virus (OYDV) is another example. Amino acid sequence of the CP of OYDV isolates from various regions of the world shows only minor variation (Kobayashi *et al.*, 1996; Shiboleth *et al.*, 1997, 2001; Tsuneyoshi *et al.*, 1998b). The mistaken serological identifications resulted, in part, from contamination of samples and extracts used for sera preparations with other latent or asymptomatic viruses (van Dijk, 1994). The lack of accurate definition systems gave rise to a confused state for garlic virus classification worldwide (Walkey, 1990).

Previous work has shown that partial or complete freedom from viruses raises garlic yields by 50% or more, mainly due to size increase of the healthy cloves and bulbs (Delecolle *et al.*, 1985; Walkey, 1990; Ohkoshi, 1991; Lot *et al.*, 1998). Virus-free propagation material is needed to produce a virus-free crop. A number of biological and serological techniques are used throughout the world for the production of such propagules (Novak, 1990; Verbeek *et al.*, 1995; Ucman *et al.*, 1998). These techniques, however, rely on support by dependable methods for the isolation and identification of garlic viruses.

In recent years, the application of molecular techniques has resulted in more accurate identification and classification of viruses in garlic (Sumi *et al.*, 1993; Nagakubo *et al.*, 1994; Yamashita *et al.*, 1995, 1996; Kobayashi *et al.*, 1996; Salomon *et al.*, 1996; Tsuneyoshi and Sumi, 1996; Shiboleth *et al.*, 1997, 2001; Tsuneyoshi *et al.*, 1998a, b; Dovas *et al.*, 2001a, b). The most commonly used methods are based on the polymerase chain reaction (PCR) technique, which, when available, markedly reduces the time needed for accurate detection and identification of viruses. These methods are equally accurate for virus identification in cultured tissues, stored bulbs and growing plants. Consequently, the process of virus elimination and maintenance of 'virus-free' lines has become more reliable and efficient.

Viruses from several taxa affect garlic, including three different potyviruses, at least two carla-type viruses, a shallot virus X (SVX), a mite-transmitted virus similar to SVX and some not yet identified latent viruses. Any system used to test for the presence of viruses should ideally be able to identify each component of the infesting population of the virus complex in the tested tissue. However, this makes the procedure very complicated and expensive. At present, specific antibodies against all the viruses that compose this complex are not yet available, nor are there specific DNA primers for the reverse-transcription polymerase chain reaction (RT-PCR) procedure.

In the absence of a single reliable technique for complete virus identification, health certification of garlic propagules should depend on a series of tests, including biological, serological and molecular methods.

In this chapter, we shall formulate a coherent general classification for the viruses infesting garlic and describe the most important of the currently available methods used to test for and eliminate these viruses so as to facilitate propagation of 'virus-free' garlic planting material.

2. Virus Diseases of Garlic

The RNA genome of plant viruses is relatively unstable and mutates quite often. The population of each specific virus (virion) is therefore a mixture of many mutants, i.e. of more than one RNA sequence. Any attempt to classify plant RNA viruses should take this point into consideration (Shiboleth *et al.*, 2001). Reciprocal false identification is also common. Hence, a recent study on shallot yellow-stripe virus (SYSV) showed that some previous classifications of potyviruses infecting shallot and Japanese bunching onion were inaccurate (van der Vlugt *et al.*, 1999). Sequence analysis of the gene coding for the viral coat protein confirmed that the former classification of those two different viruses was false and they actually belong to one species.

2.1 Potyviruses

The most common (Dovas *et al.*, 2001a) and probably the most damaging to garlic foliage and consequently to yield and quality of the bulbs are the potyviruses from the *Potyviridae* family. These include OYDV on garlic (Colour Plate 5A) and LYSV on *A. ampeloprasum* (Colour Plate 5B) (Delecolle and Lot, 1981; Delecolle *et al.*, 1985; Koch and Salomon, 1994a, b; Messiaen *et al.*, 1994; van Dijk, 1994; Barg *et al.*, 1995, 1997; Salomon *et al.*, 1996; Tsuneyoshi *et al.*, 1998b; Dovas *et al.*, 2001a, b; Shiboleth *et al.*, 2001). Garlic samples collected in Greece were infected with OYDV and LYSV at 98.5 and 83.7%, respectively (Dovas *et al.*, 2001a, b). A third potyvirus common in garlic and several other *Allium* species is the turnip mosaic virus (TuMV), which has a broad host range especially among the *Cruciferae*, where it was first discovered (Stefanac and Plese, 1980; Gera *et al.*, 1997). TuMV was reported in two wild *Allium* species of the Mediterranean basin (Stefanec and Plese, 1980) and recently in ornamental leek (*A. ampeloprasum*) in Israel (Gera *et al.*, 1997; Colour Plates 5A, B). Analysing the genes coding for CPs of OYDV, LYSV and TuMV from isolates from various parts of the world revealed a marked variation in nucleotides and in the amino acid composition (Table 13.1), thus indicating the possible development of local strains (Salomon *et al.*, 1996; Tsuneyoshi and Sumi, 1996; Tsuneyoshi *et al.*, 1998b).

Recently, an SYSV was identified in bulb onion and in Japanese bunching onion (Tsuneyoshi *et al.*, 1998b; van der Vlugt *et*

Table 13.1. Limits of detection for LYSV and OYDV by serological and molecular techniques.

Tissue	Diagnostic method used	Minimal amount of infected tissue required per reaction (μg)	Sample quantity per reaction (mg)	Sample volume per reaction (μl)	Sensitivity (infected tissue mg ml^{-1} extraction buffer)	Fold of sensitivity compared with ELISA
Leaf	ELISA	1,000	10	100	10	1
	Double-tube IC-RT-PCR	5	5	50	0.1	100
	Double-tube RT-PCR using RNA	0.02	2	1*	0.001	10,000
	Double-tube RT-PCR using LE	0.01	0.1	1	0.01	1,000
	One-step IC-RT-PCR	5	5	50	0.1	100
	One-step RT-PCR using LE	0.01	0.001	1	1	10
Bulb	Double-tube RT-PCR using BE	0.01	0.1	1	0.01	1,000
	One-step IC-RT-PCR	5,000	5	50	100	0.1
	'One-step' RT-PCR using BE	0.1	0.01	1	1	10

*1 μl of RNA extract, representing the equivalent of 2 mg of plant tissue.
LE, leaf extract; BE, bulb extract; ELISA, enzyme-linked immunosorbent assay; IC, immunocapture.

al., 1999) but not in garlic. Therefore, SYSV and SVX (Arshava *et al.*, 1995) will not be discussed here.

2.2 Carla viruses

Several carla viruses from the *Closteroviridae* were detected in garlic in many parts of the world. The effect of carla viruses in garlic, however, is less obvious than that of *Potyviridae*, although some of the carla viruses cause severe damage to garlic. The most common are the garlic common latent virus (GCLV) (van Dijk, 1993b, 1994; Tsuneyoshi *et al.*, 1998b; Dovas *et al.*, 2001a, b; Shiboleth *et al.*, 2001), shallot latent virus (SLV) (Nagakubo *et al.*, 1994; Tsuneyoshi *et al.*, 1998a), as well as the garlic mite transmitted viruses (Yamashita *et al.*, 1996). Using immunological methods, garlic virus V (GarV)-type virus was detected in garlic in Argentina and was assumed to be mite-transmitted. Application of molecular technologies to reveal the genome organization pointed to its relation to a carla or carla-like virus (Fig. 13.1; Helguera *et al.*, 1997).

2.3 Allexiviruses

A third group of viruses common in garlic are related to SVX-like viruses, of the *Allexiviruses*, a subgroup of the *Closteroviridae* (Kanyuka *et al.*, 1992; Sumi *et al.*, 1993, 1999; Arshava *et al.*, 1995; Pringle, 1998; Song *et al.*, 1998; Takaichi *et al.*, 1998; Shiboleth *et al.*, 2001). Very little research has been done on these viruses, and it is not clear if SVX actually infects garlic (Arshava *et al.*, 1995).

2.4 Mite-transmitted viruses

This group of garlic-infecting viruses has so far been classed only by the mode of transmission (van Dijk, 1991; van Dijk *et al.*,

Carla-like unclassified viruses

Shallot Virus X Group
Garlic virus A
Garlic virus B
Garlic virus C = Garlic mite-borne virus
Garlic virus D

True carla viruses

Garlic virus 1 = Garlic virus latent = Garlic mosaic virus ≠ Garlic common latent carla virus

Fig. 13.1. Classification of garlic carla viruses.

1991; van Dijk and van der Vlugt, 1994; Yamashita *et al.*, 1996). Little information is available on this group of viruses. However, a recent report proposed that some mite-transmitted viruses belong to the group of carla or carla-like viruses (Helguera *et al.*, 1997). Since the mite-transmitted viruses of garlic have not yet been completely isolated and identified, we assume for the present that some of them belong to the group of mite-transmitted potyviruses, as does the wheat streak mosaic virus (WSMV).

2.5 Nematode-transmitted viruses

This assemblage of garlic-infesting viruses has so far been grouped only by the mode of transmission of its members (Graichen, 1975). No more recent information is available.

2.6 Cumulative damage

The cumulative reduction in garlic yield caused by a mixture of viruses is very high. However, the contributing damage by individual viruses in mixtures has not yet been investigated except for that of OYDV and LYSV. Lot *et al.* (1998) compared the yields of garlic freed from the two viruses with that of standard propagation material and estimated yield loss due to virus infection at about 50%.

3. Transmission of Virus Diseases in Garlic

Viruses infecting garlic are carried over from one season to the next through the infected propagation material. In addition, there is also a continuous flux of viruses from one infected plant to another, from adjacent fields and from wild flora to cultivated fields. Mites (van Dijk *et al.*, 1991; van Dijk and van der Vlugt, 1994; Yamashita *et al.*, 1996) and nematodes (Graichen, 1975) have all been reported as vectors for garlic viruses. It has been found that virus-free garlic is quickly reinfected in open fields (Ohkoshi, 1991; Lot *et al.*, 1998), indicating that the viruses are transmitted from adjacent plots of infected garlic and/or from other plant species, such as wild *A. ampeloprasum*. The immediate candidates as vectors are arthropods. However, little information is available on the role of arthropods as vectors (Koch and Salomon, 1994b; van Dijk, 1994; van Dijk and van der Vlugt, 1994). Viruses such as those mechanically transmitted (including by wind and dust) – for example, tobacco mosaic virus (TMV) – have not yet been found in garlic.

4. Virus Detection and Identification

For many years, methods for virus isolation and identification in garlic were based on specific symptoms such as local lesions and specific symptoms on the target and test plants (Matthews, 1991). Later, more precise analytical methods were developed. Virus isolates from local lesions were propagated and used for the production of antibodies against the specific isolates, for serological assays and for electron microscopy (EM), which allows virus particles to be visually identified (Figs 13.2 and 13.3). In recent years, molecular techniques, including sequence analyses of proteins and genomes have been developed for both detection and identification of garlic viruses.

4.1 Biological methods

Earlier biological methods using test plants were only partially efficient in isolation of

(a)
(b)

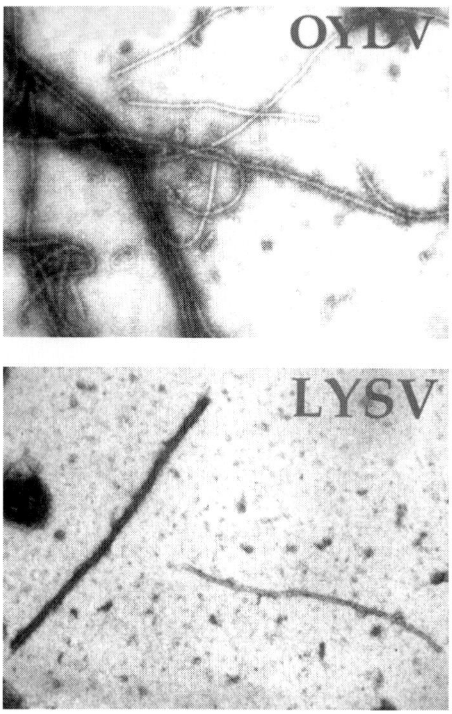

Fig. 13.2. Virus particles extracted from garlic as a mixed population treated with specific rabbit polyclonal antibody and visualized by immuno-electron microscopy. (a) With antibody against OYDV, prepared from *Escherichia coli*-expressed coat-protein gene. (b) With antibody against LYSV. In both photographs, only part of the elongated viruses reacted with the specific antibody, leaving the other virions not decorated.

(a)
(b)

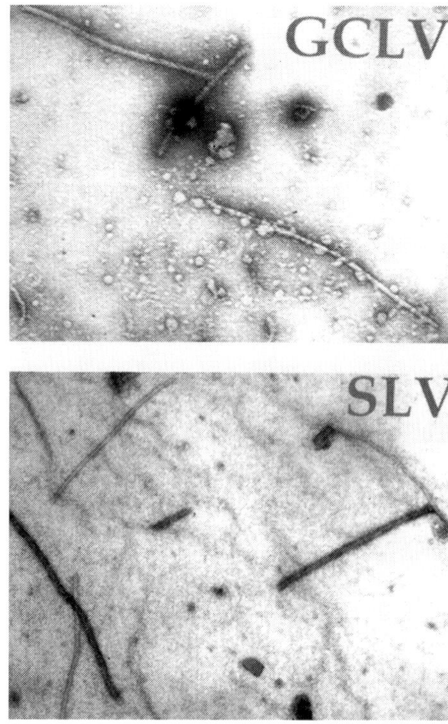

Fig. 13.3. Virus particles extracted from garlic as a mixed population were treated with specific rabbit polyclonal antibody and visualized by immuno-electron microscopy. (a) Treated with antibody against garlic carla latent virus (GCLV). (b) Treated with antibody against shallot latent virus (SLV). The binding of antibodies to GCLV was weak, shown in the weak decoration of this virus. Similarly to Fig. 13.2a,b, only the specific viruses were decorated, while the rest of the extracted virus mixture was unaffected.

individual viruses from the mixtures common in garlic. For instance, the garlic type OYDV infects only garlic, great-headed garlic (*A. ampeloprasum* L., great-headed garlic group) and leek (*A. ampeloprasum* L., leek group), but not bulb onion and other *Allium* species (van Dijk, 1993a). In the absence of hosts susceptible only to the garlic-type OYDV but not to other members of the virus complex (van Dijk, 1994), it was difficult to isolate this virus by the common biological procedures. Therefore, isolates of the garlic-type OYDV were always accompanied by other viruses (Shiboleth *et al.*, 1997). On the other hand, LYSV, the most common *Allium*-infecting potyvirus, was easily isolated using the specific *Allium* host *A. ampeloprasum* and *Chenopodium quinoae* as test plants (Delecolle and Lot, 1981; Bos, 1983; Yamashita *et al.*, 1995; Salomon *et al.*, 1996). Later, antibodies were prepared against the purified virus using LYSV coat protein expressed in bacteria (Salomon *et al.*, 1996). TuMV was also amenable to biological isolation through test plants (Gera *et al.*, 1997).

4.2 Serological methods

Immunological techniques are sensitive, specific and very accurate (Table 13.1). Specific polyclonal antisera are produced against purified virion isolates. They enable researchers to co-apply several immunological techniques (Delecolle and Lot, 1981; Conci *et al.*, 1992); the most commonly used are modified enzyme-linked immunosorbent assay (ELISA) techniques (Delecolle *et al.*, 1985; Barg *et al.*, 1994; Koch and Salomon, 1994b; Koch *et al.*, 1995a; Helguera *et al.*, 1997; Coperland, 1998). ELISA tests are inexpensive and facilitate a single-step large-scale examination of samples from meristem-tip culture or any other tissue of interest for the presence of viruses.

Antisera of high specificity can be obtained against viruses isolated from the virus mixture harboured by garlic. Among the *Potyviridae*, LYSV is most commonly detected by ELISA, and was found in 73% of the garlic clones tested around the world (van Dijk, 1994) and in 86% of the garlic bulbs analysed in Brazil (Daniels, 1999).

The frequency of OYDV incidence in the samples was only half that of LYSV (Daniels, 1999). Antisera prepared against OYDV were in most cases a mixture of antibodies against this and other viruses accompanying OYDV extracts from garlic. However, the highly specific polyclonal antibody recently induced against bacterially expressed OYDV CP enables large-scale tests of high accuracy (Dovas *et al.*, 2001a, b).

Gar-V-type virus, common in garlic in Argentina, was also detected by immunological methods (Helguera *et al.*, 1997).

4.3 Electron-microscopic visualization and the combination of serology and electron microscopy

Visualization, by decoration of isolated virions with antibodies, is now a common procedure for virus identification (Bos, 1983; Walkey, 1990; Matthews, 1991; Conci *et al.*, 1992; Koch and Salomon, 1994b; van Dijk, 1994; Dovas *et al.*, 2001a; Figs 13.2a,b, 13.3a,b). However, this procedure, although sensitive and accurate, is very expensive and depends on the availability of the EM equipment as well as of highly qualified specialists to perform it.

4.4 Molecular markers

The fact that antiserum prepared against OYDV consists of a mixture of antibodies against a number of viruses was a drawback to the value of the serological technique. This difficulty was overcome recently by molecular methods in which the genes coding for the CP of different virus species were isolated and later expressed in bacteria (Kobayashi *et al.*, 1996; Shiboleth *et al.*, 1997).

The use of specific primers based on the RT-PCR procedure produces the most sensitive and specific detection method known today (Sumi *et al.*, 1993; Nagakubo *et al.*, 1994; Kobayashi *et al.*, 1996; Salomon *et al.*, 1996; Tsuneyoshi and Sumi, 1996; Shiboleth *et al.*, 1997; Takaichi *et al.*, 1998; Tsuneyoshi *et al.*, 1998a, b; van der Vlugt *et al.*, 1999; Dovas *et al.*, 2001a, b). A recent common technique for sequencing the amino acids of an isolated protein is based on time-of-flight mass spectroscopy (TOFMS). However, the only protein easily isolated from virus-infected plant tissue is the CP. Hence, other differences in the viral genome cannot be detected by this technique.

The proper application of the RT-PCR procedure for the detection of viral RNA requires careful adjustment of the reaction conditions for the specific virus under investigation (Kobayashi *et al.*, 1996; Rosner *et al.*, 1998; Shiboleth, 1998; Dovas *et al.*, 2001a, b; Shiboleth *et al.*, 2001). The correct design of the DNA primer is essential for its accurate annealing to the viral RNA, which enables the accurate transcription and DNA multiplication of segments of the viral genome (Table 13.2).

The use of degenerated DNA primers, for sequences expected to be present in a number of viruses, can broaden the detection capacity of a single marker to a large number of viruses. However, this procedure is expected to be less efficient than the one utilizing specific primers, due to some inherent limitations. Degenerated DNA primers may either

Table 13.2. The use of GenBank information to devise, prepare and test the DNA primers that are listed in Table 13.3 (from Shiboleth *et al.*, 2001). A comparison between Israeli local virus clones to nearest GenBank accession. Local Israeli virus clones were compared to GenBank accessions with the aid of BLASTP or BLASTN (http://www.ncbi.nlm.nih.gov/).

Israeli clone	Nearest relative(s)	Amino acid identity	Nucleic acid identity
No. 7 (ORF[a] V to middle of ORF III)	GVA[b] (Ac. no. D11157) ShVX-related	Not compared	97% (compared in ORF III)
No. 11 and No. 18 (ORF V to middle of ORF III)	GVC, garlic mite-borne (Ac. no. D11159) ShVX-related	Not compared	78% in ORF V and 78% in ORF III
No. 16 (ORF VI and part of coat = ORF V)	GCLV, Ac. nos X81138 and 9 Carlavirus	97% identity	Not compared
OYDV 5–13 Ac. no. AF071226	See Fig. 13.1	Above 90% in whole coat protein gene	Not compared
LYSV FLC-CP Ac. no. AF071525	LYSV, Ac. no. D28590	Above 95% in whole coat protein gene	Not compared
TuMV W2 Ac. no. AF071526	TuMV, Ac. no. D10601	Not compared	90% in partial coat protein gene

[a]ORF, open reading frame.
[b]GVA, garlic virus A.
[c]GVC, garlic virus C.

recognize some host sequences as well, thus resulting in false positives, or may not anneal to some or all of the viral RNA, thus failing to indicate some of the infecting viruses.

A crucial factor in obtaining specific annealing is the proper reaction temperature. The two factors – specific primer and exact reaction temperature – should be determined specifically for each virus, as was done for OYDV (Kobayashi *et al.*, 1996; Shiboleth, 1998; Dovas *et al.*, 2001a, b; Shiboleth *et al.*, 2001; Fig. 13.4).

A list of various DNA primers used for the detection of OYDV in different countries

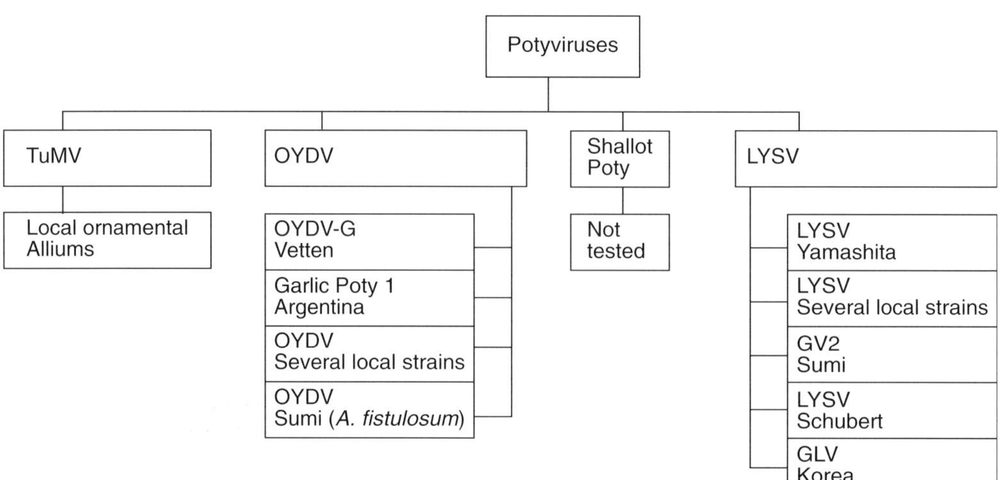

Fig. 13.4. A comparison of molecularly classified garlic potyviruses from Israel and other countries. Note that the GLV (= GMV) from Korea is identical to LYSV. In each column, all viruses listed belong to the same species. Different names were given by different authors due to lack of accurate means of virus identification.

and laboratories is presented in Table 13.3. The heterogeneity of the CP genes of OYDV, LYSV and TuMV from various sources is evident from sequence comparisons (Sumi *et al.*, 1993; Nagakubo *et al.*, 1994; Shukla *et al.*, 1994; Kobayashi *et al.*, 1996; Salomon *et al.*, 1996; Gera *et al.*, 1997; Shiboleth *et al.*, 2001). The effect of this heterogeneity on detection and identification of low-concentration virus particles, such as those present in plants grown from meristem culture, may be significant when inappropriate primers and annealing conditions are used for the RT-PCR reaction.

5. Virus Elimination Techniques

5.1 Meristem-tip culture

Embryonic cells in the garlic meristem are free (or almost free) of all infecting viruses common in the cloves and other plant tissue. Hence, plantlets regenerated from meristem tips may also be free of viruses (Novak, 1990; Ohkoshi, 1991; Wang *et al.*, 1994). However, embryonic cells in the meristem are few and are difficult to extract free from adjoining infected cells (Ma *et al.*, 1994). All meristematic cells have a similar appearance and yet the ones adjoining the actual embryonic cells may already have acquired viruses from the slightly more developed neighbouring cells, probably through newly formed plasmodesmata. Therefore, only a fraction of the plantlets grown from the cultured tips are free of viruses (Novak, 1990; Walkey, 1990; Xu *et al.*, 1994; Verbeek *et al.*, 1995; Lot *et al.*, 1998; Roberts *et al.*, 1998; Shiboleth *et al.*, 2001), and the success rate for obtaining virus-free shoots and plants is cultivar-dependent (Koch *et al.*, 1995b; Verbeek *et al.*, 1995; Shiboleth *et al.*, 2001). A high rate of regeneration was reported for some cultivars (Ucman *et al.*, 1998) and low rates for others (Koch *et al.*, 1995b). Most attempts to obtain high proportions of virus-free embryonic cells by culturing very small excised tips failed and resulted in non-viable tissues or in very low rates of propagation (plantlet formation). Therefore, further decreasing the size of sampled tissue is of no practical value, although in theory it is technically possible.

For many years, meristem-tip culture has been used worldwide for the production of 'virus-free' propagules (Peña-Iglesias and Ayuso, 1983; Bertaccini *et al.*, 1986; Walkey *et al.*, 1987; Walkey and Antill, 1989; Ma *et al.*, 1994; Messiaen *et al.*, 1994; Ravnikar *et*

Table 13.3. Primers used in OYDV cloning, expression and diagnostics.

Primer sequence	Name	Direction
5′TGAAGCATACATTGAATATA	AB2S*	Forward (5′–3′)
5′TGCTCGAAGTCAGGTTAAACGAA	GP1#3†	
5′GCTATAAAAGAGGTTCGCTATC	Q-GP1†	Forward
5′CATGCCATGGCTGGCACAGGCGAAGATGC	Nco1 F-1‡	Forward
5′CGCCATATGGCTGGCACAGGCGAAGATGC	Nde 5-13 @	Forward
5′CCGCTCGAGCATCTTAATACCAAGTAAGG	Xho 5-13 @‡	Reverse
5′TGCTGTGTGCCTCTCCGTGTCCTC	RS1§	Reverse

*AB2S is an OYDV-garlic primer, used by Dr H.J.Vetten at Braunschweig, Germany.
†Primers GP1#3 (used to create 5-13, 869 bp) and QGP1 are based on GenBank Ac. no. X89402 (Kobayashi *et al.*, 1996) and are situated in the 3′ untranslated region and nuclear inclusion body region, respectively.
‡Primers Nco1F-1 (based on clone F-1, a short homologue of 5-13) and Xho5-13 (based on clone 5-13) were used to subclone the coat protein of clone 5-13 into a pET22b(+), Plasmid, Novagen, USA (pET) expression vector.
§Primer RS1 is in a conserved potyvirus coat-protein area, based on an LYSV sequence (GV2 by Nagakubo *et al.*, 1994, GenBank Ac. no. D28590).
@ Indicates the primers used for diagnostic purposes.

al., 1994; Yun *et al.*, 1998). However, in Argentina, this method was not very effective at removing the Gar-V-type virus from garlic (Helguera *et al.*, 1997).

5.2 Thermotherapy

At temperatures above 38°C, virus multiplication in plant cells is reduced or even completely stopped. At the same time, plant meristematic cells continue to divide and replicate, though rather slowly. The slow rate of virus propagation following high-temperature treatment results in reduced amounts of virus particles available for movement into newly formed cells. Under these very stringent conditions, the newly formed tissue may be free of viruses (Conci and Nome, 1991; van Dijk, 1993a; Xu *et al.*, 1994; Verbeek *et al.*, 1995; Bruna, 1997; Ghosh *et al.*, 1997; Ucman *et al.*, 1998). The combined use of the meristem-tip culture procedure with thermotherapy increases the chance of regenerating virus-free propagules, and yet its efficiency is still genotype-dependent (Xu *et al.*, 1994; Ucman *et al.*, 1998). Moreover, the strength of the heat treatment needs to be adjusted for each cultivar.

Garlic viruses are differentially affected by thermotherapy. For example, thermotherapy was very effective in eliminating LYSV from garlic stem tips, but had no effect on OYDV (Ravnikar *et al.*, 1994; Ucman *et al.*, 1998; Shiboleth *et al.*, 2001). The application of thermotherapy should therefore be tested for each virus separately to ensure the success of the treatment.

5.3 Chemotherapy

Another approach to virus elimination from plant tissues is the use of chemicals that interfere with nucleic acid replication, thus limiting virus multiplication (Shiboleth, 1998). This method has achieved little success, since the treated tissues may suffer damage as well. Chemotherapy may also induce mutations in the cultured tissue and thus alter the horticultural traits of the propagated garlic. Furthermore, chemicals such as 5-fluorouracil (5-FU) and 6-azoguanine (6-AG) are toxic and thus dangerous both to the user and to the environment. The use of chemotherapy to reduce replication is therefore not recommended for use outside laboratory experiments.

All of the above methods are non-discriminating for a specific virus or viral groups and therefore cannot be used for the attribution of damage estimates to any single virus.

6. Analysing for Virus Presence in Meristem-tip-grown Garlic Plants

Excised meristem tips are composed of embryonic stem cells and neighbouring cells, which are in the process of differentiation and weaving intracellular cytoplasmic connections – the plasmodesmata. The embryonic stem cells are the most suitable tissue for initiation of virus-free cultures. In practice, however, excised tissues consist of stem and neighbouring cells and may harbour a few or many virus particles. Consequently, the emerging plantlets may be either virus-free or infected with any number of virus particles. A very sensitive method is therefore needed for virus detection in the very early stages of vegetative reproduction.

6.1 Biological detection methods

The sensitivity and reliability of the available detection methods determine the efficiency of the screening procedure of virus-free garlic plants from meristem-tip culture (Dovas *et al.*, 2001a, b). To guarantee freedom from viruses, the regenerated plantlets should be tested for their health status through three growing seasons. When only biological detection methods (symptom appearance and sap transmission) are available, plantlets are allowed to grow for a period sufficient for the accumulation of detectable amounts of virus particles. The tiny bulbils obtained from tissue culture are transplanted and cultivated through a second and a third season, during which viruses (if present) multiply until they reach a detectable mass (Walkey and Antill, 1989; Walkey, 1990; Messiaen *et al.*, 1994). The biological method can, in some cases, reliably be used to identify the presence of a single viral taxon, but its threshold sensitivity is very high. Other limitations were dealt with earlier in this chapter.

6.2 Serological methods

Garlic populations from meristem-tip culture consist of a mixture of virus-free and infected plants. Some of the latter contain very few virus particles, frequently below the detection threshold of the available observational methods. In such cases, viruses can be detected by the more accurate serological assays. The most common and efficient are ELISA and immuno-electron microscopy. When applied by van Dijk and co-workers (1991) to identify virus-infected plants from meristem-tip cultures, a high percentage of virus-free plants were recovered and yields improved substantially. However, high percentages of infected plants were identified in the second growth season, probably not only due to vector transmission, but also due to the threshold of virus detection by this procedure, whereby plants containing only a few virus particles were not detected in the first season.

6.3 Molecular methods

The introduction of the one- or two-step RT-PCR method for detection of viral RNA has enabled researchers to shorten the screening period markedly and with greater reliability (Dovas *et al.*, 2001a, b; Shiboleth *et al.*, 2001). The sensitivity of the two-step RT-PCR is 10^2–10^4 times higher than that of serological detection methods, such as ELISA (Shiboleth *et al.*, 2001; Table 13.1). Therefore, RT-PCR is the current preferred method for testing for the presence of viruses in meristem tips of garlic (Dovas *et al.*, 2001a, b).

The RT-PCR procedure is expensive and requires expert skills to perform it. It is therefore impractical for large-scale testing and should be applied only to a small number of plants. Early detection and elimination of infected cultures by biological or serological assays followed by RT-PCR, results in a significant shortening of time needed for the production of garlic propagules free of viruses, and at lower costs than by the latter method solely. Therefore, serological methods are used first to identify 'virus-free' plants from meristem-tip culture. This is followed by RT-PCR analysis of the small number of suspected virus-free plants, to reconfirm freedom from viruses. Consequently, clean plantlets are ready for multiplication within one season. The same screening routine is applied in the second year to produce the first certified propagation material for the production of a nucleus of virus-free material in insect-proof propagation houses. Commercial multiplication in isolated fields is the final step prior to release of certified propagules (Table 13.4).

7. Vegetative Propagation of Garlic and its Implications

Nearly all cultivated garlic clones grown today are completely sexually sterile (Etoh and Simon, Chapter 5, this volume) and therefore are propagated vegetatively from cloves. Variation is introduced mainly by natural mutations in growing plants or in tissue culture (somaclonal variation) (Novak, 1990; Koch and Salomon, 1994a) or, to a smaller extent, by induced mutations. A single outstanding plant thus becomes the origin of a new clone. However, this mode of vegetative propagation perpetuates biotic factors, especially viruses, from one generation to another.

Many growers select the largest cloves for propagation, as experience shows that they yield larger bulbs. Clove size of a given plant is affected by its position in the bulb (outer or inner whorl), by environment and field fertility and by damage from biotic factors. Selection for the largest bulbs and cloves within a given field therefore also means selection for plants in which the effects of the virus are attenuated. This practice significantly increases the cost of propagation material and of garlic production.

Elimination of the most damaging viruses from the propagation material results in yield increases, mainly due to an increase of over 50% in the size of bulbs and individual cloves due to both the reduced damage and the increased vigour of virus-free plants (Walkey and Antill, 1989; Walkey, 1990; Oh *et al.*, 1994; Verbeek *et al.*, 1995; Lot *et al.*, 1998). The consequent increase in growers' income is greater than the increase in yield due to consumers' preference for large bulbs.

8. Multiplication of Virus-free Tested Garlic from Laboratory to Crop

The propagation cycle from meristem-tip culture up to the production of a crop is the most elaborate and expensive component of the production of commercial virus–free garlic (Bhojwani *et al.*, 1982; Ohkoshi, 1991; Messiaen *et al.*, 1994; Xu *et al.*, 1994). Following the laboratory production of meristem-tip cultures, the scale-up procedure from the initial few virus-free plantlets to a controlled commercial propagation field begins with propagation in an insect-proof, 50-mesh (per inch) screen-house. Standard agricultural practice is applied throughout and, in addition, there is careful supervision and strict and careful control of possible vectors and a continuous sampling for early detection of contamination. The small-sized first-season virus-free bulblets (with two to three cloves per bulb in modern cultivars) are used for further increase in isolated fields, preferably in a cool-climate area, to reduce interaction with potential vectors and reinfestation from adjacent wild or cultivated plants.

Experience gained in Argentina, France, Spain and Japan has shown that field-grown virus-free garlic and Japanese bunching onion become heavily reinfested within three to four growing seasons in open fields (Ohkoshi, 1991; Lot *et al.*, 1998), and plants grown for propagation are no different. Reinfestation occurs rather quickly, probably by vector transmission from nearby infected commercially grown plots or from infected wild *Allium* species, as reported for LYSV (Sosa *et al.*, 1997; Lot *et al.*, 1998). For economic reasons, field propagation in isolated plots is repeated once or twice, thus obtaining maximum yields of large-sized multiclove bulbs. Continuous inspection in the propagation fields for vectors and strict pest-control management are therefore essential throughout. Random samples are tested for contamination throughout the growing season and at harvest. At each stage, the propagation material may be discarded upon detection of virus infection at a rate > 1%. This 4-year cycle of developing commercial-sized bulbs in sufficient quantities for commercial propagation is long, laborious and expensive, and yet results in a high-value product (Table 13.4).

Reinfestation nullifies the advantages expected from the virus-free propagation material, as the invading virus may be extremely virulent and the virus level may exceed that common in the original untreated garlic. This may happen due both to favourable virus propagation conditions in the vigorous plants from meristem-tip cultures and to the absence of biotic competition (Ohkoshi, 1991; Sosa *et al.*, 1997).

A number of commercial companies market certified garlic propagules. However, our

Table 13.4. Flow chart demonstrating the five steps from (1) generation of virus-free initial material to (4) commercial propagules and (5) farmers' fields. Note that, following monitoring for freedom from viruses, nuclear material propagated in an insect-proof screen-house can be further used as virus-free initial material for second, third and fourth years in the screen-house.

1st year	2nd year	3rd year	4th year	5th year
Virus elimination: generation of original virus-free initial material	Nuclear material × 2–3 multiplication	Propagation × 3–5 multiplication	Propagation × 10 multiplication	Transplanting
In vitro propagation ⇨	Screen-house ⇨	Isolated field ⇨	Isolated field ⇨	Commercial crop

tests with RT-PCR showed that many samples of this high-quality propagation material from tip culture are not completely free of viruses, but still contain very low levels of some viruses (R. Salomon, unpublished data). A California-based company, Basic Vegetable Products, produces 'virus-free' propagules of 'California Early', 'California Late' and others, and French companies such as Top Semence and Allicoop produce 'virus-free' propagules of cvs 'Messidrome', 'Germidor' and 'Printanor', which are planted commercially for several years with a marked yield advantage over the virus-infected parent lines (Sosa *et al.*, 1997; Lot *et al.*, 1998).

9. Conclusions and Future Developments

Garlic propagated from virus-free cloves shows a yield increase of 50% or more over the yield of the untreated plants (Walkey and Antill, 1989; Walkey, 1990; Lot *et al.*, 1998). The increase in yield results from more vigorous plant growth, which in turn results in larger cloves and bulbs. In economic terms, the increase in revenues may be even greater than the weight increase suggests, since the larger bulbs fetch a higher price per unit weight compared with small ones. Therefore, the use of virus-free propagation material provides an improved horticultural practice for garlic cropping, even when the price of the propagation material is higher than that of the conventional planting material.

Garlic genotypes vary markedly in their response to tissue-culture conditions. Hence, research is required to develop procedures specific to the cultivars most suitable for each region. This can be done either by public support or by commercial companies, as in the USA, Argentina, Brazil and Western Europe, where both growers and investors have profited from the introduction of virus-free propagation material.

The use of 'virus-free' propagules is spreading fast. This can only occur, however, where large plots, regional coordination and improved field management are practised as essential steps for reducing the risk of immediate reinoculation and increasing the chance of success. Growers of virus-free garlic have to adopt a complete change in management, including an absolute separation between the production of propagation material and the cultivation of the commercial crop. Many growers throughout the world use part of the commercial crop as propagation material. Adaptation to virus-free production implies that reproduction is done only by specialized growers/companies, and annual purchase of propagation material from these sources is required. Since reinfestation quickly occurs from neighbouring fields or wild plants (Sosa *et al.*, 1997), it is imperative to have regional cooperation in cultivation, sanitary controls (both of pests and of weeds, which may serve as hosts for vectors and/or garlic viruses) and management. A single, small grower or even a backyard amateur in the vicinity of a virus-free garlic production area may become a source of reinfestation and jeopardize the whole operation.

The large-sized production blocks require mechanization and the uniform crop facilitates the operations of planting, spraying, fertigation, harvest, trimming, cleaning, sorting and packing. Only big farms or regional coordination supported by farmers' organizations, regional councils or national governments can guarantee the safety of the crop and justify the investment needed for the equipment, which again needs phytosanitary attention to prevent infection by other biotic factors.

In many countries garlic is a long-established crop and local cultivars have been selected that are well adapted to the local conditions and the local markets. In addition, bulbing is dependent on day length. Thus, imported high-quality propagules of foreign cultivars may not be suitable. A more appropriate practice is to free the local clones from viruses. Furthermore, since garlic is traded between countries, imported garlic should always be inspected for diseases and pests and, if not certified for propagation, must be used only for consumption and not for planting. However, the more widespread the use of virus-free propagules becomes worldwide, the smaller will be the danger of virus spreading from one location to another.

It has yet to be determined whether freeing garlic from viruses affects quality traits, such as flavour, dry-matter content and shelf-life. The French cultivars 'Messidrome', 'Germidour' and 'Printanor', of which virus-free propagation material exists, retain all the qualities of the original cultivars.

Certified propagation material, which is not truly virus-free, is marketed in the USA. However, our tests with RT-PCR showed that many samples of this high-quality propagation material from tip culture are not completely free of viruses, but contain very low levels of some viruses, present at concentrations below the detection level of the serological methods (R. Salomon, unpublished data). This propagation material shows a superior performance over the original cultivars; however, virus propagation in it is fast and, within two seasons, the plants suffer from a high rate of virus infestation. Thus, the use of this certified propagation material is limited to one or two seasons.

Future research will have to determine the extent of damage inflicted on garlic by each individual virus and will allow the comparison of the detailed characteristics of virus-free garlic with those of the original local cultivars.

References

Arshava, N.V., Konareva, T.N., Ryabov, E.V. and Zavriev, S.K. (1995) The 42k protein of shallot virus X is expressed in infected *Allium* plants. *Molecular Biology (Moscow)* 29, 192–198 (in Russian).

Barg, E., Lesemann, D.-E., Vetten, H.J. and Green, S.K. (1994) Identification, partial characterisation and distribution of viruses infecting *Allium* crops in south and south-east Asia. *Acta Horticulturae* 358, 251–258.

Barg, E., Lesemann, D.-E., Vetten, H.J. and Schonfelder, M. (1995) Differentiation of potyviruses infecting cultivated *Allium* species. *Proceedings of the 8th Conference on Virus Diseases of Vegetables, Prague, 9–15 July 1994*, pp. 29–31.

Barg, E., Lesemann, D.-E., Vetten, H.J. and Green, S.K. (1997) Viruses of alliums and their distribution in different *Allium* crops and geographical regions. *Acta Horticulturae* 433, 607–616.

Bertaccini, A., Marani, F. and Borgia, M. (1986) Shoot-tip culture of different garlic lines for virus elimination. *Revista Ortoflorofrutticoltura Italiana* 70, 97–105.

Bhojwani, S.S., Cohen, D. and Fry, P.R. (1982) Production of virus-free garlic and field performance of micropropagated plants. *Scientia Horticulturae* 18, 39–43.

Bos, L. (1983) Viruses and virus diseases of *Allium* species. *Acta Horticulturae* 127, 11–29.

Bruna, A. (1997) Effect of thermotherapy and meristem-tip culture on production of virus-free garlic in Chile. *Acta Horticulturae* 433, 631–634.

Conci, V.C. and Nome, S.F. (1991) Virus free garlic (*Allium sativum* L.) plants obtained by thermotherapy and meristem tip culture. *Journal of Phytopathology* 132, 186–192.

Conci, V., Nome, S.F. and Milne, R.G. (1992) Filamentous viruses of garlic in Argentina. *Plant Disease* 76, 594–596.

Coperland, R. (1998) Assaying levels of plant virus by ELISA. In: Foster, G.D. and Taylor, S.C. (eds) *Methods in Molecular Biology*, Vol. 81: *Plant Virology Protocols: From Virus Isolation to Transgenic Resistance*. Humana Press, Totowa, New Jersey, pp. 455–460.

Daniels, J. (1999) Occurrence of viruses in garlic in the state of Rio Grande do Sul, Brazil. *Fitopatologia Brasileira* 24, 91–96 (in Portuguese).

Delecolle, B. and Lot, H. (1981) Garlic viruses: detection and partial characterization with immune electron microscopy of three different garlic populations with mosaic. *Agronomie* 1, 763–770 (in French).

Delecolle, B., Lot, H. and Michel, M.J. (1985) Application of ELISA for detecting onion yellow dwarf virus in garlic and shallot seeds and plants. *Phytoparasitica* 13, 266–267.

Dovas, C., Hatziloukas, E., Salomon, R., Barg, E., Shiboleth, Y.M. and Katis, N. (2001a) Incidence of viruses infecting *Allium* spp. in Greece. *European Journal of Plant Pathology* 107, 677–684.

Dovas, C., Hatziloukas, E., Salomon, R., Barg, E., Shiboleth, Y.M. and Katis, N. (2001b) Comparison of methods for virus detection in *Allium* spp. *Journal of Phytopathology* 149, 731–737.

Gera, A., Lesemann, D.-E., Cohen, J., Franck, A., Levy, S. and Salomon, R. (1997) The natural occurrence of turnip mosaic potyvirus in *Allium ampeloprasum*. *Journal of Phytopathology* 145, 289–293.

Ghosh, D.K., Ahlawat, Y.S. and Gupta, M.D. (1997) Production of virus-free garlic (*Allium sativum*) plants by thermotherapy and meristem tip culture. *Indian Journal of Agricultural Sciences* 67, 591–593.

Graichen, K. (1975) *Allium* species as natural hosts of nematode transmissible viruses. *Archiv für Phytopathologie und Pflanzenschutz* 11, 399–403.

Helguera, M., Bravo-Almonacid, F., Kobayashi, K., Rabinowicz, P.D., Conci, V. and Mentaberry, A. (1997) Immunological detection of a Gar V-type virus in Argentine garlic cultivars. *Plant Disease* 81, 1005–1010.

Kanyuka, K.V., Vishichenko, V.K., Levay, K.E., Kondrikov, D.Y., Ryabov, E.V. and Zavriev, S.K. (1992) Nucleotide sequence of shallot virus X RNA reveals a 5′-proximal cistron closely related to those of potexviruses and a unique arrangement of the 3′-proximal cistrons. *Journal of General Virology* 73, 2553–2560.

Kobayashi, K., Rabinowicz, P., Bravo-Almonacid, F., Helguera, M., Conci, V., Lot, H. and Mentaberry, A. (1996) Coat protein gene sequences of garlic and onion isolates of the onion yellow dwarf potyvirus (OYDV). *Archives of Virology* 141, 2277–2287.

Koch, M. and Salomon, R. (1994a) Improvement of garlic via somaclonal variation and virus elimination. *Acta Horticulturae* 358, 211–214.

Koch, M. and Salomon, R. (1994b) Serological detection of onion yellow dwarf virus in garlic. *Plant Disease* 78, 785–788.

Koch, M., Ta'anami, Z., Levi, S. and Salomon, R. (1995a) Testing garlic cloves and bulblets for onion yellow dwarf virus by ACP-ELISA. *Phytoparasitica* 23, 27–29.

Koch, M., Ta'anami, Z. and Salomon, R. (1995b) Improved regeneration of shoots from garlic callus. *HortScience* 30, 378.

Lot, H., Chovelon, V., Souche, S. and Delecolle, B. (1998) Effects of onion yellow dwarf and leek yellow stripe viruses on symptomatology and yield loss of three French garlic cultivars. *Plant Disease* 82, 1381–1385.

Ma, Y., Wang, H.L., Zhang, C.J. and Kang, Y.Q. (1994) High rate of virus-free plantlet regeneration via garlic scape-tip culture. *Plant Cell Reports* 14, 65–68.

Matthews, R.E.F. (1991) *Plant Virology*, 3rd edn. Academic Press, New York, 835 pp.

Messiaen, C.M., Lot, H. and Delecolle, B. (1994) Thirty years of France's experience in the production of disease-free garlic and shallot mother bulbs. *Acta Horticulturae* 358, 275–279.

Nagakubo, T., Kubo, M. and Oeda, K. (1994) Nucleotide sequence of the 3′ regions of two major viruses from mosaic diseased garlic: molecular evidence of mixed infection by a potyvirus and a carlavirus. *Phytopathology* 84, 640–645.

Novak, F.J. (1990) *Allium* tissue culture. In: Rabinowitch, H.D. and Brewster, J.L. (eds) *Onions and Allied Crops*, Vol. I. *Botany, Physiology, and Genetics*. CRC Press, Boca Raton, Florida, pp. 233–250.

Oh, D.G., Suh, H.D., Kim, K.T. and Lee, J.W. (1994) Field performance of meristem-tip-culture derived seed garlic. *Acta Horticulturae* 358, 281–284.

Ohkoshi, K. (1991) Production of virus-free plants by meristem culture vegetables and ornamental plants. In: *The Biological Control of Plant Diseases*. FFTC Book Series No. 42, ASPAC Food and Fertilizer Technology Centre, pp. 87–95.

Peña-Iglesias, A. and Ayuso, P. (1983) Characterization of Spanish garlic viruses and their elimination by *in vitro* shoot apex culture. *Acta Horticulturae* 127, 183–193.

Pringle, C.R. (1998) Virus Taxonomy – San Diego 1998. 27th Meeting of the Executive Committee of the ICTV. *Archives of Virology, Virology Division News* 143, 7.

Ravnikar, M., Plaper, I., Ucman, R. and Zel, J. (1994) Establishment of an efficient method for virus elimination in meristem cultures and regeneration of high-quality plants. In: Javornik, B., Bohanec, B. and Kreft, I. (eds) *Proceedings of the International Colloquium on the Impact of Plant Biotechnology on Agriculture, University of Ljubljana, Slovenia, 5–7 Dec. 1994*. Centre for Plant Biotechnology and Breeding, Agronomy Department, University of Ljubljana, Slovenia, pp. 97–102.

Roberts, J.D., Bebenek, K. and Kunkel, T.A. (1998) The accuracy of reverse transcriptase from HIV-1. *Science* 242, 1171–1173.

Rosner, A., Shiboleth, Y., Speigel, S., Krizbai, L. and Kolber, M. (1998) Evaluation of IC-RT-PCR for detection of prunus necrotic ringspot virus in stone fruits. In: Hadidi, A. (ed.) *Proceedings of the 17th International Symposium on Virus and Virus-like Diseases of Temperate Fruit Crops*. US Department of Agriculture, Bethesda, Maryland, 23–27 June 1997. *Acta Horticulturae* 472, 227–233.

Salomon, R., Koch, M., Levy, S. and Gal-On, A. (1996) Detection and identification of the viruses forming mixed infection in garlic. In: *Symposium Proceedings No. 65: Diagnostics in Crop Production*. British Crop Protection Council, Farnham, UK, pp. 193–198.

Shiboleth, Y.M. (1998) Molecular diagnosis of garlic (*Allium sativum* L.) viruses in Israel and evaluation of tissue culture methods for their elimination. MSc thesis, The Hebrew University of Jerusalem, Faculty of Agricultural, Food and Environmental Sciences, Rehovot, Israel.

Shiboleth, Y., Gal-On, A., Levy, S., Koch, M., Rabinowitch, H.D. and Salomon, R. (1997) Identification of viruses in garlic (*Allium sativum* L.) and closely related *Allium* species grown in Israel. In: *Proceedings of the 10th Congress of the Mediterranean Phytopathological Union, Montpellier, France, 1–5 June, 1997*, pp. 313–317.

Shiboleth, Y.M., Gal-On, A., Koch, M., Rabinowitch, H.D. and Salomon, R. (2001) Molecular characterisation of *Onion yellow dwarf virus* (OYDV) infecting garlic (*Allium sativum* L.) in Israel: thermotherapy inhibits virus elimination by meristem tip culture. *Annals of Applied Biology* 138, 187–195.

Shukla, D.D., Ward, C.W. and Brunt, A.A. (1994) *The Potyviridae*. CAB International, Wallingford, UK, 528 pp.

Song, S.I., Song, J.T., Kim, C.H., Lee, J.S. and Choi, Y.D. (1998) Molecular characterization of the garlic virus X genome. *Journal of General Virology* 79, 155–159.

Sosa, C., Muñoz, J., Navelino, P. and González, H. (1997) Evaluation of re-infection by virus in virus-free garlic 'Rosado Paraguayo' grown in Córdoba. Survey of vectors. *Acta Horticulturae* 433, 601–605.

Stefanac, Z. and Plese, N. (1980) Turnip mosaic virus in two Mediterranean *Allium* species. *Proceedings of the 5th Congress of the Mediterranean Phytopathological Union, Patras, Greece, 21–27 September 1979*, pp. 37–38.

Sumi, S., Tsuneyoshi, T. and Furutani, H. (1993) Novel rod shaped viruses isolated from garlic, *Allium sativum*, possessing a unique genome organization. *Journal of General Virology* 74, 1879–1885.

Sumi, S., Matsumi, T. and Tsuneyoshi, T. (1999) Complete nucleotide sequence of garlic viruses A and C, members of the newly ratified genus *Allexivirus*. *Archives of Virology* 144, 1819–1826.

Takaichi, M., Yamamoto, M., Nagakubo, T. and Oeda, K. (1998) Four garlic viruses identified by the reverse-transcription-polymerase chain reaction and their regional distribution in northern Japan. *Plant Disease* 82, 694–698.

Tsuneyoshi, T. and Sumi, S. (1996) Differentiation among garlic viruses in mixed infections based on RT-PCR procedures and direct tissue blotting immunoassays. *Phytopathology* 86, 253–259.

Tsuneyoshi, T., Matsumi, T., Natsuaki, K.T. and Sumi, S. (1998a) Nucleotide sequence analysis of virus isolates indicates the presence of three potyvirus species in *Allium* plants. *Archives of Virology* 143, 97–113.

Tsuneyoshi, T., Matsumi, T., Deng, T.C., Sako, I. and Sumi, S. (1998b) Differentiation of *Allium* carlaviruses isolated from different parts of the world based on the viral coat protein sequence. *Archives of Virology* 143, 1093–1107.

Ucman, R., Zel, J. and Ravnikar, M. (1998) Thermotherapy in virus elimination from garlic: influences on shoot multiplication from meristems and bulb formation *in vitro*. *Scientia Horticulturae* 73, 193–202.

van Dijk, P. (1991) Mite-borne virus isolates from cultivated *Allium* species and their classification into two new rymoviruses in the family Potyviridae. *Netherlands Journal of Plant Pathology* 97, 381–399.

van Dijk, P. (1993a) Survey and characterisation of potyviruses and their strains of *Allium* species. *Netherlands Journal of Plant Pathology* 99 (Suppl. 2), 1–48.

van Dijk, P. (1993b) Carlavirus isolates from cultivated *Allium* species represent three viruses. *Netherlands Journal of Plant Pathology* 99, 233–257.

van Dijk, P. (1994) Virus diseases of *Allium* species and prospects for their control. *Acta Horticulturae* 358, 299–306.

van Dijk, P. and van der Vlugt, R.A.A. (1994) New mite-borne virus isolates from rakkyo, shallot and wild leek species. *European Journal of Pathology* 100, 269–277.

van Dijk, P., Verbeek, M. and Bos, L. (1991) Mite borne virus isolates from cultivated *Allium* species, and their classification into two new rymoviruses in the family Potyviridae. *Netherlands Journal of Plant Pathology* 97, 381–399.

van der Vlugt, R.A.A., Steffens, P., Cuperus, C., Barg, E., Lesemann, D.E., Bos, L. and Vetten, H.J. (1999) Further evidence that shallot yellow stripe virus (SYSV) is a distinct potyvirus and reidentification of welsh onion yellow stripe virus as SYSV strain. *Phytopathology* 89, 148–155.

Verbeek, M., van Dijk, P. and van Well, P.M.A. (1995) Efficiency of eradication of four viruses from garlic (*Allium sativum*) by meristem-tip culture. *European Journal of Plant Pathology* 101, 231–239.

Walkey, D.G.A. (1990) Virus diseases. In: Rabinowitch, H.D. and Brewster, J.L. (eds) *Onions and Allied Crops*, Vol. II. *Agronomy, Biotic Interactions, Pathology, and Crop Protection*. CRC Press, Boca Raton, Florida, pp. 191–212.

Walkey, D.G.A. and Antill, D.N. (1989) Agronomic evaluation of virus free and virus infected garlic (*Allium sativum* L.). *Journal of Horticultural Science* 64, 53–60.

Walkey, D.G.A., Webb, M.J.W., Bolland, C.J. and Miller, A. (1987) Production of virus-free garlic (*Allium sativum* L.) and shallot (*A. ascalonicum* L.) by meristem-tip culture. *Journal of Horticultural Science* 62, 211–220.

Wang, H.L., Zhang, C.J. and Kang, Y.Q. (1994) High rate of virus-free plantlet regeneration via garlic scape-tip culture. *Plant Cell Reports* 14, 65–68.

Xu, P., Sun, H., Sun, R. and Yang, Y. (1994) Strategy for the use of virus-free garlic in field production. *Acta Horticulturae* 358, 307–311.

Yamashita, K., Sakai, J. and Hanada, K. (1995) Leek yellow stripe virus (LYSV) isolated from garlic and its relationship to garlic mosaic virus (GMV). *Annals of the Phytopathological Society of Japan* 61, 273–278.

Yamashita, K., Sakai, J. and Hanada, K. (1996) Characterization of a new virus from garlic (*Allium sativum* L.), garlic mite-borne mosaic virus. *Annals of the Phytopathological Society of Japan* 62, 483–489.

Yun, J.S., Hwang, S.G., Song, I.G., Lee, C.H., Yun, T., Jeong, I.M. and Park, K.Y. (1998) Mass multiplication of shoots through shoot-tip culture of garlic. *RDA Journal of Horticultural Science* 40, 14–19.

14 Sulphur Compounds in Alliums in Relation to Flavour Quality

W.M. Randle[1] and J.E. Lancaster[2]

[1]*Department of Horticulture, University of Georgia, 1111 Plant Sciences Building, Athens, GA 30602-7273, USA;* [2]*AgriFood Solutions Ltd., Voss Road, RD4, Christchurch, New Zealand*

1. Introduction

Allium species have been prized by most civilizations since antiquity. While this diverse genus has been used variously as medicine, in art or as a feature of spirituality, alliums were, and are, primarily consumed because of their unique flavours or their ability to enhance the flavours of other foods. Studies into the chemistry of *Allium* flavour began in the 1800s, but it was not until the 1940s, 1950s and 1960s that the complexity of flavour and its development among and within *Allium* species became known. During the last part of the 20th century, much has been learned about *Allium* chemistry, although questions still remain (Block, 1992). We have also begun to understand those factors that affect the quality and intensity of *Allium* flavour. These factors are almost as complex and entangled as the chemistry itself. Although we realize that flavour is the result of a multifaceted interaction among many different compounds, this review focuses on the sulphur compounds that give alliums their characteristic flavours and odours.

2. Formation of Flavour in *Allium*

2.1 *S*-alk(en)yl cysteine sulphoxide flavour precursors

The *S*-alk(en)yl cysteine sulphoxides (ACSOs), when hydrolysed by the enzyme alliinase, give rise to the flavour and pungency characteristic of the *Allium* plants. In *Allium* species, four different ACSOs have been found (Bernhard, 1970; Freeman and Whenham, 1975a; Yoo and Pike, 1998). These are (+)-*S*-methyl-L-cysteine sulphoxide (MCSO), (+)-*S*-propyl-L-cysteine sulphoxide (PCSO), *trans*-(+)-*S*-(1-propenyl)-L-cysteine sulphoxide (1-PECSO) and (+)-*S*-(2-propenyl)-L-cysteine sulphoxide (2-PECSO, also referred to as alliin) (Fig. 14.1). The sulphoxide bond can be diastereomeric, but the naturally occurring compounds are all (+) isomers.

The existence of PCSO has been disputed over the years. It was first isolated from onion by Virtanen and Matikkala (1959). Its presence and decomposition products were subsequently reported by Freeman and Whenham (1975b), Block (1992), Lancaster *et al.* (1995) and Randle *et al.* (1995). PCSO was not detected by Thomas and Parkin (1994) or Yoo and Pike (1998). Its detection, however, may be linked to the analysis method. For example, Randle *et al.* (1995) were able to detect PCSO using the methods of Thomas and Parkin (1994) if samples were eluted during high-performance liquid chromatography (HPLC) analysis, using a solvent gradient instead of constant-composition elution.

It is the quantitative and qualitative differences in these four ACSOs that give each *Allium* species its characteristic flavour. For example, the flavour and lachrymatory effect of *A. cepa* is due to the high proportion of 1-PECSO it contains, while the flavour of *A. sativum* is due to its high 2-PECSO content (Lancaster and Boland, 1990).

2.2 γ-Glutamyl peptides and the S-substituted cysteines

Twenty-four γ-glutamyl peptides, 18 of which contain sulphur, have been isolated from *Allium* species (Lancaster and Boland, 1990). γ-Glutamyl-*trans*-(+)-*S*-(1-propenyl)-cysteine sulphoxide is the major peptide

$CH_3.SO.CH_2.(NH_2).COOH$	$CH_3.CH_2.CH_2.SO.CH_2.(NH_2).COOH$
Methyl-L-cysteine sulphoxide	Propyl-L-cysteine sulphoxide
$CH_3.CH{=}CH.SO.CH_2.(NH_2).COOH$	$CH_2{=}CH.CH_2.SO.CH_2.(NH_2).COOH$
1-Propenyl-L-cysteine sulphoxide	2-Propenyl-L-cysteine sulphoxide

Fig. 14.1. The *S*-alk(en)yl cysteine sulphoxides.

component in onions (130 mg 100 g^{-1} fresh weight) (Carson, 1987), representing approximately 50% of the potential flavour and odour precursors (Whitaker, 1976). Low levels of γ-glutamyl derivatives of S-2-carboxypropyl cysteine, S-methyl and S-propenyl cysteine have also been detected in onion tissue.

In garlic the major γ-glutamyl peptides are γ-glutamyl-*trans*-(+)-*S*-(1-propenyl)-cysteine, γ-glutamyl-*S*-allyl cysteine and γ-glutamyl-*S*-methyl cysteine (Lawson, 1996). γ-Glutamyl alk(en)yl cysteine sulphoxides have not been reported in garlic. A full list of all the sulphur compounds recorded in garlic may be found in Lawson (1996).

The significance of the γ-glutamyl peptides in alliums is unclear. The presence of large quantities of γ-glutamyl peptides in dormant bulbs and seeds suggests that these peptides may function as storage sources of nitrogen and sulphur for use in sprouting or germination. For example, the loss of γ-glutamyl propenyl cysteine sulphoxide was proportional to the increase in 1-PECSO during long-term storage of onion bulbs (Kopsell *et al.*, 1999). The enzyme transpeptidase is considered to act as a hydrolase of γ-glutamyl peptides during the biosynthesis of flavour precursors (Matikkala and Virtanen, 1965a, b; Lancaster and Shaw, 1994). The γ-glutamyl peptides are not thought to be converted to flavour compounds in crushed onion, although they may contribute to flavour on cooking, due to thermal decomposition (Block, 1992). γ-Glutamyl propenyl cysteine sulphoxide and 2-carboxypropyl glutathione, however, do disappear in onion macerates (Lancaster *et al.*, 1998).

3. Localization of ACSOs

The ACSOs are found in the cytoplasm of onion cells, physically separated from alliinase (Lancaster and Collin, 1981). Analysis of 1-PECSO suggests that it is associated with the cell's endoplasmic reticulum in onion (Edwards *et al.*, 1994). Alliin (2-PECSO) is concentrated in the very abundant storage mesophyll cells of garlic cloves, with none present near the bundle-sheath cells (G.S. Ellmore, unpublished data, in Lawson, 1996).

4. Compounds Produced after Cell Lysis

When the tissues of any allium are disrupted, the enzyme alliinase hydrolyses the flavour precursors. The result is a wide range of reactive organosulphur compounds with characteristic flavour and striking bioactivity. Elucidating the chemistry of these varied and reactive sulphur compounds has been difficult: it has gradually been achieved over the 30 years since alliin was first isolated. The reader is referred to the comprehensive account of *Allium* organosulphur chemistry by Block (1992).

The first products of the reaction between alliinase and the flavour precursors are the highly reactive sulphenic acids (Fig. 14.2). The sulphenic acids condense with each other to form thiosulphinates. The thiosulphinates are responsible for the flavour of fresh onions, garlic and other alliums. These thiosulphinates participate in a cascade of non-enzymic (and possibly enzymic) rearrangements to produce thiosulphonates and sulphides and a wide range of other organosulphur compounds (Fig. 14.2). Propyl and propenyl di- and trisulphides produce the odour of cooked onions. Aged extracts of alliums develop the capaenes, compounds with multiple sulphur centres. The particular compound and its amounts depend on the conditions, i.e. temperature, the flavour precursors present in the allium and the nature of the solvent.

In garlic, the 2-propene sulphenic acid condenses to form the thiosulphinate allicin (allyl-2-propenethiosulphinate), which gives the characteristic flavour of garlic. In aged extracts of garlic, allicin can disproportionate (react with itself) to form the sulphides, thiosulphonates and the trisulphur compound called ajoene. Ajoene has notable antithrombitic activity.

In onions and other *Allium* species that contain 1-PECSO, a range of sulphur compounds is produced because of the compound's reactivity. The intermediate

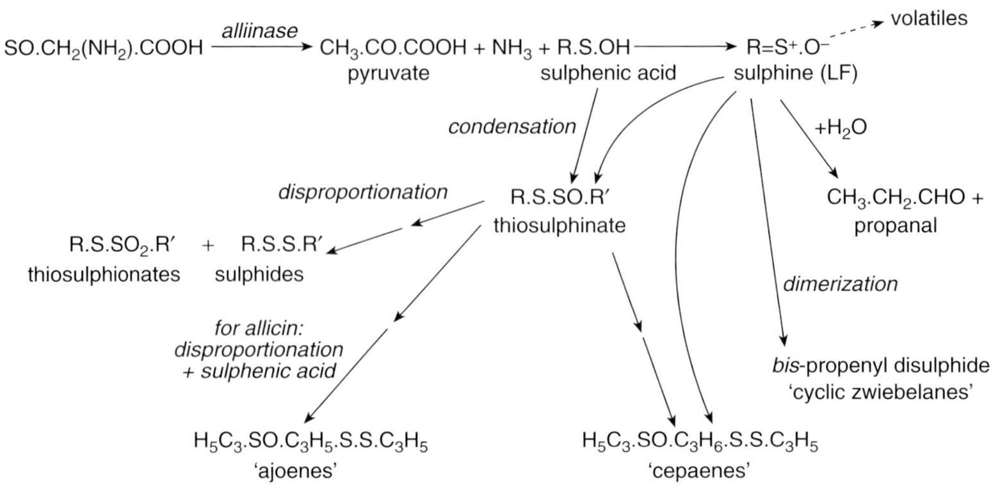

Fig. 14.2. Schematic of the main S compounds formed from the hydrolysis of ACSOs by alliinase. R, methyl, propyl, 2-propenyl(allyl) and 1-propenyl; LF, lachrymatory factor.

1-propenyl sulphenic acid rearranges almost instantly to form the sulphine propanethial-*S*-oxide. This is the lachrymatory factor (LF) of onion. The exact mechanism for the LF triggering tear production is unknown, but it is suggested that lachrymators (such as tear-gas) undergo rapid reduction by nicotinamide adenine dinucleotide phosphate (NADPH) following reception in nerve-cell membranes, triggering the tear ducts.

Most of the LF is lost to the atmosphere when onion tissue is chopped or crushed. However, the majority of the LF can be captured from onion juice if extracted in methylene chloride within 5–10 s of maceration (Kopsell, 1999). LF in solution may react in several ways. It may react with water to form propanal and inorganic sulphur. LF may also disproportionate with methyl and propyl sulphenic acids to form thiosulphinates. These produce some of the characteristic fresh onion flavour. However, propenyl-*S*(*O*)*S*-propenyl thiosulphinates are not formed from LF. LF dimerizes to form bisulphines and their derivative cyclic S–S compounds – the zwiebelanes – or the sulphinyl disulphides – the cepaenes (Block, 1992).

Whereas garlic forms mainly the thiosulphinate allicin, onion forms a wide range of unstable compounds that have differing structures and give rise to different odour perceptions. Because of this, it has been difficult to quantify the flavour content of onion reaction products in the way that has been successful with garlic. Furthermore, although the food industry has developed good processed garlic products, it has been difficult to develop products that faithfully produce the experience of fresh onion flavour.

5. Alliinase and Flavour

The official specific name for the enzyme alliinase is alliin alkyl-sulphenate-lyase (EC 4.4.1.4). The enzyme is also known as alliin lyase, *S*-alk(en)yl-L-cysteine sulphoxide lyase and cysteine sulphoxide lyase (C-S lyase). Alliinase is one of the major proteins found in *Allium*, comprising 6 and 12%, respectively, of the total soluble protein in *A. cepa* bulbs and *A. sativum* cloves (Nock and Mazelis, 1987). Alliinase was first isolated from garlic by von Stoll and Seebeck (1949).

5.1 Phylogenetic distribution of alliinase

Alliinases are most probably present in all members of the *Allium* genus (Tsuno, 1958a, b; Lancaster *et al.*, 2000a). Activity has been detected in *A. cepa, A. sativum, A. porrum* (= *A. ampeloprasum*), *A. tuberosum, A. ursinum* and *A. fistulosum* (Fujita *et al.*, 1990). Alliinase-like activity has also been reported in related genera of the *Alliaceae* and *Liliaceae*, such as *Ipheion*, *Tulbaghia* (Jacobsen *et al.*, 1968) and *Leucocoryne* (Lancaster *et al.*, 2000b). Alliinase-like activity was also reported in the South American dicotyledon *Adenocalymma alliaceum* (*Bignoniaceae*) (Apparao *et al.*, 1981). Although generally less specific in their substrate reactivity, alliin lyases have been purified from bacteria (Nomura *et al.*, 1963; Kamitani *et al.*, 1990), shiitake mushrooms (Iwami and Yasumoto, 1980), the ornamental shrub *Albizzia lophanta* (Schwimmer and Kjaer, 1960) and a variety of *Brassica* species (Hall and Smith, 1983; Ho and Mazelis, 1993; Ramirez and Whitaker, 1998).

5.2 Localization in plant tissues

Lancaster and Collin (1981) used cell-fractionation studies of protoplasts from onion bulbs to demonstrate that alliinase is compartmentalized in the vacuole. Ellmore and Feldberg (1994) used general histology and enzyme-specific antibodies to determine the distribution of alliinase within the garlic clove. Sections stained with aniline blue-black to detect general protein revealed dense deposits within the parenchymatous bundle sheaths, especially around the phloem (the small yellow spots seen when a garlic clove is cut transversely). Autofluorescence under blue light, presumably due to the pyridoxal-5′-phosphate cofactor, was only visible in bundle-sheath cells. By using an alliinase activity stain together with immunocytochemical staining with a polyclonal antibody, it was found that alliinase was concentrated in bundle-sheath cells, usually one layer thick. It was shown that the messenger RNA (mRNA) for alliinase was also localized in these cells, indicating that the alliinase is synthesized there, not transported (G.S. Ellmore, unpublished data, in Lawson, 1996). High levels in bundle sheaths place the enzyme near the phloem, where it, or related products, can be rapidly translocated during development (Ellmore and Feldberg, 1994). Using antibodies against the *A. sativum* alliinase, immunosignals were observed in the bundle-sheath cells (particularly the phloem) and guard cells of *A. tuberosum* leaves (Manabe *et al.*, 1998). Similar green autofluorescence has been observed in the vacuoles of onion guard cells. Ellmore and Feldberg (1994) suggested that this autofluorescence may also have been caused by the presence of the alliinase cofactor pyridoxal-5′-phosphate, as in garlic. The presence of alliinase in onion guard cells would situate it ideally as a defence mechanism to retard the entry of microbial pathogens through the stomata.

Rabinkov *et al.* (1994) found that the specific activity of alliinase in garlic was ten times higher in the bulb than in the leaves. Very high lyase activity was found in the roots, but there was no immunological cross-reaction with shoot alliinase, suggesting the presence of a distinct root alliin lyase.

5.3 Mode of action

Alliinase catalyses the release of the *S*-alk(en)yl sulphoxide group from the ACSO substrate. The reaction mechanism is via a pyridoxal-5′-phosphate–Schiff-base derivative, which then undergoes beta elimination (Fig. 14.3; Jansen *et al.*, 1989b). The products of this reaction, α-iminopropionic acid and the sulphenic acid, are both chemically unstable. α-Iminopropionic acid spontaneously hydrolyses to pyruvate and ammonia. The reactive sulphenic acid can combine with a range of coreactants, as described above.

Pyridoxal-5′-phosphate has been demonstrated as an essential cofactor, which also gives alliinase a characteristic absorption peak at 420 nm. Empirical measurements predict one very tightly bound pyridoxal phosphate per subunit (Tobkin and Mazelis, 1979). Pyridoxal-5′-phosphate inhibitors,

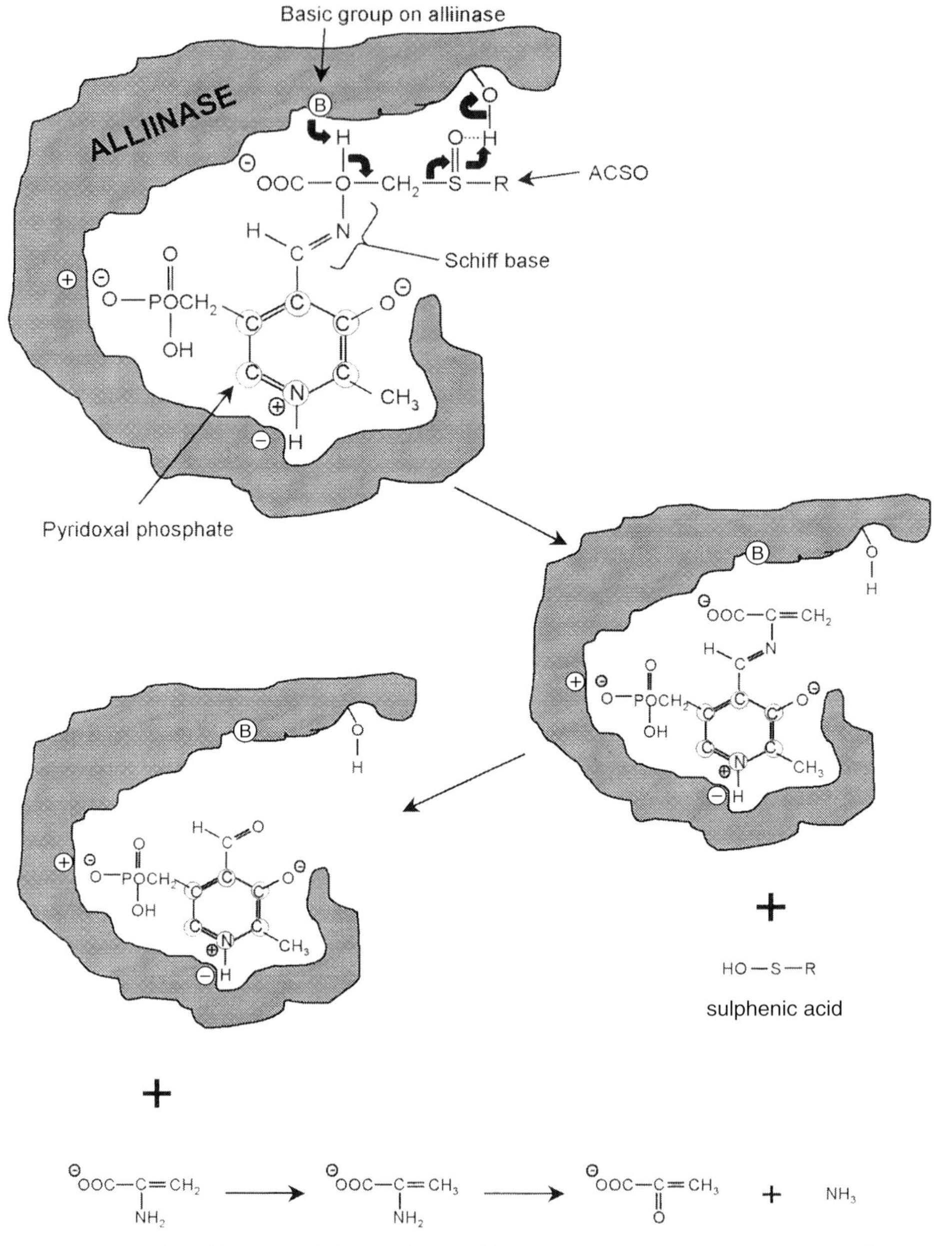

Fig. 14.3. Possible reaction mechanism of alliinase catalysed hydrolysis of ACSOs. Adapted from Block (1992) and Jansen *et al.* (1989b) and reprinted from Gilpin (1995).

such as sodium cyanide, amino-oxyacetate and amino-oxypropionate, all inhibit the activity of alliinase (Lancaster and Boland, 1990).

5.4 Chemistry/substrate specificity

The substrate specificity of *Allium* alliinase has been investigated for onion bulb

(Schwimmer, 1969; Nock and Mazelis, 1987), onion root (Lancaster *et al.*, 2000a), garlic clove (Kazaryan and Goryachenkova, 1978; Jansen *et al.*, 1989a), *A. ursinum* (wild garlic) (Landshuter *et al.*, 1994) and leek (Lohmüller *et al.*, 1994). All enzymes are active towards all of the ACSOs, even though a particular ACSO may not occur in a given allium. Examination of the Michaelis constants (K_m values) of the alliinases with a variety of substrates suggests that all of the above alliinases are similar, with alliinase having a lower affinity to MCSO than to the other substrates. Alkyl cysteines and cysteine were competitive inhibitors of alliinases (Schwimmer *et al.*, 1964; Jansen *et al.*, 1989b). It appeared that the alliinase substrate must have an aliphatic substituent on the sulphur of the L-cysteine sulphoxide and the amino group must be unsubstituted (Carson, 1987).

Onion-root alliinase differed from the bulb alliinase in having activity towards cystine (cysteine–cysteine). This alliinase thus has both C-S lyase and cystine lyase activity. In *Brassica* species C-S lyases have both cysteine sulphoxide (C-S) and cystine lyase activity (Ramirez and Whitaker, 1998), but such dual activity has not previously been reported for *Allium* alliinases. Alliinase from species other than *Allium* have activity towards a much wider range of (C-S)-containing compounds (Schwimmer and Kjaer, 1960; Nomura *et al.*, 1963; Iwami and Yasumoto, 1980; Hall and Smith, 1983; Kamitani *et al.*, 1990; Ho and Mazelis, 1993; Ramirez and Whitaker, 1998).

Work on the reaction of onion alliinase *in vivo* showed that the hydrolysis of 1-PECSO was immediate and almost 100% between 5 and 20 s after bulb maceration (Lancaster *et al.*, 1998). The hydrolysis of PCSO and MCSO was incomplete; about 50% remained even after 2 h. This study also showed a lack of quantitative relationship (non-stoichiometric) between the ACSO content of the tissue and the pyruvate produced upon maceration. Schwimmer (1969) had also shown that only 1 mol of pyruvate was produced by alliinase from 5 mol of 1-PECSO substrate. For *Tulbaghia violacea* it was shown that the C-S lyase was inactivated by an unstable precursor of pyruvate that was bound to the pyridoxal-5′-phosphate of the enzyme (Jacobsen *et al.*, 1968). These results indicate that alliinase reaction inhibition may be occurring in macerates. For onion alliinase the addition of pyridoxal-5′-phosphate cofactor enhanced the hydrolysis of the remaining MCSO and PCSO in the macerate (Lancaster *et al.*, 1998).

Block (1992) has discussed the possibility that sulphenic acid can remain bound to the alliinase via hydrogen bonding when it is attacked by a second free sulphenic acid, giving an optically active allicin. Clearly there is still much that we do not know about the mode of action of alliinase on ACSOs.

5.5 Alliinase isozymes

Evidence is growing that multiple isozymes of allium alliinase exist, with differing physical, chemical and enzymatic activities. The separation of isoforms of onion-bulb alliinase by isoelectric focusing (IEF) was reported by Nock and Mazelis (1987), although the pI of the bands was not clear. Our own experiments have shown that it is difficult to focus onion-bulb alliinase into bands, although they were present (J.E. Lancaster and M.L. Shaw, unpublished results).

Onion-root alliinase separated into two isoforms on the basis of glycosylation (Lancaster *et al.*, 2000a). Isoform 1 gave one band on IEF (pI = 9.3), whereas isoform 2 gave four bands (pI = 7.6, 7.9, 8.1 and 8.3). Leek alliinase gave two bands on IEF (pI = 7.5 and 7.6) (Landshuter *et al.*, 1994). In contrast, *A. ursinum* alliinase protein had a low pI of 4.7 (Lohmüller *et al.*, 1994). Garlic appears to have two different alliinase isoforms, one of which is specific for 1-PECSO and alliin and one that is specific for MCSO (Lawson and Hughes, 1992). The separation of these two activities and their pIs need to be determined.

5.6 Physical characterization

It has been difficult to characterize the physical state of the alliinase molecule. Alliinase can be active as a monomer in onion bulb and root (Clark *et al.*, 1998; Lancaster *et al.*, 2000a), and also as a dimer in garlic (Kazaryan and Goryachenkova, 1978), a trimer, a tetramer and even a hexamer in onion bulbs (Nock and Mazelis, 1987; Hanum *et al.*, 1995; Clark *et al.*, 1998) and a trimer in *A. ursinum* (Landshuter *et al.*, 1994) and leek (Lohmüller *et al.*, 1994). Chinese chives (*A. tuberosum*) alliinase may be active as a monomer only (Manabe *et al.*, 1998). Alliums contain lectins, particularly in the bulbs. In *A. sativum* and *A. ursinum*, alliinase has been shown to aggregate with low-molecular-mass lectins into stable, active complexes (Rabinkov *et al.*, 1995; Smeets *et al.*, 1997). This aggregation is a possible explanation for the occurrence of alliinase as multimeric forms.

Alliinase is a glycosylated enzyme in all *Allium* species except leek and *A. ursinum* (Landshuter *et al.*, 1994; Lohmüller *et al.*, 1994). The enzyme contains about 4.6% carbohydrate in *A. cepa* and 5.5% in *A. sativum* (Nock and Mazelis, 1987). On sodium dodecyl sulphate polyacrylamide gel electrophoresis (SDS-PAGE), alliinase separates into subunits of varying molecular mass, between 48 and 54 kDa (Nock and Mazelis, 1987; Landshuter *et al.*, 1994; Lohmüller *et al.*, 1994; Hanum *et al.*, 1995; Clark *et al.*, 1998; Manabe *et al.*, 1998; Lancaster *et al.*, 2000a). In *A. cepa* bulbs, deglycosylation of unequal subunits gave a single band in SDS-PAGE of size 49 kDa. Garlic alliinase contains one N-linked mannose-rich glycan (Rabinkov *et al.*, 1995). *A. cepa* root alliinase had isoforms of significantly differing glycosylation. Both contained xylose/fucose complex type N-linked glycans and, in addition, one isoform contained terminal mannose structures (Lancaster *et al.*, 2000a). It is likely that varying glycosylation between alliinases from different sources accounts for some of the heterogeneity in subunit size. The high mannose content of alliinase can account for its aggregation with mannose-specific lectins into multimeric forms.

5.6.1 Alliinase and freezing

It was generally reported that freezing onion tissue inactivated alliinase (Schwimmer and Guadagni, 1968; Whitaker, 1976). Wäfler *et al.* (1994) demonstrated that alliinase was not denatured by freezing *per se*, but by cellular processes occurring during slow freezing and thawing of onion tissue. Onion tissue retained alliinase activity when frozen in liquid N_2, stored at −80°C, homogenized in liquid N_2 and thawed in a high-salt buffer containing ethylene glycol. It was suggested that disruption of the cellular environment and localized changes in pH and ionic strength during freezing were responsible for inactivation of alliinase. Cell proteases were not thought to be involved. Alliinase of *A. ursinum*, however, was not destroyed after freezing and thawing of the native bulb tissue (Landshuter *et al.*, 1994).

5.7 Alliinase genes

Genes encoding alliinase have been isolated from bulb onion (van Damme *et al.*, 1992; Clark, 1993; Gilpin *et al.*, 1995; King *et al.*, 1998), Chinese chives (Manabe *et al.*, 1998), shallots (van Damme *et al.*, 1992) and garlic (van Damme *et al.*, 1992). Amino acid sequence was derived by codon usage from the alliinase complementary DNA (cDNA) sequence. Homology between alliinase cDNA-deduced amino acid sequences of onion (bulb and leaf), garlic and shallot was very high, at > 90% (van Damme *et al.*, 1992; Clark, 1993). Chinese chives alliinase-deduced amino acid sequence was only 66–69% homologous to other alliinase sequences, while the onion-root cDNA-deduced amino acid sequence was the most divergent, at about 50%. Southern hybridization of bulb-onion DNA with a full-length alliinase cDNA probe suggested that alliinase was encoded by a small gene family of three or four closely related members (Clark, 1993). Two alliinase loci have been mapped in bulb onion (King *et al.*, 1998).

The derived protein sequence of the coding region of the alliinase from various *Allium* sources gave a predicted mature protein of 445 amino acids (bulb onion and shallot), 448 amino acids (garlic), 447 amino acids (Chinese chives) and 453 amino acids (*A. cepa* root). The alignment of *Allium* alliinase-deduced amino acid sequences showed a consensus Asn glycosylation sequence only at Asn 146 (or Asn 143 for *A. cepa* root alliinase) (Lancaster *et al.*, 2000a). It is likely that glycosylation of site Asn 146 is necessary for alliinase activity. The derived protein sequence of the coding region of the alliinases from bulb onion, garlic and shallot contain the same four potential Asn glycosylation sites at amino acid positions 19, 146, 191 and 328. In *A. sativum*, glycosylation is only at site Asn 146 (Rabinkov *et al.*, 1995). In onion bulbs, Asn 328 was glycosylated and also Asn 146 and/or Asn 191, whereas Asn 19 was not glycosylated (U. Wäfler, M.L. Shaw and J.E. Lancaster, unpublished results).

Site-directed mutagenesis experiments in Chinese chives (Manabe *et al.*, 1998) and pyridoxal-5′-phosphate labelling studies in bulb onion (Kitamura *et al.*, 1997) have shown that a Lys in the region of 250–255 amino acids from the N terminus is essential for alliinase activity. The alignment of *Allium* alliinase-deduced amino acid sequences showed a consensus region of 35 amino acids around a highly conserved Lys 251 (Lys 248 for *A. cepa* root alliinase). The region around Lys −250 to −255 in *Allium* alliinase cDNAs is also conserved in C-S lyases for the metabolism of cysteine, homocysteine and methionine (Manabe *et al.*, 1998). However, the *A. cepa* root alliinase is the only *Allium* protein to have shown a wider substrate activity with cystine lyase as well as the cysteine sulphoxide activity.

6. Sulphur Metabolism and Flavour

6.1 Uptake and reduction of sulphur

Sulphur is one of the six macronutrients required by plants and is found in the amino acids cysteine and methionine and in a variety of metabolites. Alliums have a high sulphur content because of high concentrations of ACSOs and their metabolic intermediates. Sulphur is taken up by the roots as sulphate and transported in the vascular tissue to the leaves, where most of the sulphate assimilation and reduction to organic compounds occurs (for reviews, see Hell, 1997; Leustek and Saito, 1999).

A family of membrane transporters with specialized functions mediates sulphate uptake. The sequences of cDNAs have been cloned from seven species. Typically there is a transporter with high sulphate affinity expressed exclusively in the roots and transporters of lower affinity expressed in the leaves and the roots. The spatial pattern of this low-affinity-type transporter indicates that it must be responsible for uptake from the internal apoplastic pool of sulphate, not from the soil.

Sulphate is an inert compound that must be activated before it can be metabolized. The assimilation pathway leading from sulphate to cysteine involves at least six enzymes. Sulphate is incorporated into adenosine phosphosulphate (APS). This reaction is catalysed by the enzyme adenosine triphosphate (ATP) sulphurylase and is the sole entry point for the metabolism of sulphate. There are two ATP-sulphurylase isoforms in most plants: a major form located in the plastids and a minor form localized in the cytoplasm. The isoforms are coded by gene families. The plastid enzyme exists in both leaves and roots, and the chloroplasts in the leaf are the main site for sulphate assimilation.

Sulphate is reduced before incorporation into cysteine. Reduction is generally believed to take place in the plastids. The reaction occurs through the sequential action of two different enzymes, both localized in the plastids. APS is the first substrate and the reaction requires two electrons to produce sulphite. The enzyme responsible for this reduction of APS is still controversial (Hell, 1997; Leustek and Saito, 1999). Reaction may occur through a bound intermediate, such as glutathione, the 'APS-bound' pathway and APS sulphotransferase or via a free reductase, such as APS reductase or phosphoadenosine phosphosulphate (PAPS) reductase. The reduction of sulphite occurs via sulphite reductase and requires six electrons, donated from ferredoxin. This enzyme has been convincingly demonstrated by purification and cloning of the corresponding gene and cDNA (Bork *et al.*, 1998).

The synthesis of cysteine from serine and sulphide, from two converging pathways, represents the final step of sulphur assimilation into organic S (Fig. 14.4). The reaction is catalysed by serine acetyl transferase and *O*-acetyl serine lyase in an enzyme complex known as cysteine synthase. Unlike the other enzymes of sulphur assimilation, which are primarily localized in the plastids, cysteine synthase is found in plastids, the cytosol and mitochondria. Cysteine synthase was also localized in the vascular-bundle sheath cells (Saito, 1998). In *Arabidopsis thaliana*, *O*-acetyl serine lyase activity in the roots contributed significantly to the pool of cysteine in the plant (Barroso *et al.*, 1998). Cysteine is then available for incorporation into proteins, into glutathione – a key compound in cellular redox regulation and defence – and, for *Allium*, into the ACSOs.

6.2 ACSO biosynthesis

Figure 14.4 summarizes the proposed biosynthesis of the various *Allium* peptides and ACSOs based on the results of labelling experiments in which ^{35}S sulphate was fed to onion plants (Granroth, 1970; Lancaster and Shaw, 1989; Lawson, 1996). Addition of methacrylic acid (from valine) to glutathione gives the *S*-2-carboxypropyl derivative. Sequential hydrolysis of glycine, decarboxylation to give γ-glutamyl-*S*-1-propenyl cysteine, oxidation to γ-glutamyl-*S*-1-propenyl cysteine sulphoxide and cleavage by γ-glutamyl transpeptidase (EC 2.3.2.1) gives 1-PECSO. γ-Glutamyl transpeptidase has been shown to function as a hydrolytic enzyme in *Allium* plants (Lancaster and Shaw, 1994). Labelling experiments have established that γ-glutamyl-*S*-2-carboxypropyl cysteine is converted into 1-PECSO (Parry and Lii, 1991). Biosynthesis of 1-PECSO may also occur from the precursor to glutathione, γ-glutamyl cysteine, via similar reactions. Methylation of glutathione gives *S*-methyl glutathione, followed by conversion to MCSO. Because PCSO occurs in low amounts in onion and is not present in garlic, its biosynthesis has received less attention. Lancaster and Shaw (1989) present the possibility that it is derived from γ-glutamyl-*S*-propenyl cysteine via saturation of the double bond.

A similar biosynthetic scheme has been postulated for garlic, with an important difference (Lawson, 1996). Garlic accumulates γ-glutamyl cysteine derivatives of allyl cysteine and smaller amounts of propenyl and methyl cysteine (Lawson, 1996). γ-Glutamyl-*S*-alkenyl cysteine sulphoxides have not been found to accumulate in garlic. Thus it is suggested that, in garlic, the action of γ-glutamyl transpeptidase to cleave the glutamic acid residue precedes the action of the postulated oxidase.

Although there is good evidence that most of the biosynthesis proceeds via peptide intermediates, labelling studies with compounds other than sulphate have indicated alternative routes to ACSOs (Granroth, 1970). Alkyl thiols fed to cell cultures produced the corresponding cysteine sulphoxides (Prince *et al.*, 1997). Direct formation of ACSOs, by an unspecified route, was also suggested by Edwards *et al.* (1994). The broad substrate specificity of cysteine synthase, normally functioning to combine H_2S with *O*-acetyl serine, means that exogenous thiols can also be combined.

6.3 Regulation of sulphur metabolism

At conditions of low S, there is induction of the sulphate transporter proteins, ATP sulphurylase, APS reductase and cysteine synthase, all of which are involved in the uptake and assimilation of S into organic compounds (Hell, 1997; Saito, 1998; Leustek and Saito, 1999). Sulphur starvation induces the activity of certain enzymes (Smith *et al.*, 1997; Takahashi *et al.*, 1997, 1998; Lee and Leustek, 1998; Lappartient *et al.*, 1999). The steady-state mRNA levels of the high-affinity sulphate transporter protein in the roots increase rapidly in response to sulphur starvation. The lower-affinity form is slower

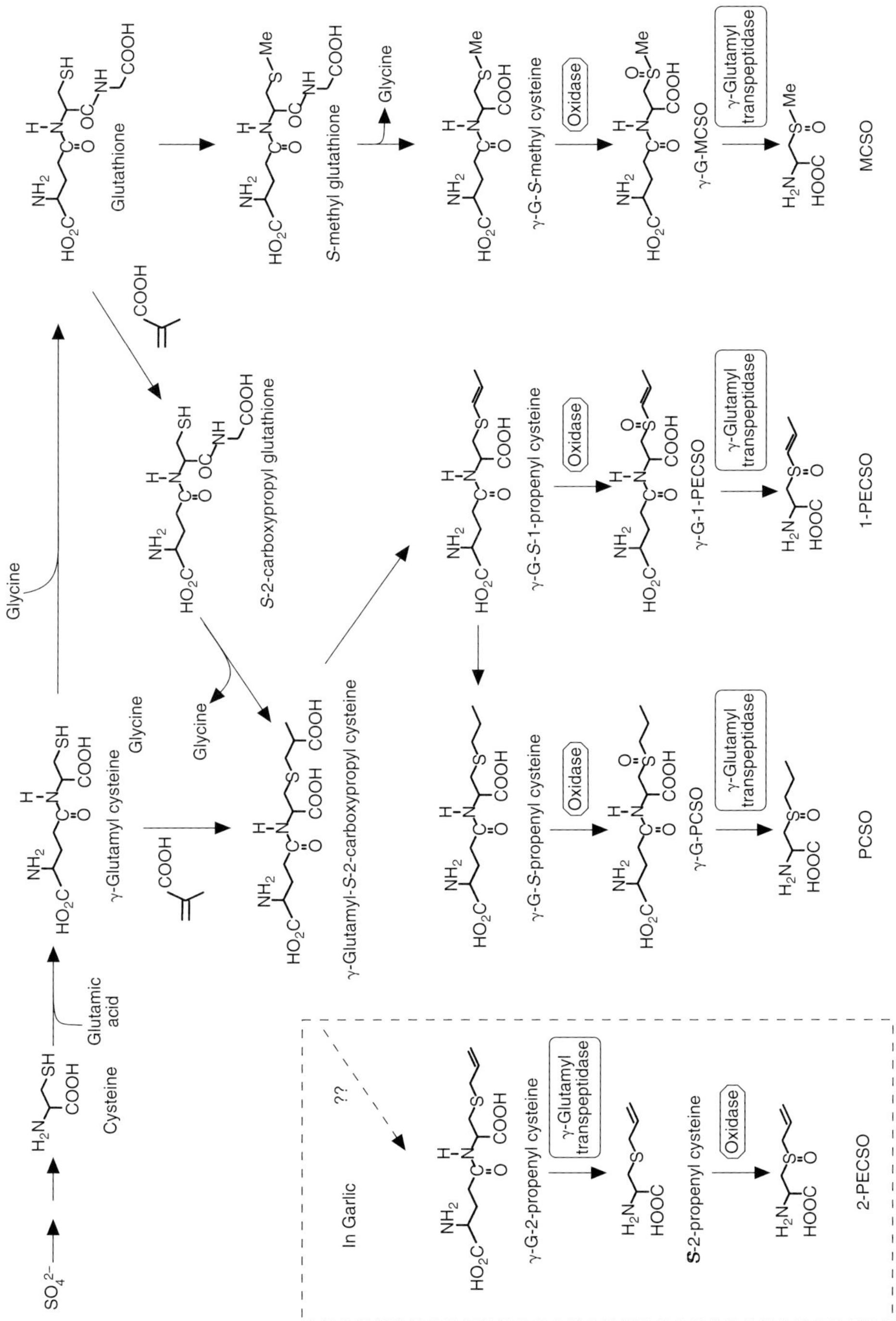

Fig. 14.4. Proposed biosynthesis of ACSOs and their intermediates.

or less responsive to sulphur starvation. In general, the activity and steady-state mRNA levels of ATP sulphurylase increase when plants are starved for sulphur. However, these changes are relatively small, being approximately twofold or less, and the regulation occurs mainly in roots. APS sulphotransferase, the postulated first sulphur reduction enzyme, is an important regulation point in sulphate assimilation. Sulphur starvation induces the accumulation of mRNA in the roots and an increase in enzyme activity. In contrast, sulphite reductase does not appear to be appreciably regulated at the mRNA level (Bork *et al.*, 1998). Cysteine synthase in the plastids of the leaves increases the level of steady-state mRNA after sulphur starvation (Takahashi *et al.*, 1997). In *Arabidopsis thaliana,* sulphur starvation was shown to increase the *O*-acetyl serine lyase in all the parts of the plant and particularly in the aerial parts (Barroso *et al.*, 1998).

When supplied to plants, reduced sulphur compounds, such as cysteine and glutathione, lower the activity of the sulphur assimilation enzymes. In general, the activity and steady-state mRNA levels of ATP sulphurylase decrease when plants are fed reduced forms of sulphur, such as cysteine or glutathione.

Two compounds have been suggested as endogenous regulators of this pathway (Leustek and Saito, 1999). Glutathione is transported through the phloem sap, and its level is markedly reduced after short-term sulphur starvation. 'Split-root' experiments supported this role for glutathione (Lappartient *et al.*, 1999). *O*-acetyl serine may act as a positive signal (Smith *et al.*, 1997) on sulphur assimilation, as evidence has shown an increase in sulphate transport-protein steady-state mRNA when this compound was fed to plants.

6.4 Remobilization of sulphur

Although the assimilation of sulphur into organic compounds is important for the growth and development of plants, much of the sulphur remains in the cell vacuole as sulphate. In oil-seed rape, it was estimated that 70–90% of the total sulphur in the middle and older leaves was sulphate and about 40% in the youngest leaves (Blake-Kalff *et al.*, 1998). During conditions of sulphur deficiency, the concentrations of all sulphur compounds decreased, but sulphate in particular acted as a sulphur source. Pulse chase experiments showed that the soluble pool of sulphur contained a small metabolically active pool of sulphur and a larger pool that is in slow equilibrium with the small pool (Sunarpi and Anderson, 1996). Remobilization of sulphur from proteins does not take place unless nitrogen is also deficient (Sunarpi and Anderson, 1997).

In onions, sulphate was estimated to be 41–48% of the total bulb sulphur (Randle *et al.*, 1999). Sulphate levels were greater in mild cultivars and with increasing levels of sulphur supply. At S-deficiency supply levels, nearly 95% of the total bulb S could be accounted for in the ACSOs and their peptide intermediates (Randle *et al.*, 1995).

Low sulphur supply decreased the levels of sulphur-containing compounds, such as ACSOs and their biosynthetic intermediates and glutathione (Randle *et al.*, 1995; Hamilton *et al.*, 1997; Leustek and Saito, 1999). Low sulphur supply has also been shown to increase the expression and activity of proteins involved in sulphur uptake and assimilation (see Section 6.3 above). Thus, when sulphur is limiting, the organic sulphur compounds are metabolized more efficiently.

The increase in alliinase activity at low sulphur levels raises the possibility that alliinase is involved endogenously in recycling ACSOs (Lancaster *et al.*, 2000c). We know that alliinase is sequestered in the vacuoles of onion cells, and hydrolyses the flavour precursors when the cells are destroyed. Alliinase may also have a role in remobilizing flavour precursors in intact cells during conditions of sulphur deprivation. In pulse chase experiments of sulphate fed to onion leaves, the specific activity of ^{35}S in 1-PECSO fell by half between days 3 and 7, providing evidence of the endogenous metabolism of this compound (Lancaster and Shaw, 1989).

Similar evidence of the loss of labelled ^{35}S from ACSOs was found by Edwards *et al.* (1994).

6.5 Defence-related regulation of sulphur

The alliinase–ACSO system has been generally regarded (although not proved) to be involved in defence against pathogens and insect attack. *Arabidopsis thaliana* contains glucosinolates, S compounds similarly construed to be involved in defence. After wounding or the application of jasmonate – an inducer of the wound response – the enzymes of S assimilation increased and glucosinolate levels themselves increased twofold. This suggests that wounded plants deliver available sulphur to synthesize defence-related substances by activating genes involved in sulphur metabolism (Harada *et al.*, 2000). It would be interesting to determine if similar responses were observed in alliums.

7. Factors Affecting Flavour Intensity and Quality

7.1 Genetic factors affecting flavour

7.1.1 Differences among cultivars

It is well known that *Allium* cultivars differ in flavour intensity. While consumers in many cultures prefer pungent cultivars, others desire cultivars that are mild and sweet. Demand for flavour quality and intensity depends on cultural preference and intended use (Jones and Mann, 1963; Rabinowitch, 1988). Through the years, using different analytical techniques, various studies have described differences in flavour quality and intensity among cultivars of onion. Platenius (1941) utilized total volatile S to separate 16 onion cultivars that ranged from 59 to 156 ppm Similarly, the volatile LF, thiopropanal *S*-oxide, was used to differentiate nine cultivars, although little separation was reported, probably because of the instability and time sensitivity of measuring this compound (Freeman and Whenham, 1975a).

Because it is easy to measure, enzymatically produced pyruvate has been used to compare cultivars in a number of studies (Schwimmer and Weston, 1961; Schwimmer and Guadagni, 1962; Bajaj *et al.*, 1980, 1990; Bedford, 1984; Randle, 1992b, c; Thomas *et al.*, 1992; Randle and Bussard, 1993a; Vavrina and Smittle, 1993; Kopsell, D.E. and Randle, 1997). Pyruvate content of cultivars has varied from 1 to 22 μmol g^{-1} fresh weight of bulb tissue. However, because pyruvate is a product from the hydrolysis of all flavour precursors, it only measures gross flavour intensity and does not differentiate for flavour quality. Total and individual ACSOs have also been used to separate onion cultivars for flavour quality and intensity (Lancaster *et al.*, 1988; Randle *et al.*, 1995; Yoo and Pike, 1998; Bacon *et al.*, 1999; Kopsell *et al.*, 1999).

Considering the complexity of S uptake, its reduction and requirement for healthy plant growth and development, its use in the flavour biosynthetic pathway and the fact that onions have been cultivated by different civilization for millennia, it is understandable that continuous variation exists for flavour intensity among onion cultivars. Cultivars differ in total plant S, and differences in flavour intensity and quality probably arise due to variability in sulphur uptake and its metabolism through the flavour biosynthetic pathway. Sixty-two onion cultivars were tested for total leaf and bulb S at two S fertility levels (Randle, 1992c). At high S fertility, leaf S ranged from 1.11 to 0.69% dry weight, while bulb S ranged from 1.03 to 0.46% dry weight. At low S fertility, leaf and bulb S were substantially lower and less variable among the tested cultivars. Poor correlations between leaf S and bulb S and between bulb S and enzymatically produced pyruvate suggested that the cultivars differed in the way S was partitioned into flavour and non-flavour compounds (Randle, 1992c; Randle and Bussard, 1993b; Randle *et al.*, 1999).

One way in which cultivars differ in partitioning S is in their ability to reduce SO_4^{-2} so that it can enter the flavour pathway. Pungent cultivars are more efficient at reducing SO_4^{-2}, whereas mild cultivars store more of the absorbed S as SO_4^{-2} in the plant's cell vacuoles, thereby excluding it from the flavour pathway (Randle *et al.*, 1999). Differences within onion for ATP sulphurylase and other enzymes responsible for SO_4^{-2} reduction, however, still need to be described.

Onion cultivars also differ in their ability to accumulate S compounds among and within the various pathways leading to flavour-precursor synthesis (Shaw *et al.*, 1989; Thomas and Parkin, 1994; Randle *et al.*, 1995; Yoo and Pike, 1998; Bacon *et al.*, 1999; Kopsell and Randle, 1999). Pungent cultivars tend to accumulate more total flavour-precursor content than milder cultivars (Randle *et al.*, 1995; Yoo and Pike, 1998; Kopsell and Randle, 1999). However, in one case, a mild cultivar accumulated more total ACSO content than more pungent cultivars (Bacon *et al.*, 1999). This was attributed to differences in environmental growing conditions among the cultivars evaluated. Pungent cultivars will also partition more sulphur into the 1-propenyl cysteine sulphoxide biosynthetic pathway, compared with milder cultivars (Shaw *et al.*, 1989; Randle *et al.*, 1995; Kopsell and Randle, 1999). Higher levels of γ-L-glutamyl-*S*-(1-propenyl)-L-cysteine sulphoxide, the penultimate peptide leading to 1-propenyl cysteine sulphoxide, and *S*-2-carboxypropyl glutathione, found early in the pathway, accumulated in higher concentration in these cultivars. γ-L-Glutamyl-*S*-(1-propenyl)-L-cysteine sulphoxide also appears to be a regulatory bottleneck in moving S through this pathway, as it accumulated up to four times the concentration of *S*-2-carboxypropyl glutathione when plants were grown in high S supply (Randle *et al.*, 1995).

Cultivars also differ in the level of *in vitro* alliinase activity (Lancaster *et al.*, 1993, 1995). While significant bulb-to-bulb variability exists, pungent cultivars had from two to three times the alliinase activity of mild cultivars. However, *in vitro* alliinase activity may not be an accurate means of classifying flavour potential in onion, because onions grown in low-sulphur environments had greater alliinase activity, albeit lower flavour potential, than onions grown in a high-sulphur environment (Lancaster *et al.*, 1995). Increased methyl cysteine sulphoxide content in relation to 1-propenyl cysteine sulphoxide and the *in vivo* alliinase activity towards these precursor changes are possible causes for this discrepancy.

7.1.2 Within-cultivar variation for flavour

While differences in mean flavour intensity or quality exist among onion cultivars, significant bulb-to-bulb variation occurs for measured flavour components. In a heterogeneous species, such as onion, inherent variability of any cultivar can be expected (Dowker, 1990). In the development of onion cultivars, the lack of active selection for flavour coupled with severe inbreeding depression, has probably contributed to this variability. Significant sample-to-sample or bulb-to-bulb variability has been described for volatile sulphur content (Platenius, 1935), enzymatically produced pyruvate and total bulb S (Randle, 1992b), alliinase activity (Lancaster *et al.*, 1993), γ-glutamyl peptides (Lancaster and Shaw, 1991) and flavour-precursor content (Lancaster and Shaw, 1991; W.M. Randle, unpublished data). In each case, the bulb-to-bulb variability highlighted the need to use multiple bulb samples when testing cultivar differences or when testing the effect of treatments on cultivar performance. Platenius (1935) found the combined tissue of ten bulbs was normally sufficient to detect a 10 ppm difference in volatile sulphur among cultivars, although increasing the sample to 20 or 25 bulbs improved his accuracy. Randle (1992b), using a statistical sampling model, determined that ten onion bulbs replicated four times would be sufficient to detect a 1 μmol pyruvate difference among treatments with 95% confidence. Significant bulb-to-bulb variability for flavour quality and intensity suggests that active selection during cultivar development is needed to improve flavour quality and consistency.

7.1.3 *The heritability of sulphur-related flavour*

There have been few studies examining the genetic basis of flavour or flavour development in *Allium*. Obtaining heritability estimates and the reporting of genetic variances are difficult in onion because of the species' complex floral morphology and susceptibility to inbreeding depression. The mere fact that cultivars differ in flavour intensity around the world and that their general flavour characteristics are passed on through successive seed cycles, however, suggests that flavour intensity is a heritable trait. When estimating heritability using controlled crosses among selected parents, values close to 1.0 indicate a high degree of association between progeny and parents, while values closer to 0 suggest that trait expression is not successfully passed from parents to progeny. However, when testing for heritability, the power of each mating design should be taken into consideration, as some tests are more stringent than others in generating heritability estimates.

When evaluating large numbers of plants, researchers have relied on rapid chemical tests to differentiate flavour intensity among individual phenotypes. Estimating enzymatically produced pyruvate (Schwimmer and Weston, 1961) has been the most widely used assay, and several laboratories have increased the speed and efficiency of the test to accommodate large sample numbers (Thomas *et al.*, 1992; Randle and Bussard, 1993b; Yoo *et al.*, 1995). Warid (1952) used parent–offspring regression to obtain a heritability estimate of 0.71 for bulb pungency in short-day onions. More recently, using generation means analysis from crosses among long-day parents and a single year's evaluation, broad-sense heritability estimates for bulb pungency ranged between 0.13 and 0.56 (Lin *et al.*, 1995). Lin and co-workers also determined that the inheritance of bulb pungency was mainly due to additive gene effects, and selection based on parental performance was therefore possible. Narrow-sense heritability estimates for bulb pungency calculated from half-sib progeny analysis among short-day onions in a single evaluation cycle were between 0.25 and 0.53 (Wall *et al.*, 1996). The corresponding realized heritability estimates from one generation of directed selection for lower pungency were between 0 and 0.51. However, the unselected population and the selected bulbs in that study were evaluated in one season and the selected population in another. Yearly environmental changes can greatly influence pungency values. In the most extensive study done to date, Simon (1995) used both generation means analysis and four- and eight-parent diallel mating designs to determine the genetic control of pungency over a 2-year evaluation period. Broad-sense heritability estimates from four long-day parental crosses ranged from 0.34 to 0.89, with significant year effects. A preponderance of additive genetic variation suggested that selection would be effective for manipulating bulb pungency. In another study, the pungency of test crosses generated in a factorial mating design tended to be equivalent to the most pungent parent when long-day open-pollinated (OP) populations were stored for 3 months and evaluated in two different years (Havey and Randle, 1996). Test crosses to Spanish OP populations were found to have the lowest pungency.

7.1.4 *Flavour progress from selection*

Generally speaking, the heritability estimates provided by the above studies indicate low to moderate ability to pass flavour characteristics from parents to progeny, suggesting that while changes in flavour intensity can be made, progress may be slow. The release of cv. 'NuMex Dulce', a selection from 'NuMex Starlite', supports this observation (Wall and Corgan, 1998). Following two cycles of recombinant selection for lower pyruvate, NuMex Dulce's average pungency was reduced from 5.2 to 4.4 μmol pyruvate g^{-1} fresh weight. Both Simon (1995) and Wall *et al.* (1996) reported progress from selection for lowering bulb pungency, with the level of progress dependent on the populations selected. Currently, breeders are actively selecting for lower pungency in programmes around the USA and in Europe and Israel. Other onion varieties released have been selected for low bulb pungency (Peterson *et al.*, 1986; Pike *et al.*, 1988a, b; Wall and Corgan, 1999).

It is often thought or assumed that soluble solids/dry-matter content and pungency are phenotypically correlated. Jones and Bisson (1934) were the first to report that low-dry-matter varieties were low in pungency while high-dry-matter types were more pungent. Platenius (1941) reported some association, but also noted that the weakest- and strongest-pungency varieties had similar dry-matter content. Similarly, Schwimmer and Guadagni (1962) and Lin *et al.* (1995) reported weak correlations between pungency and soluble solids (r = 0.57 and r = 0.50–0.42, respectively). However, the correlation is a positively biased association, in that the compounds that produce pungency also contribute to soluble solids. Moreover, exceptions can always be found in this relationship if the number of varieties evaluated is large (Bedford, 1984; Randle, 1992c), and insignificant correlations between solids and pungency have also been reported (Randle and Bussard, 1993a). Simon (1995) found that a strong positive correlation between pyruvate and soluble solids among parents and F_1 hybrids was insignificant or even became negative within and among F_3 families, and suggested that these characters might be selected independently. In fact, the biosynthetic pathways leading to flavour precursor and to soluble-carbohydrate accumulation have little in common. And, unless the genes responsible for both flavour-precursor synthesis and soluble-carbohydrate synthesis are genetically linked, it is likely that each biosynthetic system will perform independently and there will be opportunities to manipulate each system separately.

7.2 Tissue and ontogenetic factors affecting flavour

Flavour and the components of flavour are unevenly distributed in the different tissues among alliums. Total volatile sulphides from chopped tissues of five *Allium* species native to North America were higher in the foliage than in the bulbs (Saghir *et al.*, 1965). Flavour quality also differed between the tissues, since methyl and allyl sulphide radicals were found in higher concentrations in the bulbs than in the foliage. Methyl and allyl moieties produce different flavour attributes and sensory notes when consumed (Block, 1992). Boscher *et al.* (1995) reported that the concentration of the volatile sulphur moieties between foliage and bulbs differed in concentration and composition and depended on the *Allium* species examined. They also reported a poor association between the alk(en)yl cysteine sulphoxide precursors and their volatile moieties among the tissues tested.

Flavour gradients also exist within bulb tissues of different *Allium* species. Reports differ for flavour gradients in bulb onion, although the highest concentrations generally occur at the interior and base of the bulbs, while the lowest concentrations occur at the top and in the outside scales. Freeman (1975) reported a consistent increase in thiosulphinate and pyruvate concentration from the outer to the inner scales of the onion and from the outer to inner leaves of leek. Lancaster *et al.* (1986), on the other hand, reported increases in total precursor content from onion outer scales to mid-interior scales, and then decreasing precursor content in the interior scales. 1-Propenyl cysteine sulphoxide was found in higher concentration in the outer scales of bulb onion, while methyl cysteine sulphoxide occurred in higher concentration in the inner scales (Bacon *et al.*, 1999). Similarly, pyruvate concentrations increased from the outer to the inner scales in the three onion cultivars tested (Bacon *et al.*, 1999). Differences also exist within the onion bulb from top to bottom. Enzymatically produced pyruvate was highest at the base of the bulb, decreased to near the top and then increased (Randle *et al.*, 1998a). Higher pungency at the top and bottom of the bulb were the result of higher 1-propenyl cysteine sulphoxide concentrations (Bacon *et al.*, 1999). In garlic, the greatest concentrations of thiosulphinates were found in the storage leaf, with lesser amounts in the sprout and foliage leaves (Freeman, 1975).

Flavour changes with the stage of growth, since plants develop vegetatively and then translocate materials to their storage tissues. Immature onion bulbs were low in producing volatile sulphur, increased as bulbing proceeded and then decreased as bulbs matured and became dormant (Platenius and Knott, 1935). These changes were largely attributed to changes in bulb water content during the bulking and maturation processes, because the differences were minimal if expressed on a per cent dry-weight basis. Total leaf sulphur measured during early bulbing was higher than total bulb sulphur from mature plants (Randle, 1992b). Onion pungency also decreased as bulbing proceeded to maturity (Hamilton *et al.*, 1998). When *A. amplectens* and *A. anceps* plants were dormant, no differences were detected in the amount of volatile sulphides produced. But, as plants grew during the season, volatile sulphides increased. Lancaster *et al.* (1986) were the first to show that differences in flavour concentration and composition between plant tissues may be a result of ontogenetic factors as well as tissue type. In pre-bulbing onions, total flavour-precursor concentration was higher in the leaves than in the bulb scales. However, as plants began to bulb, total precursor concentrations in the bulbs started to exceed those found in the leaves. Pre-bulbing plants also had extremely low levels of γ-glutamyl peptides (Lancaster and Shaw, 1991). But, as plants bulbed, large pools of γ-glutamyl propenyl cysteine sulphoxide and, to a lesser extent, *S*-2-carboxypropyl glutathione appeared. Large increases in γ-glutamyl-*S*-allyl-L-cysteine, γ-glutamyl-*S*-(*trans*-1-propenyl)-L-cysteine and allyl cysteine sulphoxide also occurred in garlic as cloves matured during the last month of growth (Matsuura *et al.*, 1996).

7.3 Flavour changes during storage

Bulbs of onion and garlic are routinely stored for varying lengths of time before being marketed. Even though the bulbs may be dormant during this time, flavour changes have been measured in onion and garlic. However, it is difficult to draw comparisons among these studies because of dissimilarities in pre- and postharvest factors that may have affected the depth of bulb dormancy and bulb quality. Differences in storage conditions and duration also make comparisons problematic. However, flavour intensity and quality do change in storage and the changes appear to be dependent on cultivar, storage duration, depth of bulb dormancy and storage temperature.

Freeman and Whenham (1976) in the UK reported increases in enzymatically produced pyruvate for two long-storing onion cultivars in the first 210 days of storage, followed by a sharp decrease in pungency to 240 days. Temperature influenced the magnitude of the change, a fourfold increase in pungency occurring when bulbs were stored at 2°C and ambient temperatures and only a two- to threefold increase at 25°C storage. In the USA, three long-day, long-storing cultivars held at between 15 and 22°C decreased in pungency over a 4-month period (Peterson *et al.*, 1986) while, in another study, the pungency of cv. 'Spartan Banner' increased (Hanum *et al.*, 1995). At 4°C, the pungency of four long-day cultivars generally decreased over a 7-month storage period, although there were significant differences between the two years in which the study was done (Debaene *et al.*, 1999). With the poor-storing cv. 'Walla Walla', pungency more than quadrupled over a 4-week period for early-harvested bulbs, but only slightly increased over a 7-week period for mature bulbs (Mikitzel and Fellman, 1994). Poor-storing 'Granex'-type onions were evaluated under different storage conditions for flavour changes (Smittle, 1988). In every storage situation, pungency increased over a 6-month period, with the smallest increase coming from refrigerated-air and controlled-atmosphere (CA) conditions of 5% CO_2 and 3% O_2, and the largest increase (almost × 2) coming from 1°C-refrigerated storage with normal atmosphere. Later, 'Granex'-type onions were evaluated for pungency changes after June in commercial CA storage facilities in the Vidalia (Georgia, USA) growing region. Although onions from the 14 facilities represented different cultivars, different lengths of storage and different combinations of cultural and handling methods, all lots increased in pungency (Smittle, 1991). Although a simple regression equation was generated to predict pungency increases, analysis of the data revealed large deviations from the regression line, suggesting that this approach may not be straightforward.

In an attempt to determine the genetic differences among cultivars for flavour changes that occur during storage, eight cultivars were grown and stored under similar conditions (Kopsell, D.E. and Randle, 1997). Data revealed a complex pattern of flavour changes that occurred during storage and were cultivar-dependent. Some cultivars decreased in pungency and others increased, while others increased and then decreased during storage.

In garlic, bulb pungency increased from 90 to 180 days in storage, and then decreased as bulbs were stored to 300 days (Ceci *et al.*, 1991).

Flavour changes have been associated with sprouting following the loss of bulb dormancy (Lancaster and Shaw, 1991; Ceci *et al.*, 1992). When bulbs sprouted, flavour compounds were mobilized and translocated to the emerging leaves, with a corresponding increase in enzyme activity associated with the translocation. Bulb sprouting, however, occurred long after dormancy was lost. When loss of dormancy was determined by a bulb's ability to form adventitious roots, little correlation could be made between flavour changes and loss of dormancy during storage (Kopsell, D.E. and Randle, 1997; Kopsell *et al.*, 1999). For example, two short-day onion cultivars lost dormancy at a similar rate, but one increased while the other decreased in pungency. In another example, a short-day cultivar that lost dormancy during storage had a decrease in pungency from pre-storage levels similar to that of a long-day cultivar that remained completely dormant. It therefore appears that the metabolic changes which occur in the flavour pathway during storage are quite complex and are not necessarily associated with a bulb's transition through dormancy. This area needs further study, as flavour changes which occur during storage have a significant impact in the market-place.

Flavour quality has also been shown to change during onion- and garlic-bulb storage. When garlic was stored at 4°C for up to 22 weeks, significant increases in the allyl and *trans*-1-propenyl thiosulphinates were found, and at a level of about four times that of allicin (Lawson *et al.*, 1991). The increases were substantially less, however, when the bulbs were stored at 22°C. Allicin remained unchanged at both storage temperatures. Using the rearrangement products 3-vinyl-1,2-dithi-5-ene and 3-vinyl-1,2-dithi-4-ene as indicators of allicin-content behaviour during storage, similar results were obtained during 22 weeks at room-temperature storage (Ceci *et al.*, 1991). However, as storage proceeded to 43 weeks, allicin content substantially declined in garlic bulbs, while diallyl disulphide significantly increased throughout storage.

In onion, 1-propenyl cysteine sulphoxide and its decomposition products have been shown to generally increase during bulb storage. Because the products of 1-propenyl cysteine sulphoxide are responsible for the tear-inducing, heat and mouth-burn sensations, storage makes these onions taste more harsh. Similarly, Freeman and Whenham (1976) reported that thiopropanal-*S*-oxide from stored bulbs increased during the first 7 months of ambient storage, and then decreased in the 8th month. Storage at 2°C produced similar patterns, but lower levels of the lachrymator, while 6 months of 0°C storage resulted in an increase of over 50% in 1-propenyl cysteine sulphoxide in two onion cultivars in the UK (Bacon *et al.*, 1999). Short-, intermediate- and long-day cultivars that were grown and stored under similar conditions in the USA differed in the rate and amount of 1-propenyl increases (Kopsell *et al.*, 1999). One cultivar more than doubled in 1-propenyl cysteine sulphoxide content, while others only showed a 40% increase. Interestingly, the rate of 1-propenyl cysteine sulphoxide increase during storage for the seven cultivars was proportional to the loss of γ-glutamyl propenyl cysteine sulphoxide, suggesting that the activity of γ-glutamyl transpeptidase was responsible for the flavour-precursor increase throughout storage. Methyl cysteine sulphoxide, on the other hand, generally decreased during storage, although several cultivars showed little change in content. Because 1-propenyl cysteine sulphoxide hydrolysis has been shown to affect the complete hydrolysis of methyl cysteine sulphoxide (Lancaster *et al.*, 1998), the pattern of precursor change during storage has important implications for flavour quality. In fact, while alliinase activity towards 1-propenyl cysteine sulphoxide did not change during storage, little methyl cysteine sulphoxide was hydrolysed after 4 months of storage (Kopsell, 1999).

7.4 Ecological factors affecting flavour

It has long been recognized that onions grown in different areas may have distinct flavour intensities (Platenius and Knott, 1935). 'Italian Red' onions produced in Italy were milder than bulbs of the same variety grown in New York. Popular cultivars sampled within New Zealand had as much as a threefold difference in total S-related flavour-precursor content, depending on the growing location (Lancaster *et al.*, 1988). Yearly fluctuations in bulb pungency have been reported, with some differences being more than double for a particular cultivar (Platenius, 1941; Bedford, 1984; Vavrina and Smittle, 1993). Why these differences occurred obviously had a lot to do with the environment in which the onions were grown. Locations can differ in soil type, fertility, water availability, growing temperatures, solar radiation and farmer-directed management practices (Randle *et al.*, 1998b). Many of these same factors can also vary between years. Through controlled experiments, researchers have begun to isolate the specific factors that influence flavour intensity and quality.

7.4.1 Sulphur supply and flavour

Sulphur (sulphate) availability has been the most thoroughly researched, and arguably has the greatest effect on flavour intensity and quality of any ecological factor. Clearly, limiting sulphur or making it abundant to the plant must have a serious impact on the quantity of sulphur-based flavour-precursor compounds synthesized. Freeman and Mossadeghi (1970) first showed in controlled greenhouse experiments that pungency (gross flavour intensity) could be varied from almost negligible levels to high levels simply by manipulating the amount of sulphate supplied to the plant. Similar results were found for sulphate applied to garlic and *A. vineale* (Freeman and Mossadeghi, 1971). Their work also showed that a saturation point could be reached, after which additional sulphate resulted in little response in pungency. In field experiments, pungency did not respond to applied sulphur, since the element was already adequately supplied (Paterson, 1979; Hamilton *et al.*, 1998). In the Texas study by Hamilton and co-workers, soil sulphate was between 120 and 530 ppm while the irrigation water had 71 ppm sulphate. At those levels, S was well above the saturation point for sulphate and little influence on bulb pungency was found. Kumar and Sahay (1954) in India did show a pungency response to applied sulphate in field-grown onions.

For many types of onions, high flavour intensity is desired. In most cases, growing areas have adequate sulphate in the soil or sulphate is added through normal fertility practices. However, for the production of truly mild onions, the restriction of sulphate to the plant is necessary. To produce mild onions, sulphate in the soil and water should not go above 50 ppm (D.A. Smittle, University of Georgia, 1991, personal communication). As sulphate content rises above this level, it becomes increasingly difficult to produce mild onions. Because sulphate is a leachable ion, lighter soils and even sandy soils are preferred over heavy or highly organic soils for mild onion production. Since sulphur is a required element and necessary for normal plant growth and development, restricting sulphate to produce mild onions has resulted in lower bulb yields (Platenius, 1941; Kumar and Sahay, 1954; Freeman and Mossadeghi, 1970; Randle *et al.*, 1995; Hamilton *et al.*, 1997). Conversely, high levels of available sulphate can actually depress onion and garlic bulb yields (Kumar and Sahay, 1954; Randle *et al.*, 1995; Singh *et al.*, 1995; Jaggi and Dixit, 1999). Changing sulphate fertility also affects sugar and soluble-solids accumulation in onions (Randle, 1992a; Randle and Bussard, 1993a; Hamilton *et al.*, 1997). Some cultivars produce increased solids/sugar content at high sulphate fertility while others decrease their solids/sugar content. On the other hand, some cultivars, especially those with the potential to be mild, will increase solids/sugar content at low sulphate fertility (Randle and Bussard, 1993a; Hamilton *et al.*, 1997). In the dehydrator onion 'Giza 20' in Egypt, increasing sulphur fertility increased total solids content, total fructan content and bulb pungency (Bakr and Gawish, 1998). The sugars sucrose, glucose and fructose, on the other hand, decreased as sulphur fertility increased.

Onion cultivars vary in their flavour response to sulphate fertility (Randle, 1992c; Randle and Bussard, 1993a; Hamilton *et al.*, 1997). Some cultivars are greatly affected by changes in available sulphate while others show less response. Moreover, a poor correlation in the rank order of cultivars grown at high and low S fertility suggests that sulphur metabolism within the flavour pathway among onion germplasm may be quite complex (Randle, 1992c; Randle and Bussard, 1993a).

Modifying sulphate fertility also has a considerable influence on the partitioning of sulphur within the plant. With plants grown at high sulphate fertility, more sulphur is retained in the leaves during bulbing than with plants grown at low sulphate fertility (Randle *et al.*, 1993a). The difference in total leaf sulphur between plants grown at high and low sulphur levels becomes exaggerated as bulbing advances and mass translocation accompanies bulking. If leaves are allowed to senesce and dry on mature bulbs, virtually no sulphur is left in the leaves of plants grown under low sulphate fertility. As sulphate fertility increased, the amount of total bulb sulphur stored as sulphate increased from approximately 10% to almost 50% (Randle *et al.*, 1999). Randle and co-workers also observed that pungent cultivars accumulated a lower percentage of sulphate in the bulb when compared with mild cultivars. It appears that pungent cultivars have a greater metabolic requirement for sulphur and more efficiently incorporate S into and through the pathway leading to flavour-precursor accumulation. On the other hand, one mechanism that appears to account for an onion being mild is the plant's ability to partition a greater amount of the absorbed S as sulphate, thereby excluding it from the ACSO pathway.

S fertility has a pronounced affect on how organic S accumulates and is metabolized through the various peptides and precursors of the flavour pathway. Under high-S-fertility conditions, 1-propenyl cysteine sulphoxide accumulated in the highest concentration among the individual flavour precursors (Randle *et al.*, 1995). Because most onions are grown with adequate S fertility, other studies have reported that 1-propenyl cysteine sulphoxide is the major flavour precursor of onion (Block, 1992; Edwards *et al.*, 1994; Thomas and Parkin, 1994; Yoo and Pike, 1998). However, as S fertility was decreased to near-deficiency levels, methyl cysteine sulphoxide increased in concentration and became the dominant precursor (Randle *et al.*, 1995). Propyl cysteine sulphoxide, which is normally found in the lowest concentration of the individual precursors (and with some analytical methods is undetectable in onion), was found in higher concentration than 1-propenyl cysteine sulphoxide when S fertility approached deficiency levels. At low S fertility, S is efficiently metabolized through the flavour biosynthetic pathway and does not accumulate to any great extent in the peptide intermediates (Randle *et al.*, 1995). Nearly 95% of total bulb S could be accounted for in compounds of the flavour pathway. But, as S fertility increased, the peptide intermediates 2-carboxypropyl glutathione and γ-glutamyl-1-propenyl cysteine sulphoxide began to accumulate in higher concentrations. At high S fertility, less than 40% of total bulb sulphur was accounted for in compounds of the flavour pathway.

7.4.2 Other plant nutrients and flavour

NITROGEN. High nitrogen availability affected onion flavour intensity and quality (Randle, 2000). When solution N levels were varied from 0.22 to 0.97 g l^{-1} in hydroponic solutions, enzymatically produced pyruvate increased linearly but then decreased at the highest N treatment. MCSO increased as N availability increased, while 1-PECSO initially increased but then decreased at the higher N treatments. PCSO generally increased with increasing N levels. Changes in the ACSO concentrations and ratios affect sensory perceptions of onion flavour.

SELENIUM. Selenate fertility also affected onion flavour quality and intensity. Of the 16 cultivars evaluated, six had significant decreases in enzymatically produced pyruvate, although most showed a trend to decreased pungency when grown in the presence of sodium selenate (Barak and Goldman, 1997; Kopsell, D.A. and Randle, 1997). The effect of sodium selenate on flavour quality was similar to that found when onions were grown at S-stress fertility levels (Kopsell and Randle, 1999). MCSO content increased and 1-PECSO decreased in the presence of high sodium selenate fertility, although there were differences among the cultivars tested.

CALCIUM. In a field study, preplant calcium fertility had little effect on onion-bulb pungency (Randle, 1995). Only the highest calcium treatment caused pungency to be significantly higher, and the effect was attributed to the Ca application having caused a nutrient imbalance in the soil solution.

7.4.3 Temperature and flavour

Temperatures influence onion growth and development. Bulbing ceases as temperatures fall below 10°C; it reaches a maximum at around 38°C. Platenius (1941) reported that volatile sulphur compounds in onions increased with increasing temperature, although he was unable to specify whether flavour differences were due to temperature-related plant growth or to the direct effect of temperature on flavour development. Later, onions were grown at four temperatures and under two regimes: (i) for 35 days and then tested for bulb pyruvate development, and (ii) until bulb maturity and then tested for bulb pyruvate (Randle *et al.*, 1993b). In both cases, as temperatures increased from 10 to 30°C, sulphur utilization increased and bulb pungency doubled. Within these limits, the hotter the conditions, the more pungent an onion will be.

7.4.4 Water-supply and flavour

Growing onions under dry conditions will also increase bulb pungency when compared with that of onions grown under well-irrigated conditions. When onions were grown under natural rainfall or supplemented with artificial irrigation, the onions with natural rainfall produced higher volatile sulphur content compared with the onions receiving supplemental artificial irrigation (Platenius, 1941). Drier growing conditions also resulted in bulbs that had higher pungency levels, as measured by enzymatically produced pyruvate, and greater flavour intensity in taste-panel evaluations (Freeman and Mossadeghi, 1973). As bulb size was smaller under the drier conditions, it was thought that the increase in flavour strength was due to a concentration of the flavour compounds in small cells.

Water usage and sulphate uptake by onions was poorly correlated ($r = 0.09$) (W.M. Randle, unpublished data). Water usage was greatly affected by daily differences in solar radiation, while sulphate uptake was unaffected. The exact mechanism for flavour increases in water-stressed plants, however, is yet to be determined.

8. Conclusions and Future Developments

Allium flavour will continue to be of interest to researchers, producers and consumers. A key diagnostic tool for improving bulb-flavour quality and consistency will be the use of global positioning satellite systems (GPS) and precision farming techniques (Randle *et al.*, 1998b). Statistical sampling methodology and GPS technology have been effective in plotting the distribution of bulb pungency in production fields, giving producers a diagnostic tool for improving flavour quality and consistency (Colour Plate 6). Continued research into defining the attributes causing flavour variance is needed to support GPS mapping and help in its interpretation.

Improvements in the rapid measurement of onion flavour need to be made. Although pyruvate has been widely used to determine overall flavour intensity, it lacks the ability to describe the major flavour attributes. For example, the dominant sensations of heat and mouth burn are not identified in the pyruvate test, but could be tested by quantifying the LF, if a reliable and rapid method were to be developed. Bitterness is another flavour attribute that adversely affects different onion products, ranging from dehydrator types to those which are mild and sweet. The compound(s) contributing the bitter attributes have not been identified and will need to be addressed.

Finally, with advances in molecular biology, marker-assisted breeding approaches should be developed to aid in the improvement of onions with a full range of flavour quality and intensity. Active selection either on an individual bulb basis or within families would greatly reduce the bulb-to-bulb flavour variability of current cultivars. We also need better understanding of the dynamic nature of the enzymes regulating S metabolic pathways within alliums. As mentioned earlier, *Arabidopsis* has been used to improve our knowledge of the enzymes involved in early S absorption and assimilation to cysteine. These enzymes are now candidates for genetic manipulation/transformation in *Allium*. However, it would be unwise to suppress these enzymes without first addressing the enzymes involved with S metabolism in the peptides leading to precursor synthesis. Research with low-S environments has shown us that the flavour-precursor pathway is a very strong sink for available S, and suppressing S absorption could lead to S-deficiency symptoms. Conversely, amplifying S absorption without changing S metabolism through the flavour-precursor pathway could lead to substantial sulphate accumulation.

References

Apparao, M., Kjaer, A., Olsen, O., Venkata Rao, E., Rasmussen, K.W. and Sorenson, H. (1981) Alliin in the garlicky taxon *Adenocalymma alliaceum* (*Bignoniaceae*). *Phytochemistry* 20, 822–823.

Bacon, J.R., Moates, G.K., Ng, A., Rhodes, M.J.C., Smith, A.C. and Waldron, K.W. (1999) Quantitative analysis of flavour precursors and pyruvate levels in different tissues and cultivars of onion (*Allium cepa*). *Food Chemistry* 64, 257–261.

Bajaj, K.L., Kaur, G., Singh, J. and Gill, S.P.S. (1980) Chemical evaluation of some important varieties of onion (*Allium cepa* L.). *Quality of Plant Foods and Human Nutrition* 30, 117–122.

Bajaj, K.L., Kaur, G. and Chadha, M.L. (1990) Varietal variations in some important chemical constituents of onion (*Allium cepa* L.). *Tropical Science* 30, 391–395.

Bakr, A.A. and Gawish, R.A. (1998) Current studies on onion and its pungency. Part 2. Interactive effects of sulfur fertilization and drying under the aspect of solar energy utilization. *Nahrung* 42, 94–101.

Barak, P. and Goldman, I.L. (1997) Antagonistic relationship between selenate and sulfate uptake in onion (*Allium cepa*). Implications for the production of organosulfur and organoselenium compounds in plants. *Journal of Agricultural and Food Chemistry* 45, 1290–1294.

Barroso, C., Vega, J.M. and Gotor, C. (1998) The role of roots in cysteine biosynthesis by *Arabidopsis thaliana*. *Journal of Physiology and Biochemistry* 54, 189–194.

Bedford, L.V. (1984) Dry matter and pungency tests on British grown onions. *Journal of the National Institute of Agricultural Botany* 16, 581–591.

Bernhard, R.A. (1970) Chemotaxonomy, distribution studies of sulphur compounds in *Allium*. *Phytochemistry* 9, 2019–2027.

Blake-Kalff, M.M.A., Harrison, K.R., Hawkesford, M.J., Zhao, F.J. and McGrath, S.P. (1998) Distribution of sulfur within oilseed rape leaves in response to sulfur deficiency during vegetative growth. *Plant Physiology* 118, 1337–1344.

Block, E. (1992) The organosulfur chemistry of the genus *Allium* – implications for the organic chemistry of sulfur. *Angewandte Chemie, International Edition in English* 31, 1135–1178.

Bork, C., Schwenn, J.D. and Hell, R. (1998) Isolation and characterization of a gene for assimilatory sulfite reductase from *Arabidopsis thaliana*. *Gene* 21, 147–153.

Boscher, J., Auger, J., Mandon, N. and Ferary, S. (1995) Qualitative and quantitative comparison of volatile sulphides and flavour precursors in different organs of some wild and cultivated garlics. *Biochemical and Systematic Ecology* 23, 787–791.

Carson, J.F. (1987) Chemistry and biological properties of onion and garlic. *Food Review International* 3, 71–103.

Ceci, L.N., Curzio, O.A. and Pomilio, A.B. (1991) Effects of irradiation and storage on the flavor of garlic bulbs cv. 'Red'. *Journal of Food Science* 56, 44–46.

Ceci, L.N., Curzio, O.A. and Pomilio, A.B. (1992) γ-Glutamyl transpeptidase/γ-glutamyl peptidase in sprouted *Allium sativum*. *Phytochemistry* 31, 441–444.

Clark, S.A. (1993) Molecular cloning of a cDNA encoding alliinase from onion (*Allium cepa* L.). PhD dissertation, University of Canterbury, Christchurch, New Zealand.

Clark, S.A., Shaw, M.L., Every, D. and Lancaster, J.E. (1998) Physical characterization of alliinase, the flavor generating enzyme in onions. *Journal of Food Biochemistry* 22, 91–103.

Debaene, J.E.P., Goldman, I.L. and Yandell, B.S. (1999) Postharvest flux and genotype × environmental effects for onion-induced antiplatelet activity, pungency, and soluble solids in long-day onion during postharvest cold storage. *Journal of the American Society for Horticultural Science* 124, 366–372.

Dowker, B.D. (1990) Onion breeding. In: Rabinowitch, H.D. and Brewster, J.L. (eds) *Onions and Allied Crops*, Vol. I. *Botany, Physiology, and Genetics*. CRC Press, Boca Raton, Florida, pp. 215–232.

Edwards, S.J., Britton, G. and Collin, H.A. (1994) The biosynthetic pathway of the *S*-alk(en)yl-L-cysteine sulphoxides (flavor precursors) in species of *Allium*. *Plant Cell, Tissue, and Organ Culture* 38, 181–188.

Ellmore, G.S. and Feldberg, R.S. (1994) Alliin lyase localization in bundle sheaths of garlic clove (*Allium sativum*). *American Journal of Botany* 81, 89–94.

Freeman, G.G. (1975) Distribution of flavour components in onion (*Allium cepa* L.), leek (*Allium porrum*) and garlic (*Allium sativum*). *Journal of the Science of Food and Agriculture* 26, 471–481.

Freeman, G.G. and Mossadeghi, N. (1970) Effect of sulphate nutrition on flavour components of onion (*Allium cepa*). *Journal of the Science of Food and Agriculture* 21, 610–615.

Freeman, G.G. and Mossadeghi, N. (1971) Influence of sulphate nutrition on the flavour components of garlic (*Allium sativum*) and wild onion (*A. vineale*). *Journal of the Science of Food and Agriculture* 22, 330–334.

Freeman, G.G. and Mossadeghi, N. (1973) Studies on the relationship between water regime and flavour strength in water cress (*Rorippa masturium-aquaticum* [L.] Hayek), cabbage (*Brassica oleracea* var. *capitata*) and onion (*Allium cepa*). *Journal of Horticultural Science* 48, 365–378.

Freeman, G.G. and Whenham, W.J. (1975a) A rapid spectrophotometric method of determination of thiopropanal *S*-oxide (lachrymator) in onion (*Allium cepa*) and its significance in flavour studies. *Journal of the Science of Food and Agriculture* 26, 1529–1543.

Freeman, G.G. and Whenham, R.J. (1975b) A survey of volatile components of some *Allium* species in terms of *S*-alk(en)yl-L-cysteine sulphoxides present as flavour precursors. *Journal of the Science of Food and Agriculture* 26, 1869–1886.

Freeman, G.G. and Whenham, R.J. (1976) Effect of overwinter storage at three temperatures on the flavour intensity of dry bulb onions. *Journal of the Science of Food and Agriculture* 27, 37–42.

Fujita, M., Endo, M. and Sano, M.(1990) Purification and characterization of alliin lyase from Welsh onion, *Allium fistulosum* L. *Agricultural and Biological Chemistry* 54, 1077–1079.

Gilpin, B.J. (1995) Isolation and characterisation of genomic clones of alliinase from *Allium cepa* L. PhD dissertation, University of Canterbury, Christchurch, New Zealand.

Gilpin, B.J., Leung, D.W. and Lancaster, J.E. (1995) Nucleotide sequence of a nuclear clone of alliinase (accession no. L48614) from onion (PGR 95–125). *Plant Physiology* 110, 336.

Granroth, B. (1970) Biosynthesis and decomposition of cysteine derivatives in onion and other *Allium* species. *Annales Academica Scientia Fenniae Series A2* 154, 1–71.

Hall, D.I. and Smith, I.K. (1983) Partial purification and characterization of cystine lyase from cabbage (*Brassica oleracea* var. *capitata*). *Plant Physiology* 72, 654–658.

Hamilton, B.K., Pike, L.M. and Yoo, K.S. (1997) Clonal variations of pungency, sugar content, and bulb weight of onions due to sulphur nutrition. *Scientia Horticulturae* 71, 131–136.

Hamilton, B.K., Yoo, K.S. and Pike, L.M. (1998) Changes in pungency of onions by soil type, sulphur nutrition and bulb maturity. *Scientia Horticulturae* 74, 249–256.

Hanum, T., Sinha, N.K. and Cash, J.N. (1995) Characteristics of γ-glutamyl transpeptidase and alliinase on onion and their effects on the enhancement of pyruvate formation in onion macerates. *Journal of Food Biochemistry* 19, 51–65.

Harada, E., Kusano, T. and Sano, H. (2000) Differential expression of genes encoding enzymes involved in sulfur assimilation pathways in response to wounding and jasmonate in *Arabidopsis thaliana*. *Journal of Plant Physiology* 156, 272–276.

Havey, M.J. and Randle, W.M. (1996) Combining abilities for yield and bulb quality among long- and intermediate-day open-pollinated onion populations. *Journal of the American Society for Horticultural Science* 121, 604–608.

Hell, R. (1997) Molecular physiology of plant sulfur metabolism. *Planta* 202, 138–148.

Ho, M.F. and Mazelis, M. (1993) The C-S lyases of higher plants, determination of immunology by immunological procedures. *Phytochemistry* 34, 625–629.

Iwami, K. and Yasumoto, K. (1980) Alliinase-like enzymes in fruiting bodies of *Lentinus edodes*, their purification and substrate specificity. *Agricultural and Biological Chemistry* 4, 3003–3004.

Jacobsen, J.V., Yamaguchi, M., Mann, L.K., Howard, F.D. and Bernhard, R.A. (1968) An alkyl-cysteine sulfoxide lyase in *Tulbaghia violacea* and its relation to other alliinase-like enzymes. *Phytochemistry* 7, 1099–1108.

Jaggi, R.C. and Dixit, S.P. (1999) Onion (*Allium cepa*) responses to sulphur in representative vegetable growing soils of Kangra Valley of Himachal Pradesh. *Indian Journal of Agricultural Science* 69, 289–291.

Jansen, H., Müller, B. and Knobloch, K. (1989a) Characterisation of an alliin lyase preparation from garlic (*Allium sativum*). *Planta Medica* 55, 434–439.

Jansen, H., Müller, B. and Knobloch, K. (1989b) Alliin lyase from garlic, *Allium sativum*. Investigations on enzyme/substrate, enzyme/inhibitor interactions, and on a new coenzyme. *Planta Medica* 55, 440–445.

Jones, H.A. and Bisson, C.S. (1934) Moisture content of different varieties of onions. *Proceedings of the American Society for Horticultural Science* 31, 165–168.

Jones, H.A. and Mann, L.K (1963) *Onions and Their Allies. Botany, Cultivation, and Utilization*. Interscience Publications, New York, 286 pp.

Kamitani, H., Esaki, N., Tanaka, H., Imahara, H. and Soda, K. (1990) Purification and characterization of *S*-alkyl cysteine α,β-lyase from *Pseudomonas putida*. *Journal of Nutritional Science and Vitaminology* 36, 339–347.

Kazaryan, R.A. and Goryachenkova, E.V. (1978) Alliinase, purification and characterization. *Biokhimiya* 43, 1905–1913.

King, J.J., Bradeen, J.M., Bark, O., McCallum, J.A. and Havey, M.J. (1998) A low-density genetic map of onion reveals a role for tandem duplication in the evolution of an extremely large diploid genome. *Theoretical and Applied Genetics* 96, 52–62.

Kitamura, N., Shimomura, N., Iseki, J., Honma, M., Shiba, S., Tahara, S. and Mizutani, J. (1997) Cysteine, S-conjugate α-lyase activity and pyridoxal 5′-phosphate binding site of onion alliin lyase. *BioScience Biotechnology and Biochemistry* 61, 1327–1330.

Kopsell, D.A. and Randle, W.M. (1997) Short-day onion cultivars differ in bulb selenium and sulfur accumulation which can affect bulb pungency. *Euphytica* 96, 385–390.

Kopsell, D.A. and Randle, W.M. (1999) Selenium affects the *S*-alk(en)yl cysteine sulfoxides among short-day onion cultivars. *Journal of the American Society for Horticultural Science* 124, 307–311.

Kopsell, D.E. (1999) Changes of flavor components in different cultivars of *Allium cepa* L. during storage. PhD thesis, University of Georgia, Atlanta, Georgia, USA.

Kopsell, D.E. and Randle, W.M. (1997) Onion cultivars differ in pungency and bulb quality changes during storage. *HortScience* 32, 1260–1263.

Kopsell, D.E., Randle, W.M. and Eiteman, M.A. (1999) Changes in the *S*-alk(en)yl cysteine sulfoxides and their biosynthetic intermediates during onion storage. *Journal of the American Society for Horticultural Science* 124, 177–183.

Kumar, K. and Sahay, R.K. (1954) Effect of sulphur fertilization on the pungency of onion. *Current Science* 11, 368–369.

Lancaster, J.E. and Boland, M.J. (1990) Flavor biochemistry. In: Brewster, J.L. and Rabinowitch, H.D. (eds) *Onions and Allied Crops*, Vol. III. *Biochemistry, Food Science, and Minor Crops*. CRC Press, Boca Raton, Florida, pp. 33–72.

Lancaster, J.E. and Collin, H.A. (1981) Presence of alliinase in isolated vacuoles and alkyl cysteine sulphoxides in the cytoplasm of bulbs in onion (*Allium cepa*). *Plant Science Letters* 22, 169–176.

Lancaster, J.E. and Shaw, M.L. (1989) γ-Glutamyl peptides in the biosynthesis of *S*-alk(en)yl-L-cysteine sulphoxides (flavor precursors) in *Allium*. *Phytochemistry* 28, 455–460.

Lancaster, J.E. and Shaw, M.L. (1991) Metabolism of γ-glutamyl peptides during development, storage and sprouting of onion bulbs. *Phytochemistry* 30, 2857–2859.

Lancaster, J.E. and Shaw, M.L. (1994) Characterization of purified γ-glutamyl transpeptidase in onions, evidence for *in vivo* role as a peptidase. *Phytochemistry* 36, 1351–1358.

Lancaster, J.E., McCallion, B.J. and Shaw, M.L. (1986) The dynamics of the flavour precursors, the *S*-alk(en)yl-L-cysteine sulphoxides, during leaf blade and scale development in the onion (*Allium cepa*). *Physiologia Plantarum* 66, 293–297.

Lancaster, J.E., Reay, P.F., Mann, J.D., Bennett, W.D. and Sedcole, J.R. (1988) Quality in New Zealand-grown onion bulbs – a survey of chemical and physical characteristics. *New Zealand Journal of Experimental Agriculture* 16, 279–285.

Lancaster, J.E., Shaw, M.L., Clark, S., Sutton, K. and Dommisse, E. (1993) Biochemistry of onion flavor. In: *Proceedings of the 1993 National Onion Research Conference, 9–11 December 1993, Ithaca, New York, USA*. Cornell University Press, Ithaca, New York, pp. 80–84.

Lancaster, J.E., Shaw, M.L., Pither-Joyce, M.D., Farrant, J.P. and McCallum, J.A. (1995) A review of the regulation of sulfur metabolism and its effect on quality in onions and garlic. In: *Proceedings of the 1995 National Onion Research Conference*. University of Wisconsin, Madison, Wisconsin, pp. 143–155.

Lancaster, J.E., Shaw, M.L. and Randle, W.M. (1998) Differential hydrolysis of alk(en)yl cysteine sulphoxides by alliinase in onion macerates, flavour implications. *Journal of the Science of Food and Agriculture* 78, 367–372.

Lancaster, J.E., Shaw, M.L., Pither-Joyce, M.D., McCallum, J.A. and McManus, M.T. (2000a) A novel alliinase from onion roots. Biochemical characterization and cDNA cloning. *Plant Physiology* 122, 1269–1279.

Lancaster, J.E., Shaw, M.L. and Walton, E.F. (2000b) *S*-Alk(en)yl-L-cysteine sulfoxides, alliinase and aroma in *Leucocoryne*. *Phytochemistry* 55, 127–130.

Lancaster, J.E., Farrant, J. and Shaw, M.L. (2000c) Effect of sulfur fertility on alliinase, the flavor generating enzyme in onions. *Journal of Food Biochemistry* 24, 353–361.

Landshuter, J., Lohmüller, E.M. and Knobloch, K. (1994) Purification and characterization of a C-S-lyase from ramson, the wild garlic *Allium ursinum*. *Planta Medica* 60, 343–347.

Lappartient, A.G., Vidmar, J.J., Leustek, T., Glass, A.D.M. and Touraine, B. (1999) Inter-organ signaling in plants, regulation of ATP sulfurylase and sulfate transporter genes expression in roots mediated by phloem-translocated compounds. *The Plant Journal* 18, 89–95.

Lawson, L.D. (1996) The composition and chemistry of garlic cloves and processed garlic. In: Koch, H.P. and Lawson, L.D. (eds) *Garlic. The Science and Therapeutic Application of* Allium sativum *L. and Related Species*, 2nd edn. Williams and Wilkins, Baltimore, Maryland, pp. 37–107.

Lawson, L.D. and Hughes, B.G. (1992) Characterisation of the formation of allicin and other thiosulfinates from garlic. *Planta Medica* 58, 345–350.

Lawson, L.D., Wang, Z.Y.J. and Hughes, B.G. (1991) γ-Glutamyl-*S*-alkylcysteines in garlic and other *Allium* spp., precursors of age-dependent *trans*-1-propenyl thiosulfinates. *Journal of Natural Products* 54, 436–444.

Lee, S. and Leustek, T. (1998) APS kinase from *Arabidopsis thaliana*, genomic organization, expression, and kinetic analysis of the recombinant enzyme. *Biochemistry and Biophysics Research Communications* 24, 171–175.

Leustek, T. and Saito, K. (1999) Sulfate transport and assimilation in plants. *Plant Physiology* 120, 637–643.

Lin, M.-W., Watson, J.F. and Baggett, J.R. (1995) Inheritance of soluble solids and pyruvic acid content of bulb onions. *Journal of the American Society for Horticultural Science* 120, 119–122.

Lohmüller, E.M., Landshuter, J. and Knobloch, K. (1994) On the isolation and characterisation of a C-S-lyase preparation from leek, *Allium porrum*. *Planta Medica* 60, 337–342.

Manabe, T., Hasumi, A., Sugiyama, M., Yamazaki, M. and Saito, K. (1998) Alliinase (*S*-alk(en)yl-L-cysteine sulfoxide lyase) from *Allium tuberosum* (Chinese chive), purification, localization, cDNA cloning and heterologous functional expression. *European Journal of Biochemistry* 257, 21–30.

Matikkala, E.J. and Virtanen, A.I. (1965a) γ-Glutamyl peptidase (glutaminase) in germinating seeds of chives (*Allium schoenoprasum*). *Acta Chemica Scandinavica* 19, 1258–1261.

Matikkala, E.J. and Virtanen, A.I. (1965b) γ-Glutamyl peptidase in sprouting onion bulbs. *Acta Chemica Scandinavica* 19, 1261–1262.

Matsuura, H., Inagaki, M., Maeshige, K., Ide, N., Kajimura, Y. and Itakura, Y. (1996) Changes in contents of γ-glutamyl peptides and fructan during growth of *Allium sativum*. *Planta Medica* 62, 70–71.

Mikitzel, L.J. and Fellman, J.K. (1994) Flavor and quality changes in sweet onions during storage at room temperature. *Journal of Food Quality* 17, 431–445.

Nock, L.P. and Mazelis, M. (1987) The C-S lyases of higher plants. Direct comparison of the physical properties of homogeneous alliin lyase of garlic (*Allium sativum*) and onion (*Allium cepa*). *Plant Physiology* 85, 1079–1083.

Nomura, J., Nishizuka, Y. and Hayaishi, O. (1963) *S*-Alkylcysteinease, enzymatic cleavage of *S*-methyl cysteine and its sulfoxide. *Journal of Biological Chemistry* 23, 1441–1446.

Parry, R.J. and Lii, F.-L. (1991) Investigations of the biosynthesis of trans-(+)-*S*-1-propenyl-L-cysteine sulfoxide. Elucidation of the stereochemistry of the oxidative decarboxylation process. *Journal of the American Chemistry Society* 113, 4704–4706.

Paterson, D.R. (1979) *Sulfur Fertilization Effects on Onion Yield and Pungency*. Progress Report No. 3551, Texas Agricultural Experiment Station, Texas A&M University, College Station, Texas.

Peterson, C.E., Simon, P.W. and Ellerbrock, L.A. (1986) 'Sweet Sandwich' onion. *HortScience* 21, 1466–1468.

Pike, L.M., Horn, R.S. and Andersen, C.R. (1988a). 'Texas Grano 1015Y', a mild pungency, sweet, shortday onion. *HortScience* 23, 634–635.

Pike, L.M., Horn, R.S. and Andersen, C.R. (1988b) 'Texas Grano 1030Y', a late maturing, mild pungency, shortday onion. *HortScience* 23, 636–637.

Platenius, H. (1935) A method for estimating the volatile sulphur content and pungency of onions. *Journal of Agricultural Research* 51, 847–853.

Platenius, H. (1941) Factors affecting onion pungency. *Journal of Agricultural Research* 62, 371–379.

Platenius, H. and Knott, J.E. (1935) The pungency of the onion bulb as influenced by the stage of development of the plant. *Proceedings of the American Society for Horticultural Science* 33, 481–483.

Prince, C.L., Shuler, M.L. and Yamada, Y. (1997) Altering flavor profiles in onion (*Allium capa* L.) root cultures through directed biosynthesis. *Biotechnology Progress* 13, 506–510.

Rabinkov, A., Zhu, X.Z., Grafi, G. and Mirelman, D. (1994) Allin lyase (alliinase) from garlic (*Allium sativum*), biochemical characterization and cDNA cloning. *Applied Biochemistry and Biotechnology* 48, 149–171.

Rabinkov, A., Wilchek, M. and Mirelman, D. (1995) Alliinase (alliin lyase) from garlic (*Allium sativum*) is glycosylated at ASN^{146} and forms a complex with a garlic mannose-specific lectin. *Glyco-conjugate Journal* 12, 690–698.

Rabinowitch, H.D. (1988) Genetics and breeding, state-of-the-art or too slow but not too late. In: *Proceedings of the Eucarpia 4th Allium Symposium, 6–9 September*. Institute of Horticultural Research, Wellesbourne, Warwick, UK, pp. 57–69.

Ramirez, E.C. and Whitaker, J.R. (1998) Cystine lyases in plants, a comprehensive review. *Journal of Food Biochemistry* 22, 427–440.

Randle, W.M. (1992a) Sulfur nutrition affects nonstructural water-soluble carbohydrates in onion germplasm. *HortScience* 27, 52–55.

Randle, W.M. (1992b) Sampling procedures to estimate flavor potential in onion. *HortScience* 27, 1116–1117.
Randle, W.M. (1992c) Onion germplasm interacts with sulfur fertility for plant sulfur utilization and bulb pungency. *Euphytica* 59, 151–156.
Randle, W.M. (1995) Preplant calcium affects onion bulb quality and shelf-life. *HortScience* 30, 768 (abstract).
Randle, W.M. (2000) Increasing nitrogen concentration in hydroponic solutions affects onion flavor and bulb quality. *Journal of the American Society for Horticultural Science* 125, 254–259.
Randle, W.M. and Bussard, M.L. (1993a) Pungency and sugars of short-day onions as affected by sulfur nutrition. *Journal of the American Society for Horticultural Science* 118, 766–770.
Randle, W.M. and Bussard, M.L. (1993b) Streamlining onion pungency analysis. *HortScience* 28, 60.
Randle, W.M., Bussard, M.L. and Warnock, D.F. (1993a) Ontogeny and sulfur fertility affect leaf sulfur in short-day onions. *Journal of the American Society for Horticultural Science* 118, 762–765.
Randle, W.M., Bussard, M.L. and Warnock, D.F. (1993b) Temperature affects plant growth and sulfur utilization in onion (*Allium cepa*). *HortScience* 28, 467 (abstract).
Randle, W.M., Lancaster, J.E., Shaw, M.L., Sutton, K.H., Hay, R.L. and Bussard, M.L. (1995) Quantifying onion flavor compounds responding to sulfur fertility, sulfur increases levels of alk(en)yl cysteine sulfoxides and biosynthetic intermediates. *Journal of the American Society for Horticultural Science* 120, 1075–1081.
Randle, W.M., Kopsell, D.E. and Kopsell, D.A. (1998a) Considerations for implementing pungency field testing and its practical implications. In: *Proceedings of the 1998 National Onion (and Other Allium) Research Conference, 10–12 December, Sacramento, California, USA*. University of California, Davis, California, pp. 171–173.
Randle, W.M., Kopsell, D.A., Kopsell, D.E., Snyder, R.L. and Torrance, R. (1998b) Field sampling short-day onions for bulb pungency. *HortTechnology* 8, 329–332.
Randle, W.M., Kopsell, D.E., Kopsell, D.A. and Snyder, R.L. (1999) Total sulfur and sulfate accumulation in onion is affected by sulfur fertility. *Journal of Plant Nutrition* 22, 45–51.
Saghir, A.R., Mann, L.K. and Yamaguchi, M. (1965) Composition of volatiles in *Allium* as related to habitat, stage of growth and plant part. *Plant Physiology* 35, 681–685.
Saito, K. (1998) Molecular regulation and engineering of sulfur assimilation and conversion. Abstract S14 (SIII-02). *Plant and Cell Physiology* 39 (Suppl.), S5.
Schwimmer, S. (1969) Characterisation of *S*-propenyl-L-cysteine sulfoxide lyase of onion. *Archives of Biochemistry and Biophysics* 130, 312–320.
Schwimmer, S. and Guadagni, D.G. (1962) Relation between olfactory threshold concentration and pyruvic acid content of onion juice. *Journal of Food Science* 27, 94–97.
Schwimmer, S. and Guadagni, D.G. (1968) Kinetics of the enzymatic development of pyruvic acid and odour in frozen onions treated with cysteine C-S lyase. *Journal of Food Science* 33, 193–196.
Schwimmer, S. and Kjaer, A. (1960) Purification and specificity of the C-S lyase of *Albizzia lophanta*. *Biochemica et Biophysica Acta* 42, 316–324.
Schwimmer, S. and Weston, W.J. (1961) Enzymatic development of pyruvic acid in onion as a measure of pungency. *Journal of Agricultural and Food Chemistry* 9, 301–304.
Schwimmer, S., Ryan, C.A. and Wong, F. (1964) Specificity of L-cysteine sulfoxide lyase and partially competitive inhibition by *S*-alkyl-L-cysteines. *Biological Chemistry* 239, 777–782.
Shaw, M.L., Lancaster, J.E. and Lane, G.A. (1989) Quantitative analysis of the major γ-glutamyl peptides in onion bulbs (*Allium cepa*). *Journal of the Science of Food and Agriculture* 48, 459–467.
Simon, P.W. (1995) Genetic analysis of pungency and soluble solids in long-storage onions. *Euphytica* 82, 1–8.
Singh, P., Singh, V. and Malik, R.S. (1995) Effect of different doses and sources of S on yield and uptake of S by garlic. *Journal of the Indian Society of Soil Science* 43, 130–131.
Smeets, K., van Damme, E.J.M., van Leuven, F. and Peumans, W.F. (1997) Isolation and characterisation of lectins and lectin-alliinase complexes from bulbs of garlic (*Allium sativum*) and ramsons (*Allium sativum*). *Glycoconjugate Journal* 14, 331–343.
Smith, F.W., Hawkesford, M.J., Ealing, P.M., Clarkson, D.T., van den Berg, P.J., Belcher, A.R. and Warrilow, A.G. (1997) Regulation of expression of a cDNA from barley roots encoding a high affinity sulphate transporter. *The Plant Journal* 12, 875–884.
Smittle, D.A. (1988) Evaluation of storage methods for 'Granex' onions. *Journal of the American Society for Horticultural Science* 113, 877–880.

Smittle, D.A. (1991) Commercial storage of Vidalia onions. *Onion World* 7, 10–12.

Sunarpi and Anderson, J.W. (1996) Effect of sulfur nutrition on the redistribution of sulfur in vegetative soybean plants. *Plant Physiology* 112, 623–631.

Sunarpi and Anderson, J.W. (1997) Effect of nitrogen nutrition on remobilization of protein sulfur in the leaves of vegetative soybean and associated changes in soluble sulfur metabolites. *Plant Physiology* 115, 1671–1680.

Takahashi, H., Yamazaki, M., Sasakura, N., Watanabe, A., Levster, T., Engler, J.deA., Engler, G., van Montagu, M. and Saito, K. (1997) Regulation of sulfur assimilation in higher plants, a sulfate transporter induced in sulfate-starved roots plays a central role in *Arabidopsis thaliana*. *Proceedings of the National Academy of Science of the USA* 94, 11102–11107.

Takahashi, H., Watanabe, A. and Saito, K. (1998) Antisense repression of sulfate transporter in transgenic *Arabidopsis thaliana* plants. *Plant and Cell Physiology* 39 (Suppl.), abstract 55.

Thomas, D.J. and Parkin, K.L. (1994) Quantification of alk(en)yl-L-cysteine sulfoxides and related amino acids in alliums by high-performance liquid chromatography. *Journal of Agricultural and Food Chemistry* 42, 1632–1638.

Thomas, D.J., Parkin, K.L. and Simon, P.W. (1992) Development of a simple pungency indicator test for onions. *Journal of the Science of Food and Agriculture* 60, 499–504.

Tobkin, H.E. and Mazelis, M. (1979) Alliin lyase, preparation and characterization of the homogeneous enzyme from onion bulbs. *Archives of Biochemistry and Biophysics* 19, 150–157.

Tsuno, S. (1958a) Alliinase in *Allium* plants. *Bitamin* 14, 659–664.

Tsuno, S. (1958b) The nutritional value of *Allium* plants. XVII. Formation of *S*-2-(*n*-(2-methyl-4-amino-5-pyrimidyl methyl formamido)-5-hydroxy-2-pentene-3-yl) allyl disulphide with *Ipheion uniflorum*. *Bitamin* 14, 665–670.

van Damme, E.J.M., Smeets, K., Torrekens, S., van Leuven, F. and Peumans, W.J. (1992) Isolation and characterization of alliinase cDNA clones from garlic (*Allium sativum* L.) and related species. *European Journal of Biochemistry* 209, 751–757.

Vavrina, C.S. and Smittle, D.A. (1993) Evaluating sweet onion cultivars for sugar concentrations and pungency. *HortScience* 28, 804–806.

Virtanen, A.I. and Matikkala, E.J. (1959) Isolation of *S*-methyl cysteine sulphoxide and *S*-*n*-propyl cysteine sulphoxide from onion (*Allium cepa*) and the antibiotic activity of crushed onion. *Acta Chemica Scandinavica* 13, 1898–1900.

von Stoll, A. and Seebeck, E. (1949) Über die Spezifizität der Alliinase und die Synthese inmehrer dem Alliin verwandter Verbindungen. *Helvetica Chimica Acta* 32, 866–876.

Wäfler, U., Shaw, M.L. and Lancaster, J.E. (1994) Effect of freezing upon alliinase activity in onion extracts and pure enzyme preparations. *Journal of the Science of Food and Agriculture* 64, 315–318.

Wall, M.M. and Corgan, J.N. (1998) 'NuMex Dulce' onion. *HortScience* 33, 762–763.

Wall, M.M. and Corgan, J.N. (1999) 'NuMex Sweetpak' onion. *HortScience* 34, 1303–1304.

Wall, M.M., Mohammad, A. and Corgan, J.N. (1996) Heritability estimates and response to selection for the pungency and single center traits in onion. *Euphytica* 87, 133–139.

Warid, W.A. (1952) Inheritance studies in the onion. PhD dissertation, Louisiana State University, Baton Rouge, Louisiana, USA.

Whitaker, J.R. (1976) Development of flavor, odor and pungency in onion and garlic. *Advanced Food Research* 22, 73–133.

Yoo, K.S. and Pike, L.M. (1998) Determination of flavor precursor compound *S*-alk(en)yl-L-cysteine sulfoxides by an HPLC method and their distribution in *Allium* species. *Scientia Horticulturae* 75, 1–10.

Yoo, K.S., Pike, L.M. and Hamilton, B.K. (1995) A simplified pyruvic acid analysis suitable for onion breeding programs. *HortScience* 30, 1306.

15 Health and Alliums

M. Keusgen
Institute for Pharmaceutical Biology, University of Bonn, Nußallee 6, D-53115 Bonn, Germany

1. Therapeutic and Medicinal Values of Onion (*Allium cepa* L.)

1.1 Introduction: composition and chemistry

Since ancient times, onions and related species have been widely used in many parts of the world as flavouring vegetables, as well as in traditional and folk medicine. Health benefits in the treatment of many different major and minor disorders have been claimed in traditional medicine, but precise medical data are lacking to support most of these effects. In recent years, the therapeutic and medicinal values of onions have been reviewed by Augusti (1990, 1996), Koch (1994), Dorsch (1996), Koch and Lawson (1996) and Craig (1999). Therefore, the studies published during the 1990s are mainly considered in this review.

Bulbs of *A. cepa* for fresh consumption

contain mainly water (ranging between about 95 and 88%) and mono- and disaccharides (total sugars about 6%). Fructans – fructose-based polysaccharides – are also typical constituents of *Allium*. Steroid saponins are common in *Liliaceae* and closely related families and spirostanol and furostanol types were reported for *A. cepa* (Koch, 1994; Breu, 1996). During recent years, the flavonoid quercetin and related compounds have attracted special attention (Fig. 15.1; Patil and Pike, 1995; Patil *et al.*, 1995). Onion bulbs contain high levels of flavonoids. In dried red onion, free quercetin was found in concentrations up to 2.1%. Additionally, *A. cepa* contains some phenolic compounds, as well as malic, citric, succinic, fumaric and quinic acids. Vitamins, such as B_1, B_2, B_6 and biotinic, nicotinic, folic, pantothenic and ascorbic acids, are also found (Breu, 1996).

Many of the health benefits of onions and related species are attributed to organosulphur compounds, which account for 1–5% of the dry weight of the mature bulbs (Block, 1992; Koch and Lawson, 1996). The most important sulphur-containing substances are the amino acid cysteine and its derivatives, especially the S-substituted cysteine sulphoxides and the γ-glutamyl peptides. Block (1992) and Breu (1996) reviewed the complex sulphur chemistry of *Allium* (see also Randle and Lancaster, Chapter 14, this volume).

Intact plant material of onions contains the odourless cysteine sulphoxides, mainly *trans* (+)-*S*-(1-propenyl)-L-cysteine sulphoxide and (+)-*S*-methyl-L-cysteine sulphoxide in low concentrations (Fig. 15.2). The homologous propyl derivative was also reported for *A. cepa* (Breu, 1996). Additionally, these amino acid derivatives occur as the corresponding γ-glutamyl derivatives. When intact cells are disrupted, cysteine sulphoxides are rapidly converted into alk(en)ylsulphenic acid, pyruvic acid and ammonia. This reaction is catalysed by alliinase (EC 4.4.1.4) to form unstable sulphenic acids, which are rapidly converted into either the corresponding thiosulphinates or the lachrymatory factor (LF) (*Z*)-propanethial-*S*-oxide, which cause the

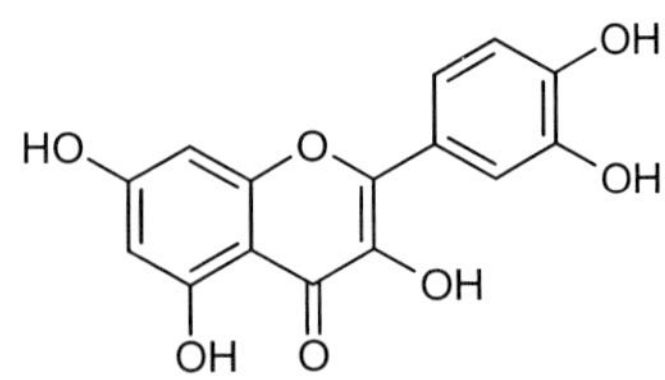

Fig. 15.1. Chemical structure of quercetin.

characteristic smell of freshly prepared onion juice (Fig. 15.2: 'primary aroma compounds'). The primary aroma compounds are also relatively unstable and quickly decompose into a variety of strong-smelling, volatile sulphur compounds. These derivatives are characteristic of processed onions, e.g. steam-distilled oils (Fig. 15.2: 'secondary aroma compounds').

1.2 Antibiotic activities

1.2.1 Crude extracts

ANTIBACTERIAL EFFECTS. The antibiotic properties of onion extracts and oils have been intensively studied in the second half of the 20th century. In traditional medicine, onions were used against different infectious diseases for many centuries. Even the Egyptian Papyrus Ebers mentioned onion-containing remedies against worms, diarrhoea, other infections and inflammatory diseases (Dorsch, 1996). The organosulphur compounds of onions and other *Allium* species, but also proteins, saponins and phenolic compounds are considered responsible for these effects. Some recent findings are summarized in Table 15.1. Onion oils and aqueous extracts were almost ineffective against Gram-negative bacteria but active against several Gram-positive bacteria (Dankert *et al.*, 1979; Elnima *et al.*, 1983; Zohri *et al.*, 1995). Onion extracts inhibit oral bacteria causing dental caries (Kim, 1997). In contrast, comparable garlic preparations were active against Gram-negative bacteria (Dankert *et al.*, 1979; Elnima *et al.*, 1983; Yoshida *et al.*, 1999a, b).

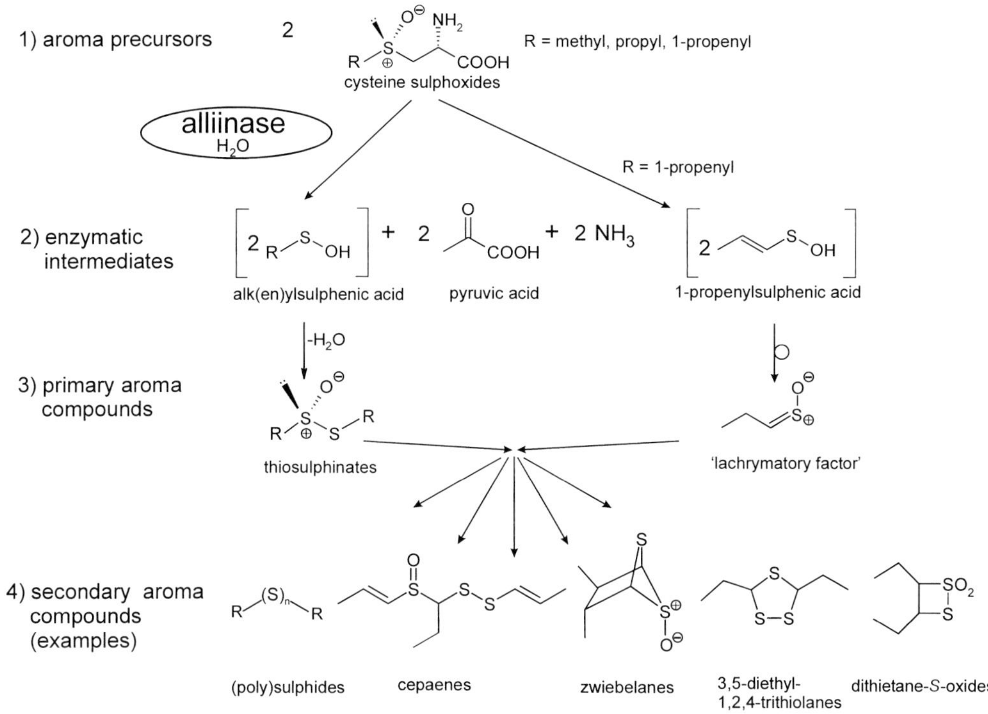

Fig. 15.2. Enzymatic cleavage of different cysteine sulphoxides, which are typical for *Allium cepa*. The reaction is catalysed by the enzyme alliinase. Alk(en)ylsulphenic acids (enzymatic intermediates) are unstable and are converted into the lachrymatory factor or thiosulphinates (primary aroma compounds). These compounds are highly reactive and result in a variety of different sulphur-containing compounds (secondary aroma compounds).

ANTI-FUNGAL AND ANTI-YEAST EFFECTS. Yin and Tsao (1999) reported the inhibitory effects of *Allium* extracts against *Aspergillus* species. Compared with bulb onion, garlic (*A. sativum*) was more effective. Treatment of the plant extracts by acetic acid and heat increased the inhibitory effect against these fungi. Other fungi are also sensitive to onion extracts. Zohri *et al.* (1995) described the inhibitory effect of onion oil against dermatophytic fungi. The best antifungal results were observed against *Microsporum canis*, *Microsporum gypseum* and *Trichophyton simii* at a concentration of 200 ppm onion oil. Onion juice was also active against yeast species (Dankert *et al.*, 1979).

1.2.2 Active ingredients

The above results were obtained with crude plant preparations. In further studies, Cammue *et al.* (1995) isolated an antimicrobial protein (Ace-AMP1) from onion seeds which was highly active against plant-pathogenic fungi at concentrations below 10 $\mu g\ ml^{-1}$. The structure of Ace-AMP1 was fully elucidated. It was demonstrated that this protein interacts with phospholipid membranes (Tassin *et al.*, 1998).

1.3 Cardiovascular effects

Recent studies have mainly focused on the antiplatelet activity of onion extracts.

Table 15.1. Effects of fresh plant material and extracts obtained from *Allium cepa*.

Effect	Author	Model
Antibiotic activities		
Antifungal activity	Yin and Tsao, 1999	*Aspergillus* species
Antibacterial activity	Kim, 1997	Oral pathogenic bacteria
Antibacterial, antifungal activity	Zohri *et al.*, 1995	Bacteria and fungi
Antibacterial, antifungal activity	Elnima *et al.*, 1983	Bacteria and fungi
Antibacterial, antifungal activity	Dankert *et al.*, 1979	Bacteria and yeast species
Plant antifungal activity	Cammue *et al.*, 1995	Plant-pathogenic fungi
Membrane interactions	Tassin *et al.*, 1998	Artificial membranes
Cardiovascular effects		
Antithrombotic activity	Bordia *et al.*, 1996	Rats
Antiplatelet activity	Ali *et al.*, 1999	Human and rabbit plasma
Antiplatelet activity	Debaene *et al.*, 1999	Human platelet-rich plasma
Antiplatelet activity	Goldman *et al.*, 1996	Human platelet-rich plasma
Antiplatelet activity	Goldman, 1996	Human platelet-rich plasma
Antiplatelet activity	Makheja and Bailey, 1990	Rabbits, human platelets
Antihyperlipidaemic activity	Kumari *et al.*, 1995	Rats
The respiratory system		
Inhibition of 5-lipoxygenase and cyclooxygenase	Wagner *et al.*, 1990	Sheep microcosms Porcine leucocytes
Antiasthmatic effects	Dorsch *et al.*, 1989	Guinea-pigs
Antiasthmatic effects	Dorsch *et al.*, 1987a, b	Guinea-pigs, humans
Metabolic diseases		
Renal lesions, diabetes	Babu and Srinivasan, 1999	Wistar rats
Antihyperglycaemic activity	Roman-Ramos *et al.*, 1995	Rabbits
Antidiabetic activity	Sheela *et al.*, 1995	Alloxan diabetic rats
Antidiabetic activity	Kumari *et al.*, 1995	Alloxan diabetic rats
Antidiabetic activity	Kumari and Augusti, 1995	Alloxan diabetic rats
Hypoglycaemic effects	Mathew and Augusti, 1975	Alloxan diabetic rabbits
Hypoglycaemic effects	Augusti *et al.*, 1974	Alloxan diabetic rabbits
Anticancer activities		
Antimutagenic effects	Ikken *et al.*, 1999	Ames test
Antimutagenic effects	Kato *et al.*, 1998	*Salmonella* strain
Protective effects	Gao *et al.*, 1999	Human case-referent study
Stomach-carcinoma protection	Dorant *et al.*, 1996	Human case-control study
Breast-cancer protection	Challier *et al.*, 1998	Human case-control study
Induction of phase II enzymes	Guyonnet *et al.*, 1999	Wistar rats
Chemopreventive activity	Siess *et al.*, 1997	Rats
Chemopreventive activity	Takada *et al.*, 1997	Ito rat-liver test
Additional activities		
Anti-inflammatory activity	Dorsch *et al.*, 1990	Human granulocytes

Reducing platelet aggregation has a preventive effect on some cardiovascular disorders, such as atherosclerosis. In addition, onion extracts have some lipid-lowering effects (Table 15.1).

The antithrombotic potential of aqueous extracts from onion and garlic was evaluated by Bordia *et al.* (1996). Extracts were administered orally and intraperitoneally to rats. A relatively low dose of the aqueous garlic extract (50 mg kg^{-1} body weight) decreased thromboxane-B_2 levels significantly, regardless of the mode of administration. Onion extracts were effective at higher concentrations (500 mg kg^{-1} body weight). Boiling the extracts before application resulted in an almost complete loss of activity. Cooking may therefore cause the decomposition of the potent antithrombotic components in alliums.

The effects of aqueous extracts of raw and boiled onion and garlic on collagen-induced platelet aggregation were studied *in vitro* (Ali *et al.*, 1999). The concentrations of onion and garlic required for a 50% inhibition of the platelet aggregation were calculated to be 90 and 7 mg ml^{-1} plasma (rabbit), respectively. Results for human plasma were similar. Boiled extracts showed a reduced inhibitory effect.

Goldman *et al.* (1996) investigated the antiplatelet activity of extracts of four bulb-onion cultivars (mild cvs 'Exhibition' and 'MSU8155B' and pungent cvs 'W434B' and 'W420B'), grown at two different locations in the USA. The highest antiplatelet activity was observed in the two pungent cultivars, i.e. onion bulbs with high sulphur content exhibited a significantly greater antiplatelet activity than those containing low levels of sulphur. Significantly greater activity was determined for three out of four cultivars grown in Oregon compared with those from Wisconsin. During postharvest cold storage at 4°C, antiplatelet activity increased by about 60% across all cultivars, reaching a maximum at 90 days of storage; however, changes in pungency were not correlated with changes in antiplatelet activity (Debaene *et al.*, 1999). Juice prepared from a flowering umbel of the onion genotype 'W420B' had antiplatelet activity 336% higher than that of juice obtained from onion bulbs (Goldman, 1996).

The active principle of onion and related species in terms of inhibition of platelet aggregation is not yet identified. However, some studies have been carried out with isolated compounds (Makheja and Bailey, 1990). Allicin, which occurs more abundantly in garlic than in onion, and adenosine, both inhibited platelet aggregation without affecting cyclo-oxygenase and lipoxygenase metabolites of arachidonic acid. The trisulphides investigated inhibited platelet aggregation, as well as thromboxane synthesis, along with the induction of lipoxygenase metabolites. The observed *in vivo* antiplatelet effects appear to be more attributable to adenosine than to allicin and alk(en)yl polysulphides of onion and garlic.

Lipid-lowering effects have been intensively studied for garlic (Table 15.2). It was also found that a typical organosulphur compound of onion and related species, namely (+)-*S*-methyl-L-cysteine sulphoxide, showed hypolipidaemic activity, as well as antidiabetic activity (Kumari *et al.*, 1995). Effects on hydroxymethylglutaryl (HMG)-coenzyme A (CoA)-reductase and on seral lipids were obtained after administration of this sulphoxide to rats at a dose of 200 mg kg^{-1} body weight over a period of 45 days. The (+)-*S*-methyl-L-cysteine sulphoxide used was isolated from *A. cepa*.

1.4 Effects on the respiratory system

One of the best investigated effects of onion is its antiasthmatic activity. In the last 20 years, onion extracts, as well as isolated or synthesized organosulphur compounds, have been studied for their potential use as medicinal drugs (Dorsch, 1996). The effects of thiosulphinates and cepaenes were investigated by *in vitro* tests. They exhibited dose-dependent inhibitory effects at 0.25 to 100 μM. Cepaenes inhibited both cyclo-oxygenase and 5-lipoxygenase by more than 75% at 10 and 1μM concentrations, respectively. These effects may be at least partly responsible for the anti-inflammatory and antiasthmatic properties of onion extracts observed *in vivo* (Wagner *et al.*, 1990).

In further studies, guinea-pigs were chosen as the model for investigating the effect of organosulphur compounds on platelet-activating factor (PAF)-induced bronchial constriction (Dorsch *et al.*, 1989; Dorsch, 1996). An oral treatment with lyophilized onion extract was active. Furthermore, a chloroform extract (at 20 mg kg^{-1} body weight) was more active than the lyophilized onion juice (at 100 mg kg^{-1} body weight). Thiosulphinates and cepaenes were identified as the active compounds (some of them had also been investigated in a previous study by Dorsch *et al.* (1987b)). Interestingly, diphenylthiosulphinate gave the best results.

Table 15.2. Effects of fresh plant material and extracts obtained from *Allium sativum*.

Effect	Author	Model
Antibiotic activities		
Antimicrobial activity	Arora and Kaur, 1999	Human pathogenic bacteria, yeasts
Antimicrobial activity	Yoshida *et al.*, 1999a	Gram-positive bacteria, yeasts
Antimicrobial activity	Yoshida *et al.*, 1999b	Bacteria, yeasts
Antibacterial activity	Jonkers *et al.*, 1999	*Helicobacter pylori*
Antibacterial activity	Sivam *et al.*, 1997	*Helicobacter pylori*
Antibacterial activity	Cellini *et al.*, 1996	*Helicobacter pylori*
Anti-*Trypanosoma* effect	Nok *et al.*, 1996	*Trypanosoma*-infected mice
Antifungal activity	Shen *et al.*, 1996	*Cryptococcus neoformans*
Antifungal activity	Venugopal and Venugopal, 1995	Dermatophytes
Antifungal activity	Pai and Platt, 1995	*Aspergillus* species
Cardiovascular activities		
Lipid-lowering effect	Oi *et al.*, 1999	Rats
Lipid-lowering effect	Isaacsohn *et al.*, 1998	Human study
Lipid-lowering effect	Ahmed and Sharma, 1997	Rats
Lipid-lowering effect	Mathew *et al.*, 1996	Rats
Lipid-lowering effect	Gebhardt and Beck, 1996	Rat hepatocytes
Lipid-lowering effect	Simons *et al.*, 1995	Human study
Lipid-lowering effect	Gebhardt, 1995	Rat hepatocytes
Lipid-lowering effect	Gebhardt *et al.*, 1994	Human and rat hepatocytes
Lipid-lowering effect	Gebhardt, 1993	Human and rat hepatocytes
Lipid-lowering effect	Jain *et al.*, 1993	Human study
Lipid-lowering effect	Holzgartner *et al.*, 1992	Human study
Lipid-lowering effect	Mader, 1990	Human study
Lipid-lowering effect, increased fibrinolysis, antiplatelet activity	Bordia *et al.*, 1998	Human study
Antiplatelet activity	Ali, 1995	Rabbit tissues
Antiplatelet activity, increased nitric oxidase activity	Das *et al.*, 1995	Placental villous tissue
Antihypertensive effect	Al Qattan *et al.*, 1999	Wistar rats
Antihypertensive effect	Ziyyat *et al.*, 1997	Human cohort study
Hypotensive activity	Melzig *et al.*, 1995	Endothelial cells
Antioxidative effect	Ide and Lau, 1999	Bovine and murine cells
Antioxidative effect	Munday *et al.*, 1999	Human study
Antiatherosclerotic effect	Koscielny *et al.*, 1999	Human study
Therapy of peripheral arterial occlusive disease	Kiesewetter *et al.*, 1993	Human study
Effects on cutaneous microcirculation	Jung *et al.*, 1991	Human study
Cardioprotective actions	Isensee *et al.*, 1993	Isolated rat heart
The respiratory system		
Improved arterial oxygenation	Abrams and Fallon, 1998	Human study
Metabolic diseases		
Antidiabetic activity	Augusti and Sheela, 1996	Rats, isolated cells
Antidiabetic activity	Sheela and Augusti, 1992	Rats
Antidiabetic activity	Swanston-Flatt *et al.*, 1990	Mice

Table 15.2. *Continued.*

Effect	Author	Model
Anticancer activities		
Induction of phase II enzymes	Munday and Munday, 1999	Rats
Induction of phase II enzymes	Guyonnet *et al.*, 1999	Wistar rats
Chemopreventive effects	Fanelli *et al.*, 1998	Chemical study
Chemopreventive effects	Dion *et al.*, 1997	*In vitro* tests
Antitumour activity	Singh and Shukla, 1998	Swiss albino mice
Antitumour activity	Riggs *et al.*, 1997	Mice
Antitumour activity	Sigounas *et al.*, 1997	Cancer cell lines
Anticlastogenic effect	Das *et al.*, 1996	Swiss albino mice
Additional activities		
Protective effect on small intestinal damage	Horie *et al.*, 1999	Rats
Adrenaline and noradrenaline increase	Oi *et al.*, 1999	Rats
Immunomodulatory effects	Liu *et al.*, 1998	Rats
Hydroxyl radical scavenger	Kourounakis and Rekka, 1991	Inactivated liver microcosms
Radiation-protective effects	Gupta, 1996	Swiss albino mice
Radiation-protective effects	Jaiswal and Bordia, 1996	Rats
Diuretic effects	Pantoja *et al.*, 1996	Rabbits

A single dose of 100 mg kg^{-1} body weight can prevent PAF-induced bronchial hyperactivity to histamine over a period of 12 h. Diphenylthiosulphinate does not occur in onion extracts, but it is closely related to native thiosulphinates.

In a human study, adult volunteers suffering from allergic bronchial asthma were treated with 100 ml of a 5% ethanol onion extract prepared from 100 g sliced onions 1 h prior to an allergen inhalation test (Dorsch *et al.*, 1987a). Immediate and late bronchial reactions were markedly reduced after onion pretreatment.

1.5 Effects on metabolic diseases

Besides the antiasthmatic effects described above, effects of onion extracts on blood glucose levels were also thoroughly investigated and reviewed by Augusti (1990). In a more recent study, the influence of diet containing onion powder on renal lesions of streptozotocin-induced diabetic Wistar rats was investigated. Babu and Srinivasan (1999) reported that dietary onion caused a beneficial modulation of the progression of renal lesions in the diabetic rats. The authors assumed that the beneficial ameliorating influence of onion on diabetic nephropathy might be mediated through the onion's ability to lower blood cholesterol levels and to reduce lipid peroxidation.

The antihyperglycaemic effect of 12 edible plants was studied on 27 healthy rabbits (Roman-Ramos *et al.*, 1995). Apart from cauliflower (*Brassica oleracea* var. *botrytis*), only onion and garlic decreased the hyperglycaemic peak, which was induced by the glucose tolerance test. In further studies, alloxan-diabetic rats were treated with *S*-methyl-L-cysteine sulphoxide (typical for many *Allium* species) and *S*-allyl-L-cysteine sulphoxide (typical for garlic) at a concentration of 200 mg kg^{-1} body weight per day (Kumari and Augusti, 1995; Kumari *et al.*, 1995; Sheela *et al.*, 1995). Both compounds significantly lowered blood glucose levels. The authors suggested that sulphoxides might prevent insulin destruction by the reduction of reducing equivalents (e.g. nicotinamide adenine dinucleotide phosphate (NADPH)) and inactivation of -SH group systems.

Cysteine sulphoxides are the principal organosulphur compounds of *Allium* species

(Fig. 15.2). However, other products resulting from alliinase conversion also showed antihyperglycaemic activity (Augusti *et al.*, 1974; Mathew and Augusti, 1975; Augusti, 1990). Ether–oil extracts from fresh onions were separated in petroleum ether-soluble and insoluble fractions, and administered to alloxan-diabetic rabbits. Both fractions exhibited a hypoglycaemic effect, but the petroleum ether-soluble fraction produced an initial hyperglycaemia. A single dose of 250 mg kg^{-1} body weight of the insoluble fraction led to a hypoglycaemia over a period of 4 h. In both fractions, alk(en)yl disulphides were the major sulphur compounds, and most of them were unsaturated. The petroleum ether-soluble fraction additionally contains thiols, which may directly interact with insulin.

1.6 Anticancer effects

In recent years, many scientists have focused their research activities on the prevention of cancer by nutrition with a diet containing a high proportion of vegetables. Reviews concerning the effects of the *Allium* species, sometimes among those of other vegetables, were published by Krishnaswamy and Polasa (1995), Block (1996), Lea (1996), Steinmetz and Potter (1996), Wargovich and Uda (1996) and Wargovich *et al.* (1996). It was shown that onions and garlic are rich in organoselenium compounds, which may help to prevent cancer. Selenium is usually fixed to sulphur-containing amino acid derivatives. Quercetin and its derivatives, which are also typical constituents of onions, are also of great interest, because of their anticarcinogenic properties (Formica and Regelson, 1995). De Vries *et al.* (1997) studied the daily intake of quercetin and related flavonoids, and the metabolism of these compounds has been intensively investigated in a clinical trial (Hollman *et al.*, 1997).

The antimutagenic effect of ethanolic extracts of onion and garlic against the mutagenicity of *N*-nitrosamines was evaluated by the Ames test (Ikken *et al.*, 1999). Onion showed a greater effect than garlic. Moreover, the mutagenicity of cooked hamburger containing ground beef was reduced by onion (Kato *et al.*, 1998). The authors suggested that the addition of onion might change the balance of the sugar content of ground beef, which effectively induces mutagenicity.

In a case–referent study in a high-epidemic area located in the Jiangsu province of China (234 cases, 234 referents), it was shown that *Allium* vegetables exhibited a protective effect against oesophageal and stomach cancer (Gao *et al.*, 1999). A frequent intake of *Allium* vegetables, such as onions, garlic, Welsh onions and Chinese chives, and also the consumption of raw vegetables and tea were inversely associated with the risk of oesophagus and stomach cancer. The study suggests that *Allium* vegetables, like other raw vegetables, might have an important protecting effect against gastrointestinal cancer diseases.

The Netherlands Cohort Study (120,852 men and women) also provides evidence for a strong inverse association between onion consumption and stomach carcinoma incidence (Dorant *et al.*, 1996). Onion consumption reduced the risk of carcinoma in the non-cardial part of the stomach. Leek and garlic supplements were not associated with stomach carcinoma risk. In a French case–control study (345 patients with primary breast carcinoma), the risk of breast cancer was shown to decrease as consumption of onion, garlic and fibre is increased (Challier *et al.*, 1998).

In a further study, the modulation of phase II drug-metabolizing enzymes, such as glutathione *S*-transferase, by organosulphur compounds from *Allium* vegetables was investigated in rat tissues (Guyonnet *et al.*, 1999). These enzymes are mainly located in the liver but were also found in the kidney and the intestine. Phase II enzymes are responsible for the detoxification of many drugs and carcinogenic compounds. Diallyl disulphide intake significantly increased activities of most phase II enzymes. But other sulphides were also active. The study suggests that diallyl disulphide could be a promising chemopreventive agent, considering its pleiotropic capacity for enzyme induction. In a similar study, the modulation by a variety of sulphides of hepatic drug-metabolizing enzymes was investigated (Siess *et al.*, 1997). Compounds containing methyl groups had little or no effect. Alkyl sulphides and diallyl disulphides showed a possible protective effect on the first step of carcinogenesis through the modulation of enzymes involved in the metabolism of carcinogenic substances.

As shown in Fig. 15.2, the sulphides described above are enzymatic degradation products of cysteine sulphoxides. The original organosulphur compounds of *Allium* species, such as *S*-methyl-L-cysteine sulphoxide and related derivatives, were examined in terms of their modifying effects on diethylnitrosamine-induced neoplasia of the liver in male rats (Takada *et al.*, 1997). In particular, *S*-methyl-L-cysteine sulphoxide and cysteine exerted significant inhibitory effects. These substances seem to be chemopreventive agents for rat hepatocarcinogenesis and their intake may be important for cancer prevention.

1.7 Further effects

Koch (1994) and Dorsch (1996) described a number of minor therapeutic effects of onion. Besides their antiasthmatic effects, onion extracts had an anti-inflammatory potential (Dorsch *et al.*, 1990). Synthetic thiosulphinates and cepaene- and/or thiosulphinate-rich onion extracts were found to inhibit *in vitro* the chemotaxis of human granulocytes, which was induced by formyl-methionine-leucine-phenylalanine. The anti-inflammatory properties of onion extracts are related, at least in part, to the inhibition of inflammatory cell influx by thiosulphinates and cepaenes.

2. Therapeutic and Medicinal Values of Garlic (*Allium sativum* L.)

2.1 Introduction and chemistry

Garlic (*A. sativum* L.) is one of the best-studied medicinal plants. As in the case of onion (*A. cepa*), the antibacterial and antiseptic properties are well known. The use of garlic was well documented by the Egyptians, Greeks and Romans. The Egyptian Papyrus Ebers mentioned 22 garlic-containing remedies, which were applied against heart problems, headache, bites, worms and tumours (Sendl, 1995). Furthermore, garlic is used in folk medicine for the prevention of stroke and atherosclerosis. Medicinal studies were often carried out with both onion and garlic. Comparative studies on the two species have already been described.

Organosulphur compounds are typically found in garlic. In contrast to onion, garlic mainly contains the *S*-(2-propenyl)-L-cysteine sulphoxide (alliin), as well as the corresponding γ-glutamyl derivative. Further cysteine sulphoxides can be detected in lower concentrations (Koch and Lawson, 1996; Keusgen, 1999). Garlic bulbs may contain up to 1.4% of the fresh weight as alliin.

When garlic tissue is damaged, the odourless alliin is converted by alliinase into the thiosulphinate allicin, which produces the typical garlic smell (Fig. 15.3). Allicin is relatively unstable and decomposes into a variety of organosulphur compounds. This reaction depends strongly on the polarity of the solvent used (Fig. 15.4). Oil extracts of garlic are characterized by a high content of vinyldithiins, mainly 2-vinyl-4-*H*-1,3-dithiin, whereas aqueous garlic juice mainly contains alkenyl sulphides, such as diallyl di- and trisulphides. In ethanolic extracts, allicin is typically converted to ajoenes. All of these breakdown products are volatile; hence the so-called 'aged garlic extracts' are nearly free of these compounds.

Fresh garlic bulbs contain about 65% water. Besides the organosulphur constituents, carbohydrates are the most abundant class of compounds present in garlic bulbs and account for about 77% of their dry weight (Koch and Lawson, 1996). The majority of the carbohydrates consist of water-soluble fructose polymers, namely fructans and fructosans.

Garlic cloves also contain saponins and sterols. The content of total saponins was determined to be 0.3–1.1% of the dry weight

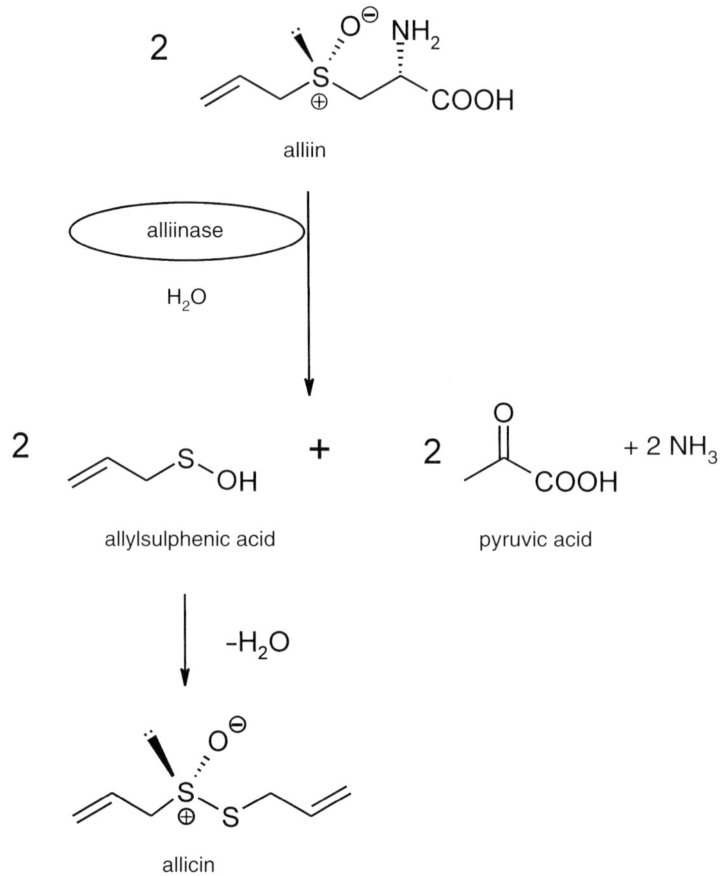

Fig. 15.3. Alliinase-catalysed cleavage of alliin into allylsulphenic acid, pyruvic acid and ammonia. Two molecules of allylsulphenic acid react spontaneously to form allicin.

of cloves. Several steroidal saponins, furostanol glycosides and spirostanol glycosides were identified. Garlic cloves contain rather large amounts of minerals and trace elements. In recent years, the content of selenium has been of special interest. Fresh garlic can show selenium concentrations of up to 70 μg in 100 g of fresh cloves. A further increase can be achieved by deliberate selenium fertilization (Ip and Lisk, 1994). In addition, vitamins, such as ascorbic acid, thiamine, riboflavin, niacin, pantothenic acid and vitamin E, are reported in garlic (Koch and Lawson, 1996). Flavonoids and phenols are found in low concentrations. Interestingly, adenosine has been reported, with values as high as 0.2–0.4 mg g^{-1} fresh weight for garlic cloves (Sendl, 1995; Koch and Lawson, 1996).

2.2 Antibiotic activities

Allicin, a typical component of freshly prepared aqueous extracts of garlic, is well known for its antibiotic activity (Farbman *et al.*, 1993; Ankri and Mirelman, 1999). In another recent study, antibacterial and antifungal activities of garlic extracts were tested (Table 15.2). A 93% bactericidal effect against *Staphylococcus epidermidis* was apparent within 1 h of incubation. The same effect

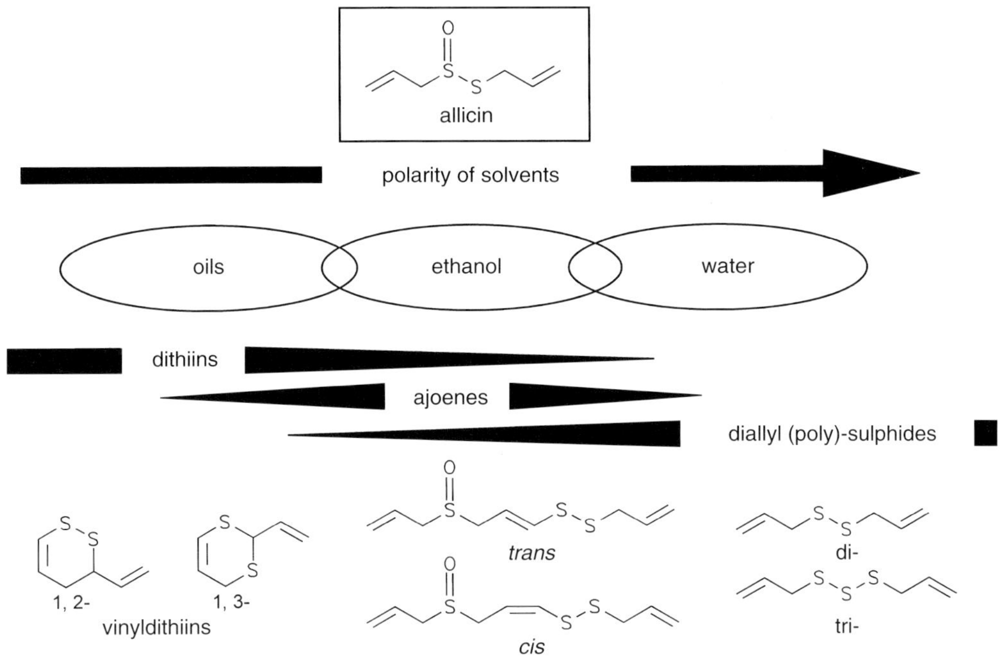

Fig. 15.4. Decomposition of allicin in the presence of solvents with different polarities. Vinyldithiins, ajoenes and diallyl sulphides are the main products. (Modified from Winkler *et al.*, 1991.)

was achieved for *Salmonella typhi* within 3 h. Yeasts were totally killed within 1 h by garlic extract. Garlic was also active against bacteria that are known to have resistance to certain antibiotics (Arora and Kaur, 1999).

Organosulphur compounds isolated from oil-macerated garlic extracts also exhibited antimicrobial activity (Yoshida *et al.*, 1999a, b). Activity against Gram-positive bacteria seems to be higher than against Gram-negative bacteria. Substances containing *S*-allyl groups were found to be more active than those containing *S*-methyl or *S*-propyl groups.

The effect of aqueous garlic extracts and commercial garlic tablets on *Helicobacter pylori* has been intensively studied (Cellini *et al.*, 1996; Sivam *et al.*, 1997; Jonkers *et al.*, 1999). *H. pylori* is a bacterium implicated in the aetiology of stomach cancer. Using garlic preparations suspended in culture medium, the minimum inhibitory concentration (MIC) *in vitro* was found to range between 2 and 17.5 mg ml^{-1} garlic or garlic tablets. Heat treatment of extracts reduced the inhibitory effect. Thiosulphinates were assumed to be the active principle of the investigated extracts. *In vitro* experiments on *H. pylori* determined that the MIC for thiosulphinates dissolved in culture solution was 40 $\mu g\ ml^{-1}$. Because of the sensitivity of *H. pylori* to garlic extracts, a high garlic intake may be related to a lower risk of stomach cancer.

An oily extract prepared from garlic pulp was found to be active against *Trypanosoma* species (Nok *et al.*, 1996). The ability of the parasites to be infective in mice was completely suppressed. Experimentally infected mice were cured of trypanosomiasis in 4 days when treated with 120 mg kg^{-1} body weight day^{-1} of a partially purified extract. Diallyl disulphide was suggested as the active principle: it may interfere with the membrane synthesis of the parasites.

Garlic extracts exhibited activity against several pathogenic fungi (Pai and Platt, 1995; Venugopal and Venugopal, 1995;

Shen *et al.*, 1996). Aqueous garlic extracts and garlic-containing preparations were tested against *Cryptococcus neoformis*, several dermatophytes and *Aspergillus* species involved in otomycosis (fungal infection of the ear). Aqueous garlic extracts and concentrated garlic oil were found to have antifungal activity. Diallyl trisulphide and other polysulphides may be the active principles of processed garlic.

2.3 Cardiovascular effects

The lipid-lowering action of garlic is one of the best investigated effects in modern phytotherapy. Exceptionally large numbers of animal and human studies have been carried out (Table 15.2). Moreover, the mechanism of the lipid-lowering – mainly cholesterol-lowering – actions has been thoroughly elucidated and is summarized in Fig. 15.5 (Gebhardt, 1993, 1995; Gebhardt *et al.*, 1994; Gebhardt and Beck, 1996).

Studies were carried out in rat and human hepatocytes. A suppression of cholesterol biosynthesis was observed, up to 30%, at concentrations of 0.5 mM allicin and ajoene. This may prevent undesired side-effects, which might be caused by a total inhibition of the cholesterol pathway. Allicin acts directly on the HMG-CoA-reductase

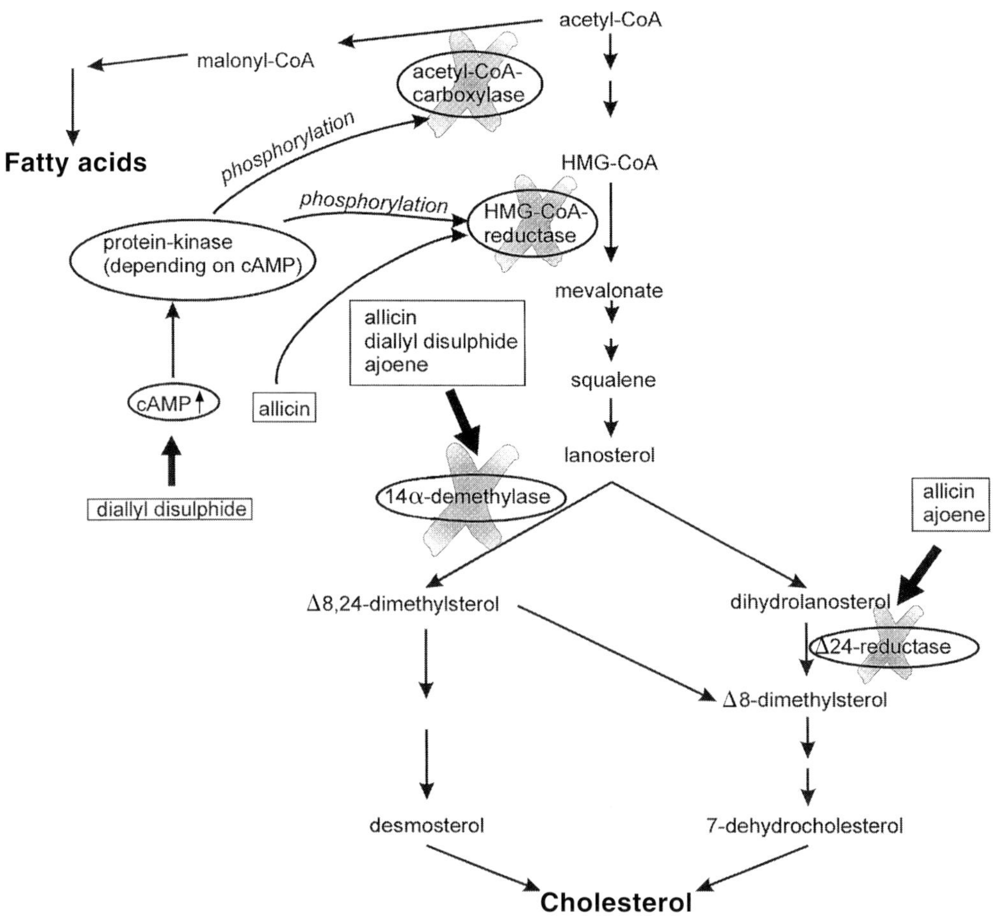

Fig. 15.5. Regulatory effects on the synthesis of fatty acids and cholesterol by different garlic compounds. Enzyme inhibition is marked by crosses.

and leads to an inhibition of the enzyme. Diallyl disulphide induces the level of cyclic adenosine monophosphate (cAMP) affecting a phosphorylation of the HMG-CoA-reductase, as well as the acetyl-CoA-carboxylase. Phosphorylated forms of both enzymes are inactive. Moreover, the two branches of cholesterol biosynthesis are inhibited by allicin, ajoene and diallyl disulphide. This demonstrates that the action of garlic compounds is rather complex and also that different garlic preparations (fresh garlic, garlic tablets and ethanolic extracts) may exhibit a lipid-lowering potential. These results were reflected in several of the studies listed in Table 15.2. Most studies gave a cholesterol-lowering effect of 10–15% for patients with hypercholesterolaemia. However, no cholesterol-lowering effect was evident in some recent studies (Simons *et al.*, 1995; Isaacsohn *et al.*, 1998). Unlike earlier studies, treatment with garlic tablets was accompanied by a strict, low-cholesterol diet. It can be assumed that garlic shows the same cholesterol-lowering effect as obtained by a nearly cholesterol-free diet. However, such a diet requires detailed information on patients as well as continuous supervision. Therefore, daily intake of garlic is a more practicable way to lower cholesterol levels that are pathologically high.

Increased serum lipid levels are one risk factor for atherosclerosis. However, reduction of antiplatelet activity, hypotension and the antioxidative effects of garlic may also contribute to the prevention of atherosclerosis. Antiplatelet activity was studied in humans and in isolated tissues. Diallyl disulphide and diallyl trisulphide were found to be the active compounds that are typically found in garlic oils (Bordia *et al.*, 1998). Raw garlic seems to be more active than boiled garlic (Ali, 1995). A dose-dependent inhibition of cyclo-oxygenase was observed in rabbit tissues treated with raw garlic. In addition, water and alcoholic extracts are potent inhibitors of platelet aggregation, which is induced by adrenaline and adenosine diphosphate (ADP). These garlic extracts showed the ability to increase nitric oxidase synthase activity intracellularly, resulting in the relaxation of blood-vessels, as well as in the inhibition of platelet aggregation (Das *et al.*, 1995). Vasodilation was also observed (Jung *et al.*, 1991).

Garlic exhibited a moderate antihypertensive effect in several studies (Melzig *et al.*, 1995; Ziyyat *et al.*, 1997; Al Qattan *et al.*, 1999). Aged garlic extract was found to inhibit cholesterol oxidation of the low-density lipoprotein fraction (LDL) of blood (Ide and Lau, 1999). Comparable results were described by Munday *et al.* (1999). Oxidized LDL has been recognized as a major factor in the initiation and progression of atherosclerosis. Continuous intake of garlic powder over 48 months significantly slowed down the increase in atherosclerotic plaque volume, or even effected a slight regression within the observational period (Koscielny *et al.*, 1999).

Moreover, a curative effect of a 12-week treatment with garlic powder was demonstrated in patients with peripheral arterial occlusive disease (Kiesewetter *et al.*, 1993). For these patients, walking distance could be significantly increased. Garlic-powder therapy also increases cutaneous microcirculation (Jung *et al.*, 1991). Experiments with isolated rat hearts showed that garlic has a cardioprotective effect, which is probably due to the free-radical-scavenging activity of garlic components and its antiarrhythmic effects (Isensee *et al.*, 1993).

2.4 Effects on the respiratory system

The antiasthmatic effects as described for onion were less marked for garlic preparations. Because of its antibiotic properties, garlic and garlic extracts have been traditionally used against infections of the respiratory system, the throat and the mouth cavity (Koch and Lawson, 1996). Abrams and Fallon (1998) describe the treatment of the hepatopulmonary syndrome with garlic. No medicinal therapy previously existed for patients with this syndrome. Standardized garlic powder was administered over a period of 6 months. Six of 15 patients responded to garlic and had less dyspnoea upon exertion. Arterial oxygenation was improved in younger patients or those with lower macroaggregated albumin-shunt fractions.

2.5 Effects on metabolic diseases

Organosulphur compounds typical for garlic were tested for their antidiabetic potential (Augusti and Sheela, 1996). Alliin isolated from garlic ameliorated the diabetic condition of alloxan-diabetic rats. It could be demonstrated *in vitro* that alliin stimulated the secretion from B cells isolated from normal rats. Furthermore, allicin at a dose of 200 mg kg^{-1} body weight significantly decreased the concentration of serum lipids and the activities of serum enzymes, such as alkaline phosphatase, acid phosphatase, lactate dehydrogenase and hepatic glucose-6-phosphatase (Sheela and Augusti, 1992).

The effect of garlic as part of a daily diet (6.25% by weight of total food intake) on blood glucose and insulin levels was studied in normal and streptozotocin-diabetic mice (Swanston-Flatt *et al.*, 1990). After 12 days of treatment with garlic, no significant effects were observed. Moreover, garlic did not significantly alter hyperglycaemia and hypoinsulinaemia of the streptozotocin-diabetic mice. This is in contrast to the previous findings, reviewed by Augusti (1990), where some studies on human volunteers treated with raw garlic and garlic oil were described. However, a slight reduction in the blood glucose levels was observed after a regular intake of at least 1 month.

2.6 Anticancer effects

The induction of phase II enzymes as a mechanism for chemoprevention has also been studied for organosulphur compounds of onion (see Section 1.6). Diallyl disulphide, a typical component of steam-distilled garlic oil and garlic juice (Fig. 15.4), seems to be the most effective compound (Guyonnet *et al.*, 1999; Munday and Munday, 1999). In the latter study, the influence of diallyl disulphide on the quinone reductase and the glutathione reductase of a number of organs was investigated. Enzymes of the forestomach, duodenum and jejunum were the most sensitive to enzyme induction caused by diallyl disulphide. A significant increase in the concentration of quinone reductase was observed at a dose of 0.3 mg kg^{-1} body weight day^{-1}. Such a dose level is close to what may be achieved by human consumption of garlic.

Several sulphur-containing compounds of garlic, mainly allyl sulphides, were tested for their chemopreventive potential (Fanelli *et al.*, 1998). Allyl sulphide, diallyl sulphide and diallyl disulphide were able to trap trichloromethyl and trichloromethylperoxyl free radicals. Furthermore, diallyl disulphide inhibited chloroform-promoted liver-microsomal lipid peroxidation. Diallyl sulphide was able to react with free radicals induced by photoactivation of peroxides. Besides these volatile sulphur compounds, alliin also showed chemopreventive properties (Dion *et al.*, 1997). Alliin and higher doses of *S*-propyl-L-cysteine sulphoxide were able to block the formation of *N*-nitrosomorpholine *in vitro*. A depressed nitrosamine formation is associated with a reduced risk for some cancers.

The anticarcinogenic activity of diallyl sulphide was tested in Swiss albino mice in the two-stage carcinogenesis test (Singh and Shukla, 1998). Skin cancers were initiated topically with a single subcarcinogenic dose of an anthracene derivative. Promotion of a skin tumour was obtained by application of a phorbol ester. Diallyl sulphide (250 μg per animal) was administered topically thrice weekly over a period of 3 weeks, 1 h before each phorbol ester treatment. Diallyl sulphide was able to reduce chemically induced mouse skin carcinogenesis.

Riggs *et al.* (1997) inoculated murine bladder carcinoma cells into mouse thighs. A garlic preparation was administered subcutaneously and orally. In both groups, tumour growth was inhibited. Mice that received 500 mg garlic 100 ml^{-1} drinking-water showed significant reductions of both tumour volume and mortality. These results suggest that garlic may be valuable for the therapy of transitional cell carcinoma of the bladder.

S-Allylmercaptocysteine, a stable sulphur

compound present in aged garlic extract, was investigated for its potential as an inhibitor of cancer-cell proliferation (Sigounas *et al.*, 1997). Two hormone-sensitive cell lines of breast and prostate cancer were very susceptible to the inhibitory influence of *S*-allylmercaptocysteine. However, the antiproliferative effect of this compound was restricted to actively growing cells.

Fresh garlic in three different concentrations (25, 50 and 100 mg kg^{-1} body weight) was investigated as a protecting agent against clastogenic effects on mouse chromosomes (Das *et al.*, 1996). Fresh garlic was administered to Swiss albino mice over a period of 60 days and frequencies of chromosomal aberrations and damaged cells induced in bone-marrow preparations were evaluated. After a dose-dependent initial enhancement for 7 days, damage and aberrations were reduced following prolonged exposure for 30 and 60 days. Therefore, daily intake of a low garlic dose for at least 30 days is assumed to give the maximum benefit as a protecting agent against the clastogenic effects of genotoxicants.

2.7 Further effects

Reuter (1995) and Koch and Lawson (1996) reviewed some additional effects of garlic and its preparations. More recently, Horie *et al.* (1999) reported on the protective effect of aged garlic extract on damage to the small intestine induced by methotrexate, an antimetabolite that is a common drug for the treatment of cancer. During treatment with methotrexate, permeability of the small intestine increased. Rats fed with aged garlic extract were almost completely protected against this damage.

Administration of alliin, diallyl disulphide and diallyl trisulphide significantly increased plasma noradrenaline and adrenaline levels in rats. In contrast, administration of disulphides without allyl residues, diallyl monosulphide and *S*-allyl-L-cysteine (desoxyalliin) did not increase adrenaline secretion (Oi *et al.*, 1999). Garlic oil and diallyl disulphide (200 mg kg^{-1} body weight) also modify the hepatic-membrane fatty acid composition of rats (Liu *et al.*, 1998). The concentration of linoleic acid was increased, whereas the concentration of arachidonic acid was decreased. Additionally, garlic oil depressed the $\Delta 6$-desaturase. Therefore, garlic oil probably shows an immunomodulatory potential.

The antioxidative properties of garlic powder and alliin were examined in a lipid peroxidation test (Kourounakis and Rekka, 1991). Garlic powder showed antioxidant activity and alliin was a very good hydroxyl radical scavenger. Radiation-protective effects in mice were described by Gupta (1996) and in rats by Jaiswal and Bordia (1996). Administration of garlic oil to Swiss albino mice resulted in prevention of radiation-induced increases of hepatic total lipids, triglycerides and phospholipids and a decrease of free fatty acids. Treatment of albino rats with alliin reduced radiation-induced mortality and showed significant protection against the tissue-damaging effects of irradiation. Furthermore, the intravenous administration of chromatographically purified fractions of garlic to rabbits elicits a dose-dependent diuretic effect, which reaches a maximum 60 min after injection and returns to normal after 90 min (Pantoja *et al.*, 1996).

3. Therapeutic Benefits of Other *Allium* Species

In addition to *A. cepa* and *A. sativum*, several studies have been carried out with shallots and with other *Allium* species (Table 15.3). It was demonstrated that a number of them contain cysteine sulphoxides in considerable amounts, as well as active alliinase (Keusgen, 1999; Krest *et al.*, 2000). Health benefits related to this class of compounds may also be claimed for these species.

Extracts prepared from shallot (*A. cepa* Aggregatum group) influenced erythrocyte shape (Tappayuthpijarn *et al.*, 1989). Hypercholesterolaemia was induced in rabbits by feeding with egg yolk. Additional feeding with 50 g of a freshly prepared shallot extract reduced the number of abnormal-shaped erythrocytes (crenation) induced by

Table 15.3. Effects of further *Allium* species.

Species	Author	Effect
A. cepa Aggreg.	Tappayuthpijarn *et al.*, 1989	Protection of erythrocyte membranes
A. cepa Aggreg.	Dankert *et al.*, 1979	Antimicrobial activity
A. cepa Aggreg.	Caldes and Prescot, 1973	Antileukaemic effect
A. chinense	Kuroda *et al.*, 1995	Inhibition of cyclic AMP phosphodiesterase and Na^+/K^+ ATPase
A. fistulosum	Chen *et al.*, 1999	Vasodilation of rat aortae
A. fistulosum	Phay *et al.*, 1999	Activity against *Fusarium oxysporum*
A. fistulosum	Fan and Chen, 1999	Activity against *Aspergillus* species
A. giganteum	Mimaki *et al.*, 1994	Inhibition of cyclic AMP phosphodiesterase
A. jesdianum	Mimaki *et al.*, 1999a	Cytostatic and cytotoxic activity
A. karataviense	Mimaki *et al.*, 1999b	Cytostatic activity
A. macleanii	Inoue *et al.*, 1995	Activity as antitumour-promoter
A. nutans	Stajner *et al.*, 1999	Antioxidant activity
A. ampeloprasum	Carotenuto *et al.*, 1997	Cytotoxic and antiproliferative activity
A. ampeloprasum	Liakopoulou-Kyriakides and Sinakos, 1992	Antiplatelet activity
A. senescens	Inoue *et al.*, 1995	Activity as antitumour-promoter
A. ursinum	Rietz *et al.*, 1993	Cardioprotective effect

A. cepa Aggreg. (shallot) in list was referred to as *A. ascalonicum* in the original papers.
A. ampeloprasum (leek) in list was referred to as *A. porrum* in the original papers.

high blood cholesterol levels. Lipid levels, however, did not decrease significantly. In a further study, an antileukaemic substance was described for shallots (Caldes and Prescott, 1973). In addition, as with bulb onion, Gram-positive bacteria were inhibited by shallot juice (Dankert *et al.*, 1979).

Extracts from rakkyo (*A. chinense*) were separated by column chromatography into several fractions (Kuroda *et al.*, 1995). A fraction containing steroidal saponins exhibited inhibitory effects against the enzymes cAMP phosphodiesterase and Na^+/K^+ adenosine triphosphatase (ATPase), at a sample concentration of 100 μg ml^{-1}. Similar activities were found for a saponin isolated from the bulbs of *A. giganteum* (Mimaki *et al.*, 1994). Furthermore, steroidal saponins isolated from *A. jesdianum* and *A. karataviense* exhibited cytostatic and cytotoxic activities against different tumour cells (Mimaki *et al.*, 1999a,b). Steroidal glycosides obtained from *A. macleanii* and *A. senescens* were evaluated for their potential as antitumour-promoter compounds (Inoue *et al.*, 1995). The glycosides described exhibited an antitumour-promoter activity, but also showed undesirable cytotoxic effects.

In a recent study, the effect of Japanese bunching onion (*A. fistulosum*) on the vascular responses in rat aortas was investigated (Chen *et al.*, 1999). Best vasodilatation was obtained with a raw green-leaf extract, whereas a boiled extract stimulated vasoconstriction. The roots of Japanese bunching onion are also of interest. A substance named fistulosin isolated from the roots showed a high activity against the fungus *Fusarium oxysporum* (Phay *et al.*, 1999). Moreover, ethanolic extracts prepared from Japanese bunching onion were tested for their inhibitory activity against growth and aflatoxin production of *Aspergillus* species (Fan and Chen, 1999). Both parameters were inhibited, depending on extract concentration and exposure times.

The antioxidant activity of leaves, bulbs and roots of *A. nutans* was intensively studied (Stajner *et al.*, 1999). All investigated plant parts exhibited antioxidant ability. Best results were obtained with extracts prepared from the leaves. Furthermore, some bioactive compounds were isolated from leek (*A. ampeloprasum*, leek group). Saponins of leek were found to be cytotoxic and exhibited a high antiproliferative activity on four different tumour cell lines *in vitro* (Carotenuto *et al.*, 1997). A peptide with a molecular weight of 1052 Da showed a strong inhibitory activity on platelet aggregation (Liakopoulou-Kyriakides and Sinakos, 1992).

The cardioprotective action of wild garlic (*A. ursinum*) on rats was studied by Rietz *et al.* (1993). Feed enriched with 2% of pulverized wild garlic leaves was administered over 8 weeks. Several cardioprotective effects were observed in isolated hearts from the treated rats. A moderate inhibition of the angiotensin-converting enzyme by wild garlic could contribute to the cardioprotective and blood-pressure-lowering action of this *Allium* species. No significant alterations in cardiac fatty acid composition could be observed. Wild garlic is used in traditional medicine against atherosclerosis and hypertonia (Koch and Lawson, 1996).

4. Conclusions

Many health benefits have been reported for different *Allium* species. *A. sativum* and *A. cepa* have been the most intensively investigated. Extracts from both species exhibited a significant antibiotic activity. However, it is not yet clear how drugs prepared from these plants can be used as modern antibiotics. Antiasthmatic, antidiabetic and a weak antiplatelet activity were proved for *A. cepa*, while *A. sativum* showed lipid-lowering effects, antiplatelet activity and antiatherosclerotic activities. The cardiovascular effects of garlic are among the best investigated of all medicinal plants.

To obtain the therapeutic effects described above, a daily intake of 50 to 100 g of *A. cepa* and 2.5 to 4 g of *A. sativum* or an adequate amount of an *Allium* preparation is recommended. As demonstrated by some case-controlled studies, a regular daily intake of both may prevent some cancer diseases, even at lower doses. *Allium* vegetables seem to lower risk for gastrointestinal cancers. It is suggested that daily intake of *A. sativum* significantly lowers the incidence of diseases that are caused by atherosclerosis. The health benefits of other *Allium* species are not yet clear, and human studies are urgently needed.

References

Abrams, G.A. and Fallon, M.B. (1998) Treatment of hepatopulmonary syndrome with *Allium sativum* L. (garlic): a pilot trial. *Journal of Clinical Gastroenterology* 27, 232–235.

Ahmed, R.S. and Sharma, S.B. (1997) Biochemical studies on combined effects of garlic (*Allium sativum* Linn.) and ginger (*Zingiber officinale* Rosc.) in albino rats. *Indian Journal of Experimental Biology* 35, 841–843.

Ali, M. (1995) Mechanism by which garlic (*Allium sativum*) inhibits cyclooxygenase activity. Effect of raw versus boiled garlic extract on the synthesis of prostanoids. *Prostaglandins, Leukotrienes and Essential Fatty Acids* 53, 397–400.

Ali, M., Bordia, T. and Mustafa, T. (1999) Effect of raw versus boiled aqueous extract of garlic and onion on platelet aggregation. *Prostaglandins, Leukotrienes and Essential Fatty Acids* 60, 43–47.

Al Qattan, K.K., Alnaqeeb, M.A. and Ali, M. (1999) The antihypertensive effect of garlic (*Allium sativum*) in the rat two-kidney-one-clip Goldblatt model. *Journal of Ethnopharmacology* 66, 217–222.

Ankri, S. and Mirelman, D. (1999) Antimicrobial properties of allicin from garlic. *Microbes and Infection* 1, 125–129.

Arora, D.S. and Kaur, J. (1999) Antimicrobial activity of spices. *International Journal of Antimicrobial Agents* 12, 257–262.

Augusti, K.T. (1990) Therapeutic and medicinal values of onions and garlic. In: Brewster, J.L. and Rabinowitch, H.D. (eds) *Onions and Allied Crops*, Vol. III. *Biochemistry, Food Science, and Minor Crops*. CRC Press, Boca Raton, Florida, pp. 94–108.

Augusti, K.T. (1996) Therapeutic values of onion (*Allium cepa* L.) and garlic (*Allium sativum* L.). *Indian Journal of Experimental Biology* 34, 634–640.

Augusti, K.T. and Sheela, C.G. (1996) Antiperoxide effect of *S*-allyl cysteine sulfoxide, an insulin secretagogue, in diabetic rats. *Experientia* 52, 115–120.

Augusti, K.T., Roy, V.C. and Semple, M. (1974) Effect of allyl propyl disulphide isolated from onion (*Allium cepa* L.) on glucose tolerance of alloxan diabetic rabbits. *Experimentia* 30, 1119–1120.

Babu, P.S. and Srinivasan, K. (1999) Renal lesions in streptozotocin-induced diabetic rats maintained on onion and capsaicin containing diets. *Journal of Nutritional Biochemistry* 10, 477–483.

Block, E. (1992) Die Organoschwefelchemie der Gattung *Allium* und ihre Bedeutung für die organische Chemie des Schwefels. *Angewandte Chemie* 104, 1158–1203.

Block, E. (1996) Recent results in the organosulphur and organoselenium chemistry of genus *Allium* and *Brassica* plants. Relevance for cancer prevention. *Advances in Experimental Medicine and Biology* 401, 155–169.

Bordia, T., Mohammed, N., Thomson, M. and Ali, M. (1996) An evaluation of garlic and onion as antithrombotic agents. *Prostaglandins, Leukotrienes and Essential Fatty Acids* 54, 183–186.

Bordia, A., Verma, S.K. and Srivastava, K.C. (1998) Effect of garlic (*Allium sativum*) on blood lipids, blood sugar, fibrinogen and fibrinolytic activity in patients with coronary artery disease. *Prostaglandins, Leukotrienes and Essential Fatty Acids* 58, 257–263.

Breu, W. (1996) *Allium cepa* L. (onion). Part 1: chemistry and analysis. *Phytomedicine* 3, 293–306.

Caldes, G. and Prescott, B. (1973) A potential antileukemic substance present in *Allium ascalonicum*. *Planta Medica* 23, 99–100.

Cammue, B.P.A., Thevissen, K., Hendriks, M., Eggermont, K., Goderis, I.J., Proost, P., van Damme, E.J.M., Osborn, R.W., Guerbette, F., Kader, J.-C. and Broekaert, W.F. (1995) A potent antimicrobial protein from onion seeds showing sequence homology to plant lipid transfer proteins. *Plant Physiology* 109, 445–455.

Carotenuto, A., Fattorusso, E., Lanzotti, V., Magno, S., De Feo, V., Carnuccio, R. and D'Acquisto, F. (1997) Porrigenins A and B, novel cytotoxic and antiproliferative sapogenins isolated from *Allium porrum*. *Journal of Natural Products* 60, 1003–1007.

Cellini, L., di Campli, E., Masulli, M., di Bartolomeo, S. and Allocati, N. (1996) Inhibition of *Helicobacter pylori* by garlic extract (*Allium sativum*). *FEMS Immunology and Medical Microbiology* 13, 273–277.

Challier, B., Perarnau, J.M. and Viel, J.F. (1998) Garlic, onion and cereal fibre as protective factors for breast cancer: a French case–control study. *European Journal of Epidemiology* 14, 737–747.

Chen, J.H., Tsai, S.J. and Chen, H.I. (1999) Welsh onion (*Allium fistulosum* L.) extracts alter vascular responses in rat aortae. *Journal of Cardiovascular Pharmacology* 33, 515–520.

Craig, W.J. (1999) Health-promoting properties of common herbs. *American Journal of Clinical Nutrition* 70, 491S-499S.

Dankert, J., Tromp, T.F., de Vries, H. and Klasen, H.J. (1979) Antimicrobial activity of crude juices of *Allium ascalonicum*, *Allium cepa* and *Allium sativum*. *Zentralblatt für Bakteriologie, Parasitenkunde, Infektionskrankheiten und Hygiene, 1, Originale A* 245, 229–239.

Das, I., Khan, N.S. and Soorana, S.R. (1995) Potent activation of nitric oxide synthase by garlic: a basis for its therapeutic applications. *Current Medical Research and Opinion* 13, 257–263.

Das, T., Choudhury, A.R., Sharma, A. and Talukder, G. (1996) Effects of crude garlic extract on mouse chromosomes *in vivo*. *Food and Chemical Toxicology* 34, 43–47.

Debaene, J.E.P., Goldman, I.L. and Yandell, B.S. (1999) Postharvest flux and genotype × environment effects for onion-induced antiplatelet activity, pungency, and soluble solids in long-day onion during postharvest cold storage. *Journal of the American Society for Horticultural Science* 124, 366–372.

de Vries, J.H., Janssen, P.L., Hollman, P.C., van Staveren, W.A. and Katan, M.B. (1997) Consumption of quercetin and kaempferol in free-living subjects eating a variety of diets. *Cancer Letters* 114, 141–144.

Dion, M.E., Agler, M. and Milner, J.A. (1997) *S*-Allyl cysteine inhibits nitrosomorpholine formation and bioactivation. *Nutrition and Cancer* 28, 1–6.

Dorant, E., van den Brandt, P.A., Goldbohm, R.A. and Sturmans, F. (1996) Consumption of onions and a reduced risk of stomach carcinoma. *Gastroenterology* 110, 12–20.

Dorsch, W. (1996) *Allium cepa* L. (onion): Part 2. Chemistry, analysis and pharmacology. *Phytomedicine* 3, 391–397.

Dorsch, W., Adelmann-Grill, B., Bayer, T., Ettl, M., Hein, G., Jaggy, H., Ring, J., Scheftner, P., Schneider, T. and Wagner, H. (1987a) Zwiebelextrakte als Asthma-Therapeutika? *Allergologie* 10, 316–326.

Dorsch, W., Ettl, M., Hein, G., Scheftner, P., Weber, J., Bayer, T. and Wagner, H. (1987b) Antiasthmatic effects of onions. Inhibition of platelet-activating factor-induced bronchial obstruction by onion oils. *International Archives of Allergy and Applied Immunology* 82, 535–536.

Dorsch, W., Scharff, J., Bayer, T. and Wagner, H. (1989) Antiasthmatic effects of onions. Prevention of platelet-activating factor induced bronchial hyperreactivity to histamine in guinea pigs by diphenylthiosulfinate. *International Archives of Allergy and Applied Immunology* 88, 228–230.

Dorsch, W., Schneider, E., Bayer, T., Breu, W. and Wagner, H. (1990) Anti-inflammatory effects of onions: inhibition of chemotaxis of human polymorphonuclear leukocytes by thiosulfinates and cepaenes. *International Archives of Allergy and Applied Immunology* 92, 39–42.

Elnima, E.I., Ahmed, S.A., Mekkawi, A.G. and Mossa, J.S. (1983) The antimicrobial activity of garlic and onion extracts. *Die Pharmazie* 38, 747–748.

Fan, J.J. and Chen, J.H. (1999) Inhibition of aflatoxin-producing fungi by Welsh onion extracts. *Journal of Food Protection* 62, 414–417.

Fanelli, S.L., Castro, G.D., de Toranzo, E.G.D. and Castro, J.A. (1998) Mechanisms of the preventive properties of some garlic components in the carbon tetrachloride promoted oxidative stress. *Research Communications in Molecular Pathology and Pharmacology* 102, 163–174.

Farbman, K.S., Barnett, E.D., Bolduc, G.R. and Klein, J.O. (1993) Antibacterial activity of garlic and onions: a historical perspective. *The Pediatric Infectious Disease Journal* 12, 163–164.

Formica, J.V. and Regelson, W. (1995) Review of the biology of quercetin and related bioflavonoids. *Food and Chemical Toxicology* 33, 1061–1080.

Gao, C.M., Takezaki, T., Ding, J.H., Li, M.S. and Tajima, K. (1999) Protective effect of *Allium* vegetables against both esophageal and stomach cancer: a simultaneous case–referent study of high-epidemic area in Jiangsu province, China. *Japanese Journal of Cancer Research* 90, 614–621.

Gebhardt, R. (1993) Multiple inhibitory effects of garlic extracts on cholesterol biosynthesis in hepatocytes. *Lipids* 28, 613–619.

Gebhardt, R. (1995) Amplification of palmitate-induced inhibition of cholesterol biosynthesis in cultured rat hepatocytes by garlic-derived organosulfur compounds. *Phytomedicine* 2, 29–34.

Gebhardt, R. and Beck, H. (1996) Differential inhibitory effects of garlic-derived organosulfur compounds on cholesterol biosynthesis in primary rat hepatocyte cultures. *Lipids* 31, 1269–1276

Gebhardt, R., Beck, H. and Wagner, K.G. (1994) Inhibition of cholesterol biosynthesis by allicin and ajoene in rat hepatocytes and HepG2 cells. *Biochimica et Biophysica Acta* 1213, 57–62.

Goldman, I.L. (1996) Elevated antiplatelet activity induced by extracts from onion umbels. *HortScience* 31, 874.

Goldman, I.L., Kopelberg, M., Debaene, J.E. and Schwartz, B.S. (1996) Antiplatelet activity in onion (*Allium cepa*) is sulfur dependent. *Thrombosis and Haemostasis* 76, 450–452.

Gupta, N.K. (1996) Modification of radiation induced changes in murine hepatic lipid profiles by garlic (*Allium sativum* Linn.) unsaturated oils. *Indian Journal of Experimental Biology* 34, 851–853.

Guyonnet, D., Siess, M.H., Le Bon, A.M. and Suschetet, M. (1999) Modulation of phase II enzymes by organosulfur compounds from *Allium* vegetables in rat tissues. *Toxicology and Applied Pharmacology* 154, 50–58.

Hollman, P.C., van Trijp, J.M., Mengelers, M.J., de Vries, J.H. and Katan, M.B. (1997) Bioavailability of the dietary antioxidant flavonol quercetin in man. *Cancer Letters* 114, 139–140.

Holzgartner, H., Schmidt, U. and Kuhn, U. (1992) Comparison of the efficacy and tolerance of a garlic preparation vs. bezafibrate. *Arzneimittel-Forschung* 42, 1473–1477.

Horie, T., Matsumoto, H., Kasagi, M., Sugiyama, A., Kikuchi, M., Karasawa, C., Awazu, S., Itakura, Y. and Fuwa, T. (1999) Protective effect of aged garlic extract on the small intestinal damage of rats induced by methotrexate administration. *Planta Medica* 65, 545–548.

Ide, N. and Lau, B.H.S. (1999) Aged garlic extract attenuates intracellular oxidative stress. *Phytomedicine* 6, 125–131.

Ikken, Y., Morales, P., Martinez, A., Marin, M.L., Haza, A.I. and Cambero, M.I. (1999) Antimutagenic effect of fruit and vegetable ethanolic extracts against *N*-nitrosamines evaluated by the Ames test. *Journal of Agricultural and Food Chemistry* 47, 3257–3264.

Inoue, T., Mimaki, Y., Sashida, Y., Nishino, A., Satomi, Y. and Nishino, H. (1995) Steroidal glycosides from *Allium macleanii* and *A. senescens*, and their inhibitory activity on tumour promotor-induced phospholipid metabolism in HeLa cells. *Phytochemistry* 40, 521–525.

Ip, C. and Lisk, D.J. (1994) Enrichment of selenium in *Allium* vegetables for cancer prevention. *Carcinogenesis* 15, 1881–1885.

Isaacsohn, J.L., Moser, M., Stein, E.A., Dudley, K., Davey, J.A., Liskov, E. and Black, H.R. (1998) Garlic powder and plasma lipids and lipoproteins: a multicenter, randomized, placebo-controlled trial. *Archives of Internal Medicine* 158, 1189–1194.

Isensee, H., Rietz, B. and Jacob, R. (1993) Cardioprotective actions of garlic (*Allium sativum*). *Arzneimittel-Forschung* 43, 94–98.

Jain, A.K., Vargas, R., Gotzkowsky, S. and McMahon, F.G. (1993) Can garlic reduce levels of serum lipids? A controlled clinical study. *American Journal of Medicine* 94, 632–635.

Jaiswal, S.K. and Bordia, A. (1996) Radio-protective effect of garlic *Allium sativum* Linn. in albino rats. *Indian Journal of Medical Sciences* 50, 231–233.

Jonkers, D., van den Broek, E., van Dooren, I., Thijs, C., Dorant, E., Hageman, G. and Stobberingh, E. (1999) Antibacterial effect of garlic and omeprazole on *Helicobacter pylori*. *Journal of Antimicrobial Chemotherapy* 43, 837–839.

Jung, E.M., Jung, F., Mrowietz, C., Kiesewetter, H., Pindur, G. and Wenzel, E. (1991) Influence of garlic powder on cutaneous microcirculation. A randomized placebo-controlled double-blind cross-over study in apparently healthy subjects. *Arzneimittel-Forschung* 41, 626–630.

Kato, T., Michikoshi, K., Minowa, Y., Maeda, Y. and Kikugawa, K. (1998) Mutagenicity of cooked hamburger is reduced by addition of onion to ground beef. *Mutation Research – Genetic Toxicology and Environmental Mutagenesis* 420, 109–114.

Keusgen, M. (1999) *Biosensorische Methoden zur quantitativen Bestimmung von Cysteinsulfoxiden*. Shaker Verlag, Aachen, Germany, 162 pp.

Kiesewetter, H., Jung, F., Jung, E.M., Blume, J., Mrowietz, C., Birk, A., Koscielny, J. and Wenzel, E. (1993) Effects of garlic coated tablets in peripheral arterial occlusive disease. *The Clinical Investigator* 71, 383–386.

Kim, J.H. (1997) Anti-bacterial action of onion (*Allium cepa* L.) extracts against oral pathogenic bacteria. *Journal of Nihon University School of Dentistry* 39, 136–141.

Koch, H.P. (1994) Die Küchenzwiebel – eine zu Unrecht vernachlässigte Arzneipflanze. *Pharmazie in unserer Zeit* 23, 333–339.

Koch, H.P. and Lawson, L.D. (1996) *Garlic. The Science and Therapeutic Application of* Allium sativum *L. and Related Species*, 2nd edn. Williams & Wilkins, Baltimore, Maryland, 329 pp.

Koscielny, J., Klussendorf, D., Latza, R., Schmitt, R., Radtke, H., Siegel, G. and Kiesewetter, H. (1999) The antiatherosclerotic effect of *Allium sativum*. *Atherosclerosis* 144, 237–249.

Kourounakis, P.N. and Rekka, E.A. (1991) Effect of active oxygen species of alliin and *Allium sativum* (garlic) powder. *Research Communications in Chemical Pathology and Pharmacology* 74, 249–252.

Krest, I., Glodek, J. and Keusgen, M. (2000) Cysteine sulfoxides and alliinase activity of some *Allium* species. *Journal of Agricultural and Food Chemistry* 48, 3753–3760.

Krishnaswamy, K. and Polasa, K. (1995) Diet, nutrition and cancer – the Indian scenario. *Indian Journal of Medical Research* 102, 200–209.

Kumari, K. and Augusti, K.T. (1995) Antidiabetic effects of *S*-methylcysteine sulfoxide on alloxan diabetes. *Planta Medica* 61, 72–74.

Kumari, K., Mathew, B.C. and Augusti, K.T. (1995) Antidiabetic and hypolipidemic effects of *S*-methyl cysteine sulfoxide isolated from *Allium cepa* Linn. *Indian Journal of Biochemistry and Biophysics* 32, 49–54.

Kuroda, M., Mimaki, Y., Kameyama, A., Sashida, Y. and Nikaido, T. (1995) Steroidal saponins from *Allium chinense* and their inhibitory activities on cyclic AMP phosphodiesterase and Na^+/K^+ ATPase. *Phytochemistry* 40, 1071–1076

Lea, M.A. (1996) Organosulfur compounds and cancer. *Advances in Experimental Medicine and Biology* 401, 147–154.

Liakopoulou-Kyriakides, M. and Sinakos, Z. (1992) A low molecular weight peptide from *Allium porrum* with inhibitory activity on platelet aggregation *in vitro*. *Biochemistry International* 28, 373–378.

Liu, C.T., Chen, H.W., Sheen, L.Y., Kung, Y.L., Chen, P.C.H. and Lii, C.K. (1998) Effect of garlic oil on hepatic arachidonic acid content and immune response in rats. *Journal of Agricultural and Food Chemistry* 46, 4642–4647.

Mader, F.H. (1990) Treatment of hyperlipidaemia with garlic-powder tablets. Evidence from the German Association of General Practitioners' multicentric placebo-controlled double-blind study. *Arzneimittel-Forschung* 40, 1111–1116.

Makheja, A.N. and Bailey, J.M. (1990) Antiplatelet constituents of garlic and onion. *Agents and Actions* 29, 360–363.

Mathew, P.T. and Augusti, K.T. (1975) Hypoglycaemic effects of onion, *Allium cepa* Linn. on diabetes mellitus – a preliminary report. *Indian Journal of Physiology and Pharmacology* 19, 213–217.

Mathew, B.C., Daniel, R.S. and Augusti, K.T. (1996) Hypolipidemic effect of garlic protein substituted for casein in diet of rats compared to those of garlic oil. *Indian Journal of Experimental Biology* 34, 337–340.

Melzig, M.F., Krause, E. and Franke, S. (1995) Inhibition of adenosine deaminase activity of aortic endothelial cells by extracts of garlic (*Allium sativum* L.). *Die Pharmazie* 50, 359–361.

Mimaki, Y., Nikaido, T., Matsumoto, K., Sashida, Y. and Ohmoto, T. (1994) New steroidal saponins from the bulbs of *Allium giganteum* exhibiting potent inhibition of cAMP phosphodiesterase activity. *Chemical and Pharmaceutical Bulletin* 42, 710–714.

Mimaki, Y., Kuroda, M., Fukasawa, T. and Sashida, Y. (1999a) Steroidal glycosides from the bulbs of *Allium jesdianum*. *Journal of Natural Products* 62, 194–197.

Mimaki, Y., Kuroda, M., Fukasawa, T. and Sashida, Y. (1999b) Steroidal saponins from the bulbs of *Allium karataviense*. *Chemical and Pharmaceutical Bulletin* 47, 738–743.

Munday, J.S., James, K.A., Fray, L.M., Kirkwood, S.W. and Thompson, K.G. (1999) Daily supplementation with aged garlic extract, but not raw garlic, protects low density lipoprotein against *in vitro* oxidation. *Atherosclerosis* 143, 399–404.

Munday, R. and Munday, C.M. (1999) Low doses of diallyl disulfide, a compound derived from garlic, increase tissue activities of quinone reductase and glutathione transferase in the gastrointestinal tract of the rat. *Nutrition and Cancer* 34, 42–48.

Nok, A.J., Williams, S. and Onyenekwe, P.C. (1996) *Allium sativum*-induced death of African trypanosomes. *Parasitology Research* 82, 634–637.

Oi, Y., Kawada, T., Shishido, C., Wada, K., Kominato, Y., Nishimura, S., Ariga, T. and Iwai, K. (1999) Allyl-containing sulfides in garlic increase uncoupling protein content in brown adipose tissue, and noradrenaline and adrenaline secretion in rats. *Journal of Nutrition* 129, 336–342.

Pai, S.T. and Platt, M.W. (1995) Antifungal effects of *Allium sativum* (garlic) extract against the *Aspergillus* species involved in otomycosis. *Letters in Applied Microbiology* 20, 14–18.

Pantoja, C.V., Norris, B.C. and Contreras, C.M. (1996) Diuretic and natriuretic effects of chromatographically purified fraction of garlic (*Allium sativum*). *Journal of Ethnopharmacology* 52, 101–105.

Patil, B.S. and Pike, L.M. (1995) Distribution of quercetin content in different rings of various coloured onion (*Allium cepa* L.) cultivars. *Journal of Horticultural Science* 70, 643–650.

Patil, B.S., Pike, L.M. and Hamilton, B.K. (1995) Changes in quercetin concentration in onion (*Allium cepa* L.) owing to location, growth, storage and soil type. *New Phytologist* 130, 349–355.

Phay, N., Higashiyama, T., Tsuji, M., Matsuura, H., Fukushi, Y., Yokota, A. and Tomita, F. (1999) An antifungal compound from roots of Welsh onion. *Phytochemistry* 52, 271–274.

Reuter, H.D. (1995) *Allium sativum* and *Allium ursinum*: Part 2. Pharmacology and medicinal application. *Phytomedicine* 2, 72–91.

Rietz, B., Isensee, H., Strobach, H., Makdessi, S. and Jacob, R. (1993) Cardioprotective actions of wild garlic (*Allium ursinum*) in ischemia and reperfusion. *Molecular and Cellular Biochemistry* 119, 143–150.

Riggs, D.R., de Haven, J.I. and Lamm, D.L. (1997) *Allium sativum* (garlic) treatment for murine transitional cell carcinoma. *Cancer* 79, 1987–1994.

Roman-Ramos, R., Flores-Saenz, J.L. and Alarcon-Aguilar, F.J. (1995) Anti-hyperglycemic effect of some edible plants. *Journal of Ethnopharmacology* 48, 25–32.

Sendl, A. (1995) *Allium sativum* and *Allium ursinum*: Part 1. Chemistry analysis, history, botany. *Phytomedicine* 1, 323–339.

Sheela, C.G. and Augusti, K.T. (1992) Antidiabetic effects of *S*-allyl cysteine sulphoxide isolated from garlic *Allium sativum* Linn. *Indian Journal of Experimental Biology* 30, 523–526.

Sheela, C.G., Kumud, K. and Augusti, K.T. (1995) Anti-diabetic effects of onion and garlic sulfoxide amino acids in rats. *Planta Medica* 61, 356–357.

Shen, J., Davis, L.E., Wallace, J.M., Cai, Y. and Lawson, L.D. (1996) Enhanced diallyl trisulfide has *in vitro* synergy with amphotericin B against *Cryptococcus neoformans*. *Planta Medica* 62, 415–418.

Siess, M.H., Le Bon, A.M., Canivenc-Lavier, M.C. and Suschetet, M. (1997) Modification of hepatic drug metabolizing enzymes in rats treated with alkyl sulfides. *Cancer Letters* 120, 195–201.

Sigounas, G., Hooker, J., Anagnostou, A. and Steiner, M. (1997) *S*-Allylmercaptocysteine inhibits cell proliferation and reduces the viability of erythroleukemia, breast, and prostate cancer cell lines. *Nutrition and Cancer* 27, 186–191.

Simons, L.A., Balasubramaniam, S., von Konigsmark, M., Parfitt, A., Simons, J. and Peters, W. (1995) On the effect of garlic on plasma lipids and lipoproteins in mild hypercholesterolaemia. *Atherosclerosis* 113, 219–225.

Singh, A. and Shukla, Y. (1998) Antitumor activity of diallyl sulfide in two-stage mouse skin model of carcinogenesis. *Biomedical and Environmental Sciences* 11, 258–263.

Sivam, G.P., Lampe, J.W., Ulness, B., Swanzy, S.R. and Potter, J.D. (1997) *Helicobacter pylori – in vitro* susceptibility to garlic (*Allium sativum*) extract. *Nutrition and Cancer* 27, 118–121.

Stajner, D., Milic, N. and Canadanovic-Brunet, J. (1999) An investigation into the antioxidant activity of *Allium nutans* L. *Phytotherapy Research* 13, 333–336.

Steinmetz, K.A. and Potter, J.D. (1996) Vegetables, fruit, and cancer prevention: a review. *Journal of the American Dietetic Association* 96, 1027–1039.

Swanston-Flatt, S.K., Day, C., Bailey, C.J. and Flatt, P.R. (1990) Traditional plant treatments for diabetes. Studies in normal and streptozotocin diabetic mice. *Diabetologia* 33, 462–464.

Takada, N., Yano, Y., Wanibuchi, H., Otani, S. and Fukushima, S. (1997) *S*-methylcysteine and cysteine are inhibitors of induction of glutathione *S*-transferase placental form-positive foci during initiation and promotion phase of rat hepatocarcinogenesis. *Japanese Journal of Cancer Research* 88, 435–442.

Tappayuthpijarn, P., Dejatiwongse, Q., Hincheranan, T. and Suriyant, P.N. (1989) Effect of *Allium ascalonicum* on erythrocyte shape in induced hypercholesterolemia rabbits. *Journal of the Medical Association of Thailand* 72, 448–451.

Tassin, S., Broekaert, W.F., Mario, D., Acland, D.P., Ptak, M., Vovelle, F. and Sodano, P. (1998) Solution structure of Ace-AMP1, a potent antimicrobial protein extracted from onion seeds. Structural analogies with plant nonspecific lipid transfer proteins. *Biochemistry* 17, 3623–3637.

Venugopal, P.V. and Venugopal, T.V. (1995) Antidermatophytic activity of garlic (*Allium sativum*) *in vitro*. *International Journal of Dermatology* 34, 278–279.

Wagner, H., Dorsch, W., Bayer, T., Breu, W. and Willer, F. (1990) Antiasthmatic effects of onions: inhibition of 5-lipoxygenase and cyclooxygenase *in vitro* by thiosulfinates and 'Cepaenes'. *Prostaglandins, Leukotrienes and Essential Fatty Acids* 39, 59–62.

Wargovich, M.J. and Uda, N. (1996) *Allium* vegetables and the potential for chemoprevention of cancer. *Advances in Experimental Medicine and Biology* 401, 171–177.

Wargovich, M.J., Uda, N., Woods, C., Velaso, M. and McKee, K. (1996) Allium vegetables: their role in the prevention of cancer. *Biochemical Society Transactions* 24, 811–814.

Winkler, G., Iberl, B. and Knobloch, K. (1991) Neue Ergebnisse über Allicin-Folgeprodukte in lipophilen Medien. *Deutsche Apotheker Zeitung* 131 (Suppl. 24), 6–8.

Yin, M.C. and Tsao, S.M. (1999) Inhibitory effect of seven *Allium* plants upon three *Aspergillus* species. *International Journal of Food Microbiology* 49, 49–56.

Yoshida, H., Katsuzaki, M., Ohta, R., Ishikawa, K., Fukuda, H., Fujino, T. and Suzuki, A. (1999a) An organosulfur compound isolated from oil-macerated garlic extract, and its antimicrobial effect. *Bioscience, Biotechnology and Biochemistry* 63, 588–590.

Yoshida, H., Katsuzaki, M., Ohta, R., Ishikawa, K., Fukuda, H., Fujino, T. and Suzuki, A. (1999b) Antimicrobial activity of the thiosulfinates isolated from oil-macerated garlic extract. *Bioscience, Biotechnology and Biochemistry* 63, 591–594.

Ziyyat, A., Legssyer, A., Mekhfi, H., Dassouli, A., Serhrouchni, M. and Benjelloun, W. (1997) Phytotherapy of hypertension and diabetes in oriental Morocco. *Journal of Ethnopharmacology* 58, 45–54.

Zohri, A.N., Abdel-Gawad, K. and Saber, S. (1995) Antibacterial, antidermatophytic and antitoxigenic activities of onion (*Allium cepa* L.) oil. *Microbiological Research* 150, 167–172.

16 Onions in the Tropics: Cultivars and Country Reports

L. Currah
Currah Consultancy, 14 Eton Road, Stratford-upon-Avon CV37 7EJ, UK

1 Introduction

A high proportion of the world's onions is grown and consumed in the tropics. India alone produces 4.8 million metric tonnes (t) per year currently, one-tenth of the world's total of 47.8 million t for 2000 (FAO, 2001). Other major producers are Iran, Pakistan and Brazil, with up to 1 million t year^{-1} each. Mexico increased its production of green onions to nearly 1 million t in 1999 (FAO, 2000). The low onion yields (less than 10 t ha^{-1}) from many tropical countries are striking (Tables 16.1–16.3).

Onion growing is expanding in the tropics and yields are rising in some countries; new cultivars are being released with improved disease and heat tolerance; neglected genetic material is being used to create new open-pollinated (OP) cultivars and hybrids, and integrated crop-management practices are starting to be adopted. Genetic conservation

Table 16.1. Production of onions in the world, Asia, Australia and other major producing countries in 2000: areas, average national yields and total production (from FAO, 2001).

Country	Area under onions (ha)	Average yield (t ha^{-1})	Total production ('000 t)
World	2,705,030	17.66	47,781
Asia			
Bangladesh	34,454	4.0	138
China, PR	575,820	21.2	12,187*
India	480,600	11.4	5,467
Indonesia	79,949	10.1	805
Iran	65,000	27.7	1,800
Iraq	8,400	3.5	30
Israel	3,000	30.0	90
Myanmar	50,000	10.1	507
Oman	700	13.4	9
Pakistan	109,760	15.0	1,648
Philippines	12,500	7.2	90
Qatar	170	26.5	5
Saudi Arabia	14,000	15.0	210
Sri Lanka	10,740	9.8	105
Thailand	18,700	16.0	300
Vietnam	62,000	3.0	184
Yemen	5,273	14.5	77
Australia			
Australia	5,600	39.1	219
Other countries for comparison			
Korea, Rep.	16,131	58.0	936
Netherlands	19,682	38.9	766
Spain	23,000	43.5	1,000
USA	70,170	47.6	3,337

* Evidence from Chinese sources suggests that only about 10% of this figure may actually refer to dry bulb onions (Xu *et al.*, 1994).

of tropical onions is beginning, though not yet in a systematic way. Studies of the short-day (SD) cultivars grown in the tropics are providing new information on their physiological responses to day length and to moderate to high temperatures (Wiles, 1989; Chanda, 1996; Mettananda and Fordham, 1997, 1999; Wickramasinghe *et al.*, 2000). Reports on tropical shallots have appeared (e.g. David *et al.*, 1998) and seed of new SD shallots is becoming commercially available. In onion socio-economics, the functioning of an important onion trade route from Niger to Côte d'Ivoire in West Africa has now been described (David and Moustier, 1996, 1998).

In this chapter, I aim to pinpoint some important information sources, to summarize the types of SD onion cultivars grown in the tropics and to indicate recent country reports, grouped by regions. SD onions from the USA and from Israel are described briefly, since they are some of the most productive cultivars grown in the tropics. This chapter aims to consolidate information on onions in tropical countries for readers who are new to the topic.

2. Reviews, Surveys and International Reports

El Baradi (1971) of the Royal Tropical Institute, Amsterdam, wrote the first truly international account of onions in the

Table 16.2. Production of onions in Africa in 2000: areas, average national yields and total production (from FAO, 2001).

Country	Area under onions (ha)	Average yield (t ha^{-1})	Total production ('000 t)
Algeria	28,000	13.6	380
Benin	801	14.4	12
Botswana	60	15.0	1
Cameroon	7,300	5.8	42
Chad	700	20.0	14
Congo, Dem. Rep.	9,500	5.6	53
Egypt	40,000	25.0	1,000
Ethiopia	4,200	9.5	40
Ghana	5,000	7.7	39
Kenya	4,000	5.0	20
Libya	9,500	19.0	180
Madagascar	680	8.5	6
Malawi	2,500	7.2	18
Mali	2,224	29.0	65
Mauritius	350	25.7	9
Morocco	28,000	18.2	510
Mozambique	1,200	4.8	6
Niger	7,500	24.0	180
Nigeria	–	–	596
Senegal	3,000	21.7	65
South Africa	19,000	21.1	400
Sudan	8,000	7.1	57
Tanzania	18,000	3.0	54
Tunisia	8,900	14.9	133
Uganda	31,000	4.1	127
Zambia	1,800	15.0	27
Zimbabwe	160	15.0	2

Table 16.3. Production of onions in Central and South America and the Caribbean in 2000: areas, average national yields and total production (from FAO, 2001).

Country	Area under onions (ha)	Average yield (t ha^{-1})	Total production ('000 t)
Bolivia	6,600	7.6	50
Brazil	65,366	16.5	1,078
Colombia	14,000	17.7	248
Costa Rica	660	22.7	15
Cuba	2,377	6.6	16
Dominican Rep.	4,060	9.7	39
El Salvador	450	8.3	4
Guatemala	5,000	6.4	32
Haiti	800	5.0	4
Honduras	2,000	4.9	10
Jamaica	181	9.1	2
Mexico	8,250	12.1	100
Nicaragua	2,500	2.4	6
Panama	500	17.8	9
Paraguay	5,000	6.8	34
Peru	15,000	24.4	367
Venezuela	7,986	22.6	181

tropics. A key review of onion storage in the tropics appeared soon after, in 1972 (Thompson *et al.*, 1972). In the late 1980s, Uzo and Currah (1990) reviewed the agronomy of tropical onions, mainly based on West African work. Currah and Proctor (1990) attempted a more global review of literature, including the results of an international survey of onions and tropical storage technology by the Natural Resources Institute (NRI) in the UK. A report covering the first 5 years of an international cultivar trials project appeared in 1997 (Currah *et al.*, 1997), and work on genotype × environment interactions of the cultivars in the trials is in progress (A.J.R. Godfrey, New Zealand, 2000, personal communication).

The last 10 years of the 20th century saw a marked expansion in the availability of information on onions in the tropics. International symposia were held in Bangkok in 1993 (Midmore, 1994); in Mendoza, Argentina, in 1994 (Burba and Galmarini, 1997); and in Adelaide in 1997 (Armstrong, 2001). Regional onion workshops in Maradi, Niger (FAO, 1992; de Bon, 1993), and Nairobi, Kenya (Rabinowitch *et al.*, 1997) generated numerous country reports. These meetings stimulated greater international cooperation between researchers.

In India, the National Horticultural Research and Development Foundation (NHRDF) was set up in 1977 to promote onion and garlic culture and has an extensive publication list. The National Research Centre for Onion and Garlic, near Pune, India, was established recently and has now published its first annual report and several advisory works.

The *Onion Newsletter for the Tropics*, published by the NRI in the UK (1989 to 1996) and the *Allium Information Newsletter* (published annually by the US Department of Agriculture, University of Wisconsin, 1990 onwards) include many reports from onion researchers in the tropics.

Two major international schemes are important sources of information. The Asian Vegetable Research and Development Center (AVRDC) in Taiwan adopted the three bulbing alliums for genetic improvement in 1992. The Center's researchers have collected and assessed many lines of onions, shallots and garlic. A breeding programme aims to improve the resistance of onions to hot and humid tropical climates (Pathak, 1997). In West Africa, the Food and Agriculture Organization (FAO)'s Coopération Régionale pour le Développement des Productions Horticoles en Afrique (RADHORT) network promotes information exchange on market-garden crops within ten countries. A regular liaison bulletin includes recommended national lists for onions (FAO *et al.*, 2000; Table 16.4).

A manual by Brice *et al.* (1997) provides information for selecting onion storage methods in the tropics, taking into account the local climate and the investment levels possible. Practical advice is given on structures, equipment and the maintenance of onion stores.

3. Onion Cultivars Grown in the Tropics and Country Reports

3.1 An overview of the diversity of short-day onions

All onions are physiologically 'long-day' plants, but the mechanism that controls onion bulbing is really a phytochrome response to the length of the night. Therefore, in the so-called 'short-day' onions that are grown in the tropics, bulbing is in fact induced in response to night lengths, which are relatively long, at around 12 h. Intermediate-day (ID) and long-day (LD) cultivars grown at higher latitudes are induced to form bulbs by nights that are relatively shorter (i.e. nights of 11–8 h, corresponding to days of 13–16 h). I will use the convention of referring to 'short-day' onions here.

Most onions grown in the tropics are of the short-day (SD) type, but many different kinds of SD onion exist (Currah and Proctor, 1990). One way to classify them is by the amount of homogeneity which the populations have attained through selection and breeding: from the most genetically varied, the landraces, through OP cultivars to the most advanced hybrids. In the tropics, a complete range of these stages can be found.

Landrace onions are well adapted to local conditions and may have useful characters, such as long storage ability (Rouamba *et al.*, 2001) and disease tolerance (for example, through the rapid leaf-wax replacement found in some Brazilian onions (e.g. Ferreira and da Costa, 1983)). However, it is common for them to vary greatly in size, shape and colour. The next stage is represented by well-established selections, such as 'Poona Red' (India), 'Early Lockyer Brown' (Australia) and 'Red Creole' (Louisiana, USA). These are traded as named cultivars and may be maintained by seed companies rather than by farmers. They represent landraces that have been deliberately stabilized, and they often provide useful basic materials for further selection and breeding. Examples of more highly bred or selected commercial OP cultivars are 'Early Red/Moulin Rouge' (Hazera Genetics, Israel), 'Red Star PVP' (Petoseed, USA), 'Claret' (Bejo, Holland) and 'Agrifound Light Red' (NHRDF, India). This class of cultivars is being actively developed by introductions and crossing.

Hybrids form the next group. Established hybrids (mostly of the 'Granex' type) were bred in the USA from the 1950s onwards and some have been adopted as standard in some tropical production zones: for example, cvs 'Dessex' (Sunseeds, USA), 'Granex 429' and 'Granex 33' (Asgrow, USA). A wide range of more recent SD hybrids has been developed, principally for the southern production regions of the USA and Mexico, e.g. cvs 'Mercedes' (Peto) and 'Rio Raji Red' (Rio Colorado, USA). From Israel, a range of hybrids has been created which combine high productivity with improved storage performance (e.g. cvs 'Arad', 'Galil', 'Sivan', the new releases 'RAM 735', 'RAM 781' and others, from collaboration between onion breeders at the Faculty of Agricultural, Food and Environmental Quality of the Hebrew University of Jerusalem and Hazera Genetics). In Australia, the seed company Arthur Yates has recently produced a range of hybrids based on local onion material: cvs 'Predator', 'Cavalier', 'Centurion' and 'Gladiator'. Some of these are being grown in the subtropical production area near Brisbane. These cultivars incorporate better storage qualities through the addition of genetic material from LD onions from further south in Australia, the long-keeping 'Creamgolds'.

During the later 1990s, a varied range of new OP and hybrid cultivars based on distinctively 'tropical' onions began to be released: for example, Bejo's new range based on Indian 'Bombay Red' ('Orient', 'Capri', 'Flint', 'Flare') and Asgrow's 'Serrana' from Brazil and 'Red Kano' hybrid from West African source material. These onion cultivars represent a novel departure for the international seed companies and recognize the tastes of consumers within the tropics, many of whom require a regular supply of medium to small, pungent onions.

SD onions have been maintained as landraces and OP cultivars over wide geographical zones, compared with the more localized ID and LD cultivars. Today there is still a broad range of genetic variation within the SD onion gene pool, as shown by Bark and Havey (1995) in studies using restriction fragment length polymorphisms (RFLPs). Genetic selection and crossing in SD onions is accelerating. But, at the same time, the disappearance of traditional landraces is a threat to onion genetic diversity: rapid action is needed to safeguard this diversity for the future (Astley, 1990; Rouamba and Currah, 1998).

The following sections present a brief account of the onion cultivars grown in the tropics, where they originated and how they have been developed in recent years. Starting from the Indian subcontinent, we will follow them around the world in a roughly westerly direction, finishing in eastern Asia, where bulb onions (as distinct from shallots) are a comparatively recent crop. Tables 16.4 to 16.7 list onion cultivars reported from tropical countries. Table 16.8 lists some of the SD cultivars grown in the tropics by US, Israeli and Dutch seed companies. Country reports will also be indicated.

Table 16.4. Onion cultivars grown in tropical and subtropical Asia.

Countries	Cultivars	Source
Bangladesh	Faridpur Bhati (multiplier), Taherpuri (Rajshahi)	Currah and Proctor, 1990
	Kalasnagari, Zhitka, Salta	Rahim and Siddique, 1991
China, PR	(1991, Fuzhou) Fuzhou Red Skin, Texas Early Grano; (1991, Hubei) Red Skin, White Skin	NRI survey, 1991
	(In Yangtse River basin) Shanghai Red, Nanjing Yellow, Yangzhou Red; (In S) Yunnan Red, Kunming Purple, Minhou Purple	Xu *et al.*, 1994
India	N-53, N-780, N-2–4-1, Nasik Red, N-257–9-1, Pusa Red, Bellery Red, Punjab Selection, Rose onion, Pusa Ratnar, Patna Red, Poona Red, White Patna, Red Globe, Large Red, Agrifound Light Red, Agrifound Dark Red, Agrifound Rose, Agrifound White, Early Grano, N-404, N-207–1, shallots and multiplier onions	Currah and Proctor, 1990
	Arka Niketan, Pusa Madhavi, Hisar-2, Arka Kalyan, Pusa White Round, Pusa Flat White, Punjab-48, Udaipur 102; Multiplier cvs: Co-3, Co-4, Agrifound Red	Pandita, 1994
	Punjab Red Round, Punjab Naroya, Punjab Selection	Daljeet Singh, 1997
	Baswant 780, Arka Bindu, Udaipur 103, Kalianpur Red Round, Pusa White Flat, Pusa White Round; Multiplier cv: MDU 1	V. Gowda (pers.), 2000
Indonesia	Shallot varieties: Ampenan, Cloja, Bima, Bima Kuning, Bauji, Balijo, Sumenep, Bawang Lampung, Betawi Cipanas, Maja Kuning	Currah and Proctor, 1990
	Sudapurna, Menteng, Kantong	Permadi, 1994
Iran	Red Azarshahr, White Kashan, Texas Early Grano, Yellow Sweet Spanish	NRI survey, 1990
	(Red) Tarom, Ray, Isfahan, (White) Qum	Saffarian, 1994
Israel	Beit Alpha, Ori, Grano, Ben Shemen, Moab	Currah and Proctor, 1990
Myanmar	Shwephalar, Baungsauk, Shwephalar Hteikmauk, Sint-th, Mindon (shallot)	NRI survey, 1991
Nepal	Red Creole, Nasik Red, Mallajh	L. Currah (pers.), 1991
Oman	Red Creole, Texas Yellow Grano, Local White, Crystal Wax	NRI survey, 1989
Pakistan	Red Tunic, Texas Early Grano, Local White, Phulkara, Faisalabad Early, Desi Early Red, Swat	Currah and Proctor, 1990
	Chaltan, Sariab Red	Banaras and Khurshid, 1993
	Hazara	T. Brown (pers.), 2000
Philippines	Red Pinoy (ex Red Creole)	S. Groot (pers.), 1989
	Shallot, multiplier onions, Red Globe, Red Creole, Yellow Granex	Currah and Proctor, 1990
	Dingras Red Globe, Banaras (multiplier type)	Lopez and Anit, 1994
Saudi Arabia	US Grano and Granex hybrids	T. Will (pers.), 1994
Sri Lanka	Poona Red, Local red shallot, Vedala Vengayam Kalpitiya (K-1)	Currah and Proctor, 1990
	Pusa Red, Bombay Red, N-53, Rampur Red; shallots: Jaffna local and Vethalan	Sangakkara, 1994
Taiwan	Granex 429, Texas Early Grano, Superex, Equanex	Currah and Proctor, 1990
Thailand	Yellow Granex	Currah and Proctor, 1990
	Dessex, Granex 429; Shallots: Sisaket, Chiangmai, Bang Chang	Sukonthasing and Parnutat, 1994
Yemen	Texas Early Grano, Pusa Red, Red Creole, Bombay Red, Golden Creole, Baftaim (Bafteem)	Currah and Proctor, 1990

Currah and Proctor (1990) give information on yields and storage; sources noted 'NRI survey' are replies received after the NRI bulletin had been prepared for publication. pers., personal communication/ personal observation.

Table 16.5. Onion cultivars grown in Africa.

Countries	Cultivars	Source of information
Angola	Red Creole, Texas Grano	NRI survey, 1991
Benin	Ayo Massu, De Malanville, Du Niger, Jaune d'Espagne Rouge d'Espagne, Blanc d'Espagne	NRI survey, 1991
	Violet de Galmi, Violet de Zaria, shallot	NRI survey, 1994
Botswana	Granex, Pyramid, Texas Grano, Bon Accord	Currah and Proctor, 1990
Burkina Faso	Violet de Galmi, Violet de Garango, Violet de Koudougou	Currah and Proctor, 1990
	Violet de Soumarana, Blanc de Tarna, Jaune Hatif de Valence, Texas Early Grano 502, Red Creole C-5, Superex	RADHORT, 1992
Cameroon	Violet de Maroua, Violet de Garoua	RADHORT, 1998
	Goudami	P. Adama (pers.), 1994
Cape Verde	Violet de Galmi, Texas Early Yellow Grano 502 PRR, Excel 986 PRR, Jaune Hatif de Valence, shallots	RADHORT, 1992
Chad	D'Abéché, Du Chari Violacé, D'Amsilep, Dabra, De Binder, De Ngama, Texas Grano, Jaune de Valence, Jaune Paille des Vertus, Espagnol, Crystal Wax, Violet de Galmi	RADHORT, 1992
Côte d'Ivoire	Early Texas Yellow Grano, Texas Early Grano 502, Violet de Galmi, Red Creole, Brown Spanish, Rouge de Tana, Espagnol, Jaune Paille des Vertus, Boldor, Paris, Printanier Parisien, Yaakouri	RADHORT, 1992
	Shallots	David *et al.*, 1998
Egypt	Giza 6, Beheri, Giza 20	Currah and Proctor, 1990
Ethiopia	Adama Red, Mermiru Brown, Red Creole, shallots	Currah and Proctor, 1990
	Melkam (ex Pusa Red)	Aklilu and Dessalegne, 1999
Ghana	Bawku Red, Texas Grano, Red Creole	Currah and Proctor, 1990
Guinea	Texas Early Grano, Yellow Bermuda, Blanc de Galmi, Violet de Galmi	RADHORT, 1992
Guinea Bissau	Violet de Galmi, Red Creole, Texas Grano	RADHORT, 1992
Kenya	Red Creole, Tropicana, Red Creole C-5, Bombay Red, Texas Grano	Kimani and Mbatia, 1993
Mali	Shallots (several local varieties), Violet de Galmi, Blanc de Galmi	RADHORT, 1998
Mauritania	Violet de Galmi, Early Texas Grano, Red Creole, Blanc Hatif de Paris, Jaune d'Espagne, Blanc de Soumarana, Boldor, Timor, Jaune Hatif de Valence, Caraibe	RADHORT, 1992
	IRAT-69 CNRADA	RADHORT, 1998
Mauritius	Rouge local (Potiah), Gatchi, Rodrigues onion, Red Creole, Yellow Texas	Currah and Proctor, 1990
Niger	Violet de Galmi, Blanc de Galmi, Blanc de Soumarana Blanc de Galmi × White Creole	RADHORT, 1992
	Violet de Madoua, Blanc de Tarna	Baudoin *et al.*, 1994
	Violet de Zaria, Violet de Soumarana	RADHORT, 1998
Nigeria	Kano Red, White, Gindin Tasa, Wuyan Bijimi, Wuyan Makorowa	Currah and Proctor, 1990
	Zaria Red	RADHORT, 1998
Senegal	Jaune Hâtif de Valence, Violet de Galmi, Early Texas Grano 502 PRR, Yaakaar, Red Creole, Jaune de l'Espagne	RADHORT, 1992
	Rouge d'Amposta	L. Currah (pers.), 1994
Sierra Leone	Texas Grano, Red Creole	Currah and Proctor, 1990
Sudan	Nasi Red, Saggai Red, Dongola Yellow, Dongola White, Wad Ramli, Shundi Yellow, Hilalia, Kunnur	Currah and Proctor, 1990

Continued.

Table 16.5. *Continued.*

Countries	Cultivars	Source of information
Swaziland	Texas Grano, De Wildt, Pyramid, Granex	Currah and Proctor, 1990
Tanzania	Red Creole, Red Bombay, Texas Grano, shallots	Currah and Proctor, 1990
	Kakhi	Mtaita and Msuya, 1994
	Ex-Duluti-ARTZ	Mulungu *et al.*, 1998
Togo	Local yellow, Red Creole, Texas Grano	Currah and Proctor, 1990
Tunisia	Early White, Ultra Red, Aarbi	Currah and Proctor, 1990
Uganda	Red Creole, Tropicana, Bombay Red, Burgundy Red, Yellow Creole, Texas Grano, shallots	Namirembe-Ssonkko *et al.*, 1997
Zambia	Henry's Special, Extra Early Premium, Yellow Granex, Texas Early Grano	Currah and Proctor, 1990
	Pusa Red, Red Creole, shallots	Mingochi and Luchen, 1997
Zimbabwe	Dessex, Early Premium, Pyramid, Texas Grano, Gold Rush, Hojem	Currah and Proctor, 1990
	Cape Flat	Msika and Jackson, 1997

See footnotes to Table 16.4.

3.2 Onions in southern Asia

The Indian subcontinent seems likely to have been the first tropical area reached by the onion after it left its region of origin in Central Asia. Probably onions were carried south with nomadic peoples and then cultivated by agricultural settlers, at roughly the same time that other onion stocks were being developed in the Middle East and the Mediterranean basin (Fritsch and Friesen, Chapter 1, this volume). From the northern Indian plains, they could easily have been disseminated into more southerly parts of the subcontinent.

Many different ecological and latitude zones exist within southern Asia, and many different cultivars of pungent red and white onions have been developed over the centuries in India, Pakistan, Bangladesh and Sri Lanka. Onions grown in the southern parts of this region now differ considerably from those grown in the north.

3.2.1 India

For India, a good account of growing methods was published by NHRDF (undated) and the situation in the 1980s, illustrated by commonly grown cultivars of the time, was described by Pandey (1990). Information on onions grown today and on the markets for Indian exports comes from Pandey and Bhonde (1999).

Southern Indian onions tend to be smaller than northern cultivars. For example, the 'Rose' onion of Bangalore, grown for export, is 25–35 mm in diameter and is of the type known as the 'small onion'. Some southern Indian cultivars are of the multiplier type, with several bulblets – for example the 'Co' numbered cultivars (Co-1, 2, 3, 4) grown in Tamil Nadu (NHRDF, undated): these can mature in as little as 65 days. Locally they are referred to as 'Podisu', 'Mutlore' or 'Natu' onions (Pandey and Bhonde, 1999).

Cultivars grown in more northerly states, such as Maharashtra, Gujarat and the Punjab, produce medium-sized (40–65 mm diameter), flat to globe-shaped onions, which are usually highly pungent and with some internal doubling, known generically as 'big onions'. Cultivars are adapted either to the main onion season, the dry Rabi (winter, about 11–12 h photoperiod) or the wetter Kharif (summer, about 13 h photoperiod) production season. The cultivars grown in the winter are harvested from February to June. The cool Rabi growing season ends with rapidly rising temperatures in the spring, before the onset of the monsoon; the main crop for storage and export matures at this time. Smaller volumes are produced during the Kharif season, when the climate is wet; the onions produced are usually consumed fresh, and are adapted to bulbing during periods of

Table 16.6. Onion cultivars grown in tropical and subtropical Americas and the Caribbean.

Countries	Cultivars	Source
Argentina	Morada INTA	Currah and Proctor, 1990
	Navideña INTA	Galmarini *et al.*, 1995
	Angaco INTA, Ancasti INTA, Blanca Chata INTA, Morotí INTA, Valcatorce INTA, Valuno INTA	Galmarini, 1992
	Tontal INTA, Cobriza INTA, Antartica INTA	Galmarini, 1997
	Refinta 20	Galmarini, 2000
Barbados	Golden, Texas Early Grano 502, Texas Yellow Grano, Robust, Granex	Currah and Proctor, 1990
	Galil, Arad	F. Chandler (pers.), 1998
Belize	Yellow Granex, Red Creole	Currah and Proctor, 1990
Brazil	Baia Periforme Precoce, Granex, Piro Ouro, Texas Grano, Pera Norte, Crioula, Texas Grano 502, Norte-14, Jubileu, IPA varieties	Currah and Proctor, 1990
	Granex 33, Granex 429	L. Currah (pers.), 1994
	Alfa Tropical, Beta Cristal, Conquista, Roxa de Barreiro	Lopes Leite *et al.,* 2000
Chile	Calderana or Copiapina, Texas Grano 502, Torontina or Pascuina, Valenciana	Currah and Proctor, 1990
Colombia	Texas Early Grano 502 PRR, Yellow Granex, Red Creole, Red Creole C-5, White Creole Red Bermuda, Red Burgundy, New Mexico Yellow Grano, Granex 429, Ocañera	Currah and Proctor, 1990
Costa Rica	Yellow Bermuda, Granex, Dessex	Currah and Proctor, 1990
Ecuador	Paiteña or Colorada (shallot type), Blanca	
	Texas Grano 502, Granex, Red Creole, Calred, Red Burgundy	NRI survey, 1989
Honduras	Yellow Granex, Texas Grano 502, Red Creole	Ramírez and Kline, 1992
Jamaica	Texas Early Grano, Red Creole, Granex, New Mexico Yellow Grano, New Mexico White Grano, El Toro	Currah and Proctor, 1990
Martinique	Yellow Granex	Currah and Proctor, 1990
Mexico	White Cojumatlan, Purple Cojumatlan, La Chona, Santa Cruz, Copandaro, San Elias; Early Supreme, Contessa PVP, White Granex, Robust, Rio Unico, Rio Grande	Pérez Moreno *et al.,* 1996
Panama	Granex 33, Granex 429, Granex Yellow, Dessex, Texas Grano 502, Red Granex	Currah and Proctor, 1990
	Gladalan Brown, Contessa, Granex 2000, Equanex, Yellow Granex PRR	Sánchez and Serrano, 1994
Peru	Roja Arequipeña, Red Creole, Roja Italiana, Roja Americana, Regal, Texas Early Grano, Crystal White	NRI survey, 1991
Venezuela	Texas Early Grano 502, Yellow Granex, Excel 986, New Mexico Yellow Grano, Islena Amarilla, Islena Roja, Red Burgundy, Criolla, Granex 33, Granex 429	Currah and Proctor, 1990
	Brownsville, Texas Grano 438, Red Creole, White Creole	Díaz, 1993
	Canaria Dulce, Lara, Americana	D. Delgado (pers.), 1999

See footnotes to Table 16.4.

Table 16.7. Onion cultivars grown in Australia and Oceania.

Countries	Cultivars	Source
Australia	Early Lockyer Brown, Early Lockyer White, Golden Brown, Gladalan Brown, Gladalan White	Currah and Proctor, 1990
	Lockyer Gold, Brownsville	New World Seeds Catalogue, 1992
	Wallon Brown	L. Currah (pers.), 1997
Fiji	Texas Early Grano, Awahia, Tropired	Hampton, 1975
	Superex, Tropi Red	Currah and Proctor, 1990
Hawaii	Texas Yellow Grano, Granex	Currah and Proctor, 1990
New Caledonia	Early Lockyer Brown, Golden Brown, Gladalan Brown	Daly, 1996
Papua New Guinea	Gladalan Brown, Awahia, Superex	Currah and Proctor, 1990
	Texas Early Grano, Yellow Granex, Red Creole	Wiles, 1994
	Rio Enrique, Tropic Brown, Dessex, Pira Ouro, Rio Bravo	Sowei, 1995

See footnotes to Table 16.4

Table 16.8. Short-day onion cultivars sold by some companies in the USA, Israel and The Netherlands.

Company	Cultivars	Type
Asgrow	Brownsville PVP	Early Texas Grano
	Contessa PVP	White Granex
	Granex 33	Granex hybrid
	Granex 429	Granex hybrid
	Houston PVP	Texas Grano
	Marquesa PVP	Granex hybrid
	Redbone PVP	Red Creole selection
	Red Creole	Red Creole
	Red Kano	W African hybrid
	Regia PVP	Early Spanish
	Riviera	Spanish ID hybrid
	Serrana PVP	Brazilian selection
	Texas Grano 438 PVP	Texas Grano
	Texas Grano 502 PRR	Texas Grano
	Utopia	Spanish ID hybrid
	XP 6700	Granex hybrid
	XP 6712	Granex hybrid
	Encino PVP	Texas Grano
	La Joya PVP	White Grano
Bejo	Claret	Red Creole
	Orient	Bombay Red hybrid
	Capri	Bombay Red hybrid
	Flint	Bombay Red hybrid
	Flare	Bombay Red hybrid
	Solist	Bombay Red
	White Hawk	White Grano
	Domingo	Granex hybrid
	Liberty	Texas Grano
	Rox	Shallot hybrid
	Tropix	Shallot selection
Hazera	Arad	Granex storage hybrid
	Ben Shemen	US Spanish selection
	Deko	White dehydrator
	Moulin Rouge/Early Red	Red Grano
	El Ad	Granex storage hybrid
	Eytan	Egyptian (ID) selection
	Galil/Grandstand	Granex storage hybrid
	Grano 2000	Grano shaped hybrid

Table 16.8. *Continued.*

Company	Cultivars	Type
Hazera	Hazera Yellow Granex	Granex hybrid
	HA-95	Granex storage hybrid
	HA-489	Granex storage hybrid
	HA-688/Jaguar	Granex storage hybrid
	HA-944	Granex storage hybrid
	Jenin	Egyptian (ID) selection
	Ori	V. Early Texas Grano
	RAM 710	Granex storage hybrid
	RAM 735	Granex storage hybrid
	Red Synthetic/Ofir	Red Creole
	Sivan	Pink Granex storage hybrid
Peto Seed	Aspen	White dehydrator
	Cadillac	Granex hybrid
	Candy	US Spanish early int. hybrid
	Creole Red PRR PVP	Red Creole
	Chula Vista	Granex hybrid
	Equanex	Granex hybrid
	Jaguar	Granex hybrid
	Linda Vista	Granex hybrid
	Mercedes	Granex hybrid
	Savannah Sweet	Granex hybrid
	Primavera	Granex hybrid
	PSR 11390	White dehydrator
	PSX 13889	Granex hybrid
	Red Comet	Red Creole cross?
	Red Star PVP	Red Creole selection
	PS 8392	Granex hybrid
	RS 392	Granex hybrid
	PS 492	Granex hybrid
Rio Colorado	Don Victor	Granex hybrid
	Excalibur	Granex hybrid
	Mr Max	Granex hybrid
	Nikita	Grano hybrid
	NuMex BR-1	Early Grano
	Ringer Grano Improved	Early Grano
	Rio Bravo	Early Granex hybrid
	Rio Blanco Grande	White Grano
	Rio Enrique	Grano hybrid
	Rio Hondo	Grano hybrid
	Rio Raji Red	Red Grano hybrid
	Rio Redondo	White Granex hybrid
	Rio Santiago	Red Granex hybrid
	Rio Selecto	Grano hybrid
	Rio Zorro	Grano hybrid
	Sweet Dixie	Granex hybrid
Sunseeds	Colossal PVP	Early Grano
	Dehydrator no. 3	White dehydrator
	Dessex	Granex hybrid
	Primero	White dehydrator
	Red Creole PRR PVP	Red Creole
	Regal PVP	Red Grano
	Ringer Grano PVP	Early Grano
	Robust	White Granex hybrid
	Rojo	Red Granex hybrid
	White Creole PRR PVP	White dehydrator
	Yellow Granex Improved PRR	Granex hybrid
	Red Granex	Red Granex hybrid
	Early Supreme	Granex hybrid

decreasing day lengths. The Rangda or late Kharif season follows this.

Two thousand accessions of *Allium*, mostly landraces, were collected in India during the 1980s (Thomas and Dabas, 1986). Improved selections have been made since the 1950s (Pandita, 1994; Daljeet Singh, 1997). The older onion cultivars, named after their districts of origin, include 'Nasik Red' and 'Bombay White'. 'Pusa Red' was an early named cultivar developed by a government agency and was released in 1975 (Pandita, 1994). Onion breeding takes place at several government institutions and Agricultural Universities. In Maharashtra, an important region for the production of 'big onions' for export, the NHRDF developed cvs 'Agrifound Light Red' (Rabi) and 'Agrifound Dark Red' (Kharif); other members of the series are 'Agrifound Red' ('small onion') and 'Agrifound White' (dehydrator). The Indian Institute of Horticultural Research (IIHR) in Bangalore, Karnataka, produced the 'Arka' series of improved cultivars. Multiplier onions were bred from local material at Tamil Nadu Agricultural University (Vadivelu and Muthukrishnan, 1982). Pandita (1994) and Kalloo (1998) listed the important cultivars released from state-supported sources (Table 16.4). A descriptive list provided by Dr Veera Gowda (2000, personal communication) of IIHR was also used in compiling this table. Imported cultivars 'Early Grano' (yellow) and 'Spanish Brown' (an ID cultivar for northerly high-altitude areas) are grown to a limited extent in India and there is increasing interest in dehydration varieties.

Seed productivity of 30 Indian cultivars was assessed by Padule *et al.* (1996). Joint ventures in breeding and seed production have been set up during the past 10 years, for example, the Bejo-Sheetal Seed Company. Pathak and Gowda (1994) described the start of hybrid onion breeding in India and Pathak (1999) updated this account recently. Seed supplies of improved Indian-bred cultivars are not yet adequate to meet demand (Lallan Singh, 1998) and their maintenance presents practical problems (V. Gowda, IIHR, India, 2000, personal communication).

Kalloo's (1998) review of vegetable-crop advances in India mentioned the increasing importance of biofertilizers and of the possibilities for organic production methods for Indian conditions (see Bosch Serra and Currah, Chapter 9, this volume). Bhonde (1998) reviewed onion postharvest and storage in India; better designs of ambient stores are bringing about an improvement in onion storage life

3.2.2 Sri Lanka

Sangakkara (1994) outlined the allium situation in the island. By weight, about three times more shallots and multiplier onions are produced than bulb onions; most supplies of bulbs are imported from India. Some of the shallots grown bolt readily, while others never produce flowers in Sri Lanka conditions. The short growing period imposed by high temperatures and the relatively short between-monsoon intervals in the south of India and in Sri Lanka probably contribute to the greater use of multiplier onion and shallots in these regions (see Section 4, below). Most of the Sri Lankan allium production is in the drier Yala season; yields are lower in the wet or Maha season (trials by K.A. Mettananda and by E.R.S.P. Edirimanna, in Currah *et al.*, 1997).

Yields of all alliaceous crops tend to be low in Sri Lanka. 'K-1', also called 'Kalpitiya Selection', is a new cultivar of onion bred for local production in recent years from Indian and Ethiopian onion materials (Kuruppuarachchi, 1992).

3.2.3 Pakistan

Pakistan is an important onion-producing country where the supply is not always equal to year-round demand: in some years it may be a net exporter of onions (U.K. Baloch, in Currah and Proctor, 1990), while onions are imported in times of shortage. The major cropping season is from October to February and the minor season from January to May; the national supply comes from distinct production areas at different times of year. Several local red cultivars are grown, adapted to the various provinces and seasons. Average yields are now about 18 t ha^{-1} (Table 16.1). A brief account of the onion situation in Pakistan by Banaras and Khurshid (1993) listed promising cultivars (Table 16.4). The pink Rabi cultivar 'Phulkara', which stores well, and a red Kharif type, 'Hazara', are grown in Sindh province to the east of Karachi, while onions are also produced in Balochistan, in the Punjab (during the winter) and in the area north of Lahore (T. Brown, North South Seeds, Pakistan, 2000, personal communication). In the north, the 'Swat-1' red onion was selected in the Mingora area and has an ID response.

3.2.4 Bangladesh

Because of the wet climate and short growing season, only about 15% of the onion needs of Bangladesh are supplied from within the country. Cv. 'Faridpur Bhatti' is an older multiplier-onion type, while cv. 'Taherpuri' is a fairly small-bulbed red onion, which is quick to mature, with good storage ability. Onion research, seed-production problems and cultivars grown were reported by Rahim and Siddique (1990, 1991; see Table 16.4). About 30% of the country's onions are produced from sets (Rahim *et al.*, 1992) and most of the rest from transplants. Hossain and Islam (1994) reviewed the country situation and emphasized the need for better seed production and onion storage methods.

3.2.5 Nepal

Both shallots and bulb onions are traditional crops, and in some parts of Nepal and in Himachal Pradesh state in India they are included in production systems for the hill regions (Regmi, 1994; Arya and Bakashi, 1999). 'Red Creole'-type onions were introduced comparatively recently; they bolt less than Indian cultivars in the mid-hill regions (around 1500 m in altitude), and seed is produced in rain-shadow areas. The pink-skinned local cultivar 'Mallajh' is maintained near Pokhara. Farmers find that it needs fewer inputs to produce a crop, compared with imported cultivars (Regmi, 1994). Research aims to introduce higher-yielding cultivars and improve seed production (Jaiswal *et al.*, 1997). Budathoki (1997) devised a system of onion-set production, using low- and high-altitude growing areas to extend the production season. Regular reports on *Allium* research are published from Lumle and Pakhribas Agricultural Research Stations (e.g. Jaiswal and Subedi, 1996; Bhattarai and Subedi, 1998).

3.3 South-western Asia

The countries of south-western Asia are subtropical rather than tropical, but I include them here since this region forms an important bridge between Asia and Africa and the Mediterranean.

3.3.1 Iran

Many different local onion types are still grown in Iran (Saffarian, 1994; Table 16.4), an exporting country that provides India and Pakistan with onions when they have scarcities. Ramin (1999) showed that the local cv. 'Dorcheh' had better storage properties than imported 'Grano' type onions in high- and low-temperature conditions.

3.3.2 The Arabian peninsula

Saudi Arabia and the Emirates are expanding as onion producers. On irrigated land in Saudi Arabia, total onion production rose to 210,000 t in 1999 (FAO, 2000). Cultivars are mainly modern US 'Granex' hybrids, and cold stores are being built for onions. The Emirates act as onion-trading nations, with high imports and exports. From Oman, investigations on intercropping onions with legumes (faba beans and chickpeas) are reported (Ghobashi and El-Aweel, 1999).

In Yemen, cultivars of Sudanese origin are being used for breeding (Mohamedali, 1992a, b; Table 16.4). The Yemeni selection 'Bafteem' has so far given the best yields and storage quality in trials (G.H. Mohamedali, in Currah *et al.*, 1997). The Indian cv. 'Pusa Red' is maintained locally and seed of 'Bombay Red' is imported from India. Bulbs stored under unventilated conditions at high temperatures are damaged by several pathogens, including bacteria (Maude *et al.*, 1991). Experimental bins intermittently ventilated with ambient air allowed onions to be stored longer, compared with the local practice of stacking sacks on pallets (Brice *et al.*, 1995; see also Gubb and MacTavish, Chapter 10, this volume).

3.3.3 Israel

Israel today is a dynamic onion-breeding centre, which has taken onion cultivars from various sources to develop new SD and ID material suitable for the advanced country's onion culture and for the tropics (Table 16.8). Modern onion-breeding methods are being applied. Early, bolting-resistant selections from US OP 'Grano' types were made (e.g. cvs 'Ori', 'Yodalef') and an extensive range of new SD hybrids was created by incorporating genes from LD onions with better skins and storage properties. This produced high-yielding cultivars, with better skin quality and dormancy than the US 'Grano' and 'Granex' types (Rabinowitch and Peters, 1991). Cultivars now popular in the tropics are the light brown hybrids 'Galil', 'Arad', 'RAM 735', 'RAM 781' (now named 'Ada') and others (Table 16.8). The pinkish-brown cv. 'Sivan' is preferred in tropical countries where red onions are popular. The Israeli SD cultivars from Hazera Genetics give good yields and store well in ambient conditions in Thailand (Peters *et al.*, 1989), in Kenya (Kariuki and Kimani, 1997a) and in Zimbabwe (Msika *et al.*, 1994; Msika and Jackson, 1997). Some are being adopted as standard in the tropics (e.g. 'Galil', marketed as 'Grandstand', in Barbados (F. Chandler, Barbados, 1998, personal communication)).

3.4 North-eastern Africa

Mediterranean onions today present many variants for colours, shapes, sizes and pungency differences and also, crucially, in day-length responses, since they can be produced at several different seasons of the year (see Bosch Serra and Currah, Chapter 9, this volume).

3.4.1 Egypt

Egyptian onions, such as cvs 'Beheri', 'Giza-6' and 'Giza-20', are used for export. They are medium–large, globe-shaped reddish-brown onions, which store well under dry hot conditions. Much onion research takes place in Egypt, but there are few reviews that summarize the work. Yields of onions in Egypt have been raised substantially in recent years by improved management in the newly reclaimed lands. However, Egyptian growers benefit little from breeding efforts outside their own country, due to attempts to preserve the local gene pool from contamination (H.D. Rabinowitch, Israel, 2000, personal communication). Abdallah (1998) found that transplant production was significantly improved by soil solarization, and Satour *et al.* (1989) showed the benefit of this soil-disinfecting method in the onion-production field. Bahnasawy *et al.* (1998) measured the effects of diurnal variations in temperature and relative humidity on onions inside storage heaps, as a step towards a modelling approach for storage improvement.

3.4.2 Sudan

In Sudan, many local landraces are still maintained. Most are red onions but they include some yellow, white and mixed populations (Table 16.5; Mohamedali, 1994). Some probably originated in Egypt. There are regional types from Dongola, the Khartoum region and Zalingi, and 'For' landrace onions. Storage investigations under local conditions were reported by Musa *et al.* in 1994. Hayden (1990), Hayden and Maude (1992) and Hayden *et al.* (1994a, b) published detailed studies of the biology of black mould on onions in Sudan and in the UK, showing the part played by seeds in transmitting the disease. El-Nagerabi and Ahmed (2001) are continuing these studies.

3.4.3 *Ethiopia*

Shallots are the traditional alliaceous crop of the Ethiopian highlands, but in the 1980s, Sudanese onion cultivars were selected for growing there. The most popular of these is cv. 'Adama Red' (Currah, 1985; Jackson, 1987). Recently, a new cultivar 'Melkam', was selected from the Indian cv. 'Pusa Red', suitable for lowland irrigated production (Aklilu and Dessalegne, 1999). A recent country report was given by Aklilu (1997). Trials have been made on seed production in modern shallot cultivars (Aklilu, 1998) to help farmers to keep shallots free from viruses and to reduce production costs.

3.5 Eastern and southern Africa

The highland countries of East and southern Africa lack the traditional bulb-onion cultivars grown in West Africa and Sudan. Red onions are favoured in Angola and Mozambique, while yellow/brown onions are preferred in Botswana, Zimbabwe and Zambia. Shallots are produced on a large scale in Uganda. In Kenya, red onions are more in demand and fetch better prices than brown ones. In Kenya, Tanzania and Uganda, the usual onion cultivars on offer are the imported cvs 'Red Creole', 'Bombay Red' and 'Texas Early Grano', together with the old 'Red Creole' hybrid 'Tropicana'. In Tanzania, a selection, 'Ex-Duluti-ARTZ', from 'Bombay Red' was reported by Mulungu *et al.* (1998).

'Bombay Red' types are valued for their earliness in East Africa, while 'Red Creole' can be grown in the cooler high-altitude sites without bolting. 'Red Creole C-5' is somewhat tolerant to purple blotch (*Alternaria porri*) and shows superior field keeping ability. Recently, modern hybrids from the Hazera Genetics range have been taken up by growers in Kenya.

The main growing areas in these countries tend to be at high altitudes, where the constant day length and cool temperatures may result in thick-necked bulbs that fail to mature fully. The bulbs are harvested when prices are favourable: the tops may be cut off while still green and the onions sold straight away. Foliar diseases, such as purple blotch (*A. porri*), are common; the highland areas are therefore useful when selecting for disease resistance. Kimani (1997) gave a country report for Kenya, listing the main production areas. Kimani *et al.* (1994) and Kariuki and Kimani (1997b) reported on the possibilities for producing onion seed in Kenya, including vernalization studies.

At meetings in the 1990s, country reports were presented on Malawi (Kwapata and Maliro, 1997); on Uganda (Namirembe-Ssonkko *et al.*, 1997); on Tanzania (Mtaita and Msuya, 1994; Msuya, 1997); and on Zambia (Mingochi and Luchen, 1997). In southern Africa, large-scale farmers grow South African long-storing cultivars, such as 'Pyramid', or hybrids, such as cv. 'Dessex'. The Indian cv. 'Pusa Red' has been grown and maintained in Zambia for some years. In Zimbabwe, small-scale farmers grow 'Texas Early Grano' from South African seed and other OP cultivars developed in northern South Africa. The starting material for these seems to have been some flat yellow cultivars imported long ago into the Cape area, probably by the Dutch. The onion material introduced into South Africa has been subjected to strong selection pressure for bulbing in the tropics ('tropicalization'), including culture under warmer conditions and shorter days in the northern province of Transvaal. Joubert (1986) described the development of four ID–SD cultivars in South Africa (Table 16.5). The flat brown cv. 'Pyramid' is particularly suitable for set production. Cv. 'Texas Early Grano', maintained and selected in South Africa, responds to shorter day lengths and also stores better, compared with the original US stocks (R.L. Msika, Zimbabwe, 1991, unpublished data).

The country situation in Zimbabwe and trials of new cultivars were described by Msika *et al.* (1994) and Msika and Jackson (1997). Some farmers in southern African countries use storage barns with hot-air ventilation, adapted from tobacco-drying technology; this enables them to extend the onion-marketing season considerably (Currah and Proctor, 1990). Seed-production experiments comparing natural and artificial vernalization of bulbs on a range of cultivars were described by Msika *et al.* (1997); in general, 'Creole' cultivars responded better than 'Grano' types.

3.6 West Africa

Onions are a traditional crop of inland countries of the Sahel, such as Niger (Table 16.2). In Mali, shallots are a speciality of the Dogon highlands and, in Guinea, they are grown in the Fouta Djalon mountains. Nabos (1976) described how onions from Niger were selected and stabilized. Cvs 'Violet de Galmi', 'IRAT-69', 'Blanc de Galmi' and 'Blanc de Soumarana' were developed in the 1960s. 'Violet de Galmi' is now well known in West Africa and has a good reputation with consumers (David and Moustier, 1996). It is a fairly large, pungent, purplish-brown, flattened-globe onion with good storage life under hot dry conditions. Several national institutions maintain their own strains of it (Mauritania, Senegal, Côte d'Ivoire and Burkina Faso). White variants of the local onions have possibilities for dehydration, though their dry-matter content is lower than that of US dehydration cultivars.

During the 1990s, the Dakar-based network RADHORT, under the FAO, promoted cultivar trials in the participating countries of West Africa (Baudoin *et al.*, 1994). De Bon (1993) listed many landraces (Table 16.5), and FAO (1992) published individual country reports for Burkina Faso, Cape Verde, Chad, Côte d'Ivoire, Guinea, Guinea-Bissau, Mali, Niger, Mauritania and Senegal. Accounts of two Senegal production areas describe how onion production is managed locally (de Bon *et al.*, 1991, 1992). Cultivars for successive seasons from the earliest to the latest are 'Violet de Galmi' and 'Early Texas Grano', and then 'Jaune Hâtif de Valence' and 'Jaune Géant d'Espagne'. The latest cultivar to be sown is 'Rouge d'Amposta' for summer production. The seed company Tropicasem has developed cvs 'Noflaye' and 'Rouge de Tana', the former selected against bolting, from 'Violet de Galmi' material (C. Duranton, Technisem, France, 1999, personal communication). Recent developments in onion breeding in the region were described by Rouamba in FAO *et al.* (2000).

Rouamba *et al.* (1993) in Burkina Faso collected 41 onion accessions from ten countries of the region. Three distinct groups could be distinguished among 19 entries by agromorphological and isozyme markers (Rouamba *et al.*, 1994). In recent multisite trials on ten regional lines in comparison with 'Early Texas Grano', all the local onions stored far longer than 'Grano', and lines from Nigeria (2NA) and Burkina (12BF) were the highest-yielding of the local onions (Rouamba *et al.*, 2001). Thirty-eight West African onion lines are now stored in a gene bank in Burkina Faso.

An early account of traditional growing practices in Nigeria was given by Inyang (1966). Denton and Ojeifo (1990) summarized production methods in northern Nigeria, and Bednarz and Kadams (1989) described seed production. Cv. 'Kano Red', a regional OP cultivar from northern Nigeria, usually shows high rates of bolting when in trials in other climates with a cool growing period (Currah *et al.*, 1997).

In Ghana, a red local cultivar, 'Bawku', is produced in the north-east (Sinnadurai, 1970; Sinnadurai and Abu, 1977), while near the coast, shallots are grown on sandy soils (Sinnadurai, 1973). In Côte d'Ivoire, L. Fondio collected 20 lines of shallots during the 1990s. Some have potential for seed reproduction and for supplying city demand when bulb-onion prices are high (David *et al.*, 1998). In Guinea, the shallots grown in the Fouta Djalon highlands were collected and described during the 1990s by the national research organization, Institut de Recherche Agronomique de Guinée (S. Soumah, IRAG, Guinea, 1995, personal communication).

Within West Africa, there is a lively trade in onions from inland producing countries to the large cities of the coast. Niger is estimated to produce 200,000 t of onions per year. An oligopoly of specialized merchants with links to farmers at one end of the marketing chain and to onion traders in cities, such as Abidjan, Lagos or Accra, controls this traffic. David and Moustier (1996, 1998) described how the marketing networks are organized, the roles of the numerous agents in the marketing chain and their relations with the ten or so 'chiefs' who ultimately control it (David and Moustier, 1998). Perceptions of 'quality' by growers, merchants and consumers differ: for example, growers sell onions by the sack, not by weight, so they like to produce large-sized onions; merchants know that medium-sized onions travel better and stay dormant longer (and also fill the sacks better); consumers prefer medium-sized or smaller onions (David and Moustier, 1996).

3.7 Short-day onions from the USA grown in the tropics

An account of SD onions in the tropics must mention the SD onions developed in the USA, which are cultivated in many tropical countries. SD cultivars of onions used for overwinter production on the Mediterranean coast of Spain are the source of many cultivars developed in the USA during the 20th century, notably the 'Grano' and 'Granex' types. All *A. cepa* onions were introduced to the Americas from the Old World (Havey, 1991).

The earliest SD onions grown in the USA may have been the pungent 'Creoles' (discussed later). The Bermuda cultivars were brought to Texas in the 1890s; they are flat, low-pungency cultivars, originally from the Canary Islands. Later, selections were made in New Mexico of 'Grano' material based on the Spanish cv. 'Babosa'; this juicy, early-summer onion was the parent material of the popular OP cultivar 'Texas Early Grano 502', released in 1944 (see Bosch Serra and Currah, Chapter 9, this volume). The website at http://aggie-horticulture.tamu.edu/plantanswers/publications/onions/onionhis.html gives the history of these onions.

'Grano' onions are mild, with low dry-matter content. The vigorous plants produce top-shaped bulbs with thin skin, thick juicy scales and little internal doubling. Their rather elastic day-length response and strong growth are the keys to their success in tropical locations. Cultivars of this type include 'Texas Early Grano 502 PRR', 'Brownsville', 'Houston' and 'Texas Grano 438'. PRR designates cultivars selected for improved tolerance to pink root rot (*Pyrenochaeta terrestris*), a soil-borne fungal disease that attacks onions and other crops under hot conditions.

Crossing 'Grano' cultivars with onions of the pink-root-rot-tolerant cv. 'Bermuda' gave the 'Granex' onions, with thick flat, or flattened globe-shaped bulbs and vigorous leaf growth. Good quality features of 'Granex' onions are their high productivity, single-centred bulbs and thin necks at maturity. However, US 'Grano' and 'Granex' cultivars have short storage life even under cool conditions and are easily bruised by careless handling. Under ambient tropical conditions without forced ventilation, their storage life is short (on average less than 2 months) (Currah and Proctor, 1990; Peters *et al.*, 1994). The 'Granex' range includes popular hybrids such as 'Dessex', 'Henry's Special', 'Granex 33' and 'Granex 429', grown in many tropical countries.

Newer SD hybrids, including white and red as well as yellow/brown cultivars, have been produced by several US companies. Red hybrids include 'Red Granex', 'Rio Raji Red' and 'Rojo'; examples of whites are 'Robust', 'Rio Redondo' and 'La Joya'.

Many 'Grano' and 'Granex' onions are grown in Mexico and Central America and in tropical and subtropical South America. Some US companies arrange onion-seed production of their hybrid cultivars in Chile, Argentina and South Africa. Companies that offer ranges of US-bred SD 'Grano'/'Granex' onions include Seminis Vegetable Seeds (the Peto and Asgrow companies), Rio Colorado and Sunseeds. (See Table 16.8.)

3.8 Why Creole onions are widely grown in the tropics

'Red Creole' is a traditional onion of Louisiana, USA; it probably originated from the western Mediterranean, though it also has similarities to West African onions. Magruder *et al.* in 1941 recorded that it had been grown near New Orleans for over 100 years. Creole onions are now grown in countries such as Nepal and in East and West Africa (Tables 16.4 and 16.5). They grow in some of the more difficult tropical climates, produce reasonable yields, have tolerance to purple blotch and fulfil needs for moderate-sized, pungent onions. Creole onions are fairly slow-growing, with more numerous foliage leaves, higher leaf waxiness (presumably providing better tolerance to hot conditions as well as to attack by biotic factors), greater pungency and more internal scales compared with the 'Grano'/'Granex' cultivars. The mature bulbs have several tough dry skins. White and yellow colour variants exist, but red Creoles are the most widely grown. Good transport resistance, long dormancy and reasonable storage life under hot conditions are other useful characters for the tropics. Creole selections with improved pink-root tolerance exist, for example 'Red Creole PRR PVP' from Sunseed and 'Creole Red PRR' from Peto. White Creoles have provided the basis for the development of some SD dehydration onions, as all Creole onions have relatively high dry-matter content. The average storage life quoted for 'Creole' cultivars in the NRI survey was 4–5 months (Currah and Proctor, 1990).

Cultivars developed from the 'Creole' group include 'Red Synthetic', marketed as 'Ofir' (Hazera Genetics, Israel), 'Red Star PVP' (Peto), 'Claret' (Bejo) and 'Red Pinoy' (East-West Seeds, Philippines). After a period of relative neglect, the Creoles are now being improved for tropical markets.

3.9 Onions in tropical America and the Caribbean

In the French Antilles, the local shallots and other vegetative alliums were described by Messiaen (1992). In Cuba, scientists worked with the group from the Institut für Pflanzengenetik und Kulturpflanzenforschung (IPK), Gatersleben, Germany, to collect and describe the local alliums (Esquivel and Hammer, 1992); they reported on several introductions, including *A. canadense, A. chinense, A. tuberosum* and another unidentified edible species. Programmes of improvement based on locally maintained 'Creole' types of onion were started (see Table 16.6) and there has been considerable research on seed production (e.g. Prats Pérez *et al.*, 1996). In the Dominican Republic, Hazera cultivars from Israel are used for commercial bulb production (R. Peters, Hazera, Israel, 1998, personal communication).

A development programme for onions in Barbados was described in 1990 by Small and Chandler, at a time when the 'blast' disease (*Xanthomonas campestris*) (see Mark *et al.*, Chapter 11, this volume) was severely damaging the crop. Cv. 'Grandstand' (also called 'Galil') from Israel has now been adopted as a standard cultivar, since it is highly productive and also stores and travels well (F. Chandler, Barbados, 1997, personal communication).

Central America and Mexico had a large number of local onion landraces at one time (Jones and Mann, 1963). Some regional Central American cultivars are now being displaced by cultivars from the USA. The United States Agency for International Development encouraged sweet-onion production for export from Central America during the 1980s and 1990s (Cerna *et al.*, 1993; Gaskell, 1993). With the coming of the North American Free Trade Agreement (NAFTA), more onions started to be produced for export in Mexico. Green salad onions are supplied to Europe; some are *A. cepa* salad cultivars and others *A. fistulosum*. Jones and Mann (1963) were impressed by the year-round production of the white cv. 'Cojumatlan' on the Mexican highland plateau. US seed companies are now breeding new white cultivars for Mexico, as well as yellow cultivars for export to the USA. A brief description of the Mexican situation for onion and garlic by Pérez Moreno *et al.* (1996) mentions some of the local cultivars that are still being grown (Table 16.6).

The Honduras situation was outlined by Ramírez and Kline (1992). US 'Granex' and 'Grano' cultivars are largely grown, although local red onions were still being produced during the early 1990s. Bin-storage studies were described by Medlicott *et al.* (1995). The method was less successful here than in Yemen (Brice *et al.*, 1995), since the soft 'Grano'/'Granex' onions were easily crushed in deep bins. In wet weather the air was too humid to ventilate the bins, leading to sprouting and rotting. However, ventilated bin storage extended onion storage life during drier weather.

Medina (1980) described onion cultivation in Guatemala, at a time when heavy use of chemicals was still being advocated. Many countries in the Americas are now moving towards a more environmentally friendly approach to onion production. In Panama, local advisers devised solar heaters, which can be used to dry and cure onions to lengthen their storage life (Sánchez and Serrano, 1994).

In Colombia, extension agencies published a guide to onion and garlic growing (López-Avila, 1996), with advice on integrated pest management (IPM). The local 'Ocañera' onion of the highland regions produces a rather small pinkish to reddish onion and forms doubles easily, compared with modern cultivars of the bulb onion. Somewhat similar multiplier onions, usually sold with their leafy tops, are produced in the highlands of Ecuador, where they are called 'Paiteña' or 'Colorada' (Currah and Proctor, 1990).

In Venezuela, a guide to onion and garlic production was published by the extension organization Fundación Servicio para el Agricultor (FUSAGRI) in 1986. Díaz (1993) summarized the country situation briefly, listing the main cultivars grown (Table 16.6). Growers in the Quíbor production region produce onions on heavy clay soils, using irrigation methods introduced by immigrants from the Canary Islands during the 1950s. Cultivars are chosen for their suitability for the different times of year, taking advantage of the subtle difference in day-length response within the 'Grano'/'Granex' group. In Quíbor, at 9°S, these cultivars are used at successive planting dates: 'Canaria Dulce', 'Superex' (an SD hybrid bred by Takii, Japan), 'Lara', 'Ringer Grano', 'Granex 429', 'Henry's Special' and 'Texas Early Grano 502 PRR'; at the hottest period of the year, the ID cvs 'Candy' and 'Utopia' are grown. Cv. 'Texas Grano 438' can form bulbs all year round, whereas others, such as the early cultivars 'Canaria Dulce' and 'Lara', used for overwinter production, bulb prematurely if grown during the hottest times of year (D. Delgado, Venezuela, 1999, personal communication).

There is little local seed production in the tropical parts of the Americas to the north of Peru, with its red 'Arequipa' onions (Anculle Arenas and Delgado de la Flor, 1997) or the Bahia/Pernambuco production region in north-east Brazil. The central American region is therefore mainly dependent on imported seed at present. Peru provides sweet onions from the highlands which are marketed in the USA during the off-season for Georgia's popular Vidalia onions.

In Brazil, introductions from Europe into the southerly parts of the country were developed there into the 'Crioula' and 'Baia Periforme' types of bulb onion (Lopes Leite *et al.*, 2000). Selection of these onion types for more tropical zones started during the 1960s. In São Paulo State, selection for a shorter day-length response and against anthracnose disease (*Colletotrichum gloeosporioides*) (de Costa and de Melo, 1984) resulted in commercial cultivars, such as 'Serrana'. Material from the São Paulo programme was taken to Pernambuco State at about 9°S and selected again for shorter-day bulbing and resistance to constantly hot conditions. The chosen bulbs were forced into flower by cold treatment, and further selections were made over several cycles, using stratified mass selection (Menezes *et al.*, 1979). For Bahia and Pernambuco, cultivars with improved qualities for long-distance transport were needed, as the traditional cv. 'Amarela Chata das Canarias' was not sufficiently robust. De França *et al.* (1997) summarized progress by the Empresa Pernambucana de Pesquisa Agropecuária (IPA), the state agricultural research organization, in breeding new cultivars. The most popular of the IPA series so far are 'Pera IPA-4' and 'Composto IPA-6'; 'Mutuali IPA-8' and 'Franciscana IPA-10' are red cultivars with potential for other parts of the tropics (Ghana: L. Abbey, 1997, personal communication; Mauritius: P. Hanoomanjee, 1996, personal communication). The latest yellow-skinned cultivars to be released in the 1990s were 'Belém IPA-10' and 'ValeOuro IPA-11'. These onions are mostly high globe in shape, sometimes wider towards the base, in contrast with the typical 'Grano' bulb, which is widest near the top.

In trials in the São Francisco Valley, the most productive cultivars were 'Texas Grano' and 'Granex', with over 50 t ha^{-1}. Several Brazilian cultivars, including 'IPA-11' and 'IPA-10', produced bulbs of 50–90 mm in diameter, the size preferred by local consumers (Costa *et al.*, 1999), but with lower yields than the US cultivars. Evidence for the improved resistance of local cultivars to onion anthracnose was recently reported (Assunção *et al.*, 1999). In Texas, cv. 'IPA-3' has shown superior resistance to the thrips *Frankliniella occidentalis* compared with cv. 'TG 1015Y' (Hamilton *et al.*, 1999).

Two distinctive types of *A. cepa* grown in Pernambuco are the dark red multiplier onions, produced all year round at the village of Salgueiro (from locally produced seed), and the green shallots, grown around Vitoria, to supply the coastal city of Recife.

In the most northerly parts of Argentina, around Catamarca and Santiago del Estero, SD onions are grown in the cool winter period. Imported hybrid cultivars are grown as well as the local SD cv. 'Valencianita'. Though very distant from the main national market in Buenos Aires, the area has potential for exporting into the neighbouring countries (Fernández, 1998). The red cultivar 'Morada INTA' is grown in the state of Corrientes and has good storage properties (Lenscak *et al.*, 1998).

A Latin-American genetic-resource conservation network for the allium crops was set up during the 1990s (Galmarini, 1996) and aims to safeguard as many as possible of the region's locally developed onion cultivars.

3.10 Australia

Australia covers a wide range of latitudes, with temperate to tropical onion-production areas. Cultivars grown in the tropical and subtropical zones are listed in Table 16.7. Inland from Brisbane in Queensland at about 27°S is the Lockyer Valley, home to the 'Early Lockyer' and 'Gladalan' cultivars. Recent selections are 'Golden Brown' (derived from the the Japanese cv. 'Senshyu') and 'Wallon Brown' from local material. The 'Lockyer' cultivars are fairly small, globular and very early to bulb in tropical climates. The 'Gladalan' cultivars are slightly later, high-yielding, deep-globe-shaped bulbs with good skin quality; they resemble brown Spanish summer storage onions, but with a shorter day-length response. They are among some of the most productive onions outside the 'Grano'/ 'Granex' group in trials in the tropics, with consistent yields across a range of climates (Currah *et al.*, 1997). Hybrids with improved storage quality are now available (Table 16.7). Farther north in Queensland, a potential production area exists near the town of Emerald, where large sweet onions might be grown for export (D.J. Midmore, Australia, 1997, personal communication). New Caledonia, at similar latitudes to parts of Queensland, has an emerging onion industry (Daly, 1996).

The traditional export onions of southern Australia, the 'Creamgold' types, need a longer day length in order to bulb, but have been reported to do well in some subtropical areas, such as the Indian Punjab (Daljeet Singh, Bathinda, India, 1997, personal communication).

Wiles (1994) described the recently established onion industry in Papua New Guinea where bulb onions can be produced around the lowland area near Port Moresby and also in some highland areas. Sowei (1995) gave the results of trials to identify cultivars for this new industry (Table 16.7).

3.11 South-East and eastern Asia

The remaining countries of South-East Asia are mostly those in which large-bulbed onions are comparatively recent introductions, as until recently, shallots and other vegetatively propagated onions, such as *Allium* × *wakegi*, were traditionally grown (Table 16.4). In some of these countries, 'Granex'-type onions are now produced to supply the Japanese market: in Thailand, the Philippines and Taiwan for example. Trials in north-eastern Thailand showed the superior storage quality of the Israeli 'Granex'-type cultivars in comparison with US cultivars (Peters *et al.*, 1994). In trials in Taiwan, two local yellow onion cultivars, 'Tainung' and 'Tainan', performed better in ambient storage than a range of SD material from other parts of the world, although the highest yields came from 'Granex' types (N.C. Chen, in Currah *et al.*, 1997). Consumers in South-East Asian countries often prefer pungent red onions, shallots or leafy garlic for their own consumption. The onions and other alliums grown in this region were described by Midmore (1994).

3.11.1 Taiwan

In Taiwan, Lin (1994) explained the comparatively recent development of bulb-onion production for export and noted that special leafy garlic cultivars are grown for eating green. The AVRDC in Taiwan is working on the bulb alliums; disease and pest resistance, heat tolerance (Pathak *et al.*, 1996) and storage qualities are being targeted for improvement.

3.11.2 Indonesia

Shallots and wakegi onions (*Allium* × *wakegi*), a cross between *A. cepa* and *A. fistulosum*, are mostly grown in this country, but many bulb onions are imported; import substitution would be desirable (Permadi, 1994). Possible improvements are to develop true seed of shallot (Permadi, 1993; see also Rabinowitch and Kamenetsky, Chapter 17, this volume) and to move towards more environmentally friendly disease- and pest-control methods. Dibyantoro (1998) suggested that entomopathogenic viruses might be used, in combination with sex-pheromone traps, for the control of insects on shallots, with a potential reduction in insecticide spraying of 85%. Breeding to introduce resistance to anthracnose in shallots is in progress (Wietsma *et al.*, 1998) and molecular studies on the Indonesian alliums have recently been reported (Arifin *et al.*, 2000). Progress in breeding resistance to beet armyworm (*Spodoptera exigua*), an important pest in Indonesia, is reported from The Netherlands (Zheng, 2000).

3.11.3 Myanmar, Malaysia and Singapore

Thein (1994) described the onions grown in Myanmar (Burma), where local cultivars are produced in large quantities (Tables 16.1 and 16.4). Vimala *et al.* (1994) gave a country report for Malaysia, where few allium crops are grown. The country is one of the major onion importers in Asia. Here and in Singapore, small red onions are popular and many of those produced in southern India are destined for these markets. Production of shallots and *A. fistulosum* could probably be expanded and agronomic experiments are being reported (Leong and Salbiah, 2001).

3.11.4 The Philippines

Lopez and Anit (1994) summarized the situation in the Philippines, another country where small, red, multiple-bulbed onions (e.g. cv. 'Batanes') are grown for local consumption and 'Granex'-type onions for export. Cv. 'Red Pinoy' was selected locally from 'Red Creole' by the East-West Seed Company; improvements include higher proportions of single-centred bulbs. A workshop held in 1992 allowed an exchange of ideas on onion postharvest (Bautista *et al.*, 1992). Azucena (1993) made a socio-economic study on onion production in the Philippines, including the marketing situation.

3.11.5 Thailand

Sukonthasing and Parnutat (1994) reported on allium growing in Thailand. Red shallots are produced in several southern areas for local use and for export to Indonesia, and 'Granex'-type onions are produced near Chiang Mai. Thailand exports these onions to Japan. Supermarkets in Bangkok offer a wide variety of alliaceous products including baby leeks, Chinese chives (both green and blanched), flower buds and stalks of Chinese chives and shallots, as well as bulbs of onions, shallots and garlic. Onion production in Thailand is restricted to the dry season. The summer monsoon rains (April to September) prohibit onion cultivation. During the summer, the supply of bulb onion is from storage only, and the popular 'Granex' does not store well, resulting at times in rocketing onion prices (Peters *et al.*, 1994).

3.11.6 China

The allium situation in mainland China as a whole was summarized by Xu *et al.* (1994), who listed the onions produced in the tropical south of the country (Kunming, Gansu and Yunnan), where they are grown as an overwintered crop. The bulb onion was introduced into China within the 20th century. *A. fistulosum* is still the more popular food crop and constitutes the largest area under alliums; calculations based on the figures on allium production supplied by Xu *et al.* (1994) seem to show that only about 10% of the allium area in the country is used to produce bulb onions. Du (1994) summarized the *Allium* genetic-resources situation in China and gave a historical account of allium culture there.

3.11.7 Japan

Japan lies outside the tropics, but the Takii Seed Company has bred SD onions for the tropics: they market a popular brown globe cultivar with high yields and good skin quality, cv. 'Superex', though its storage capability is rather short (R. Peters, H.D. Rabinowitch and T. Kowithayakorn, Israel and Thailand, 2000, personal communication). This cultivar has performed well across a range of tropical climates in trials (Currah *et al.*, 1997). Further SD cultivars are being developed.

4. Shallots and Multiplier Onions in the Tropics

Many countries in the equatorial tropics grow shallots rather than onions. Bednarz (1994) theorized that, when *A. cepa* is near the margins of its range of adaptability (because of either cold or excessive heat), vegetatively propagated forms tend to be selected. Although bulb onions can be grown on the equator at high altitudes, as in East Africa, most growers of lowland bulbing alliums on the coasts of West and Central Africa, Sri Lanka, Malaysia, Thailand, Indonesia and the Philippines grow shallots or multiplier onions (*A. cepa* Aggregatum group), which produce clusters of small bulbs. Some shallot clones flower profusely and the flower stalks may be sold as edible floral buds.

Shallots and multiplier onions have some advantages: they can be maintained vegetatively, and they produce an economic yield within 2–3 months. Abbey *et al.* (1998) in Ghana showed that shallots could lose up to 50% of their foliage with little effect on yields.

An international collection of tropical shallots has been made at the AVRDC, Taiwan (Pathak, 1994, 1997). True seed of SD shallot cvs 'Rox' and 'Tropix' is now available from Bejo Seeds in The Netherlands, and of cvs of the RAM series, such as 'RAM-7411' and 'RAM-7419', from Hazera Genetics, Israel.

5. The Future for Onions in the Tropics

It seems certain that demand for onions will continue to rise in the tropics and that, as better and more adapted cultivars are released and new technology is adopted (for example, drip fertigation), productivity will improve. The move towards more environmentally friendly production methods is being taken seriously in some countries. Predictive methods that will enable pest- and disease-prevention measures to be applied on a rational basis are being developed (e.g. for downy mildew in Queensland: FitzGerald and O'Brien, 1994; for thrips in Brazil: Gonçalves, 1998; see also Lorbeer *et al.*, Chapter 12, this volume). There is a greater appreciation now of the value of organic manures, not only for their nutrient content but also for their biotic components, which may have the potential to combat soil-borne diseases.

Dangers that may lie ahead include the possibility of thrips-borne viruses increasing in major production areas. One tospovirus (iris yellow spot) has already been reported from India (Kumar and Rawal, 1999) and from north-east Brazil (Nagata *et al.*, 1999; Pozzer *et al.*, 1999), where it is said to be capable of causing complete loss of the crop. The dangers from seed-borne diseases too have probably been underestimated in the past (Romeiro *et al.*, 1993; Boff, 1996). Growers need to be vigilant and seek help if new diseases and pests appear in the crop.

Acknowledgements

I thank my innumerable colleagues in tropical countries for their contributions to this chapter. I am grateful to the NRI, UK, for supporting my work from 1988 to 1995 and to Horticulture Research International (HRI), Wellesbourne, UK, for allowing me to use the HRI library. I also thank Professor Haim Rabinowitch and Dr Ross Peters for their help in developing my understanding of the topic.

References

Abbey, L., Kanton, R. and Braimah, H. (1998) Susceptibility of shallots to the timing and severity of leaf damage. *Journal of Horticultural Science and Biotechnology* 73, 803–805.

Abdallah, M.M.F. (1998) Improving vegetable transplants using soil solarization. II. Onion '*Allium cepa*'. *Annals of Agricultural Science (Cairo), Special Issue*, 3, 831–843.

Aklilu, S. (1997) Onion research and production in Ethiopia. *Acta Horticulturae* 433, 95–97.

Aklilu, S. (1998) Prospect of growing shallots from true seed under irrigation in the Rift Valley. *AgriTopia* 13(3/4), 8–9.

Aklilu, S. and Dessalegne, L. (1999) Melkam – new onion cultivar for the lowland irrigated production. *AgriTopia* 14, 6.

Anculle Arenas, A. and Delgado de la Flor, F. (1997) Caracterización y evaluación de la cebolla (*Allium cepa* var. *cepa*) 'roja arequipeña'. *Acta Horticulturae* 433, 171–177.

Arifin, N.S., Ozaki, Y. and Okubo, H. (2000) Genetic diversity in Indonesian shallot (*Allium cepa* var. *ascalonicum*) and *Allium* × *wakegi* revealed by RAPD markers and origin of *A.* × *wakegi* indentified by RFLP analyses of amplified chloroplast genes. *Euphytica* 111, 23–31.

Armstrong, J. (ed.) (2001) *Proceedings of the Second International Symposium on Edible* Alliaceae, *10–13 November 1997, Adelaide, Australia. Acta Horticulturae* 555, 304 pp.

Arya, P.S. and Bakashi, B.R. (1999) Onion based cropping system studies under mid-hill conditions of Himachal Pradesh. *Advances in Horticulture and Forestry* 6, 79–85.

Assunção, I.P., Coelho, R.S.B., Lima, G.S.deA., Lima, J.A.S. and Tavares, S.C.C.deH. (1999) Reação de cultivares de cebola a isolados de *Colletotrichum gloeosporioides* coletados na região do Submédio São Francisco. *Summa Phytopathologica* 25(3), 205–209.

Astley, D. (1990) Conservation of genetic resources. In: Rabinowitch, H.D. and Brewster, J.L. (eds) *Onions and Allied Crops*, Vol. I. *Botany, Physiology, and Genetics*. CRC Press, Boca Raton, Florida, pp. 177–198.

Azucena, C.F. (1993) Socio-economic considerations in onion production and handling in the Philippines: opportunities for technology transfer. *Onion Newsletter for the Tropics* 5, 10–16.

Bahnasawy, A.H., Ghaly, A.E., El-Haddad, Z.A. and El-Ansawy, M.Y. (1998) Evaluating the current system of onion storage in Egypt. In: *Proceedings of the Northeastern Agricultural and Biological Engineering Conference, Halifax, Nova Scotia, Canada, 2–5 August 1998*. American Society of Agricultural Engineers, St Joseph, Minnesota, pp. 1–30.

Banaras, M. and Khurshid, H. (1993) Onion production and research status in Pakistan. *Onion Newsletter for the Tropics* 5, 24–25.

Bark, O.H. and Havey, M.J. (1995) Similarities and relationships among open-pollinated populations of the bulb onion as estimated by nuclear RFLPs. *Theoretical and Applied Genetics* 90, 607–614.

Baudoin, W., Bâ, M.L. and Jeangille, P. (1994) Onion production and constraints in the Sahelian countries of Africa. *Acta Horticulturae* 358, 37–42.

Bautista, O.K., Nuevo, P.A. and Masilungan, G.D. (eds) (1992) *Onion Handling and Marketing in the Philippines. Proceedings of a Workshop held at Los Baños, 2 October 1992*. Postharvest Horticulture Training and Research Centre, Los Baños, Philippines, 56 pp.

Bednarz, F. (1994) A generalized picture of the onion (*Allium cepa* L.). *Acta Horticulturae* 358, 321–324.

Bednarz, F. and Kadams, A. (1989) The traditional technology of onion (*Allium cepa* L.) seed production in Nigeria and possible ways of its improvement. *Beitrage zur Tropischen Landwirtschaft und Veterinarmedizin* 27, 179–185.

Bhattarai, S.P. and Subedi, P.P. (1998) *Effect of Curing and Storage Methods on Post Harvest Losses of Bulb Onion in the Low Hills*. Working Paper No. 98/20, Lumle Agricultural Research Centre, Lumle, Pokhara, Nepal, 5 pp.

Bhonde, S.R. (1998) Storage of onion and salient features of post-harvest technology. *News Letter, NHRDF* 18(1), 10–15.

Boff, P. (1996) Seed infection by *Colletotrichum gloeosporioides* F. Sp. *cepae* is the main cause of recent onion bulb rot in South Brazil. *Onion Newsletter for the Tropics* 7, 16–17.

Brice, J., Currah, L., Malins, A. and Bancroft, R. (1997) *Onion Storage in the Tropics: A Practical Guide to Methods of Storage and their Selection.* Natural Resources Institute, University of Greenwich, Chatham, UK, 120 pp.

Brice, J.R., Bisbrown, A.J.K. and Curd, L. (1995) Onion storage trials at high ambient temperatures in the Republic of Yemen. *Journal of Agricultural Engineering Research* 62, 185–192.

Budathoki, K. (1997) The effects of environmental conditions on the growth and development of onions (*Allium cepa* L.) in Nepal. PhD thesis, Department of Agriculture and Horticulture, Wye College, University of London, UK.

Burba, J.L. and Galmarini, C.R. (eds) (1997) *Proceedings of the First International Symposium on Edible Alliaceae, 14–18 March 1994, Mendoza, Argentina. Acta Horticulturae* 433, 652 pp.

Cerna, O., Kline, S., Kline, W. and Ramírez, D. (1993) *Guía sobre Producción de Cebolla para Exportación.* Fondación Hondureña de Investigación Agrícola, San Pedro Sula, Honduras, 59 pp.

Chanda, C.T. (1996) Manipulation and prediction of onion flower initiation in tropical onions. MSc thesis, Department of Agriculture and Horticulture, University of Nottingham, UK.

Costa, N.D., de Resende, G.M. and Souza Dias, R.deC. (1999) *Produtividade de cultivares de cebola no vale do São Francisco*. Comunicado Tecnico No. 80, CPATSA, Petrolina, Pernambuco, Brazil, 5 pp.

Currah, L. (1985) Review of three onion improvement schemes in the tropics. *Tropical Agriculture (Trinidad)* 62, 131–136.

Currah, L. and Proctor, F.J. (1990) *Onions in Tropical Regions*. Bulletin 35, Natural Resources Institute, Chatham, UK, 232 + xiii pp.

Currah, L., Green, S. and Orchard, J. (1997) *International Collaborative Short-day Onion Trials, 1990–1995*. Supplement to Onion Newsletter for the Tropics, Natural Resources Institute, University of Greenwich. 230 pp. (Obtainable on diskette from L. Currah.)

da Costa, C.P. and de Melo, I.S. (1984) Progreso genético obtido com dois ciclos de seleção massal em cebola visando resistência a *Colletotrichum gloeosporioides* Penz. *Horticultura Brasileira* 2, 21–23.

Daljeet Singh (1997) Onion improvement in India. *Acta Horticulturae* 433, 75–81.

Daly, P. (1996) La production de l'oignon en Nouvelle-Calédonie: un souci de qualité des opérateurs de la filière. *Fruits* 51, 367–374.

David, O. and Moustier, P. (1996) La formation de la qualité dans la filière oignon en Afrique de l'Ouest. *Fruits* 51, 347–358.

David, O. and Moustier, P. (1998) 'Lorsque le gros maigrit, le maigre meurt': l'organisation du commerce de l'oignon en Afrique de l'Ouest. *Autrepart* 6, 105–122.

David, O., Fondio, L. and Moustier, P. (1998) Place des échalotes ivoiriennes sur le marché d'Abidjan par rapport aux oignons importés. *Fruits* 53, 127–140.

de Bon, H. (1993) *Commercialisation, culture et multiplication de l'oignon en Afrique soudano-sahelienne.* Technical Report, GCP/RAF/244/BEL FAO, Dakar, Senegal, 88 pp.

de Bon, H., Faye, F. and Pages, J. (1991) La production d'oignon dans le Gandiolais, Sénégal. *Onion Newsletter for the Tropics* 3, 5–7.

de Bon, H., Pages, J. and Bâ, M.L. (1992) La culture de l'oignon dans la moyenne vallée du fleuve Sénégal. *Onion Newsletter for the Tropics* 4, 6–10.

de França, J.G.E., Candeia, J.A., de Menezes, J.T., Maranhão, E.A.deA., Menezes, D. and Wanderley, L.J.daG. (1997) Development of short-day yellow onion for tropical environments of the Brazilian northeast. *Acta Horticulturae* 433, 285–289.

Denton, L. and Ojeifo, I.M. (1990) Onion production practices and their improvement in Nigeria. *Onion Newsletter for the Tropics* 2, 10–13.

Díaz, R. (1993) Onion production in Venezuela. *Onion Newsletter for the Tropics* 5, 16–18.

Dibyantoro, A.L.H. (1998) Pesticide residues on some vegetables, and reductions possible by integrated pest management. *ACIAR Proceedings Series* 85, 375–379.

Du, W.F. (1994) Resources, research, and utilization of Alliums in China. *Acta Horticulturae* 358, 143–146.

El Baradi, T. (1971) Onion growing in the tropics. *Tropical Abstracts* 26, 285–291.

El-Nagerabi, S.A.F. and Ahmed, A.H.M. (2001) The effect of black mould (*Aspergillus niger*) on two Sudanese cultivars of onion. *Tropical Science* 41, 95–99.

Esquivel, M. and Hammer, K. (1992) The cultivated species of the genus *Allium* in Cuba. In: Hanelt, P., Hammer, K. and Knüpffer, H. (eds) *The Genus* Allium *– Taxonomic Problems and Genetic Resources. Proceedings of an International Symposium held at Gatersleben, Germany, 11–13 June 1991.* IPK, Gatersleben, Germany, pp. 43–48.

FAO (1992) *Compte Rendu du 3me Atelier Technique – problématique de la production et de la multiplication de l'oignon en Afrique de l'Ouest. Maradi, Republic of Niger, 9–11 March 1992.* GCP/RAF/244/BEL, FAO, Dakar, Senegal, 61pp.

FAO (2000 and 2001) Agrostat database, updated annually: http://apps1.fao.org/servlet/

FAO, Silva, N., Rouamba, A., Seck, P.A. and Coly, E.V. (2000) A recent experience for co-operative horticultural research and development in Africa (RADHORT). *Acta Horticulturae* 524, 217–234.

Fernández, F. (1998) El cultivo del cebolla en Santiago del Estero. In: *Papers of the III Jornadas de Actualización del Cultivo de Cebolla, San Juan, Argentina, 22–23 October 1998.* INTA Estación Experimental San Juan, San Juan, Argentina, pp. 3–4.

Ferreira, P.V. and da Costa, C.P. (1983) Comportamento varietal de cebola (*Allium cepa* L.) do grupo ceroso em relação à velocidade de reposição de cera foliar. *Revista Brasileira do Genetica* 6, 709–717.

FitzGerald, S.M. and O'Brien, R.G. (1994) Validation of 'Downcast' in the prediction of sporulation–infection periods of *Peronospora destructor* in the Lockyer Valley. *Australian Journal of Experimental Agriculture* 34, 537–539.

FUSAGRI (1986) *Cebolla y Ajo.* Serie Petroleo y Agricultura No. 9, Fundación Servicio para el Agricultor, Venezuela, 89 pp.

Galmarini, C. (1996) Establishment of the Latin-American network for genetic resources of onion. *Onion Newsletter for the Tropics* 7, 10–12.

Galmarini, C.R. (1992) Recursos geneticos de cebolla (*Allium cepa* L.): importancia, recolección, conservación y uso en Argentina. *Onion Newsletter for the Tropics* 4, 26–31.

Galmarini, C.R. (ed.) (1997) *Manual del cultivo de la cebolla.* Instituto Nacional de Tecnología Agropecuária, Centro Regional Cuyo, Argentina, 128 pp.

Galmarini, C.R. (2000) Onion cultivars released by La Consulta Experiment Station, INTA, Argentina. *HortScience* 35, 1360–1362.

Gaskell, M.L. (1993) A strategy for export-orientated onion production from Central America. *Proceedings of the Interamerican Society for Tropical Horticulture* 36, 67–72.

Ghobashi, A.A. and El-Aweel, M.A.T. (1999) Productivity of local onion and garlic cultivars grown in monoculture and in association with faba bean and chick pea in Oman. *Assiut Journal of Agricultural Sciences* 30(3), 47–64.

Gonçalves, P.A.S. (1998) Determinação do nivel de dano econônico de tripes em cebola. *Horticultura Brasileira* 16, 128–131.

Hamilton, B.K., Pike, L.M., Sparks, A.N., Bender, D.A., Jones, R.W., Candeia, J. and de França, G. (1999) Heritability of thrips resistance in the 'IPA-3' onion cultivar in South Texas. *Euphytica* 109, 117–122.

Hampton, R.E. (1975) Onion production in Fiji. *Fiji Agricultural Journal* 37, 17–21.

Havey, M.J. (1991) History of releases from the onion breeding program of the United States Department of Agriculture. *Allium Improvement Newsletter* 1, 50–54.

Hayden, N.J. (1990) Investigation of the biology, epidemiology and control of black mould (*Aspergillus niger*) on onions (*Allium cepa*). PhD thesis, University of Birmingham, UK.

Hayden, N.J. and Maude, R.B. (1992) The role of seed-borne *Aspergillus niger* in transmission of black mould of onion. *Plant Pathology* 41, 573–581.

Hayden, N.J., Maude, R.B. and Proctor, F.J. (1994a) Studies on the biology of black mould (*Aspergillus niger*) on temperate and tropical onions. 1. A comparison of sources of the disease in temperate and tropical field crops. *Plant Pathology* 43, 562–569.

Hayden, N.J., Maude, R.B. and Proctor, F.J. (1994b) Studies on the biology of black mould (*Aspergillus niger*) on temperate and tropical onions. 2. The effect of treatments on the control of seedborne *A. niger*. *Plant Pathology* 43, 570–578.

Hossain, A.K.M.A. and Islam, M.J. (1994) Status of *Allium* production in Bangladesh. *Acta Horticulturae* 358, 33–36.

Inyang, A.O. (1966) Onion cultivation in northern Nigeria. *Samaru Agricultural Newsletter* 8(5), 60–66.

Jackson, T.H. (1987) Ethiopian horticulture – a progress report. *Professional Horticulture* 1, 48–53.

Jaiswal, J.P. and Subedi, P.P. (1996) *Normal and Off-season Onion Varietal Trials Conducted at Outreach Research Sites in 1994/95*. Working Paper No. 96/14, Lumle Regional Agricultural Research Centre, Lumle, Pokhara, Nepal, iii + 22 pp.

Jaiswal, J.P., Bhattarai, S.P. and Subedi, P.P. (1997) *Findings of Varietal Trials on Normal Season Onion and Seed Production Study on New Elite Onion Cultivar Conducted at Outreach Research Sites during 1996/97*. Working paper No. 97/49, Lumle Agricultural Research Centre, Lumle, Pokhara, Nepal, iii + 8 pp.

Jones, H.A. and Mann, L.K. (1963) *Onions and their Allies*. Leonard Hill, London, 286 pp.

Joubert, T.G.laG. (1986) *Onion Cultivars*, reprinted from *Farming in South Africa* as *Onions, Leeks and Garlic*. C.1/1986, Department of Agriculture and Water Supply, Pretoria, 4 pp.

Kalloo, G. (1998) Vegetable research in India. *Indian Journal of Agricultural Sciences* 68, 515–526.

Kariuki, J.W. and Kimani, P.M. (1997a) Yield and storage potential of onion cultivars in Kenya. In: Rabinowitch, H.D., Kimani, P.M. and Peters, R. (eds) *Proceedings of the First Eastern Africa Regional Alliums Workshop, 21–22 September 1994, Nairobi, Kenya*. MASHAV/CINADCO, Tel Aviv, Israel, pp. 50–56.

Kariuki, J.W. and Kimani, P.M. (1997b) Potential of onion seed production in Kenya. In: Rabinowitch, H.D., Kimani, P.M. and Peters, R. (eds) *Proceedings of the First Eastern Africa Regional Alliums Workshop, 21–22 September 1994, Nairobi, Kenya*. MASHAV/CINADCO, Tel Aviv, Israel, pp. 87–93.

Kimani, P.M. (1997) Onion production and research in Kenya: an overview. In: Rabinowitch, H.D., Kimani, P.M. and Peters, R. (eds) *Proceedings of the First Eastern Africa Regional Alliums Workshop, 21–22 September 1994, Nairobi, Kenya*. MASHAV/CINADCO, Tel Aviv, Israel, pp. 7–13.

Kimani, P.M. and Mbatia, O.L.E. (1993) Production and marketing of onions in Kenya: status, problems and potential. *Onion Newsletter for the Tropics* 5, 18–23.

Kimani, P.M., Kariuki, J.L.W., Peters, R. and Rabinowitch, H.D. (1994) Potential of onion seed production in a tropical environment. *Acta Horticulturae* 358, 341–348.

Kumar, N.K.K. and Rawal, R.D. (1999) Onion thrips, *Thrips tabaci*, a vector of onion tospovirus. *Insect Environment* 5(2), 52.

Kuruppuarachchi, D.S.P. (1992) True seed production of 'large onion' at Kalpitiya in the north western dry zone of Sri Lanka. *Onion Newsletter for the Tropics* 4, 36–39.

Kwapata, M.B. and Maliro, M.F. (1997) Status and potential production of *Allium cepa* in Malawi. In: Rabinowitch, H.D., Kimani, P.M. and Peters, R. (eds) *Proceedings of the First Eastern Africa Regional Alliums Workshop, 21–22 September 1994, Nairobi, Kenya*. MASHAV/CINADCO, Tel Aviv, Israel, pp. 18–21.

Lallan Singh (1998) Onion seed production – problems and remedial measures. *News Letter, NHRDF* 18(1), 1–9.

Lenscak, M.P., Colombo, M.delH., Pletsch, R. and de Vallejos, V.C. (1998) El cultivo de cebolla en la Provincia de Corrientes. In: *Papers of the III Jornadas de Actualización del Cultivo de Cebolla, San Juan, Argentina, 22–23 October 1998*. INTA Estación Experimental San Juan, San Juan, Argentina, p. 2.

Leong, A.C. and Salbiah, H. (2001) Potential of shallot and onion production in the hot humid lowlands of Malaysia. In: *Proceedings of the 3rd International Symposium on Edible Alliaceae, University of Georgia, Athens, Georgia, USA, 30 October–3 November 2000*. University of Georgia, Athens, Georgia, pp. 181–184.

Lin, C.H. (1994) The progress and problems of *Allium* crops in Taiwan. *Acta Horticulturae* 358, 53–60.

Lopes Leite, D., Galmarini, C.R. and Havey, M.J. (2000) Cytoplasms of elite open-pollinated onions from Argentina and Brazil. *Allium Improvement Newsletter* 9, 1–4.

Lopez, E.L. and Anit, E.A. (1994) Allium production in the Philippines. *Acta Horticulturae* 358, 61–70.

López-Avila, A. (ed.) (1996) *El Cultivo del Ajo y las Cebollas en Colombia*. Corpoica-ICA-Pronatta, Bogotá, Colombia, 124 pp.

Magruder, R., Webster, R.E., Jones, H.A., Randall, T.E., Snyder, G.B., Brown, H.D., Hawthorn, L.R. and Wilson, A.L. (1941) *Descriptions of Types of Principal American Varieties of Onions*. Miscellaneous Publication No. 435, USDA, Washington, DC, 87 pp.

Maude, R.B., Lyons, N.F., Curd, L., El Muallem, A.B. and Bamakrama, H. (1991) Disease problems of onions in the Republic of Yemen. *Onion Newsletter for the Tropics* 3, 34–38.

Medina, J.N. (1980) Cultivo de la cebolla. *Camara del Agro* 2(2), 16–18.

Medlicott, A., Brice, J., Salgado, T. and Ramírez, D. (1995) Forced ambient air storage of different onion cultivars. *HortTechnology* 5, 52–57.

Menezes, D., Wanderley, L.J.daG., de Queiroz, M.A. and de Melo, P.C.T. (1979) Eficiência da seleção massal na adaptação de populaçoes de cebola (*Allium cepa* L.) ao cultura de verão, no submédio São Francisco. *Pesquisa Agropecuária Pernambucana* 3, 113–118.

Messiaen, C.-M. (1992) The botanical identity of *Allium* spp. grown in the French West Indies under the name of 'cives'. *Onion Newsletter for the Tropics* 4, 48–49.

Mettananda, K.A. and Fordham, R. (1997) The effects of 12 and 16 hour daylength treatments on the onset of bulbing in 21 onion cultivars (*Allium cepa* L.) and its application to screening germplasm for use in the tropics. *Journal of Horticultural Science* 72, 981–988.

Mettananda, K.A. and Fordham, R. (1999) The effects of plant size and leaf number on the bulbing of tropical short-day onion cultivars (*Allium cepa* L.) under controlled environments in the United Kingdom and tropical field conditions in Sri Lanka. *Journal of Horticultural Science and Biotechnology* 74, 622–631.

Midmore, D.J. (ed.) (1994) *Proceedings of the First International Symposium on Alliums for the Tropics, Bangkok and Chiang Mai, 15–19 February 1993. Acta Horticulturae* 358, 433 pp.

Mingochi, D.S. and Luchen, S.W.S. (1997) The status and prospects of onion production and research in Zambia. In: Rabinowitch, H.D., Kimani, P.M. and Peters, R. (eds) *Proceedings of the First Eastern Africa Regional Alliums Workshop, 21–22 September 1994, Nairobi, Kenya*. MASHAV/CINADCO, Tel Aviv, Israel, pp. 33–41.

Mohamedali, G.H. (1992a) Onion breeding strategies in the Republic of Yemen. *Onion Newsletter for the Tropics* 4, 31–33.

Mohamedali, G.H. (1992b) Onion seed production project in the Republic of Yemen. *Onion Newsletter for the Tropics* 4, 33–35.

Mohamedali, G.H. (1994) Onion breeding prospects and achievements in the arid tropics of northern Sudan. *Acta Horticulturae* 358, 231–234.

Msika, R.L. and Jackson, J.E. (1997) Onion production and research in Zimbabwe: country report. In: Rabinowitch, H.D., Kimani, P.M. and Peters, R. (eds) *Proceedings of the First Eastern Africa Regional Alliums Workshop, 21–22 September 1994, Nairobi, Kenya*. MASHAV/CINADCO, Tel Aviv, Israel, pp. 42–48.

Msika, R.L., Jackson, J.E. and Currah, L. (1994). Selection of onion cultivars for yield, earliness in cropping, and storage potential in Zimbabwe. *Acta Horticulturae* 358, 235–238.

Msika, R.L., Jackson, J.E. and Currah, L. (1997) Onion seed production trials in the highveld of Zimbabwe. *Acta Horticulturae* 433, 337–344.

Msuya, D.G. (1997) Farmers' experience, research input and prospects in alliums production in Tanzania. In: Rabinowitch, H.D., Kimani, P.M. and Peters, R. (eds) *Proceedings of the First Eastern Africa Regional Alliums Workshop, 21–22 September 1994, Nairobi, Kenya*. MASHAV/CINADCO, Tel Aviv, Israel, pp. 29–32.

Mtaita, T.A. and Msuya, D. (1994) Current status of allium production in Tanzania. *Acta Horticulturae* 358, 71–74.

Mulungu, L.S., Reuben, S.O.W.M., Nchimbi-Msolla, S., Misangu, R.N., Mbilinyi, L.B. and Macha, M.M. (1998) Performance of nine exotic and local onion (*Allium cepa* L.) genotypes grown under a dry season tropical condition at Morogoro, Tanzania: 1. Yield and its components. *South African Journal of Science* 94, 451–454.

Musa, S.K., Abdalla, Y.M., Haimoura, E. and Suleiman, Y. (1994) Improvement of onion storage in the Sudan. *Tropical Science* 34, 185–190.

Nabos, J. (1976) L'amélioration de l'oignon (*Allium cepa* L.) au Niger. *Agronomie Tropicale* 31, 387–397.

Nagata, T., Almeida, A.C.L., Resende, R.deO. and de Avila, A.C. (1999) The identification of the vector species of Iris Yellow Spot tospovirus occurring on onion in Brazil. *Plant Disease* 83, 399.

Namirembe-Ssonkko, R., Ssekyewa, C. and Akemo, C. (1997) The status of onion production and research in Uganda and prospects for future work. In: Rabinowitch, H.D., Kimani, P.M. and Peters, R. (eds) *Proceedings of the First Eastern Africa Regional Alliums Workshop, 21–22 September 1994, Nairobi, Kenya*. MASHAV/CINADCO, Tel Aviv, Israel, pp. 22–28.

NHRDF (National Horticultural Research and Development Foundation, India, formerly AADF) (undated) *Onion Production in India*. NHRDF, Nasik, India, 72 pp.

Padule, D.N., Kotecha, P.M. and Lohate, S.R. (1996) Seed production potential of different genotypes of onion. *Onion Newsletter for the Tropics* 7, 38–41.

Pandey, U.B. (1990) *AADF. The Past Twelve Years (1977–1989)*. AADF Information Bulletin No. 2, AADF (now NHRDF), Nasik, India, 50 pp.

Pandey, U.B. and Bhonde, S.R. (1999) Onion export. *NHRDF News Letter* 19(4), 1–6.

Pandita, M.L. (1994) Status of *Allium* production and research in India. *Acta Horticulturae* 358, 79–86.

Pathak, C.S. (1994). *Allium* improvement for the tropics: problems and AVRDC strategy. *Acta Horticulturae* 358, 23–29.

Pathak, C.S. (1997) Allium crop situation in Asia. *Acta Horticulturae* 433, 53–74.

Pathak, C.S. (1999) Hybrid seed production in onion. *Journal of New Seeds* 1(3/4), 89–108.

Pathak, C.S. and Gowda, R.V. (1994) Breeding for the development of onion hybrids in India: problems and prospects. *Acta Horticulturae* 433, 239–242.

Pathak, C.S., Ko, S.S. and Cherng, S.J. (1996) Breeding for summer stress tolerance in onions. *TVIS Newsletter* 1(1), 19.

Pérez Moreno, L., Salinas Glez, J.G. and Sánchez Pérez, J.R. (1996) Onion (*Allium cepa* L.) and garlic (*A. sativum* L.) production in Mexico. *Onion Newsletter for the Tropics* 7, 65–66.

Permadi, A.H. (1993) Growing shallot from true seed – research results and problems. *Onion Newsletter for the Tropics* 5, 35–39.

Permadi, A.H. (1994) Allium production and research status in Indonesia. *Acta Horticulturae* 358, 87–93.

Peters, R., Rabinowitch, H.D., Kowithayakorn, T. and Kuruppuruachchi, D.S.P. (1989) Selecting for productivity and good storage in short-day onions. *Onion Newsletter for the Tropics* 1, 12–15.

Peters, R., Kowithayakorn, T., Chalard, T. and Rabinowitch, H.D. (1994) The effect of date of harvest on shelf life of onions stored by hanging from leaves. *Acta Horticulturae* 358, 365–368.

Pozzer, L., Bezerra, I.C., Kormelink, R., Prins, M., Peters, D., Resende, R.deO. and de Avila, A.C. (1999) Characterization of a tospovirus isolate of Iris Yellow Spot tospovirus occurring on onion in Brazil. *Plant Disease* 83, 345–350.

Prats Pérez, A., Muñoz de Con, L. and Fundora Mayor, Z. (1996) Influence of onion bulb size and its locality of origin on seed yield. *Onion Newsletter for the Tropics* 7, 25–32.

Rabinowitch, H.D. and Peters, R. (1991) Onion breeding in Israel – a historical account. *Allium Improvement Newsletter* 1, 42–45.

Rabinowitch, H.D., Kimani, P.M. and Peters, R. (eds) (1997) *Proceedings of the First Eastern Africa Regional Alliums Workshop, Nairobi, Kenya, 21–22 September 1994*. MASHAV/CINADCO, Tel Aviv, Israel, xvii + 131pp.

RADHORT (1992) Listes variétales par pays. In: *Bulletin de Liaison* 4. GCP/RAF/244/BEL, FAO, Dakar, Senegal, pp. 100–288.

RADHORT (1998) Selection and breeding of onions (*Allium cepa* L.) in West Africa: progress made and prospects for the future. In: *Bulletin de Liaison* 14. GCP/RAF/244/BEL, FAO, Dakar, Senegal, p. 62.

Rahim, M.A. and Siddique, M.A. (1990) Research on onion in Bangladesh. *Onion Newsletter for the Tropics* 2, 5–10.

Rahim, M.A and Siddique, M.A. (1991) Onion seed production situation in Bangladesh – problems, prospects and research. *Onion Newsletter for the Tropics* 3, 39–41.

Rahim, M.A., Hakim, M.A., Begum, A. and Islam, M.S. (1992) Scope for increasing the total yield and fulfilling the demand for onions during the period of shortage in Bangladesh through the bulb-to-bulb (set) method of production. *Onion Newsletter for the Tropics* 4, 4–6.

Ramin, A.A. (1999) Storage potential of bulb onions (*Allium cepa* L.) under high temperatures. *Journal of Horticultural Science and Biotechnology* 74, 181–186.

Ramírez, D. and Kline, W. (1992) Cultura de la cebolla en Honduras: problemas y investigación. *Onion Newsletter for the Tropics* 4, 10–14.

Regmi, S.K. (1994) Onion production, storage, and seed production in Nepal. *Onion Newsletter for the Tropics* 6, 7–10.

Romeiro, R.daS., de Oliveira, J.R. and Melo, L.M.M. (1993) Etiologia e transmissão pela semente de uma bacteriose da cebola. *Revista Brasileira de Sementes* 15, 221–226.

Rouamba, A. and Currah, L. (1998) Collections of short-day onion germplasm in West Africa: a survey. *Genetic Resources and Crop Evolution* 45, 81–85.

Rouamba, A., Ricroch, A. and Sarr, A. (1993) Collecting onion germplasm in West Africa. *FAO/IBPGR Plant Genetic Resources Newsletter* 94/95, 15–17.

Rouamba, A., Ricroch, A., Sandmeier, M., Robert, T. and Sarr, A. (1994) Evaluation of genetic resources of onion (*Allium cepa* L.) in West Africa. *Acta Horticulturae* 358, 173–179.

Rouamba, A., Gbene, R.H., Bâ, D., Dembele, D., Ricroch, A. and Currah, L. (2001) Agronomic and physiological evaluation of some regional populations of onion (*Allium cepa* L.) in field and storage trials in West Africa. *Tropical Science* 41, 78–84.

Saffarian, A. (1994) Onion production and its constraints in Iran. *Acta Horticulturae* 358, 95–100.

Sánchez, E. and Serrano, C.E. (1994) *Manual del Cultivo de la Cebolla para las Tierras Altas de Chiriquí.* IDIAP, Panamá, 42 pp.

Sangakkara, U.R. (1994) Production of alliums in Sri Lanka: prospects and constraints. *Acta Horticulturae* 358, 101–106.

Satour, M.M., Grinstein, A., Rabinowitch, H.D., Abdel-Rahim, M.F., Katan, J., El-Yamani, T. and Radwan, A. (1989) Soil solarization in onion fields in Egypt and Israel: short- and long-term effects. *Acta Horticulturae* 255, 151–159.

Sinnadurai, S. (1970) A note of the bulbing and flowering habits of the 'Bawku' onion. *Tropical Agriculture (Trinidad)* 47, 77–79.

Sinnadurai, S. (1973) Shallot farming in Ghana. *Economic Botany* 27, 438–441.

Sinnadurai, S. and Abu, J.F. (1977) Onion farming in Ghana. *Economic Botany* 31, 312–314.

Small, W. and Chandler, F. (1990) The Onion Development Programme in Barbados. *Onion Newsletter for the Tropics* 2, 27–31.

Sowei, J.W. (1995) Onion cultivar selection for the lowlands of Central Province in Papua New Guinea. *Papua New Guinea Journal of Agriculture, Forestry and Fisheries* 38, 76–83.

Sukonthasing, S. and Parnutat, C. (1994) The status of alliums in Thailand. *Acta Horticulturae* 358, 107–110.

Thein, U.A. (1994) Status of onion cultivation in Myanmar. *Acta Horticulturae* 358, 111–114.

Thomas, T.A. and Dabas, B.S. (1986) Genetic resources of onion in India. In: *Proceedings of the 2nd International* Allium *Conference, Strasbourg, France, July 1986*, pp. 5–13.

Thompson, A.K., Booth, R.H. and Proctor, F.J. (1972) Onion storage in the tropics. *Tropical Science* 14, 19–34.

Uzo, J.O. and Currah, L. (1990) Cultural systems and agronomic practices in tropical climates. In: Rabinowitch, H.D. and Brewster, J.L. (eds) *Onions and Allied Crops*, Vol. II. *Agronomy, Biotic Interactions, Pathology, and Crop Protection.* CRC Press, Boca Raton, Florida, pp. 49–62.

Vadivelu, B. and Muthukrishnan, C.R. (1982) CO4 onion – a high yielding, good storing hybrid. *South Indian Horticulture* 30, 142–143.

Vimala, P., Zainab, R.S., Zaharah, A. and Salbiah, H. (1994) Status of alliums in Malaysia. *Acta Horticulturae* 358, 115–122.

Wickramasinghe, U.L., Wright, C.J. and Currah, L. (2000) Bulbing responses of two cultivars of red tropical onions to photoperiod, light integral and temperature under controlled growth conditions. *Journal of Horticultural Science and Biotechnology* 75, 304–311.

Wietsma, W., Grubben, G., Henken, B., Zheng, S., Putrasamedja, S., Permadi, A.H., Sofiari, E., Krens, F., Jacobsen, E. and Kik, C. (1998) Development of pest and disease resistant shallot cultivars by means of breeding and biotechnology. *Indonesian Agricultural Research and Development Journal* 20(4), 78–82.

Wiles, G.C. (1989) The effect of light and temperature on bulb initiation and development in tropical cultivars of onion (*Allium cepa* L.). PhD thesis, Department of Horticulture, Wye College, University of London, UK.

Wiles, G.C. (1994) Onion production in Papua New Guinea. *Acta Horticulturae* 358, 123–126.

Xu, P., Sun, H., Sun, R. and Yang, Y. (1994) Allium production and status in China. *Acta Horticulturae* 358, 127–131.

Zheng, S.-J. (2000) Towards onions and shallots (*Allium cepa* L.) resistant to beet armyworm (*Spodoptera exigua* Hübner) by transgenesis and conventional breeding. PhD thesis, Landbouwuniversiteit, Wageningen, The Netherlands, 146 pp.

17 Shallot (*Allium cepa*, Aggregatum Group)

H.D. Rabinowitch[1] and R. Kamenetsky[2]

[1]*Institute of Plant Science and Genetics in Agriculture, The Hebrew University of Jerusalem, Faculty of Agricultural, Food and Environmental Quality Sciences, PO Box 12, Rehovot 76100, Israel;* [2]*Department of Ornamental Horticulture, The Volcani Center, Bet Dagan 50250, Israel*

1. Introduction

On a global scale, shallot (*Allium cepa* L. Aggregatum group) is a minor alliaceous crop. However, in South-East Asia – for example, Indonesia, Sri Lanka and Thailand – as well as in some African countries, such as Uganda, Ethiopia and Côte d'Ivoire, where onion seed is hard to produce, where onion culture is difficult and also where the growing season is too short for the production of bulb onion, the vegetatively propagated shallot is cultivated as an important substitute for bulb onion (Currah and Proctor, 1990; Grubben, 1994; Brice *et al.*, 1997; David *et al.*, 1998; Currah, Chapter 16, this volume). Some tropical clones of shallot flower more readily than

(eds H.D. Rabinowitch and L. Currah)

those from temperate climates (Currah and Proctor, 1990). In many south-eastern Asian countries and elsewhere, the green shallot inflorescences are harvested just after the scape reaches its final length (with the green spathe still closed), and the edible floral buds are used as salad onions (Colour Plate 7A).

Additional advantages of tropical and subtropical shallots are tolerance to the hot and humid tropical climate, better tolerance to pests and diseases, and longer storage life than standard short-day onions, such as 'Granex' and 'Superex' (for details see Currah, Chapter 16, this volume).

Many of these genotypes are also preferred to bulb onions by consumers, for their good culinary qualities, such as high pungency (Grubben, 1994). In addition, because of their unique flavour (Aura, 1963; Messiaen *et al.*, 1993), a number of intermediate- (ID) and long-day (LD) shallot clones are popular in many European countries (Messiaen *et al.*, 1993; Cohat and Le Nard, 1998), in the USA (Jones and Mann, 1963) and in South America, e.g. Argentina (Galmarini, 1997).

The majority of shallot genotypes are clonally propagated, even where seed production is possible, to maintain the unique quality traits and population homogeneity of the highly heterozygous plants (A. Hadi, cited by Currah and Proctor, 1990).

Vegetative propagation, however, suffers from some major disadvantages. The most important ones are the low rate of propagation (propagating material constitutes about 40% of the total production costs in Indonesia (Adiyoga and Soetiarso, 1997)); the high costs associated with the large amount of storage space required; losses during storage due to decay and sprouting; perpetuation of soil-borne diseases and pests; and the need for hand-labour prior to and at planting. In addition, under vegetative propagation there is no 'cleansing' sexual cycle to eliminate viruses from the vegetative tissues. Therefore, there is a gradual or fast increase in virus contamination, with a subsequent decrease in yield (Walkey, 1990). As in garlic, viruses in shallot can be eliminated by meristem-tip culture (Chovelon *et al.*, 1989; Lapitan *et al.*, 1991; Fletcher *et al.*, 1998). Unlike garlic, however, many shallot cultivars can be induced to flower relatively easily, and seed propagation offers a fast, cheap and complete system for virus elimination (Messiaen *et al.*, 1993; Grubben, 1994). In addition, shallot fertility enables selection for superior breeding lines (Messiaen, 1989, 1993; Messiaen *et al.*, 1993; Grubben, 1994) and, with the introduction of male sterility (Yamashita and Tashiro, 1999) or by crossing with an appropriate male-sterile common onion (Messiaen, 1989), hybrid cultivars can be produced. Indeed, several shallot hybrids have recently been released by breeders in Israel (Faculty of Agricultural, Food and Environmental Quality Sciences of the Hebrew University of Jerusalem, in collaboration with Hazera Genetics) and in Holland (Bejo Seed Company). Attempts are being made in the Philippines (Duqueza and Eugenio, 1973), Indonesia (Permadi, 1993), Ghana and some other places to free local material from pests and viruses by leaving selected plants in the field to flower during a second season. The offspring so produced result from cross-pollination between the selected plants, thus forming virus-free, heterozygous semi-synthetic cultivars.

2. Taxonomy

For many years, the name *Allium ascalonicum* was mistakenly used in the literature for shallots, as the name was first given to a distinct wild *Allium* species from the Near East (Hanelt, 1990). However, as early as 1956, J. Helm (cited by Hanelt, 1990) related shallot to the *A. cepa* taxon. Taking into consideration the modes of propagation and growth forms, Helm (1956) classified *A. cepa* into four botanical varieties: var. *cepa* (common onion); var. *viviparum* (top onion); var. *aggregatum* (potato or multiplier onion); and the shallot-like var. *cepiforme*. Later, Jones and Mann (1963) subdivided this taxonomic alliance into three horticultural groups: (i) the common onion group (bulb onion with large single bulbs, mostly without topsets, multiplication mainly from seed); (ii) the aggregatum group (shallot, potato onion and ever-ready onion, with many lateral bulbs, mostly without topsets, multiplication almost exclusively vegetative); and (iii) the proliferum group (top onion, produces poorly developed or depressed, aggregate bulbs, mostly with topsets and usually sterile).

In 1990, Hanelt subdivided the large *A. cepa* species into two groups: Common Onion (synonyms: *A. cepa* L. var. *cepa*; *A. cepa* L. ssp. *cepa* and ssp. *australe* Trofim.) and Aggregatum group (synonyms: *A. ascalonicum* auct. non Strand; *A. cepa* ssp. *orientale* Kazak.; *A. cepa* var. *ascalonicum* Baker). Messiaen *et al.* (1993) named the shallot *A. cepa* var. *aggregatum*.

The fertile shallot intercrosses freely with bulb onion to produce fertile offspring (Atkin, 1953; Hanelt, 1990; Messiaen *et al.*, 1993) and the two plants exhibit a strong cytological (Kalkman, 1984a, b) and morphological resemblance (Colour Plate 7B). Hence, it is proposed that both plants belong to one botanical species, *A. cepa* (see Fritsch and Friesen, Chapter 1, and Klaas and Friesen, Chapter 8, this volume). We therefore prefer to name the shallot *A. cepa* L. Aggregatum group.

One notable exception to the above classification, from the southern and eastern parts of France, has to be made for the so-called 'grey shallot' cv. 'Grise de la Drôme', which is also cultivated in Argentina. A detailed description of the grey shallot was given by Messiaen *et al.* (1993, 1994). The bulbs are covered with several skins adhering to each other, which form a grey-coloured 'shell'. The roots are thick and do not die back during bulbing. The grey shallot also differs from the standard shallot by the light green colour of the leaves, and it rarely blooms. The bulbs are highly esteemed for their unique flavour. Based on the scape and umbel morphology, Messiaen *et al.* (1993, 1994) identified this unique genotype as *A. oschaninii*. A large-scale isozyme screening led Maaß (1996) to confirm that the grey shallot does not belong with *A. cepa*, but rather with either *A. oschaninii* or *A. vavilovii*. Friesen and Klaas (1998) used both genomic *in situ* hybridization (GISH) and random amplified polymorphic DNA (RAPD) for molecular identification of this plant. They concluded that most chromosomes of the 'Grise de la Drôme' genome originated from *A. oschaninii* with one and a half chromosome arms from either *A. cepa* or *A. vavilovii* (Fritsch and Friesen, Chapter 1 and Klaas and Friesen, Chapter 8, this volume). Unlike other shallot clones, the population of 'Grise de la Drôme' is heterogeneous, but the level of variation has not yet been studied. A homogeneous clone, 'Griselle', was selected from 'Grise de la Drôme' and, following meristem-tip culture and careful propagation in an insect-proof house, the newly selected clone was released by the Institut National de la Recherche Agronomique (INRA), France (R. Kahane, INRA, France, 2000, personal communication).

Another plant, called a shallot in the USA, originated from a cross between the bulbous shallot and Japanese bunching onion, *A. fistulosum*. Chromosome doubling resulted in an amphidiploid plant, which in turn was back-crossed to shallot. The resulting white offspring is grown mainly in the state of Louisiana. It is a heat-tolerant, prolific and cluster-producing plant, which has only a short dormancy (Jones and Mann, 1963; Messiaen *et al.*, 1993; Kik, Chapter 4, this volume). The best-known cultivar is 'Delta Giant', which in the USA is known as shallot (Jones and Mann, 1963). Clearly, this is quite different from the European and tropical shallot, which is currently denoted by *A. cepa* Aggregatum group, and we shall not deal with it further here.

3. Morphology

Morphologically, a shallot bulb (synonyms: set, bulblet, bulbil) is very similar to the bulb of the common onion. A mature bulb consists of a compressed stem axis or basal plate, storage leaf-bases of the outer leaves, which have lost their blades, and bladeless 'true scales'. In the centre of each bulb, there are a few leaf buds, which under favourable conditions sprout when dormancy ends. Unlike the modern bulb onion, a typical shallot bulb contains a number of laterals in the axils of the inner leaves. All sets formed from a single propagule usually remain attached to the original basal plate, thus forming a cluster of sets (Jones and Mann, 1963; Cohat, 1982; Currah and Proctor, 1990; Messiaen *et al.*, 1993; Krontal *et al.*, 1998). A cluster may contain five to 15 (Messiaen *et al.*, 1993) or two or three to 35 lateral buds (tropical shallots (H.D. Rabinowitch and R. Kamenetsky, personal observation)). Occasionally, when the season is short, bulbing occurs prior to the separation of the laterals. In such cases, all laterals of a common origin remain hidden under the cover of a common skin, and separation occurs in storage or even at the beginning of the second growing season.

The foliage and the inflorescence of shallots are usually smaller than those of the bulb onion. However, root morphology, the unifacial, hollow, slightly flattened tubular leaves, the hollow scape, the terminal inflorescence and the flowers are similar to those of common onion (Jones and Mann, 1963). Messiaen *et al.* (1993) reported that flowering in French shallot clones is irregular, that the inflorescence may bear some topsets and that some genotypes, e.g. an unnamed cultivar from the West Indies cultivated in France (possibly an interspecific hybrid (L. Currah, UK, 2000, personal communication)) do not bloom.

4. Developmental Morphology and Physiology

Shallot physiology has rarely been studied and the factors leading to flowering and the developmental biology of lateral shoots have only recently been described (Krontal *et al.*, 1998, 2000).

4.1 Seed, seed germination and seedlings

With a few exceptions, shallot is currently propagated vegetatively in most parts of the world. In Israel, 100% of the shallot grown commercially is propagated from seed. Krontal *et al.* (1998) were the first to provide a scientific description of seedling development, based on material derived from a nameless Thai landrace, Israeli Genebank accession no. 66–1004. Seeds of tropical shallot are smaller than those of bulb onion, the 1000-seed weights being on average roughly 2–3 g and 3–4 g, respectively. Seeds have no dormancy and readily germinate when moisture is available. The black seed-coat is crinkled and the seed is irregular in shape, like that of onion.

Similarly to other alliums, on germination the base of the cotyledon elongates quickly, thus forcing the root and the tiny basal plate out of the seed-coat. The primary root grows downwards, and the crooked cotyledon grows from its base in the opposite direction, while the cotyledon tip remains attached to the seed-coat. On seedling emergence, the knee (hook) breaks out of the soil surface, and immediately thereafter the cotyledon turns green and unbends. The short-lived primary root dies quickly and short non-branched roots are continuously produced from the basal part of the true stem, to form a shallow but dense root system. The first foliage leaf elongates within the cotyledon sheath and emerges through a pore in the side of the cotyledon. During the first growing season of seedlings, several true leaves, composed of a sheath and a cylindrical lamina, are formed.

4.2 Leaf development

The formation of a new leaf primordium begins as a protrusion at the apex, located on the top surface of the basal plate inside the bulb (Krontal *et al.*, 1998). The apical dome then grows as a crescent that subsequently develops into a partly lifted, complete ring (Fig. 17.1A). The first foliage leaf elongates within the cotyledon and emerges through a pore in its sheath. The next foliage primordium differentiates in the shoot apex but on the opposite side to the previous one, to form an alternate tubular unifacial leaf. Each successive leaf thus develops within the sheath of the next older one and emerges as described above, in a

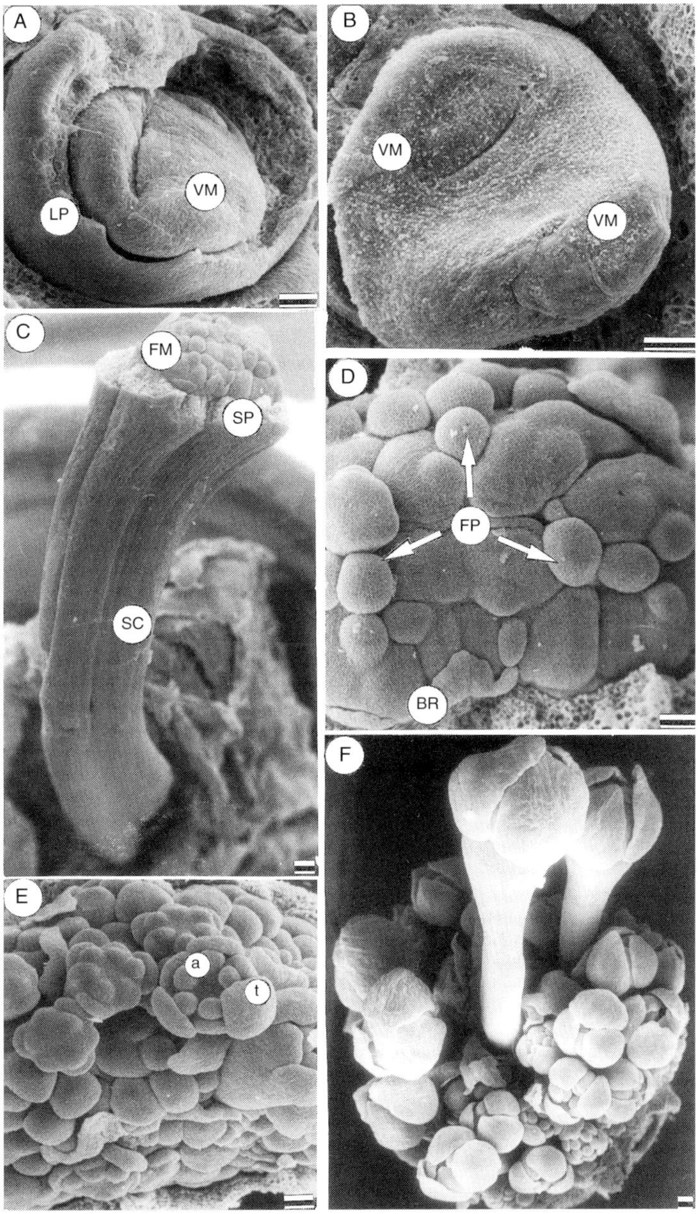

Fig. 17.1. Scanning electron photomicrographs of developing shallot apex during its vegetative and initial reproductive phases. Bar = 0.1 mm. (A) Initiation of leaf primordia in the apical vegetative meristem (VM), older leaf primordium (LP) removed. Apex size is 0.6 mm in diameter. (B) Lateral initiation in a shallot seedling: the apex is divided into two meristematic centres (VM) of differentiation. (C) Scape (SC) is 10 mm long. Formation of floral primordia in a 1 mm diameter floral meristem (FM). Spathe (SP) removed. (D) Differentiation of four centres of development in a reproductive meristem, with spathe removed. Initiation of leaf-like bracts (BR) in the inflorescence are visible. First flower primordia (FP) are initiated. (E) Flat developing inflorescence is 2 mm in diameter. Floral differentiation is visible in older flowers, while younger flowers still appear as meristematic domes. In older flowers, tepals (t) and anthers (a) are visible. The outer whorl of floral parts forms first. (F) As a result of pedicel growth, the inflorescence becomes hemispherical in shape. Different stages of floral initiation are visible. Spathe and bracts removed. Scape is 150 mm long.

way that is very similar to the development of leaves in the bulb onion (Heath and Holdsworth, 1948) and other alliaceous crops (Brewster, 1994). Hence, throughout plant growth and development, older leaves envelop the younger ones. Consequently, scales (former leaf-bases and bladeless leaf initials) within each bulb are arranged in a concentric order, as in the bulb onion (Figs 17.1B and 17.2; Jones and Mann, 1963; De Mason, 1990; Brewster, 1994). The upper parts of the sheaths form the pseudostem.

Superficially, the structure of the shallot leaf is similar to that of bulb onion. However, to the best of our knowledge, no description is available of shallot leaf anatomy. The whitish, green-veined leaf-base (syn. leaf sheath) forms a hollow tube, open at the top, while the green leaf blade is also hollow, but the tip is closed. As in the bulb onion (van Kampen, 1970), under non-inductive conditions of short days and intermediate temperatures, shallot plants can theoretically produce an unlimited number of leaves, as long as the plants are kept healthy and nutrient and water supplies last. Under field conditions and when photoperiod is long, however, development and leaf production cease with the onset of bulbing.

Fig. 17.2. Diagrammatic representation of shallot structure: ➔, vegetative growing point; O, Inflorescence; ▽, foliage leaf or leaf primordium; ▽, dry protective skin formed from a bladed leaf.

Temperature and genotypic traits have a marked effect on leaf development within the bulb during storage (Krontal *et al.*, 2000). Maximum leaf elongation was recorded in stored bulbs at 10°C (Fig. 17.3), and sprouting did not occur in bulbs of the accession no. 66–1004 stored at 30°C.

4.3 Branching and lateral formation

4.3.1 Morphology and development

Shallot branching results from the loss of the apical dominance. As in chives (*A. schoenoprasum*), where lateral initiation occurs after the development of every two or three leaves (Poulson, 1990), the initiation of the first lateral shoot in the apex of shallot is already evident in seedlings after the differentiation of the third leaf (Krontal *et al.*, 1998). At this point, the shallot apical meristem subdivides into two sections (Fig. 17.1B). Each of the newly formed branches produces new leaves and develops into a lateral shoot. This process continues and the

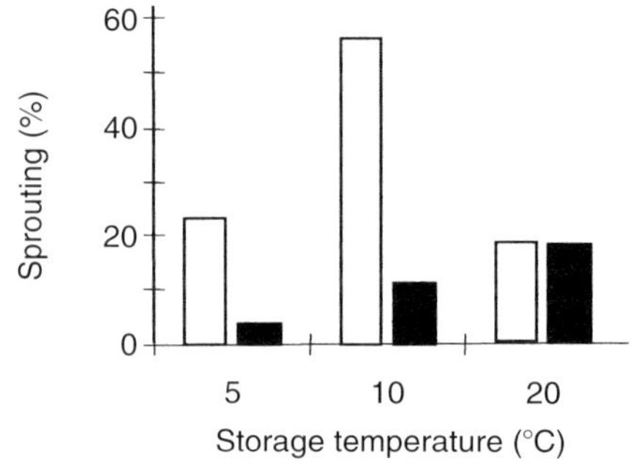

Fig. 17.3. Effect of storage temperature and genotype on sprouting of shallot bulbs in storage. Bulbs were stored at 5, 10, 20 and 30°C for 21 days and at 30°C for 42 and 56 days. No sprouting occurred at 30°C. □, accession no. 66–1004; ■, accession no. 977–1011. (From Krontal *et al.*, 2000, with permission.)

consequent growth of growing points results in bulbs with multiple centres. Initially, the side-shoots are tightly connected to each other and exchange of phloem sap is common. Later, older shoots become independent entities without any vascular exchange, thus explaining the independent flowering of each side-shoot (Cottignies *et al.*, 1999). The sheath of the older leaf in whose axil it originated surrounds the new lateral. Towards maturation and during storage, however, when the older leaf sheaths senesce and decay or dry up, the laterals form visible side-shoots, i.e. a cluster of multihearted sets (Colour Plate 7B; Messiaen *et al.*, 1993; Sumiati, 1994) attached by a common base.

Lateral shoots proceed to initiate and develop leaves in the same way that the main apex does. This habit of shallot branching differs from that of the bulb onion, where often 13 or more leaves are produced prior to the first doubling (Eto, 1956; Rabinowitch, 1979). Additionally, in the bulb onion, each lateral growing point develops into a dormant adventitious bud between the original shoot apex and the youngest leaf (Brewster, 1994) whereas, in shallot, lateral shoots and inflorescences develop simultaneously with the main apex and the primary umbel (Krontal *et al.*, 1998).

4.3.2 Environmental effects

Cohat (1982) and Cohat and Tromeur (1986) suggested that the main factors contributing to the size of the cluster are the weight of the grandmother propagule and the field conditions during the development of the mother bulb. They stated that, in vegetatively propagated shallot, the initials of the following season's laterals already exist in the mature propagules and that the number of sets per cluster cannot be manipulated by storage temperature or planting distance (see also Messiaen *et al.*, 1993). Abbey and Fordham (1997) working with tropical hybrid shallots, cv. 'Tropix', reported that water stress had no effect on the number of bulbils per plant, and Abbey *et al.* (1998) provided evidence that, in 4-week-old 'Tropix' plants, but not in older ones, the number of lateral buds increased with the severity of damage to the foliage, but the number of bulbils per plant was significantly lower. It would be interesting to examine the effect of defoliation treatment on the number of lateral shoots in the offspring.

Storage treatments had no effect on the number of laterals in tropical shallots. However, in contrast with this finding, growing temperatures markedly affected vegetative development (Krontal *et al.*, 2000). Zaharah *et al.* (1994) reported that, in Malaysia, fertilization with palm-oil mill effluents resulted in an increase in the number of shallot bulbs per unit area and in their size. They did not, however, indicate whether this increase in number of bulbs resulted from an increase in the number of laterals or if more of the bulbils produced reached a marketable size.

Regardless of preplanting treatment, all plants of accession no. 66–1004 grown at 17/9°C (day/night) in a phytotron in Israel developed smaller leaf mass and lower fresh weight than those at 29/21°C. While the number of laterals per plant under the latter regime was significantly higher in number (averaging 25–29), they were also smaller and thinner than those produced at 17/9°C (five to seven in number) (Fig. 17.4; Krontal *et al.*, 2000). G. Neupane (Wye College, UK, 1993, unpublished data) studied the effect of temperature and day length on cvs 'G102' and 'G106' (Bejo, Holland) as well as the effect of plant density on bulbing. High-density (nine plants per pot) bulbing distinctly promoted earliness as compared with less crowded plants (L. Currah, UK, 2001, personal communication).

4.4 Florogenesis and flowering physiology

4.4.1 Transition from vegetative to generative stage: physiological age

When grown from seed, the transition of accession no. 66–1004 from the vegetative to the reproductive phase occurred at the

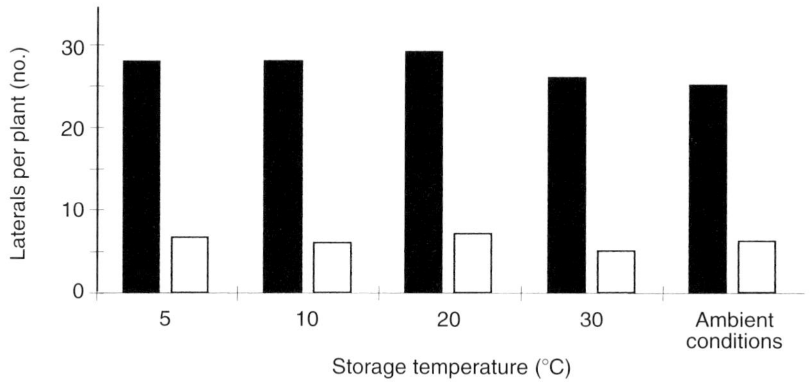

Fig. 17.4. Effect of storage and growth temperature on number of laterals per plant in shallot accession 66–1004. Following storage at ambient conditions or at constant temperature of 5, 10, 20 or 30°C for 28 days, the clusters were separated into single sets. Each set was planted in a single pot and placed in a phytotron chamber at 17/9 (□) or 29/21°C (■), day/night, respectively, 10 h photoperiod. (From Krontal *et al.*, 2000, with permission.)

physiological age of six leaves (including leaf primordia) (Fig. 17.2; Krontal *et al.*, 1998). In comparison, a critical physiological age of 12 (range 10–14) leaves (including leaf buds) has been reported as being required for most of the studied bulb-onion genotypes before they become receptive to floral induction, and some bolting-resistant cultivars, e.g. 'Senshyu Semi-Globe Yellow', have a markedly longer juvenile period (for detailed reviews see Rabinowitch, 1985, 1990; Brewster, 1987, 1994; Kamenetsky and Rabinowitch, Chapter 2, this volume).

In our studies, growth temperatures had a marked effect on the transition of seedlings from the juvenile to the generative phase. At 17/9°C (day/night) in a phytotron, shallot plants of accession no. 66–1004 remained vegetative for 160 days, during which the seedlings developed a large vegetative mass of 13–16 leaves before the first inflorescence became visible (Fig. 17.5A,C). All plants flowered and produced normal scapes. At 26/18°C, the first inflorescence became visible only 180 days after emergence, but only half of the seedling population with 18 or more leaves actually bloomed (Fig. 17.5B). In this phytotron chamber, two out of ten plants developed normal inflorescences, three produced malformed umbels (Fig. 17.5D,E), and five remained vegetative throughout. Plants exposed to high temperatures of 29/21°C remained vegetative and withered and died after producing about 20 leaves per propagule (Krontal *et al.*, 2000).

Sowing date and field temperatures also affected flowering time and the number of bolted seedlings of accession no. 66–1004 (Fig. 17.6; Krontal *et al.*, 2000). In a field experiment in Israel, early-sown plants (1 and 15 October) had their first outwardly visible scapes at the end of January, 50% bloom was recorded in the middle of February and all plants produced more than three scapes per cluster. Plants from the latest sowing (30 October) reached 50% bloom only in the middle of March, and maximum blooming rate was 55%, with only one scape per bulb cluster. In all cases, tops were down in May and bulb ripening occurred immediately thereafter. It was obvious that, as in bulb onion, flowering in field-grown shallot was initiated by the cold of autumn and winter (Krontal *et al.*, 2000). As autumn progresses, temperatures become gradually lower, light intensity weakens (due to cloudiness) and days become shorter, reaching a minimum in December. Early-sown shallots were therefore exposed to more favourable growth conditions and thus had a faster growth rate than those from the latest

Fig. 17. 5. Effect of growth temperature on shallot development. (A) Flowering shallot. Plants were grown in a 17/9°C (day/night) phytotron chamber. (B) Shallot plants in a 29/21°C (day/night) phytotron chamber. Note the lavish foliage and the poor flowering. (C) A normal shallot inflorescence at the early stages of bloom. Growth conditions as in (A). (D) A mixed inflorescence containing flowers and topsets. Growth conditions as in (B). (E) A completely transformed inflorescence. The head consists solely of topsets. Growth conditions as in (B).

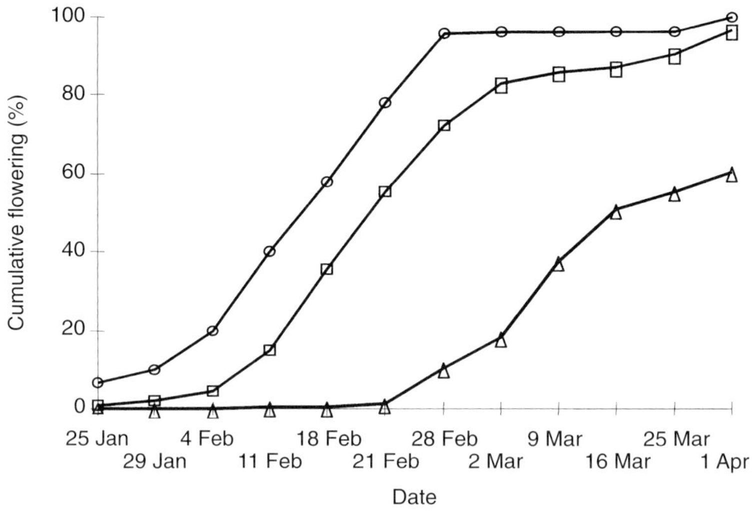

Fig. 17.6. Effect of sowing date on floral development in shallot accession no. 66–1004. Seeds were sown on 1 (O), 15 (□) and 30 (Δ) October at the experimental station in Rehovot, Israel. The first scape/cluster was counted; data points show cumulative percentages of flowering sets per cluster. (From Krontal *et al.*, 2000, with permission.)

sowing date. Consequently, the former accumulated more mass, and were able to produce earlier and more auxiliary buds, lateral shoots and inflorescences than the ones from the late-sown plants.

The fact that, like the bulb onion, post-juvenile shallots exhibit an increase in sensitivity to cold induction may explain the high percentage of bolting in plants from early sowing/planting and/or from large bulbs, and consequently their higher losses in terms of bulb yield and quality. Alternatively, when aimed at hybrid seed production, large sets and early planting are preferred, and optimal 'nicking' (synchronization of flowering) between male-sterile plants and pollinators is therefore possible by proper manipulation of planting time and set size.

4.4.2 Developmental morphology and florogenesis

The few studies on the flowering process in shallot indicate a morphological resemblance to the bulb onion (Messiaen *et al.*, 1993; Krontal *et al.*, 1998). During the transition of the seedlings from vegetative to reproductive stage, there is a change from monopodial to sympodial growth, as described by De Mason (1990) in the bulb onion. The reproductive transition of shallot, however, has no inhibitory effect on further development of auxiliary vegetative meristems.

The shallot inflorescence is an umbel-like flower arrangement, made of many flower clusters, all of which arise from a common meristem. Krontal *et al.* (1998) studied shallot floral development in detail, and showed that, after the transition of the apical meristem from vegetative to reproductive, the spathe develops quickly to envelop the apex. Later, the apical meristem of the shallot swells to become hemispherical and, following the initial elongation of the scape, the apex subdivides into four centres of differentiation (Fig. 17.1C, D), in which the floral primordia develop unevenly in a helical order. De Mason (1990) stated that the onion scape elongates and extends beyond the leaves only after the inflorescence is formed. In shallot, however, several flower clusters can be clearly distinguished in each of the four centres of differentiation when the flower stalk reaches only 50 mm in length. At this time, membranous leaf-like bracts appear both in the centre and at the periphery of the inflorescence (Fig. 17.1D). Further development is rather slow and organ differentiation begins only when the scape is 75–100 mm long. At this stage, the spathe is 12–15 mm long and the inflorescence is *c.* 2 mm in diameter; older flower primordia are clearly visible to the naked eye, while younger ones still appear as undifferentiated meristematic domes on the base of the inflorescence (Fig. 17.1E). Each centre of development is covered by thin membranous bracts and contains six to seven developing flower clusters, each containing five to ten flower buds, arranged in a spiral order within the cluster.

In the bulb onion, the protandrous flowers consist of five whorls of three organs each. In order of initiation and differentiation, these are the outer and inner perianth lobes, the outer and inner stamens and three carpels united into one pistil. The perianth lobe and the subtended stamen appear to arise simultaneously from a single primordium (Jones and Emsweller, 1936; Esau, 1965; De Mason, 1990). Floral morphology in shallot is very similar; however, no clear direction of primordia differentiation in individual shallot flowers could be distinguished.

When the shallot scape reaches 15 cm in length, the originally flat inflorescence becomes spherical. The number and size of the differentiated flowers increase with time, and new undifferentiated domes become visible at the base of the inflorescence (Fig. 17.1F). Flower formation is almost complete when the stalk is 30 cm long. At this time, pedicels of older flowers are 1.6 mm long and the gynoecium segments are visible in most flowers. The carpels develop as three protruding areas within the inner anthers, and they meet at the heart of the flowers to form the trilocular ovary. Simultaneously, the anthers – 0.3 mm in diameter – reach their typical shape and the apex of the fused ovary develops into a style, which is too short to be functional when the flower first opens and continues to elongate after anthesis, becoming receptive a few days later.

Under Israeli conditions, the flower stalk of the shallots studied reaches a length of 35–50 cm before the spathe opens, whereas onion scapes at a similar physiological stage reach 1–1.8 m in length. At blooming, all pedicels become almost equal in length; hence flower clusters can no longer be recognized in the shallot inflorescence.

4.4.3 Environmental regulation of flowering processes

Hartsema (1961) reported that floral initiation and development occur only rarely in stored shallot bulbs (no name) and take place mainly in growing plants. However, Messiaen *et al.* (1993) reported that storage of mother bulbs of cv. 'Half-long Jersey' at low temperatures, but not at subzero, promoted bolting and that long storage at high temperatures suppressed bolting, as in the bulb onion (for review, see Rabinowitch, 1990; Kamenetsky and Rabinowitch, Chapter 2, this volume). Flowering may be induced in bulb onion (Rabinowitch, 1990) and in shallot (Krontal *et al.*, 1998, 2000; Kamenetsky and Rabinowitch, Chapter 2, this volume) both during the growing period and during storage of bulbs, provided the plants have passed the juvenile phase. Following cold induction in storage, high temperatures during the growing period may inhibit floral development in both onion (Rabinowitch, 1990) and shallot (Krontal *et al.*, 2000) plants and cause the reversal of the process, whereas low and intermediate temperatures during both phases affect the time of flowering and promote the completion of the process. In Israel, storage at 5, 7 and 10°C resulted in fast bolting in the field and longer cold storage was more effective in promoting bolting, whereas storage at 30°C resulted in delayed scape emergence (as compared with plants from low and intermediate storage temperatures) (Fig. 17.7). Most plants stored at 5–20°C went on to bloom within the first 4–6 weeks after planting in the phytotron at 17/9°C (day/night), but high temperatures during growth (29/21°C) suppressed the inflorescences already initiated during storage (Fig. 17.8). In France, long storage of cv.

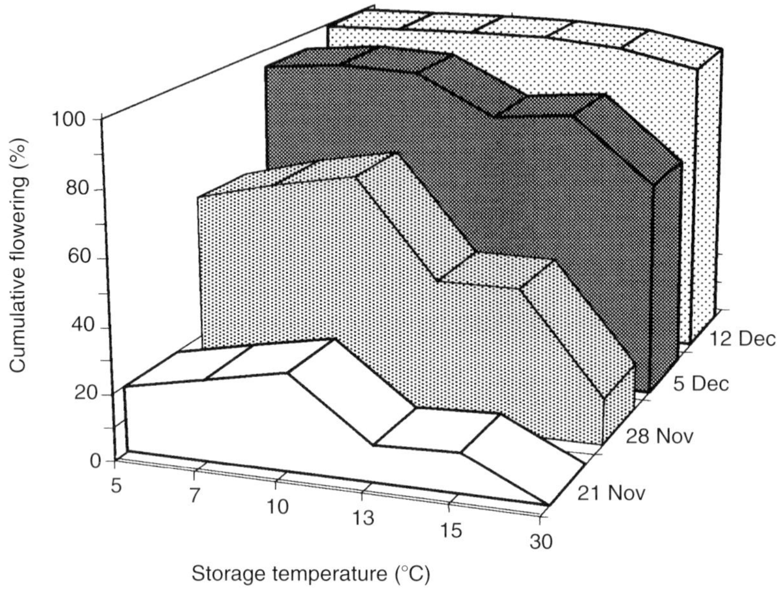

Fig. 17.7. Effect of storage at 5, 7, 10, 13, 15 or 30°C for 12 and 24 days on floral development of large shallot in the field in Israel (data are means from two accessions and from 12 and 24 days storage) (from Krontal *et al.*, 2000, with permission).

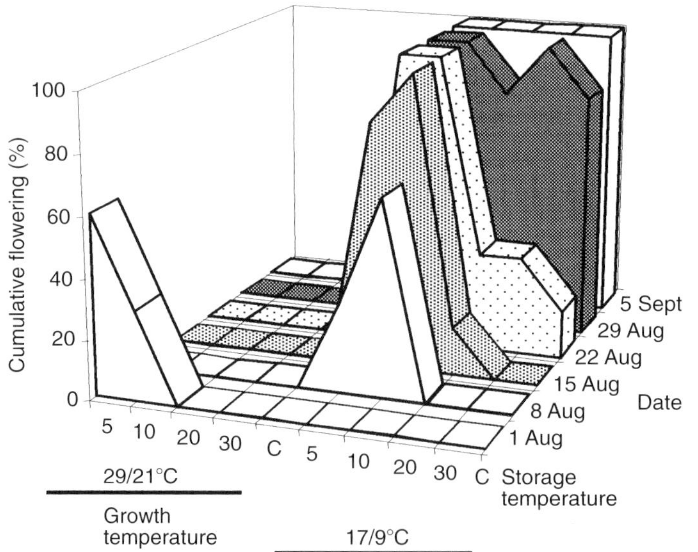

Fig. 17.8. Effect of storage and growth temperature on floral development of shallot. Large bulbs (> 20 mm) were stored at ambient conditions or at 5, 10, 20 and 30°C for 28 days and planted in the phytotron at 17/9 and 29/21°C, day/night, respectively, day length 10 h. (From Krontal *et al.*, 2000, with permission.)

'Mikor' at 30°C resulted in lower flowering percentage as compared with the control (Cottignies *et al.*, 1997).

Similar results were obtained in the tropics. Cool storage for 70–90 days at 5–15°C induced flowering in Ghanaian shallots (Sinnadurai and Amuti, 1971), and similar results were obtained in Indonesia with local clones after 4 weeks at 4–9°C (S. Prasodjo, cited by Currah and Proctor, 1990).

It is clear that storage conditions and bulb size (Fig. 17.9) affect both the time and amount of bolting and that genotypes vary significantly in their response to cold induction (see Section 4.4.4, below). Photoperiod may also be involved in floral development and long days are essential for floral scape elongation (Abbey and Fordham, 1998). Cold temperatures promoted and high temperatures delayed flowering, with an optimum between 5 and 10°C, in both field and phytotron experiments in Israel (Krontal *et al.*, 2000). Plants from large bulbs produced their inflorescences earlier and faster than those from smaller ones.

For bulb or seed production, respectively, either reduced bolting or lavish flowering are the aims. Environmental effects may be used, in combination with propagule size and genotype, to reduce the risk of bolting in the shallot bulb-production field and to promote and time flowering (nicking) in the fields of hybrid-seed production.

4.4.4 Genetic variability

The three shallot accessions (accession no. 66–1004 from Thailand and accession nos 977–1011 and 977–1009 from Nepal) tested by Krontal *et al.* (2000) in Israel responded differently to cold induction in the field. Plants of 977–1011 did not bloom, whereas plants of 977–1009 and 64–1004 bolted when exposed to cold treatment. In Brittany, France, the two cvs 'Mikor' and 'Jermor' varied markedly in their response

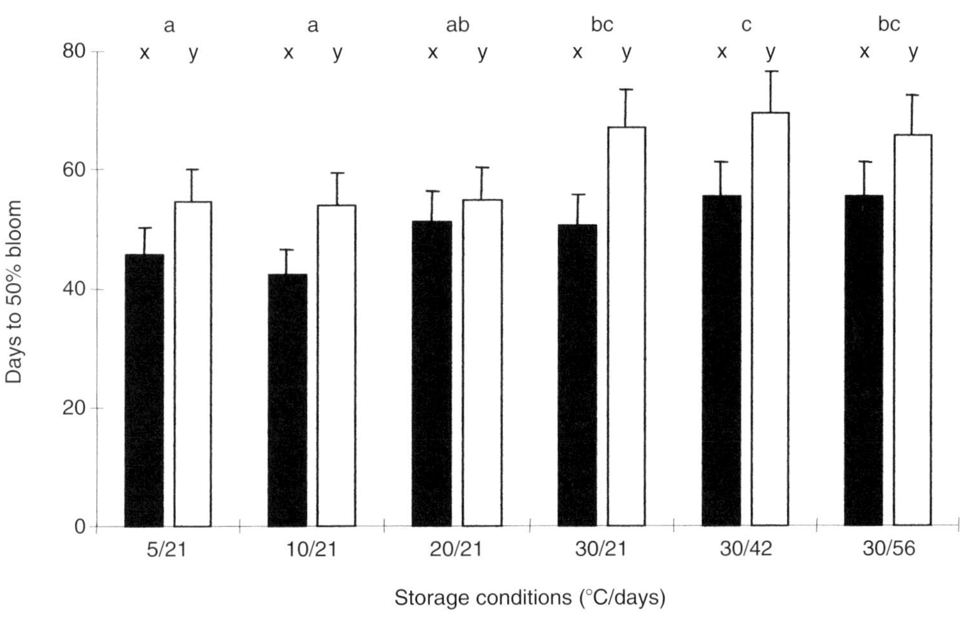

Fig. 17.9. Effect of shallot propagule size on the time from planting to bloom. Small (□) and large (■) shallot sets were incubated at 5, 10, 20 and 30°C for 21 days, and at 30°C for 42 and 56 days, prior to planting in the field. Comparisons were made between large and small propagules (x, y) and among storage conditions (a, b, c). Means followed by the same letters are not different, $P = 0.05$, Duncan's multiple range test. (From Krontal *et al.*, 2000, with permission.)

to storage and field temperatures. When 'Jermor' bulbs were planted in mid-October after storage either at ambient conditions or at 30°C, they very rarely flowered. Under the same conditions, 'Mikor' flowered abundantly and each cluster bore numerous floral buds (Cottignies *et al.*, 1997). The marked genotypic differences indicate that there is room for selection for genotypes less susceptible to bolting. Since lateral shoots and scapes compete for the same assimilates, it is expected that breeders will consciously select against easy-bolting genotypes, much as they do with the bulb onion.

4.4.5 A comparison between the shallot and the bulb onion

The shallots studied by our group differed considerably from the bulb onion in a number of developmental processes related to both branching and florogenesis, but also exhibited some great similarities. The main points are as follows.

1. Physiological age to branching: bulb onions are selected for single centres and minimum doubling (Rabinowitch, 1979), whereas shallots are distinguished by the production of lateral shoots; doubling in bulb-onion and shallot seedlings begins at the physiological ages of 13 and three leaves, respectively.
2. Minimum physiological age for floral initiation: the minimum length of the juvenile phase in bulb onion and shallot lasts for 10–14 and six leaves, respectively, and in both plants flowering depends on cold induction, and sensitivity increases with physiological age.
3. Plant mass: in both shallot and bulb onion, plants from larger sets flower more readily than those from small ones (Brewster, 1994; Krontal *et al.*, 2000).
4. Cold induction: both onion and shallot require cold induction for flowering; provided they have passed the juvenile stage, the two crop plants respond to cold induction both in storage and in the field, with an optimum between 5 and 10°C.
5. Heat inhibitory effect: high temperatures in the production field may suppress the inflorescence already initiated during storage of the bulb onion (Thompson and Smith, 1938; Heath, 1943a, b; Heath and Mathur, 1944; Aoba, 1960) and in shallot (Krontal *et al.*, 2000).
6. Inflorescence development: in the bulb onion, 'only when the inflorescence is formed does the true stem elongate through the leaf sheaths and extend beyond the other parts of the plant' (De Mason, 1990) whereas, in shallot, flower differentiation commences only after the initial elongation of the scape, and flower development is complete only when the stalk reaches 30 cm in length.
7. Individual flower: the floral morphology of onion and shallot is rather similar; however, De Mason (1990) noted that the outer whorl of onion flowers arises in a clockwise direction, while the inner one arises in an anticlockwise direction; no clear direction of primordia differentiation in individual shallot flowers could be distinguished (Krontal *et al.*, 1998).
8. Auxiliary buds: in the bulb onion, inflorescence development is rarely accompanied by the active growth of new auxiliary shoots; in shallot, simultaneous development of the terminal (principal) inflorescence with auxiliary (lateral) vegetative shoots and additional inflorescences is rather common.

Regardless of the above differences, the similarities in morphology, in many physiological processes, in cytology and at molecular level reaffirm earlier conclusions that the two plants belong to the same botanical species (Kalkmann, 1984b; Hanelt, 1990; Arifin and Okubo, 1996; Le Thierry D'Ennequin *et al.*, 1996; Maaß, 1996; Klaas, 1998; see also Fritsch and Friesen, Chapter 1, Havey, Chapter 3, and Klaas and Friesen, Chapter 8, this volume).

4.5 Bulb development

Bulb formation in shallot occurs in response to long photoperiods and relatively high temperatures (Jenkins, 1954), and different cultivars can be distinguished by the minimum day length they need to induce them to bulb (Arifin and Okubo, 1996). The young shallot foliage responds to the photoperiodic stimulus (Abbey *et al.*, 1998) and the induction of bulb development does not involve cell division. The process commences with the elongation and swelling of both leaf sheaths and young leaf buds of each individual lateral shoot, probably due to cell expansion, as in the bulb onion (Brewster, 1990, 1994; De Mason, 1990). In the bulb onion, the bulbing process results in inhibition of further leaf initiation and degeneration of the blades of the developing leafbuds, thus converting the latter into storage scales or 'true scales' (Heath and Holdsworth, 1948). It is assumed that similar processes occur in the bulbing shallot. The swelling of leaf-bases and scales soon becomes visible, and bulbing is typically characterized by an increase in the bulbing ratio, i.e. the ratio between maximum bulb and minimum pseudostem diameter. In shallot, due to its branching habit and the presence of clustered sets, this parameter is easier to assess than the 'leaf ratio' (Heath and Hollies, 1965; Brewster, 1990, 1994), which is frequently used by physiologists to determine the onset of onion bulbing. As each of the clustered sets increases in size, the carbohydrate reserves of the outermost leaf-bases are translocated to the inner developing scales, and the former develop into dry skins with a range of clonal-dependent colours from reddish-purple to brown/yellow (Dahlen, 1995; Arifin *et al.*, 1999a) or white (selections available in our collection). The accelerated transport of assimilates from the blades to the leaf-bases is accompanied by the senescence of the shoot-borne roots, and the reduced water supply results in loss of leaf turgidity. Consequently, the pseudostem of each of the clustered sets becomes hollow and loses its physical strength, due to the arrest of development of new leaves and loss of leaf turgor. The outcome of the two processes is the collapse of the green foliage. However, translocation of assimilates from the blades to the sheaths and storage scales continues as long as the leaf blades are still green. Leaves wither from top to bottom and from the outside (oldest) towards the centre (youngest). The process ends with the senescence and death of the youngest green leaf and entrance into the dormant phase.

A single shallot bulb contains several shoot initials (Colour Plate 7B, Fig. 17.2) that resemble those of doubled onions (Colour Plate 7B; Rabinowitch, 1979), and each bulblet is covered with one to three protective skins. Dormancy lasts between 2.5 and 4 months (Sinnadurai and Amuti, 1971; Currah and Proctor, 1990; Messiaen *et al.*, 1993) or 5 months at 27–32°C (van der Meer, 1990, in Grubben, 1994). Messiaen *et al.* (1993) reported that storage of the French shallot cultivar 'Half-long Jersey' at −2°C resulted in longer dormancy than that at room temperature or at constant 9 or 30°C, and Grubben (1994) stated that storage of planting material at high temperatures in Indonesia promoted sprouting after replanting. Little is published of the developmental response of grey shallot to storage temperatures. However, Messiaen *et al.* (1993) reported that the bulbs remain dormant for a longer period at 7 than at 12°C.

5. Agronomy

Vegetative propagation has until now been dominant in shallot culture throughout the world (Jones and Mann, 1963; Currah and Proctor, 1990; Messiaen *et al.*, 1993; Wietsma *et al.*, 1998). When grown from sets, the shallot-growing season is relatively short, thus enabling production where onion from seed cannot produce economic yields of commercially acceptable sized bulbs. However, when grown from seed, hybrid shallots with strong heterosis have a fast growth rate and bear high yields already after 3–4 months of growth (H.D. Rabinowitch and R. Kamenetsky, personal experience).

Management and agrotechniques for shallot in the tropics were dealt with in detail by Currah and Proctor (1990), Abbey and Fordham (1998); and Wietsema *et al.* (1998). In more temperate lands, similar practices are used in Europe, the USA and Argentina, where sets are transplanted on raised beds at 25–40 plants m^{-2}. Yields of vegetatively propagated shallots range between 5 and 30 t ha^{-1} in Indonesia (Subijanto, 1988) and 20 and 30 t ha^{-1} in Holland and France. Our experience with direct seeding or transplanting 1.5-month-old seedlings of our F_1 hybrids gave yield means of 30–50 t ha^{-1} in commercial plots in Israel. High yields of shallots grown from seeds were also obtained in Belgium (Vanparys, 1999a, b, 2000) and Ethiopia (open-pollinated selections: Aklilu, 1998).

Our results show that cultivation at high temperature leads to an increased number of side-shoots as compared with cultivation in mild conditions (Fig. 17.4). It is therefore tempting to produce sets for propagation in areas where temperatures are high. However, as for garlic, the risk of virus contamination is markedly higher in warm conditions and thus such a practice should be avoided.

In Indonesia, shallot regrowth after planting is enhanced by cutting the tops off the bulbs. This can save the farmer time, as dormancy is a factor that prevents shallot crops from being grown in rapid succession and long storage also increases the risks of losses from pests and from desiccation (Arifin *et al.*, 1999b).

6. Storage

Dry shallot bulbs are sold either fresh or from storage. Shallot clones vary considerably in storage life, with a range of 2 to 9 months (Currah and Proctor, 1990), and storage temperature and genetic traits are the main factors that influence storage life (Currah and Proctor, 1990; Messiaen *et al.*, 1993; Grubben, 1994). In Thailand, high N level in the stored bulbs was found to be associated with short keeping, with premature harvest when carried out before leaf wilting and with poor postharvest handling (Ruaysoongnern, 1994). Storage diseases, early sprouting and shrivelling seem to be the main limiting factors for long keeping of shallots in tropical and subtropical countries. Bulb onions and most shallots store well at low (~0°C) and high (roughly 25–~30°C) temperatures (Komochi, 1990; Krontal *et al.*, 2000; Gubb and MacTavish, Chapter 10, this volume). However, shallots can be stored for long periods under ambient conditions in the tropics, over 5 months in some trials (Currah and Proctor, 1990; van der Meer, 1990, in Grubben, 1994). Storage in shaded heaps in the field (Colour Plate 7C) or in open sheds (Colour Plate 7D) under ambient conditions is common in the tropics, in Israel and in other places. In temperate climates, such as in Holland and France, cold storage (at 1–2°C) of long-keeping cultivars is common for extended periods, with minimum losses (Cohat and Le Nard, 1998). Brice *et al.* (1997) compiled recommendations for methods of storage and their selection for onions, which can also be applied for improving shallot storage conditions.

7. Diseases and Pests

Loss of shallot yield from pests and diseases is common all over the world, and chemical treatment is the major means currently used to reduce damage (Anon., 1986; Suhardi, 1996). However, good agricultural practices can be used to partially control losses. Practices that are essential for high-quality long-keeping yields include crop rotation; drip irrigation (which is preferred over sprinkler irrigation to maintain low air humidity); proper spacing, to allow free passage of air so as to reduce the relative humidity of the air; proper harvesting and curing practices (for details on bacterial contamination at harvest, see Brewster, 1994; Mark *et al.*, Chapter 11, this volume); well-ventilated (Colour Plate 7C,D) or cold storage; and proper sanitation. No information is published on resistance/tolerance to pests in shallots, but some landraces show better field tolerance to some foliage diseases than the bulb onion, and differences in resistance between cultivars are noticeable (Currah and Proctor, 1990).

Shallots are susceptible to a number of airborne and soil-borne fungi, as well as to insects, nematodes, bacteria and viruses (Messiaen *et al.*, 1993, 1994). The main diseases and pests are: anthracnose (*Colletotrichum gloeosporioides*), basal rot

(*Fusarium oxysporum*), downy mildew (*Peronospora destructor*), moulds (*Aspergillus niger*, *Penicillium corymbiferum*, *Penicillium cyclopium*), neck rot (*Botrytis allii*), onion blast (*Botrytis squamosa*), pink root (*Pyrenochaeta terrestris*), purple blotch (*Alternaria porri*), smudge (*Colletotrichum circinans*), white rot (*Sclerotium cepivorum*), *Sclerotium rolfsii*, nematodes (*Ditylenchus dipsaci*), thrips (*Thrips tabaci*), beet armyworm (*Spodoptera exigua*) and other *Spodoptera* sp. caterpillars, as well as a number of virus diseases (Jones and Mann, 1963; Currah and Proctor, 1990; Messiaen *et al.*, 1993; Brewster, 1994; Grubben, 1994; Wietsma *et al.*, 1998; Kuruppu, 1999). Virus infection results in heavy losses (van Dijk, 1992). In comparison to *A. cepa* shallots, the grey shallot selected from 'Grise de la Drôme', cv. 'Griselle' is highly susceptible to white rot and to *Botrytis* spp. and downy mildew (*Peronospora destructor*) (Messiaen *et al.*, 1993, 1994). Tolerance/resistance to purple blotch was reported for red shallot and the bulb onion 'Red Creole' from Ethiopia (Currah and Proctor, 1990). Crosses between Jersey-type shallots and *A. roylei*, which could confer resistance to downy mildew and *B. squamosa*, are being made experimentally in France (Cohat and Le Nard, 1998).

The vegetative propagation habit of the shallot crop perpetuates diseases and pests. Meristem-tip culture or seed propagation provide the only means to free the crop from viruses (e.g. Lapitan *et al.*, 1991; Aklilu, 1998), and hot-water treatment is used to free shallot sets from nematodes (Anon., 1972; Bridge, 1975).

The cross-pollinated, vegetatively propagated shallot is considered a minor *Allium* crop. Until recently (when seed-propagated hybrid cultivars gained momentum), little effort was invested in breeding shallot, and less so for resistance to diseases specific to tropical conditions. However, recently, a Dutch–Indonesian group (Wietsma *et al.*, 1998) launched new breeding work, using classical and modern technologies for the introduction of genes for tolerance/resistance to anthracnose (sources for tolerance identified in a local Indonesian clone 'Sumenep') (Suhardi, 1993). Wietsma *et al.* (1998) screened a large gene pool of wild relatives, and found sources of resistance in *A. altaicum*, *A. fistulosum* and *A. galanthum*. The tolerance of 'Sumenep', however, was not confirmed (Galvan *et al.*, 1997).

8. Abiotic Stress

Shallot plants in the field may suffer from too high light intensity and saline conditions (Ko *et al.*, 1993), from high temperatures (Krontal *et al.*, 2000), from water stress (Abbey and Fordham, 1997, 1998) and from mechanical damage to the foliage (Abbey *et al.*, 1998). In general, any kind of stress in shallot usually shows up as tip burn or leaf-tip dieback (L. Currah, UK, 2001, personal communication). Propagule size, the number of lateral buds and planting date affect the number of sets per plant, size and yield (Cohat, 1982; Cohat and Tromeur, 1986; Messiaen *et al.*, 1993; Ryu *et al.*, 1998; Suh and Ryu, 1998).

9. Composition and Quality

Standards for quality grades of shallot bulbs were issued by the US Department of Agriculture (USDA) (Anon., 1946). Shallot bulbs are usually smaller and more highly flavoured than those of the single-hearted bulb onion. Shallots contain higher levels of fats and soluble solids, including sugars, than bulb onion (16–33% vs. 7–15% dry weight, respectively) (Currah and Proctor, 1990; Messiaen *et al.*, 1993) which, together with sulphur-containing compounds, make shallot an essential component in gourmet cooking. In cv. 'Griselle', the dry-matter content is as high as 30% (Cohat and Le Nard, 1998).

The dry matter of shallot consists of 70–85% carbohydrates, mainly fructans, glucose, fructose and sucrose. As in the bulb onion, cell-wall components, such as cellulose and pectins, contribute to 10–15% of the carbohydrate fraction (Messiaen *et al.*, 1993). However, dry-matter content was lower and pyruvate content was higher in seed-grown vs. set-propagated shallot (Bufler, 1998).

Little research has been done on shallot pigmentation. It is reasonable to assume that, much like the bulb onion, the red shallot contains anthocyanins (glucosides of cyanidin) (Joslyn and Peterson, 1958) and the yellow colour is largely of the flavonol quercetin (Kuroda and Umeda, 1951; Hermann, 1958). Recent studies in Japan by Arifin *et al.* (1999a) on Indonesian shallots confirmed this.

10. Nutraceutical Traits

For a detailed review on the health benefits of shallot and other alliums, see the review by Keusgen (Chapter 15, this volume). Beneficial sulphur-containing compounds are common in most alliums (Lancaster and Boland, 1990), including shallots (Messiaen *et al.*, 1993). It is therefore not surprising that shallot extracts exhibit therapeutic features. Feeding rabbits with freshly prepared shallot extract reduced the number of abnormal-shaped erythrocytes caused by high levels of cholesterol in the blood (Tappayuthpijarn *et al.*, 1989). Additionally, an antileukaemic substance was described for shallots (Caldes and Prescott, 1973) and Gram-positive bacteria were inhibited by shallot juice (Dankert *et al.*, 1979). Shallots may also contain high levels of antibiotic factors, due to their high dry matter and pungency (Currah and Proctor, 1990).

11. Conclusions

There are many strong morphological and developmental similarities between the bulb onion and the shallot. Nevertheless, a number of distinct physiological and developmental differences have been observed between the two alliaceous crops.

Regardless of its minor economic importance, it is expected that, in the near future, breakthroughs in molecular biology and genetic engineering of the bulb onion will markedly contribute to shallot improvement, due to the close relationship between the two crops. In addition, it seems likely that seed propagation will become increasingly popular and modern F_1 hybrids with improved performance will replace the traditional landraces, except perhaps in France, where growers have strongly defended their national vegetatively produced cultivars as being the only genuine horticultural shallots (L. Currah, UK, 2000, personal communication). However, even in France, the future potential of seed-grown shallots is now being recognized (Cohat and Le Nard, 1998).

While local clones are still available, it is important to mobilize international resources for a systematic collection of these genetic treasures. The available diversity is important not only for the secondary crop of shallot but also for the most important member of the *A. cepa* complex, the bulb onion. Long storage, tolerance to some biotic and abiotic stresses, nutraceutical and culinarily advantageous characteristics and other quality traits (skins, pigmentation and more) are only a few of the traits that could be used to improve the bulb onion with no need to cross species barriers.

References

Abbey, L. and Fordham, R. (1997) Growth and yield sensitivity of pre-bulbing and bulbing shallots to water-stress. *Crop Research (India)* 14, 307–313.

Abbey, L. and Fordham, R. (1998) Abiotic stress affects shallot growth performance. *Crop Research (India)* 16, 66–69.

Abbey, L., Kanton, R. and Braimah, H. (1998) Susceptibility of shallots to the timing and severity of leaf damage. *Journal of Horticultural Science and Biotechnology* 73, 803–805.

Adiyoga, W. and Soetiarso, T.A. (1997) The study of comparative advantage and economic incentive in shallot production. *Indonesian Journal of Horticulture* 7, 614–621.

Aklilu, S. (1998) Prospects of growing shallots from true seed under irrigation in the Rift Valley. *AgriTopia* 13 (3/4), 8–9.

Anon. (1946) *United States Standards for Grades of Bunched Shallots*. USDA, Washington, DC, pp. 189–191.

Anon. (1972) *Hot Water Treatment of Plant Material.* Bulletin No. 201, UK Ministry of Agriculture, Fisheries and Food, HMSO, London, 46 pp.
Anon. (1986) *Pest Control in Tropical Onions.* Tropical Development and Research Institute, London, 109 pp.
Aoba, T. (1960) The influence of the storage temperature for onion bulbs on their seed production. *Journal of the Japanese Society for Horticultural Science* 29, 135–141 (in Japanese, with English summary).
Arifin, N.S. and Okubo, H. (1996) Geographical distribution of allozyme patterns in shallot (*Allium cepa* var. *ascalonicum* Backer) and wakegi onion (*A.* × *wakegi*, Araki). *Euphytica* 91, 305–313.
Arifin, N.S., Miyajima, I. and Okubo, H. (1999a) Variation in pigments in the bulbs of shallot *Allium cepa* var. *ascalonicum* and *Allium* × *wakegi*. *Journal of the Faculty of Agriculture, Kyushu University* 43, 303–308.
Arifin, N.S., Okubo, H. and Miho, N. (1999b) Dormancy in shallot (*Allium cepa ascalonicum*) and *Allium* × *wakegi* bulbs and its breaking by scale cutting. *Journal of the Faculty of Agriculture, Kyushu University* 43, 309–315.
Atkin, J.D. (1953) Genetic and cytological studies of the hybrid *Allium cepa* × *A. ascalonicum*. PhD thesis, University of California, Davis, California, USA.
Aura, K. (1963) Studies of the vegetatively propagated onions cultivated in Finland, with special reference to flowering and storage. *Annales Agriculturae Fenniae* 2 (Suppl. 5), 1–74.
Brewster, J.L. (1987) Vernalization of the onion – a quantitative approach. In: Atherton, J.G. (ed.) *The Manipulation of Flowering*. Butterworths, London, pp. 171–183.
Brewster, J.L. (1990) Physiology of crop growth and bulbing. In: Rabinowitch, H.D. and Brewster, J.L. (eds) *Onions and Allied Crops*, Vol. I. *Botany, Physiology, and Genetics*. CRC Press, Boca Raton, Florida, pp. 53–88.
Brewster, J.L. (1994) *Onions and Other Vegetable Alliums*. CAB International, Wallingford, UK, 236 pp.
Brice, J., Currah, L., Malins, A. and Bancroft, R. (1997) *Onion Storage in the Tropics. A Practical Guide to Methods of Storage and Their Selection*. Natural Resources Institute, University of Greenwich, Chatham, UK, 120 pp.
Bridge, J. (1975) Hot water treatment to control plant parasitic nematodes of tropical crops. *Mededelingen van de Faculteit Landbouwwetenschappen, Rijksuniversiteit Gent* 40, 249–259.
Bufler, G. (1998) Quality comparison for seed and plant shallot cultivars. *Gemüse (München)* 34, 460–462 (in German).
Caldes, G. and Prescott, B. (1973) A potential antileukemic substance present in *Allium ascalonicum*. *Planta Medica* 23, 99–100.
Chovelon, V., Leroux, J.P. and Doré, C. (1989) Disease and pest control in garlic and shallot: meristem culture and virus free varieties. In: *Cinquantenaire de la culture* in vitro *chez les végétaux. Xme Colloque de la Section Française de l'International Association for Plant Tissue Culture. Station de Génétique et d'Amélioration des Plantes, Centre de Recherches Agronomiques, Versailles, France, 24–25 October 1989.* Colloques de l'INRA No. 51, Editions INRA, Paris, pp. 141–150.
Cohat, J. (1982) Influence du calibre des bulbes de semence d'échalote sur leur taux de multiplication et leur rendement. *Pépiniéristes, Horticulteurs, Maraîchers, Revue Horticole* 231, 21–24.
Cohat, J. and Le Nard, M. (1998) L'échalote en France: variétés et amélioration génétique. *Pépiniéristes, Horticulteurs, Maraîchers, Revue Horticole* 392, 53–56.
Cohat, J. and Tromeur, C. (1986) Influence du poids et de la densité de plantation des bulbes d'échalote sur les caractéristiques de la récolte et la prolificité des bulbes-fils. *Agronomie* 6, 85–90.
Cottignies, A., Cohat, J., Le Nard, M. and Hourmant, A. (1997) Cycle cultural et floraison de l'échalote, *Allium cepa* L. var. *aggregatum* (cvs Mikor et Jermor). *Acta Botanica Gallica* 144, 209–216.
Cottignies, A., Cohat, J., le Floc'h, G., Delpierre, N. and Le Nard, M. (1999) L'unité biologique du bulbe d'échalote au cours du temps. *Acta Botanica Gallica* 146, 169–178.
Currah, L. and Proctor, F.J. (1990) *Onions in Tropical Regions*. Bulletin 35, Natural Resources Institute, Chatham Maritime, Kent, UK, 232 pp.
Dahlen, M. (1995) *A Cook's Guide to Chinese Vegetables*. The Guidebook Company, Hong Kong, 22 pp.
Dankert, J., Tromp, T.F., de Vries, H. and Klasen, H.J. (1979) Antimicrobial activity of crude juices of *Allium ascalonicum, Allium cepa* and *Allium sativum. Zentralblatt für Bakteriologie, Parasitenkunde, Infektionskrankheiten und Hygiene, 1, Originale A* 245, 229–239.
David, O., Fondio, L. and Moustier, P. (1998) The position of Côte d'Ivoire shallots on the Abidjan market in relation to imported onions. *Fruits* 53, 127–140 (in French).

De Mason, D.A. (1990) Morphology and anatomy of *Allium*. In: Rabinowitch, H.D. and Brewster, J.L. (eds) *Onions and Allied Crops*, Vol. I. *Botany, Physiology, and Genetics*. CRC Press, Boca Raton, Florida, pp. 27–51.

Duqueza, L.C. and Eugenio, T.S. (1973) The effect of fertilizer and different planting materials on the production of bulb and seed of Batanes, a native variety of onion. *Philippines Journal of Plant Industry* 38(3/4), 25–41.

Esau, K. (1965) *Plant Anatomy*, 2nd edn. John Wiley, London, 767 pp.

Eto, K. (1956) Studies of bolting onion. I. Relation between the flower-bud formation and the bulb division. *Journal of the Japanese Society for Horticultural Science* 25, 187–193 (in Japanese, with English summary).

Fletcher, P.J., Fletcher, J.D. and Lewthwaite, S.L. (1998) *In vitro* elimination of onion yellow dwarf and shallot latent viruses in shallots (*Allium cepa* var. *ascalonicum* L.). *New Zealand Journal of Crop and Horticultural Science* 26, 23–26.

Friesen, N. and Klaas, M. (1998) Origin of some minor vegetatively propagated *Allium* crops studied with RAPD and GISH. *Genetic Resources and Crop Evolution* 45, 511–523.

Galmarini, C.R. (1997) Allium crop situation in Argentina. *Acta Horticulturae* 433, 35–52.

Galvan, G.A., Wietsma, W.A., Putrasemedja, S., Permadi, A.H. and Kik, C. (1997) Screening for resistance to anthracnose (*Colletotrichum gloeosporioides* Penz.) in *Allium cepa* and its wild relatives. *Euphytica* 95, 173–178.

Grubben, G.J.H. (1994) Constraints for shallot, garlic, and Welsh onion in Indonesia: a case study on the evolution of *Allium* crops in the equatorial tropics. *Acta Horticulturae* 358, 333–339.

Hanelt, P. (1990) Taxonomy, evolution and history. In: Rabinowitch, H.D. and Brewster, J.L. (eds) *Onions and Allied Crops*, Vol. I. *Botany, Physiology and Genetics*. CRC Press, Boca Raton, Florida, pp. 1–26.

Hartsema, A.M. (1961) Influence of temperatures on flower formation and flowering of bulbous and tuberous plants. In: Ruhland, W. (ed.) *Encyclopedia of Plant Physiology*. Springer-Verlag, Berlin, Germany, pp. 4–123.

Heath, O.V.S. (1943a) Studies in the physiology of the onion plant. I. An investigation of factors concerned in the flowering ('bolting') of onion grown from sets and its prevention. Part 1. Production and storage of onion sets, and field results. *Annals of Applied Biology* 30, 208–220.

Heath, O.V.S. (1943b) Studies in the physiology of the onion plant. I. An investigation of factors concerned in the flowering ('bolting') of onions grown from sets and its prevention. II. Effects of day length and temperature on onion grown from sets, and general discussion. *Annals of Applied Biology* 30, 308–319.

Heath, O.V.S. and Holdsworth, M. (1948) Morphogenic factors as exemplified by the onion plant. *Symposia of the Society for Experimental Biology* 2, 326–350.

Heath, O.V.S. and Hollies, M.A. (1965) Studies in the physiology of the onion plant. VI. A sensitive morphological test for bulbing and its use in detecting bulb development in sterile culture. *Journal of Experimental Botany* 16, 128–144.

Heath, O.V.S. and Mathur, P.B. (1944) Studies in the physiology of the onion plant. II. Inflorescence initiation and development, and other changes in the internal morphology of onion sets, as influenced by temperature and day length. *Annals of Applied Biology* 31, 173–187.

Helm, J. (1956) Die zu Würz – und Speisezwecken kultivierten Arten der Gattung Allium L. *Kulturpflanze* 4, 130–180.

Hermann, K. (1958) Über die Flavonole und Phenole der Zweibel (*Allium cepa* L.) *Archiv der Pharmazie (Berlin)* 291, 238–247.

Jenkins, J.M. (1954) Some effects of different day length and temperatures upon bulb formation in shallots. *Proceedings of the American Society for Horticultural Science* 64, 311–314.

Jones, H.A. and Emsweller, S.L. (1936) Development of the flower and macrogametophyte of *Allium cepa*. *Hilgardia* 10, 415–428.

Jones, H.A. and Mann, L.K. (1963) *Onions and Their Allies. Botany, Cultivation, and Utilization*. Interscience Publishers, New York, 286 pp.

Joslyn, M.A. and Peterson, R.G. (1958) Reddening of white onion bulb purees. *Journal of Agricultural and Food Chemistry* 6, 754–765.

Kalkman, E.R. (1984a) Cytotaxonomic studies in the genus *Allium*. Usefulness of C-banding for description and classification. In: *Proceedings of Eucarpia 3rd* Allium *Symposium*. Institute for Horticultural Plant Breeding (IVT), Wageningen, The Netherlands, pp. 74–77.

Kalkman, E.R. (1984b) Analysis of the C-banded karyotype of *Allium cepa* L. Standard system of nomenclature and polymorphism. *Genetica* 65, 141–148.

Klaas, M. (1998) Applications and impact of molecular markers on evolutionary and diversity studies in the genus *Allium*. *Plant Breeding* 117, 297–308.

Ko, K.-D., Park, S.-K. and Lee, E.-H. (1993) Effect of shading, medium and ionic strength on the growth of hydroponically grown shallot (*Allium ascalonicum* L.) in summer season. *RDA Journal of Agricultural Science* 35, 381–385 (in Korean, with English summary).

Komochi, S. (1990) Bulb dormancy and storage physiology. In: Rabinowitch, H.D. and Brewster, J.L. (eds) *Onions and Allied Crops*, Vol. I. Botany, *Physiology, and Genetics*. CRC Press, Boca Raton, Florida, pp. 89–111.

Krontal, Y., Kamenetsky, R. and Rabinowitch, H.D. (1998) Lateral development and florogenesis of a tropical shallot: a comparison with bulb onion. *International Journal of Plant Science* 159, 57–64.

Krontal, Y., Kamenetsky, R. and Rabinowitch, H.D. (2000) Flowering physiology and some vegetative traits of short-day shallot – a comparison with bulb onion. *Journal of Horticultural Science and Biotechnology* 75, 35–41.

Kuroda, C. and Umeda, M. (1951) The pigments and the related compounds in the outer skins of onion bulb. *Journal of the Scientific Research Institute, Tokyo* 45, 17–22.

Kuruppu, P.U. (1999) First report of *Fusarium oxysporum* causing a leaf twisting disease on *Allium cepa* var. *ascalonicum* in Sri Lanka. *Plant Disease* 83, 695 (abstract).

Lancaster, J.E. and Boland, M.J. (1990) Flavor biochemistry. In: Brewster, J.L. and Rabinowitch, H.D. (eds) *Onions and Allied Crops*, Vol. III. *Biochemistry, Food Science, and Minor Crops*. CRC Press, Boca Raton, Florida, pp. 33–72.

Lapitan, V.P.C., Pateña, L.F. and Rosario, T.L. (1991) *In vitro* system of producing shallot (*Allium cepa* var. group aggregatum) planting material. *Philippines Journal of Crop Science* 16(3), 95–101.

Le Thierry D'Ennequin, M., Panaud, O., Robert, T. and Ricroch, A. (1996) Assessment of genetic relationships among sexual and asexual forms of *Allium cepa* using morphological traits and RAPD markers. *Heredity* 78, 403–409.

Maaß, H.I. (1996) About the origin of the French grey shallot. *Genetic Resources and Crop Evolution* 43, 291–292.

Messiaen, C.M. (1989) *Le Potager Tropical*. Presses Universitaires de France, Paris, 580 pp.

Messiaen, C.M. (1993) The possible use of F_1 hybrids between shallot clones as a starting point for several vegetative generations. *Onion Newsletter for the Tropics* 4, 49–51.

Messiaen, C.M., Cohat, J., Leroux, J.P., Pichon, M. and Beyries, A. (1993) *Les Allium alimentaires reproduits par voie végétative*. INRA, Paris, 228 pp. (in French, English language booklet, 42 pp.).

Messiaen, C.M., Lot, H. and Delecolle, B. (1994) Thirty years of France's experience in the production of disease-free garlic and shallot mother bulbs. *Acta Horticulturae* 358, 275–279.

Permadi, A.H. (1993) Growing shallot from true seed – research results and problems. *Onion Newsletter for the Tropics* 5, 35–39.

Poulson, N. (1990) Chives *Allium schoenoprasum* L. In: Brewster, J.L. and Rabinowitch, H.D. (eds) *Onions and Allied Crops*, Vol. III. *Biochemistry, Food Science, and Minor Crops*. CRC Press, Boca Raton, Florida, pp. 213–250.

Rabinowitch, H.D. (1979) Doubling of onion bulbs as affected by size and planting date of sets. *Annals of Applied Biology* 93, 63–65.

Rabinowitch, H.D. (1985) Onions and other edible *Alliums*. In: Halevy, A.H. (ed.) *Handbook of Flowering*, Vol. 1. CRC Press, Boca Raton, Florida, pp. 398–409.

Rabinowitch, H.D. (1990) Physiology of flowering. In: Rabinowitch, H.D. and Brewster, J.L. (eds) *Onions and Allied Crops*, Vol. I. *Botany, Physiology, and Genetics*. CRC Press, Boca Raton, Florida, pp. 113–134.

Ruaysoongnern, S. (1994) Management factors affecting keeping quality of shallot in Sisaket, Northeastern Thailand. *Acta Horticulturae* 358, 375–381.

Ryu, Y.-W., Suh, J.-K., Hwang, H.-J., Ha, I.-J. and Kim, W.-I. (1998) Effect of bulb size at planting on the growth and yield of shallot (*Allium cepa* var. *ascalonicum* Baker). *RDA Journal of Horticultural Science* 40, 105–108 (in Korean, with English summary).

Sinnadurai, S. and Amuti, S.K. (1971) Dormancy of shallots in Ghana. *Experimental Agriculture* 7, 17–20.

Subijanto (1988) Research and development to increase export potential of horticultural products from Indonesia. *Indonesian Agricultural Research and Development Journal* 10, 87–94.

Suh, J.-K. and Ryu, Y.-W. (1998) Effect of planting date under spring and autumn culture on the growth and yield of shallot (*Allium cepa* var. *ascalonicum* Baker). *RDA Journal of Horticultural Science* 40, 98–104 (in Korean, with English summary).

Suhardi, H.A. (1993) Anthracnose on shallot (*Allium cepa* group *aggregatum*) in Java. *Onion Newsletter for the Tropics* 5, 48–50.

Suhardi, H.A. (1996) Effect of planting date and fungicide applications on the intensity of anthracnose on shallot. *Indonesian Journal of Horticulture* 6, 172–180.

Sumiati, E. (1994). Response of shallot and garlic to different altitudes. *Acta Horticulturae* 358, 395–400.

Tappayuthpijarn, P., Dejatiwongse, Q., Hincheranan, T. and Suriyant, P.N. (1989) Effect of *Allium ascalonicum* on erythrocyte shape in induced hypercholesterolemia rabbits. *Journal of the Medical Association of Thailand* 72, 448–451.

Thompson, H.C. and Smith, O. (1938) *Seedstalk and Bulb Development in the Onion* (Allium cepa *L.).* Bulletin No. 708, Cornell University Agricultural Experimental Station, Ithaca, New York, 21 pp.

van Dijk, P. (1992) *Virus Diseases of Garlic, Shallot, and Welsh Onion in Java, and Prospects for their Control. Report of a Consultancy Study*. CPRO-DLO, Wageningen, The Netherlands, 72 pp. (in Dutch).

van Kampen, J. (1970) *Shortening the Breeding Cycle in Onions.* Mededelingen Proefstation voor de Groenteteelt in de Vollegrond, No. 51, University of Agriculture, Wageningen, The Netherlands, 72 pp. (in Dutch).

Vanparys, L. (1999a) *Shallots from Seed.* Mededeling No. 404, Provinciaal Onderzoek Voorlichtingscentrum voor Land en Tuinbouw, Beitem Roeselare, 4 pp. (in Dutch).

Vanparys, L. (1999b) Shallots from seed. Cultivation of shallots from seed: mirage does fine! *Proeftuinnieuws* 9(4), 31–32 (in Dutch).

Vanparys, L. (2000) Cultivation of seed shallots. *Proeftuinnieuws* 10(9), Belgium, 14–15 (in Dutch).

Walkey, D. (1990) Virus diseases. In: Rabinowitch, H.D. and Brewster, J.L. (eds) *Onions and Allied Crops*, Vol. II. *Agronomy, Biotic Interactions, Pathology, and Crop Protection*. CRC Press, Boca Raton, Florida, pp. 191–212.

Wietsma, W., Grubben, G., Henken, B., Zheng, S., Putrasemedjo, S., Hadi Anggoro, A., Sofiari, E., Krens, F., Jacobsen, E. and Kik, C. (1998) Development of pest and disease resistant shallot cultivars by means of breeding and biotechnology. *Indonesian Agricultural Research and Development Journal* 20(4), 78–82.

Yamashita, K. and Tashiro, Y. (1999) Possibility of developing male-sterile line of shallot (*Allium cepa* L. Aggregatum group) with cytoplasm from *A. galanthum* Kar. et Kir. *Journal of the Japanese Society for Horticultural Science* 68, 256–262.

Zaharah, A., Vimala, P., Siti Zainab, R. and Salbiah, H. (1994) Response of onion and shallot to organic fertilization on bris (Rudua series) soil, in Malaysia. *Acta Horticulturae* 358, 429–432.

18 Leek: Advances in Agronomy and Breeding

H. De Clercq and E. Van Bockstaele
Department of Plant Genetics and Breeding (DvP), Centre for Agricultural Research-Ghent (CLO-Gent), Caritasstraat 21, 9090 Melle, Belgium

1. Botany

1.1 Origin and distribution

Leeks have been cultivated from very early times. The garden leek was a popular vegetable in the ancient Near East when the Egyptians built their pyramids – for example, that of Cheops, 2500 BC. Leek was an important vegetable for the Greeks and Romans, and its use later spread throughout medieval Europe (Silvertand, 1996).

Leek (French: poireau; Dutch: prei; German: lauch, porree; Spanish: puerro; Italian: porro; Latin: *Allium ampeloprasum* L. leek group), unlike onion, is indifferent to day length and the same genotype can be grown and produce an economic yield over a wide range of latitudes. Leeks are well adapted to cool conditions and are harvested throughout the winter in the maritime countries of Western Europe, and can be grown in subtropical countries, such as Senegal, if irrigation water is available (De Clercq, 1981).

In the European Union, leek is cultivated as an outdoor vegetable on about 30,000 ha, as compared with 104,500, 36,000 and 2300 ha for onion, garlic and shallot, respectively (Eurostat, 1999); leek production in Europe is detailed in Table 18.1. In recent years, leeks have been becoming more popular in the USA and elsewhere.

1.2 Taxonomy and cultonomy

Leek is a member of the genus *Allium* (family: *Alliaceae*), which comprises about 700 species spread over the northern hemisphere (Klaas, 1998; Fritsch and Friesen, Chapter 1, this volume). Morphological, anatomical, karyological and serological studies, together with the life cycle, spread and ecology of alliums, have demanded an infrageneric classification (Hanelt *et al.*, 1992), in which the genus is divided into six subgenera. The most important are the subgenera *Rhizirideum*, with onion and chives, and *Allium*, with leek, garlic, kurrat, pearl-onion and great-headed garlic (Table 18.2; Fritsch and Friesen, Chapter 1, this volume).

There is still some disagreement about the botanical nomenclature of leek. Traditionally leek had the name *Allium porrum* L., given by Linnaeus and still used in botanical floras from Belgium and The Netherlands (van der Meijden, 1996; Lambinon *et al.*, 1998). In recent years, some taxonomists have preferred *Allium ampeloprasum* L. as the scientific name for leek (Hanelt, 1996; Silvertand, 1996; Buiteveld, 1998; Klaas, 1998). Others use the name *Allium ampeloprasum* var. *porrum* (Hanelt, 1990; Smith and Crowther, 1995; Khazanehdari and Jones, 1997; De Clercq *et al.*, 1999). We shall use the Fritsch and Friesen (Chapter 1, this volume) nomenclature, which is *A. ampeloprasum*, leek group, for consistency within this book.

In addition to classical taxonomy with Linnaean names, there is increasing interest in a cultonomic approach for the classification of cultivated plants. A culton is a systematic group of cultivated plants based on one or more user criteria (Hetterscheid and Brandenburg, 1995). The culton or cultivar-group system provides the necessary flexibility to describe a very varied crop and can be easily adjusted to new developments and needs. In the leek taxon, the following cultivar groups are distinct: *Allium* Swiss Giant Leek Group, *Allium* Autumn Giant Leek Group, *Allium* Blue Green Autumn Leek Group, *Allium* Winter Giant Leek Group and *Allium* Blue Green Winter Leek Group (Hetterscheid *et al.*, 1999). The Turkish or Bulgarian Giant Leek, the Egyptian Kurrat, the Tarée Irani, the Poireau Perpétuel, the Prei Anak and other cultivated varieties will eventually be listed using the cultonomic approach.

1.3 Biology and physiology

Leek is a slow-growing monocotyledonous species. For many centuries, plants have been selected for the development of concentric, ensheathing leaf-bases, together with folded immature leaf blades at the centre, which form long edible pseudostems. Different types occur, varying in the length and slenderness of the pseudostem. Turkish and Bulgarian types have long, thin pseudostems (Fig. 18.1a), whereas those from Western Europe have shorter, thicker ones (van der Meer and Hanelt, 1990; Fig. 18.1b). All types of leeks freely intercross with each other, and the wide genetic

Table 18.1. Leek production area (ha) in EU countries (Eurostat, 1999).

Country	Area (ha)	Country	Area (ha)
Belgium	5700	Ireland	<100
Denmark	400	Italy	1000
Finland	<100	Netherlands	3700
France	8300	Spain	2400
Germany	2400	Sweden	<100
Greece	1800	UK	2600

Table 18.2. Taxonomic and cultonomic approach to the classification of leek *(Allium porrum* L.).

Taxonomy (the Gatersleben approach) (Hanelt *et al.*, 1992)				Cultonomy (Hetterscheid *et al.*, 1999)
Family	*Alliaceae*			
Genus	*Allium*			
Subgenus 1	*Bromatorrhiza*	3 sections	9 species	
Subgenus 2	*Rhizirideum*	15 sections	170 species	
Section	*Schoenoprasum*	*A. schoenoprasum* L. (chives)		
Section	*Cepa*	*A. ascalonicum* L. (shallot)		Group I True Shallot Group II Shallot Group III Echalion (Onion)
		A. cepa L. (onion)		Group IV Onion or IVa Yellow-Brown IVb Red IVc White IVd Silverskin Onion
		A. fistulosum L. (bunching onion)		Group V Welsh Onion Group VI Japanese Bunching
Subgenus 3	*Allium*	5 sections	280 species	
Section	*Allium*	*A. ampeloprasum* var. *porrum* (leek) var. *kurrat* (kurrat) var. *sectivum* (pearl onion)		Group VII Swiss Giant Leek Group VIII Autumn Giant Leek Group IX Blue Green Autumn Leek Group X Winter Giant Leek Group XI Blue Green Winter Leek
		A. sativum L. (garlic)		
Subgenus 4	*Melanocrommyum*	12 sections	110 species	
Subgenus 5	*Caloscordum*	1 section	2 species	
Subgenus 6	*Amerallium*	11 sections	130 species	

variation for seasonal adaptation between cultivars enables the year-round cultivation of the crop.

The objective in leek culture is the production of shoots of marketable size before the leek plants bolt. In temperate Europe, premature bolting may be a problem of very early plantings (Wurr *et al.*, 1999) and normal bolting occurs in late-spring-harvested crops. In most countries, leek plants are transplanted after a nursery period of about 12 weeks (Fig. 18.2a,b). This method permits the rejection of weak-growing plants. However, in the UK, direct drilling is more frequent.

Allium ampeloprasum is a biennial species, which does not normally produce bulbs. However, after flowering, bulblets/bulbils sometimes form in the leaf axil at the base of the flower stalk. Some genotypes produce topsets on the flowering umbel (Schweisguth, 1970). Cutting off flower buds at an early stage of development can induce topset formation in the umbel (Fig. 18.3a,b).

(a)

(b)

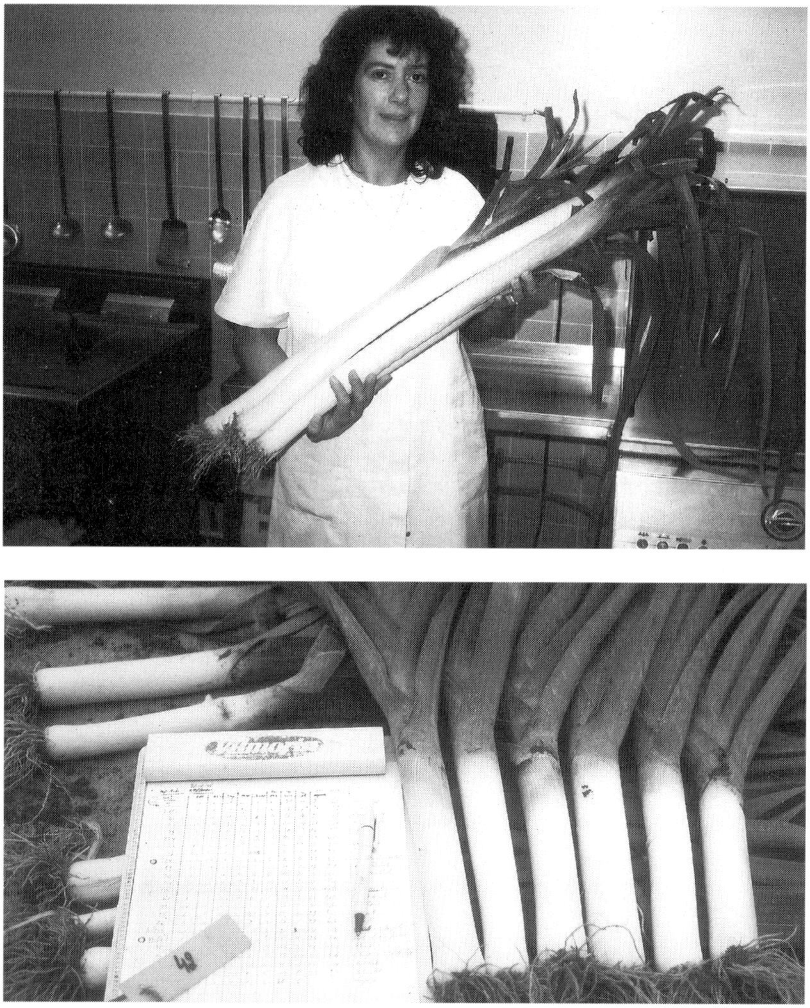

Fig. 18.1. Different types of leek (*Allium ampeloprasum* L.). Long pseudostems (a) are typical of summer leek, while a short stem is typical for the winter leek type (b).

Leek needs vernalization for flower induction, and blooms in Europe during midsummer when days are long (from about 10 June to 20 July). The globose umbels contain hundreds of flowers, which vary in colour from light pink to dark purple. They are insect-pollinated and, although a protandrous flowering mechanism exists within individual flowers, 10–30% of self-pollination occurs in open-pollinated varieties (Berninger and Buret, 1967). High temperatures (30/20°C) during pollination, fertilization and seed ripening result in less seed, but the seeds ripen faster than at lower temperatures (20/10°C) (Gray *et al.*, 1992).

2. Agronomy

In keeping with the comparatively restricted geographical area where leeks are popular with consumers, sources of information on leek agronomy are mostly from Western Europe and Poland. A comprehensive guide to leek growing was published in France (de Bohec *et al.*, 1993) and this covers many

(a)

(b)

Fig. 18.2. An indoor nursery of leek plants (a) and an outdoor production field on ridges (b).

agronomic topics, including growing leeks for different markets, the costs of production and the seasons for leeks in different parts of the country. In The Netherlands and Belgium, regular reports on agronomic advances appear in print from research stations (e.g. Callens, 1999), and a Netherlands growers' guide is also available (de Kraker and Bosch, 1993). Belgium gives particularly good coverage of regular variety-trial reports from research stations (e.g. Vanparys, 1998). Control of *Thrips tabaci* on leeks has received a lot of attention in recent years: monitoring methods have been developed and the course of epidemics has been traced. In Poland, with its harsher winter climate, methods of plant raising and transplanting using protection from the weather have been reported (e.g. Kolota and Adamczewska-Sowinska, 1995). In this section, we will discuss a selection of recent references on agronomic topics.

2.1 Traditional cultivation methods

Leek can be grown in practically all soil types that have a relatively open texture. The culture of leeks requires a long growing season and they do well in cool, wet weather.

(a)

(b)

Fig. 18.3. Leek seed production (a) and induction of bulbil formation on the umbel (b).

Compared with other *Allium* crops, leek is very tolerant to cold weather, although the optimum temperature for vegetative growth is around 20°C. Local landraces, adapted to the different climates and market demands, have been developed in many European countries from Bulgaria to Ireland and in other parts of the world (e.g. Middle East).

In most countries, leek is sown at high density in seed-beds and is transplanted after 12 weeks when it has reached pencil thickness. Leek transplants are planted into 15–22 cm deep planting holes, often at 12 × 50 cm or 10 × 65 cm spacings.

Harvesting can be performed manually behind a lifting machine or mechanically by a lifter-harvester (Fig. 18.4a,b), after the plants are undercut, usually by a vibrating knife. Following lifting, the outer leaves are removed, the remaining leaves are shortened and the plants are washed or brushed, graded for length and diameter and packed into boxes. Leeks are sometimes sold loose and sometimes prepacked in trays with

(a)

(b)

Fig. 18.4. Harvest of leek can be performed manually behind a lifting plough (a) or entirely mechanically with a lifter-harvester (b).

plastic covers or in plastic bags. The requirements for prepacking leeks include uniform lengths of the white portion of the pseudostem. New products, such as 'baby leeks', are also appearing in European markets.

In Europe about 90% of the leek crop is sold on the fresh market and 10% is processed by the industry. Some processed leeks are used for freezing, some are freeze-dried and some are used to prepare ready-cooked dishes. The requirements for these different uses are discussed by de Bohec *et al.* (1993).

There is an increasing demand for the development of 'organic' leek seed, as for other vegetable crops, since the current derogation of this requirement for the production of 'organic' vegetables in Europe will cease at the end of 2003.

2.2 Advances in agronomy

2.2.1. Seed quality and uniformity of the crop

Most of the well-known commercial cultivars of leek have until recently been open-pollinated. Consequently, one of the major problems with this crop, besides vulnerability to pests and diseases, has been poor uniformity. As a mostly transplanted crop, selection of the plant material at planting takes time, but at harvest a considerable percentage of the remaining plants still have to be rejected. For direct drilling, the requirements for phenotypic uniformity and for seed quality are even higher.

Many efforts have been made to improve leek seed quality. Gray and Steckel (1986a) found that selfed seed was inferior in weight, with a higher coefficient of variation, and gave lower germination and more variable seedling weights, compared with plants from outcrossed seed. Furthermore, they agreed with Benjamin (1984) that only about 10% of the variation in seedling weight could be accounted for by variation in seed weight. Field emergence could be improved by seed production at high density (97 plants instead of 11 plants m^{-2}): close planting probably increases the percentage crossing and advances the seed-crop harvest date (Gray and Steckel, 1986b; Gray *et al.*, 1992).

In our experiments in Belgium, agronomic aspects, such as emergence, plant growth and yield, were well correlated with individual seed weight. Initially, using seeds cleaned in a standard way by threshing and air-separation, we found a weak correlation between plant growth and individual seed weight (Table 18.3). However, yield was strongly correlated with plant weight at transplanting and also with shaft diameter at harvest. In a further experiment, leek seeds were harvested and cleaned by hand to retain even the smallest seeds. Here we found a strong correlation between seed weight and transplant weight (Table 18.4). These results emphasize the importance of adequate seed cleaning and grading and, following this, the selection of large and uniform seedlings at transplanting for improved crop uniformity in leek.

A further improvement in germination performance and field uniformity can be achieved by seed priming, in which controlled hydration of seeds permits pregermination metabolic events to take place without radicle emergence. The process engineering of leek seeds was developed, comprising osmotic priming, washing, fluidized-bed drying (heated air is blown up from underneath through a layer of seeds to promote rapid drying while they are floating in the air) and film coating: this has been proven feasible (Bujalski *et al.*, 1991). The superiority of the processed seeds is usually reflected in improved germination, rapid and uniform emergence in the field and improved early plant growth compared with untreated seed.

Table 18.3. Correlation coefficients of growth characteristics and seed weight of leek after industrial cleaning of the seed (from H. De Clercq, D. Peusens, I. Roldán-Ruiz and E. Van Bockstaele, unpublished).

	Germination	Seedling growth	Transplant weight	Adult weight	Shaft thickness
Seed weight	0.009	0.204 *	0.137	0.050	0.052
Germination		0.446 **	0.319 **	0.290 *	0.289 *
Seedling growth			0.348 **	0.123 *	0.117 *
Transplant weight				0.749 **	0.743 **
Adult weight					0.976 **

*, Correlation is significant ($P < 0.05$); **, correlation is significant ($P < 0.01$).

Table 18.4. Correlation coefficients of growth characteristics and seed weight of leek after hand-cleaning of the seed (with retention of small seeds) (from H. De Clercq, D. Peusens, I. Roldán-Ruiz and E. Van Bockstaele, unpublished).

Trait	Seedling length	Transplant weight	Transplant length
Seed weight	0.526 **	0.248 **	0.148 **
Seedling length		0.346 **	0.336 **
Transplant weight			0.707 **

**, Correlation is significant ($P < 0.01$).

Priming is used commercially in the UK: for example, all the leek seed sold by the Elsom Seed Company is primed, using the drum priming process patented by Rowse (1996) (R. Dobbs, UK, 2000, personal communication.). Other priming processes may be used elsewhere in the seed trade. Treated seed cannot be stored for as long as unprimed seed but, provided it is sown in the same season, this is not a problem. Further studies must define optimum overall conditions and assess the long-term storage life of treated seeds. (See also Section 3.2 below on F_1 hybrids.)

2.2.2 Year-round production and plant raising under protected conditions

The large number of leek cultivars now available makes it possible to sow the crop in Western Europe from December to June. Harvesting takes place from June to May in the next year. The cultivation of leek in Europe is divided, according to the time of harvest, into three main periods, i.e. summer, autumn and winter. The 'winter' leek can be harvested until early April and kept in cold stores for a few weeks (see Section 2.2.8 below). In order to fill the supply gap in the northern regions during May and June (when leeks normally bolt), leeks are imported from southern Europe. In Belgium a special method of cultivation, called 'Oude Jonkman' or 'Stekprei', is sometimes applied to fill the gap between the winter and summer production seasons (Vanparys, 1991). Under this method, sowing takes place in July and August, transplanting in October and November and harvesting in May and June. Only winter-hardy, good-growing but slow-bolting cultivars are adapted to this cultural technique.

Protected nurseries in Poland, where winters are more severe than those in Western Europe, use fleece (non-woven polypropylene fabric, about 17 g m^{-2}) to cover the growing seedlings and thus improve earliness and yields (Rumpel *et al.*, 1995), while early sowing and field protection of block-raised plants with perforated plastic covers also gave better crops of leeks (Kolota and Adamczewska-Sowinska, 1996). Studies on plant-raising methods have shown the possibilities of using multiple block-raised transplants by sowing directly into compressed-soil blocks (Kolota and Adamczewska-Sowinska, 1995). Field uniformity can be improved by grading bare-root seedlings and planting the different size-grades separately (Embrechts, 1996). De Bohec and Fouyer (1999) summarized plant-raising methods currently used in France.

Leek production using hydroponics has been studied in Belgium (Tongaram *et al.*, 1994a, b). In Hebei Province, China, summer leek production in greenhouses has been reported (Chen *et al.*, 1998). Yields equivalent to 60 t ha^{-1} were achieved in this system.

2.2.3. Direct drilling and weed control

In France, Belgium and The Netherlands, leeks are seldom sown in the production field, but, in the UK, direct drilling is common. Direct sowing has some disadvantages: there is no opportunity to select for transplant uniformity, weeds must be controlled throughout crop life and the plants must be 'earthed up' to produce long white shafts.

Weed control is done either mechanically or chemically by herbicides, such as chlorprofam (in Chlor-IPC, Shell) or propachlor (in Ramrod, Monsanto) (de Kraker and Bosch, 1993). More recently, methabenzthiazuron (in Tribunil, Bayer) and cyanazin (in Bladex, Shell) were very efficient weedkillers in leek fields. Currently, these products are being tested for their safety and efficiency on nursery plots (PCG, 2000).

Non-chemical weed-control methods are being studied in Denmark, where they have also been developed for onions (Melander *et al.*, 1999; see also Bosch Serra and Currah, Chapter 9, this volume).

2.2.4 Fertilizer and water studies

In The Netherlands, 7-year studies on integrated production methods for leeks (Kroonen-Backbier and Rovers, 1998) resulted in significant reductions in fertilizer use. The requirements for P, K and MgO were substantially lower than those recommended earlier in Holland, while targets for N can be tailored to suit the crop's requirements at different growth stages. While there were considerable savings on fertilizers and pesticides, the cost of a mulch with cereal straw, used to reduce *Phytophthora porri* damage, increased production costs in the integrated system. Yields were 15–20% lower than those from conventional systems, with lower crop quality, particularly in autumn leeks (Kroonen-Backbier and Rovers, 1998).

In France, a 4-year study was used to develop a method to calculate the N needs of leek plants, depending on growth stages and the N status of the soil. A quick test is used to determine nitrate-N levels in the soil, while a graphical model, called ZENIT, can be used to support decision-making (Berry and Thicoïpe, 1998).

Several recent studies have added to our understanding of the fertilizer requirements of leeks: these include work by some research groups to reduce the leaching of nitrogen from leek fields when grown following green-manure crops (Ekbladh, 1995); the demonstrated benefits of ammonium polyphosphate compared with triple superphosphate (Vanaerde, 1998); and the development in Belgium of the controlled-uptake long-term ammonium nutrition (CULTAN) system which involves the injection of concentrated urea ammonium sulphate into the soil between the rows (de Rooster and Spiessens, 1998). The CULTAN system gave crop-yield increases of 8–10% compared with the other forms of N applied, and also improved the storage quality of the leeks.

A model of daily plant growth and N uptake was developed by Kuzyakov *et al.* (1996). It was based on the relative growth rate of the crop and made use of the environmental factors radiation, air temperature, soil moisture and mineral N content of the soil. The model could be used to simulate leek growth.

Mulching and irrigation requirements for summer-grown leeks have been investigated in Belgium. Raising transplants under fleece, followed by mulching and drip irrigation after transplanting, produced somewhat heavier plants than those from unprotected soil (Benoit and Ceustermans, 1999a). Recommendations on water requirements for Belgium were for 8 l m^{-2} $week^{-1}$ for the first 6 weeks after transplanting, increasing up to 40 l m^{-2} $week^{-1}$ during the next 2 months for leeks to be harvested in October. Different methods of calculating water requirements were compared: either water was added to compensate for calculated evapotranspiration, or rainfall was supplemented to give a total of 40 l of water m^{-2} $week^{-1}$. By August, mulched leeks were significantly larger than those in open ground, but, by September, the difference was much smaller. The supplementary-irrigation plan also gave higher-weight leeks by August compared with the evapotranspiration formula, though these gains were also more important early in the growing season than at later stages. Less water was needed by the crop in the cool autumn, from September onwards.

2.2.5 Leek physiology and growth modelling

Several groups have studied the physiology of leek growth, particularly to understand

and prevent bolting, and some have modelled various aspects (for N-fertilizer responses, see previous section). Hay and Kemp (1992) in Scotland formulated a simple model based on primordium development. They found that, during the main period of leaf production of autumn leeks, successive leaves initiated, developed and senesced at equal intervals of accumulated temperature/thermal time. The plastochron interval was represented by 92 day-degrees and the number of fully expanded leaves remained constant. A model of canopy expansion was based on the temperature relations of primordium initiation and additional data on leaf expansion and leaf dimensions. Leaf area indices (LAIs) computed by this model were verified against the thermal-time course of the LAI previously observed (Hay and Kemp, 1992).

In Germany, Wiebe (1994) found that leeks have an obligatory vernalization requirement, whereas the effects of the photoperiod were quantitative. The juvenile phase of leek growth ended when the seedling had five visible leaves or weighed about 2 g. The optimal vernalization temperature was estimated to be about 5°C and the range at which induction of flowering was possible ranged from 0 to 18°C. Devernalization took place at temperatures over 18°C. Later, Wiebe (1995) compared the response of 9-week-old seedlings of three cultivars to cold treatments at 2–14°C for 3–6 weeks, followed by 2 weeks under glass at 15 or 22°C, prior to planting outdoors in May. Optimal temperatures for bolting induction were 2–8°C, and some plants were susceptible to just 3 weeks of vernalization. The author recommended that leek plants should be raised at temperatures of 16–18°C from the point when the plants have four to five leaves, in order to reduce bolting. Thirteen cultivars were ranked for their bolting susceptibility.

In the UK, Wurr *et al.* (1999) looked at the growth patterns of the early cv. 'Prelina' in two locations over 3 years. Dates of sowing, nursery temperatures and transplant type were some of the factors examined. Nursery temperatures had no effect on marketable yields or bolting percentages. Outdoor conditions early in the season had the most influence on bolting; hence transplants from early sowing or large seedlings from peat blocks were the most likely to bolt. A temperature of 7°C was the most effective for vernalization. Delaying transplanting until the outside temperature was sufficiently high to discourage bolting was suggested. Models to predict the rate of increase of leek diameter and the rate of flower stalk extension were developed, to help to identify the optimal harvest time (Wurr *et al.*, 1999).

2.2.6 Integrated pest control, forecasting for disease and pest attack in leek

THRIPS. The major pest on leeks is the thrips (*T. tabaci* Lindeman). These 2 mm long insects hide between the inner leaf blades, where they feed on cell fluids. The green leaves lose their colour as the empty surface cells form thousands of highly visible grey spots. The economic damage due to quality loss is serious.

Thrips damage is most severe when plants are water-stressed in hot, dry weather. In these conditions leaf expansion is slow and the increase in thrips number is fast. At 30°C, it takes only 11 days for the insect to develop from egg to adult (Edelson and Magaro, 1988). By means of blue sticky insect traps, the migration of the thrips can be monitored (see Lorbeer *et al.*, Chapter 12, this volume). Edelson and Magaro (1988) developed a thrips-forecasting model based on a day-degree model: the minimum temperature for development is 11.5°C. It takes 95.4 day-degrees from egg to larva, and another 132.8 day-degrees from hatched larva to adult. Under Belgian conditions, a first small flight of thrips is registered in June; the second and third flights in August and in September are the more important (Callens, 1999). Threshold studies revealed the need to start control treatments at a count of one or two thrips per plant. Any attempt to economize on pesticide by delayed application until three to five insects per plant are present results in severe damage later in the season. Integrated pest control includes forecasting the thrips attacks so that three to five treatments can be sufficient to control the pest (PCG, 2000).

In France, thrips-monitoring studies have been made since the early 1990s, using variously coloured sticky traps (e.g. Villeneuve *et al.*, 1997). Recent conclusions are that weekly pesticide applications during August, based on a threshold of one thrips per leek plant, may be sufficient to control the pest, and that both leeks and onions have considerable potential for recovery from thrips attack (Villeneuve *et al.*, 1999).

In Germany, the main infestation period was from July to October, with peak numbers in August, and the impact, biology and control of thrips on leeks have been studied intensively. Richter *et al.* (1999) concluded that monitoring could be used to estimate when thrips-control measures should be employed, thus avoiding excessive insecticide use, and that thrips damage has probably been overestimated in the past. The data collected on crop losses here also showed that leeks can recover well from thrips damage after peak infection time has passed (Richter *et al.*, 1999).

Given the low damage threshold for thrips, much research has been done to control this pest efficiently. The better chemical products are carbamates, including methiocarb (in Mesurol, Bayer) and furathiocarb (in Deltanet, Ciba-Geigy). Some phosphorus compounds, such as acefaat (in Orthene, Tomen France) and malathion, and also pyrethrins, may have protective effects. Novel pesticides for use against thrips are being tested. In The Netherlands, seed-coatings with fipronil were effective in protecting seedlings, with no apparent phytotoxicity (Ester *et al.*, 1997).

Another approach suitable for 'organic' production is intercropping with legumes or other plants to discourage thrips from feeding in large numbers on leeks. Den Belder and Elderson (1998) studied the feasibility of intercropping in pot and field experiments in The Netherlands. Intercropping with clover led to reduced thrips populations on leeks, even when the legume was trimmed. Theunissen and Schelling (1998) reported that intercropping leeks with clover (*Trifolium fragiferum*), either throughout the field or in between rows, suppressed both larval and adult thrips populations. A possible explanation for this type of pest suppression is discussed in a review by Finch and Collier (2000), which, though it refers to pests of cruciferous crops, may have implications for future work on the pests of leeks.

Mulching with coloured plastic has also been investigated as a means of discouraging thrips on leeks. These studies have sometimes been combined with those on water use (Benoit and Ceustermans, 1999a, b, c). For example, Benoit and Ceustermans (1999c) monitored during August–September and found that blue/black plastic mulch with the blue colour facing upwards resulted in lower thrips numbers per plant (1.5 compared with 10.9 in bare plots), resulting in a 13% increase in yield compared with unmulched plots.

LEEK RUST. Leek rust (*Puccinia porri* G. Wint., syn. *P. allii* F. Rudolphi) causes severe damage on European leeks. As the crop is now cultivated all year round, the uredo stage of leek rust is present throughout the year. During winter, low temperatures inhibit the formation of uredosori (the bodies that produce uredospores, one of a possible five types of rust spores). As soon as the temperature increases in spring, the epidemic of leek rust starts again. The disease develops most frequently under conditions of high humidity and low rainfall, while immersion of the spores in water reduces their viability. The highest infection efficiency occurs at 100% relative humidity (RH) at 10–15°C, and temperatures above 24°C and below 10°C inhibit infection.

The economic threshold for leek rust is low, as all leaves are prone to damage and leaf removal is not practical. A regular spray schedule with protectant fungicides (e.g. maneb or zineb) should give adequate protection (Schwartz and Mohan, 1995). Spraying fenpropimorph (in Corbel, Basf and Schering), either alone or in mixtures with maneb, provides a good control. Compounds of the triazole group – tebuconazole (in Horizon, Bayer) and epoxiconazole (in Opus, Basf) – are also effective;

treatments with propiconazole (in Tilt, Liro) resulted in outstanding control. If necessary, curative chemical treatment must be used to control leek rust, until better tolerance or resistance is developed in commercial leeks (see Section 3.3.2).

De Jong *et al.* (1995) developed a model to simulate chemical control of leek rust in The Netherlands, and concluded that spraying can be reduced only if fields are isolated from other leek crops and if the planting material is disease-free. It is also important to avoid temporal overlapping between leek crops in nearby fields. Without these conditions, heavy use of sprays will be unavoidable. Novel studies are being conducted to develop a disease-forecasting sytem for leek rust, based on the integration of models that describe the latent period of *P. porri* and spore production (T. Gilles, UK, 2001, personal communication).

WHITE-TIP DISEASE. White-tip disease, the most important leek disease in Europe during the winter, is caused by *Phytophthora porri* Foister. Infected leaves show papery-white local lesions, sometimes surrounded by dark-green waterlogged zones. Sporangia can develop in wet lesions and may release 10–30 zoospores, while oospores are formed when the leaves dry up and may survive for a long period. Harvest losses may be severe: in some cases, total crop loss is reported (Smilde, 1996).

Smilde (1996) studied the epidemic curves of white-tip (*P. porri*) in the production fields of leek and their correlation with rainfall. The correlation between rainfall and disease was relatively high at the onset of an epidemic and relatively low later in the season. Apparently, rainfall is necessary for initial infection with white-tip, but less important for subsequent autoinfection. Because white-tip is considered to be a soil-borne disease, recent epidemiological studies not only focus on temperature and humidity but are also obtaining basic information on the frequency and intensity of rainfall and susceptibility of the soil to splashing (related to soil texture and how well the soil particles are aggregated together) (PCG, 2000).

A two-step chemical control method can be used: during September and October maneb is used preventively; when the first symptoms are visible, more systemic products such as benalaxyl (in Galben, ICI) or metalaxyl (in Ridomil, Liro), are used curatively.

2.2.7 Reducing labour requirements for leek cultivation

The biggest reduction in labour requirement during the last century in leeks, as well as in other crops, was due to chemical weed control. Direct drilling has economic advantages compared with transplant production, but also some disadvantages (see Section 2.2.3).

Under a direct-drilling system, high-quality seed ('hybrids' are increasingly used) is sown with a pneumatic or belt seed drill. Transplanting leeks is laborious, and several mechanical systems are in use: plants transplanted on a flat field need to be earthed up during growth, whereas planting on a bed system or on ridges gives the opportunity for deeper planting, which allows the production of a longer white shaft. For each planting system, suitable machinery is available.

Mature leek plants are harvested by a lifter-harvester (Fig. 18.4b), sometimes equipped with a leaf cutter and cleaning brushes, so that the leeks can be transported immediately from the field to the packhouse. In most packhouses, the leeks are rinsed, the leaf blades are trimmed and then the plants are graded in washing–peeling tunnels into different diameter classes for the fresh market (Fig. 18.5).

2.2.8 Storage of leek and leek quality

Experiments on leek storage conditions were reported from Poland (Grzegorzewska and Bakowski, 1996), where 5 kg crates holding leeks upright were found most suitable and a storage temperature of −1.5°C was more effective than one of 0°C. In France, a temperature of −2°C is recommended for leeks being stored for 2 months or more (de Bohec *et al.*, 1993). In

Fig. 18.5. Postharvest cleaning of leeks can be done in a washing–peeling tunnel.

Scandinavia, stores at −1 to 0°C and at 95% RH allow leeks to be kept for 8 weeks.

Under optimal controlled-atmosphere (CA) conditions at about 10% CO_2 and 1% O_2, the duration of storage can be even longer (Hoftun, 1978). In Belgium, CA at 1–2% O_2 and 5–10% CO_2 gave good results. Young leek plants are more difficult to store, due to their faster respiration, than mature plants (ones that have reached optimal marketing size). Besides respiration, the amounts of reserve compounds in the cells and the wax layers coating the leaves influence leek storability (A. Schenk, Belgium, 2000, personal communication).

The composition of alliums was reviewed by Fenwick and Hanley (1990). The major storage tissues of leek are the leaf sheaths, which are normally 1–2% lower in dry matter (DM) than those of bulb onion: about 11% DM in our tests. DM constituents are 70–85% storage carbohydrates (mostly fructans), 10–20% proteins and about 1% lipids and ash. The flavour compounds in alliums are sulphur-containing non-protein amino acids, with a common general structure of cysteine sulphoxide, but with differences in their chemical R groups between the major allium crops. Besides methyl, leek contains mainly propyl as the R group, whereas onion and shallot contain mostly propenyl, and garlic contains mostly allyl R groups.

Several studies have recently reported on quality aspects of leeks. In Italy, for example, the timing of harvesting of autumn/winter leeks was studied in relation to the concentration of volatile flavour compounds. For late-April-sown leeks, the best organoleptic quality, in four cultivars, was measured in December–January, although higher yields were obtained only later, in March (Ferrari *et al.*, 1999). Further studies on the actual flavour compounds present in the white part of the leeks showed that these were at their highest concentrations during the recommended December harvest period (di Cesare *et al.*, 1999). In Denmark, Sørensen *et al.* (1995) found that restricting the water-supply to leek crops resulted in increased dietary fibres, vitamin C, nitrate, protein, Ca, Mg and Mn. The nutritional quality of the leeks improved over time, since the concentration of K rose and there was a reduction in nitrate content.

Brunsgaard *et al.* (1997) compared the effect of a range of N levels on leek quality in dietary experiments with rats: protein content increased with N applications, while during the autumn the protein content tended to fall; the total biological food value rose over time from September to November.

3. Genetics and Breeding

3.1 Genetics of leek

Plants of the *A. ampeloprasum* complex have a variable degree of ploidy: $2n$ = 24, 32, 40, 48 or 56. However, all commercial leek cultivars probably have the tetraploid state ($2n = 4x = 32$) (van der Meer and Hanelt, 1990). Several authors consider that leek is an autotetraploid (Berninger and Buret, 1967; Schweisguth, 1970; Stack and Roelofs, 1996), but others suggest that leek is an allotetraploid, originating from three related wild parents (Koul and Gohil, 1970; Khazanehdari *et al.*, 1995).

During meiosis, leek chromosomes show 71% quadrivalents in prophase I, while in metaphase I most of the quadrivalents transpose to bivalents (Jones *et al.*, 1996). In these bivalents the chiasmata are confined to the pericentromeric one-third of the chromosome (Levan, 1940). Several authors have suggested that this strong chiasma localization has survival value, because it is a bivalentizing mechanism, reducing the frequency of unbalanced gametes (Stack, 1993; Khazanehdari and Jones, 1997; Kik, Chapter 4, this volume). A consequence of chiasma localization may be the suppression of recombination in the distal two-thirds of the chromosomes. Some cytologists speculated that many genes in leek are inherited as tightly linked 'supergenes' (Ved Brat, 1965) or 'gene blocks' (Gohil, 1984). This block inheritance, if it exists, may severely limit the possibilities for leek breeding, as desired alleles may be linked with one or more of the frequent recessive deficiency alleles (Berninger and Buret, 1967) or with one or more of the many genes from wild relatives (Smilde *et al.*, 1997). However, leek chromosomes actually pair along their whole length in prophase I (Levan, 1940; Khazanehdari *et al.*, 1995). Therefore points of recombination may be distributed in a more random fashion than appears from the chiasma positions at the more condensed state of the chromosome in metaphase I. Points of recombination can be seen in pachytene as recombination nodules (RNs) (Sherman and Stack, 1995). In leek, Stack and Roelofs (1996) found that 1.5% of the RNs were outside the proximal two-thirds of the chromosomes. The positions of RNs and centromeres could not be determined simultaneously. Therefore, the authors could not exclude the possibility that most RNs are confined to the proximal one-third of the chromosomes, and consequently the speculation about block inheritance in large blocks, comprising two-thirds of the genome (Gohil, 1984) could not be rejected.

More recently, amplified fragment length polymorphism (AFLP) technology (Vos *et al.*,

1995) has provided an opportunity to generate many molecular markers without prior sequence information from the genome. AFLP markers reveal a high polymorphic rate in many crops, such as ryegrasses (*Lolium* spp.) (Roldán-Ruiz *et al.*, 2000) and azaleas (*Rhododendron* spp.) (de Riek *et al.*, 1999). Leek is classified among the species with a large genome (24.1–26.8 pg per 2C) (Labani and Elkington, 1987; Arumuganathan and Earle, 1991). EcoRI/MseI AFLP markers represent both hyper- and hypomethylated regions of the chromosomes. Therefore, they may be distributed randomly over the whole chromosome (they identify active and non-active genes, which are supposedly spread at random throughout the chromosomes) (Boivin *et al.*, 1999). Indeed, in the bulb onion, these molecular markers appear to be evenly distributed over the linkage map (van Heusden *et al.*, 2000), while, in barley, AFLPs reveal suppressed recombination near centromeres (Qi *et al.*, 1998).

Extremely strong suppression in large parts of the genome, as is supposedly the case in leek, should therefore be detectable with relatively few AFLP markers. Smilde *et al.* (1999) showed that AFLP markers in leek are not inherited in large linkage blocks, in spite of the chiasma concentration mainly in the proximal one-third and RNs occurring mainly in the proximal two-thirds of the chromosome. Therefore, it now appears that exceptionally strong linkages in two-thirds of the genome do not limit recombination and therefore should not impede the improvement of leek via classical breeding procedures.

3.2 Breeding history

The early leek cultivars were actually landraces and were highly variable in agronomic and morphological traits: they were often named for their localities of origin. They were well adapted to local conditions and were propagated by mass selection. Leek growers harvested their own seeds from selected plants, which were chosen for superior phenotypic characteristics.

It seems obvious that selection for winter-hardiness took place in the cool temperate zone and for early and fast-growing landraces in southern European and Mediterranean countries with less severe winters (Silvertand, 1996). However, little early written evidence is available on leek landraces. Some Greek and Roman writers mention the distinction between *porrum capitatum* (leek) and *porrum sectilis* (chives), but it was De Combes (1752) who, for the first time, mentioned two leek landraces in France: 'Le Long' and 'Le Court'. A very famous landrace, 'Musselburgh' or 'Scotch Flag', from near Edinburgh in Scotland, has been known since the early 19th century. Vilmorin-Andrieux (1856) described several leek cultivars and also mentioned a long-shafted winter leek ('Poireau long d'hiver de Paris') as a selection out of 'Le Long'. 'Le Court' was also known as 'Gros du midi', 'Broad Flag', 'London Flag' or 'American Flag'. Another winter-hardy selection, cultivated since 1874, was derived from the famous french 'Gros court de Rouen' and known in France, Belgium, Britain and The Netherlands as the king of the leeks: 'De Carentan' or 'The Monstrous Carentan' (Gibault, 1912; Bois, 1927). Both the 'Musselburgh' and the 'Carentan' were also popular in the USA at the beginning of the 20th century (Cox and Starr, 1927). During this time, local varieties, such as 'Elephant', an autumn leek, 'Dutch Brabander Winter' or 'Flanders Winter leek' and 'de Liège' or 'Luikse Winter', were selected.

Leek breeding was reviewed by Currah (1986) and by Pink (1992), and we will mainly review later work. During the second half of the 20th century, numerous new cultivars were developed. Winter-hardiness, long shafts in winter varieties, erectness of the leaves and dark leaf colour were the desirable traits. From the 1960s onwards, modern methods based on selection of family breeding lines replaced mass selection and cv. 'Alaska' was the first modern leek release. This was followed by a number of modern cultivars, developed through selection. Within the UK and Belgium these replaced most of the older landraces, causing severe genetic erosion. In Belgium there are still some landraces of leek, but they are fast disappearing and being replaced by commercial cultivars (Fig. 18.6; De Clercq *et al.*, 1999).

Nowadays F_1 hybrids are gaining popularity among growers, due to their higher yields and improved uniformity compared with open-pollinated cultivars. Hybrid (in fact, 'near-hybrid', as none of the male parents are true inbreds) seed production utilizes a naturally occurring male-sterile clone (Havey, Chapter 3, this volume) discovered in the UK by Smith and Crowther (1995). The first male parents used in hybrid seed production were not purposely bred, but commercial varieties were tested for their combining ability with the male-sterile lines. Therefore, yield improvement in the near-hybrids seems to be, in part, due to the elimination of all inbred plants derived from self-pollination (which are common in open-pollinated cultivars), together with heterotic effects at some loci (Smith and Crowther, 1995). The cultivar 'Carlton F_1' was the first commercial hybrid leek produced by this means, and is being followed by many others, where deliberately selected improved male parents are now being used.

Another approach for leek improvement may be to transform the genome of leek, either by *Agrobacterium*, which has been unsuccessful until now, or by 'biolistics' (Wang, 1996). By cotransforming the calli with *barnase/barstar* genes (Mariani *et al.*, 1990), the transformation frequency was significantly improved, and Wang (1996) obtained nine transgenic plant lines using this method.

A cheaper system for large-scale hybrid seed production would be one based on cytoplasmic male sterility (CMS), which until now has not been found in leek, either naturally or after induction. In recent years, significant progress has been made in the development of protoplast to plant systems for *Allium* species (Buiteveld and Creemers-Molenaar, 1994). In the future, this system may be used in CMS introgression from onion or chives to leek. With that aim, Peterka *et al.* (1997) generated an interspecific hybrid between S-cytoplasmic onion and leek as an initial step to transfer CMS from onion to leek, and Buiteveld (1998) developed a method for somatic hybridization between leek and onion by protoplast fusion.

3.3 Current breeding goals

3.3.1 Breeding for uniformity

AGRICULTURAL AND MOLECULAR ASPECTS OF UNIFORMITY. Leek is an outbreeding species with up to 20% self-fertilization. Much of the variation within open-pollinated cultivars is explained by the strong inbreeding depression that occurs in selfed plants (Berninger and Buret, 1967). After one selfed generation, plant weight declined by 26–62% compared with the average for the starting population (Gagnebin and Bonnet, 1979). Testing I_1 and I_2 lines, Schweisguth (1970) recorded an average yield decrease of 35% (between 12 and 54%) for each of the inbred generations, but achieved higher uniformity for plant habit and leaf colour. In contrast, Smith and Crowther (1995) showed that yields of I_1 and I_2 lines decreased by 28 and 32%, respectively, with no reduction in the coefficients of variation within populations/families for several measured traits. Our own experiments with I_1 and I_2 populations show that I_1 yields decreased by about 60–80% and I_2 yields by an additional 27%, with increased coefficients of variation for plant weight and shaft length (Table 18.5, Fig. 18.7).

Khazanehdari and Jones (1997) found frequencies of aneuploids of 4.3–8.4% in each of the four leek cultivars used in their trial, and aneuploidy can also increase variability in an autotetraploid crop.

Dominant AFLP markers can be used to determine the relationship between yield and cross-fertilization. Smilde *et al.* (1999) used primer combinations with seven selective bases and showed that in leek these markers are not inherited in large linkage blocks. We used AFLP markers on four parents, which were combined in two pair-crosses of two individual plants, and their offspring. Two of the six primer combinations

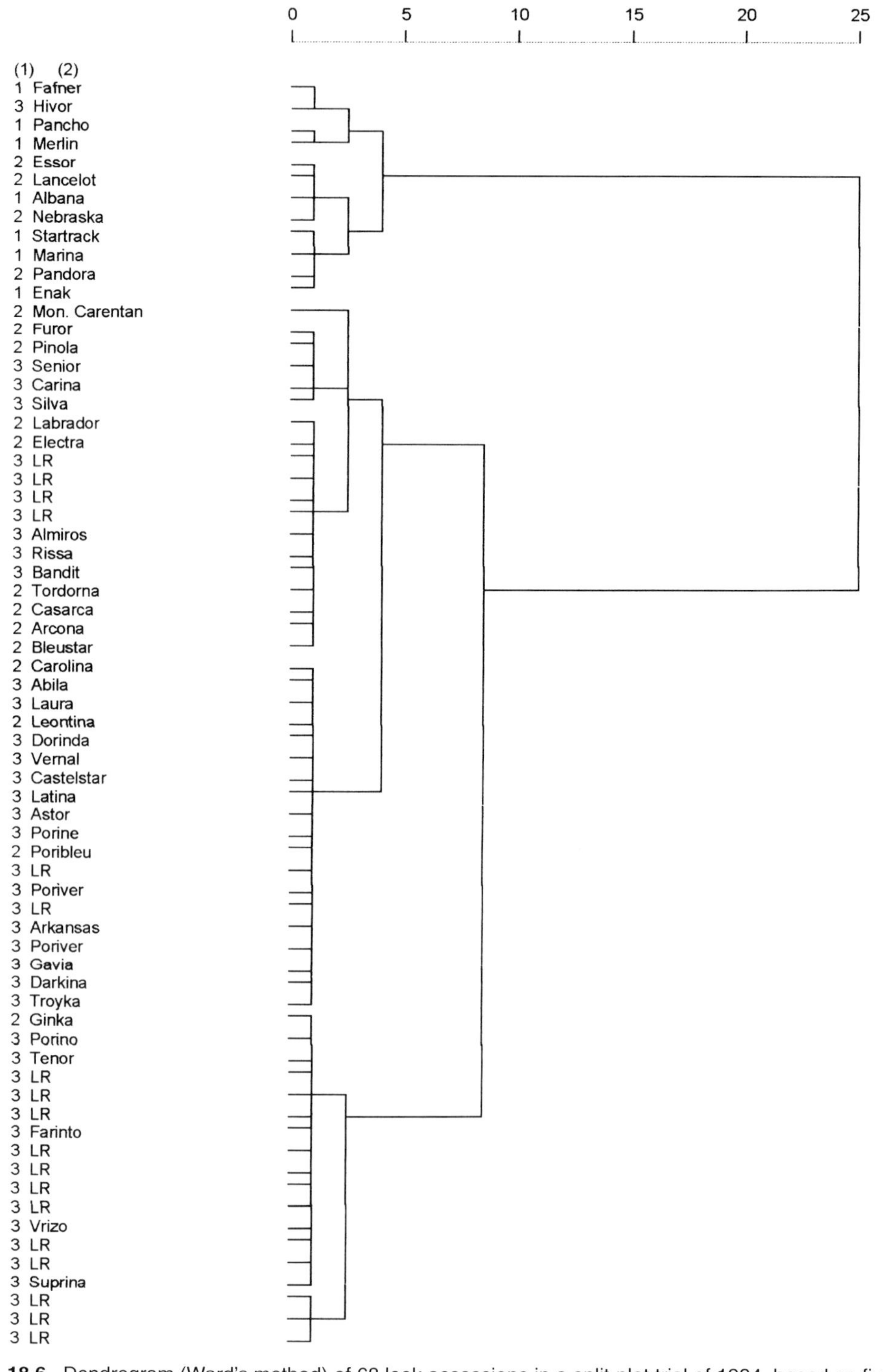

Fig. 18.6. Dendrogram (Ward's method) of 68 leek accessions in a split-plot trial of 1994, based on five parameters: leaf colour, shaft length, senescence, yield in September and yield in February. (1) Leek type in catalogue: 1 = summer-, 2 = autumn- and 3 = winter-type leek; (2) cultivar names. LR indicates leek of Belgian origin (different landraces). (From De Clercq *et al.*, 1999.)

Table 18.5. A comparison between an open-pollinated leek (Landrace Y) and the first (L152) and second (L567) self-pollinated populations: seed weight (10^{-3} g), germination (1–5), transplant weight (g), harvest weight (g) and shaft length (mm). Data show means and coefficients of variation (CV%). (From H. De Clercq, D. Peusens, I. Roldán-Ruiz and E. Van Bockstaele, unpublished).

	Landrace Y		L152		L567	
Trait	Mean	CV%	Mean	CV%	Mean	CV%
Seed weight	2.28	22.5	4.04	15.5	3.97	14.3
Germination	3.11	39.9	2.70	41.5	3.16	44.9
Transplant weight	1.71	88.9	1.05	96.2	0.85	111.8
Harvest weight	45.57	61.9	7.43	76.6	5.28	97.0
Shaft length	16.95	8.8	15.72	7.8	15.36	14.1

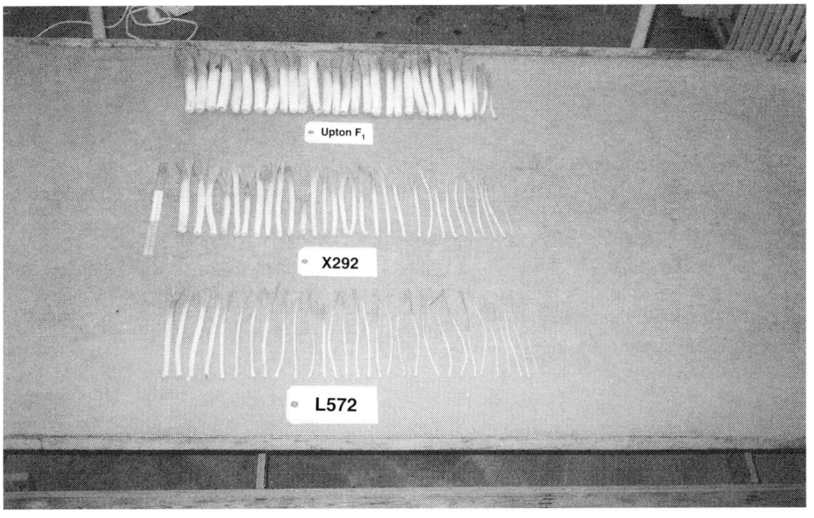

Fig. 18.7. Inbreeding depression in leek after self-pollination (L572) compared with pair–crossing (X292) and controlled cross-pollination (Upton F_1).

with seven selective bases gave a number of exclusive fragments for each parent. These markers were used to screen the progeny genotypes, and this indicated that high-yielding plants were generally heterozygous (D. Peusens, I. Roldán-Ruiz, H. De Clercq and E. Van Bockstaele, unpublished data).

FAMILY SELECTION, CLONAL PROPAGATION AND INBREEDING IN LEEK. The problem of poor uniformity can only partly be solved by traditional breeding methods, such as mass and family selection. Clonal propagation of selected plants and inbreeding in leek are currently receiving more attention. Hybrid seed, however, seems to be the most powerful tool for cultivar improvement.

Family selection is the basic breeding method used for leek improvement, using the system of half-sib families (i.e. same mother, father unknown). We studied the heritability of the percentage of weak plants in populations developed using this scheme. In eight half-sib families created from among the offspring of cv. 'Metro', the average percentage of weak plants in the second and third generations was stable at 17%. In other cases, cultivars 'SW 8026' and 'Bulgina' originally had 24 and 9.4% weak plants, respectively: this was markedly lowered in the first generation of selection to only 7.6 and 5.0% weak plants, respectively. Hence, improvement in uniformity seems possible through family selection.

Leek clones can be propagated by topsets (Schweisguth, 1970; Debergh and Standaert-de Metsenaere, 1976) or, *in vitro*, using the basal plate (Doré, 1988), the floral stem (Lavrijsen *et al.*, 1993) or the umbel (Baumunk-Wende, 1989) as explants. These clones show phenotypic uniformity. To increase phenotypic homogeneity in cultivars derived from them, while retaining heterozygosity in the new populations, we are developing polycross-based cultivars (synthetic varieties) by cross-pollination between clones propagated either by topsets or *in vitro*. The resulting S_1 synthetic cultivars derived from the same clonal mother plants are nearly identical in vegetative and reproductive traits, such as upright habit, leaf colour, flowering time, length of scape and anthocyanin pigmentation of the tepals. Selection between clones is more reliable than attempting to select individual plants from segregating populations. The parental clones are propagated vegetatively, thus guaranteeing their perpetuation.

Leek suffers from strong inbreeding depression, which may be associated with a high frequency of lethal genes. This means that there is only a low probability of developing productive inbred lines. Using the single-seed descent (SSD) method, Smith and Crowther (1995) demonstrated the magnitude of deleterious inbreeding effect in leek and concluded that only the first generation of inbred lines should be used as parents in the production of hybrids. Over several years, we recorded a substantial variation between inbred genotypes: some produced only a few seeds with very low germination and poor growth potential, while others produced hundreds or thousands of seeds with vigorous plants.

CONTROLLED CROSS-POLLINATION AND HYBRID SEED PRODUCTION. Schweisguth (1970) selected among the offspring of two individual leek plants those with vigorous growth. He then crossed two selected plants from the different parents and found that the hybrid progeny were more vigorous than the most vigorous parent. Kampe (1980) emasculated individual flowers and hand-pollinated them to produce experimental hybrids that demonstrated hybrid vigour for both yield and diameter of shaft, but not for shaft length. Smith and Crowther (1995) found plants with natural forms of nuclear male sterility by searching in field populations. These male-sterile (MS) plants were propagated vegetatively and cross-pollinated to produce 'near-hybrids'. In the same period (1990–1995), Silvertand in The Netherlands discovered a few (0.02%) MS plants in seed-production fields. He increased the percentage of MS plants by full-sib pollination of the MS plant with fertile plants from the same half-sib population (Silvertand, 1996). None of the offspring populations was 100% MS, which means that none were found to be 'cytoplasmically male-sterile' (CMS). Based on segregation studies, monogenic inheritance could be postulated for the genetic control of the MS trait in most of the progenies (Schweisguth, 1970; Rauber, 1989; Silvertand, 1996).

The appearance of MS leek flowers differs from that of fertile flowers: they can be either more narrow and more closed or more open and larger than a normal flower. The tepal colour is usually purplish-white, but some sterile flowers are white or whitish-yellow. Inside immature, unopened MS flower buds, the anthers already have a shrunken appearance, compared with the swollen-looking unopened anthers of fertile flower buds. Sterile anthers can even be confused with anthers in older normal flowers which have already shed their pollen and degenerated while the tepals have closed up again. We noticed that the first few flowers of a male-fertile leek umbel sometimes appear to be sterile, while the later-opening flowers may all be normal and fertile. Male-sterility is an exceptional state, and not all MS plants identified are useful for breeding, since some are too weak to propagate or give no or only very few seeds after cross-pollination.

Using emasculation and hand-pollination, we made four experimental 'near-hybrids', which had a 17% higher germination rate, a doubled seedling weight and a tripled weight at harvest compared with offspring from self-pollination of the same plants. The gain in uniformity of the 'near-hybrids' is reflected by a 50% decrease in the standard deviation of the harvest weight in comparison with that of the I_1 inbreds (Table 18.6, Fig. 18.8).

3.3.2 Breeding for resistance

Inherited resistance can provide reliable protection from pests and disease and reduce production costs and pollution. Whenever available, resistance in the crop is better for both producers and consumers compared with chemical control. Consumers are deeply concerned about health and the safety of foods. Therefore growers nowadays try to limit the use of fertilizers and pesticides, in order to reduce pollution and to protect human health and the environment. In several western countries, governments are beginning to support research into organic farming. In future, both conventional and organic farming will increasingly require resistant cultivars.

THRIPS RESISTANCE. In onion, resistant cultivars produce leaves that are widely separated on the pseudostem, thus minimizing the shelter for thrips between the leaves (Jones and Mann, 1963). Leaf position and age in cucumber have a significant effect on thrips reproduction, whereas plant age has not. In general, thrips reproduction was highest on young cucumber leaves (de Kogel *et al.*, 1997). Some resistant chrysanthemum cultivars produce a low content of the amino acids that are essential for the moulting of the pupae of the insect. Some resistant clover varieties act in a comparable way, thus limiting thrips numbers (J. Harrewijn, PRI, The Netherlands, 1997, personal communication).

Theiler and Buser (1996) worked on clones and inbreds of leek and found that progenies from somewhat tolerant plants showed a significantly lower attack by thrips than those from highly susceptible plants. They concluded that there is a genetic basis for thrips resistance in leek. We found only a slight difference in susceptibility among currently available cultivars. Within one cultivar, individual plants are sometimes more severely damaged by thrips than others and show silver-coloured leaves. Hence, individual plant selection may be worth pursuing.

RUST RESISTANCE. Dixon (1976) suggested that some cultivars are more susceptible to rust (*Puccinia allii*) than others. Studying the reaction of leek cultivars to infection by rust, clear differences were found in sporulation intensity, infection frequency and latent period. These parameters can serve for describing variation in host response (Uma and Taylor, 1991). Inoculation tests indicated 'fast-' and 'slow-rusting' cultivars. Further epidemic studies were done by de Jong (1995), and by de Jong and de Bree (1995). Plant age was important but differed in its effects: some cultivars were more susceptible as young plants (cv. 'Albana'), others as old plants (cv. 'Cortina'). Other factors influencing the progress of the leek-rust

Table 18.6. Plant characteristics of 'near-hybrid' leeks made by hand-pollination after emasculation, in comparison with their I_1 inbreds produced at the same time (mean of four genotypes) (From H. De Clercq, D. Peusens, I. Roldán-Ruiz and E. Van Bockstaele, unpublished).

Genotype	Germination	Transplant weight (g)	Damping-off in field	Adult weight (g)
'Near-hybrids'				
Mean	80%	10.9	6%	289
% STDEV*	8	38	4	33
I_1 inbreds				
Mean	63%	5.7	23%	109
% STDEV	30	47	13	64

* Standard deviation.

(a)

(b)

Fig. 18.8. Four progenies of controlled cross and self-fertilization in leek after emasculation and hand-pollination of two genotypes. (a) X241 F_1 emasculated and cross-pollinated by X242 L_1; (b) the reverse situation: X242 F_1 crossed and X241 L_1 self-fertilized.

epidemic were weather conditions (e.g. wind direction, wind speed and the growing conditions of the leek crop). Clarkson *et al.* (1996) found that optimal conditions for epidemics in the UK occur during August and September.

In our experiments, some F_1s between a susceptible and tolerant plant had an intermediate sensitivity, while others showed a maternal effect of sensitivity which slowed down in the G_2 (= F_2) generation (Fig. 18.9). In the UK, experiments have also

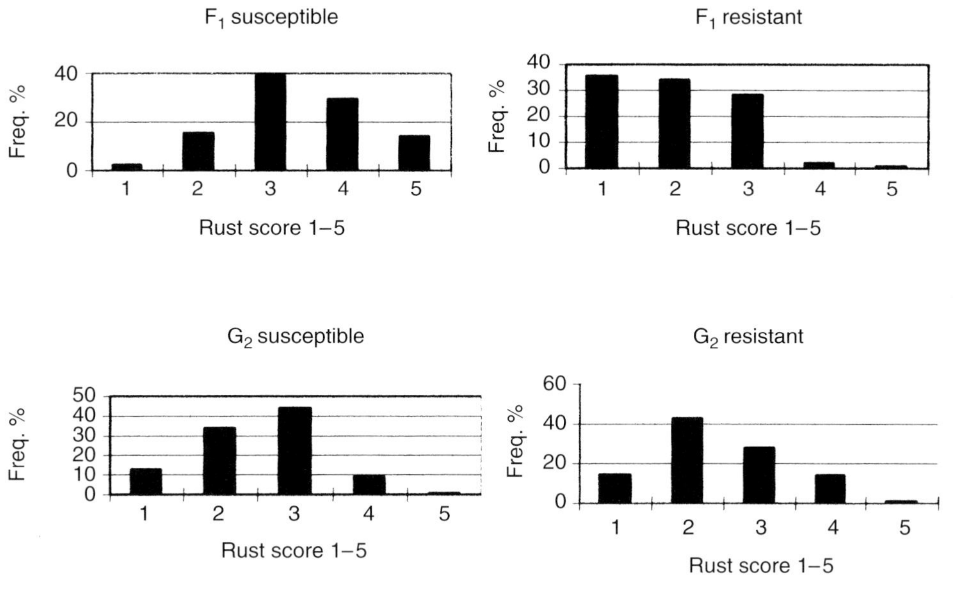

Fig. 18.9. Infection rate of leek rust in the field in October 1999. Two F_1 populations compared with their respective G_2 populations at DvP-CLO Experimental Station, Belgium.

been done on methods to define partial resistance to rust in leeks (Smith *et al.*, 2000). Although 16 commercial leek cultivars did not show great differences in resistance, a wider range of resistance levels appeared when half-sib and inbred progenies from the cultivars were compared. Therefore, selection within currently available cultivars is recommended as a viable short- to medium-term strategy for breeders wishing to improve rust resistance in leeks (Smith *et al.*, 2000). Smith *et al.* (2001) also crossed a range of related species with leeks in search of potential donors of leek-rust tolerance. They regard the species *A. commutatum* as the most promising crossable species and breeding has reached the F_2 stage: work is continuing.

WHITE-TIP RESISTANCE. Field and glasshouse experiments, with and without zoospore inoculations, revealed that leeks have a genetically based sensitivity to *Phytophthora porri* (Smilde *et al.*, 1995). Even within common cultivars, enough variation for resistance was found for breeding to start, without the need for backcrossing to sources of higher resistance. A recently developed screening method is based on inoculation by 24 h immersion of leek seedlings at the three- to six-leaf stage in a suspension of *c.* 100 zoospores ml^{-1}. This non-destructive test can be used to select young tolerant plants shortly before planting in a selection field (Smilde *et al.*, 1997). Smilde and co-authors studied the genetic basis of resistance to *P. porri* in winter leek cv. 'Carina' and some landraces. They used five classes to define resistance/tolerance. In cv. 'Carina' they found a partial resistance, thought to be controlled by a few recessive genes with independent quantitative expression. There was no significant response to selection for susceptibility. In experimental crosses of 'Carina' with landraces, they found resistance to white tip corresponding to a few loci with dominant genes (in accession CGN 873243) or additive polygenes (in accession PI 368351). Smilde *et al.* (1997) concluded that partial resistance to *P. porri* can be selected within one generation and that combining dominant resistance genes in leek hybrids is a possible strategy for the future. It is predicted that within 10–20 years *P. porri*-resistant cultivars may become available.

4. Conclusions

It is appropriate to point out in conclusion some scientific topics for investigation in the near future. It seems very worthwhile to continue to explore the centres of diversity of leek and its genetic relatives. Conservation *in vivo* or in seed banks of leeks and their related species will help to maintain enough genetic diversity for the future, now that hybrid breeding in leek is founded on very few genotypes. By means of DNA markers, further taxonomic and cultonomic studies can clarify relationships and distinctness between taxa of alliums and groups of leek. Physiological studies will define the optimal application of fertilizers and water-supply in order to achieve economic production of high-yielding leeks, while respecting the safety of food production and care for the environment. Research on integrated pest control methods in leek with application of more 'biological' products and warnings by local forecasting networks must help to reduce pesticide residues. Leek physiology and biochemistry need to be explored further in order to improve postharvest treatments, storage and food-processing methods. Marker-assisted breeding will become more important, not only in genetic studies related to the type or the uniformity of a leek crop, but also in relation to the desirable quality traits of leek – for example, better pest and disease tolerance, higher N-use efficiency, seed vigour and plant-growth adaptation. All this research will help the leek to become a worldwide crop.

References

Arumuganathan, K. and Earle, E.D. (1991) Nuclear DNA content of some important plant species. *Plant Molecular Biology Reporter* 9, 208–218.

Baumunk-Wende, E. (1989) Application of tissue culture in leek breeding (*Allium porrum* L.). *Gartenbauwissenschaft* 54, 20–24.

Benjamin, L.R. (1984) The relative importance of some different sources of plant weight variation in drilled and transplanted leeks. *Journal of Agricultural Science (Cambridge)* 103, 527–537.

Benoit, F. and Ceustermans, N. (1999a) Prei. Bodemmulch en druppelbevloeiing bij prei. *Proeftuinnieuws* 8(9), 36–37.

Benoit, F. and Ceustermans, N. (1999b) Prei. Witte bodembedekking en irrigatie hij prei: wat zijn mogelijkheden? *Proeftuinnieuws* 9(7), 13–14.

Benoit, F. and Ceustermans, N. (1999c) Prei. Gekleurde bodemmulch en Thripsproblematiek bij prei. *Proeftuinnieuws* 9(12), 29–30.

Berninger, E. and Buret, P. (1967) Étude des déficients chlorophylliens chez deux espèces cultivées du genre *Allium*: l'oignon *A. cepa* L. et le poireau *A. porrum*. *Annales de l'Amélioration des Plantes* 17, 175–194.

Berry, D. and Thicoïpe, J.P. (1998) La fertilization azotée du poireau d'automne. *Infos (Paris)* 140, 39–42.

Bois, D. (1927) *Les Plantes alimentaires chez tous les peuples*. Lechevalier, Paris, pp. 507–509.

Boivin, K., Deu, M., Rami, J.F., Trouche, G. and Hamon, P. (1999) Towards a saturated sorghum map using RFLP and AFLP markers. *Theoretical and Applied Genetics* 98, 320–328.

Brunsgaard, G., Sørensen, J.N., Kaack, K. and Eggum, B.O. (1997) Protein quality and energy density of leek (*Allium porrum* L.) as influenced by water and nitrogen supply and plant age at harvest. *Journal of the Science of Food and Agriculture* 74, 237–243.

Buiteveld, J. (1998) Regeneration and interspecific somatic hybridization in *Allium* for transfer of cytoplasmic male sterility to leek. PhD thesis, Wageningen Agricultural University, The Netherlands.

Buiteveld, J. and Creemers-Molenaar, J. (1994) Plant regeneration from protoplasts isolated from suspension cultures of leek (*Allium ampeloprasum* L.). *Plant Science* 100, 203–210.

Bujalski, N., Nienow, A.W., Petch, G.M., Drew, R.L.K. and Maude, R.B. (1991) The process engineering of leek seeds: a feasibility study. *Seed Science and Technology* 20, 129–139.

Callens, D. (1999) *Prei, overzicht van het onderzoek 1998*. Onderzoeks- en Voorlichtingscentrum voor Land- en Tuinbouw West-Vlaanderen, Rumbeke, Belgium, 120 pp.

Chen, G.-L., Dong, J.-Z. and Zhang, G.-H. (1998) Efficient cultivation of greenhouse leeks. *China Vegetables* 4, 49–50 (in Chinese).

Clarkson, J.P., Davies, J. and Kennedy, R. (1996) Insights on leek rust. *The Grower* 125(23), 25.

Cox, J.F. and Starr, G.E. (1927) *Seed Production and Marketing*. Wiley and Sons, New York and London, pp. 372–373.

Currah, L. (1986) Leek breeding: a review. *Journal of Horticultural Science* 61, 407–415.

Debergh, P. and Standaert-de Metsenaere, R. (1976) Neoformation of bulbils in *Allium porrum* L. cultured *in vitro*. *Scientia Horticulturae* 5, 11–12.

de Bohec, J. and Fouyer, L. (1999) Poireau: l'élevage du plant. *Infos (Paris)* 148, 35–39.

de Bohec, J., Erard, P. and Leteinturier, J. (1993) *Le Poireau. Guide Technique*. CTIFL, Paris, 185 pp.

De Clercq, H. (1981) *Note Concernant une Prospection sur le Maraîchage ou Sine-Saloum*. Centre pour le Développement de l'Horticulture, Cambérène, Dakar, Senegal, 35 pp.

De Clercq, H., Baert, J. and Van Bockstaele, E. (1999) Breeding potential of Belgian landraces of leek (*Allium ampeloprasum* L. var. *porrum*). *Euphytica* 106, 101–109.

De Combes (1752) *L'École du Jardin Potager*, Vol. II. Boudet, Paris, pp. 396–404.

de Jong, P.D. (1995) Growth of leek rust epidemics in time in three cultivars during the early stage of the epidemic. *European Journal of Plant Pathology* 101, 139–148.

de Jong, P.D. and de Bree, J. (1995) Analysis of the spatial distribution of rust infected leek plants with the black-white join-count statistics. *European Journal of Plant Pathology* 101, 133–137.

de Jong, P.D., Daamen, R.A. and Rabbinge, R. (1995) The reduction of chemical control of leek rust, a simulation study. *European Journal of Plant Pathology* 101, 687–693.

de Kogel, W.J., Balkema-Boomstra, A., van der Hoek, M., Zijlstra, S. and Mollema, C. (1997) Resistance to western flower thrips in greenhouse cucumber: effect of leaf position and plant age on thrips reproduction. *Euphytica* 94, 63–67.

de Kraker, J. and Bosch, H. (1993) *Teelt van Prei*. Teelthandleiding no. 56, Proefstation voor de Akkerbouw en de Groenteteelt in de Vollegrond (PAGV), Lelystad, The Netherlands, 112 pp.

den Belder, E. and Elderson, J. (1998) Suitability of leek for *Thrips tabaci* is reduced by intercropping with clover. *Proceedings of the Section Experimental and Applied Entomology of the Netherlands Entomological Society* 9, 123–127.

De Riek, J., Dendauw, J., Mertens, M., De Loose, M., Heursel, J. and Van Bockstaele, E. (1999) Validation of criteria for the selection of AFLP markers to assess the genetic variation of a breeder's collection of evergreen azaleas. *Theoretical and Applied Genetics* 99, 1155–1165.

De Rooster, L. and Spiessens, K. (1998) Prei. Stikstofsbemestingsproef: Nitrophoska Stabil en Cultan vergeleken met Nitrophoska Perfekt. *Proeftuinnieuws* 8(5), 20–23.

di Cesare, L.F., Giombelli, R., Senesi, E. and Ferrari, V. (1999) Influenza del periodo di raccolta sulla qualità di alcune cultivar di porro coltivate in Italia. *Alimentari Industrie* 38, 667–673.

Dixon, G.R. (1976) Observations on the incidence of leek rust in NIAB trials. *Journal of the National Institute of Agricultural Botany* 14, 100–104.

Doré, C. (1988) Multiplication végétative et conservation *in vitro* chez le poireau (*Allium porrum* L.). *Agronomie* 8, 509–511.

Edelson, J.V. and Magaro, J.J. (1988) Development of onion thrips, *Thrips tabaci* Lind., as a function of temperature. *The Southwestern Entomologist* 13, 171–176.

Ekbladh, G. (1995) N effects of organic manures on leeks. Influence of raised beds and mulching on N availability. In: Kristensen, K., Stopes, C., Kolster, P., Granstedt, A. and Hodges, D. (eds) *Nitrogen Leaching in Organic Agriculture. Proceedings of an International Workshop*. Royal Veterinary and Agricultural University, Copenhagen, pp. 157–171.

Embrechts, A.J.M. (1996) *Sortering van plantmateriaal bij prei*. Publicatie – Proefstation voor de Akkerbouw en de Groenteteelt in de Vollegrond No. 81B, Lelystad, The Netherlands, pp. 59–69.

Ester, A., de Vogel, R. and Bouma, E. (1997) Controlling *Thrips tabaci* (Lind.) in leek by film-coating seeds with insecticides. *Crop Protection* 16, 673–677.

Eurostat (1999) *Crop Production, Yearly Statistics*, Theme 5, no. 2. Publication Office of the European Community, Luxembourg.

Fenwick, R.G. and Hanley, A.B. (1990) Chemical composition. In: Brewster, J.L. and Rabinowitch, H.D. (eds) *Onions and Allied Crops*, Vol. III. *Biochemistry, Food Science, and Minor Crops*. CRC Press, Boca Raton, Florida, pp. 17–31.

Ferrari, V., di Cesare, L. and Schiavi, M. (1999) Valutazione di quattro cultivar di porro a raccolta autunno-vernina. *Informatore Agrario* 55(2), 55–58.

Finch, S. and Collier, R.H. (2000) Host-plant selection by insects – a theory based on 'appropriate/inappropriate landings' by pest insects of cruciferous crops. *Entomologia Experimentalis et Applicata* 96, 91–102.

Gagnebin, F. and Bonnet, J.C. (1979) Quelques considérations sur la culture et l'amélioration du poireau. *Revue Horticole Suisse* 52, 112–116.

Gibault, M.G. (1912) *Histoire des Légumes*. Librairie Horticole, Paris, pp. 167–172.

Gohil, R.N. (1984) Extent of recombination possible in the cultivated leek. In: *Proceedings of EUCARPIA 3rd* Allium *Symposium*. IVT, Wageningen, The Netherlands, pp. 99–105.

Gray, D. and Steckel, J.R.A. (1986a) Self- and open-pollination as factor influencing seed quality in leek. *Annals of Applied Biology* 108, 167–170.

Gray, D. and Steckel, J.R.A. (1986b) The effect of seed-crop plant density, transplant size, harvest date and seed grading on leek (*Allium porrum* L.) seed quality. *Journal of Horticultural Science* 61, 315–323.

Gray, D., Steckel, J.R.A. and Hands, L.J. (1992) Leek (*Allium porrum* L.) seed development and germination. *Seed Science Research* 2(2), 89–95.

Grzegorzewska, M. and Bakowski, J. (1996) The effect of cultivar and storage conditions on the quality and storage durability of leeks. *Biuletyn Warzywniczy* 45, 77–89 (in Polish).

Hanelt, P. (1990) Taxonomy, evolution, and history. In: Rabinowitch, H.D. and Brewster, J.L. (eds) *Onions and Allied Crops*, Vol. I. *Botany, Physiology, and Genetics*. CRC Press, Boca Raton, Florida, pp. 1–26.

Hanelt, P. (1996) Proposal to conserve the name *Allium ampeloprasum* against *A. porrum* (*Liliaceae*). *Taxon* 45, 691–692.

Hanelt, P., Schultze-Motel, J., Fritsch, R., Kruse, J., Maaß, H.I., Ohle, H. and Pistrick, K. (1992) Infrageneric grouping of *Allium*, the Gatersleben approach. In: Hanelt, P., Hammer, K. and Knüpffer, H. (eds) *The Genus* Allium, *Taxonomic Problems and Genetic Resources*. *Proceedings of an International Symposium held at Gatersleben*. IPK, Gatersleben, Germany, pp. 107–123.

Hay, R.K.M. and Kemp, D.R. (1992) The prediction of leaf canopy expansion in the leek from a simple model dependent on primordial development. *Annals of Applied Biology* 120, 537–545.

Hetterscheid, W.L.A. and Brandenburg, W.A. (1995) Culton versus taxon: conceptual issues in cultivated plant systematics. *Taxon* 44, 161–175.

Hetterscheid, W.L.A., van Ettekoven, C., van den Berg, R.G. and Brandenburg, W.A. (1999) Cultonomy in statutory registration exemplified by *Allium* L. crops. *Plant Varieties and Seeds* 12, 149–160.

Hoftun, H. (1978) Storage of leeks. III. Storage in controlled atmospheres. *Meldinger fra Norges Landbrukshøgskole* 57(38), 1–14 (in Norwegian).

Jones, G.H., Khazanehdari, K.A. and Ford-Lloyd, B.V. (1996) Meiosis in the leek (*Allium porrum* L.) revisited. II. Metaphase I observations. *Heredity* 76, 186–191.

Jones, H.A. and Mann, L.K. (1963) *Onions and Their Allies*. Leonard Hill, London, 268 pp.

Kampe, R. (1980) Untersuchungen zum Ausmass von Hybrideffekten bei Porree. *Archiv für Züchtungsforschung* 10, 123–131.

Khazanehdari, K.A. and Jones, G.H. (1997) The causes and consequences of meiotic irregularity in the leek (*Allium ampeloprasum* ssp. *porrum*): implications for quality and uniformity. *Euphytica* 93, 313–319.

Khazanehdari, K.A., Jones, G.H. and Ford-Lloyd, B.V. (1995) Meiosis in *Allium porrum* L. (the leek) revisited. I. Prophase I pairing. *Chromosome Research* 3, 433–439.

Klaas, M. (1998) Applications and impact of molecular markers on evolutionary and diversity studies in the genus *Allium*. *Plant Breeding* 117, 297–308.

Kolota, E. and Adamczewska-Sowinska, K. (1995) The effect of seedling raising and planting methods in early production of leeks. *Folia Horticulturae* 7(1), 3–13.

Kolota, E. and Adamczewska-Sowinska, K. (1996) Yield of early leeks in relation to agronomic variables. *Folia Horticulturae* 8(1), 19–28.

Koul, A.K. and Gohil, R.N. (1970) Cytology of the tetraploid *Allium ampeloprasum* with chiasma localization. *Chromosoma* 29, 12–19.

Kroonen-Backbier, B.M.A. and Rovers, J.A.J.M. (1998) Geintegreerde preiteelt beidt perspectief. *PAV-Bulletin Vollegrondsgroenteteelt* August, 18–23.

Kuzyakov, Y., Rühlmann, J., Gutezeit, B. and Geyer, B. (1996) Modelling on the growth and N uptake of leek and broccoli. *Acta Horticulturae* 428, 181–191.

Labani, R.M. and Elkington, T.T. (1987) Nuclear DNA variation in the genus *Allium* L. (*Liliaceae*). *Heredity* 59, 119–128.

Lambinon, J., de Langhe, J.-E., Delvosalle, L. and Duvigneaud, J. (1998) *Flora van België, het Groothertogdom Luxemburg, Noord-Frankrijk en de aangrenzende gebieden*. Meise, Nationale Plantentuin van België, Brussels, 1091 pp.

Lavrijsen, P.J.M., Silvertand, B.C.H.J. and van Harten, A.M. (1993) Bloemsteelmethode opent deur naar hybriderassen. *Prophyta* 3, 28–30.

Levan, A. (1940) Meiosis of *Allium porrum*, a tetraploid species with chiasma localisation. *Hereditas* 26, 454–462.

Mariani, C., Beuckeleer, M.D., Truettner, J., Leemans, J. and Goldberg, R.B. (1990) Induction of male sterility in plants by a chimaeric ribonuclease gene. *Nature* 347, 737–741.

Melander, B., Korsgaard, M. and Willumsen, J. (1999) Resultater og erfaringer med ukrudtsbekaempelse i okologiske frilandsgronsage. In: *16th Danish Plant Protection Conference. Crop Protection in Organic Farming. Pests and Diseases. DJF Rapport, Markbrug* 10, 85–95.

PCG (Proefcentrum voor de Groenteteelt Oost-Vlaanderen) (2000) *Jaarverslag 1999*. Kruishoutem, Belgium, 308 pp.

Peterka, H., Budahn, H. and Schrader, O. (1997) Interspecific hybrids between onion (*Allium cepa* L.) with S-cytoplasm and leek (*Allium ampeloprasum* L.). *Theoretical and Applied Genetics* 94, 383–389.

Pink, D.A.C. (1992) Leek *Allium ampeloprasum* L. In: Kalloo, G. and Bergh, B.O. (eds) *Genetic Improvement of Vegetable Crops*. Pergamon Press, Oxford, pp. 29–33.

Qi, X., Stam, P. and Lindhout, P. (1998) The use of locus-specific AFLP markers to construct a high-density molecular map in barley. *Theoretical and Applied Genetics* 96, 376–384.

Rauber, M. (1989) Untersuchungen zur männlichen Sterilität und *in vitro* Massenvermehrung bei *Allium schoenoprasum* L. und *Allium ampeloprasum* L. var. *porrum*. PhD thesis, Universität Hannover, Germany.

Richter, E., Hommes, M. and Krauthausen, J.H. (1999) Investigations on the supervised control of *Thrips tabaci* in leek and onion crops. In: Finch, S., Hartfield, C. and Brunel, E. (eds) *Integrated Control in Field Vegetable Crops. Bulletin OILB/SROP* 22(5), 61–72.

Roldán-Ruiz, I., Dendauw, J., Van Bockstaele, E., Depicker, A. and De Loose, M. (2000) AFLP markers reveal high polymorphic rates in ryegrasses (*Lolium* spp.). *Molecular Breeding* 6, 125–134.

Rowse, H.R. (1996) Drum priming – an environmentally friendly way of improving seed performance. *Journal of the Royal Agricultural Society of England* 157, 77–83.

Rumpel, J., Grudzien, K. and Borkowski, J. (1995) Wintering some vegetables under fleece. In: *Materialy ogólnopolskiej konferencji naukowej Nauka Praktyce Ogradniczej z okazji XXV-lecia Wydzialu Ogrodniczego Akademii Rolniczej w Lublinie*. Wydzial Ogrodniczy Akademia Rolnicza w Lublinie, Poland, pp. 687–690 (in Polish).

Schwartz, H.F. and Mohan, S.K. (1995) *Compendium of Onion and Garlic Diseases*. The American Phytopathological Society, St Paul, Minnesota, 54 pp.

Schweisguth, B. (1970) Études préliminaires à l'amélioration du poireau *A. porrum* L. Proposition d'une méthode d'amélioration. *Annales de l'Amélioration des Plantes* 20, 215–231.

Sherman, J.D. and Stack, S.M. (1995) Two-dimensional spreads of synaptonemal complexes from solanaceous plants. VI. High-resolution recombination nodule map for tomato (*Lycopersicon esculentum*). *Genetics* 141, 683–708.

Silvertand, B. (1996) Induction, maintenance and utilisation of male sterility in leek (*Allium ampeloprasum* L.). PhD thesis, Wageningen Agricultural University, The Netherlands.

Smilde, W.D. (1996) *Phytophthora* in leek: epidemiology and resistance. PhD thesis, Wageningen Agricultural University, The Netherlands.

Smilde, W.D., van Nes, M. and Reinink, K. (1995) Resistance to *Phytophthora porri* in leek and some of its wild relatives. *Euphytica* 83, 131–138.

Smilde, W.D., van Nes, M., Reinink, K. and Kik, C. (1997) Genetical studies of resistance to *Phytophthora porri* in *Allium porrum*. *Euphytica* 93, 345–352.

Smilde, W.D., van Heusden, A.W. and Kik, C. (1999) AFLPs in leek (*Allium porrum*) are not inherited in large linkage blocks. *Euphytica* 110, 127–132.

Smith, B.M. and Crowther, T.C. (1995) Inbreeding depression and single cross hybrids in leeks (*Allium ampeloprasum* ssp. *porrum*). *Euphytica* 86, 87–94.

Smith, B.M., Crowther, T.C., Clarkson, J.P. and Trueman, L. (2000) Partial resistance to rust (*Puccinia allii*) in cultivated leek (*Allium ampeloprasum* ssp. *porrum*): estimation and improvement. *Annals of Applied Biology* 137, 43–51.

Smith, B.M., Crowther, T.C., Treu, R., Trueman, L., McClenaghan, E.R., Astley, D. and Holmes, D. (2002) Wild species of *Allium* sect. *allium* and their value for introducing rust (*Puccinia allii*) resistance into cultivated leek (*Allium ampeloprasum* ssp. *porrum*). *Genetic Resources and Crop Evolution* (in press).

Sørensen, J.N., Johansen, A.S. and Kaack, K. (1995) Marketable and nutritional quality of leeks as affected by water and nitrogen supply and plant age at harvest. *Journal of the Science of Food and Agriculture* 68, 367–373.

Stack, S.M. (1993) Diploidization in autotetraploid *Allium porrum* by restricted crossing over. *American Journal of Botany* 80, 78–79.

Stack, S.M. and Roelofs, D. (1996) Localized chiasmata and meiotic nodules in the tetraploid onion *Allium porrum*. *Genome* 39, 770–783.

Theiler, R. and Buser, H.P. (1996) Resistenzzüchtung beim Lauch. Klonen-Pflanzen im Vergleich zu Sämlingen. *Der Gemüsebau/Le Maraîcher* 59(5), 4–6.

Theunissen, J. and Schelling, G. (1998) Infestation of leek by *Thrips tabaci* as related to spatial and temporal patterns of undersowing. *BioControl* 43, 107–119.

Tongaram, D., Schrevens, E., de Rijck, G. and de Proft, M. (1994a) Comparison of plant supporting systems and varieties for the hydroponic cultivation of leek. *Acta Horticulturae* 358, 401–405.

Tongaram, D., Schrevens, E. and de Rijck, G. (1994b) The optimization of the composition of the nutrient solution for hydroponics leek cropping. *Acta Horticulturae* 358, 407–413.

Uma, N.U. and Taylor, G.S. (1991) Reaction of leek cultivars to infection by *Puccinia allii*. *Plant Pathology* 40, 221–225.

Vanaerde, H. (1998) Prei. Ammoniumpolyfosfaat: een wondermiddel? *Proeftuinnieuws* 8(3), 35–36.

van der Meer, Q.P. and Hanelt, P. (1990) Leek *Allium ampeloprasum* L. In: Brewster, J.L. and Rabinowitch, H.D. (eds) *Onions and Allied Crops*, Vol. III. *Biochemistry, Food Science, and Minor Crops*. CRC Press, Boca Raton, Florida, pp. 179–196.

van der Meijden, R. (1996) *Heukels' Flora van Nederland*. Wolters-Noordhoff, Groningen, The Netherlands, 676 pp.

van Heusden, A.W., van Ooijen, J.W., Vrielink-Van Ginkel, M., Verbeek, W.H.J., Wietsma, W.A. and Kik, C. (2000) A genetic map of an interspecific cross in *Allium* based on amplified fragment length polymorphism (AFLP^{TM}) markers. *Theoretical and Applied Genetics* 100, 118–126.

Vanparys, L. (1991) Teelt van 'Oude Jonkman' of 'Stekprei'. *Revue de l'Agriculture* 44, 7–19.

Vanparys, L. (1998) *Cultivaronderzoek in de winterteelt van prei*. Mededeling – Provinciaal Onderzoek- en Voorlichtingscentrum voor Land- en Tuinbouw, No. 399, Beitem-Roeselare, Belgium, 4 pp.

Ved Brat, S. (1965) Genetic systems in *Allium* III. Meiosis and breeding systems. *Heredity* 20, 325–338.

Villeneuve, F., Bosc, J.P., Letouze, P. and Levalet, M. (1997) Activité de vols de *Thrips tabaci* en parcelles de poireaux et possibilités de lutte raisonnée. In: *Quatrième Conférence Internationale sur les Ravageurs en Agriculture*, 6–8 January 1997, le Corum, Montpellier, France. Vol. 2. Association Nationale pour la Protection des Plantes, Paris, pp. 563–572.

Villeneuve, F., Thicoïpe, J.P., Legrand, M. and Bosc, J.P. (1999) Peut-on raisonner les interventions contre le Thrips sur poireau? Quelles sont les stratégies? *Phytoma* 519, 32–37.

Vilmorin-Andrieux (1856) *Plantes potagères*. Vilmorin-Andrieux, Paris.

Vos, P., Hogers, R., Bleeker, M., Reijans, M., van de Lee, T., Hornes, M., Frijters, A., Pot, J., Peleman, J., Kuiper, M. and Zabeau, M. (1995) AFLP^{TM}: a new technique for DNA fingerprinting. *Nucleic Acids Research* 23(21), 4407–4414.

Wang, H. (1996) Genetic engineering. Male sterility in leek (*Allium porrum* L.). PhD thesis, Universiteit Gent, Faculteit Landbouwkundige en toegepaste biologische wetenschappen, Belgium.

Wiebe, H.J. (1994) Effects of temperature and daylength on bolting of leek. *Scientia Horticulturae* 59, 177–185.

Wiebe, H.J. (1995) Ursachen für das vorzeitige Schossen von Poree. *Gemüse (München)* 31, 689–700.

Wurr, D.C.E., Fellows, J.R., Hambidge, A.J. and Fuller, M.P. (1999) Growth, development and bolting of early leeks in the UK. *Journal of Horticultural Science and Biotechnology* 74, 140–146.

19 Ornamental Alliums

R. Kamenetsky[1] and R.M. Fritsch[2]
[1]*Department of Ornamental Horticulture, The Volcani Center, Bet Dagan 50250, Israel;* [2]*Institut für Pflanzengenetik und Kulturpflanzenforschung, Corrensstrasse 3, D-06466 Gatersleben, Germany*

1. Introduction

Up to the mid-19th century, only a small number of *Allium* species were cultivated as ornamental plants. This situation changed when Eduard Regel (1887) and other botanists described a number of spectacular species from Central Asia. Most of these taxa were introduced in the 1870s and 1880s via the Imperial Botanical Garden of St Petersburg, Russia, and later through other European gardens (Rümpler, 1882). Other magnificent *Allium* species were collected in the 1880s by British exploration trips in

 Allium Crop Science: Recent Advances (eds H.D. Rabinowitch and L. Currah)

South-West Asia and shipped to Europe for conservation and utilization (Dadd, 1987).

Recently, *Allium* species have become popular in rock gardens, herbaceous beds and perennial borders, especially for spring and summer blooming. These plants have also been introduced as commercial cut flowers for outdoor cultivation and forcing in greenhouses (*International Checklist*, 1991; De Hertogh and Zimmer, 1993; Bijl, 1995).

The ornamental value of the most popular species is based on their striking multiflowered inflorescences (e.g. *A. giganteum, A. aflatunense, A. karataviense*). However, some decorative alliums have only a few large flowers per umbel (e.g. *A. oreophilum, A. moly, A. insubricum*). The visual display is sometimes accompanied by a sweetish odour (e.g. *A. cyrilli* and *A. darwasicum*) or by spectacular foliage (e.g. *A. karataviense* and *A. macleanii*).

Commonly, blooming of individual umbels lasts 1 or 2 weeks, but several species (e.g. *A. giganteum*) may retain their ornamental traits for up to 1 month.

The Netherlands is the world's largest commercial producer of ornamental *Allium* bulbs. From 1995/96 to 1998/99, the annual area of production of ornamental *Allium* bulbs in The Netherlands increased by 33%, from 85 to 113 ha, and included 40 species and cultivars (PVS/BKD, 1998/99). A few species are currently produced by commercial growers in Israel, France, Japan and, recently, Latvia (Ruksans, 1998). Production is continuously increasing.

2. Botanical Classification, Morphology and Geographical Distribution

Allium is a very diverse genus, consisting of about 700 species and presenting a vast array of morphological traits (for details, see Fritsch and Friesen, Chapter 1, this volume; Table 19.1).

2.1 Subgenus *Melanocrommyum*

The largest number of ornamental species and cultivars belong to the subgenus *Melanocrommyum* (the main part of section *Molium* in the old sense of Regel and other authors), which includes about 130 species (Fritsch, 1992a). This large and diverse group is spread through the semi-deserts, deserts and mountainous steppes from the Canary Islands to Kazakhstan, China and Pakistan, with the centre of diversity in the eastern Mediterranean area and south-west and Central Asia (Fritsch, 1992a). It includes many endemic species, with life cycles adapted to the place of origin (Pistrick, 1992; Kamenetsky, 1994; Kamenetsky and Japarova, 1997). The plants possess true tunicated bulbs, with very thick outer scales and thinner inner storage scales (Fig. 19.1a). The mostly broad and flat leaves vary greatly in number and lack clearly visible sheath parts. The root system is annual and superficial and is well adapted for efficient absorption of water and minerals during the short growing period (Kamenetsky, 1992). The biomorphological type of the *Melanocrommyum* species has been named the 'melanoallium' type (Pastor and Valdes, 1985) or the 'extreme geoephemeroid' (Kamenetsky, 1992).

The floral stem varies in height from short (10–20 cm, *A. oreophilum*) to very long (180 cm, *A. giganteum, A. stipitatum*). Most species have dense and attractively coloured inflorescences (e.g. *A. aflatunense, A. giganteum*), but some have loose umbels with stiff, unequal pedicels and star-shaped flowers (e.g. *A. cristophii, A. schubertii*). Most species of this group contain only a very low level of cysteine sulphoxides (the source of volatile sulphur compounds) and are therefore almost odourless (M. Keusgen, Quedlinburg, 2000, personal communication).

2.2 Subgenus *Allium*

The subgenus *Allium*, the largest of the genus, is most diverse in the Mediterranean area, Asia Minor and Central Asia (Mathew, 1996). Ornamental *A. flavum, A. atroviolaceum, A. ampeloprasum* and *A. sphaerocephalon* are grown commercially. These species possess true ovoid or subglobose bulbs, with one, two or a few tunicated scales

Table 19.1. Ornamental alliums – popular species with ornamental potential. This list is a selection from about 300 species presented in horticultural catalogues and books on herbaceous and bulbous plants. For more species and detailed information, see also Bailey and Bailey (1976), Stearn and Campbell (1986), Huxley *et al.* (1991), Davies (1992) and Griffiths (1994). Most species are bulbous; rhizomatous habit is indicated. Economically important species are underlined.

No.	Species name	Origin	Specific characters	Flowering period and use
Subgenus *Melanocrommyum*				
1	<u>*Allium aflatunense*</u>* hort.		Incorrect name for *A. hollandicum*	
2	*Allium akaka* Gmel.	NW Iran, N Iraq, E Turkey, Transcaucasus	Leaves 1–2, broad, thick, scape 5–15 cm, umbel loose, flowers cream to pinkish, tepals recurved, after bloom starry	June Rock gardens
3	*Allium albopilosum* Wright		Synonym of *A. cristophii*	
4	*Allium alexeianum* Regel	Central Asia, Zeravshan and Turkestan ranges	Leaves 1–3, elliptic blue-green, scape 10–25 cm, head loose, pedicels unequal, flowers starlike, tepals greenish-white, purple midvein, finally stiff and spiky	May–June Rock gardens, dry bouquets
5	<u>*Allium altissimum*</u>* Regel	Central Asia, Pamir-Alai ranges	Leaves hairless, 3–6 cm wide, scape 90–120 cm, head denser, flowers deep carmine-violet; similar to *A. stipitatum*	May–June Herbaceous beds, cut flowers, dry bouquets
6	*Allium aschersonianum* Barbey	Israel, Lebanon, Jordan	Leaves linear-lanceolate, scape (30) 60–80 cm, umbel semi-globose, flowers red-purple	February–April Herbaceous beds, cut flowers, potted plants, forcing
7	*Allium atropurpureum* W. et K. Syn. *A. nigrum* var. *atropurpureum*	SE Europe	Leaves smooth, lanceolate, scape 40–100 cm, umbel finally semi-globose, dense, flowers starlike, tepals narrow, light to deep purple	June Herbaceous beds and borders, cut flowers
8	*Allium backhousianum* Regel Syn. *A. gulczense*	Central Asia, Alai range	Leaves large, broad, scape 120–140 cm, heads dense, flowers slightly greenish- or yellowish-white, tepals up to 15 mm, narrow triangular	May–June Solitary plants for herbaceous beds, cut flowers
9	*Allium cardiostemon* Fisch. et Mey.	N Iran, E Turkey, Transcaucasus	Leaves narrowly linear, scape 30–50 cm, heads dense, flowers small, blackish-purple, filaments purple with heart-shaped dilated basal part	June Rock gardens and borders

Continued.

Table 19.1. *Continued.*

No.	Species name	Origin	Specific characters	Flowering period and use
10	*Allium caspium* (Pallas) M. Bieb.	Caspian deserts and Central Asia to Pakistan	Leaves large, glaucous, scape stout, 20–35 cm, head large, loose, pedicels unequal, flowers pinkish, triangular–campanulate, filaments pink to lilac, very long exserted	June Well drained and dry places of rock gardens, excellent for dry bouquets
11	*Allium christophii*		Incorrect spelling of *A. cristophii*	
12	*Allium cristophii* Trautv. Syn. *A. albopilosum*	SW Asia, Kopetdag range	Leaves linear, sparsely to densely long white-hairy, scape 60–90 cm, head very loose, flowers large, starlike, tepals silvery pink to deep purple, lanceolate, after bloom stiff and spiky	June–July Striking in herbaceous borders and beds, excellent for dry bouquets
13	*Allium cupuliferum*[†] Regel	Central Asia (Uzbekistan), Nura-Tau range	Leaves 1–2, lanceolate, bluish green, flat on the scree, scape flexuous, 20–30 cm, pedicels upwards curved, unequal, umbel broadly fascicular, loose, flowers funnel-shaped, deep pink	May Rock gardens, dry borders, cut flowers
14	*Allium cyrilli* Ten.	NE Mediterranean area	Leaves lanceolate, smooth, scape 30–80 cm, umbel fascicular to semi-globose, flowers whitish to yellowish-green or pinkish, petals narrow with conspicuous midvein, filaments long, fleshy, conical	Mid-May to mid-June Small plantings for herbaceous beds, peculiar cut flowers with strong smell like carnations
15	*Allium darwasicum* Regel	Central Asia, S Pamir Alai, N Hindu-Kush ranges	Leaves coarse, narrowly lanceolate, scape 30–60 cm, umbel fascicular to semi-globose, dense, flowers funnel-shaped, dirty white or greenish yellow with darker midvein	Mid-May to end June Dry rock gardens and herbaceous borders, cut flower with pleasant smell like hyacinths
16	*Allium decipiens*[‡] Fisch.	N Caucasus to Volga area	Leaves narrow lanceolate, scape smooth, 30–50 cm, heads semi-globose to subglobose, flowers bright pinkish-violet	May–June Herbaceous borders and beds, cut flowers
17	*Allium elatum* Regel		Synonym of *A. macleanii*	
18	*Allium fetisowii* Regel	N Tien-Shan ranges	Leaves lanceolate, smooth, scape 40–50 cm, heads smaller, dense, flowers deep pink to pinkish-lilac, inner filament bases widened with narrow side-teeth	May–June Small plantings in herbaceous borders, cut flowers

19	<u>*Allium giganteum*</u> Regel	S Pamir–Alai and N Hindu-Kush	Leaves large, dull, scape dull, smooth, 100–150 cm, head extremely dense becoming 2–3 times larger during bloom, pedicels unequal, flowers broadly cup-shaped to star-like, pink to purple	June–July Attractive solitary plants in herbaceous beds, cut flowers Long-lasting bloom
20	*Allium gulczense* O. Fedtsch.		Synonym of *A. backhousianum* *A. gultschense* is incorrect spelling form of *A. gulczense*	
21	*Allium hirtifolium* Boiss.		Synonym of *A. stipitatum*	
22	<u>*Allium hollandicum*</u> R.M. Fritsch Incorrectly named *A. aflatunense**	Wild origin not known yet, distributed from Holland	Leaves lanceolate, ribbed, canaliculate, up to 6 cm broad, scape basally ribbed, 40–90 (120) cm, heads subglobose, moderately dense, pedicels equal, flowers pink to deep violet-purple, rarely white	(April–) May Borders and herbaceous beds, excellent cut flower
23	<u>*Allium jesdianum*</u>§ Boiss. et Buhse Syn. *A. angustitepalum, A. rosenbachianum*	Central Iran, S Pamir–Alai to N Hindu-Kush ranges	Leaves narrowly lanceolate, scape basally ribbed, 40–80 (100) cm, head subglobose, moderately dense, flowers starlike, pink to violet, upper filament parts conspicuously white	May (–beginning of June) Herbaceous beds and borders, best as small plantings, excellent cut flowers, dry bouquets
24	<u>*Allium karataviense*</u> Regel	Central Asia, Karatau to S Pamir–Alai ranges	Leaves 1–2, thick, bluish-green, reddish flushed, scape flexuous 20–30 cm, head semi-globose finally globose, flowers cream to meat-coloured, midvein darker, capsules deeply incised	May–June (July) Rock gardens, herbaceous borders, potted plants, fruiting umbels conspicuous in bouquets
25	*Allium lipskyanum*† Vved.	Central Asia, S Pamir–Alai ranges	Leaves narrow, canaliculate, scape 30–50 (80) cm, umbel initially very densely fascicular, later a very loose ovate head, flowers campanulate, pinkish-carmine with darker midvein	May Herbaceous beds and borders, conspicuous in dry bouquets
26	*Allium macleanii* Baker Syn. *A. elatum*	S Pamir–Alai range to Hindu-Kush	Like *A. giganteum* but scape shorter, leaves shorter and glossy, heads less dense, flowers pink to lilac, tepals cuspidate, filaments shorter exserted	May–June Herbaceous beds, cut flowers
27	*Allium multibulbosum* Jacq.		Synonym of *A. nigrum*	
28	*Allium nevskianum* Vved. ex Wendelbo	Central Asia, S Pamir–Alai ranges	Leaves long-ovate, thick, bluish-green, scape flexuous, 5–20 cm, head large, loose, flowers reddish to purple, midvein dark, tepals weak and crumpled after bloom	May–June Rock gardens, dry bouquets

Continued.

Table 19.1. *Continued.*

No.	Species name	Origin	Specific characters	Flowering period and use
29	*Allium nigrum* L. Syn. *A. multibulbosum*	Mediterranean area	Leaves linear-lanceolate, smooth, scape 40–90 cm, umbel broad-fascicular, dense, flowers greenish-white to pink with greenish midvein, ovaries broad, red to purple or greenish-black	(March) April–June Herbaceous beds and borders, cut flowers, forcing
30	*Allium nigrum* var. *atropurpureum* (W. et K.) Vis.		Synonym of *A. atropurpureum*	
31	*Allium oreophilum* C.A. Mey Syn. *A. ostrowskianum*	Caucasus to Tien-Shan and Tarbagatai	Leaves 2, linear, green to greyish-violet, longer than the flexuous 5–20 cm scape, umbel loose, initially fascicular later subglobose, flowers large campanulate, pink to deep brownish purple	June Rock gardens and borders, dry bouquets
32	*Allium ostrowskianum* Regel		Synonym of *A. oreophilum*	
33	*Allium protensum* Wendelbo formerly merged with *A. schubertii*	S Tajikistan, N Afghanistan, Kazakh deserts	Leaves large, smooth, scape 30–40 cm, head large, loose, pedicels strongly unequal, flowers starlike, tepals cream, stiff and starry after bloom, midvein brown to purple, broad	June Rock gardens and borders, excellent for dry bouquets
34	*Allium regelii* Trautv.	Kopet-Dag to SW Hindu-Kush	Leaves narrowly lanceolate, scape (20) 50–80 cm, inflorescence composed of (1) 2–6 superposed whorls, pedicels unequal, flowers few, campanulate, tepals bright lilac to purple, midvein dark	End May to mid-June Herbaceous beds and borders, cut flowers
35	*Allium rosenbachianum* hort§		Synonym of *A. rosenorum*	
36	*Allium rosenorum*§ R.M. Fritsch Syn. *A. rosenbachianum, A. jesdianum*	Central Asia, Zeravshan and Hissar ranges	Leaves many, linear-lanceolate, channelled, whole scape densely ribbed, 60–90 (120) cm, head dense, flowers shining pink, upper part of pedicels light pink	Mid-May to beginning of June Small plantings in herbaceous beds and borders, excellent for cut flowers
37	*Allium schubertii* Zucc. Name also misapplied to *A. protensum*	Near East and N Africa	Leaves large, smooth, scape 40–50 cm, head very large and loose, pedicels very unequal, flowers starlike, pink, after bloom crumpled, midvein inconspicuous	June Herbaceous beds, cut flowers, forcing, potted plants, dry bouquets

38	<u>*Allium stipitatum*</u>* Regel Syn. *A. hirtifolium*	Central Asia to Afghanistan and Iran	Leaves large, densely to sparsely hairy, or hairless with rough margin, scape 100–150 cm, completely smooth (only ribbed when dry), heads large, semi- to subglobose, flowers starlike, pink to purple, tepals lanceolate, reflexed, midvein darker	May–June Herbaceous borders and beds, cut flowers, forcing, dry bouquets
Subgenus *Amerallium*				
39	*Allium acuminatum* Hook. Syn. *A. murrayanum*	Pacific North America	Leaves 1–2, scape 10–30 cm, umbel with few bright pink to reddish-purple flowers	May–July Herbaceous beds, rock gardens
40	*Allium amplectens* Torr.	N America, Wyoming to N California	Leaves narrow, scape 20–50 cm, umbel subglobose, many white to pinkish flowers	(April) May–June Rock gardens
41	*Allium bulgaricum*[‖] (Janka) Prodán		Synonym of *A. siculum*	
42	*Allium cernuum* Roth	Canada to Mexico	Rhizomatous, leaves linear, scape 2-angled, (15) 30–60 cm, spathe recurved, umbel broad, loose, pedicels pendent, after bloom bent upwards, flowers pink to white, anthers strongly exserted	July–August Cooler spots of rock gardens and borders, dry bouquets
43	*Allium dioscoridis*[‖] Sibth. et Sm.		Synonym of *A. siculum*	
44	*Allium drummondii* Regel	N America, Nebraska to Texas	Rhizomatous, leaves narrow linear, scape 10–30 cm, umbel rather loose, fascicular, flowers pink to white	June Rock gardens and herbaceous borders
45	*Allium falcifolium* J.D. Hook	North coast ranges of Oregon and California	Leaves linear, falcate, scape flat, winged, 5–12 cm, umbels fairly dense, flowers large, deep pink to white, tepals recurved	May–June Rock gardens and herbaceous borders
46	*Allium fragrans* Vent.		Synonym of *Nothoscordum borbonicum*	
47	*Allium inodorum* Aiton		Synonym of *Nothoscordum borbonicum*	
48	*Allium insubricum* Boiss. et Reut.	S Alps	Rhizomatous, differs from *A. narcissiflorum* by scape 15–30 cm, umbel permanently pendent with fewer purple flowers	June (–July) Rock gardens and borders

Continued.

Table 19.1. *Continued.*

No.	Species name	Origin	Specific characters	Flowering period and use
49	*Allium luteum* hort.		Synonym of *A. moly* Also offered as *A. moly* 'Luteum'	
50	*Allium macranthum* Baker	Himalaya to SW China	Small tufts, fleshy roots, leaves linear, scape angled, 25–40 cm, umbel lax, flowers 7–12, large, pendent, deep purple, ovary tipped by conical outgrowths	July–August Borders and cool spots
51	*Allium moly* L. Syn. *A. luteum*	Mediterranean area	Leaves lanceolate, scape 15–35 cm, umbel loose, fascicular to semi-globose, flowers large, funnel-shaped, tepals yellow with green midvein	June Rock gardens, herbaceous beds and borders, cut flowers, forcing
52	*Allium murrayanum* Regel		Synonym of *A. acuminatum*	
53	*Allium neapolitanum* Cyr. Syn. *A. cowanii*	Mediterranean area	Leaves linear-lanceolate, keeled, scape 3-angled, 20–35 cm, umbel loose, fascicular to semi-globose, flowers cup-shaped, pure white	(February–) May Rock gardens, herbaceous beds, cut flowers, forcing
54	*Allium paradoxum* (M. Bieb.) G. Don var. *normale* Stearn	Transcaucasus, Kopet-Dag and Elburz ranges	Leaves 1–2, deep green, flat, keeled, scape 20–30 cm, edged, flowers 2–5, large, pendent, pure white	May Herbaceous borders, rock gardens, cut flowers, pot plants
55	*Allium siculum*[‖] Ucria Syn. *Nectaroscordum siculum*, *A. bulgaricum*, *A. dioscoridis*	N and E Mediterranean area	Leaves narrow, strongly keeled, yellowish green, scape 60–90 (120) cm, umbel very loose, fascicular, flowers large, pedicels pendent, unequal, tepals very broad, greenish with red and yellowish midvein	End May–mid-June Herbaceous beds, cut flowers, dry bouquets All plant parts with pungent smell
56	*Allium tripedale*[‖] Trautv. Syn. *Nectaroscordum tripedale*	SW Asia	Like *A. siculum* but pedicels shorter, flowers white with pinkish flush, earlier flowering	May Herbaceous beds, cut flowers, dry bouquets
57	*Allium triquetrum* L.	SW Europe	Leaves linear, strongly keeled, scape 15–25 cm, 3-angled, umbel fascicular, flowers few, pendent, white, midvein broad, green	February–April Herbaceous beds, cut flowers, forcing, pot plants
58	*Allium unifolium* Kell.	N America, Oregon, California	Bulbs on rhizome, leaves 2–3, flat, keeled, scape 40–50 cm, umbel fascicular, flowers few, rather large, bell-shaped, light to deep pink or violet	May–June Herbaceous beds and borders, cut flowers, forcing, pot plants

59	*Allium ursinum* L.	Europe and Asia	Leaves keeled, broadly lanceolate, long petiolate, scape angled, 30–50 cm, umbel loose, fascicular to semi-globose, flowers several, starlike, pure white, anthers included	May Shady borders, landscaping of shady places All plant parts with garlic smell
60	*Allium zebdanense* L.	Lebanon	Leaves narrow, flat, scape 20–40 cm, umbels fascicular, flowers few, large, white, campanulate	May Herbaceous borders, rock gardens, pot plants
61	*Nectaroscordum bulgaricum* Janka, *N. dioscoridis* Sibth. et Sm., *N. siculum* (Ucria) Lindl.		Synonyms of *A. siculum*	
62	*Nectaroscordum tripedale* (Trautv.) Traub		Synonym of *A. tripedale*	
63	*Nothoscordum bivalve* (L.) Britton	N America, S Central and SE USA	Leaves straight upright, yellowish-green, very narrowly linear, channelled, scape 15–25 cm, umbel narrow, flowers few, yellowish, funnel-shaped	May–July (–October) Herbaceous borders
64	*Nothoscordum borbonicum* (Ait.) Ravenna Syn. *Allium fragrans, N. fragrans* (Vent.) Kunth, *N. inodorum* (Ait.) Nichols., *N. gracile* sensu Stearn	Subtrop. Mexico and adjacent areas, naturalized worldwide	Leaves narrowly linear, flat, glaucous, scape 20–60 cm, umbel narrow, flowers several, white, funnel-shaped, fragrant, tepals often lilac flushed, midvein dark	Year-round, in temperate zones April–October Herbaceous borders in warm regions A noxious weed under frost-free conditions
Subgenus *Allium*				
65	*Allium ampeloprasum* L.	SW Europe to Near East	Leaves flat linear, keeled, scape 40–100 (180) cm, head globose, very dense, flowers papillose, violet to purple or white, anthers and long threadlike side-teeth of inner filaments exserted	May–June (July) Herbaceous beds, cut flowers, forcing
66	*Allium atroviolaceum* Boiss.	Crimea, Caucasus to Central Asia	Leaves flat, keeled, scape 60–120 (180) cm, head very dense, globose, flowers small, tuberculate, blackish-violet to blackish-purple, anthers and elongated side teeth of inner filaments exserted	July Small plantings for dry borders and herbaceous beds

Continued.

Table 19.1. *Continued.*

No.	Species name	Origin	Specific characters	Flowering period and use
67	*Allium azureum* Ledeb.		Synonym of *A. caeruleum*	
68	*Allium caeruleum* Pallas Syn. *A. azureum*	SW Siberia and Central Asia to Caucasus	Leaves 3-angled, smooth, scape 30–60 (90) cm, very dense heads of intensely blue, small flowers, filaments with 2 short basal side-teeth	June–July Hot and dry borders and beds, cut flowers, forcing, dry bouquets
69	*Allium caesium* Schrenk	Central Asia to S Siberia	Like *A. caeruleum* but leaves semi-cylindrical, coarse, scape 30–80 cm, flowers bluish-grey to sky-blue, 2 teeth in the upper filament parts beside the anthers	June Dry borders and rock gardens, cut flowers, dry bouquets
70	*Allium callimischon* Link	S Greece, Crete	Scape 10 cm (tall form 35 cm), covered by leaf sheaths, umbel lax, flowers few, small, whitish or pale pink (ssp. *haemostictum* Stearn, flowers dark-red-spotted)	September–October Hot dry spots of rock gardens
71	*Allium carinatum* L. ssp. *pulchellum* (G. Don) Bonnier et Layens Syn. *A. pulchellum*	Central Europe and Mediterranean area to Caucasus	Leaves narrowly linear, ribbed, scape 20–60 cm, spathe with 2 long appendages, umbel fascicular, loose, pedicels pendent, after bloom upright, flowers ovate, pink to violet, anthers long, exserted	June–July (August) Rock gardens and herbaceous borders
72	*Allium coeruleum*		Incorrect spelling of *A. caeruleum*	
73	*Allium flavum* L.	S and SE Europe to Asia Minor	Leaf blades semi-cylindrical, glaucous or deep green, scape 15–60 cm, umbel lax, pedicels unequal, spathe with 2 strongly elongated tips, flowers small, pendent, lemon yellow to golden, filaments long exserted	May–June Rock gardens and sunny borders
74	*Allium flavum* var. *minus* Boiss. Syn. var. *nanum* hort., var. *pumilum* hort., also named 'Minor'	Throughout the area of the common form	Scape only 5–12 cm	May–June Rock gardens
75	*Allium jajlae* Vved.		Synonym of *A. scorodoprasum* ssp. *jajlae*	
76	*Allium rotundum* L.		Synonym of *A. scorodoprasum* ssp. *rotundum*	

77	*Allium scorodoprasum* L. (ssp. *jajlae* (Vved.) Stearn and ssp. *rotundum* (L.) Stearn only), Syn. *A. jajlae*, *A. rotundum*	Central Europe to Turkey and Caucasus	Leaves flat, keeled, long sheaths, scape 25–80 cm, head dense, ovate to globose, flowers small, tepals papillose along the keel, dark brownish to purplish-red (ssp. *rotundum*) or pink to violet (ssp. *jajlae*)	Mid-May–June (ssp. *jajlae*) or mid-June–July) Herbaceous beds and borders, cut flowers, dry inflorescences with fruits are decorative
78	*Allium sphaerocephalon* L.	S and SE Europe to Asia Minor and Near East	Leaves semi-cylindrical, scape 30–80 cm, heads ovate, very dense, pedicels unequal, flowers deep pink to purple or brown-red, anthers and cylindrical side teeth of broadened inner filament bases exserted	June–July Alpine gardens and borders, cut flowers, forcing
Subgenus *Rhizirideum*				
79	*Allium albidum* Fischer ex Bieb. ssp. *albidum* and ssp. *caucasicum* (Regel) Stearn) Syn. *A. flavescens*	Eastern European steppes, slopes of Crimea and Caucasus	Rhizomatous, leaves narrow linear, flat, scape 10–30 cm, umbel semi-globose, lax, flowers yellowish-cream (ssp. *albidum*), or umbel denser, flowers whitish ± reddish flushed (ssp. *caucasicum*)	June–July Herbaceous borders, rock gardens
80	*Allium altyncolicum* Friesen Syn. *A. ledebourianum* hort.	S Siberia, shore of Lake Teletskoe	Rhizomatous, scape 40–60 cm, leaves thick, umbel dense, pedicels green, flowers large, glossy, filaments roughly as long as tepals	June Herbaceous beds and borders
81	*Allium amabile* Stapf	S China	Small rhizomes, thick storage roots, leaves threadlike, scape 10–20 cm, umbel loose, flowers few, large, narrow, deep pink to red	July–August Rock gardens and borders
82	*Allium angulosum* L.	Europe to Siberia	Rhizomatous, leaves narrowly linear, below sharply edged, scape two-angled, 20–45 cm, umbel semi-globose, flowers pink, tepals as long as filaments	July–August Rock gardens and herbaceous borders
83	*Allium barsczewskii* Lipsky	Central Asia, Pamir–Alai and Tien-Shan ranges	Rhizomatous, bulb tunics reticulate, leaves narrow linear, scape 30–60 cm, umbels fascicular, flowers few, campanulate, white or pink to carmine	June–July Borders and rock gardens, cut flowers

Continued.

Table 19.1. *Continued.*

No.	Species name	Origin	Specific characters	Flowering period and use
84	*Allium beesianum* W.W. Sm.	E Himalaya	Leaves linear, keeled, scape 20–45 cm, umbel small, fascicular, pendent, flowers few, narrowly ovate, light blue, up to 13 mm long	July–August Rock gardens and borders
85	*Allium cyaneum* Regel Syn. *A. purdomii*	NW Himalaya	Rhizomatous, leaves threadlike, scape thin, flexuous, 15–25 cm, umbel few-flowered, fascicular to semi-globose, flowers cobalt-blue, anthers long exserted	July–August Rock gardens and borders
86	*Allium cyathophorum* Bur. et Franch. var. *farreri* (Stearn) Stearn Syn. *A. farreri*	W China	Densely rhizomatous, roots thick, leaves narrowly linear, keeled, scape 2-angled, 20–30 cm, umbel fascicular, initially pendent, flowers long, deep pink to purple-violet, filaments cup-shaped, united	(May) June–July Rock gardens and not too dry borders May become invasive by self-seeding
87	*Allium farreri* Stearn		Synonym of *A. cyathophorum* var. *farreri*	
88	*Allium flavescens* Bess.		Synonym of *A. albidum* ssp. *albidum*	
89	*Allium kansuense* Regel		Synonym of *A. sikkimense*	
90	*Allium ledebourianum* Roem. et Schult. Name also misapplied to *A. altyncolicum*	S Siberia, S Altai range	Rhizomatous, scape 70–90 cm, umbels fascicular, pedicels blackish, flowers bluish-lilac, filaments exserted	June Herbaceous beds and borders
91	*Allium lusitanicum* Lam. Syn. *A. senescens* ssp. *montanum*, *A. senescens* of European authors	West and Central Europe	Rhizomatous, like *A. angulosum* but leaf blades without sharp angles below, umbel dense, flowers deep pink to purplish, filaments 1/5 longer than tepals	August–October Herbaceous borders and rock gardens, cut flowers
92	*Allium mairei* Lév. Syn. *A. yunnanense*	SW China	Rhizomatous, leaves threadlike, scape 10–40 cm, umbels with few large flowers, tepals pink or white with red spots	August–September Rock gardens and borders
93	*Allium narcissiflorum* Vill.	S and W Alps	Rhizomatous, leaves linear, scape 20–40 cm, umbel fascicular, initially pendent, flowers few, very large, pink to carmine	July–August Rock gardens, herbaceous borders, pot plants

94	*Allium nutans* L.	S Siberia and E Kazakhstan	Rhizomatous, like *A. senescens* but leaves wider and spread out, scape strongly 2-angled, umbel pendent before bloom, flowers whitish to bluish lilac or purplish, inner filament bases widened with 2 rounded teeth	August Herbaceous borders and beds
95	*Allium obliquum* L.	SW Siberia, E Europe, Romania	Bulbs large, ovate, leaves many, flat, keeled, with telescope-like sheaths, stem 60–120 cm, head dense, small, flowers yellow, anthers long, exserted	June–July Herbaceous beds and borders
96	*Allium odorum* L.		Synonym of *A. ramosum*	
97	*Allium purdomii* W.W. Sm		Synonym of *A. cyaneum*	
98	*Allium ramosum*¶ L. Syn. *A. tuberosum*	Mongolia, N China, E Asia	Rhizomatous, bulb tunics reticulate, leaves narrow, scape 2-angled, 50–80 cm, umbel dense, semi-globose, flowers white, tepals 5–9 mm, midvein brownish	July–October Herbaceous beds and borders, cut flowers Flowers with sweet smell
99	*Allium schoenoprasum* L. Syn. *A. sibiricum*	Europe, Asia, N America	Rhizomatous, leaves thin, cylindrical, hollow, scape 10–50 cm, hollow, heads dense, subglobose, flowers white or pink to carmine, midvein darker, tepals twice the length of filaments; polymorphous species, many ornamental varieties were selected	May–June, second bloom in late summer possible Herbaceous beds, rock gardens, cut flowers
100	*Allium senescens* L. Formerly merged with *A. lusitanicum*	Central and S Siberia	Rhizomatous, leaves linear, erect, scape 30–60 cm, heads dense, globose, flowers purplish or lilac, inner filament bases without teeth	June–August Borders and herbaceous beds, cut flowers
101	*Allium sibiricum* L.		Synonym of *A. schoenoprasum,* Robust arctic-montane form	
102	*Allium sikkimense* Baker Syn. *A. tibeticum, A. kansuense*	Himalaya to W China	Like *A. cyaneum* but plants larger, leaves broader, flat, flowers larger, filaments shorter than tepals	July–August Herbaceous borders
103	*Allium tibeticum* Rendle		Synonym of *A. sikkimense*	
104	*Allium tuberosum*¶ Rottl. ex Spreng		Synonym of *A. ramosum*	

Continued.

Table 19.1. *Continued.*

No.	Species name	Origin	Specific characters	Flowering period and use
105	*Allium victorialis* L.	Temperate Europe and Asia	Rhizomatous, leaves lanceolate, basally petiolate, scape angled, 30–60 cm, head dense, flowers cream to yellowish, filaments long, exserted	June–August Herbaceous beds, damp soil
106	*Allium yunnanense* Diels		Synonym of *A. mairei*	

* According to the modern *Allium* taxonomy (Gregory *et al.*, 1998), *Allium hollandicum* is a correct scientific name for the economically important species commonly named by horticulturists *Allium aflatunense.* This species is also closely related to *A. altissimum* and to the very polymorphous *A. stipitatum.*

§ Scientific name *Allium rosenbachianum* (hort.) is incorrectly applied by horticulturists to *Allium jesdianum* Boiss. et Buhse, *A. angustitepalum* Wendelbo or *A. rosenorum* R.M. Fritsch. True *Allium rosenbachianum* Regel, with lanceolate leaves, smooth scape and larger and less dense flower-heads, is found in Central Asia.

† *Allium lipskyanum* is generally sold under the name *A. cupuliferum.*

‡ South Siberian species, such as *A. tulipifolium*, are sometimes included in *A. decipiens.*

¶ *Allium ramosum* (early-flowering, large tepals) and *A. tuberosum* (late-flowering, small tepals) are connected by a wide range of transitional forms.

‖ The former genus *Nectaroscordum* belongs to genus *Allium* and is very close to subgenus *Amerallium* (see Fritsch and Friesen, Chapter 1, this volume).

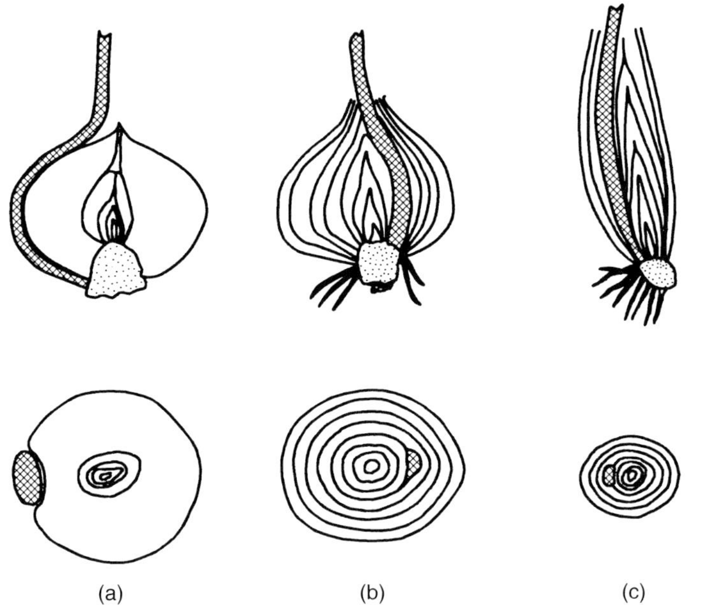

Fig. 19.1. Morphological structure of underground organs in the genus *Allium*. (a) *A. altissimum* Regel, subgenus *Melanocrommyum*; (b) *A. caesium* Schrenk, subgenus *Allium*; (c) *A. hymenorrhizum* Ledeb., subgenus *Rhizirideum*. (From Kamenetsky, 1996, with permission.)

(Fig. 19.1b), and annual root systems. The leaves are commonly narrow, thin or semi-cylindrical, with elongated above-ground leaf sheaths (Hanelt *et al.*, 1992). The floral stem may be any length from rather short (about 5 cm, *A. flavum* var. *minus*) to rather tall (150 cm, *A. atroviolaceum*, *A. ampeloprasum*). The inflorescence is either dense and drumstick-like (section *Allium*) or loose and spectacular, with colourful campanulate flowers on nodding pedicels (section *Codonoprasum*).

2.3 Subgenus *Rhizirideum*

The subgenus *Rhizirideum*, in the wide sense, groups together the members of several different evolutionary lines with well-developed rhizomes and mostly small and elongated false bulbs, which consist of more than two moderately thickened leaf-bases (Fig. 19.1c; see also Fritsch and Friesen, Chapter 1, this volume). Taxonomically, this grouping is artificial, but it is useful for horticultural purposes and will therefore be used here in this sense.

The leaves are narrow and often flat, sometimes cylindrical or semicylindrical (Cheremushkina, 1992; Hanelt *et al.*, 1992), and the root system is perennial (Fritsch, 1992b; Kamenetsky, 1992). Most species are adapted to humid or moderately dry conditions, and they grow in all altitude belts of Europe, Asia and North America. *A. barsczewskii*, *A. lusitanicum*, *A. nutans*, *A. obliquum* and others represent the ornamental potential of this subgenus.

2.4 Subgenus *Amerallium*

The subgenus *Amerallium* contains plants with diverse morphology. *A. moly*, *A. neapolitanum*, *A. siculum* and *A. unifolium* are common in the Mediterranean basin and in North America, and are adapted to a broad range of differing ecological conditions, from hot dry deserts to very humid conditions in dense forests (Mann, 1960; Gregory *et al.*, 1998). Some species possess rhizomes, poorly developed bulbs and short leaf sheaths (e.g. *A. cernuum*). Others produce distinct bulbs and broad leaves similar to those of the subgenus *Melanocrommyum* (e.g. *A. moly*), or very narrow leaves, as in the subgenus *Allium* (e.g. *A. unifolium*).

3. Horticultural Traits

Of the ornamental alliums, the most popular commercial cultivars are either selections or hybrids between species from the *Melanocrommyum* or *Amerallium* groups (Table 19.2; Wendelbo, 1967; Davies, 1992; Dubouzet *et al.*, 1992; De Hertogh and Zimmer, 1993; Bijl, 1995; Friesen *et al.*, 1997).

Flowering of ornamental alliums occurs in May–June in the northern-hemisphere temperate zone (Tables 19.1, 19.2). After flowering, the roots and above-ground part dry off, and the bulb enters a 'rest' period, which lasts for 3–5 months. In horticultural practice, bulbs are kept in storage, cleaned, handled and shipped to growers during the 'rest' period. *Melanocrommyum* species with large bulbs, such as *A. aflatunense, A. giganteum* and *A. macleanii*, are used in ornamental gardening and also for cut-flower production in greenhouses and open fields; some may be used as potted plants (*A. karataviense, A. cristophii, A. oreophilum, A. unifolium*; Table 19.1). Early-flowering small-bulb species, such as *A. oreophilum, A. moly* and *A. roseum*, are popular in European and American gardens (Davies, 1992; De Hertogh and Zimmer, 1993).

Rhizomatous species (e.g. *A. tuberosum, A. lusitanicum, A. nutans*) have rich foliage, which, in Europe, remains green from spring to autumn, and they produce many small and attractive inflorescences, which continue to flower through the summer and, in many cases, dry off only in the autumn. In spite of the fact that most of these species are traditional ornamentals in temperate climates, with evident potential for gardening, they are not yet widely used in ornamental horticulture in western countries.

4. Growth, Development and Flowering

A complete growth cycle of *Allium* species begins with seed germination, continues with a juvenile period of vegetative growth, which lasts from 1 to 5 years, and ends with the generative period and senescence.

4.1 Seed germination and juvenile period

In many ornamental species, mature seeds enter a dormant phase. Temperature is the principal factor affecting dormancy release and seed germination in *Allium* species. In general, germination is affected by temperature level and duration of temperature treatment (Dalezkaya and Nikiforova, 1984; Specht and Keller, 1997; Kamenetsky and Gutterman, 2000; Table 19.3).

The optimum temperature for seed germination of some ornamental alliums is closely related to the climatic conditions of their natural habitats. Thus, the germination temperatures of species from the Mediterranean basin, steppe or continental climatic zones in Central Asia correspond with cool-, cold- and warm-germinating types, respectively (Aoba, 1967; Table 19.3). However, many species show special ecological adaptations with regard to germination. For instance, *A. rothii* (subgenus *Melanocrommyum*) from the Negev Desert, Israel, where the winter is mild, germinates best within 14–28 days of wetting at 15°C (Gutterman *et al.*, 1995). Druselmann (1992) reported that the epigeal mode of germination dominates in *Allium* species, while only a few species which have adapted to humid conditions (e.g. *A. ursinum* and *A. victorialis*) show the hypogeal mode of germination.

Two morphological types of seedlings have been distinguished among *Allium* species with epigeal germination: the unspecialized and variable *A. cepa*-type, which

Table 19.2. Popular ornamental *Allium* cultivars. Economically important cultivars are underlined.

Cultivar name	Origin and classification	Specific characters, flowering period and use
Subgenus *Melanocrommyum*		
'Album'	*A. hollandicum* (possibly selected from *A. jesdianum* or *A. rosenbachianum*, see Table 19.1, footnotes * and §)	Scape ribbed 100 cm, white slightly purplish-tinged flowers May
'Album'	*A. stipitatum*	Scape smooth 100–120 cm, flowers pure white May–June
'Beau Regarde'	*A. cristophii* × *A. giganteum*	Scape strong 90 cm, head large, flowers lilac blue
'Colanda'	*A. rosenorum*	Scape 100 cm, tepals inside purple, outside dark violet
'Firmament'	*A. atropurpureum* × *A. cristophii*	Scape 80 cm, large umbels, flowers deep purple
'<u>Gladiator</u>'	*A. hollandicum* × *A. stipitatum*	Leaves large, somewhat hairy, scape 70–120 cm, head very dense, broadly globose, flowers purple; sterile May–June Herbaceous beds, cut flowers, forcing
'<u>Globemaster</u>'	*A. cristophii* × *A. macleanii*	Leaves large, bright green, glossy, scape glossy, 80–100 cm, heads large, flowers violet June Solitary plants in herbaceous borders, cut flowers, forcing; long-lasting bloom
'Globus'	*A. karataviense* × *A. stipitatum*	Leaves hairless, glaucous, scape 40–50 cm, head large, initially semi-globose, flowers pinkish May–June Herbaceous beds and borders, potting; long-lasting bloom
'His Excellency'	*A. macleanii*	Leaves slowly withering during bloom, scape 90 cm, flowers pinkish-violet
'John Dix'	*A. giganteum* × *A. cristophii*	Scape 100 cm, flowers mauve, paler to the centre
'Lucy Ball'	*A. macleanii* × *A. hollandicum*	Leaves yellowish-green, scape ribbed, 80–100 cm, head dense, flowers lilac–purple; sterile May Herbaceous borders, cut flowers, forcing

Continued.

Table 19.2. *Continued.*

Cultivar name	Origin and classification	Specific characters, flowering period and use
'Mars'	*A. stipitatum*	Leaves glossy, yellowish green, scape 100–120 cm, head large, flowers bright violet May–June
'Michael Hoog'	*A. rosenorum* (*A. jesdianum* or *A. rosenbachianum*, see Table 19.1, footnotes * and §)	Leaves narrow, ribbed, scape wholly ribbed, 100–120 cm, head large, moderately dense, flowers intensely pinkish-purple Beginning of May
'Mont Blanc'	*A. stipitatum*	Scape 100 cm, head large, flowers true white, anthers white May
'Mother of Pearl'	*A. hollandicum*	Scape 80–100 cm, flowers violet–purple
'Mount Everest'	*A. stipitatum*	Scape 100 cm, flowers white, slightly greenish, anthers yellow May
'Per Wendelbo'	*A. jesdianum*	Leaves lanceolate, scape 120 cm, head large, flowers deep purple, upper filament parts white May
'Purple King'	*A. rosenorum* (*A. jesdianum* or *A. rosenbachianum*, see Table 19.1, footnotes * and §)	Scape 70 cm, head dense, flowers dark purple End of May Herbaceous beds and borders, excellent cut flowers
'Purple Sensation'	*A. hollandicum*	Scape basally ribbed, 60–80 cm, head moderately dense, flowers deep purple Mid- to end of May Herbaceous beds and borders, excellent cut flowers, forcing
'Purple Surprise'	*A. hollandicum* Selection from cv. 'Purple Sensation'	Scapes 100 cm, flowers brighter, bloom later End of May Herbaceous beds, cut flowers, forcing
'Red Globe'	*A. karataviense*	Leaves narrower, scape 20–40 cm, head larger, flowers red to purple
'Rien Poortvliet'	*A. hollandicum* × *A. stipitatum* Selection from cv. 'Gladiator'	Earlier flowering, scape 90 cm, flowers amethyst-violet
'Rosy Giant'	*A. giganteum*	Scape 100 cm, flowers amaranth–rose edged darker
'White Giant'	*A. stipitatum*	Same characters as 'Mount Everest', probably identical May

Table 19.2. *Continued.*

Cultivar name	Origin and classification	Specific characters, flowering period and use
Subgenus *Amerallium*		
'Cowanii'	*A. neapolitanum* Syn. *A. cowanii* Lindl.	Early flowering, scape 30–40 cm, flowers pure white
'Grandiflorum'	*A. neapolitanum*	Flowers larger than the type
'Jeannine'	*A. moly*	Robust and large, flowers golden yellow June
Subgenus *Allium*		
'Minor'	*A. flavum* var. *minus*	May–June
Subgenus *Rhizirideum*		
'Album'	*A. schoenoprasum*	Scape 30 cm, flowers white with green midvein
'Superbum'	*A. nutans*	Scape taller, heads very dense, flowers purple August

Table 19.3. Mean optimum temperature for seed germination of ornamental *Allium* taxa.

Subgenus	Main geographical distribution	Period of optimal germination	Optimum temperature (°C)
Melanocrommyum	Steppes and semi-deserts of SW to Central Asia	2–7 months [1, 3–5]	4–5 [1, 4, 5]
Allium	Mediterranean basin	2–8 weeks [1, 4]	10–20 [1, 4]
Rhizirideum	Temperate climatic zones	4–30 days [2, 4]	15–26 [2, 4]

[1] Aoba (1967); [2] Durdyev (1981); [3] Dalezkaya and Nikiforova (1984); [4] Specht and Keller (1997); [5] Kamenetsky and Gutterman (2000).

dominates in the genus, and the specialized *A. karataviense*-type, which occurs only among species of the subgenus *Melanocrommyum*. In both types, the cotyledon emerges from the seed and pushes the embryonal rootlet downward. One to 2 days later, the upper part of the cotyledon develops an inverted-U-shaped bend (loop; knee), which is pushed upwards through the soil surface by the elongation of the two sides of the cotyledon. Later, in the *A. cepa*-type, the first foliage leaf elongates within the sheath of the cotyledon and emerges through a slit in its side. During the first season, one to several primary leaves, composed of a sheath and a cylindrical lamina, are formed, and several branching adventitious roots replace the primary root (De Mason, 1990). In the *A. karataviense*-type, the epigeal part of the cotyledon remains green for several weeks without any other leaf formation. This part is the sole assimilating organ during the whole season, and often reaches more than 10 cm in length (Druselmann, 1992). Neither adventitious nor lateral roots are formed. In *A. rosenbachianum* (= *A. rosenorum*), the cotyledon develops normally at 5–25°C, but elongates more rapidly at 20–25°C than at 5°C (Aoba, 1968). At the end of the season, the storage leaves develop at the underground growing point and form a small bulb at a depth of

3–20 cm. This mechanism allows fast elongation of the subterranean part of the seedling and formation of a bulblet at a depth sufficient to protect it from desiccation during the dry hot summer (Glimcher, 1951; Kamenetsky, 1994).

The length of the juvenile stage of the *Allium* plant ranges from a few months in *Rhizirideum* species to several years in *Melanocrommyum* species. During this phase, the leaf form changes from the thread-like cotyledon to the species-specific final form, and the size of the bulb increases. The juvenile apical meristem produces only leaves and cannot be induced to bloom (Baitulin *et al.*, 1986; Kamenetsky, 1994; R.M. Fritsch, personal observations).

4.2 Annual growth rhythm

Most ornamental *Allium* species originate in the temperate zone and require a warm–cool–warm annual thermoperiodic cycle. The major difference among the species and among the cultivars is in cold requirement and cold-hardiness of the bulbs. Alliums range from very cold-resistant (*A. aflatunense, A. giganteum*) to cold-susceptible (*A. neapolitanum, A. triquetrum*) (De Hertogh and Zimmer, 1993). This range is exploited in horticulture, and different *Allium* species can be used for various purposes in a wide range of climatic zones, including arid regions.

There is significant variation among *Allium* spp. with regard to their annual life cycle and morphogenesis. Most rhizomatous species produce false bulbs, which are made of leaf sheaths. These plants have no dormancy or rest period; they form new leaves and renewal bulbs throughout the year and low winter temperatures only slow down these processes (Fig. 19.2a,b; Cheremushkina, 1992; Pistrick, 1992; Kamenetsky, 1996). Typical rhizomatous species, such as *A. nutans* or *A. senescens*, can form 20–22 leaves during 1 year (Baitulin *et al.*, 1986). Formation of the generative shoot in these species occurs in the spring and is followed by flowering in the summer. Some of them undergo two or three flowering cycles during one summer, during which they develop one to several complete leaves and a few prophylls per flower scape (Fig. 19.3a,b; Cheremushkina, 1985; Kruse, 1992).

The structure of bulbous species is characterized by a sequence of different leaf types, with photosynthesis, storage and protective functions. The intensity of intrabulb ramification ranges from numerous axillary daughter bulbs (*A. moly, A. oreophilum, A. ampeloprasum, A. scorodoprasum*; Figs 19.3c,d, 19.4) to only one or two bulbs (*A. aflatunense, A. macleanii, A. flavum*). In some *Melanocrommyum* species, there is no intrabulb ramification (*A. aschersonianum, A. rothii*; Fig. 19.3e).

During seed maturation, bulbous species lose their roots and above-ground parts and enter a 'rest' period, which allows them to survive unfavourable environmental conditions. Development of new vegetative and generative organs continues inside the 'resting' subterranean bulbs (Fig. 19.2c–e). The nutrient reserves supply the energy and material demands for building new cells and for resprouting (Aoba, 1970; Baitulin *et al.*, 1986). The duration of this 'rest' period depends on the degree of natural adaptation to environmental conditions. Thus, *A. karataviense* from Central Asian semi-deserts and *A. rothii* from the Israeli desert remain at 'rest' for about 4 and 6 months, respectively (Kamenetsky, 1996).

4.3 Floral development

The critical size for flowering is species-dependent and ranges from 3–5 cm in bulb circumference for *A. caeruleum, A. neapolitanum* and *A. unifolium*, through 12–14 cm for *A. aflatunense, A. cristophii* and *A. karataviense*, to 20–22 cm for *A. giganteum* and for the cultivars 'Gladiator', 'Globemaster' and 'Violet Beauty' (De Hertogh and Zimmer, 1993). The capacity for flower initiation is not necessarily related to the amount of reserves, therefore, the physiological significance of bulb size is not completely clear. The size of the apical meristem has been suggested as one of the determining factors for normal flower

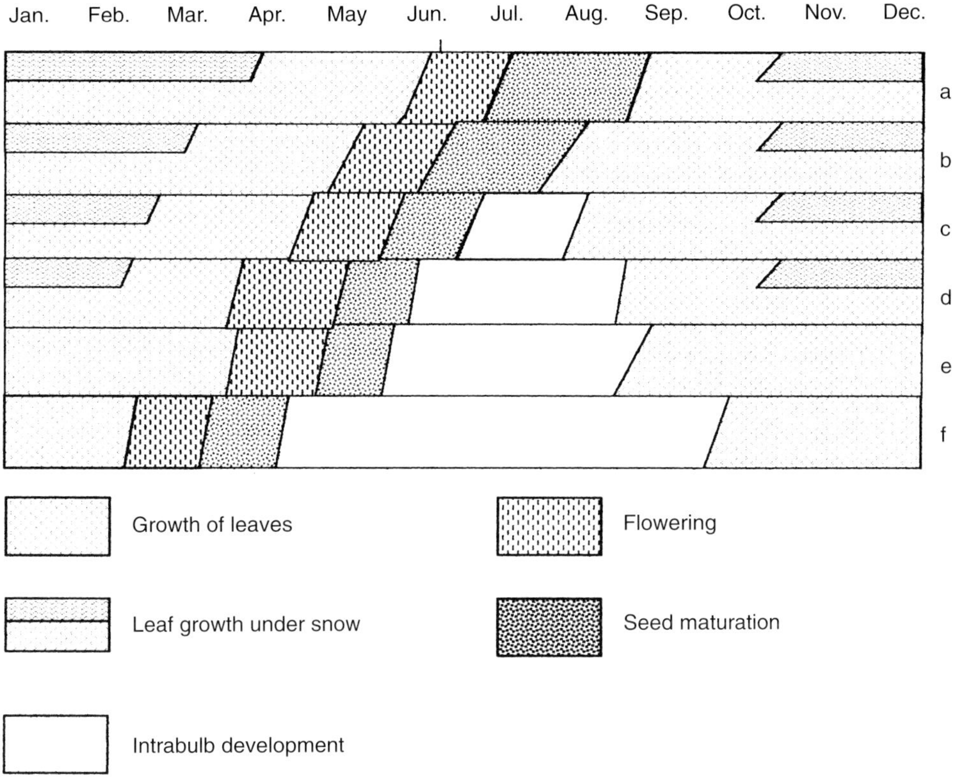

Fig. 19.2. Annual life cycles of some *Allium* species. (a) *A. nutans* (mountains, Siberia); (b) *A. pskemense* (mountains, Kazakhstan); (c) *A. caeruleum* (steppe, Russia); (d) *A. karataviense* (semi-desert, Kazakhstan); (e) *A. ampeloprasum* (semi-desert, Mediterranean); (f) *A. rothii* (desert, Israel). (From Kamenetsky, 1996, with permission.)

differentiation (Halevy, 1990; Le Nard and De Hertogh, 1993).

In addition to physiological age and genetics, environmental conditions, especially temperature, affect annual development (Table 19.4; see also Kamenetsky and Rabinowitch, Chapter 2, this volume). In general, the importance of temperature for growth and development in geophytes has been clearly established in the horticultural literature (Hartsema, 1961; Halevy, 1990; Le Nard and De Hertogh, 1993).

Floral initiation and differentiation have been studied for only a few ornamental *Allium* species. Thus, floral development has been described for the rhizomatous *A. odorum* (= *A. ramosum*) (Weber, 1929), the bulbous *A. sphaerocephalon* (subgenus *Allium*) (Berghoef and Zevenbergen, 1992) and several species from the subgenus *Amerallium* (Mann, 1959). Light and scanning electron microscopy of florogenesis in species from the subgenus *Melanocrommyum* revealed that all branches of the umbel-like inflorescence arise from a common meristem during the 'rest' period. In each inflorescence, many stages of floral development occur simultaneously (Kamenetsky, 1994, 1997; Kamenetsky and Japarova, 1997).

Floral initiation and differentiation of *Melanocrommyum* species proceeds during the 'rest' period at relatively high temperatures, but further leaf and floral stalk elongation demands a prolonged period of low temperatures (Dosser, 1980; Zimmer and Renken, 1984; De Hertogh and Zimmer, 1993;

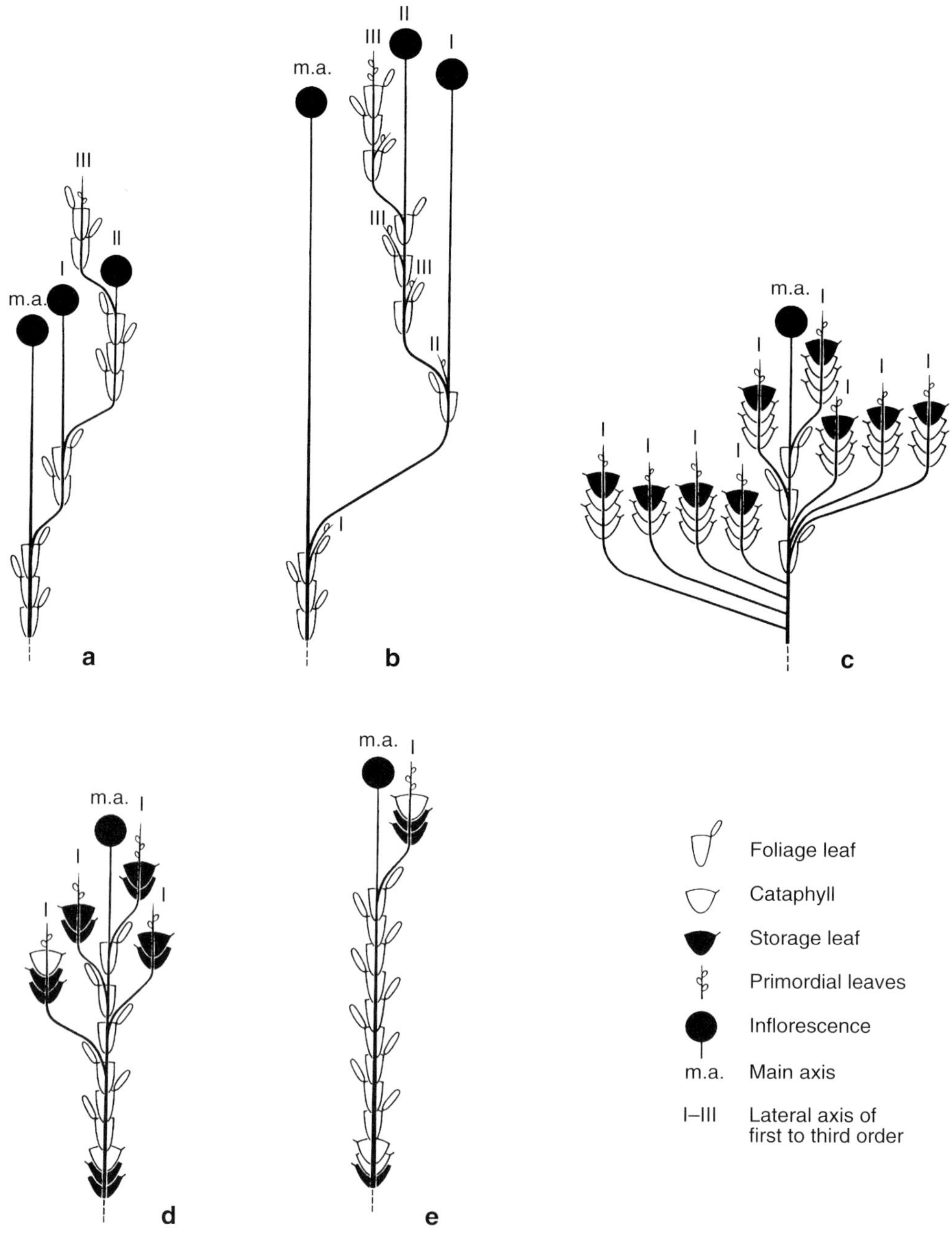

Fig. 19.3. Diagrams of growth and branching inside the bulb of *Allium* species. (a) *A. tuberosum*; (b) *A. cyathophorum*; (c) *A. moly*; (d) *A. stipitatum*; (e) *A. cristophii*. (From Kruse, 1992, with permission.)

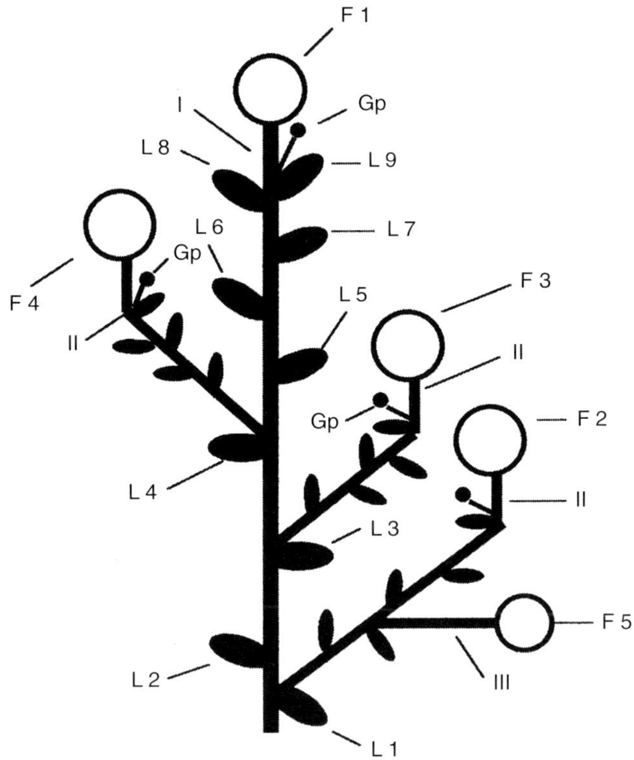

Fig. 19.4. Diagram of branching system in *Allium unifolium*. I, main axis; II, secondary axis; III, tertiary axis; F1, F2, F3, F4, F5, inflorescences; L1, L2, L3, L4, L5, L6, L7, L8, L9, leaves; Gp, growing point. (From Kodaira *et al.*, 1996, with permission.)

Zemah *et al.*, 1999, 2001). However, *A. aschersonianum*, *A. nigrum* and *A. rothii* from the Mediterranean basin remain vegetative during the summer 'rest' period, and floral differentiation occurs in their bulbs in autumn. These species sprout and develop normal leaves and inflorescences during the winter and require no cold induction for floral development and stalk elongation (Table 19.4; Kamenetsky, 1994; Kamenetsky *et al.*, 2000).

Similarly, flower initiation and differentiation of the species from the subgenus *Amerallium* occur at mild (9–17°C) temperatures during autumn storage or even after planting (Table 19.4). Lower storage temperatures before planting (2–9°C) accelerate scape emergence and blooming but, at the same time, result in decreases in flowering percentage and scape length (Kodaira *et al.*, 1991a, b, 1996; Maeda *et al.*, 1994; van Leeuwen and van der Weijden, 1994).

The species from the subgenus *Allium* form inflorescences only after the initiation of several leaf primordia during the vegetative growth stage, whereas storage temperatures before bulb planting have no conspicuous effect on flower initiation and differentiation (Table 19.4). After sprouting and leaf formation, these species require mild (17–20°C) temperatures and long days for successful blooming (Berghoef and Zevenbergen, 1992; van Leeuwen and van der Weijden, 1994).

Experimental data on environmental effects on the florogenesis of the rhizomatous species from the subgenus *Rhizirideum* are very limited. The process of flower

Table 19.4. The environmental requirements for floral induction and development of the *Allium* species from different origins and taxonomic groups.

Species	Origin	Environmental conditions for flowering			Remarks
		Initiation	Floral development	Stalk elongation	
Subgenus *Melanocrommyum* *A. aflatunense*[2, 11, 12] (= *A. hollandicum*) *A. karataviense*[3, 5] *A. altissimum*[5] *A. oreophilum*[9] *A. cristophii* (= *A. christophii*)[2]	Central and South-west Asia	Warm temperatures during growth of parent plant or summer 'rest' period (May–July)	Mild/warm temperatures during 'rest' period and dry autumn (August–November)	Long storage at low temperatures (4–5°C for 16 weeks[2,12]; 8°C for 24 weeks[11]) Moderate growth temperatures (17–23°C day/ 9–15°C night[11])	After differentiation of leaf primordia, the apical meristem may remain inactive for 6–10 weeks[5] Autumn storage at 15°C accelerates floral stalk emergence but reduces flowering percentage[9] No effect of day length on florogenesis[11]
A. rothii[4] *A. aschersonianum*[6]	Mediterranean, semi-arid zone	Warm temperatures (25–30°C), during summer 'rest' period (June–August)	Mild temperatures (20–25°C) after 'rest' period, during autumn storage (September–October)	Water supply and mild temperatures (15–25°C) after planting (December–January)	Post-harvest storage at mild temperature (20–25°C) accelerates intrabulb floral development and further flowering[6] Preplanting storage at low temperatures (9–13°C) may cause flower malformations[6]
Subgenus *Amerallium* *A. neapolitanum*[7, 9, 10] (= *A. cowanii*) *A. unifolium*[8, 9, 10] *A. moly*[9] *A. roseum*[8, 9, 10]	Mediterranean areas, California	Autumn storage at mild temperatures (9–17°C, *A. unifolium*[8]) or at room temperatures (*A. cowanii*[10])	Mild temperatures during autumn preplanting storage (9–17°C[10]) or during growth (10–20°C[8])	Winter storage at 9 or 13°C for 10 weeks (*A. roseum*[10]) or 4°C for 12 weeks (*A. moly*[9]) Water supply and mild temperatures during growth (10–20°C) (February-April)[9]	Bulbs stored at 15–20°C during summer bloomed earlier that those stored at 25–30°C[7] Storage at lower temperatures (2, 5 and 9°C) accelerates flowering but reduces percentage of flowering plants[9, 10] and scape length[10] Growth at >20°C reduces scape length[8]

Subgenus *Allium*					
A. caeruleum[5] (= *A. caesium*) *A. sphaerocephalon*[2,9] *A. ampeloprasum*[2]	Central Asia, Europe, Mediterranean	Autumn storage at 17°C (*A. caeruleum*[10]) Storage temperatures (2–21°C) do not affect floral initiation, which occurs only after planting (*A. sphaerocephalon*[1])	Autumn storage at 17°C (*A. caeruleum*[10])	Winter storage at 4°C for 12 weeks (*A. caeruleum*[9]) Mild temperatures (17–20°C[1]) and long day during growth[1,2,9]	Autumn storage at 2, 5 and 9°C reduces flowering percentage and flower quality (*A. caeruleum*[10]) Storage-temperature treatments affect flowering percentage but do not influence the process of scape emergence and bloom High growth temperatures and short day keep the plants in the vegetative stage (*A. sphaerocephalon*[1])

[1]Berghoef and Zevenbergen, 1992; [2] De Hertogh and Zimmer, 1993; [3] Dosser, 1980; [4]Kamenetsky, 1994; [5]Kamenetsky and Japarova, 1997; [6]Kamenetsky *et al.*, 2000; [7]Kodaira *et al.*, 1991a; [8]Kodaira *et al.*, 1996; [9]Maeda *et al.*, 1994; [10] van Leeuwen and van der Weijden, 1994; [11]Zemah *et al.*, 2001; [12]Zimmer *et al.*, 1985.

initiation and development in *A. nutans, A. senescens* and *A. galanthum* is relatively short: the apical meristem turns to the generative stage in the spring, just before floral-stalk elongation and summer flowering (Cheremushkina, 1985).

The effect of photoperiod on the flowering of ornamental *Alliums* has not been extensively studied. Long days accelerate flowering of *A. ampeloprasum, A. moly* and *A. roseum* (De Hertogh and Zimmer, 1993; Maeda *et al.*, 1994) and provide an essential condition for flower initiation in *A. sphaerocephalon* (Berghoef and Zevenbergen, 1992).

For all species studied, only the appropriate sequence of environmental conditions, especially temperatures, at the various physiological stages of plant development leads to normal flowering. In general, this sequence of environmental factors reflects the adaptation of *Allium* species to the specific conditions of their natural habitats and has to be taken into consideration during the growing and forcing of *Allium* plants.

4.4 Postharvest storage of cut flowers

When alliums are used as cut flowers, they can be cut when the florets are approximately 50% open, and kept under cold conditions (0–2°C) for up to 2 weeks before selling (Bijl, 1977; Mevel, 1983; De Hertogh, 1996).

4.5 Bulb development

Experimental data on the effects of environment on bulb production are rather limited. Zimmer and Renken (1984) and Zimmer *et al.* (1985) studied the effect of temperature on the subsequent bulb growth of *Melanocrommyum* species. They found that *A. aflatunense* (= *A. hollandicum*) developed one or two renewal buds inside the parent bulb, at the base of the scape, simultaneously with floral differentiation and before the low-temperature period. By mid-November, the developing buds reached 2 mm and their growth was inhibited by low-temperature treatment. After the low-temperature requirement was met, the buds enlarged rapidly and, during the growing period, they replaced the old bulb and served for the continuation of the parent plant.

Storage conditions conducive to scape elongation also resulted in the formation of a renewal bulb and a few daughter bulbs in *A. aflatunense*. However, the development of the renewal bulb required less cold treatment than floral development: the heaviest daughter bulbs were obtained after storage at 9°C for 8 weeks followed by 4°C for 8 weeks, before planting. Although cold storage at 4°C for 16 weeks resulted in the longest scapes, the renewal bulbs were smaller than the parent bulb (Zemah *et al.*, 2001). Plants of *A. cristophii* that had been stored for 24 weeks at 8°C produced the highest fresh weight of bulbs within 16 weeks after planting (Zimmer *et al.*, 1985).

In *A. neapolitanum* and *A. atropurpureum*, high growth temperatures resulted in the development of numerous daughter bulbs with relatively low fresh weight (Zimmer and Weckeck, 1989). This effect has also been observed in Israel, where high growing temperatures resulted in the formation of many small daughter bulbs of the *Melanocrommyum* species from Central Asia (R. Kamenetsky, personal observations).

Growing conditions for the cultivation of *Allium* species may also determine bulb and flower development for the next season. For instance, bulbs of *A. unifolium* were grown during their first season in a plastic house and outdoors in Kagoshima (Japan). Although the bulbs of both groups were stored and replanted under the same conditions, after planting in the next season the first group flowered 1 month earlier (Kodaira *et al.*, 1996).

These findings suggest that, as in many other geophytes (Le Nard and De Hertogh, 1993; Brewster, 1994), the processes of bulb and floral development in ornamental *Allium* species are linked. Only accurate fulfilment of the environmental requirements during the growth of the parent plants and during the subsequent bulb storage and growth periods will lead to successful flowering and bulb production.

4.6 Postharvest storage of bulbs

After cleaning and grading, *Allium* bulbs have to be stored and transported in well-ventilated ethylene-free containers at 20–23°C, except for *A. giganteum*, which has to be stored at 25–28°C (De Hertogh and Zimmer, 1993).

5. Propagation

Commercially, ornamental *Alliums* are propagated both from seed and vegetatively.

5.1 Propagation from seed

Seeds of rhizomatous species have no dormancy and can be sown late in the spring. They germinate quickly and grow throughout the summer to produce plants of marketable size in the autumn. Some species are able to produce flowers in the first season, but most bloom in the second year of development (R.M. Fritsch, unpublished data).

Seeds of bulbous species are sown in Europe shortly after harvest or in the autumn. Germination occurs in the spring, after seed exposure to low winter temperatures. In Israel, seeds of the *Melanocrommyum* species from Central Asia have to be wetted at low temperatures for at least 8–10 weeks before germination (R. Kamenetsky, unpublished data). After seed germination, the visible growth period (presence of a green leaf) lasts only 6–8 weeks in the first year and is followed by a 'rest' period for the newly produced bulbs (see above, Section 4). The juvenile period lasts for 2–3 or 3–5 years in species with small or large bulbs, respectively. The above-ground vegetative growth period lasts about 12–15 weeks per season (R.M. Fritsch and J. Kruse, unpublished data).

In The Netherlands, the field is lightly fertilized and mulched with straw after seeding. The bulbs are harvested 2 years later. After grading, bulbs of non-flowering sizes are replanted and the commercial sizes are marketed (De Hertogh and Zimmer, 1993). In Israel, plastic containers are used for seed germination and first-year bulb production of *A. nigrum* and *A. aschersonianum*. After harvest in May, small (0.5–1.0 cm in diameter) bulbs are stored at 20–25°C until October and are then replanted in the field for 2–3 years' cultivation (R. Kamenetsky, unpublished data).

5.2 Propagation from bulbs

Allium species, like most ornamental geophytes, are propagated vegetatively, from axillary bulbs, bulblets on stolons, division of rhizomes and topsets (Kamenetsky, 1993). The most common method for vegetative propagation of rhizomatous species is by rhizome division (Davies, 1992). Juvenile and generative plants differ in their ability to produce laterals; thus, juvenile plants of *A. senescens* produce only one or two laterals, whereas adult generative plants annually form six to 12 laterals (Cheremushkina, 1985).

Bulbous species vary in the propagation rate of their axillary daughter bulbs. Thus, *A. moly*, *A. rosenorum*, *A. stipitatum*, several species of the subgenus *Allium* (section *Allium*) and commercial strains from The Netherlands produce many daughter bulbs, whereas multiplication of most wild plants of subgenus *Allium* (section *Codonoprasum*) and of subgenus *Melanocrommyum* (e.g. *A. oreophilum*, *A. giganteum*, *A. macleanii* and other ornamental species) is rather low. This process can, however, be enhanced by suitable growth conditions (fertilization, optimal water-supply, growing temperatures between 20 and 30°C (R.M. Fritsch, personal observation)). Some species from the subgenus *Melanocrommyum* (e.g. *A. aschersonianum* and *A. rothii*) form only one renewal bulb each year to replace the parent bulb (Kamenetsky, 1993, 1994; Kamenetsky *et al.*, 2000).

Allium species can be multiplied artificially by autumn scaling, as demonstrated for *A. cristophii* and *A. giganteum* (Alkema, 1976) and for *A. aschersonianum* (D. Sandler-Ziv, Volcani Center, Israel, 2000, personal communication).

For some species, a factor limiting either their increased use or the rapid introduction of new germplasm is their low rate of natural multiplication. Therefore, systems for rapid multiplication, especially tissue-culture methods, should be developed for these species and cultivars.

Protocols for rapid propagation via tissue culture have been developed for *A. aflatunense*, *A. ampeloprasum* and *A. aschersonianum* (Ziv *et al.*, 1983; Evenor *et al.*, 1997; H. Lilien-Kipnis, Israel, 1999, personal communication). The highest regeneration rate of *A. aflatunense* was obtained from the explants derived from the developing inflorescence, which resulted in hundreds of plantlets and bulblets from a single bulb within a few months. No results of experiments on further development and hardening of the plantlets and their flowering ability were reported.

6. Pests and Diseases

The reaction of ornamental alliums to pests and diseases was reviewed by Davies (1992). Under conditions of high air humidity, many *Allium* species may suffer severely from downy mildew (*Peronospora destructor*) especially if the infection begins early in the summer. The damaged leaves and scapes may suffer damage and die, and the surviving storage organs are weak and prone to storage decay. Observations on 13 *Allium* species and cultivars from the subgenus *Melanocrommyum* in Israel indicated complete susceptibility to downy mildew and to pink root (*Pyrenochaeta terrestris*) in all tested plants (Kik *et al.*, 1999). The same ornamental alliums were tested for tolerance/resistance to two soil-borne *Sclerotium* species: one accession of *A. stipitatum* was found to be resistant to *S. cepivorum* and *S. perniciosum*, whereas the other species were tolerant to these diseases (Kik *et al.*, 1999).

In Europe, most bulbous species are susceptible to *Botrytis cinerea*, which infects plants when the soil is too damp during summer, or if damaged bulbs are stored under high air humidity. *Penicillium* infection is quite common in bulbs stored at low temperatures prior to forcing (Davies, 1992).

Eelworms (nematodes) may become a problem if alliums are repeatedly grown in the same plot; therefore, crop rotation is strongly advised. Warm-water treatment may be helpful against nematodes, as reported for *A. oreophilum* (De Hertogh and Zimmer, 1993). Bulbs of several *Allium* species are prone to damage by wireworms, onion fly and other insects. The wounds caused by these pests often provide entrance for fungal and bacterial diseases. The major sucking insects, which may also transmit viruses, are aphids and thrips (De Hertogh and Zimmer, 1993).

Viral infections that accompany commercial cultivation and bulb production are sometimes very severe and may cause floral malformations and plant death (Davies, 1992; R.M. Fritsch, unpublished data). Onion yellow dwarf virus (OYDV), which causes mosaic symptoms, and tobacco rattle virus (TRV), which causes chlorotic venation, are the two main viruses infecting ornamental alliums (De Hertogh and Zimmer, 1993).

7. Agronomic Practices

Fertile loamy soils are the most suitable for growing and multiplying ornamental alliums. Planting depth depends on species and bulb size and ranges from 5 to 30 cm. Contractile roots, developed by many species, may move the bulb to the right depth, so that small bulbs may descend by 5 cm during one vegetation period (R.M. Fritsch, personal observations). Therefore, the soil (growing medium) should be properly prepared to a depth of 25–30 cm or more.

Rhizomatous species are planted shallowly and the rhizomes, which are covered with 1–2 cm of soil, grow obliquely or vertically to position the plant at the right depth.

Most ornamental alliums grow well in sunny places with moderate moisture in the spring and drier conditions in the summer. Several taxa from humid areas (*A. cyathophorum* and *A. cyaneum*) require high air and soil humidity; others (*A. ursinum*) are adapted to moderate light intensity and may suffer damage under high light intensity or in permanent deep shadow.

When grown as perennial plants, species from arid habitats (*A. karataviense, A. cristophii*) need water-supply only as long as the complete leaves are green; later, during the summer, they should be kept under dry hot conditions to enhance normal inflorescence development for the following year. Nevertheless, most rhizomatous species that flower through the summer need a permanent water-supply.

Most species benefit from mineral and organic-matter supplements to ensure optimal growth during their vegetative development. In the case of perennial cultivation, regular replanting every third year may be necessary to reduce the plant stand and to promote growth (*A. stipitatum, A. ramosum, A. senescens*). Fertilization with well-composted manure and 500 kg ha^{-1} of N : P : K at 7–14–28 in the autumn and 500 kg ha^{-1} of 12–10–18 during the winter are recommended for bulb production (De Hertogh and Zimmer, 1993). Several perennial species, such as *A. flavum*, are rather short-lived and should be resown every 3 or 4 years (Davies, 1992; R.M. Fritsch, unpublished data).

8. Breeding Goals and Future Developments

Several characteristics are essential for the successful cultivation of ornamental alliums and to ensure that they are attractive to consumers. These characteristics include the persistence of green foliage through the blooming period and resistance or tolerance to pests and diseases and to extreme environmental conditions (e.g. low and high temperatures, salinity). Successful commercialization requires variety in the flower and foliage colours, as well as in the sizes and shapes of the inflorescence and the scape. Extended life, pleasant fragrance, resistance to shipment and a long production season are also important traits (De Hertogh and Zimmer, 1993). For cut-flower types, upright scapes longer than 60 cm are needed, and vase life should be 10 days or more. Cut flowers should be easily forced, to fit market demands, and should be capable of being stored and shipped after cutting (Bijl, 1977). For these requirements, precise knowledge of florogenesis, flowering physiology and environmental requirements is essential.

For the development of potted plants, which flower lavishly for a long period of time, a short and straight scape is needed and the presence of attractive and long-lasting foliage is especially important.

Special efforts should be invested in the development of propagation systems: seed propagation necessitates improved knowledge of the specific conditions for seed storage; further development of aseptic techniques in tissue culture is needed for successful introduction of new *Allium* varieties as commercial crops, as well as for elimination of viral infection and diseases in the existing ornamental alliums.

The natural variability of *Allium* species provides a tremendous potential for a wide range of different ornamental crops. Therefore, systematic collection, preservation and evaluation of wild species are needed to maintain this natural treasure and to expand the ornamental potential of the species. Specialized *Allium* collections exist today in Germany, Israel, England and the USA, and many botanical gardens keep vegetatively propagated plants.

9. Concluding Remarks

Ornamental alliums have become popular for gardens and also as cut flowers and potted plants. Further collection and preservation of wild *Allium* species and their evaluation for ornamental traits could achieve expansion of the available selection of these crops. For effective breeding and introduction of new cultivars into ornamental horticulture, the development of efficient propagation techniques, especially tissue culture, is needed. Investigations into florogenesis, flowering physiology and storage requirements of the prospective species, especially of the new bulbous cultivars and rhizomatous species, will extend our knowledge and provide new techniques for plant cultivation and forcing. Special attention has to be given to disease resistance of new ornamental species and to their tolerance to a wide range of climatic conditions.

Acknowledgements

We thank Dr J. Kruse (Gatersleben) for permission to use an unpublished manuscript in German on ornamental alliums. Thanks are also due to the Israeli Flower Board and the Institute für Pflanzengenetik und Kulturpflanzenforschung (IPK), Gatersleben, who helped us to cover the expenses for publication of colour photographs (Colour Plates 8–13).

References

Alkema, H.Y. (1976) Vegetatieve vermeerdering van *Allium* species. *Weekblad voor Bloembollencultuur* 86, 981–982.

Aoba, T. (1967) Effects of different temperatures on seed germination of garden ornamentals in *Allium*. *Journal of the Japanese Society for Horticultural Science* 36, 333–338 (in Japanese).

Aoba, T. (1968) Studies on propagation of *Allium rosenbachianum* Regel. II. Process of bulb formation in seedling. *Journal of the Japanese Society for Horticultural Science* 37, 166–171 (in Japanese).

Aoba, T. (1970) Effect of low temperature on the bulb or corm formation in some ornamental plants. *Journal of the Japanese Society for Horticultural Science* 39, 369–374 (in Japanese).

Bailey, L.H. and Bailey, E.Z. (1976) *Hortus Third. A Concise Dictionary of Plants Cultivated in the United States and Canada*. Macmillan and Collier Macmillan, New York and London, pp. 47–54.

Baitulin, I.O., Rakhimbaev, I.R. and Kamenezkaja, I.I. (1986) *Introdukcija i morfogenez dikorastushchikh lukov Kazakhstana*. Nauka, Alma-Ata, Kazakhstan, 154 pp.

Berghoef, J. and Zevenbergen, A.P. (1992) Effects of environmental conditions on flower initiation and development of *Allium sphaerocephalon* L. *Acta Horticulturae* 325, 91–96.

Bijl, J. (1977) Diverse alliums zijn uitstekende snijbloemen, zeer houdbaar op water en dikwijls goed bestand tegen het 'cellen'. *Weekblad voor Bloembollencultuur* 87, 1025–1028.

Bijl, J.R. (1995) *Allium* – flowering onions. *Herbertia* 50, 88–94.

Brewster, J.L. (1994) *Onions and other Vegetable Alliums*. CAB International, Wallingford, UK, 236 pp.

Cheremushkina,V.A. (1985) Osobennosty ritma sezonnogo razvitija i varianty malogo jiznennogo zikla kornevishnikh lukov (Seasonal development rhythm and variants of the minor life cycle in rhizomatous onions). *Bjulleten Moskovskogo Obshchestva Ispitatelei Prirody* 90(4), 96–106.

Cheremushkina, V.A. (1992) Evolution of life forms of species in subgenus *Rhizirideum* (Koch) Wendelbo, genus *Allium* L. In: Hanelt, P., Hammer, K. and Knüpffer, H. (eds) *The Genus* Allium – *Taxonomic Problems and Genetic Resources. Proceedings of an International Symposium held at Gatersleben, 11–13 June 1991*. IPK, Gatersleben, Germany, pp. 27–34.

Dadd, R. (1987) The discovery and introduction of *Allium giganteum*. *Kew Magazine* 4, 91–96.

Dalezkaya, T.V. and Nikiforova, V.N. (1984) Study of seed germination in some *Allium* species. In: Nekrasov, V.I. (ed.) *Ecological Problems of Seed Production in Introduced Forms. Abstracts of Papers from the 7th All-Union Conference*. Zinatne, Riga, Latvia, pp. 24–25 (in Russian).

Davies, D. (1992) *Alliums. The Ornamental Onions*. Batsford, London, 168 pp.

De Hertogh, A.A. (1996) *Holland Bulb Forcer's Guide*, 5th edn. International Flower Bulb Centre, Hillegom, The Netherlands, 365 pp.

De Hertogh, A.A. and Zimmer, K. (1993) *Allium* – ornamental species. In: De Hertogh, A.A. and Le Nard, M. (eds) *The Physiology of Flower Bulbs*. Elsevier, Amsterdam, pp. 187–200.

De Mason, D.A. (1990) Morphology and anatomy of *Allium*. In: Rabinowitch, H.D. and Brewster, J.L. (eds) *Onions and Allied Crops*, Vol. I. *Botany, Physiology, and Genetics*. CRC Press, Boca Raton, Florida, pp. 27–51.

Dosser, A. (1980) *Allium karataviense* – a lonely little onion in a petunia patch. *North Carolina Flower Growers Bulletin* 24, 11–12.

Druselmann, S. (1992) Vergleichende Untersuchungen an Vertretern der *Alliaceae* Agardh. 1. Morphologie der Keimpflanzen der Gattung *Allium* L. *Flora* 186, 37–52.

Dubouzet, J.G., Arisumi, K.I., Etoh, T., Matsuo, E. and Sakata, Y. (1992) Studies on the development of new ornamental *Allium* through interspecific hybridization. I. Evaluation of commercial ornamental *Allium* for adaptation to Kagoshima conditions. *Journal of the Japanese Society for Horticultural Science* 61, 659–664.

Durdyev, B.D. (1981) Biological features of onion wild relatives from Central Asia. Onions in the flora of USSR and Turkmenia. *Introduction and Ecology of Plants* 7, 231–250 (in Russian).

Evenor, D., Levi-Nissim, A., Afgin, L., Lilien-Kipnis, H. and Watad, A.A. (1997) Regeneration of plantlets and bulblets from explants and callus of *Allium aflatunense* cultivars and selection from indigenous Israeli *Allium ampeloprasum*. *Acta Horticulturae* 430(1), pp. 325–330.

Friesen, N., Fritsch, R. and Bachmann, K. (1997) Hybrid origin of some ornamentals of *Allium* subgenus *Melanocrommyum* verified with GISH and RAPD. *Theoretical and Applied Genetics* 95, 1229–1238.

Fritsch, R.M. (1992a) Infra-subgeneric grouping in subgenus *Melanocrommyum* (Webb et Berth.) Rouy. In: Hanelt, P., Hammer, K. and Knüpffer, H. (eds) *The Genus* Allium *– Taxonomic Problems and Genetic Resources. Proceedings of an International Symposium held at Gatersleben, 11–13 June, 1991*. IPK, Gatersleben, Germany, pp. 67–75.

Fritsch, R.M. (1992b) Zur Wurzelanatomie in der Gattung *Allium* L. (*Alliaceae*). *Beiträge zur Biologie der Pflanzen* 67, 129–160.

Glimcher, J. (1951) On the biology of germination in the *Melanocrommyum* section of the genus *Allium*. *Palestine Journal of Botany* 5, 115–118.

Gregory, M., Fritsch, R.M., Friesen, N.W., Khassanov, F.O. and McNeal, D.W. (1998) *Nomenclator Alliorum.* Allium *Names and Synonyms – a World Guide.* Royal Botanic Gardens, Kew, UK, 83 pp.

Griffiths, M. (1994) *Index of Garden Plants*. Timber Press, Portland, Oregon, pp. 43–47.

Gutterman, Y., Kamenetsky, R. and van Rooyen, M. (1995) A comparative study of seed germination of two *Allium* species from different habitats in the Negev Desert highlands. *Journal of Arid Environments* 29, 303–315.

Halevy, A.H. (1990) Recent advances in control of flowering and growth habit of geophytes. *Acta Horticulturae* 266, 35–42.

Hanelt, P., Schultze-Motel, J., Fritsch, R., Kruse, J., Maaß, H.I., Ohle, H. and Pistrick, K. (1992) Infrageneric grouping of *Allium* – the Gatersleben approach. In: Hanelt, P., Hammer, K. and Knüpffer, H. (eds) *The Genus* Allium *– Taxonomic Problems and Genetic Resources. Proceedings of an International Symposium held at Gatersleben, 11–13 June, 1991*. IPK, Gatersleben, Germany, pp. 107–123.

Hartsema, A.M. (1961) Influence of temperatures on flower formation and flowering of bulbous and tuberous plants. In: Ruhland, W. (ed.) *Handbuch der Pflanzenphysiologie,* Vol. 16. Springer-Verlag, Berlin, pp. 123–167.

Huxley, A., Griffiths, M. and Levy, M. (eds) (1991) *Dictionary of Gardening,* Vol. 1. Macmillan, London, pp. 107–115.

International Checklist for Hyacinths and Miscellaneous Bulbs (1991) Royal General Bulbgrowers Association, Hillegom, The Netherlands, pp. 1–14.

Kamenetsky, R. (1992) Morphological types and root systems as indicators of evolutionary pathways in the genus *Allium.* In: Hanelt, P., Hammer, K. and Knüpffer, H. (eds) *The Genus* Allium *– Taxonomic Problems and Genetic Resources. Proceedings of an International Symposium held at Gatersleben, 11–13 June, 1991*. IPK, Gatersleben, Germany, pp. 129–135.

Kamenetsky, R. (1993) Vegetative propagation of species of genus *Allium* L. *Water Science and Technology* 27, 511–517.

Kamenetsky, R. (1994) Life cycle, flower initiation and propagation of the desert geophyte *Allium rothii*. *International Journal of Plant Science* 155, 597–605.

Kamenetsky, R. (1996) Life cycle and morphological features of *Allium* L. species in connection with geographical distribution. *Bocconea* 5, 251–257.

Kamenetsky, R. (1997) Inflorescence of *Allium* species (subgenus *Melanocrommyum*): structure and development. *Acta Horticulturae* 430, 141–146.

Kamenetsky, R. and Gutterman, Y. (2000) Germination strategies of some *Allium* species of the subgenus *Melanocrommyum* from the arid zone of Central Asia. *Journal of Arid Environments* 45, 61–72.

Kamenetsky, R. and Japarova, N. (1997) Relationship between annual cycle and floral development of three *Allium* species from subgenus *Melanocrommyum*. *Journal of Arid Environments* 35, 473–485.

Kamenetsky, R., Gilad, Z. and Rabinowitch, E. (2000) *Development of* A. aschersonianum *from Israeli Flora as New Ornamental Crop for Cut Flower and Bulb Production*. Final Report, The Foundation of the Chief Scientist of the Ministry of Agriculture of Israel, Bet Dagan, 22 pp. (in Hebrew).

Kik, C., Kamenetsky, R. and Rabinowitch, H.D. (1999) *Development of high quality* Allium *species for cut flower production. Final Report Dutch–Israeli Agricultural Research Programme (DIARP)*, Wageningen, The Netherlands, 40 pp.

Kodaira, E., Mori, G. and Imanishi, H. (1991a) Effect of storage temperatures on the flowering of *Allium cowanii*. *Journal of the Japanese Society for Horticultural Science* 60, 566–567.

Kodaira, E., Mori, G. and Imanishi, H. (1991b) Effect of storage temperatures on the flowering of *A. unifolium*. *Journal of the Japanese Society for Horticultural Science* 61, 602–603.

Kodaira, E., Mori, G., Takeuchi, M. and Imanishi, H. (1996) Effects of temperature on the growth and flowering of *Allium unifolium* Kellogg. *Journal of the Japanese Society for Horticultural Science* 65, 373–380.

Kruse, J. (1992) Growth form characters and their variation in *Allium* L. In: Hanelt, P., Hammer, K. and Knüpffer, H. (eds) *The Genus* Allium *– Taxonomic Problems and Genetic Resources. Proceedings of an International Symposium held at Gatersleben, 11–13 June, 1991*. IPK, Gatersleben, Germany, pp. 173–179.

Le Nard, M. and De Hertogh, A.A. (1993) Bulb growth and development and flowering. In: De Hertogh, A.A. and Le Nard, M. (eds) *The Physiology of Flowering Bulbs*. Elsevier, Amsterdam, pp. 29–43.

Maeda, M., Dubouzet, J.G., Arisumi, K.I., Etoh, T. and Sakata, Y. (1994) Effects of cold storage and staggered planting in forcing culture of spring-flowering *Allium* species. *Journal of the Japanese Society for Horticultural Science* 63, 629–638.

Mann, L. (1959) The *Allium* inflorescence: some species of the section *Molium*. *American Journal of Botany* 46, 730–739.

Mann, L. (1960) Bulb organization in *Allium*: some species of the section *Molium*. *American Journal of Botany* 47, 765–771.

Mathew, B. (1996) *A Review of* Allium *Section* Allium. Royal Botanical Gardens, Kew, UK, 176 pp.

Mevel, A. (1983) Deux bulbeuses en vue: *Allium* et *Liatris*. *Horticulture Française* 150, 21–24.

Pastor, J. and Valdes, B. (1985) Bulb structure in some species of *Allium* (*Liliaceae*) of the Iberian Peninsula. *Annales Musei Goulandris* 7, 249–261.

Pistrick, K. (1992) Phenological variability in the genus *Allium* L. In: Hanelt, P., Hammer, K. and Knüpffer, H. (eds) *The Genus* Allium *– Taxonomic Problems and Genetic Resources. Proceedings of an International Symposium held at Gatersleben, 11–13 June, 1991*. IPK, Gatersleben, Germany, pp. 243–249.

PVS/BKD (1998/99) *Bloembollen – Voorjaarsbloeiers. Beplante oppervlakten, 1995/96 tot en met 1998/99*. Productschap Tuinbouw, The Hague, and Bloembollen-keuringsdienst, Lisse, The Netherlands, pp. 97–99.

Regel, E. (1887) *Allii* species Asiae Centralis in Asia Media a Turcomania desertisque Araliensibus et Caspicis usque ad Mongoliam crescentes. *Acta Horti Petropolitani* 10, 278–362.

Ruksans, J. (1998) *Catalogue of Bulb Nursery*. Rozula, Latvia, 44 pp.

Rümpler, T. (1882) *Die schönblühenden Zwiebelgewächse, ihre Kultur im Garten, Gewächshaus und Zimmer.* Paul Parey, Berlin, 460 pp.

Specht, C.E. and Keller, E.R.J. (1997) Temperature requirements for seed germination in the species of the genus *Allium* L. *Genetic Resources and Crop Evolution* 44, 509–517.

Stearn, W.T. and Campbell, E. (1986) *Allium* Linnaeus. In: Walters, S.M., Brady, A., Brickell, C.D., Cullen, J., Green, P.S., Lewis, J., Matthews, V.A., Webb, D.A., Yeo, P.F. and Alexander, J.C.M. (eds) *The European Garden Flora*. Vol. 1, *Pteridophyta, Gymnospermae, Angiospermae – Monocotyledons*, Part I. Cambridge University Press, Cambridge, pp. 233–246.

van Leeuwen, P.J. and van der Weijden, J.A. (1994) Vervroegen beperkt mogelijk voor enkele soorten. *Vakblad voor de Bloemisterij* 29, 28–29.

Weber, E. (1929) Entwicklungsgeschichtliche Untersuchungen über die Gattung *Allium*. *Botanisches Archiv* 25, 1–44.

Wendelbo, P. (1967) The genus *Allium*. A lily group discussion. *The Royal Horticultural Society Lily Year Book* 30. London, pp. 86–100.

Zemah, H., Bendel, P., Rabinowitch, H. and Kamenetsky, R. (1999) Visualization of morphological structure and water status during storage of *Allium aflatunense* bulbs by NMR imaging. *Plant Science* 147, 65–73.

Zemah, H., Rabinowitch, H.D. and Kamenetsky, R. (2001) Florogenesis and the effect of temperatures on the development of *Allium aflatunense*. *Journal of Horticultural Science and Biotechnology* 76, 507–513.

Zimmer, K. and Renken, M. (1984) Untersuchungen an *Allium aflatunense*. Gute Blüte nur nach Kuhlbehandlung. *Deutscher Gartenbau* 38, 2004–2008.

Zimmer, K. and Weckeck, K. (1989) Effect of temperature on some ornamental alliums. *Acta Horticulturae* 246, 131–135.

Zimmer, K., Walingen, M. and Gebauer, B. (1985) Untersuchungen an *Allium christophii*. *Deutscher Gartenbau* 39, 2206–2209.

Ziv, M., Hertz, N. and Biran, Y. (1983) Vegetative reproduction of *Allium ampeloprasum* L. *in vivo* and *in vitro*. *Israel Journal of Botany* 32, 1–9.

Index

Note: page numbers in **bold** indicate tables or figures.